Aviation Maintenance Handbook–Powerplant

2023

U.S. Department of Transportation

Federal Aviation Administration

Flight Standards Service

Preface

The Aviation Maintenance Technician Handbook–Powerplant (FAA-H-8083-32B) is one of a series of three handbooks for persons preparing for certification as a powerplant mechanic. It is intended that this handbook provide the basic information on principles, fundamentals, and technical procedures in the subject matter areas relating to the powerplant rating. It is designed to aid students enrolled in a formal course of instruction, as well as the individual who is studying on his or her own. Since the knowledge requirements for the airframe and powerplant ratings closely parallel each other in some subject areas, the chapters which discuss fire protection systems and electrical systems contain some material which is also duplicated in the Aviation Maintenance Technician Handbook–Airframe (FAA-H-8083-31B).

This handbook contains an explanation of the units that make up each of the systems that bring fuel, air, and ignition together in an aircraft engine for combustion. It also contains information on engine construction features, lubrication systems, exhaust systems, cooling systems, cylinder removal and replacement, compression checks, and valve adjustments. Because there are so many different types of aircraft in use today, it is reasonable to expect that differences exist in airframe components and systems. To avoid undue repetition, the practice of using representative systems and units is carried out throughout the handbook. Subject matter treatment is from a generalized point of view and should be supplemented by reference to manufacturer's manuals or other textbooks if more detail is desired. This handbook is not intended to replace, substitute for, or supersede official regulations or the manufacturer's instructions. Occasionally the word "must" or similar language is used where the desired action is deemed critical. The use of such language is not intended to add to, interpret, or relieve a duty imposed by Title 14 of the Code of Federal Regulations (14 CFR).

The subject of Human Factors is contained in the Aviation Maintenance Technician Handbook—General (FAA-H-8083-30) (as revised).

This handbook is available for download, in PDF format, from www.faa.gov.

This handbook is published by the United States Department of Transportation, Federal Aviation Administration, Airman Testing Standards Branch, AFS-630, P.O. Box 25082, Oklahoma City, OK 73125.

Comments regarding this publication should be emailed to AFS630comments@faa.gov.

Acknowledgments

The Aviation Maintenance Technician Handbook–Powerplant (FAA-H-8083-32B) was produced by the Federal Aviation Administration (FAA). The FAA wishes to acknowledge the following contributors:

Mr. Tom Wild for images used throughout this handbook

Free Images Live (www.freeimageslive.co.uk) for image used in Chapter 1

Mr. Stephen Sweet (www.stephensweet.com) for image used in Chapter 1

Mr. Omar Filipovic (www.glasair-owners.com) for image used in Chapter 1

Mr. Warren Lane (Atomic Metalsmith, Inc.) for image used in Chapter 1

Pratt & Whitney for images used in Chapters 2, 3, 6, 7, and 8

Teledyne Continental Motors (www.genuinecontinental.aero) for images used in Chapters 2, 3, and 11

Aircraft Tool Supply Company (www.aircraft-tool.com) for images used in Chapter 4

Chief Aircraft (www.chiefaircraft.com) for images used in Chapter 4

DeltaHawk Engines, Inc. (www.deltahawkengines.com) for image used in Chapter 6

Mr. Felix Gottwald for image used in Chapter 7

Mr. Stephen Christopher (www.schristo.com) for images used in Chapter 8

Mr. Yunjin Lee for images used in Chapter 9

Mr. Marco Leerentveld (www.flightillusion.com) for image used in Chapter 10

Aeromax Aviation, LLC (www.aeromaxaviation.com) for images used in Chapter 11

Avid Aircraft (www.avidflyeraircraft.com) for image used in Chapter 11

Flight and Safety Design (www.eco1aircraft.com) for image used in Chapter 11

Great Plains Aircraft Supply Co., Inc. (www.greatplainsas.com) for image used in Chapter 11

Lycoming Engines (www.lycoming.textron.com) for image used in Chapter 11

Revmaster LLC Aviation (revmasteraviation.com) for images used in Chapter 11

Rotech Research Canada, Ltd. (www.rotec.com) for images used in Chapter 11

Additional appreciation is extended to Mr. Gary E. Hoyle, Dean of Students, Pittsburgh Institute of Aeronautics; Mr. Tom Wild, Purdue University; Dr. Ronald Sterkenburg, Associate Professor of the Department of Aviation Technology, Purdue University; for their technical support and input.

Table of Contents

Chapter 1
Aircraft Engines
General Requirements 1-1
    Power & Weight 1-2
    Fuel Economy 1-2
    Durability & Reliability 1-2
    Operating Flexibility 1-3
    Compactness 1-3
    Powerplant Selection 1-3
Types of Engines 1-3
    Inline Engines 1-3
    Opposed or O-Type Engines 1-4
    V-Type Engines 1-4
    Radial Engines 1-4
Reciprocating Engines 1-5
    Design & Construction 1-5
    Crankcase Section 1-5
    Accessory Section 1-7
    Accessory Gear Trains 1-7
Crankshafts 1-7
    Crankshaft Balance 1-8
    Dynamic Dampers 1-8
Connecting Rods 1-9
    Master-and-Articulated Rod Assembly 1-9
    Knuckle Pins 1-10
    Plain-Type Connecting Rods 1-11
    Fork-and-Blade Rod Assembly 1-11
Pistons 1-11
    Piston Construction 1-12
    Piston Pin 1-13
Piston Rings 1-13
    Piston Ring Construction 1-13
    Compression Ring 1-13
    Oil Control Rings 1-13
    Oil Scraper Ring 1-13
Cylinders 1-14
    Cylinder Heads 1-14
    Cylinder Barrels 1-15
    Cylinder Numbering 1-15
Firing Order 1-16
    Single-Row Radial Engines 1-16
    Double-Row Radial Engines 1-16
Valves 1-16
    Valve Construction 1-17
Valve Operating Mechanism 1-18
    Cam Rings 1-18

Camshaft 1-20
Tappet Assembly 1-20
Solid Lifters/Tappets 1-20
Hydraulic Valve Tappets/Lifters 1-20
Push Rod 1-21
Rocker Arms 1-21
Valve Springs 1-21
Bearings 1-22
Plain Bearings 1-22
Ball Bearings 1-22
Roller Bearings 1-22
Propeller Reduction Gearing 1-23
Propeller Shafts 1-23
Reciprocating Engine Operating Principles 1-25
Operating Cycles 1-25
Four-Stroke Cycle 1-25
Intake Stroke 1-26
Compression Stroke 1-27
Power Stroke 1-27
Exhaust Stroke 1-27
Two-Stroke Cycle 1-27
Rotary Cycle 1-27
Diesel Cycle 1-27
Reciprocating Engine Power & Efficiencies 1-27
Work 1-27
Horsepower 1-28
Piston Displacement 1-28
Area of a Circle 1-28
Compression Ratio 1-29
Indicated Horsepower 1-30
Brake Horsepower 1-31
Friction Horsepower 1-32
Friction & Brake Mean Effective Pressures 1-33
Thrust Horsepower 1-33
Efficiencies 1-33
Thermal Efficiency 1-33
Mechanical Efficiency 1-35
Volumetric Efficiency 1-35
Propulsive Efficiency 1-36
Gas Turbine Engines 1-36
Types & Construction 1-36
Air Entrance 1-37
Accessory Section 1-38
Compressor Section 1-38
Compressor Types 1-38
Centrifugal-Flow Compressors 1-38
Axial-Flow Compressor 1-40
Diffuser 1-43
Combustion Section 1-43
Turbine Section 1-46
Exhaust Section 1-51

Gas Turbine Engine Bearings & Seals 1-52
Turboprop Engines 1-54
Turboshaft Engines 1-55
Turbofan Engines 1-55
Turbine Engine Operating Principles 1-56
Thrust 1-58
Gas Turbine Engine Performance 1-59
    Ram Recovery 1-60

Chapter 2
Engine Fuel & Fuel Metering Systems
Fuel System Requirements 2-1
    Vapor Lock 2-1
Basic Fuel System 2-2
Fuel Metering Devices for Reciprocating Engines 2-2
    Air-Fuel Mixtures 2-3
Carburetion Principles 2-5
    Venturi Principles 2-5
    Application of Venturi Principle to Carburetor 2-5
    Metering & Discharge of Fuel 2-6
Carburetor Systems 2-7
Carburetor Types 2-8
Carburetor Icing 2-8
Float-Type Carburetors 2-9
    Float Chamber Mechanism System 2-10
    Main Metering System 2-10
    Idling System 2-12
    Mixture Control System 2-12
    Accelerating System 2-13
    Economizer System 2-13
Pressure Injection Carburetors 2-14
    Typical Injection Carburetor 2-15
    Throttle Body 2-15
    Regulator Unit 2-15
    Fuel Control Unit 2-17
Automatic Mixture Control (AMC) 2-18
Stromberg PS Carburetor 2-19
    Accelerating Pump 2-21
    Manual Mixture Control 2-21
Fuel-Injection Systems 2-21
    Bendix/Precision Fuel-Injection System 2-21
    Fuel Injector 2-21
    Airflow Section 2-21
    Regulator Section 2-22
    Fuel Metering Section 2-23
    Flow Divider 2-23
    Fuel Discharge Nozzles 2-24
    Continental/TCM Fuel-Injection System 2-24
    Fuel-Injection Pump 2-25
    Air-Fuel Control Unit 2-25
    Fuel Control Assembly 2-26

Fuel Manifold Valve 2-26
Fuel Discharge Nozzle 2-27
Carburetor Maintenance 2-27
Carburetor Removal 2-27
Installation of Carburetor 2-29
Rigging Carburetor Controls 2-29
Adjusting Idle Mixtures 2-29
Idle Speed Adjustment 2-30
Fuel System Inspection & Maintenance 2-31
Complete System 2-31
Fuel Tanks 2-32
Lines & Fittings 2-32
Selector Valves 2-32
Pumps 2-32
Main Line Strainers 2-32
Fuel Quantity Gauges 2-32
Fuel Pressure Gauge 2-32
Pressure Warning Signal 2-32
Water Injection Systems for Reciprocating Engines 2-33
Turbine Engine Fuel System—General Requirements 2-33
Turbine Fuel Controls 2-33
Hydromechanical Fuel Control 2-34
Hydromechanical/Electronic Fuel Control 2-34
FADEC Fuel Control Systems 2-36
FADEC for an Auxiliary Power Unit 2-36
FADEC Fuel Control Propulsion Engine 2-36
Fuel System Operation 2-40
Water Injection System 2-40
Fuel Control Maintenance 2-40
Engine Fuel System Components 2-41
Main Fuel Pumps (Engine Driven) 2-41
Fuel Heater 2-42
Fuel Filters 2-42
Fuel Spray Nozzles & Fuel Manifolds 2-43
Simplex Fuel Nozzle 2-44
Duplex Fuel Nozzle 2-44
Airblast Nozzles 2-44
Flow Divider 2-45
Fuel Pressurizing & Dump Valves 2-45
Combustion Drain Valves 2-46
Fuel Quantity Indicating Units 2-46

Chapter 3
Induction & Exhaust Systems

Reciprocating Engine Induction Systems 3-1
Basic Carburetor Induction System 3-1
Induction System Icing 3-3
Induction System Filtering 3-4
Induction System Inspection & Maintenance 3-4
Extinguishing Engine Fires 3-4
Induction System Troubleshooting 3-5

Supercharged Induction Systems 3-5
   Internally Driven Superchargers 3-6
   Turbosuperchargers 3-7
   Normalizer Turbocharger 3-9
   Ground-Boosted Turbosupercharger System 3-9
   A Typical Turbosupercharger System 3-10
   Turbocharger Controllers & System Descriptions 3-14
   Variable Absolute Pressure Controller (VAPC) 3-14
   Sloped Controller 3-15
   Absolute Pressure Controller 3-15
   Turbocharger System Troubleshooting 3-15
Turbine Engine Inlet Systems 3-16
   Divided-Entrance Duct...3-17
   Variable-Geometry Duct 3-17
   Compressor Inlet Screens 3-18
   Bellmouth Compressor Inlets 3-19
   Turboprop & Turboshaft Compressor Inlets 3-19
   Turbofan Engine Inlet Sections 3-19
Reciprocating Engine Exhaust Systems 3-21
   Radial Engine Exhaust Collector Ring System..3-22
   Manifold & Augmentor Exhaust Assembly 3-22
   Reciprocating Engine Exhaust System Maintenance Practices 3-22
   Exhaust System Inspection 3-22
   Muffler & Heat Exchanger Failures 3-23
   Exhaust Manifold & Stack Failures 3-24
   Internal Muffler Failures 3-24
Exhaust Systems with Turbocharger...3-24
   Augmentor Exhaust System 3-24
   Exhaust System Repairs 3-24
   Turbine Engine Exhaust Nozzles..3-25
   Convergent Exhaust Nozzle 3-26
   Convergent-Divergent Exhaust Nozzle...3-26
Thrust Reversers..3-27
Afterburning/Thrust Augmentation 3-27
Thrust Vectoring 3-29
Engine Noise Suppression 3-29
Turbine Engine Emissions..3-31

Chapter 4
Engine Ignition & Electrical Systems

Reciprocating Engine Ignition Systems..4-1
Magneto-Ignition System Operating Principles 4-1
   High-Tension Magneto System Theory of Operation..4-2
      Magnetic Circuit 4-2
      Primary Electrical Circuit...4-3
      Secondary Electrical Circuit 4-6
   Magneto & Distributor Venting 4-7
   Ignition Harness 4-7
   Ignition Switches..4-8
   Single & Dual High-Tension System Magnetos 4-10

Magneto Mounting Systems 4-10
    High- & Low-Tension Magneto Systems 4-11
Types of DC Generators 4-12
    Series Wound DC Generators 4-12
        Parallel (Shunt) Wound DC Generators...4-12
        Compound Wound DC Generators...4-13
    Generator Ratings 4-13
    DC Generator Maintenance 4-13
FADEC System Description 4-14
    Low-Voltage Harness...4-14
    Electronic Control Unit (ECU) 4-15
    PowerLink Ignition System 4-16
    Engine Indicating & Crew Alerting System (EICAS) 4-16
Auxiliary Ignition Units 4-17
    Booster Coil 4-18
    Impulse Coupling 4-19
    High-Tension Retard Breaker Vibrator 4-21
    Low-Tension Retard Breaker Vibrator 4-23
Spark Plugs 4-24
    Reciprocating Engine Ignition System Maintenance & Inspection 4-26
Magneto-Ignition Timing Devices 4-26
    Built-In Engine Timing Reference Marks 4-26
    Timing Discs 4-27
    Piston Position Indicators 4-27
    Timing Lights...4-28
Checking the Internal Timing of a Magneto...4-29
    High-Tension Magneto E-Gap Setting (Bench Timing) 4-29
    Timing the High-tension Magneto to the Engine 4-30
    Performing Ignition System Checks 4-31
    Ignition Switch Check 4-31
    Maintenance & Inspection of Ignition Leads 4-31
    Replacement of Ignition Harness..4-33
    Checking Ignition Induction Vibrator Systems 4-33
Spark Plug Inspection & Maintenance 4-33
    Carbon Fouling of Spark Plugs 4-33
    Oil Fouling of Spark Plugs 4-34
    Lead Fouling of Spark Plugs 4-34
    Graphite Fouling of Spark Plugs...4-35
    Gap Erosion of Spark Plugs 4-35
    Spark Plug Removal 4-35
    Spark Plug Reconditioning Service 4-36
    Inspection Prior to Installation..4-37
    Spark Plug Installation 4-38
    Spark Plug Lead Installation 4-38
    Breaker Point Inspection..4-39
    Dielectric Inspection..4-41
    Ignition Harness Maintenance 4-41
    High-Tension Ignition Harness Faults 4-41
    Harness Testing 4-42
Turbine Engine Ignition Systems..4-43
    Capacitor Discharge Exciter Unit 4-44

Igniter Plugs 4-45
Turbine Ignition System Inspection & Maintenance 4-45
    Inspection 4-45
    Check System Operation 4-46
    Repair 4-46
Removal, Maintenance, & Installation of Ignition System Components 4-46
    Ignition System Leads 4-46
    Igniter Plugs 4-47
Powerplant Electrical Systems 4-47
    Wire Size 4-48
    Factors Affecting the Selection of Wire Size 4-48
    Factors Affecting Selection of Conductor Material 4-50
    Voltage Drop in Aircraft Wire & Cable 4-50
    Conductor Insulation 4-52
    Identifying Wire & Cable 4-53
    Electrical Wiring Installation 4-54
    Wire Groups & Bundles 4-54
    Twisting Wires 4-54
    Spliced Connections in Wire Bundles 4-54
    Slack in Wiring Bundles 4-54
    Bend Radii 4-55
    Routing & Installation 4-55
    Protection Against Chafing 4-56
    Protection Against High Temperature 4-56
    Protection Against Solvents & Fluids 4-56
    Protection of Wires in Wheel Well Area 4-57
    Routing Precautions 4-57
    Installation of Cable Clamps 4-57
Lacing & Tying Wire Bundles 4-58
    Single-Cord Lacing 4-58
    Double-Cord Lacing 4-59
    Lacing Branch-Offs 4-59
    Tying 4-59
Cutting Wire & Cable 4-60
Stripping Wire & Cable 4-61
    Solderless Terminals & Splices 4-61
    Copper Wire Terminals 4-62
    Crimping Tools 4-62
    Aluminum Wire Terminals 4-63
    Splicing Copper Wires Using Preinsulated Wires 4-63
Emergency Splicing Repairs 4-63
    Splicing with Solder & Potting Compound 4-63
Connecting Terminal Lugs to Terminal Blocks 4-64
Bonding & Grounding 4-65
    General Bonding & Grounding Procedures 4-65
Connectors 4-67
    Types of Connectors 4-67
    Connector Identification 4-68
    Installation of Connectors 4-69
Conduit 4-69
Electrical Equipment Installation 4-70

Electrical Load Limits..4-70

Controlling or Monitoring the Electrical Load 4-70

Circuit Protection Devices..4-70

Switches 4-71

Relays...4-71

Chapter 5
Engine Starting Systems

Introduction 5-1

Reciprocating Engine Starting Systems...5-1

Inertia Starters 5-1

Direct Cranking Electric Starter 5-2

Direct Cranking Electric Starting System for Large Reciprocating Engines.............................5-3

Direct Cranking Electric Starting System for Small Aircraft..5-6

Reciprocating Engine Starting System Maintenance Practices...5-6

Troubleshooting Small Aircraft Starting Systems..5-7

Gas Turbine Engine Starters 5-8

Electric Starting Systems & Starter Generator Starting System 5-10

Troubleshooting a Starter Generator Starting System...5-12

Air Turbine Starters 5-12

Air Turbine Starter Troubleshooting Guide 5-16

Chapter 6
Lubrication & Cooling Systems

Principles of Engine Lubrication 6-1

Types of Friction 6-1

Functions of Engine Oil...6-1

Requirements & Characteristics of Reciprocating Engine Lubricants...6-2

Viscosity...6-2

Viscosity Index...6-2

Flash Point & Fire Point 6-3

Cloud Point & Pour Point 6-3

Specific Gravity..6-3

Reciprocating Engine Lubrication Systems...6-4

Combination Splash & Pressure Lubrication 6-4

Lubrication System Requirements 6-4

Dry Sump Oil Systems...6-4

Oil Tanks 6-4

Oil Pump 6-6

Oil Filters..6-7

Oil Pressure Regulating Valve 6-7

Oil Pressure Gauge 6-8

Oil Temperature Indicator 6-9

Oil Cooler...6-9

Oil Cooler Flow Control Valve 6-9

Surge Protection Valves 6-10

Airflow Controls...6-10

Dry Sump Lubrication System Operation 6-11

Wet-Sump Lubrication System Operation 6-13

Lubrication System Maintenance Practices 6-13

Oil Tank..6-13

Oil Cooler 6-14
Oil Temperature Bulbs 6-15
Pressure & Scavenge Oil Screens 6-15
Oil Pressure Relief Valve 6-16
Recommendations for Changing Oil 6-17
Draining Oil 6-17
Oil & Filter Change & Screen Cleaning 6-17
Oil Filter Removal Canister Type Housing 6-17
Oil Filter/Screen Content Inspection 6-17
Assembly of & Installation of Oil Filters 6-18
Troubleshooting Oil Systems 6-18
Requirements for Turbine Engine Lubricants 6-18
Turbine Oil Health & Safety Precautions 6-19
Spectrometric Oil Analysis Program 6-20
Typical Wear Metals & Additives 6-20
Turbine Engine Lubrication Systems 6-20
Turbine Lubrication System Components 6-21
Oil Tank 6-21
Oil Pump 6-22
Turbine Oil Filters 6-23
Oil Pressure Regulating Valve 6-24
Oil Pressure Relief Valve 6-24
Oil Jets 6-25
Lubrication System Instrumentation 6-25
Lubrication System Breather Systems (Vents) 6-25
Lubrication System Check Valve 6-26
Lubrication System Thermostatic Bypass Valves 6-27
Air-Oil Coolers 6-27
Fuel-Oil Coolers 6-27
Deoiler 6-28
Magnetic Chip Detectors 6-28
Typical Dry-Sump Pressure Regulated Turbine Lubrication System 6-28
Pressure System 6-28
Scavenge System 6-28
Breather Pressurizing System 6-29
Typical Dry-Sump Variable Pressure Lubrication System 6-29
Pressure Subsystem 6-29
Scavenger Subsystem 6-30
Breather Subsystem 6-30
Turbine Engine Wet-Sump Lubrication System 6-30
Turbine Engine Oil System Maintenance 6-31
Engine Cooling Systems 6-32
Reciprocating Engine Cooling Systems 6-32
Reciprocating Engine Cooling System Maintenance 6-34
Maintenance of Engine Cowling 6-34
Engine Cylinder Cooling Fin Inspection 6-36
Cylinder Baffle & Deflector System Inspection 6-36
Cylinder Temperature Indicating Systems 6-37
Exhaust Gas Temperature Indicating Systems 6-38
Turbine Engine Cooling 6-38
Accessory Zone Cooling 6-38

Turbine Engine Insulation Blankets 6-40

Chapter 7
Propellers

General 7-1
Basic Propeller Principles 7-1
Propeller Aerodynamic Process 7-2
    Aerodynamic Factors 7-4
    Propeller Controls & Instruments 7-5
Propeller Location 7-5
    Tractor Propeller 7-5
    Pusher Propellers 7-6
Types of Propellers 7-6
    Fixed-Pitch Propeller 7-6
    Test Club Propeller 7-6
    Ground-Adjustable Propeller 7-6
    Controllable-Pitch Propeller 7-6
    Constant-Speed Propellers 7-7
    Feathering Propellers 7-8
    Reverse-Pitch Propellers 7-8
Propeller Governor 7-8
    Governor Mechanism 7-9
    Underspeed Condition 7-9
    Overspeed Condition 7-11
    On-Speed Condition 7-11
    Governor System Operation 7-11
Propellers Used on General Aviation Aircraft 7-12
    Fixed-Pitch Wooden Propellers 7-12
    Metal Fixed-Pitch Propellers 7-13
Constant-Speed Propellers 7-14
    Hartzell Constant-Speed, Nonfeathering 7-14
    Constant-Speed Feathering Propeller 7-15
    Unfeathering 7-16
Propeller Auxiliary Systems 7-17
    Ice Control Systems 7-17
        Anti-Icing Systems 7-17
        Deicing Systems 7-18
    Propeller Synchronization & Synchrophasing 7-19
    Autofeathering System 7-20
Propeller Inspection & Maintenance 7-20
    Wood Propeller Inspection 7-21
    Metal Propeller Inspection 7-21
    Aluminum Propeller Inspection 7-21
    Composite Propeller Inspection 7-21
Propeller Vibration 7-22
    Blade Tracking 7-22
    Checking & Adjusting Propeller Blade Angles 7-23
    Universal Propeller Protractor 7-23
Propeller Balancing 7-24
    Static Balancing 7-24
    Dynamic Balancing 7-25

    Balancing Procedure 7-25
  Propeller Removal & Installation 7-26
    Removal 7-26
    Installation..7-27
  Servicing Propellers 7-27
    Cleaning Propeller Blades 7-27
    Charging the Propeller Air Dome 7-27
    Propeller Lubrication...7-27
  Propeller Overhaul 7-29
    The Hub 7-29
    Prop Reassembly 7-29
  Troubleshooting Propellers 7-30
    Hunting & Surging..7-30
    Engine Speed Varies with Flight Attitude (Airspeed) 7-30
    Failure to Feather or Feathers Slowly 7-30
  Turboprop Engines & Propeller Control Systems..7-30
    Reduction Gear Assembly 7-31
    Turbo-Propeller Assembly...7-31
  Pratt & Whitney PT6 Hartzell Propeller System 7-31
  Hamilton Standard Hydromatic Propellers 7-34
    Principles of Operation 7-36
    Feathering Operation...7-37
    Unfeathering Operation 7-38
    Setting the Propeller Governor 7-40

Chapter 8
Engine Removal & Replacement

  Introduction 8-1
  Reasons for Removal of Reciprocating Engines..8-1
    Engine or Component Lifespan Exceeded 8-1
    Engine Sudden Stoppage or Propeller Strike...8-1
    Sudden Reduction in Speed 8-1
    Metal Particles in the Oil 8-2
    Spectrometric Oil Analysis Engine Inspection Program..8-2
    Turbine Engine Condition Monitoring Programs 8-2
    Engine Operational Problems...8-2
  General Procedures for Engine Removal & Installation 8-3
    Preparation of Engines for Installation..8-3
    QECA Buildup Method for Changing of Engines 8-3
    Depreservation of an Engine 8-4
    Inspection & Depreservation of Accessories 8-5
  Inspection & Replacement of Powerplant External Units & Systems..............................8-5
  Preparing the Engine for Removal 8-6
    Draining the Engine 8-6
    Electrical Disconnects...8-6
    Disconnection of Engine Controls...8-7
    Disconnection of Lines..8-8
    Other Disconnections 8-8
  Removing the Engine 8-8
    Hoisting the Engine...8-9
  Hoisting & Mounting the Engine for Installation 8-9

Connections & Adjustments 8-10
Preparation of Engine for Ground & Flight Testing 8-12
   Pre-Oiling 8-12
   Fuel System Bleeding 8-13
Propeller Check 8-13
Checks & Adjustments After Engine Runup & Operation 8-13
Rigging, Inspections, & Adjustments 8-13
   Rigging Power Controls 8-14
   Adjusting the Fuel Control 8-14
Turboprop Powerplant Removal & Installation 8-15
Reciprocating Helicopter Engine & QECA 8-15
   Removal of Helicopter QECA 8-16
   Installation, Rigging, & Adjustment of Helicopter QECA 8-16
   Testing the Engine Installation 8-16
Engine Mounts 8-16
   Mounts for Reciprocating Engines 8-16
   Mounts for Turbofan Engines 8-17
   Turbine Vibration Isolation Engine Mounts 8-17
Preservation & Storage of Engines 8-18
   Corrosion-Preventive Materials 8-18
   Corrosion-Preventive Compounds 8-18
   Dehydrating Agents 8-18
Engine Preservation & Return to Service 8-19
Engine Shipping Containers 8-21
Inspection of Stored Engines 8-21
Preservation & Depreservation of Gas Turbine Engines 8-22

Chapter 9
Engine Fire Protection Systems
Introduction 9-1
   Components 9-1
   Engine Fire Detection Systems 9-2
      Thermal Switch System 9-2
      Thermocouple Systems 9-2
      Optical Fire Detection Systems 9-3
      Pneumatic Thermal Fire Detection 9-3
      Continuous-Loop Detector Systems 9-3
   Fire Zones 9-7
Engine Fire Extinguishing System 9-7
   Fire Extinguishing Agents 9-7
   Turbine Engine Ground Fire Protection 9-8
   Containers 9-8
   Discharge Valves 9-8
   Pressure Indication 9-8
   Two-Way Check Valve 9-9
   Discharge Indicators 9-9
      Thermal Discharge Indicator (Red Disc) 9-9
      Yellow Disc Discharge Indicator 9-9
   Fire Switch 9-9
   Warning Systems 9-10
Fire Detection System Maintenance 9-10

Fire Detection System Troubleshooting 9-11

Fire Extinguisher System Maintenance Practices 9-12

Boeing 777 Aircraft Fire Detection & Extinguishing System 9-13

    Overheat Detection 9-13

    Fire Detection 9-13

    Nacelle Temperature Recording 9-13

    Continuous Fault Monitoring 9-13

    Single/Dual Loop Operation 9-14

    System Test 9-14

    Boeing 777 Fire Extinguisher System 9-14

        Fire Extinguisher Containers 9-14

    Squib 9-16

    Engine Fire Switches 9-16

    Engine Fire Operation 9-17

APU Fire Detection & Extinguishing System 9-17

    APU Fire Warning 9-18

    Fire Bottle Discharge 9-18

Chapter 10
Engine Maintenance & Operation

Reciprocating Engine Overhaul 10-1

    Top Overhaul 10-1

    Major Overhaul & Major Repairs 10-1

General Overhaul Procedures 10-1

Receiving Inspection 10-2

Disassembly 10-2

Inspection Process 10-2

Visual Inspection 10-3

    Cylinder Head 10-4

    Piston, Valve Train, & Piston Pin 10-5

    Crankshaft & Connecting Rods 10-5

Cleaning 10-5

    Degreasing 10-5

    Removing Hard Carbon 10-5

Structural Inspection 10-6

    Dye Penetrant Inspection 10-6

    Eddy Current Inspection 10-7

    Ultrasonic Inspection 10-7

        Pulse-Echo 10-7

        Through Transmission 10-7

        Resonance 10-7

    Magnetic Particle Inspection 10-7

    X-ray 10-7

Dimensional Inspection 10-7

    Cylinder Barrel 10-7

    Rocker Arms & Shafts 10-8

    Crankshaft 10-9

    Checking Alignment 10-9

    Repair & Replacement 10-10

    Cylinder Assembly Reconditioning 10-11

    Piston & Piston Pins 10-11

     Valves & Valve Springs 10-11

     Refacing Valve Seats 10-13

     Valve Reconditioning 10-14

     Valve Lapping & Leak Testing 10-17

     Piston Repairs 10-17

     Cylinder Grinding & Honing 10-17

   Reassembly 10-19

   Installation & Testing 10-19

     Testing Reciprocating Engines 10-19

     Test Cell Requirements 10-20

     Engine Instruments 10-20

     Carburetor Air Temperature (CAT) Indicator 10-21

     Fuel Pressure Indicator 10-21

     Oil Pressure Indicator 10-22

     Oil Temperature Indicator 10-22

     Fuel-Flow Meter 10-22

     Manifold Pressure Indicator 10-23

     Tachometer Indicator 10-23

     Cylinder Head Temperature Indicator 10-23

     Torquemeter 10-24

     Warning Systems 10-24

   Reciprocating Engine Operation 10-24

     Engine Instruments 10-24

     Engine Starting 10-25

     Pre-Oiling 10-25

     Hydraulic Lock 10-25

     Engine Warm-Up 10-25

     Ground Check 10-26

     Fuel Pressure & Oil Pressure Check 10-27

     Propeller Pitch Check 10-27

     Power Check 10-27

     Idle Speed & Idle Mixture Checks 10-28

     Engine Stopping 10-28

   Basic Engine Operating Principles 10-29

     Combustion Process 10-29

     Detonation 10-29

     Pre-Ignition 10-30

     Backfiring 10-30

     Afterfiring 10-31

   Factors Affecting Engine Operation 10-31

     Compression 10-31

     Fuel Metering 10-31

     Idle Mixture 10-33

     Induction Manifold 10-33

     Operational Effect of Valve Clearance 10-33

   Engine Troubleshooting 10-35

     Valve Blow-By 10-39

   Cylinder Compression Tests 10-39

     Differential Pressure Tester 10-39

     Cylinder Replacement 10-41

   Cylinder Removal 10-41

Cylinder Installation 10-42
Cold Cylinder Check 10-43
Turbine Engine Maintenance 10-44
  Compressor Section 10-45
  Inspection & Cleaning 10-45
  Causes of Blade Damage 10-45
  Blending & Replacement 10-47
Combustion Section Inspection 10-47
  Marking Materials for Combustion Section Parts 10-49
  Inspection & Repair of Combustion Chambers 10-49
  Fuel Nozzle & Support Assemblies 10-50
  Turbine Disc Inspection 10-50
  Turbine Blade Inspection 10-50
  Turbine Blade Replacement Procedure 10-51
  Turbine Nozzle Inlet Guide Vane Inspection 10-52
  Clearances 10-52
  Exhaust Section 10-54
Engine Ratings 10-54
Turbine Engine Instruments 10-54
  Engine Pressure Ratio Indicator 10-54
  Torquemeter (Turboprop Engines) 10-55
  Tachometer 10-55
  Exhaust Gas Temperature Indicator (EGT) 10-55
  Fuel-Flow Indicator 10-55
  Engine Oil Pressure Indicator 10-55
  Engine Oil Temperature Indicator 10-55
Turbine Engine Operation 10-57
  Ground Operation Engine Fire 10-57
  Engine Checks 10-57
  Checking Takeoff Thrust 10-57
  Ambient Conditions 10-58
Engine Shutdown 10-58
Troubleshooting Turbine Engines 10-59
Turboprop Operation 10-59
  Troubleshooting Procedures for Turboprop Engines 10-59
Turbine Engine Calibration & Testing 10-59
  Turbine Engine Analyzer Uses 10-59
  Analyzer Safety Precautions 10-61
  Continuity Check of Aircraft EGT Circuit 10-63
  Functional Check of Aircraft EGT Circuit 10-63
  EGT Indicator Check 10-64
  Resistance & Insulation Check 10-64
  Tachometer Check 10-64
Troubleshooting EGT System 10-64
  One or More Inoperative Thermocouples in Engine Parallel Harness 10-64
  Engine Thermocouples Out of Calibration 10-64
  EGT Circuit Error 10-65
  Resistance of Circuit Out of Tolerance 10-65
  Shorts to Ground/Shorts Between Leads 10-65
Troubleshooting Aircraft Tachometer System 10-65

Chapter 11
Light-Sport Aircraft Engines

Engine General Requirements 11-1
Personnel Authorized to Perform Inspection & Maintenance on Light-Sport Engines 11-2
   Authorized Personnel That Meet FAA Regulations...11-3
Types of Light-Sport & Experimental Engines...11-3
   Light-Sport Aircraft Engines 11-3
   Two-Cycle, Two Cylinder Rotax Engine 11-3
   Rotax 447 UL Single Capacitor Discharge Ignition (SCDI) & Rotax 503 UL Dual Capacitor Discharge
   Ignition (DCDI) 11-3
      Rotax 582 UL DCDI 11-4
   Description of Systems for Two-Stroke Engines...11-4
      Cooling System of Rotax 447 UL SCDI & Rotax 503 UL DCDI 11-4
      Cooling System of the Rotax 582 UL DCDI 11-4
   Lubrication Systems 11-4
      Oil Injection Lubrication of Rotax 503 UL DCDE & 582 UL DCDI 11-4
   Electric System 11-5
   Fuel System 11-5
      Fuel-Oil Mixing Procedure 11-5
Opposed Light-Sport, Experimental, & Certificated Engines 11-5
   Rotax 912/914 11-5
   Description of Systems 11-6
      Cooling System..11-6
      Fuel System..11-6
      Lubrication System 11-7
      Electric System 11-8
      Turbocharger & Control System 11-8
   HKS 700T Engine 11-9
   Jabiru Light-Sport Engines 11-10
      Jabiru 2200 Aircraft Engine 11-11
      Aeromax Aviation 100 (IFB) Aircraft Engine 11-11
Direct Drive VW Engines...11-12
   Revmaster R-2300 Engine 11-12
   Great Plains Aircraft Volkswagen (VW)Conversions 11-14
   Teledyne Continental 0-200 Engine 11-15
   Lycoming 0-233 Series Light-Sport Aircraft Engine 11-15
General Maintenance Practices on Light-Sport Rotax Engines 11-16
Maintenance Schedule Procedures & Maintenance Checklist...11-16
   Carburetor Synchronization 11-17
   Pneumatic Synchronization 11-18
   Idle Speed Adjustment...11-19
   Optimizing Engine Running 11-19
   Checking the Carburetor Actuation..11-19
Lubrication System...11-19
   Oil Level Check..11-19
   Oil Change 11-20
   Cleaning the Oil Tank...11-20
   Inspecting the Magnetic Plug..11-21
   Checking the Propeller Gearbox..11-21
   Checking the Friction Torque in Free Rotation 11-21

    Daily Maintenance Checks 11-21

Pre-flight Checks 11-22

Troubleshooting & Abnormal Operation 11-22

    Troubleshooting 11-23

        Engine Keeps Running With Ignition OFF 11-23

        Knocking Under Load 11-23

    Abnormal Operation 11-23

        Exceeding the Maximum Admissible Engine Speed 11-23

        Exceeding Maximum Admissible Cylinder Head Temperature 11-23

        Exceeding Maximum Admissible Exhaust Gas Temperature........................ 11-23

Engine Preservation 11-23

General Maintenance Practices for the Light-Sport Jabiru Engines 11-23

    Engine & Engine Compartment Inspection 11-23

    Lubrication System 11-24

    Carburetor Adjustment & Checks 11-24

    Spark Plugs 11-24

    Exhaust System 11-24

    Head Bolts 11-24

    Tachometer & Sender 11-25

Engine Inspection Charts 11-25

Glossary **G-1**

Index **I-1**

Chapter 1
Aircraft Engines

General Requirements

Aircraft require thrust to produce enough speed for the wings to provide lift or enough thrust to overcome the weight of the aircraft for vertical takeoff. For an aircraft to remain in level flight, thrust must be provided that is equal to and in the opposite direction of the aircraft drag. This thrust, or propulsive force, is provided by a suitable type of aircraft heat engine. All heat engines have in common the ability to convert heat energy into mechanical energy by the flow of some fluid mass (generally air) through the engine. In all cases, the heat energy is released at a point in the cycle where the working pressure is high relative to atmospheric pressure.

The propulsive force is obtained by the displacement of a working fluid (again, atmospheric air). This air is not necessarily the same air used within the engine. By displacing air in a direction opposite to that in which the aircraft is propelled, thrust can be developed. This is an application of Newton's third law of motion. It states that for every action there is an equal and opposite reaction. So, as air is being displaced to the rear of the aircraft the aircraft is moved forward by this principle. One misinterpretation of this principle is air is pushing against the air behind the aircraft making it move forward. This is not true. Rockets in space have no air to push against, yet, they can produce thrust by using Newton's third law. Atmospheric air is the principal fluid used for propulsion in every type of aircraft powerplant except the rocket, in which the total combustion gases are accelerated and displaced. The rocket must provide all the fuel and oxygen for combustion and does not depend on atmospheric air. A rocket carries its own oxidizer rather than using ambient air for combustion. It discharges the gaseous byproducts of combustion through the exhaust nozzle at an extremely high velocity (action) and it is propelled in the other direction (reaction).

The propellers of aircraft powered by reciprocating or turboprop engines accelerate a large mass of air at a relatively lower velocity by turning a propeller. The same amount of thrust can be generated by accelerating a small mass of air to a very high velocity. The working fluid (air) used for the propulsive force is a different quantity of air than that used within the engine to produce the mechanical energy to turn the propeller.

Turbojets, ramjets, and pulse jets are examples of engines that accelerate a smaller quantity of air through a large velocity change. They use the same working fluid for propulsive force that is used within the engine. One problem with these types of engines is the noise made by the high velocity air exiting the engine. The term turbojet was used to describe any gas turbine engine, but with the differences in gas turbines used in aircraft, this term is used to describe a type of gas turbine that passes all the gases through the core of the engine directly.

Turbojets, ramjets, and pulse jets have very little to no use in modern aircraft due to noise and fuel consumption. Small general aviation aircraft use mostly horizontally opposed reciprocating piston engines. While some aircraft still use radial reciprocating piston engines, their use is very limited. Many aircraft use a form of the gas turbine engine to produce power for thrust. These engines are normally the turboprop, turboshaft, turbofan, and a few turbojet engines. "Turbojet" is the former term for any turbine engine. Now that there are so many different types of turbine engines, the term used to describe most turbine engines is "gas turbine engine." All four of the previously mentioned engines belong to the gas turbine family.

All aircraft engines must meet certain general requirements of efficiency, economy, and reliability. Besides being economical in fuel consumption, an aircraft engine must be economical in the cost of original procurement and the cost of maintenance; and it must meet exacting requirements of efficiency and low weight-to-horsepower ratio. It must be capable of sustained high-power output with no sacrifice in reliability; it must also have the durability to operate for long periods of time between overhauls. It needs to be as compact as possible yet have easy accessibility for maintenance. It is required to be as vibration free as possible and be able to cover a wide range of power output at various speeds and altitudes.

These requirements dictate the use of ignition systems that deliver the firing impulse to the spark plugs at the proper time in all kinds of weather and under other adverse conditions. Engine fuel delivery systems provide metered fuel at the correct proportion of air-fuel ingested by the engine regardless of the attitude, altitude, or type of weather in which the engine is operated. The engine needs a type of oil system that delivers oil under the proper pressure to lubricate and cool all of the operating parts of the engine when it is running. Also, it must have a system of damping units to damp out the vibrations of the engine when it is operating.

Power & Weight

The useful output of all aircraft powerplants is thrust, the force which propels the aircraft. Since the reciprocating engine is rated in brake horsepower (bhp), the gas turbine engine is rated in thrust horsepower (thp):

$$\text{Thp} = \frac{\text{thrust} \times \text{aircraft speed (mph)}}{375 \text{ mile-pounds per hour}}$$

The value of 375 mile-pounds per hour is derived from the basic horsepower formula as follows:

$$1 \text{ hp} = 33,000 \text{ ft-lb per minute}$$

$$33,000 \times 60 = 1,980,000 \text{ ft-lb per hour}$$

$$\frac{1,980,000}{5,280 \text{ ft in a mile}} = 375 \text{ mile-pounds per hour}$$

One horsepower equals 33,000 ft-lb per minute or 375 mile-pounds per hour. Under static conditions, thrust is figured as equivalent to approximately 2.6 pounds per hour.

If a gas turbine is producing 4,000 pounds of thrust and the aircraft in which the engine is installed is traveling at 500 mph, the thp is:

$$\frac{4,000 \times 500}{375} = 5,333.33 \text{ thp}$$

It is necessary to calculate the horsepower for each speed of an aircraft, since the horsepower varies with speed. Therefore, it is not practical to try to rate or compare the output of a turbine engine on a horsepower basis. The aircraft engine operates at a relatively high percentage of its maximum power output throughout its service life. The aircraft engine is at full power output whenever a takeoff is made. It may hold this power for a period of time up to the limits set by the manufacturer. The engine is seldom held at a maximum power for more than 2 minutes, and usually not that long. Within a few seconds after lift-off, the power is reduced to a power that is used for climbing and that can be maintained for longer periods of time. After the aircraft has climbed to cruising altitude, the power of the engine(s) is further reduced to a cruise power

which can be maintained for the duration of the flight. If the weight of an engine per brake horsepower (called the specific weight of the engine) is decreased, the useful load that an aircraft can carry and the performance of the aircraft obviously are increased. Every excess pound of weight carried by an aircraft engine reduces its performance. Tremendous improvement in reducing the weight of the aircraft engine through improved design and metallurgy has resulted in reciprocating engines with a much improved power-to-weight ratio (specific weight).

Fuel Economy

The basic parameter for describing the fuel economy of aircraft engines is usually specific fuel consumption. Specific fuel consumption for gas turbines is the fuel flow measured in (lb/hr) divided by thrust (lb), and for reciprocating engines the fuel flow (lb/hr) divided by brake horsepower. These are called thrust-specific fuel consumption and brake-specific fuel consumption, respectively. Equivalent specific fuel consumption is used for the turboprop engine and is the fuel flow in pounds per hour divided by a turboprop's equivalent shaft horsepower. Comparisons can be made between the various engines on a specific fuel consumption basis. At low speed, the reciprocating and turboprop engines have better economy than the pure turbojet or turbofan engines. However, at high speed, because of losses in propeller efficiency, the reciprocating or turboprop engine's efficiency becomes limited above 400 mph less than that of the turbofan.

Durability & Reliability

Durability and reliability are usually considered identical factors since it is difficult to mention one without including the other. Simply put, reliability is measured as the mean time between failures, while durability is measured as the mean time between overhauls.

More specifically, an aircraft engine is reliable when it can perform at the specified ratings in widely varying flight attitudes and in extreme weather conditions. Standards of powerplant reliability are agreed upon by the Federal Aviation Administration (FAA), the engine manufacturer, and the airframe manufacturer. The engine manufacturer ensures the reliability of the product by design, research, and testing. Close control of manufacturing and assembly procedures is maintained, and each engine is tested before it leaves the factory.

Durability is the amount of engine life obtained while maintaining the desired reliability. The fact that an engine has successfully completed its type or proof test indicates that it can be operated in a normal manner over a long period before requiring overhaul. However, no definite time interval between overhauls is specified or implied in the engine rating. The time between overhauls (TBO) varies with the operating conditions, such as engine temperatures, amount of time the engine is operated at high-power settings, and the maintenance received. Recommended TBOs are specified by the engine manufacturer.

Reliability and durability are built into the engine by the manufacturer, but the continued reliability of the engine is determined by the maintenance, overhaul, and operating personnel. Careful maintenance and overhaul methods, thorough periodical and preflight inspections, and strict observance of the operating limits established by the engine manufacturer make engine failure a rare occurrence.

Operating Flexibility

Operating flexibility is the ability of an engine to run smoothly and give desired performance at all speeds from idling to full-power output. The aircraft engine must also function efficiently through all the variations in atmospheric conditions encountered in widespread operations.

Compactness

To affect proper streamlining and balancing of an aircraft, the shape and size of the engine must be as compact as possible. In single-engine aircraft, the shape and size of the engine also affect the view of the pilot, making a smaller engine better from this standpoint, in addition to reducing the drag created by a large frontal area.

Weight limitations, naturally, are closely related to the compactness requirement. The more elongated and spread out an engine is, the more difficult it becomes to keep the specific weight within the allowable limits.

Powerplant Selection

Engine specific weight and specific fuel consumption were discussed in the previous paragraphs, but for certain design requirements, the final powerplant selection may be based on factors other than those that can be discussed from an analytical point of view. For that reason, a general discussion of powerplant selection follows.

For aircraft whose cruising speed does not exceed 250 mph, the reciprocating engine is the usual choice of powerplant. When economy is required in the low speed range, the conventional reciprocating engine is chosen because of its excellent efficiency and relatively low cost. When high altitude performance is required, the turbo-supercharged reciprocating engine may be chosen because it is capable of maintaining rated power to a high altitude (above 30,000 feet). Gas turbine engines operate most economically at high altitudes. Although in most cases the gas turbine engine provides superior performance, the cost of gas turbine

engines is a limiting factor. In the range of cruising speed of 180 to 350 mph, the turboprop engine performs very well. It develops more power per pound of weight than does the reciprocating engine, thus allowing a greater fuel load or payload for engines of a given power. From 350 mph up to Mach .8–.9, turbofan engines are generally used for airline operations. Aircraft intended to operate at Mach 1 or higher are powered by pure turbojet engines/afterburning (augmented) engines, or low-bypass turbofan engines.

Types of Engines

Aircraft engines can be classified by several methods. They can be classed by operating cycles, cylinder arrangement, or the method of thrust production. All are heat engines that convert fuel into heat energy that is converted to mechanical energy to produce thrust. Most of the current aircraft engines are of the internal combustion type because the combustion process takes place inside the engine. Aircraft engines come in many different types, such as gas turbine based, reciprocating piston, rotary, two or four cycle, spark ignition, diesel, and air or water cooled. Reciprocating and gas turbine engines also have subdivisions based on the type of cylinder arrangement (piston) and speed range (gas turbine).

Many types of reciprocating engines have been designed. However, manufacturers have developed some designs that are used more commonly than others and are, therefore, recognized as conventional. Reciprocating engines may be classified according to the cylinder arrangement (inline, V-type, radial, and opposed) or according to the method of cooling (liquid cooled or air cooled). Actually, all piston engines are cooled by transferring excess heat to the surrounding air. In air-cooled engines, this heat transfer is direct from the cylinders to the air. Therefore, it is necessary to provide thin metal fins on the cylinders of an air-cooled engine in order to have increased surface for sufficient heat transfer. Most reciprocating aircraft engines are air cooled although a few high powered engines use an efficient liquid-cooling system. In liquid-cooled engines, the heat is transferred from the cylinders to the coolant, which is then sent through tubing and cooled within a radiator placed in the airstream. The coolant radiator must be large enough to cool the liquid efficiently. The main problem with liquid cooling is the added weight of coolant, heat exchanger (radiator), and tubing to connect the components. Liquid cooled engines do allow high power to be obtained from the engine safely.

Inline Engines

An inline engine generally has an even number of cylinders, although some three-cylinder engines have been constructed. This engine may be either liquid cooled or air cooled and has only one crank shaft, which is located either above or below the cylinders. If the engine is designed to operate with the cylinders below the crankshaft, it is called an inverted engine.

The inline engine has a small frontal area and is better adapted to streamlining. When mounted with the cylinders in an inverted position, it offers the added advantages of a shorter landing gear and greater pilot visibility. With increase in engine size, the air cooled, inline type offers additional problems to provide proper cooling; therefore, this type of engine is confined to low- and medium-horsepower engines used in very old light aircraft.

Opposed or O-Type Engines

The opposed-type engine has two banks of cylinders directly opposite each other with a crankshaft in the center. *[Figure 1-1]* The pistons of both cylinder banks are connected to the single crankshaft. Although the engine can be either liquid cooled or air cooled, the air-cooled version is used predominantly in aviation. It is generally mounted with the cylinders in a horizontal position. The opposed-type engine has a low weight-to-horsepower ratio, and its narrow silhouette makes it ideal for horizontal installation on the aircraft wings (twin engine applications). Another advantage is its low vibration characteristics.

V-Type Engines

In V-type engines, the cylinders are arranged in two inline banks generally set 60° apart. Most of the engines have 12 cylinders, which are either liquid cooled or air cooled. The engines are designated by a V followed by a dash and the piston displacement in cubic inches. For example, V-1710. This type of engine was used mostly during the Second World War and its use is mostly limited to older aircraft.

Radial Engines

The radial engine consists of a row, or rows, of cylinders arranged radially about a central crankcase. *[Figure 1-2]* This type of engine has proven to be very rugged and dependable. The number of cylinders which make up a row may be three,

Figure 1-2. *Radial engine.*

five, seven, or nine. Some radial engines have two rows of seven or nine cylinders arranged radially about the crankcase, one in front of the other. These are called double-row radials. *[Figure 1-3]* One type of radial engine has four rows of cylinders with seven cylinders in each row for a total of 28 cylinders. Radial engines are still used in some older cargo airplanes, war birds, and crop spray airplanes. Although many of these engines still exist, their use is limited. The single-row, nine-cylinder radial engine is of relatively simple construction, having a one-piece nose and a two-section main crankcase. The larger twin-row engines are of slightly more complex construction than the single row engines. For example, the crankcase of the Wright R-3350 engine is composed of the crankcase front section, four crankcase main sections (front main, front center, rear center, and

Figure 1-1. *A typical four-cylinder opposed engine.*

Figure 1-3. *Double row radials.*

rear main), rear cam and tappet housing, supercharger front housing, supercharger rear housing, and supercharger rear housing cover. Pratt and Whitney engines of comparable size incorporate the same basic sections, although the construction and the nomenclature differ considerably.

Reciprocating Engines

Design & Construction

The basic major components of a reciprocating engine are the crankcase, cylinders, pistons, connecting rods, valves, valve-operating mechanism, and crankshaft. In the head of each cylinder are the valves and spark plugs. One of the valves is in a passage leading from the induction system; the other is in a passage leading to the exhaust system. Inside each cylinder is a movable piston connected to a crankshaft by a connecting rod. *Figure 1-4* illustrates the basic parts of a reciprocating engine.

Crankcase Section

The foundation of an engine is the crankcase. It contains the bearings and bearing supports in which the crankshaft revolves. Besides supporting itself, the crankcase must provide a tight enclosure for the lubricating oil and must support various external and internal mechanisms of the engine. It also provides support for attachment of the cylinder assemblies, and the powerplant to the aircraft. It must be sufficiently rigid and strong to prevent misalignment of the crankshaft and its bearings. Cast or forged aluminum alloy is generally used for crankcase construction because it is light and strong. The crankcase is subjected to many variations of mechanical loads and other forces. Since the cylinders are fastened to the crankcase, the tremendous forces placed on the cylinder tend to pull the cylinder off the crankcase. The unbalanced centrifugal and inertia forces of the crankshaft acting through the main bearings subject the crankcase to bending moments which change continuously in direction and magnitude. The crankcase must have sufficient stiffness to withstand these bending moments without major deflections. *[Figure 1-5]*

If the engine is equipped with a propeller reduction gear, the front or drive end is subjected to additional forces. In addition to the thrust forces developed by the propeller under high power output, there are severe centrifugal and gyroscopic forces applied to the crankcase due to sudden changes in the direction of flight, such as those occurring during maneuvers of the airplane. Gyroscopic forces are particularly severe when a heavy propeller is installed. To absorb centrifugal loads, a large centrifugal bearing is used in the nose section.

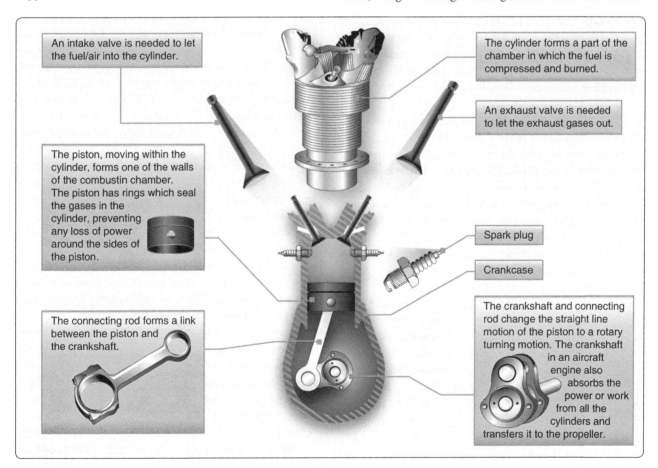

Figure 1-4. *Basic parts of a reciprocating engine.*

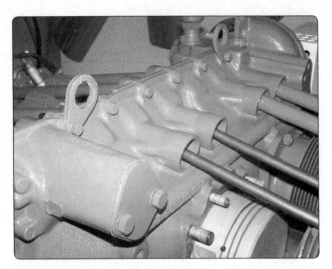

Figure 1-5. *The crankcase.*

The shape of the nose or front of the crankcase section varies considerably. In general, it is either tapered or round. Depending upon the type of reciprocating engine, the nose or front area of the crankcase varies somewhat. If the propeller is driven directly by the crankshaft, less area is needed for this component of the engine. The crankcases used on engines having opposed or inline cylinder arrangements vary in form for the different types of engines, but in general they are approximately cylindrical. One or more sides are surfaced to serve as a base to which the cylinders are attached by means of cap screws, bolts, or studs. These accurately machined surfaces are frequently referred to as cylinder pads.

If the propeller is driven by reduction gearing (gears that slow down the speed of the propeller less than the engine), more area is required to house the reduction gears. A tapered nose section is used quite frequently on direct-drive, low-powered engines, because extra space is not required to house the propeller reduction gears. Crankcase nose sections are usually cast of either aluminum alloy or magnesium. The crankcase nose section on engines that develop from 1,000 to 2,500 hp is usually larger to house reduction gears and sometimes ribbed to get as much strength as possible.

The governor is used to control propeller speed and blade angle. The mounting of the propeller governor varies. On some engines, it is located on the rear section, although this complicates the installation, especially if the propeller is operated or controlled by oil pressure, because of the distance between the governor and propeller. Where hydraulically operated propellers are used, it is good practice to mount the governor on the nose section as close to the propeller as possible to reduce the length of the oil passages. The governor is then driven either from gear teeth on the periphery of the bell gear or by some other suitable means. This basic

arrangement is also used for turboprops.

On some of the larger radial engines, a small chamber is located on the bottom of the nose section to collect the oil. This is called the nose section oil sump. Since the nose section transmits many varied forces to the main crankcase or power section, it must be secured properly to transmit the loads efficiently.

The machined surfaces on which the cylinders are mounted are called cylinder pads. They are provided with a suitable means of retaining or fastening the cylinders to the crankcase. The general practice in securing the cylinder flange to the pad is to mount studs in threaded holes in the crankcase. The inner portion of the cylinder pads are sometimes chamfered or tapered to permit the installation of a large rubber O-ring around the cylinder skirt, which effectively seals the joint between the cylinder and the crankcase pads against oil leakage.

Because oil is thrown about the crankcase, especially on inverted inline and radial-type engines, the cylinder skirts extend a considerable distance into the crankcase sections to reduce the flow of oil into the inverted cylinders. The piston and ring assemblies must be arranged so that they throw out the oil splashed directly into them.

Mounting lugs are spaced about the periphery of the rear of the crankcase or the diffuser section of a radial engine. These are used to attach the engine assembly to the engine mount or framework provided for attaching the powerplant to the fuselage of single-engine aircraft or to the wing nacelle structure of multiengine aircraft. The mounting lugs may be either integral with the crankcase or diffuser section or detachable, as in the case of flexible or dynamic engine mounts.

The mounting arrangement supports the entire powerplant including the propeller, and therefore is designed to provide ample strength for rapid maneuvers or other loadings. Because of the elongation and contraction of the cylinders, the intake pipes which carry the mixture from the diffuser chamber through the intake valve ports are arranged to provide a slip joint which must be leak proof. The atmospheric pressure on the outside of the case of an un-supercharged engine is higher than on the inside, especially when the engine is operating at idling speed. If the engine is equipped with a supercharger and operated at full throttle, the pressure is considerably higher on the inside than on the outside of the case. If the slip joint connection has a slight leakage, the engine may idle fast due to a slight leaning of the mixture. If the leak is quite large, it may not idle at all. At open throttle, a small leak probably would not be noticeable in operation of the engine, but the slight leaning of the air-fuel mixture might cause detonation or damage to the valves and valve seats. On some radial engines, the intake pipe has considerable length and on some inline engines, the intake pipe is at right angles to the cylinders. In these cases, flexibility of the intake pipe or its arrangement eliminates the need for a slip joint. In any case, the engine induction system must be arranged so that it does not leak air and change the desired air-fuel ratio.

Accessory Section

The accessory (rear) section usually is of cast construction and the material may be either aluminum alloy, which is used most widely, or magnesium, which has been used to some extent. On some engines, it is cast in one piece and provided with means for mounting the accessories, such as magnetos, carburetors, fuel, oil, vacuum pumps, starter, generator, tachometer drive, etc., in the various locations required to facilitate accessibility. Other adaptations consist of an aluminum alloy casting and a separate cast magnesium cover plate on which the accessory mounts are arranged. Accessory drive shafts are mounted in suitable drive arrangements that are carried out to the accessory mounting pads. In this manner, the various gear ratios can be arranged to give the proper drive speed to magnetos, pumps, and other accessories to obtain correct timing or functioning.

Accessory Gear Trains

Gear trains, containing both spur- and bevel-type gears, are used in the different types of engines for driving engine components and accessories. Spur-type gears are generally used to drive the heavier loaded accessories or those requiring the least play or backlash in the gear train. Bevel gears permit angular location of short stub shafts leading to the various accessory mounting pads. On opposed, reciprocating engines, the accessory gear trains are usually simple arrangements. Many of these engines use simple gear trains to drive the engine's accessories at the proper speeds.

Crankshafts

The crankshaft is carried in a position parallel to the longitudinal axis of the crankcase and is generally supported by a main bearing between each throw. The crankshaft main bearings must be supported rigidly in the crankcase. This usually is accomplished by means of transverse webs in the crankcase, one for each main bearing. The webs form an integral part of the structure and, in addition to supporting the main bearings, add to the strength of the entire case.

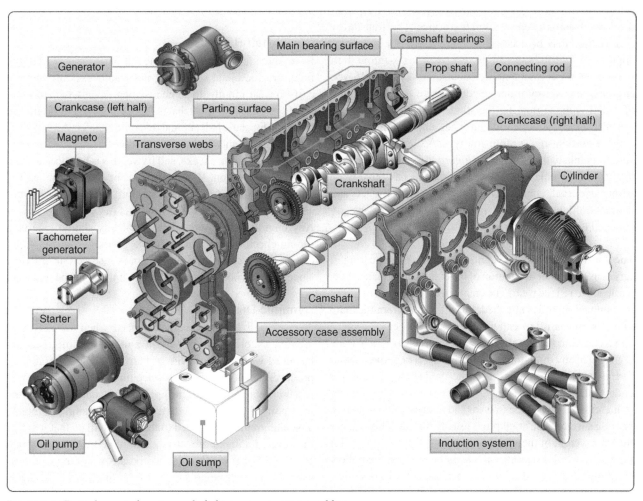

Figure 1-6. *Typical opposed engine exploded into component assemblies.*

The crankcase is divided into two sections in a longitudinal plane. This division may be in the plane of the crankshaft so that one-half of the main bearing (and sometimes camshaft bearings) are carried in one section of the case and the other half in the opposite section. [Figure 1-6] Another method is to divide the case in such a manner that the main bearings are secured to only one section of the case on which the cylinders are attached, thereby providing means of removing a section of the crankcase for inspection without disturbing the bearing adjustment.

The crankshaft is the backbone of the reciprocating engine. It is subjected to most of the forces developed by the engine. Its main purpose is to transform the reciprocating motion of the piston and connecting rod into rotary motion for rotation of the propeller. The crankshaft, as the name implies, is a shaft composed of one or more cranks located at specified points along its length. The cranks, or throws, are formed by forging offsets into a shaft before it is machined. Since crankshafts must be very strong, they generally are forged from a very strong alloy, such as chromium-nickel-molybdenum steel.

A crankshaft may be of single-piece or multipiece construction. Figure 1-7 shows two representative types of solid crankshafts used in aircraft engines. The four-throw construction may be used either on four-cylinder horizontal opposed or four-cylinder inline engines. The six-throw shaft is used on six-cylinder inline engines, 12-cylinder V-type engines, and six-cylinder opposed engines. Crankshafts of radial engines may be the single-throw, two-throw, or four-throw type, depending on whether the engine is the single-row, twin-row, or four-row type. A single-throw radial engine crankshaft is shown in Figure 1-8. No matter how many throws it may have, each crankshaft has three main parts—a journal, crankpin, and crank cheek. Flyweights and dampers, although not a true part of a crankshaft, are usually attached to it to reduce engine vibration.

The journal is supported by, and rotates in, a main bearing. It serves as the center of rotation of the crankshaft. It is surface-hardened to reduce wear. The crankpin is the section to which the connecting rod is attached. It is off-center from the main journals and is often called the throw. Two crank cheeks and a crankpin make a throw. When a force is applied to the crankpin in any direction other than parallel or perpendicular to and through the center line of the crankshaft, it causes the crankshaft to rotate. The outer surface is hardened by nitriding to increase its resistance to wear and to provide the required bearing surface. The crankpin is usually hollow. This reduces the total weight of the crankshaft and provides a passage for the transfer of lubricating oil. On early engines, the hollow crankpin also served as a chamber for collecting sludge, carbon deposits, and other foreign material. Centrifugal force

threw these substances to the outside of the chamber and kept them from reaching the connecting-rod bearing surface. Due to the use of ashless dispersant oils, newer engines no longer use sludge chambers. On some engines, a passage is drilled in the crank cheek to allow oil from the hollow crankshaft to be sprayed on the cylinder walls. The crank cheek connects the crankpin to the main journal. In some designs, the cheek extends beyond the journal and carries a flyweight to balance the crankshaft. The crank cheek must be of sturdy construction to obtain the required rigidity between the crankpin and the journal.

In all cases, the type of crankshaft and the number of crankpins must correspond with the cylinder arrangement of the engine. The position of the cranks on the crankshaft in relation to the other cranks of the same shaft is expressed in degrees.

The simplest crankshaft is the single-throw or 360° type. This type is used in a single-row radial engine. It can be constructed in one or two pieces. Two main bearings (one on each end) are provided when this type of crankshaft is used. The double-throw or 180° crankshaft is used on double-row radial engines. In the radial-type engine, one throw is provided for each row of cylinders.

Crankshaft Balance

Excessive vibration in an engine not only results in fatigue failure of the metal structures, but also causes the moving parts to wear rapidly. In some instances, excessive vibration is caused by a crankshaft that is not balanced. Crankshafts are balanced for static balance and dynamic balance. A crankshaft is statically balanced when the weight of the entire assembly of crankpins, crank cheeks, and flyweights is balanced around the axis of rotation. When checked for static balance, it is placed on two knife edges. If the shaft tends to turn toward any one position during the test, it is out of static balance. Any engine to be overhauled completely should receive a runout check of its crankshaft as a first step. Any question concerning crankshaft replacement is resolved at this time since a shaft whose runout is beyond limits must be replaced.

Dynamic Dampers

A crankshaft is dynamically balanced when all the forces created by crankshaft rotation and power impulses are balanced within themselves so that little or no vibration is produced when the engine is operating. To reduce vibration to a minimum during engine operation, dynamic dampers are incorporated on the crankshaft. A dynamic damper is merely a pendulum that is fastened to the crankshaft so that it is free to move in a small arc. It is incorporated in the flyweight assembly. Some crankshafts incorporate two or more of these assemblies, each being attached to a different crank cheek. The distance the pendulum moves and, thus, its vibrating

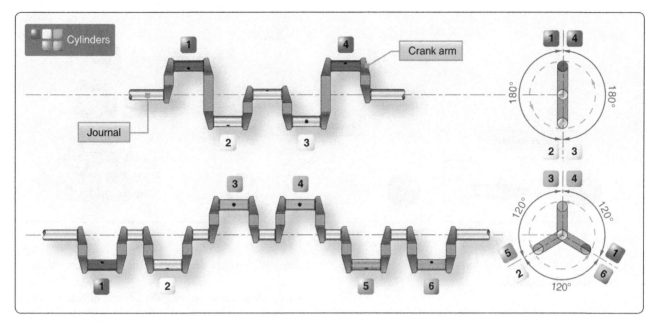

Figure 1-7. *Solid types of crankshafts.*

frequency corresponds to the frequency of the power impulses of the engine. When the vibration frequency of the crankshaft occurs, the pendulum oscillates out of time with the crankshaft vibration, thus reducing vibration to a minimum.

The construction of the dynamic damper used in one engine consists of a movable slotted-steel flyweight attached to the

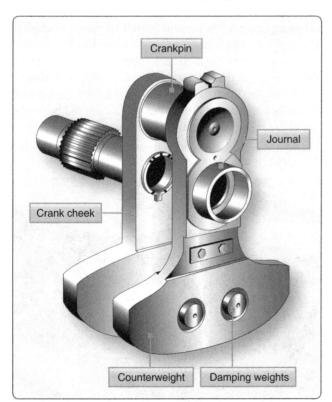

Figure 1-8. *A single throw radial engine crankshaft.*

crank cheek. Two spool-shaped steel pins extend into the slot and pass through oversized holes in the flyweight and crank cheek. The difference in the diameter between the pins and the holes provides a pendulum effect. An analogy of the functioning of a dynamic damper is shown in *Figure 1-9*.

Connecting Rods

The connecting rod is the link that transmits forces between the piston and the crankshaft. *[Figure 1-10]* Connecting rods must be strong enough to remain rigid under load and yet be light enough to reduce the inertia forces that are produced when the rod and piston stop, change direction, and start again at the end of each stroke.

There are four types of connecting-rod assemblies *[Figure 1-11]*:

1. Plain.
2. Fork and blade.
3. Master and articulated.
4. Split-type.

Master-and-Articulated Rod Assembly

The master-and-articulated rod assembly is commonly used in radial engines. In a radial engine, the piston in one cylinder in each row is connected to the crankshaft by a master rod. All other pistons in the row are connected to the master rod by articulated rods. In an 18-cylinder engine, which has two rows of cylinders, there are two master rods and 16 articulated rods. The articulated rods are constructed of forged steel alloy in either the I- or H-shape, denoting the cross-sectional shape. Bronze bushings are pressed into the

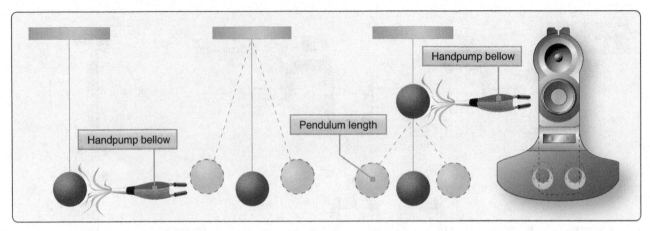

Figure 1-9. *Principles of a dynamic damper.*

bores in each end of the articulated rod to provide knuckle-pin and piston-pin bearings.

The master rod serves as the connecting link between the piston pin and the crankpin. The crankpin end, or the big end, contains the crankpin or master rod bearing. Flanges around the big end provide for the attachment of the articulated rods. The articulated rods are attached to the master rod by knuckle pins, which are pressed into holes in the master rod flanges during assembly. A plain bearing, usually called a piston-pin bushing, is installed in the piston end of the master rod to receive the piston pin.

When a crankshaft of the split-spline or split-clamp type is employed, a one-piece master rod is used. The master and articulated rods are assembled and then installed on the crankpin; the crankshaft sections are then joined together. In engines that use the one-piece type of crankshaft, the big end of the master rod is split, as is the master rod bearing. The main part of the master rod is installed on the crankpin; then the bearing cap is set in place and bolted to the master rod. The centers of the knuckle pins do not coincide with the center of the crankpin. Thus, while the crankpin center describes a true circle for each revolution of the crankshaft, the centers of the knuckle pins describe an elliptical path. *[Figure 1-12]* The elliptical paths are symmetrical about a center line through the master rod cylinder. It can be seen that the major diameters of the ellipses are not the same. Thus, the link rods have varying degrees of angularity relative to the center of the crank throw.

Because of the varying angularity of the link rods and the elliptical motion of the knuckle pins, all pistons do not move an equal amount in each cylinder for a given number of degrees of crank throw movement. This variation in piston position between cylinders can have considerable effect on engine operation. To minimize the effect of these factors on valve and ignition timing, the knuckle pin holes in the master rod flange are not equidistant from the center of the crankpin, thereby offsetting to an extent the effect of the link rod angularity.

Another method of minimizing the adverse effects on engine operation is to use a compensated magneto. In this magneto the breaker cam has a number of lobes equal to the number of cylinders on the engine. To compensate for the variation in piston position due to link rod angularity, the breaker cam lobes are ground with uneven spacing. This allows the breaker contacts to open when the piston is in the correct firing position. This is further outlined during the discussion on ignition timing in Chapter 4, Engine Ignition & Electrical Systems.

Knuckle Pins

The knuckle pins are of solid construction except for the oil passages drilled in the pins, which lubricate the knuckle pin bushings. These pins may be installed by pressing into holes in the master rod flanges so that they are prevented from turning in the master rod. Knuckle pins may also be installed with a loose fit so that they can turn in the master rod flange holes, and also turn in the articulating rod bushings. These are called full-floating knuckle pins. In either type of installation, a lock plate on each side retains the knuckle pin and prevents a lateral movement.

Figure 1-10. *A connecting rod between the piston and crankshaft.*

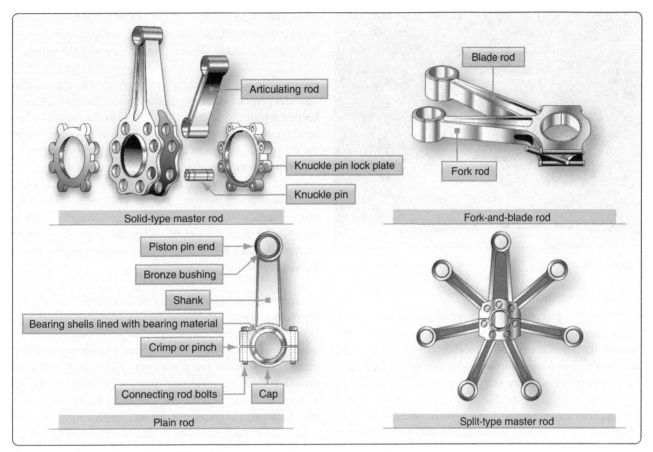

Figure 1-11. *Connecting rod assemblies.*

Figure 1-12. *Elliptical travel path of knuckle pins in an articulated rod assembly.*

Plain-Type Connecting Rods

Plain-type connecting rods are used in inline and opposed engines. The end of the rod attached to the crankpin is fitted with a cap and a two-piece bearing. The bearing cap is held on the end of the rod by bolts or studs. To maintain proper fit and balance, connecting rods should always be replaced in the same cylinder and in the same relative position.

Fork-and-Blade Rod Assembly

The fork-and-blade rod assembly is used primarily in V-type engines. The forked rod is split at the crankpin end to allow space for the blade rod to fit between the prongs. A single two-piece bearing is used on the crankshaft end of the rod. This type of connecting rod is not used much on modern engines.

Pistons

The piston of a reciprocating engine is a cylindrical member which moves back and forth within a steel cylinder. *[Figure 1-13]* The piston acts as a moving wall within the combustion chamber. As the piston moves down in the cylinder, it draws in the air-fuel mixture. As it moves upward, it compresses the charge, ignition occurs, and the expanding gases force the piston downward. This force is transmitted to the crankshaft through the connecting rod. On the return upward stroke, the piston forces the exhaust gases from the cylinder and the cycle repeats.

Figure 1-13. *A piston.*

Piston Construction

The majority of aircraft engine pistons are machined from aluminum alloy forgings. Grooves are machined in the outside surface of the piston to receive the piston rings, and cooling fins are provided on the inside of the piston for greater heat transfer to the engine oil.

Pistons may be either the trunk type or the slipper type. *[Figure 1-14]* Slipper-type pistons are not used in modern, high-powered engines because they do not provide adequate

strength or wear resistance. The top of the piston, or head, may be flat, convex, or concave. Recesses may be machined in the piston head to prevent interference with the valves.

Modern engines use cam ground pistons that are a larger diameter perpendicular to the piston pin. This larger diameter keeps the piston straight in the cylinder as the engine warms up from initial startup. As the piston heats up during warm up, the part of the piston in line with the pin has more mass and expands more making the piston completely round. At low temperatures, the piston is oval shaped and, when it warms to operating temperature, it becomes round. This process reduces the tendency of the piston to cock or slap in the cylinder during warm up. When the engine reaches its normal operating temperature, the piston assumes the correct dimensions in the cylinder.

As many as six grooves may be machined around the piston to accommodate the compression rings and oil rings. *[Figure 1-15]* The compression rings are installed in the three uppermost grooves; the oil control rings are installed immediately above the piston pin. The piston is usually drilled at the oil control ring grooves to allow surplus oil scraped from the cylinder walls by the oil control rings to pass back into the crankcase. An oil scraper ring is installed at the base of the piston wall or skirt to prevent excessive oil consumption. The portions of the piston walls that lie between

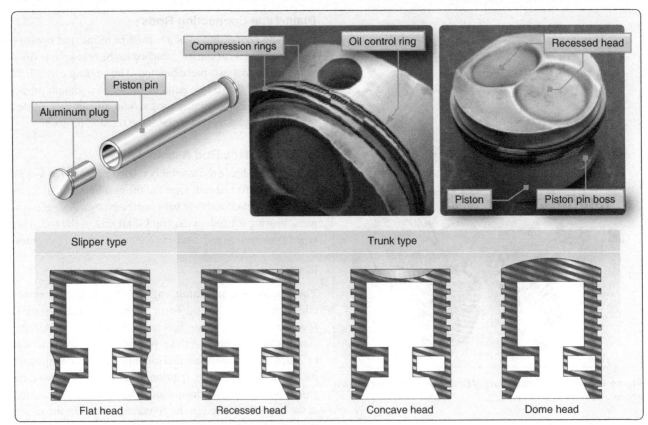

Figure 1-14. *Piston assembly and types of pistons.*

ring grooves are called the ring lands. In addition to acting as a guide for the piston head, the piston skirt incorporates the piston-pin bosses. The piston-pin bosses are of heavy construction to enable the heavy load on the piston head to be transferred to the piston pin.

Piston Pin

The piston pin joins the piston to the connecting rod. It is machined in the form of a tube from a nickel steel alloy forging, casehardened, and ground. The piston pin is sometimes called a wristpin because of the similarity between the relative motions of the piston and the articulated rod and that of the human arm. The piston pin used in modern aircraft engines is the full-floating type, so called because the pin is free to rotate in both the piston and in the connecting rod piston-pin bearing. The piston pin must be held in place to prevent the pin ends from scoring the cylinder walls. A plug of relatively soft aluminum in the pin end provides a good bearing surface against the cylinder wall.

Piston Rings

The piston rings prevent leakage of gas pressure from the combustion chamber and reduce to a minimum the seepage of oil into the combustion chamber. *[Figure 1-15]* The rings fit into the piston grooves but spring out to press against the cylinder walls; when properly lubricated, the rings form an effective gas seal.

Piston Ring Construction

Most piston rings are made of high-grade cast iron. *[Figure 1-14]* After the rings are made, they are ground to

Figure 1-15. *Machined rings around a piston.*

the cross-section desired. Then they are split so that they can be slipped over the outside of the piston and into the ring grooves that are machined in the piston wall. Since their purpose is to seal the clearance between the piston and the cylinder wall, they must fit the cylinder wall snugly enough to provide a gastight fit. They must exert equal pressure at all points on the cylinder wall and must make a gastight fit against the sides of the ring grooves.

Gray cast iron is most often used in making piston rings. In some engines, chrome-plated mild steel piston rings are used in the top compression ring groove because these rings can better withstand the high temperatures present at this point. Chrome rings must be used with steel cylinder walls. Never use chrome rings on chrome cylinders.

Compression Ring

The purpose of the compression rings is to prevent the escape of combustion gases past the piston during engine operation. They are placed in the ring grooves immediately below the piston head. The number of compression rings used on each piston is determined by the type of engine and its design, although most aircraft engines use two compression rings plus one or more oil control rings.

The cross-section of the ring is either rectangular or wedge shaped with a tapered face. The tapered face presents a narrow bearing edge to the cylinder wall, which helps to reduce friction and provide better sealing.

Oil Control Rings

Oil control rings are placed in the grooves immediately below the compression rings and above the piston pin bores. There may be one or more oil control rings per piston; two rings may be installed in the same groove, or they may be installed in separate grooves. Oil control rings regulate the thickness of the oil film on the cylinder wall. If too much oil enters the combustion chamber, it burns and leaves a thick coating of carbon on the combustion chamber walls, the piston head, the spark plugs, and the valve heads. This carbon can cause the valves and piston rings to stick if it enters the ring grooves or valve guides. In addition, the carbon can cause spark plug misfiring as well as detonation, pre-ignition, or excessive oil consumption. To allow the surplus oil to return to the crankcase, holes are drilled in the bottom of the oil control piston ring grooves or in the lands next to these grooves.

Oil Scraper Ring

The oil scraper ring usually has a beveled face and is installed in the groove at the bottom of the piston skirt. The ring is installed with the scraping edge away from the piston head or in the reverse position, depending upon cylinder position and the engine series. In the reverse position, the scraper ring retains the surplus oil above the ring on the upward piston stroke, and this oil is returned to the crankcase by the oil control rings on the downward stroke.

Cylinders

The portion of the engine in which the power is developed is called the cylinder. *[Figure 1-16]* The cylinder provides a combustion chamber where the burning and expansion of gases take place, and it houses the piston and the connecting rod. There are four major factors that need to be considered in the design and construction of the cylinder assembly. It must:

1. Be strong enough to withstand the internal pressures developed during engine operation.

2. Be constructed of a lightweight metal to keep down engine weight.

3. Have good heat-conducting properties for efficient cooling.

4. Be comparatively easy and inexpensive to manufacture, inspect, and maintain.

The cylinder head of an air cooled engine is generally made of aluminum alloy because aluminum alloy is a good conductor of heat and its light weight reduces the overall engine weight. Cylinder heads are forged or die-cast for greater strength. The inner shape of a cylinder head is generally semispherical. The semispherical shape is stronger than conventionalist design and aids in a more rapid and thorough scavenging of the exhaust gases.

The cylinder used in the air cooled engine is the overhead valve type. *[Figure 1-17]* Each cylinder is an assembly of two major parts: cylinder head and cylinder barrel. At assembly, the cylinder head is expanded by heating and then screwed down on the cylinder barrel, which has been chilled. When the head cools and contracts and the barrel warms up and expands, a gastight joint results. The majority of the cylinders used are constructed in this manner using an aluminum head and a steel barrel. *[Figure 1-18]*

Cylinder Heads

The purpose of the cylinder head is to provide a place for combustion of the air-fuel mixture and to give the cylinder more heat conductivity for adequate cooling. The air-fuel mixture is ignited by the spark in the combustion chamber and commences burning as the piston travels toward top dead center (top of its travel) on the compression stroke. The ignited charge is rapidly expanding at this time, and pressure is increasing so that, as the piston travels through the top dead center position, it is driven downward on the power stroke.

Figure 1-16. *An example of an engine cylinder.*

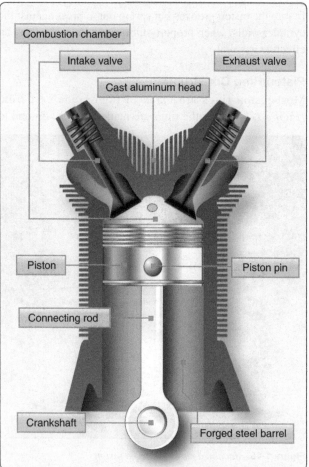

Figure 1-17. *Cutaway view of the cylinder assembly.*

The intake and exhaust valve ports are located in the cylinder head along with the spark plugs and the intake and exhaust valve actuating mechanisms.

After the cylinder head is cast, the spark plug bushings, valve guides, rocker arm bushings, and valve seats are installed in the cylinder head. Spark plug openings may be fitted with bronze or steel bushings that are shrunk and screwed into the openings. Stainless steel Heli-Coil spark plug inserts are used in many engines currently manufactured. Bronze or steel valve guides are usually shrunk or screwed into drilled openings in the cylinder head to provide guides for the valve stems. These are generally located at an angle to the center line of the cylinder. The valve seats are circular rings of hardened metal that protect the relatively soft metal of the cylinder head from the hammering action of the valves (as they open and close) and from the exhaust gases.

The cylinder heads of air cooled engines are subjected to extreme temperatures; it is therefore necessary to provide adequate cooling fin area and to use metals that conduct heat rapidly. Cylinder heads of air cooled engines are usually cast or forged. Aluminum alloy is used in the construction for a number of reasons. It is well adapted for casting or for the machining of deep, closely spaced fins, and it is more resistant than most metals to the corrosive attack of tetraethyl lead in gasoline. The greatest improvement in air cooling has resulted from reducing the thickness of the fins and increasing their depth. In this way, the fin area has been increased in modern engines. Cooling fins taper from 0.090" at the base to 0.060" at the tip end. Because of the difference in temperature in the various sections of the cylinder head, it is necessary to provide more cooling-fin area on some sections than on others. The exhaust valve region is the hottest part of the internal surface; therefore, more fin area is provided around the outside of the cylinder in this section.

Figure 1-18. *The aluminum head and steel barrel of a cylinder.*

Cylinder Barrels

The cylinder barrel in which the piston operates must be made of a high-strength material, usually steel. It must be as light as possible yet have the proper characteristics for operating under high temperatures. It must be made of a good bearing material and have high tensile strength. The cylinder barrel is made of a steel alloy forging with the inner surface hardened to resist wear of the piston and the piston rings which bear against it. This hardening is usually done by exposing the steel to ammonia or cyanide gas while the steel is very hot. The steel soaks up nitrogen from the gas, which forms iron nitrides on the exposed surface. As a result of this process, the metal is said to be nitrided. This nitriding only penetrates into the barrel surface a few thousands of an inch. As the cylinder barrels wear due to use, they can be repaired by chroming. This is a process that plates chromium on the surface of the cylinder barrel and brings it back to new standard dimensions. Chromium-plated cylinders should use cast iron rings. Honing the cylinder walls is a process that brings it to the correct dimensions and provides crosshatch pattern for seating the piston rings during engine break-in. Some engine cylinder barrels are choked at the top, or they are smaller in diameter to allow for heat expansion and wear.

In some instances, the barrel has threads on the outside surface at one end so that it can be screwed into the cylinder head. The cooling fins are machined as an integral part of the barrel and have limits on repair and service.

Cylinder Numbering

Occasionally, it is necessary to refer to the left or right side of the engine or to a particular cylinder. Therefore, it is necessary to know the engine directions and how cylinders of an engine are numbered. The propeller shaft end of the engine is always the front end, and the accessory end is the rear end, regardless of how the engine is mounted in an aircraft. When referring to the right side or left side of an engine, always assume the view is from the rear or accessory end. As seen from this position, crankshaft rotation is referred to as either clockwise or counterclockwise.

Inline and V-type engine cylinders are usually numbered from the rear. In V-engines, the cylinder banks are known as the right bank and the left bank, as viewed from the accessory end. *[Figure 1-19]* The cylinder numbering of the opposed engine shown begins with the right rear as No. 1 and the left rear as No. 2. The one forward of No. 1 is No. 3; the one forward of No. 2 is No. 4, and so on. The numbering of opposed engine cylinders is by no means standard. Some manufacturers number their cylinders from the rear and others from the front of the engine. Always refer to the appropriate engine manual to determine the numbering system used by that manufacturer.

Single-row radial engine cylinders are numbered clockwise when viewed from the rear. Cylinder No. 1 is the top cylinder. In double-row engines, the same system is used. The No. 1 cylinder is the top one in the rear row. No. 2 cylinder is the first one clockwise from No. 1, but No. 2 is in the front row. No. 3 cylinder is the next one clockwise to No. 2 but is in the rear row. Thus, all odd-numbered cylinders are in the rear row, and all even-numbered cylinders are in the front row.

Firing Order

The firing order of an engine is the sequence in which the power event occurs in the different cylinders. The firing order is designed to provide for balance and to eliminate vibration to the greatest extent possible. In radial engines, the firing order must follow a special pattern since the firing impulses must follow the motion of the crank throw during its rotation. In inline engines, the firing orders may vary somewhat, yet most orders are arranged so that the firing of cylinders is evenly distributed along the crankshaft. Six-cylinder inline engines generally have a firing order of 1-5-3-6-2-4. Cylinder firing order in opposed engines can usually be listed in pairs of cylinders, as each pair fires across the center main bearing. The firing order of six-cylinder opposed engines is 1-4-5-2-3-6. The firing order of one model four-cylinder opposed engine is 1-4-2-3, but on another model, it is 1-3-2-4.

Single-Row Radial Engines

On a single-row radial engine, all the odd-numbered cylinders fire in numerical succession; then, the even numbered cylinders fire in numerical succession. On a five-cylinder radial engine, for example, the firing order is 1-3-5-2-4, and on a seven-cylinder radial engine it is 1-3-5-7-2-4-6. The firing order of a nine-cylinder radial engine is 1-3-5-7-9-2-4-6-8.

Double-Row Radial Engines

On a double-row radial engine, the firing order is somewhat complicated. The firing order is arranged with the firing impulse occurring in a cylinder in one row and then in a cylinder in the other row; therefore, two cylinders in the same row never fire in succession.

An easy method for computing the firing order of a 14-cylinder, double-row radial engine is to start with any number from 1 to 14 and add 9 or subtract 5 (these are called the firing order numbers), whichever gives an answer between 1 and 14, inclusive. For example, starting with 8, 9 cannot be added since the answer would then be more than 14; therefore, subtract 5 from 8 to get 3, add 9 to 3 to get 12, subtract 5 from 12 to get 7, subtract 5 from 7 to get 2, and so on.

The firing order numbers of an 18-cylinder, double-row radial engine are 11 and 7; that is, begin with any number from 1 to 18 and add 11 or subtract 7. For example, beginning with 1, add 11 to get 12; 11 cannot be added to 12 because the total would be more than 18, so subtract 7 to get 5, add 11 to 5 to get 16, subtract 7 from 16 to get 9, subtract 7 from 9 to get 2, add 11 to 2 to get 13, and continue this process for 18 cylinders.

Valves

The air-fuel mixture enters the cylinders through the intake valve ports, and burned gases are expelled through the exhaust valve ports. The head of each valve opens and closes these cylinder ports. The valves used in aircraft engines are the conventional poppet type. The valves are also typed by their shape and are called either mushroom or tulip because of their resemblance to the shape of these plants. *Figure 1-20* illustrates various shapes and types of these valves.

Figure 1-19. *Numbering of engine cylinders.*

Valve Construction

The valves in the cylinders of an aircraft engine are subjected to high temperatures, corrosion, and operating stresses; thus, the metal alloy in the valves must be able to resist all these factors. Because intake valves operate at lower temperatures than exhaust valves, they can be made of chromic-nickel steel. Exhaust valves are usually made of nichrome, silchrome, or cobalt-chromium steel because these materials are much more heat resistant.

The valve head has a ground face that forms a seal against the ground valve seat in the cylinder head when the valve is closed. The face of the valve is usually ground to an angle of either 30° or 45°. In some engines, the intake-valve face is ground to an angle of 30°, and the exhaust-valve face is ground to a 45° angle. Valve faces are often made more durable by the application of a material called stellite. About 1/16 inch of this alloy is welded to the valve face and ground to the correct angle. Stellite is resistant to high-temperature corrosion and also withstands the shock and wear associated with valve operation. Some engine manufacturers use a nichrome facing on the valves. This serves the same purpose as the stellite material.

The valve stem acts as a pilot for the valve head and rides in the valve guide installed in the cylinder head for this purpose. *[Figure 1-21]* The valve stem is surface hardened to resist wear. The neck is the part that forms the junction between the head and the stem. The tip of the valve is hardened to withstand the hammering of the valve rocker arm as it opens the valve. A machined groove on the stem near the tip receives the split-ring stem keys. These stem keys form a lock ring to hold the valve spring retaining washer in place. *[Figure 1-22]*

Some intake and exhaust valve stems are hollow and partially filled with metallic sodium. This material is used because it is an excellent heat conductor. The sodium melts at approximately 208 °F and the reciprocating motion of the valve circulates the liquid sodium, allowing it to carry away heat from the valve head to the valve stem where it is dissipated through the valve guide to the cylinder head and the cooling fins. Thus, the operating temperature of the valve may be reduced as much as 300° to 400 °F. Under no circumstances should a sodium-filled valve be cut open

Figure 1-21. *View of valve guide installed on a cylinder head.*

Figure 1-22. *Stem keys forming a lock ring to hold valve spring retaining washers in place.*

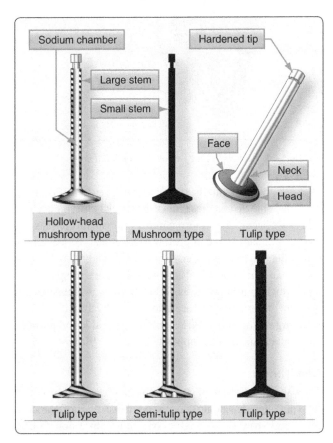

Figure 1-20. *Various valve types.*

or subjected to treatment which may cause it to rupture. Exposure of the sodium in these valves to the outside air results in fire or explosion with possible personal injury.

The most commonly used intake valves have solid stems, and the head is either flat or tulip shaped. Intake valves for low-power engines are usually flat headed. In some engines, the intake valve may be the tulip type and have a smaller stem than the exhaust valve or it may be similar to the exhaust valve but have a solid stem and head. Although these valves are similar, they are not interchangeable since the faces of the valves are constructed of different material. The intake valve usually has a flat milled on the tip to identify it.

Valve Operating Mechanism

For a reciprocating engine to operate properly, each valve must open at the proper time, stay open for the required length of time, and close at the proper time. Intake valves are opened just before the piston reaches top dead center, and exhaust valves remain open after top dead center. At a particular instant, therefore, both valves are open at the same time (end of the exhaust stroke and beginning of the intake stroke). This valve overlap permits better volumetric efficiency and lowers the cylinder operating temperature. This timing of the valves is controlled by the valve-operating mechanism and is referred to as the valve timing.

The valve lift (distance that the valve is lifted off its seat) and the valve duration (length of time the valve is held open) are both determined by the shape of the cam lobes. Typical cam lobes are illustrated in *Figure 1-23*. The portion of the lobe that gently starts the valve operating mechanism moving is called a ramp, or step. The ramp is machined on each side of

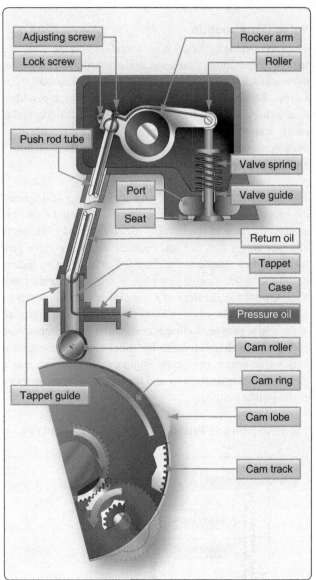

Figure 1-24. *Valve operating mechanism (radial engine).*

the cam lobe to permit the rocker arm to be eased into contact with the valve tip and thus reduce the shock load which would otherwise occur. The valve operating mechanism consists of a cam ring or camshaft equipped with lobes that work against a cam roller or a cam follower. *[Figures 1-24 and 1-25]* The cam follower pushes a push rod and ball socket, actuating a rocker arm, which in turn opens the valve. Springs, which slip over the stem of the valves and are held in place by the valve-spring retaining washer and stem key, close each valve and push the valve mechanism in the opposite direction. *[Figure 1-26]*

Cam Rings

The valve mechanism of a radial engine is operated by one or two cam rings, depending upon the number of rows of cylinders. In a single-row radial engine, one ring with a double cam track is used. One track operates the intake valves, the other

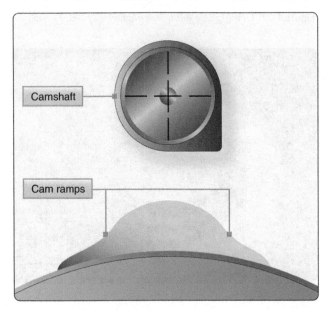

Figure 1-23. *Typical cam lobes.*

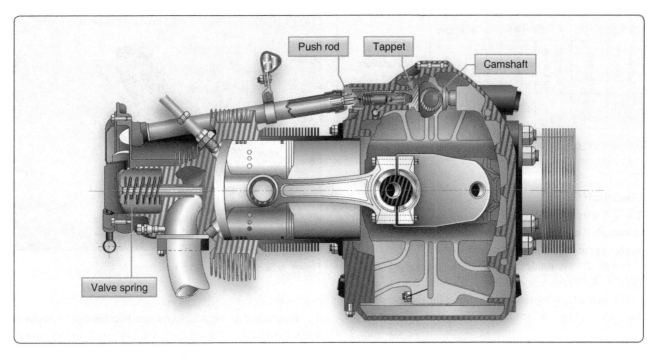

Figure 1-25. *Valve operating mechanism (opposed engine).*

operates the exhaust valves. The cam ring is a circular piece of steel with a series of cams or lobes on the outer surface. The surface of these lobes and the space between them (on which the cam rollers ride) is known as the cam track. As the cam ring revolves, the lobes cause the cam roller to raise the tappet in the tappet guide, thereby transmitting the force through the push rod and rocker arm to open the valve. In a single-row radial engine, the cam ring is usually located between the propeller reduction gearing and the front end of the power section. In a twin-row radial engine, a second cam for the operation of the valves in the rear row is installed between the rear end of the power section and the supercharger section.

The cam ring is mounted concentrically with the crankshaft and is driven by the crankshaft at a reduced rate of speed through the cam intermediate drive gear assembly. The cam ring has two parallel sets of lobes spaced around the outer periphery, one set (cam track) for the intake valves and the other for the exhaust valves. The cam rings used may have four or five lobes on both the intake and the exhaust tracks. The timing of the valve events is determined by the spacing of these lobes and the speed and direction at which the cam rings are driven in relation to the speed and direction of the crankshaft. The method of driving the cam varies on different makes of engines. The cam ring can be designed with teeth on either the inside or outside periphery. If the reduction gear meshes with the teeth on the outside of the ring, the cam turns in the direction of rotation of the crankshaft. If the ring is driven from the inside, the cam turns in the opposite direction from the crankshaft. *[Figure 1-24]*

A four-lobe cam may be used on either a seven-cylinder or nine-cylinder engine. *[Figure 1-27]* On the seven cylinder, it rotates in the same direction as the crankshaft, and on the nine cylinder, opposite the crankshaft rotation. On the nine-cylinder engine, the spacing between cylinders is 40° and the firing order is 1-3-5-7-9-2-4-6-8. This means that there is a space of 80° between firing impulses. The spacing on the four lobes of the cam ring is 90°, which is greater than the spacing between impulses. Therefore, to obtain proper relation of valve operations and firing order, it is necessary to drive the cam opposite the crankshaft rotation. Using the four-lobe cam on the seven-cylinder engine, the spacing between the firing of the cylinders is greater than the spacing of the cam lobes. Therefore, it is necessary for the cam to rotate in the same direction as the crankshaft.

Figure 1-26. *A typical set of valve springs used to dampen oscillations. Multiple springs are used to protect against breakage.*

5 Cylinders		7 Cylinders		9 Cylinders		Direction of Rotation
Number of Lobes	Speed	Number of Lobes	Speed	Number of Lobes	Speed	
3	1/6	4	1/8	5	1/10	with crankshaft
2	1/4	3	1/6	4	1/8	opposite crankshaft

Figure 1-27. *Radial engines, cam ring table.*

Camshaft

The valve mechanism of an opposed engine is operated by a camshaft. The camshaft is driven by a gear that mates with another gear attached to the crankshaft. *[Figure 1-28]* The camshaft always rotates at one-half the crankshaft speed. As the camshaft revolves, the lobes cause the tappet assembly to rise in the tappet guide, transmitting the force through the push rod and rocker arm to open the valve. *[Figure 1-29]*

Tappet Assembly

The tappet assembly consists of:

1. A cylindrical tappet, which slides in and out in a tappet guide installed in one of the crankcase sections around the cam ring;

2. A tappet roller, which follows the contour of the cam ring and lobes;

3. A tappet ball socket or push rod socket; and

4. A tappet spring.

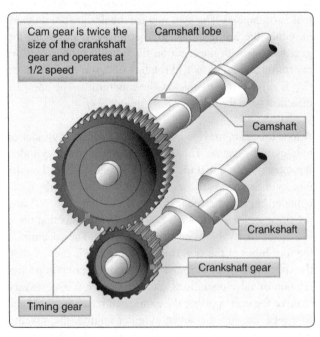

Figure 1-28. *Cam drive mechanism opposed-type aircraft engine.*

Figure 1-29. *Cam load on lifter body.*

The function of the tappet assembly is to convert the rotational movement of the cam lobe into reciprocating motion and to transmit this motion to the push rod, rocker arm, and then to the valve tip, opening the valve at the proper time. The purpose of the tappet spring is to take up the clearance between the rocker arm and the valve tip to reduce the shock load when the valve is opened. A hole is drilled through the tappet to allow engine oil to flow to the hollow push rods to lubricate the rocker assemblies.

Solid Lifters/Tappets

Solid lifters or cam followers generally require the valve clearance to be adjusted manually by adjusting a screw and lock nut. Valve clearance is needed to assure that the valve has enough clearance in the valve train to close completely. This adjustment or inspection was a continuous maintenance item until hydraulic lifters were used.

Hydraulic Valve Tappets/Lifters

Some aircraft engines incorporate hydraulic tappets that automatically keep the valve clearance at zero, eliminating the necessity for any valve clearance adjustment mechanism. A typical hydraulic tappet (zero-lash valve lifter) is shown in *Figure 1-30.*

When the engine valve is closed, the face of the tappet body (cam follower) is on the base circle or back of the cam. *[Figure 1-30]* The light plunger spring lifts the hydraulic plunger so that its outer end contacts the push rod socket, exerting a light pressure against it, thus eliminating any clearance in the valve linkage. As the plunger moves outward, the ball check valve moves off its seat. Oil from the supply chamber, which is directly connected with the engine lubrication system, flows in and fills the pressure chamber. As the camshaft rotates, the cam pushes the tappet body and the

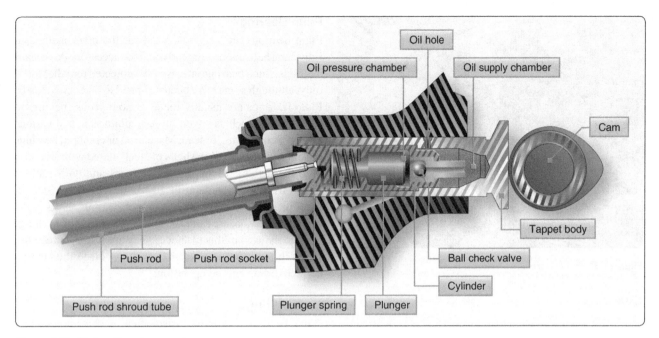

Figure 1-30. *Hydraulic valve tappets.*

hydraulic lifter cylinder outward. This action forces the ball check valve onto its seat; thus, the body of oil trapped in the pressure chamber acts as a cushion. During the interval when the engine valve is off its seat, a predetermined leakage occurs between plunger and cylinder bore, which compensates for any expansion or contraction in the valve train. Immediately after the engine valve closes, the amount of oil required to fill the pressure chamber flows in from the supply chamber, preparing for another cycle of operation.

Hydraulic valve lifters are normally adjusted at the time of overhaul. They are assembled dry (no lubrication), clearances checked, and adjustments are usually made by using push rods of different lengths. A minimum and maximum valve clearance is established. Any measurement between these extremes is acceptable, but approximately half way between the extremes is desired. Hydraulic valve lifters require less maintenance, are better lubricated, and operate more quietly than the screw adjustment type.

Push Rod

The push rod, tubular in form, transmits the lifting force from the valve tappet to the rocker arm. A hardened-steel ball is pressed over or into each end of the tube. One ball end fits into the socket of the rocker arm. In some instances, the balls are on the tappet and rocker arm, and the sockets are on the push rod. The tubular form is employed because of its lightness and strength. It permits the engine lubricating oil under pressure to pass through the hollow rod and the drilled ball ends to lubricate the ball ends, rocker-arm bearing, and valve-stem guide. The push rod is enclosed in a tubular housing that extends from the crankcase to the cylinder head, referred to as push rod tubes.

Rocker Arms

The rocker arms transmit the lifting force from the cams to the valves. *[Figure 1-31]* Rocker arm assemblies are supported by a plain, roller, or ball bearing, or a combination of these, which serves as a pivot. Generally, one end of the arm bears against the push rod and the other bears on the valve stem. One end of the rocker arm is sometimes slotted to accommodate a steel roller. The opposite end is constructed with either a threaded split clamp and locking bolt or a tapped hole. The arm may have an adjusting screw, for adjusting the clearance between the rocker arm and the valve stem tip. The screw can be adjusted to the specified clearance to make certain that the valve closes fully.

Valve Springs

Each valve is closed by two or three helical springs. If a single spring were used, it would vibrate or surge at certain speeds. To eliminate this difficulty, two or more springs (one inside the other) are installed on each valve. Each spring vibrates at a different engine speed and rapid damping out of all spring-surge vibrations during engine operation results. Two or more springs also reduce danger of weakness and possible failure by breakage due to heat and metal fatigue. The springs are held in place by split locks installed in the recess of the valve spring upper retainer or washer, and engage a groove machined into the valve stem. The functions of the valve springs are to close the valve and to hold the valve securely on the valve seat.

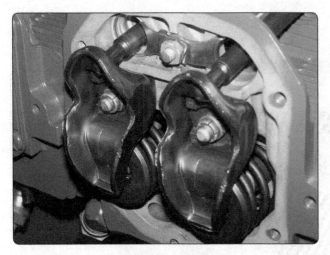

Figure 1-31. *Rocker arms.*

Bearings

A bearing is any surface which supports, or is supported by, another surface. A good bearing must be composed of material that is strong enough to withstand the pressure imposed on it and should permit the other surface to move with a minimum of friction and wear. The parts must be held in position within very close tolerances to provide efficient and quiet operation, and yet allow freedom of motion. Lubricated bearings of many types are used to accomplish this, and at the same time reduce friction of moving parts so that power loss is not excessive and to prevent high oil temperatures caused by failed or failing bearings. Bearings are required to take radial loads, thrust loads, or a combination of the two. An example of a radial load would be a rotating shaft being held or contained in one position on a radial plane. Thrust load would be the rotating shaft being contained from moving axially along the shafts axis. These radial and thrust loads are illustrated in *Figure 1-32*. There are two ways in which bearing surfaces move in relation to each other. One is by the sliding movement of one metal against the other (sliding friction), and the second is for one surface to roll over the other (rolling friction). The three different types of bearings in general use are plain, roller, and ball. *[Figure 1-33]*

Plain Bearings

Plain bearings are generally used for the crankshaft, cam ring, camshaft, connecting rods, and the accessory drive shaft bearings. Such bearings are usually subjected to radial loads only, although some have been designed to take thrust loads. Plain bearings are usually made of nonferrous (having no iron) metals, such as silver, bronze, aluminum, and various alloys of copper, tin, or lead. Master rod or crankpin bearings in some engines are thin shells of steel, plated with silver on both the inside and the outside surfaces and with lead-tin plated over the silver on the inside surface only. Smaller bearings, such as those used to support various shafts in the accessory section, are called bushings. Porous Oilite bushings are widely used in this instance. They are impregnated with oil so that the heat of friction brings the oil to the bearing surface during engine operation.

Ball Bearings

A ball bearing assembly consists of grooved inner and outer races, one or more sets of balls, in bearings designed for disassembly, and a bearing retainer. They are used for shaft bearings and rocker arm bearings in some reciprocating engines. Special deep-groove ball bearings are used to transmit propeller thrust and radial loads to the engine nose section of radial engines. Since this type of bearing can accept both radial and thrust loads, it is used in gas turbine engines to support one end of a shaft (radial loads) and to keep the shaft from moving axially (thrust loads).

Roller Bearings

Roller bearings are made in many types and shapes, but the two types generally used in the aircraft engine are the straight roller and the tapered roller bearings. Straight roller bearings are used where the bearing is subjected to radial loads only. In tapered roller bearings, the inner- and outer-race bearing surfaces are cone-shaped. Such bearings withstand both radial and thrust loads. Straight roller bearings are used in high power reciprocating aircraft engines for the crankshaft main bearings. They are also used in gas turbine applications where radial loads are high. Generally, a rotating shaft in a gas turbine engine is supported by a deep-groove ball bearing

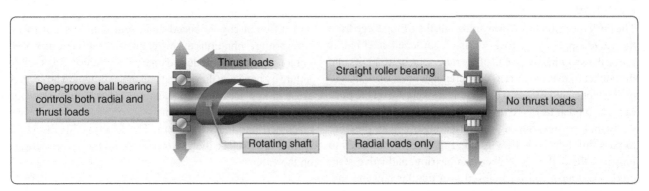

Figure 1-32. *Radial and thrust loads.*

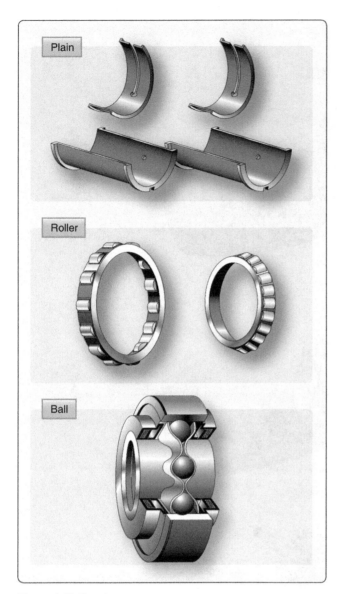

Figure 1-33. *Bearings.*

(radial and thrust loads) on one end and a straight roller bearing (radial loads only) on the other end.

Propeller Reduction Gearing

The increased brake horsepower delivered by a high horsepower engine results partly from increased crankshaft rpm. It is therefore necessary to provide reduction gears to limit the propeller rotation speed to a value at which efficient operation is obtained. Whenever the speed of the blade tips approaches the speed of sound, the efficiency of the propeller decreases rapidly. Reduction gearing for engines allows the engine to operate at a higher rpm, developing more power while slowing down the propeller rpm. This prevents the propeller efficiency from decreasing. Since reduction gearing must withstand extremely high stresses, the gears are machined from steel forgings. Many types of reduction gearing systems

are in use. The three types most commonly used are spur planetary, bevel planetary, and spur and pinion. [*Figure 1-34*]

The spur planetary reduction gearing consists of a large driving gear or sun gear splined (and sometimes shrunk) to the crankshaft, a large stationary gear, called a bell gear, and a set of small spur planetary pinion gears mounted on a carrier ring. The ring is fastened to the propeller shaft and the planetary gears mesh with both the sun gear and the stationary bell or ring gear. The stationary gear is bolted or splined to the front section housing. When the engine is operating, the sun gear rotates. Because the planetary gears are meshed with this ring, they also must rotate. Since they also mesh with the stationary gear, they walk or roll around it as they rotate, and the ring in which they are mounted rotates the propeller shaft in the same direction as the crankshaft but at a reduced speed.

In some engines, the bell gear is mounted on the propeller shaft, and the planetary pinion gear cage is held stationary. The sun gear is splined to the crankshaft and acts as a driving gear. In such an arrangement, the propeller travels at a reduced speed but in opposite direction to the crankshaft.

In the bevel planetary reduction gearing system, the driving gear is machined with beveled external teeth and is attached to the crankshaft. A set of mating bevel pinion gears is mounted in a cage attached to the end of the propeller shaft. The pinion gears are driven by the drive gear and walk around the stationary gear, which is bolted or splined to the front section housing. The thrust of the bevel pinion gears is absorbed by a thrust ball bearing of special design. The drive and the fixed gears are generally supported by heavy-duty ball bearings. This type of planetary reduction assembly is more compact than the other one described and, therefore, can be used where a smaller propeller gear step-down is desired. In the case of gas turbine turboprop engines, more than one stage of reduction gearing is used due to the high output speeds of the engine. Several types of lower powered engines can use the spur and pinion reduction gear arrangement.

Propeller Shafts

Propeller shafts may be of three major types: tapered, splined, or flanged. Tapered shafts are identified by taper numbers. Splined and flanged shafts are identified by SAE numbers. The propeller shaft of most low power output engines is forged as part of the crankshaft. It is tapered, and a milled slot is provided so that the propeller hub can be keyed to the shaft. The keyway and key index of the propeller are in relation to the No. 1 cylinder top dead center. The end of the shaft is threaded to receive the propeller retaining nut. Tapered propeller shafts are common on older and smaller engines.

The propeller shaft of high-output radial engines is generally

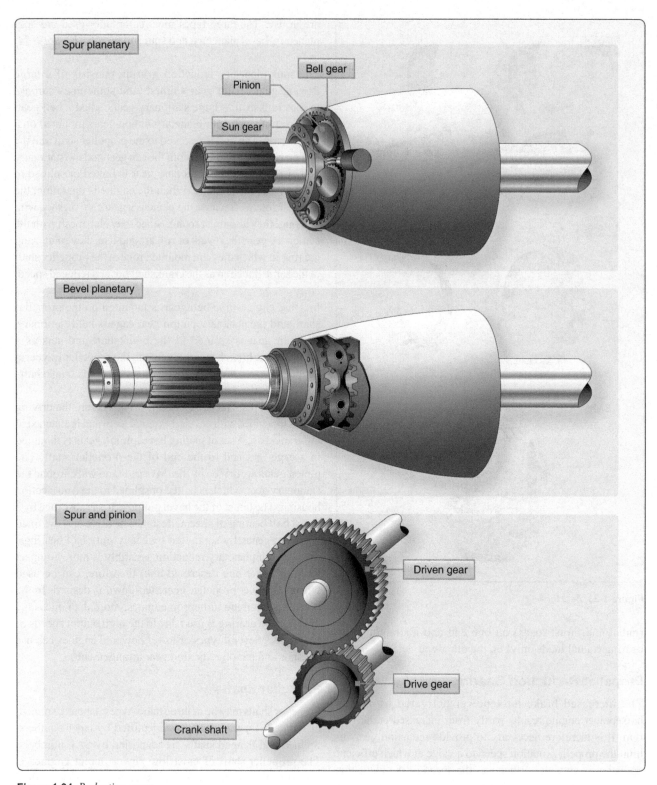

Figure 1-34. *Reduction gears.*

splined. It is threaded on one end for a propeller hub nut. The thrust bearing, which absorbs propeller thrust, is located around the shaft and transmits the thrust to the nose section housing. The shaft is threaded for attaching the thrust-bearing retaining nut. On the portion protruding from the housing (between the two sets of threads), splines are located to receive the splined propeller hub. The shaft is generally machined from a steel-alloy forging throughout its length. The propeller shaft may be connected by reduction gearing to the engine crankshaft, but in smaller engines the propeller shaft is simply an extension of the engine crankshaft. To turn the propeller shaft, the engine crankshaft must revolve.

Flanged propeller shafts are used on most modern reciprocating and turboprop engines. One end of the shaft is flanged with drilled holes to accept the propeller mounting bolts. The installation may be a short shaft with internal threading to accept the distributor valve to be used with a controllable propeller. The flanged propeller shaft is a very common installation on most propeller driven aircraft.

Reciprocating Engine Operating Principles

The relationships between pressure, volume, and temperature of gases are the basic principles of engine operation. An internal combustion engine is a device for converting heat energy into mechanical energy. Gasoline is vaporized and mixed with air, forced or drawn into a cylinder, compressed by a piston, and then ignited by an electric spark. The conversion of the resultant heat energy into mechanical energy and then into work is accomplished in the cylinder. *Figure 1-35* illustrates the various engine components necessary to accomplish this conversion and also presents the principal terms used to indicate engine operation.

The operating cycle of an internal combustion reciprocating engine includes the series of events required to induct, compress, ignite, and burn, causing expansion of the air-fuel charge in the cylinder and to scavenge or exhaust the byproducts of the combustion process. When the compressed mixture is ignited, the resultant gases of combustion expand very rapidly and force the piston to move away from the cylinder head. This downward motion of the piston, acting on the crankshaft through the connecting rod, is converted to a circular or rotary motion by the crankshaft. A valve in the top or head of the cylinder opens to allow the burned gases to escape, and the momentum of the crankshaft and the propeller forces the piston back up in the cylinder where it is ready for the next event in the cycle. Another valve in the cylinder head then opens to let in a fresh charge of the air-fuel mixture. The valve allowing for the escape of the burning exhaust gases is called the exhaust valve, and the valve which lets in the fresh charge of the air-fuel mixture is called the intake valve. These valves are opened and closed mechanically at the proper times by the valve-operating mechanism. Therefore, the order of the five events of a four stroke cycle engine are intake, compression, ignition, power, and exhaust.

The bore of a cylinder is its inside diameter. The stroke is the distance the piston moves from one end of the cylinder to the other, specifically from top dead center (TDC) to bottom dead center (BDC), or vice versa. *[Figure 1-35]*

Operating Cycles

There are several operating cycles in use:

1. Four stroke.
2. Two stroke.
3. Rotary.
4. Diesel.

Four-Stroke Cycle

The vast majority of certified aircraft reciprocating engines operate on the four-stroke cycle, sometimes called the Otto cycle after its originator, a German physicist. The four-stroke cycle engine has many advantages for use in aircraft. One advantage is that it lends itself readily to high performance through supercharging.

In this type of engine, four strokes are required to complete the required series of events or operating cycle of each cylinder. *[Figure 1-36]* Two complete revolutions of the crankshaft (720°) are required for the four strokes; thus, each cylinder in an engine of this type fires once in every two revolutions of the crankshaft. In the following discussion of the four-stroke cycle engine operation, note that the timing of the ignition and

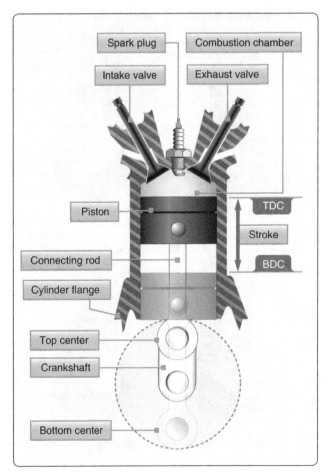

Figure 1-35. *Components and terminology of engine operation.*

the valve events vary considerably in different engines. Many factors influence the timing of a specific engine, and it is most important that the engine manufacturer's recommendations in this respect be followed in maintenance and overhaul. The timing of the valve and ignition events is always specified in degrees of crankshaft travel. It should be remembered that a certain amount of crankshaft travel is required to open a valve fully; therefore, the specified timing represents the start of opening rather than the full-open position of the valve. An example valve timing chart can be seen in *Figure 1-37*.

Intake Stroke

During the intake stroke, the piston is pulled downward in the cylinder by the rotation of the crankshaft. This reduces the pressure in the cylinder and causes air under atmospheric pressure to flow through the carburetor, which meters the correct amount of fuel. The air-fuel mixture passes through the intake pipes and intake valves into the cylinders. The quantity or weight of the air-fuel charge depends upon the degree of throttle opening.

The intake valve is opened considerably before the piston

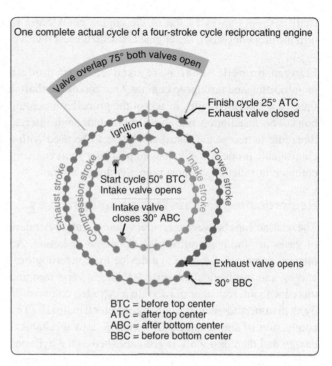

Figure 1-37. *Valve timing chart.*

reaches TDC on the exhaust stroke, in order to induce a greater quantity of the air-fuel charge into the cylinder and thus increase the horsepower. The distance the valve may be opened before TDC, however, is limited by several factors, such as the possibility that hot gases remaining in the cylinder from the previous cycle may flash back into the intake pipe and the induction system.

In all high-power aircraft engines, both the intake and the exhaust valves are off the valve seats at TDC at the start of the intake stroke. As mentioned above, the intake valve opens before TDC on the exhaust stroke (valve lead), and the closing of the exhaust valve is delayed considerably after the piston has passed TDC and has started the intake stroke (valve lag). This timing is called valve overlap and is designed to aid in cooling the cylinder internally by circulating the cool incoming air-fuel mixture, to increase the amount of the air-fuel mixture induced into the cylinder, and to aid in scavenging the byproducts of combustion from the cylinder.

The intake valve is timed to close about 50° to 75° past BDC on the compression stroke, depending upon the specific engine, to allow the momentum of the incoming gases to charge the cylinder more completely. Because of the comparatively large volume of the cylinder above the piston when the piston is near BDC, the slight upward travel of the piston during this time does not have a great effect on the incoming flow of gases. This late timing can be carried too far because the gases may be forced back through the intake valve and defeat the purpose of the late closing.

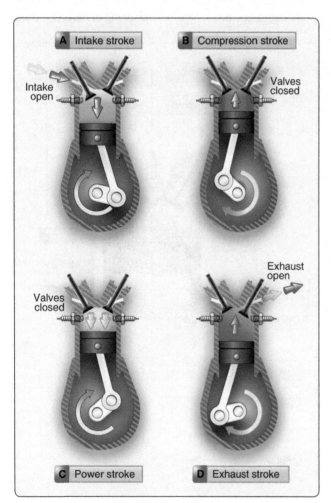

Figure 1-36. *Four stroke cycle.*

Compression Stroke

After the intake valve is closed, the continued upward travel of the piston compresses the air-fuel mixture to obtain the desired burning and expansion characteristics. The charge is fired by means of an electric spark as the piston approaches TDC. The time of ignition varies from 20° to 35° before TDC, depending upon the requirements of the specific engine to ensure complete combustion of the charge by the time the piston is slightly past the TDC position.

Many factors affect ignition timing, and the engine manufacturer has expended considerable time in research and testing to determine the best setting. All engines incorporate devices for adjusting the ignition timing, and it is most important that the ignition system be timed according to the engine manufacturer's recommendations.

Power Stroke

As the piston moves through the TDC position at the end of the compression stroke and starts down on the power stroke, it is pushed downward by the rapid expansion of the burning gases within the cylinder head with a force that can be greater than 15 tons (30,000 psi) at maximum power output of the engine. The temperature of these burning gases may be between 3,000 °F and 4,000 °F. As the piston is forced downward during the power stroke by the pressure of the burning gases exerted upon it, the downward movement of the connecting rod is changed to rotary movement by the crankshaft. Then, the rotary movement is transmitted to the propeller shaft to drive the propeller. As the burning gases are expanded, the temperature drops to within safe limits before the exhaust gases flow out through the exhaust port.

The timing of the exhaust valve opening is determined by, among other considerations, the desirability of using as much of the expansive force as possible and of scavenging the cylinder as completely and rapidly as possible. The valve is opened considerably before BDC on the power stroke (on some engines at 50° and 75° before BDC) while there is still some pressure in the cylinder. This timing is used so that the pressure can force the gases out of the exhaust port as soon as possible. This process frees the cylinder of waste heat after the desired expansion has been obtained and avoids overheating the cylinder and the piston. Thorough scavenging is very important, because any exhaust products remaining in the cylinder dilute the incoming air-fuel charge at the start of the next cycle.

Exhaust Stroke

As the piston travels through BDC at the completion of the power stroke and starts upward on the exhaust stroke, it begins to push the burned exhaust gases out the exhaust port.

The speed of the exhaust gases leaving the cylinder creates a low pressure in the cylinder. This low or reduced pressure speeds the flow of the fresh air-fuel charge into the cylinder as the intake valve is beginning to open. The intake valve opening is timed to occur at 8° to 55° before TDC on the exhaust stroke on various engines.

Two-Stroke Cycle

The two-stroke-cycle engine has re-emerged and is being used in ultra-light, light sport, and many experimental aircraft. As the name implies, two-stroke cycle engines require only one upstroke and one down stroke of the piston to complete the required series of events in the cylinder. Thus, the engine completes the operating cycle in one revolution of the crankshaft. The intake and exhaust functions are accomplished during the same stroke. These engines can be either air or water cooled and generally require a gear reduction housing between the engine and propeller.

Rotary Cycle

The rotary cycle has a three-sided rotor that turns inside an elliptical housing, completing three of the four cycles for each revolution. These engines can be single rotor or multi-rotor and can be air or water cooled. They are used mostly with experimental and light aircraft. Vibration characteristics are also very low for this type of engine.

Diesel Cycle

The diesel cycle depends on high compression pressures to provide for the ignition of the air-fuel charge in the cylinder. As air is drawn in the cylinder, it is compressed by a piston and, at maximum pressure, fuel is sprayed in the cylinder. At this point, the high pressure and temperature in the cylinder causes the fuel to burn increasing the internal pressure of the cylinder. This drives the piston down, turning or driving the crankshaft. Water and air cooled engines that can operate on JET A fuel (kerosene) use a version of the diesel cycle. There are many types of diesel cycles in use including two-stroke and four-stroke diesels.

Reciprocating Engine Power & Efficiencies

All aircraft engines are rated according to their ability to do work and produce power. This section presents an explanation of work and power and how they are calculated. Also discussed are the various efficiencies that govern the power output of a reciprocating engine.

Work

A physicist defines work as force times distance. Work done by a force acting on a body is equal to the magnitude of the force multiplied by the distance through which the force acts.

$$\text{Work (W)} = \text{Force (F)} \times \text{Distance (D)}$$

Work is measured by several standards. The most common unit is called foot-pound (ft-lb). If a one-pound mass is raised one foot, one ft-lb of work has been performed. The greater the mass is and/or the greater the distance is, the greater the work performed.

Horsepower

The common unit of mechanical power is the horsepower (hp). Late in the 18th century, James Watt, the inventor of the steam engine, found that an English workhorse could work at the rate of 550 ft-lb per second, or 33,000 ft-lb per minute, for a reasonable length of time. From his observations came the unit of horsepower, which is the standard unit of mechanical power in the English system of measurement. To calculate the hp rating of an engine, divide the power developed in ft-lb per minute by 33,000, or the power in ft-lb per second by 550.

$$\text{One hp} = \frac{\text{ft-lb per min}}{33{,}000}$$
$$\text{or}$$
$$\frac{\text{ft-lb per sec}}{550}$$

As stated above, work is the product of force and distance, and power is work per unit of time. Consequently, if a 33,000-lb weight is lifted through a vertical distance of 1 foot in 1 minute, the power expended is 33,000 ft-lb per minute, or exactly 1 hp.

Work is performed not only when a force is applied for lifting; force may be applied in any direction. If a 100-lb weight is dragged along the ground, a force is still being applied to perform work, although the direction of the resulting motion is approximately horizontal. The amount of this force would depend upon the roughness of the ground.

If the weight were attached to a spring scale graduated in pounds, then dragged by pulling on the scale handle, the amount of force required could be measured. Assume that the force required is 90 lb, and the 100-lb weight is dragged 660 feet in 2 minutes. The amount of work performed in the 2 minutes is 59,400 ft-lb or 29,700 ft-lb per minute. Since 1 hp is 33,000 ft-lb per minute, the hp expended in this case is 29,700 divided by 33,000, or 0.9 hp.

Piston Displacement

When other factors remain equal, the greater the piston displacement, the greater the maximum horsepower an engine is capable of developing. When a piston moves from BDC to TDC, it displaces a specific volume. The volume displaced by the piston is known as piston displacement and is expressed in cubic inches for most American-made engines and cubic centimeters for others.

The piston displacement of one cylinder may be obtained by multiplying the area of the cross-section of the cylinder by the total distance the piston moves in the cylinder in one stroke. For multicylinder engines, this product is multiplied by the number of cylinders to get the total piston displacement of the engine.

Since the volume (V) of a geometric cylinder equals the area (A) of the base multiplied by the height (h), it is expressed mathematically as follows:

$$V = A \times h$$

The area of the base is the area of the cross-section of the cylinder.

Area of a Circle

To find the area of a circle, it is necessary to use a number called pi (π). This number represents the ratio of the circumference to the diameter of any circle. Pi cannot be stated exactly because it is a never-ending decimal. It is 3.1416 expressed to four decimal places, which is accurate enough for most computations.

The area of a circle, as in a rectangle or triangle, must be expressed in square units. The distance that is one-half the diameter of a circle is known as the radius. The area of any circle is found by squaring the radius (r) and multiplying by π. The formula is as follows:

$$A = \pi r^2$$

The radius of a circle is equal to ½ the diameter:

$$r = \frac{d}{2}$$

Example
Compute the piston displacement of the PWA 14 cylinder engine having a cylinder with a 5.5 inch diameter and a 5.5 inch stroke. Formulas required are:

$$r = \frac{d}{2}$$
$$A = \pi r^2$$
$$V = A \times h$$
$$\text{Total } V = V \times n \text{ (number of cylinders)}$$

Substitute values into these formulas and complete the calculation.

$$r = \frac{d}{2} = \frac{5.5 \text{ inches (in)}}{2} = 2.75 \text{ in}$$

$A = \pi r^2 = 3.1416 \,(2.75 \text{ in} \times 2.75 \text{ in})$

$A = 3.1416 \times 7.5625 \text{ square inches (in}^2) = 23.7584 \text{ in}^2$

$V = A \times h = 23.7584 \text{ in}^2 \times 5.5 \text{ in} = 130.6712 \text{ cubic inches (in}^3)$

$\text{Total } V = V \times n = 130.6712 \text{ in}^3 \times 14$

$\text{Total } V = 1829.3968 \text{ in}^3$

Rounded off to the next whole number, total piston displacement equals 1,829 cubic inches.

Another method of calculating the piston displacement uses the diameter of the piston instead of the radius in the formula for the area of the base.

$$A = \tfrac{1}{4}\,(\pi)(d^2)$$

$\text{Substituting } A = \tfrac{1}{4} \times 3.1416 \times 5.5 \text{ in} \times 5.5 \text{ in}$

$A = 0.7854 \times 30.25 \text{ in}^2$

$A = 23.758 \text{ in}^2$

From this point on, the calculations are identical to the preceding example.

Compression Ratio

All internal combustion engines must compress the air-fuel mixture to receive a reasonable amount of work from each power stroke. The air-fuel charge in the cylinder can be compared to a coil spring in that the more it is compressed, the more work it is potentially capable of doing.

The compression ratio of an engine is a comparison of the volume of space in a cylinder when the piston is at the bottom of the stroke to the volume of space when the piston is at the top of the stroke. *[Figure 1-38]* This comparison is expressed as a ratio, hence the term compression ratio. Compression ratio is a controlling factor in the maximum horsepower developed by an engine, but it is limited by present day fuel grades and the high engine speeds and manifold pressures required for takeoff. For example, if there are 140 cubic inches of space in the cylinder when the piston is at the bottom and there are 20 cubic inches of space when the piston is at the top of the stroke, the compression ratio would be 140 to 20. If this ratio is expressed in fraction form, it would be 140/20 or 7 to 1, usually represented as 7:1.

The limitations placed on compression ratios, manifold pressure, and the manifold pressure's effect on compression

pressures has a major effect on engine operation. Manifold pressure is the average absolute pressure of the air or air-fuel charge in the intake manifold and is measured in units of inches of mercury ("Hg). Manifold pressure is dependent on engine speed (throttle setting) and the degree supercharging. The operation of the supercharger increases the weight of the charge entering the cylinder. When a true supercharger is used with the aircraft engine, the manifold pressure may be considerably higher than the pressure of the outside atmosphere. The advantage of this condition is that a greater amount of charge is forced into a given cylinder volume, and a greater output of horsepower results.

Compression ratio and manifold pressure determine the pressure in the cylinder in that portion of the operating cycle when both valves are closed. The pressure of the charge before compression is determined by manifold pressure, while the pressure at the height of compression (just prior to ignition) is determined by manifold pressure times the compression ratio. For example, if an engine were operating at a manifold pressure of 30 "Hg with a compression ratio of 7:1, the pressure at the instant before ignition would be approximately 210 "Hg. However, at a manifold pressure of 60 "Hg, the pressure would be 420 "Hg.

Without going into great detail, it has been shown that the compression event magnifies the effect of varying the manifold pressure, and the magnitude of both affects the pressure of the fuel charge just before the instant of ignition. If the pressure at this time becomes too high, pre-ignition or detonation occur and produce overheating. Pre-ignition is when the fuel air charge starts to burn before the spark plug fires. Detonation occurs when the fuel air charge is ignited by the spark plug, but instead of burning at a controlled rate, it explodes causing cylinder temperatures and pressures to spike very quickly. If this condition exists for very long, the engine can be damaged or destroyed.

One of the reasons for using engines with high compression ratios is to obtain long-range fuel economy, to convert more heat energy into useful work than is done in engines of low compression ratio. Since more heat of the charge is converted into useful work, less heat is absorbed by the cylinder walls. This factor promotes cooler engine operation, which in turn increases the thermal efficiency. Here again, a compromise is needed between the demand for fuel economy and the demand for maximum horsepower without detonation. Some manufacturers of high compression engines suppress detonation at high manifold pressures by using high octane fuel and limiting maximum manifold pressure.

Indicated Horsepower

The indicated horsepower produced by an engine is the horsepower calculated from the indicated mean effective pressure and the other factors which affect the power output of an engine. Indicated horsepower is the power developed in the combustion chambers without reference to friction losses within the engine. This horsepower is calculated as a function of the actual cylinder pressure recorded during engine operation.

To facilitate the indicated horsepower calculations, a mechanical indicating device, such as is attached to the engine cylinder, captures the actual pressure existing in the cylinder during the complete operating cycle. This pressure variation can be represented by the kind of graph shown in *Figure 1-39*. Notice that the cylinder pressure rises on the compression stroke, reaches a peak after top center, and decreases as the piston moves down on the power stroke. Since the cylinder pressure varies during the operating cycle, an average pressure (line AB) is computed. This average pressure, if applied steadily during the time of the power stroke, would do the same amount of work as the varying pressure during the same period. This average pressure is known as indicated mean effective pressure and is included in the indicated horsepower calculation with other engine specifications. If the characteristics and the indicated mean effective pressure of an engine are known, it is possible to calculate the indicated horsepower rating.

The indicated horsepower for a four-stroke cycle engine can be calculated from the following formula, in which the letter symbols in the numerator are arranged to spell the word "PLANK" to assist in memorizing the formula:

$$\text{Indicated horsepower} = \frac{\text{PLANK}}{33,000}$$

Where:

P = Indicated mean effective pressure, in psi
L = Length of the stroke, in feet or in fractions of a foot
A = Area of the piston head or cross-sectional area of the cylinder, in square inches
N = Number of power strokes per minute: $\frac{\text{rpm}}{2}$
K = Number of cylinders

In the formula above, the area of the piston multiplied by the indicated mean effective pressure gives the force acting on the piston in pounds. This force multiplied by the length of the stroke in feet gives the work performed in one power stroke, which, multiplied by the number of power strokes per minute, gives the number of ft-lb per minute of work produced by one cylinder. Multiplying this result by the number of cylinders in the engine gives the amount of work performed, in ft-lb, by the engine. Since hp is defined as work done at the rate of 33,000 ft-lb per minute, the total number of ft-lb of work performed by the engine is divided by 33,000 to find the indicated horsepower.

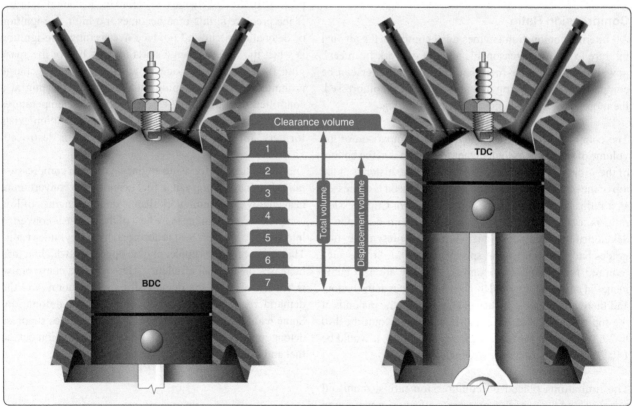

Figure 1-38. *Compression ratio.*

Example

Given:

P = 1.65 lb/in^2

L = 0.5 ft.

A = 5.5 inches

N = 1,500

K = 12

Indicated hp = $\frac{PLANK}{33,000 \text{ ft-lb/min}}$

Find indicated hp.

A is found by using the equation:

A = $\frac{1}{4}\pi D^2$

A = $\frac{1}{4}$ x 3.1416 x 5.5 in x 5.5 in

= 23.76 in^2

N is found by multiplying the rpm by $\frac{1}{2}$:

N = $\frac{1}{2}$ x 3,000 = 1,500 rpm

Now, substituting in the formula:

Indicated hp = $\frac{1.65 \text{ lb/in}^2 \text{ x } 0.5 \text{ in x } 23.76 \text{ in}^2 \text{ x } 1,500 \text{ rpm x } 12}{33,000 \text{ ft lb/min}}$

Indicated hp = 1,069.123

Brake Horsepower

The indicated horsepower calculation discussed in the preceding paragraph is the theoretical power of a frictionless engine. The total horsepower lost in overcoming friction must be subtracted from the indicated horsepower to arrive at the actual horsepower delivered to the propeller. The power delivered to the propeller for useful work is known as brake horsepower (bhp). The difference between indicated and brake horsepower is known as friction horsepower, which is the horsepower required to overcome mechanical losses, such as the pumping action of the pistons, the friction of the pistons, and the friction of all other moving parts.

The measurement of an engine's bhp involves the measurement of a quantity known as torque or twisting moment. Torque is the product of a force and the distance of the force from the axis about which it acts, or

Torque = force x distance
(at right angles to the force)

Torque is a measure of load and is properly expressed in pound-inches (lb-in) or pound-feet (lb-ft). Torque should not be confused with work, which is expressed in inch-pounds (in-lb) or foot-pounds (ft-lb).

There are numerous devices for measuring torque, such as a dynamometer or a torque meter. One very simple type of device that can be used to demonstrate torque calculations is the Prony brake. *[Figure 1-40]* All of these torque-measuring devices are usable to calculate power output of an engine on a test stand. It consists essentially of a hinged collar, or brake, which can be clamped to a drum splined to the propeller shaft. The collar and drum form a friction brake, which can be adjusted by a wheel. An arm of a known length is rigidly attached to or is a part of the hinged collar and terminates at a point that rests on a set of scales. As the propeller shaft rotates, it tends to carry the hinged collar of the brake with it and is prevented from doing so only by the arm that rests on the scale. The scale indicates the force necessary to arrest the motion of the arm. If the resulting force registered on the scale is multiplied by the length of the arm, the resulting product is the torque exerted by the rotating shaft. For example, if the scale registers 200 pounds and the length of the arm is 3.18 feet, the torque exerted by the shaft is:

200 lb x 3.18 ft = 636 lb-ft

Once the torque is known, the work done per revolution of the propeller shaft can be computed without difficulty by the equation:

Work per revolution = 2π x torque

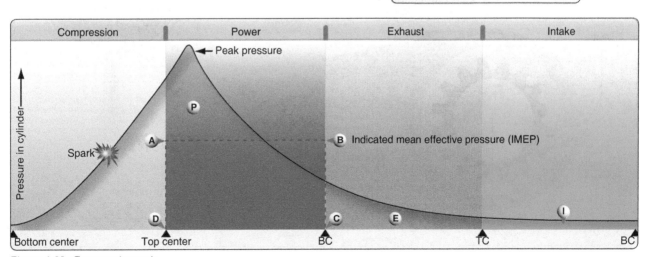

Figure 1-39. *Compression ratio.*

If work per revolution is multiplied by the rpm, the result is work per minute, or power. If the work is expressed in ft-lb per minute, this quantity is divided by 33,000. The result is the brake horsepower of the shaft.

$$\text{Power} = \text{Work per revolution} \times \text{rpm}$$

$$\text{bhp} = \frac{\text{Work per revolution} \times \text{rpm}}{33,000}$$

$$\frac{2\pi r \times \text{force on scales (lb)} \times \text{length of arm (ft)} \times \text{rpm}}{33,000}$$

Example

Given:

Force on scales	= 200 lb
Length of arm	= 3.18 ft
rpm	= 3,000
π	= 3.1416

Find bhp substituting in equation:

$$\text{bhp} = \frac{6.2832 \times 200 \times 3.18 \times 3,000}{33,000}$$
$$= 363.2$$
$$= 363$$

As long as the friction between the brake collar and propeller shaft drum is great enough to impose an appreciable load on the engine, but is not great enough to stop the engine, it is not necessary to know the amount of friction between the collar and drum to compute the bhp. If there were no load imposed, there would be no torque to measure, and the engine would "run away." If the imposed load is so great that the engine stalls, there may be considerable torque to measure, but there is no rpm. In either case, it is impossible to measure the bhp of the engine. However, if a reasonable amount of friction exists between the brake drum and the collar and the load is then increased, the tendency of the propeller shaft to carry the collar and arm about with it becomes greater, thus imposing a greater force upon the scales. As long as the torque increase is proportional to the rpm decrease, the horsepower delivered at the shaft remains unchanged. This can be seen from the equation in which $2\pi r$ and 33,000 are constants and torque and rpm are variables. If the change in rpm is inversely proportional to the change in torque, their product remains unchanged, and bhp remains unchanged. This is important. It shows that horsepower is the function of both torque and rpm, and can be changed by changing either torque, rpm, or both.

Friction Horsepower

Friction horsepower is the indicated horsepower minus brake horsepower. It is the horsepower used by an engine in overcoming the friction of moving parts, drawing in fuel, expelling exhaust, driving oil and fuel pumps, and other engine accessories. On modern aircraft engines, this power loss through friction may be as high as 10 to 15 percent of

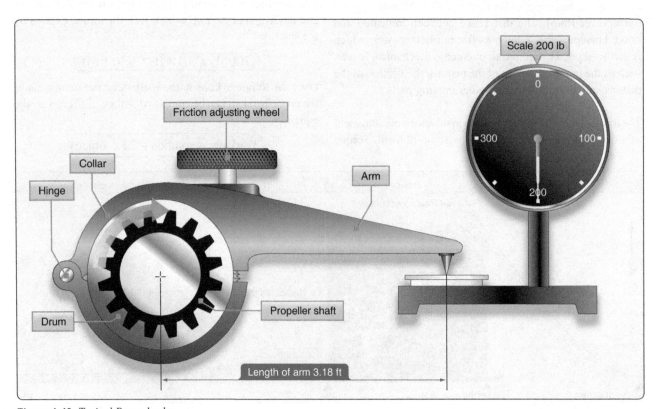

Figure 1-40. *Typical Prony brake.*

the indicated horsepower.

Friction & Brake Mean Effective Pressures

The indicated mean effective pressure (IMEP), discussed previously, is the average pressure produced in the combustion chamber during the operating cycle and is an expression of the theoretical, frictionless power known as indicated horsepower. In addition to completely disregarding power lost to friction, indicated horsepower gives no indication of how much actual power is delivered to the propeller shaft for doing useful work. However, it is related to actual pressures that occur in the cylinder and can be used as a measure of these pressures.

To compute the friction loss and net power output, the indicated horsepower of a cylinder may be thought of as two separate powers, each producing a different effect. The first power overcomes internal friction, and the horsepower thus consumed is known as friction horsepower. The second power, known as brake horsepower, produces useful work at the propeller. That portion of IMEP that produces brake horsepower is called brake mean effective pressure (BMEP). The remaining pressure used to overcome internal friction is called friction mean effective pressure (FMEP). *[Figure 1-41]* IMEP is a useful expression of total cylinder power output, but is not a real physical quantity; likewise, FMEP and BMEP are theoretical but useful expressions of friction losses and net power output.

Although BMEP and FMEP have no real existence in the cylinder, they provide a convenient means of representing pressure limits or rating engine performance throughout its entire operating range. There is an operating relationship between IMEP, BMEP, and FMEP.

One of the basic limitations placed on engine operation is the pressure developed in the cylinder during combustion. In the discussion of compression ratios and indicated mean effective pressure, it was found that, within limits, increased pressure resulted in increased power. It was also noted that if the cylinder pressure were not controlled within close limits, it would impose dangerous internal loads that might result in engine failure. Therefore, it is important to have a means of determining these cylinder pressures as a protective measure and for efficient application of power.

If the bhp is known, the BMEP can be computed by means of the following equation:

$$BMEP = \frac{bhp \times 33,000}{LANK}$$

Given:

bhp = 1,000

Stroke = 6 in

Bore = 5.5 in

rpm = 3,000

Number of cycles = 12

Find BMEP:

Find length of stroke (in ft):

L = 6 in = 0.5 ft

Find area of cylinder bore:

$A = \frac{1}{4}\pi r D^2$

A = ¼ × 3.1416 × 5.5 in × 5.5 in

$A = 23.76\ in^2$

Find number of power strokes per min:

N = ½ × rpm

N = ½ × 3,000

N = 1,500

Then, substituting in the equation:

$$BMEP = \frac{1,000\ bhp \times 33,000\ ft\text{-}lb/min}{0.5\ ft \times 23.76\ in^2 \times 1,500\ strokes/min \times 12}$$

$$= 154.32\ lb/in^2$$

Thrust Horsepower

Thrust horsepower can be considered the result of the engine and the propeller working together. If a propeller could be designed to be 100 percent efficient, the thrust and the bph would be the same. However, the efficiency of the propeller varies with the engine speed, attitude, altitude, temperature, and airspeed. Thus, the ratio of the thrust horsepower and the bhp delivered to the propeller shaft will never be equal. For example, if an engine develops 1,000 bhp, and it is used with a propeller having 85 percent efficiency, the thrust horsepower of that engine-propeller combination is 85 percent of 1,000 or 850 thrust hp. Of the four types of horsepower discussed, it is the thrust horsepower that determines the performance of the engine-propeller combination.

Efficiencies

Thermal Efficiency

Any study of engines and power involves consideration of heat as the source of power. The heat produced by the burning of gasoline in the cylinders causes a rapid expansion of the gases in the cylinder, and this, in turn, moves the pistons

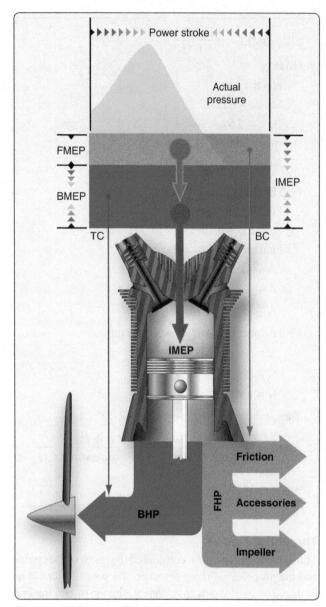

Figure 1-41. *Power and pressure.*

and creates mechanical energy. It has long been known that mechanical work can be converted into heat and that a given amount of heat contains the energy equivalent of a certain amount of mechanical work. Heat and work are theoretically interchangeable and bear a fixed relation to each other. Heat can therefore be measured in work units (for example, ft-lb) as well as in heat units. The British thermal unit (BTU) of heat is the quantity of heat required to raise the temperature of 1 pound of water by 1 °F. It is equivalent to 778 ft-lb of mechanical work. A pound of petroleum fuel, when burned with enough air to consume it completely, gives up about 20,000 BTU, the equivalent of 15,560,000 ft-lb of mechanical work. These quantities express the heat energy of the fuel in heat and work units, respectively.

The ratio of useful work done by an engine to the heat energy of the fuel it uses, expressed in work or heat units, is called the thermal efficiency of the engine. If two similar engines use equal amounts of fuel, the engine that converts into work the greater part of the energy in the fuel (higher thermal efficiency) delivers the greater amount of power. Furthermore, the engine that has the higher thermal efficiency has less waste heat to dispose of to the valves, cylinders, pistons, and cooling system of the engine. A high thermal efficiency also means low specific fuel consumption and, therefore, less fuel for a flight of a given distance at a given power. Thus, the practical importance of a high thermal efficiency is threefold, and it constitutes one of the most desirable features in the performance of an aircraft engine.

Of the total heat produced, 25 to 30 percent is utilized for power output, 15 to 20 percent is lost in cooling (heat radiated from cylinder head fins), 5 to 10 percent is lost in overcoming friction of moving parts; and 40 to 45 percent is lost through the exhaust. Anything that increases the heat content going into mechanical work on the piston, which reduces the friction and pumping losses, or which reduces the quantity of unburned fuel or the heat lost to the engine parts, increases the thermal efficiency.

The portion of the total heat of combustion that is turned into mechanical work depends to a great extent upon the compression ratio. The compression ratio is the ratio of the piston displacement plus combustion chamber space to the combustion chamber space, as mentioned earlier. Other things being equal, the higher the compression ratio is, the larger is the proportion of the heat energy of combustion turned into useful work at the crankshaft. On the other hand, increasing the compression ratio increases the cylinder head temperature. This is a limiting factor because the extremely high temperature created by high compression ratios causes the material in the cylinder to deteriorate rapidly and the fuel to detonate instead of burning at a controlled rate.

The thermal efficiency of an engine may be based on either bhp or indicated horsepower (ihp) and is represented by the following formula:

$$\text{Indicated thermal efficiency} = \frac{\text{ihp} \times 33,000}{\text{weight of fuel burned/min.} \times \text{heat value} \times 778}$$

The formula for brake thermal efficiency is the same as shown above, except the value for bhp is inserted instead of the value for ihp.

Example

An engine delivers 85 bhp for a period of 1 hour and during that time consumes 50 pounds of fuel. Assuming the fuel has a heat content of 18,800 BTU per pound, find the thermal efficiency of the engine:

$$\frac{8.5 \text{ ihp} \times 33,000}{0.833 \times 18,800 \text{ BTU} \times 778} = \frac{2,805,000}{12,184,569}$$

Brake thermal efficiency = 0.23 or 23 percent

Reciprocating engines are only about 34 percent thermally efficient; that is, they transform only about 34 percent of the total heat potential of the burning fuel into mechanical energy. The remainder of the heat is lost through the exhaust gases, the cooling system, and the friction within the engine. Thermal distribution in a reciprocating engine is illustrated in *Figure 1-42.*

Mechanical Efficiency

Mechanical efficiency is the ratio that shows how much of the power developed by the expanding gases in the cylinder is actually delivered to the output shaft. It is a comparison between the bhp and the ihp. It can be expressed by the following formula:

$$\text{Mechanical efficiency} = \frac{\text{bhp}}{\text{ihp}}$$

Brake horsepower is the useful power delivered to the propeller shaft. Indicated horsepower is the total hp developed in the cylinders. The difference between the two is friction horsepower (fhp), the power lost in overcoming friction. The factor that has the greatest effect on mechanical efficiency is the friction within the engine itself. The friction between moving parts in an engine remains practically constant throughout an engine's speed range. Therefore, the mechanical efficiency of an engine is highest when the engine is running at the rpm at which maximum bhp is developed. Mechanical efficiency of the average aircraft reciprocating engine approaches 90 percent.

Volumetric Efficiency

Volumetric efficiency is a ratio expressed in terms of percentages. It is a comparison of the volume of air-fuel charge (corrected for temperature and pressure) inducted into the cylinders to the total piston displacement of the engine. Various factors cause departure from a 100 percent volumetric efficiency. The pistons of a naturally aspirated engine displace the same volume each time they travel from top center to bottom center of the cylinders. The amount of charge that fills this volume on the intake stroke depends on the existing pressure and temperature of the surrounding atmosphere. Therefore, to find the volumetric efficiency of an engine, standards for atmospheric pressure and temperature had to be established. The U.S. standard atmosphere was established in 1958, and provides the necessary pressure and temperature values to calculate volumetric efficiency.

The standard sea level temperature is 59 °F, or 15 °C. At this temperature, the pressure of one atmosphere is 14.69 lb/in², and this pressure supports a column of mercury (Hg) 29.92 inches high, or 29.92 "Hg. These standard sea level conditions determine a standard density, and if the engine draws in a volume of charge of this density exactly equal to its piston displacement, it is said to be operating at 100 percent volumetric efficiency. An engine drawing in less volume than this has a volumetric efficiency lower than 100 percent. An engine equipped with true supercharging (boost above 30.00 "Hg) may have a volumetric efficiency greater than 100 percent. The equation for volumetric efficiency is as follows:

$$\text{Volumetric efficiency} = \frac{\text{Volume of charge (corrected for temperature and pressure)}}{\text{Piston displacement}}$$

Many factors decrease volumetric efficiency, including:

- Part-throttle operation;
- Long intake pipes of small diameter;
- Sharp bends in the induction system;

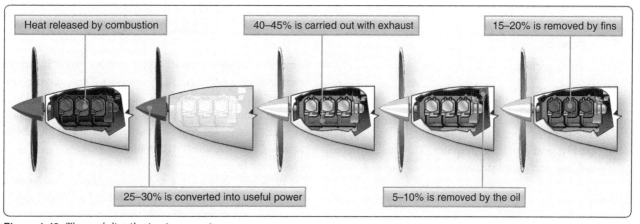

Figure 1-42. *Thermal distribution in an engine.*

- Carburetor air temperature too high;

- Cylinder-head temperature too high;

- Incomplete scavenging; and

- Improper valve timing.

Propulsive Efficiency

A propeller is used with an engine to provide thrust. The engine supplies bhp through a rotating shaft, and the propeller absorbs the bhp and converts it into thrust hp. In this conversion, some power is wasted. Since the efficiency of any machine is the ratio of useful power output to the power input, propulsive efficiency (in this case, propeller efficiency) is the ratio of thrust hp to bhp. On the average, thrust hp constitutes approximately 80 percent of the bhp. The other 20 percent is lost in friction and slippage. Controlling the blade angle of the propeller is the best method of obtaining maximum propulsive efficiency for all conditions encountered in flight.

During takeoff, when the aircraft is moving at low speeds and when maximum power and thrust are required, a low propeller blade angle gives maximum thrust. For high-speed flying or diving, the blade angle is increased to obtain maximum thrust and efficiency. The constant-speed propeller is used to give required thrust at maximum efficiency for all flight conditions.

Gas Turbine Engines

Types & Construction

In a reciprocating engine, the functions of intake, compression, combustion, and exhaust all take place in the same combustion chamber. Consequently, each must have exclusive occupancy of the chamber during its respective part of the combustion cycle. A significant feature of the gas turbine engine is that separate sections are devoted to each function, and all functions are performed simultaneously without interruption.

A typical gas turbine engine consists of:

1. An air inlet,

2. Compressor section,

3. Combustion section,

4. Turbine section,

5. Exhaust section,

6. Accessory section, and

7. The systems necessary for starting, lubrication, fuel supply, and auxiliary purposes, such as anti-icing, cooling, and pressurization.

The major components of all gas turbine engines are basically the same; however, the nomenclature of the component parts of various engines currently in use varies slightly due to the difference in each manufacturer's terminology. These differences are reflected in the applicable maintenance manuals. One of the greatest single factors influencing the construction features of any gas turbine engine is the type of compressor or compressors for which the engine is designed.

Four types of gas turbine engines are used to propel and power aircraft. They are the turbofan, turboprop, turboshaft, and turbojet. The term "turbojet" was used to describe any gas turbine engine used in aircraft. As gas turbine technology evolved, these other engine types were developed to take the place of the pure turbojet engine. The turbojet engine has problems with noise and fuel consumption in the speed range that airliners fly (.8 Mach). Due to these problems, use of pure turbojet engines is very limited. So, almost all airliner-type aircraft use a turbofan engine. It was developed to turn a large fan or set of fans at the front of the engine and produces about 80 percent of the thrust from the engine. This engine was quieter and had better fuel consumption in this speed range. Turbofan engines have more than one shaft in the engine; many are two-shaft engines. This means that there are two sets of compressors and turbines that drive them. These two-shafted engines use two spools (a spool is a compressor and a shaft and turbines that drive that compressor). In a two-spool engine, there is a high-pressure spool and a low-pressure spool. The low-pressure spool generally contains the fan(s) and the turbine stages it takes to drive them. The high-pressure spool is the high-pressure compressor, shaft, and turbines. This spool makes up the core of the engine, and this is where the combustion section is located.

Turbofan engines can be low bypass or high bypass. The amount of air that is bypassed around the core of the engine determines the bypass ratio. As can be seen in *Figure 1-43*, the air generally driven by the fan does not pass through the internal working core of the engine. The amount of air flow in lb/sec from the fan bypass to the core flow of the engine is the bypass ratio.

$$\text{Bypass ratio} = \frac{100 \text{ lb/sec flow fan}}{20 \text{ lb/sec flowcore}} = 5\text{:}1 \text{ bypass ratio}$$

Some low-bypass turbofan engines are used in speed ranges above .8 Mach (military aircraft). These engines use augmenters or afterburners to increase thrust. By adding more fuel nozzles and a flame holder in the exhaust system extra fuel can be sprayed and burned which can give large increases in thrust for short amounts of time.

The turboprop engine is a gas turbine engine that turns a propeller through a speed reduction gear box. This type of engine is most efficient in the 300 to 400 mph speed range and can use shorter runways that other aircraft. Approximately 80 to 85 percent of the energy developed by the gas turbine engine is used to drive the propeller. The rest of the available energy exits the exhaust as thrust. By adding the horsepower developed by the engine shaft and the horsepower in the exiting thrust, the answer is equivalent shaft horsepower.

With regard to aircraft, the turboshaft engine is a gas turbine engine made to transfer horsepower to a shaft that turns a helicopter transmission or is an onboard auxiliary power unit (APU). An APU is used on turbine-powered aircraft to provide electrical power and bleed air on the ground and a backup generator in flight. Turboshaft engines can come in many different styles, shapes, and horsepower ranges.

Air Entrance

The air entrance is designed to conduct incoming air to the compressor with a minimum energy loss resulting from drag or ram pressure loss; that is, the flow of air into the compressor should be free of turbulence to achieve maximum operating efficiency. Proper inlet design contributes materially to aircraft performance by increasing the ratio of compressor discharge pressure to duct inlet pressure.

This is also referred to as the compressor pressure ratio. This ratio is the outlet pressure divided by the inlet pressure. The amount of air passing through the engine is dependent upon three factors:

1. The compressor speed (rpm).
2. The forward speed of the aircraft.
3. The density of the ambient (surrounding) air.

Turbine inlet type is dictated by the type of gas turbine engine. A high-bypass turbofan engine inlet is completely different from a turboprop or turboshaft inlet. Large gas turbine-powered aircraft almost always have a turbofan engine. The inlet on this type of engine is bolted to the front (A flange) of the engine. These engines are mounted on the wings, or nacelles, on the aft fuselage, and a few are in the vertical fin. A typical turbofan inlet can be seen in *Figure 1-44*. Since on most modern turbofan engines the huge fan is the first part of the aircraft the incoming air comes into contact with, icing protection must be provided. This prevents chunks of ice from forming on the leading edge of the inlet, breaking loose, and damaging the fan. Warm air is bled from the engine's compressor and is ducted through the inlet to prevent ice from forming. If inlet guide vanes are used to straighten the air flow, then they also have anti-icing air flowing through them. The inlet also contains some sound-reducing materials that absorb the fan noise and make the engine quieter.

Turboprops and turboshafts can use an inlet screen to help filter out ice or debris from entering the engine. A deflector vane and a heated inlet lip are used to prevent ice from forming and allowing large chunks to enter the engine.

On military aircraft, the divided entrance permits the use of very short ducts with a resultant small pressure drop through skin friction. Military aircraft can fly at speeds above Mach 1, but the airflow through the engine must always stay below Mach 1. Supersonic air flow in the engine would destroy the engine. By using convergent and divergent shaped ducts, the air flow is controlled and dropped to subsonic speeds before entering the engine. Supersonic inlets are used to slow the incoming engine air to less than Mach 1 before it enters the engine.

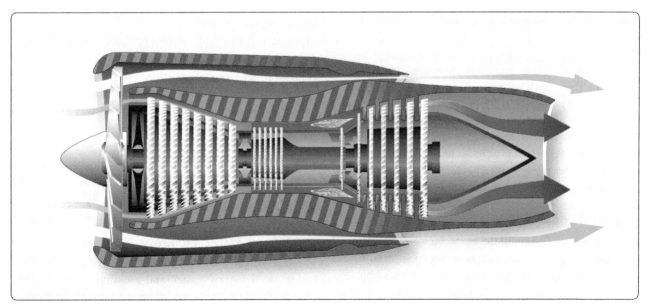

Figure 1-43. *Turbofan engine with separate nozzles, fan, and core.*

Accessory Section

The accessory section of the gas turbine engine has various functions. The primary function is to provide space for the mounting of accessories necessary for operation and control of the engine. Generally, it also includes accessories concerned with the aircraft, such as electric generators and hydraulic pumps. Secondary functions include acting as an oil reservoir and/or oil sump and housing the accessory drive gears and reduction gears.

The arrangement and driving of accessories has always been a major problem on gas turbine engines. Driven accessories on turbofans are usually mounted on the accessory gear box, which is on the bottom of the engine. The location of the accessory gear box varies somewhat, but most turboprops and turboshafts have the accessory cases mounted to the back section of the engine.

The components of the accessory section of all gas turbine engines have essentially the same purpose, even though they often differ quite extensively in construction details and nomenclature.

The basic elements of the accessory section are:

1. The accessory case, which has machined mounting pads for the engine-driven accessories, and

2. The gear train, which is housed within the accessory case.

The accessory case may be designed to act as an oil reservoir. If an oil tank is utilized, a sump is usually provided below the front bearing support for the drainage and scavenging of oil used to lubricate bearings and drive gears. The accessory case is also provided with adequate tubing or cored passages for spraying lubricating oil on the gear train and supporting bearings.

The gear train is driven by the engine high-pressure compressor through an accessory drive shaft (tower shaft) gear coupling, which splines with a gear box gear and the high-pressure compressor. The reduction gearing within the case provides suitable drive speeds for each engine accessory or component. Because the rotor operating rpm is so high, the accessory reduction gear ratios are relatively high. The accessory drives are supported by ball bearings assembled in the mounting pad bores of the accessory case. *[Figure 1-45]*

Compressor Section

The compressor section of the gas turbine engine has many functions. Its primary function is to supply air in sufficient quantity to satisfy the requirements of the combustion burners. Specifically, to fulfill its purpose, the compressor must increase the pressure of the mass of air received from the air inlet duct, and then discharge it to the burners in the quantity and at the pressures required.

A secondary function of the compressor is to supply bleed-air for various purposes in the engine and aircraft. The bleed-air is taken from any of the various pressure stages of the compressor. The exact location of the bleed ports is, of course, dependent on the pressure or temperature required for a particular job. The ports are small openings in the compressor case adjacent to the particular stage from which the air is to be bled; thus, varying degrees of pressure are available simply by tapping into the appropriate stage. Air is often bled from the final or highest pressure stage since, at this point, pressure and air temperature are at a maximum. At times it may be necessary to cool this high-pressure air. If it is used for cabin pressurization or other purposes to which excess heat would be uncomfortable or detrimental, the air is sent through an air conditioning unit before it enters the cabin.

Bleed air is utilized in a wide variety of ways. Some of the current applications of bleed air are:

1. Cabin pressurization, heating, and cooling;

2. Deicing and anti-icing equipment;

3. Pneumatic starting of engines; and

4. Auxiliary drive units (ADU).

Another function of the compressor bleed air is for the operation of vortex dissipaters. The vortex dissipater supplies a high-velocity stream of compressor bleed air blown from a nozzle into an area where vortices are likely to form. Vortex dissipaters destroy the vortices that would otherwise suck debris from the ground into engines mounted in pods that are low to the ground.

Compressor Types

The two principal types of compressors currently being used in gas turbine aircraft engines are centrifugal flow and axial flow. The centrifugal-flow compressor achieves its purpose by picking up the entering air and accelerating it outwardly by centrifugal action. The axial-flow compressor compresses air while the air continues in its original direction of flow, thus avoiding the energy loss caused by turns. The components of each of these two types of compressor have their individual functions in the compression of air for the combustion section. A stage in a compressor is considered to be a rise in pressure.

Centrifugal-Flow Compressors

The centrifugal-flow compressor consists of an impeller (rotor), a diffuser (stator), and a compressor manifold. *[Figure 1-46]* Centrifugal compressors have a high pressure rise per stage that can be around 8:1. Generally centrifugal

Figure 1-44. *Typical turbofan inlet.*

compressors are limited to two stages due to efficiency concerns. The two main functional elements are the impeller and the diffuser. Although the diffuser is a separate unit and is placed inside and bolted to the manifold, the entire assembly (diffuser and manifold) is often referred to as the diffuser. For clarification during compressor familiarization, the units are treated individually. The impeller is usually made from forged aluminum alloy, heat treated, machined, and smoothed for minimum flow restriction and turbulence.

In most types, the impeller is fabricated from a single forging. This type impeller is shown in *Figure 1-46*. The impeller, whose function is to pick up and accelerate the air outwardly to the diffuser, may be either of two types—single entry or double entry. The principal differences between the two types of impellers are size and ducting arrangement. The double-entry type has a smaller diameter but is usually operated at a higher rotational speed to assure sufficient airflow. The single-entry impeller, shown in *Figure 1-47*, permits convenient ducting directly to the impeller eye (inducer vanes) as opposed to the more complicated ducting necessary to reach the rear side of the double-entry type. Although slightly more efficient in receiving air, the single-entry impeller must be large in diameter to deliver the same quantity of air as the double-entry type. This, of course, increases the overall diameter of the engine.

Included in the ducting for double-entry compressor engines is the plenum chamber. This chamber is necessary for a double-entry compressor because the air must enter the engine at almost right angles to the engine axis. Therefore, in order to give a positive flow, the air must surround the engine compressor at a positive pressure before entering

the compressor. Included in some installations as necessary parts of the plenum chamber are the auxiliary air-intake doors (blow-in doors). These blow-in doors admit air to the engine compartment during ground operation, when air requirements for the engine are in excess of the airflow through the inlet ducts. The doors are held closed by spring action when the engine is not operating. During operation, however, the doors open automatically whenever engine compartment pressure drops below atmospheric pressure. During takeoff and flight, ram air pressure in the engine compartment aids the springs in holding the doors closed.

The diffuser is an annular chamber provided with a number of vanes forming a series of divergent passages into the manifold. The diffuser vanes direct the flow of air from the impeller to the manifold at an angle designed to retain the maximum amount of energy imparted by the impeller. They also deliver the air to the manifold at a velocity and pressure satisfactory for use in the combustion chambers. Refer to *Figure 1-46A* and note the arrow indicating the path of airflow through the diffuser, then through the manifold.

The compressor manifold shown in *Figure 1-46A* diverts the flow of air from the diffuser, which is an integral part of the manifold, into the combustion chambers. The manifold has one outlet port for each chamber so that the air is evenly divided. A compressor outlet elbow is bolted to each of the outlet ports. These air outlets are constructed in the form of ducts and are known by a variety of names, such as air outlet ducts, outlet elbows, or combustion chamber inlet ducts. Regardless of the terminology used, these outlet ducts perform a very important part of the diffusion process; that is, they change the radial direction of the airflow to an axial direction, in which the diffusion process is completed after the turn. To help the elbows perform this function in an efficient manner, turning vanes (cascade vanes) are sometimes fitted inside the elbows. These vanes reduce air pressure losses by

Figure 1-45. *Typical turboprop accessory case.*

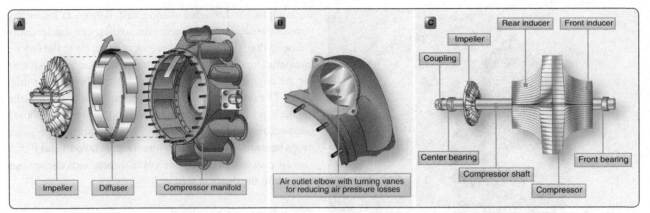

Figure 1-46. *(A) Components of a centrifugal-flow compressor; (B) Air outlet elbow with turning vanes for reducing air pressure losses; (C) Components of a double-entry centrifugal-flow compressor.*

Figure 1-47. *Single-entry impeller.*

presenting a smooth, turning surface. *[Figure 1-46B]*

Axial-Flow Compressor

The axial flow compressor is a combination of an engine compressor and high-pressure turbine that drives it using a connecting drive shaft. In a single-spool engine, the high-pressure turbine drives the entire compressor. In a dual-spool engine, the compressor and high-pressure turbine are both split into two segments. Each compressor segment is driven by its corresponding turbine using two separate drive shafts, with one inside the other. The first stage turbine drives the N2 compressor.

The axial-flow compressor has two main elements: a rotor and a stator. The rotor has blades fixed on a spindle. These blades impel air rearward in the same manner as a propeller because of their angle and airfoil contour. The rotor, turning at high speed, takes in air at the compressor inlet and impels it through a series of stages. From inlet to exit, the air flows along an axial path and is compressed at a ratio of approximately 1.25:1 per stage. The action of the

rotor increases the compression of the air at each stage and accelerates it rearward through several stages. With this increased velocity, energy is transferred from the compressor to the air in the form of velocity energy. The stator blades act as diffusers at each stage, partially converting high velocity to pressure. Each consecutive pair of rotor and stator blades constitutes a pressure stage. The number of rows of blades (stages) is determined by the amount of air and total pressure rise required. Compressor pressure ratio increases with the number of compression stages. Most engines utilize up to 16 stages and more.

The stator has rows of vanes, which are in turn attached inside an enclosing case. The stator vanes, which are stationary, project radially toward the rotor axis and fit closely on either side of each stage of the rotor blades. In some cases, the compressor case, into which the stator vanes are fitted, is horizontally divided into halves. Either the upper or lower half may be removed for inspection or maintenance of rotor and stator blades.

The function of the stator vanes is to receive air from the air inlet duct or from each preceding stage and increase the pressure of the air and deliver it to the next stage at the correct velocity and pressure. They also control the direction of air to each rotor stage to obtain the maximum possible compressor blade efficiency. Shown in *Figure 1-48* are the rotor and stator elements of a typical axial-flow compressor. The first stage rotor blades can be preceded by an inlet guide vane assembly that can be fixed or variable.

The guide vanes direct the airflow into the first stage rotor blades at the proper angle and impart a swirling motion to the air entering the compressor. This preswirl, in the direction of engine rotation, improves the aerodynamic characteristics of the compressor by reducing drag on the first stage rotor blades. The inlet guide vanes are curved steel vanes usually welded to steel inner and outer shrouds.

At the discharge end of the compressor, the stator vanes are constructed to straighten the airflow to eliminate turbulence. These vanes are called straightening vanes or the outlet vane assembly. The casings of axial-flow compressors not only support the stator vanes and provide the outer wall of the axial path the air follows, but they also provide the means for extracting compressor air for various purposes. The stator vanes are usually made of steel with corrosion- and erosion-resistant qualities. Quite frequently, they are shrouded (enclosed) by a band of suitable material to simplify the fastening problem. The vanes are welded into the shrouds, and the outer shroud is secured to the compressor housing inner wall by radial retaining screws.

The rotor blades are usually made of stainless steel with the latter stages being made of titanium. The design of blade attachment to the rotor disc rims varies, but they are commonly fitted into discs by either bulb-type or fir-tree methods. *[Figure 1-49]* The blades are then locked into place by differing methods. Compressor blade tips are reduced

in thickness by cutouts, referred to as blade profiles. These profiles prevent serious damage to the blade or housing should the blades contact the compressor housing. This condition can occur if rotor blades become excessively loose or if rotor support is reduced by a malfunctioning bearing. Even though blade profiles greatly reduce such possibilities, occasionally a blade may break under stress of rubbing and cause considerable damage to compressor blades and stator vane assemblies. The blades vary in length from entry to discharge because the annular working space (drum to casing) is reduced progressively toward the rear by the decrease in the casing diameter. *[Figure 1-50]* This feature provides for a fairly constant velocity through the compressor, which helps to keep the flow of air constant.

The rotor features either drum-type or disc-type construction. The drum-type rotor consists of rings that are flanged to fit one against the other, wherein the entire assembly can then be held together by through bolts. This type of construction is satisfactory for low-speed compressors where centrifugal stresses are low. The disc-type rotor consists of a series of discs machined from aluminum forgings, shrunk over a steel shaft, with rotor blades dovetailed into the disc rims. Another method of rotor construction is to machine the discs and shaft from a single aluminum forging, and then to bolt steel stub shafts on the front and rear of the assembly to provide bearing support surfaces and splines for joining the turbine shaft. The drum-type and disc-type rotors are illustrated in *Figures 1-50* and *1-51*, respectively.

The combination of the compressor stages and turbine stages on a common shaft is an engine referred to as an engine spool. The common shaft is provided by joining the turbine and compressor shafts by a suitable method. The engine's spool is supported by bearings, which are seated in suitable bearing housings.

As mentioned earlier, there are two configurations of the axial compressor currently in use: the single rotor/spool and the dual rotor/spool, sometimes referred to as solid spool and split spool (two spool, dual spool).

One version of the solid-spool (one spool) compressor uses variable inlet guide vanes. Also, the first few rows of stator vanes are variable. The main difference between variable inlet guide vane (VIGV) and a variable stator vane (VSV) is their position with regard to the rotor blades. VIGV are in front of the rotor blades, and VSV are behind the rotor blades. The angles of the inlet guide vanes and the first several stages of the stator vanes are can be variable. During operation, air enters the front of the engine and is directed into the compressor at the proper angle by the variable inlet guide and directed by the VSV. The air is compressed and forced into the combustion

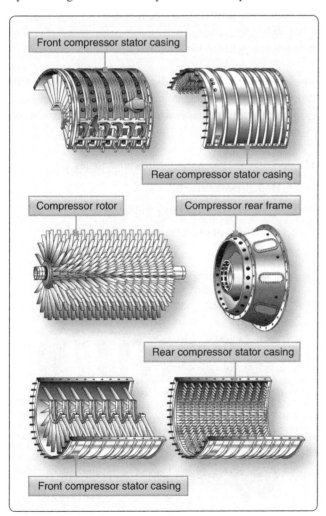

Figure 1-48. *Rotor and stator elements of a typical axial-flow compressor.*

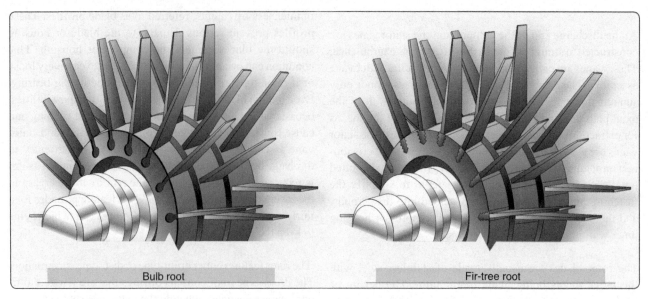

Figure 1-49. *Common designs of compressor blade attachment to the rotor disc.*

section. A fuel nozzle that extends into each combustion liner atomizes the fuel for combustion. These variables are controlled in direct relation to the amount of power the engine is required to produce by the power lever position.

Most turbofan engines are of the split-spool compressor type. Most large turbofan engines use a large fan with a few stages of compression called the low-pressure spool. These turbofans incorporate two compressors with their respective turbines and interconnecting shafts, which form two physically independent rotor systems. Many dual rotor systems have rotors turning in opposite directions and with no mechanical connection to each other. The second spool, referred to as the high-pressure spool and is the compressor for the gas generator and core of the engine, supplies air to the combustion section of the engine.

The advantages and disadvantages of both types of compressors are included in the following list. Even though each type has advantages and disadvantages, each has its use by type and size of engine.

The centrifugal-flow compressor's advantages are:

- High pressure rise per stage,

- Efficiency over wide rotational speed range,

- Simplicity of manufacture and low cost,

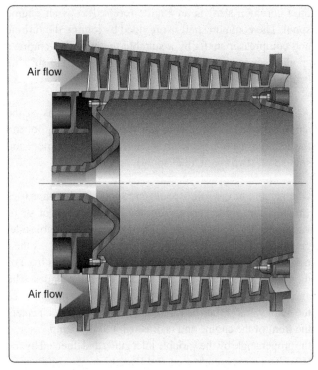

Figure 1-50. *Drum type compressor rotor.*

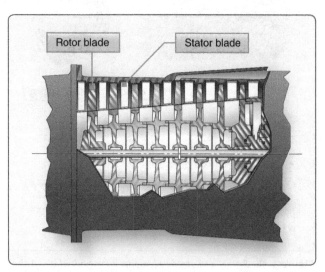

Figure 1-51. *Disc-type compressor rotor.*

- Low weight, and
- Low starting power requirements.

The centrifugal-flow compressor's disadvantages are:

- Its large frontal area for a given airflow and
- Losses in turns between stages.

The axial-flow compressor's advantages are:

- High peak efficiencies;
- Small frontal area for given airflow;
- Straight-through flow, allowing high ram efficiency; and
- Increased pressure rise by increasing number of stages, with negligible losses.

The axial-flow compressor's disadvantages are:

- Good efficiencies over only narrow rotational speed range,
- Difficulty of manufacture and high cost,
- Relatively high weight, and
- High starting power requirements (partially overcome by split compressors).

Diffuser

The diffuser is the divergent section of the engine after the compressor and before the combustion section. It has the all-important function of reducing high-velocity compressor discharge air to increased pressure at a slower velocity. This prepares the air for entry into the flame burning area of the combustion section at a lower velocity so that the flame of combustion can burn continuously. If the air passed through the flame area at a high velocity, it could extinguish the flame.

Combustion Section

The combustion section houses the combustion process, which raises the temperature of the air passing through the engine. This process releases energy contained in the air-fuel mixture. The major part of this energy is required at the turbine or turbine stages to drive the compressor. About ⅔ of the energy is used to drive the gas generator compressor. The remaining energy passes through the remaining turbine stages that absorb more of the energy to drive the fan, output shaft, or propeller. Only the pure turbojet allows the air to create all the thrust or propulsion by exiting the rear of the engine in the form of a high-velocity jet. These other engine types have some jet velocity out the rear of the engine but most of the thrust or power is generated by the additional turbine stages driving a large fan, propeller, or helicopter rotor blades.

The primary function of the combustion section is, of course, to burn the air-fuel mixture, thereby adding heat energy to the air. To do this efficiently, the combustion chamber must:

- Provide the means for proper mixing of the fuel and air to assure good combustion,
- Burn this mixture efficiently,
- Cool the hot combustion products to a temperature that the turbine inlet guide vanes/blades can withstand under operating conditions, and
- Deliver the hot gases to the turbine section.

The location of the combustion section is directly between the compressor and the turbine sections. The combustion chambers are always arranged coaxially with the compressor and turbine regardless of type, since the chambers must be in a through-flow position to function efficiently. All combustion chambers contain the same basic elements:

1. Casing.
2. Perforated inner liner.
3. Fuel injection system.
4. Some means for initial ignition.
5. Fuel drainage system to drain off unburned fuel after engine shutdown.

There are currently four basic types of combustion chambers, variations within type being in detail only. These types are:

1. Can-type.
2. Can-annular type.
3. Annular type.
4. Reverse-flow type.

The can-type combustion chamber is typical of the type used on turboshaft and APUs. *[Figure 1-52]* Each of the can-type combustion chambers consists of an outer case or housing, within which there is a perforated stainless steel (highly heat resistant) combustion chamber liner or inner liner. *[Figure 1-53]* The outer case is removed to facilitate liner replacement.

Older engines with several combustion cans had each can with interconnector (flame propagation) tube, which was a necessary part of the can-type combustion chambers. Since each can is a separate burner operating independently of the other cans, there must be some way to spread combustion during the initial starting operation. This is accomplished by interconnecting all the chambers. As the flame is started by the spark igniter plugs in two of the lower chambers, it passes through the tubes and ignites the combustible mixture in the adjacent chamber and continues until all the chambers

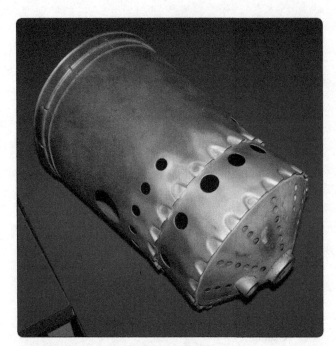

Figure 1-52. *Can-type combustion chamber.*

Figure 1-53. *Inside view of a combustion chamber liner.*

are burning.

The flame tubes vary in construction details from one engine to another, although the basic components are almost identical. *[Figure 1-54]* The spark igniters previously mentioned are normally two in number and are located in two of the can-type combustion chambers.

Another very important requirement in the construction of combustion chambers is providing the means for draining unburned fuel. This drainage prevents gum deposits in the fuel manifold, nozzles, and combustion chambers. These deposits

are caused by the residue left when the fuel evaporates. Probably most important is the danger of afterfire if the fuel is allowed to accumulate after shutdown. If the fuel is not drained, a great possibility exists that, at the next starting attempt, the excess fuel in the combustion chamber will ignite and exhaust gas temperature will exceed safe operating limits.

The liners of the can-type combustors have perforations of various sizes and shapes, each hole having a specific purpose and effect on flame propagation within the liner. *[Figure 1-52]* The air entering the combustion chamber is divided by the proper holes, louvers, and slots into two main streams—primary and secondary air. The primary or combustion air is directed inside the liner at the front end, where it mixes with the fuel and is burned. Secondary or cooling air passes between the outer casing and the liner and joins the combustion gases through larger holes toward the rear of the liner, cooling the combustion gases from about 3,500 °F to near 1,500 °F. To aid in atomization of the fuel, holes are provided around the fuel nozzle in the dome or inlet end of the can-type combustor liner. Louvers are also provided along the axial length of the liners to direct a cooling layer of air along the inside wall of the liner. This layer of air also tends to control the flame pattern by keeping it centered in the liner, thereby preventing burning of the liner walls. *Figure 1-55* illustrates the annular combustion chamber liner.

Some provision is always made in the combustion chamber case for installation of a fuel nozzle. The fuel nozzle delivers the fuel into the liner in a finely atomized spray. The more the spray is atomized, the more rapid and efficient the burning process is.

Two types of fuel nozzle currently being used in the various types of combustion chambers are the simplex nozzle and the duplex nozzle. The construction features of these nozzles are covered in greater detail in Chapter 2, Engine Fuel & Fuel Metering Systems.

The spark igniter plugs of the annular combustion chamber are the same basic type used in the can-type combustion chambers, although construction details may vary. There are usually two igniters mounted on the boss provided on each of the chamber housings. The igniters must be long enough to protrude from the housing into the combustion chamber.

The burners are interconnected by projecting flame tubes which facilitate the engine-starting process as mentioned previously in the can-type combustion chamber familiarization. The flame tubes function identically to those previously discussed, differing only in construction details.

The can-annular combustion chamber is not used in modern

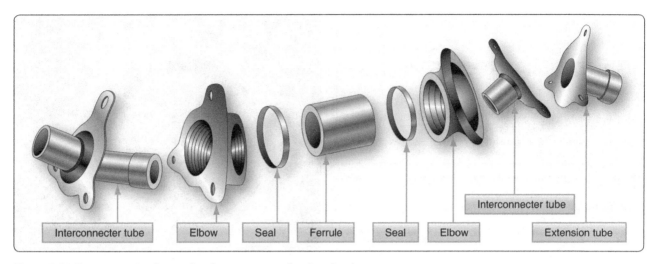

Figure 1-54. *Interconnecting flame tubes for can-type combustion chambers.*

engines. The forward face of each chamber presents six apertures, which align with the six fuel nozzles of the corresponding fuel nozzle cluster. *[Figure 1-56]* These nozzles are the dual-orifice (duplex) type requiring the use of a flow-divider (pressurizing valve), as mentioned in the can-type combustion chamber discussion. Around each nozzle are preswirl vanes for imparting a swirling motion to the fuel spray, which results in better atomization of the fuel, better burning, and efficiency. The swirl vanes function to provide two effects imperative to proper flame propagation:

1. High flame speed—better mixing of air and fuel, ensuring spontaneous burning.

2. Low air velocity axially—swirling eliminates overly rapid flame movement axially.

The swirl vanes greatly aid flame propagation, since a high degree of turbulence in the early combustion and cooling stages is desirable. The vigorous mechanical mixing of the fuel vapor with the primary air is necessary, since mixing by diffusion alone is too slow. This same mechanical mixing is also established by other means, such as placing coarse

screens in the diffuser outlet, as is the case in most axial-flow engines.

The can-annular combustion chambers also must have the required fuel drain valves located in two or more of the bottom chambers, assuring proper drainage and elimination of residual fuel burning at the next start.

The flow of air through the holes and louvers of the can-annular chambers, is almost identical with the flow through other types of burners. *[Figure 1-56]* Special baffling is used to swirl the combustion airflow and to give it turbulence. *Figure 1-57* shows the flow of combustion air, metal cooling air, and the diluent or gas cooling air. The air flow direction

Figure 1-55. *Annular combustion chamber liner.*

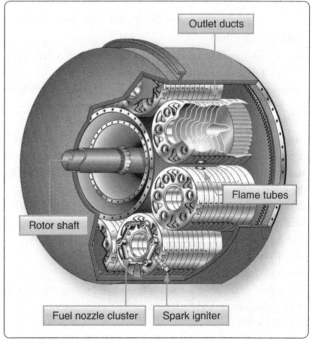

Figure 1-56. *Can-annular combustion chamber components and arrangement.*

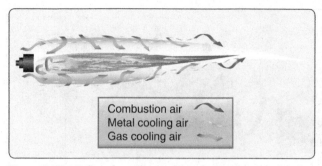

Combustion air
Metal cooling air
Gas cooling air

Figure 1-57. *Airflow through a can-annular combustion chamber.*

is indicated by the arrows.

The basic components of an annular combustion chamber are a housing and a liner, as in the can type. The liner consists of an undivided circular shroud extending all the way around the outside of the turbine shaft housing. The chamber is constructed of heat-resistant materials, which are sometimes coated with thermal barrier materials, such as ceramic materials. The annular combustion chamber is illustrated in *Figure 1-58*. Modern turbine engines usually have an annular combustion chamber. As can be seen in *Figure 1-59*, the annular combustion chamber also uses louvers and holes to prevent the flame from contacting the side of the combustion chamber.

A reverse-flow combustor is a type of combustor in which the air from the compressor enters the combustor outer case and reverses its direction as it flows into the inner liner. It again reverses its direction before it flows through the turbine. Reverse-flow combustors are used where engine length is critical.

Turbine Section

The turbine transforms a portion of the kinetic (velocity) energy of the exhaust gases into mechanical energy to drive the gas generator compressor and accessories. The sole purpose of the gas generator turbine is to absorb approximately 60 to 70 percent of the total pressure energy from the exhaust gases. The exact amount of energy absorption at the turbine is determined by the load the turbine is driving (i.e., compressor size and type, number of accessories, and the load applied by the other turbine stages). These turbine stages can be used to drive a low-pressure compressor (fan), propeller, and shaft. The turbine section of a gas turbine engine is located aft, or downstream, of the combustion chamber. Specifically, it is directly behind the combustion chamber outlet.

The turbine assembly consists of two basic elements: turbine inlet guide vanes and turbine disc. *[Figures 1-60 and 1-61]* The stator element is known by a variety of names, of which turbine inlet nozzle vanes, turbine inlet guide vanes, and nozzle diaphragm are three of the most commonly used.

Figure 1-58. *Annular combustion chamber with ceramic coating.*

Figure 1-59. *Combustion chamber louvers and holes.*

Figure 1-60. *Turbine inlet guide vanes.*

The turbine inlet nozzle vanes are located directly aft of the combustion chambers and immediately forward of the turbine wheel. This is the highest or hottest temperature that comes in contact with metal components in the engine. The turbine inlet temperature must be controlled, or damage will occur to the turbine inlet vanes.

After the combustion chamber has introduced the heat energy into the mass airflow and delivered it evenly to the turbine

Figure 1-61. *Turbine disc.*

inlet nozzles, the nozzles must prepare the mass air flow to drive the turbine rotor. The stationary vanes of the turbine inlet nozzles are contoured and set at such an angle that they form a number of small nozzles discharging gas at extremely high speed; thus, the nozzle converts a varying portion of the heat and pressure energy to velocity energy that can then be converted to mechanical energy through the turbine blades.

There are three types of turbine blades: the impulse turbine blade, reaction turbine blade, and the reaction-impulse turbine blade. The impulse turbine blade is also referred to as a bucket. This is because as the stream of air strikes the center of the blade it changes the direction of the energy as it causes the blades to rotate the disc and rotor shaft. The turbine nozzle guide vanes can usually be adjusted during engine overhaul and assembly in order to increase the efficiency of the air stream striking the blades or buckets of the turbine. *[Figure 1-62]*

Reaction turbine blades cause the disc to rotate by the aerodynamic action of the airstream directed to flow past the blade at a particular angle in order to develop the most efficient power from the turbine engine. *[Figure 1-62]*

The reaction-impulse turbine blade combines the action of both the impulse and reaction blades designs. The blade has more of the bucket shape of the impulse blade at the blade root and it also has more of an airfoil shape of the reaction blade on the second half of the blade toward the outer end of the blade.

The second purpose of the turbine inlet nozzle is to deflect the gases to a specific angle in the direction of turbine wheel rotation. Since the gas flow from the nozzle must enter the turbine blade passageway while it is still rotating, it is essential to aim the gas in the general direction of turbine rotation.

The turbine inlet nozzle assembly consists of an inner

shroud and an outer shroud between which the nozzle vanes are fixed. The number and size of inlet vanes employed vary with different types and sizes of engines. *Figure 1-63* illustrates typical turbine inlet nozzles featuring loose and welded vanes. The vanes of the turbine inlet nozzle may be assembled between the outer and inner shrouds or rings in a variety of ways. Although the actual elements may vary slightly in configuration and construction features, there is one characteristic peculiar to all turbine inlet nozzles: the nozzle vanes must be constructed to allow thermal expansion. Otherwise, there would be severe distortion or warping of the metal components because of rapid temperature changes.

The thermal expansion of turbine nozzles is accomplished by one of several methods. One method necessitates loose assembly of the supporting inner and outer vane shrouds. *[Figure 1-63A]*

Each vane fits into a contoured slot in the shrouds, which conforms to the airfoil shape of the vane. These slots are slightly larger than the vanes to give a loose fit. For further support, the inner and outer shrouds are encased by inner and outer support rings, which provide increased strength and rigidity. These support rings also facilitate removal of the nozzle vanes as a unit. Without the rings, the vanes could fall out as the shrouds were removed.

Another method of thermal expansion construction is to fit the vanes into inner and outer shrouds; however, in this method the vanes are welded or riveted into position. *[Figure 1-63B]* Some means must be provided to allow thermal expansion; therefore, either the inner or the outer shroud ring is cut into segments. The saw cuts separating the segments allow sufficient expansion to prevent stress and warping of the vanes.

The rotor element of the turbine section consists essentially of a shaft and a wheel. *[Figure 1-64]* The turbine wheel is a dynamically balanced unit consisting of blades attached to a rotating disc. The disc, in turn, is attached to the main power-transmitting shaft of the engine. The exhaust gases leaving the turbine inlet nozzle vanes act on the blades of the turbine wheel, causing the assembly to rotate at a very high rate of speed. The high rotational speed imposes severe centrifugal loads on the turbine wheel, and at the same time the elevated temperatures result in a lowering of the strength of the material. Consequently, the engine speed and temperature must be controlled to keep turbine operation within safe limits.

The turbine disc is referred to as such without blades. When the turbine blades are installed, the disc then becomes the turbine wheel. The disc acts as an anchoring component for the turbine blades. Since the disc is bolted or welded to the

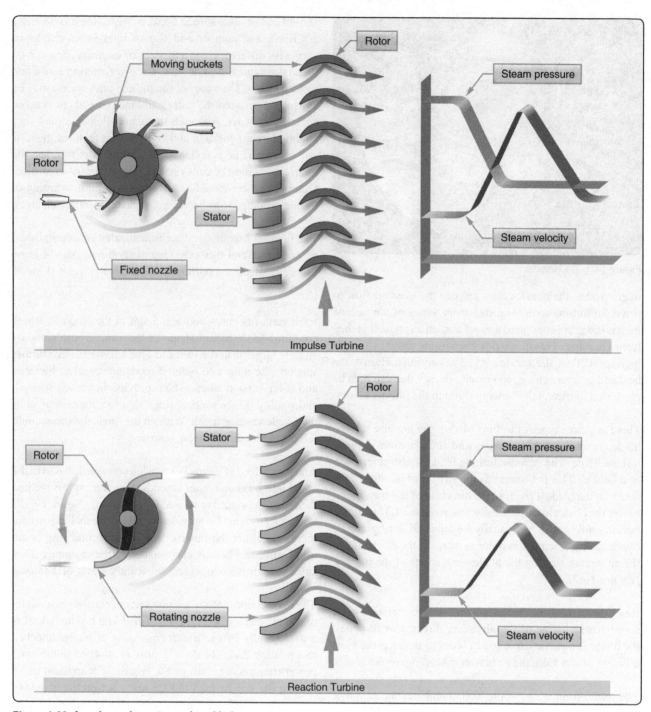

Figure 1-62. *Impulse and reaction turbine blades.*

shaft, the blades can transmit to the rotor shaft the energy they extract from the exhaust gases.

The disc rim is exposed to the hot gases passing through the blades and absorbs considerable heat from these gases. In addition, the rim also absorbs heat from the turbine blades by conduction. Hence, disc rim temperatures are normally high and well above the temperatures of the more remote inner portion of the disc. As a result of these temperature gradients, thermal stresses are added to the rotational stresses.

Additionally, turbine blades are generally more susceptible to operating damage than compressor blades due to the exposure of high temperatures. There are various methods to relieve, at least partially, the aforementioned stresses. One such method is to bleed cooling air back onto the face of the disc.

Another method of relieving the thermal stresses of the disc is incidental to blade installation. A series of grooves or notches, conforming to the blade root design, are broached in the rim of the disc. These grooves allow attachment of the

Figure 1-63. *Typical turbine nozzle vane assemblies.*

A. Turbine nozzle vane assembly with loose-fitting vanes

B. Turbine nozzle vane assembly with welded vanes

turbine blades to the disc; at the same time, space is provided by the notches for thermal expansion of the disc. Sufficient clearance exists between the blade root and the notch to permit movement of the turbine blade when the disc is cold. During engine operation, expansion of the disc decreases the clearance. This causes the blade root to fit tightly in the disc rim.

The turbine shaft is usually fabricated from alloy steel. *[Figure 1-64]* It must be capable of absorbing the high torque loads that are exerted on it.

The methods of connecting the shaft to the turbine disc vary. In one method, the shaft is welded to the disc, which has a butt or protrusion provided for the joint. Another method is by bolting. This method requires that the shaft have a hub that fits a machined surface on the disc face. Then, the bolts are inserted through holes in the shaft hub and anchored in tapped holes in the disc. Of the two connection methods, bolting is more common.

The turbine shaft must have some means for attachment to the compressor rotor hub. This is usually accomplished by a spline cut on the forward end of the shaft. The spline fits into a coupling device between the compressor and turbine shafts. If a coupling is not used, the splined end of the turbine shaft may fit into a splined recess in the compressor rotor hub. This splined coupling arrangement is used almost exclusively with centrifugal compressor engines, while axial compressor engines may use either of these described methods.

There are various ways of attaching turbine blades, some similar to compressor blade attachment. The most satisfactory method utilizes the fir-tree design. *[Figure 1-65]*

The blades are retained in their respective grooves by a variety of methods, the more common of which are peening, welding, lock tabs, and riveting. *Figure 1-66* shows a typical turbine wheel using rivets for blade retention.

The peening method of blade retention is used frequently in various ways. One of the most common applications of peening requires a small notch to be ground in the edge of the blade fir-tree root prior to the blade installation. After the blade is inserted into the disc, the notch is filled by the disc metal, which is "flowed" into it by a small punch-mark made in the disc adjacent to the notch. The tool used for this job is similar to a center punch.

Another method of blade retention is to construct the root of the blade so that it contains all the elements necessary for its retention. This method uses the blade root as a stop made on one end of the root so that the blade can be inserted and removed in one direction only, while on the opposite end is a tang. This tang is bent to secure the blade in the disc.

Turbine blades may be either forged or cast, depending on the composition of the alloys. Most blades are precision cast and finish ground to the desired shape. Many turbine blades are cast as a single crystal, which gives the blades better strength and heat properties. Heat barrier coating, such as ceramic coating, and air flow cooling help keep the turbine blades and inlet nozzles cooler. This allows the exhaust temperature to be raised, increasing the efficiency of the engine. *Figure 1-67*

Figure 1-64. *Rotor elements of the turbine assembly.*

shows a turbine blade with air holes for cooling purposes.

Most turbines are open at the outer perimeter of the blades; however, a second type called the shrouded turbine is sometimes used. The shrouded turbine blades, in effect, form a band around the outer perimeter of the turbine wheel. This improves efficiency and vibration characteristics and permits lighter stage weights. On the other hand, it limits turbine speed and requires more blades. *[Figure 1-68]*

In turbine rotor construction, it occasionally becomes necessary to utilize turbines of more than one stage. A single turbine wheel often cannot absorb enough power from the exhaust gases to drive the components dependent on the turbine for rotative power; thus, it is necessary to add additional turbine stages.

A turbine stage consists of a row of stationary vanes or nozzles, followed by a row of rotating blades. In some models of turboprop engine, as many as five turbine stages have been utilized successfully. It should be remembered that, regardless of the number of wheels necessary for driving engine components, there is always a turbine nozzle preceding each wheel.

As was brought out in the preceding discussion of turbine stages, the occasional use of more than one turbine wheel is warranted in cases of heavy rotational loads. It should also be pointed out that the same loads that necessitate multistage turbines often make it advantageous to incorporate multiple compressor rotors.

In the single-stage rotor turbine, the power is developed by one turbine rotor, and all engine-driven parts are driven by

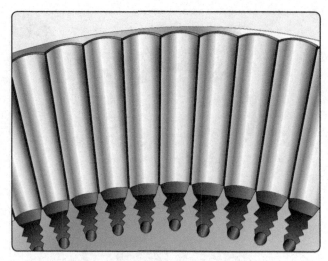

Figure 1-66. *Rivet method of turbine blade retention.*

this single wheel. *[Figure 1-69]* This arrangement is used on engines where the need for low weight and compactness predominates. This is the simplest version of the pure turbojet engine.

A multistage turbine is shown in *Figure 1-70.* In multiple spool engines, each spool has its own set of turbine stages. Each set of turbine stages turns the compressor attached to it. Most turbofan engines have two spools: low pressure (fan shaft a few stages of compression and the turbine to drive it) and high pressure (high pressure compressor shaft and high pressure turbine). *[Figure 1-71]*

The remaining element to be discussed concerning turbine familiarization is the turbine casing or housing. The turbine casing encloses the turbine wheel and the nozzle vane assembly, and at the same time gives either direct or indirect support to the stator elements of the turbine section. It always has flanges provided front and rear for bolting the assembly to the combustion chamber housing and the exhaust cone assembly, respectively. A turbine casing is illustrated in

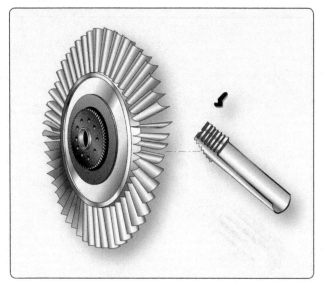

Figure 1-65. *Turbine blade with fir-tree design and lock-tab method of blade retention.*

Figure 1-67. *Turbine blade with cooling holes.*

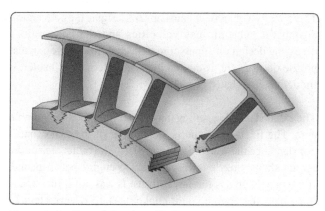

Figure 1-68. *Shrouded turbine blades.*

Figure 1-72.

Exhaust Section

The exhaust section of the gas turbine engine consists of several components. Although the components have individual purposes, they also have one common function: they must direct the flow of hot gases rearward in such a manner as to prevent turbulence and, at the same time, impart a high final or exit velocity to the gases. In performing the various functions, each of the components affects the flow of gases in different ways. The exhaust section is located directly behind the turbine section and ends when the gases are ejected at the rear in the form of high-velocity exhaust gases. The components of the exhaust section include the exhaust cone, tailpipe (if required), and the exhaust nozzle. The exhaust cone collects the exhaust gases discharged from the turbine section and gradually converts them into a solid flow of gases. In performing this, the velocity of the gases is decreased slightly and the pressure increased. This is due to the diverging passage between the outer duct and the inner cone; that is, the annular area between the two units increases rearward. The exhaust cone assembly consists of an outer shell or duct, an inner cone, three or four radial hollow struts or fins, and the necessary number of tie rods to aid the struts in supporting the inner cone from the outer duct.

The outer shell or duct is usually made of stainless steel and is attached to the rear flange of the turbine case. This element collects the exhaust gases and delivers them directly to the exhaust nozzle. The duct must be constructed to include such features as a predetermined number of thermocouple bosses for installing exhaust temperature thermocouples, and there must also be insertion holes for the supporting tie rods. In some cases, tie rods are not used for supporting the inner cone. If such is the case, the hollow struts provide the sole support of the inner cone, the struts being spot-welded in position to the inside surface of the duct and to the inner cone, respectively. *[Figure 1-73]* The radial struts actually have a twofold function. They not only support the inner cone in the

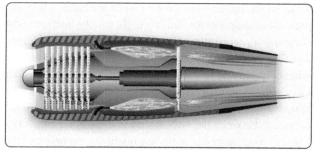

Figure 1-69. *Single-stage rotor turbine.*

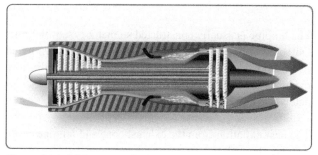

Figure 1-70. *Multirotor turbine.*

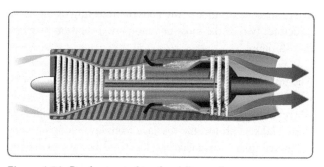

Figure 1-71. *Dual-rotor turbine for split-spool compressor.*

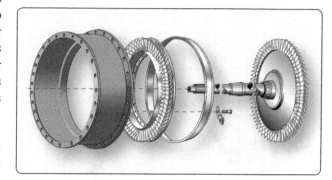

Figure 1-72. *Turbine casing assembly.*

exhaust duct, but they also perform the important function of straightening the swirling exhaust gases that would otherwise leave the turbine at an angle of approximately 45°.

The centrally located inner cone fits rather closely against the rear face of the turbine disc, preventing turbulence of the gases as they leave the turbine wheel. The cone is supported by the radial struts. In some configurations, a small hole is located in the exit tip of the cone. This hole allows cooling air to be circulated from the aft end of the cone, where the pressure of the gases is relatively high, into the interior of the cone and consequently against the face of the turbine wheel. The flow of air is positive, since the air pressure at the turbine wheel is relatively low due to rotation of the wheel; thus, air circulation is assured. The gases used for cooling the turbine wheel return to the main path of flow by passing through the clearance between the turbine disc and the inner cone. The exhaust cone assembly is the terminating component of the basic engine. The remaining component (the exhaust nozzle) is usually considered an airframe component.

The tailpipe is usually constructed so that it is semiflexible. On some tailpipes, a bellows arrangement is incorporated in its construction, allowing movement in installation, maintenance, and in thermal expansion. This eliminates stress and warping which would otherwise be present.

The heat radiation from the exhaust cone and tailpipe could damage the airframe components surrounding these units. For this reason, some means of insulation had to be devised. There are several suitable methods of protecting the fuselage structure; two of the most common are insulation blankets and shrouds.

The insulation blanket, illustrated in *Figures 1-74* and *1-75*, consists of several layers of aluminum foil, each separated by a layer of fiberglass or some other suitable material. Although these blankets protect the fuselage from heat radiation, they are used primarily to reduce heat loss from the exhaust system. The reduction of heat loss improves engine performance.

There are two types of exhaust nozzle designs: the converging design for subsonic gas velocities and the converging-diverging design for supersonic gas velocities. These exhaust nozzle designs are discussed in greater detail in Chapter 3, Induction & Exhaust Systems.

The exhaust nozzle opening may be of either fixed or variable area. The fixed-area type is the simpler of the two exhaust nozzles since there are no moving parts. The outlet area of the fixed exhaust nozzle is very critical to engine performance. If the nozzle area is too large, thrust is wasted; if the area is too small, the engine could choke or stall. A variable-area exhaust nozzle is used when an augmenter or afterburner is used due to the increased mass of flow when the afterburner is activated. It must increase its open area when the afterburner is selected. When the afterburner is off, the exhaust nozzle closes to a smaller area of opening.

Gas Turbine Engine Bearings & Seals

The main bearings have the critical function of supporting the main engine rotor. The number of bearings necessary for proper engine support is, for the most part, determined by the length and weight of the engine rotor. The length and weight are directly affected by the type of compressor used in the engine. Naturally, a two-spool compressor requires more bearing support. The minimum number of bearings required to support one shaft is one deep groove ball bearing (thrust and radial loads) and one straight roller bearing (radial load only). Sometimes, it is necessary to use more than one roller bearing if the shaft is subject to vibration or its length is excessive. The gas turbine rotors are supported by ball and roller bearings, which are antifriction bearings. *[Figure 1-76]* Many newer engines use hydraulic bearings, in which the outside race is surrounded by a thin film of oil. This reduces vibrations transmitted to the engine.

In general, antifriction bearings are preferred largely because they:

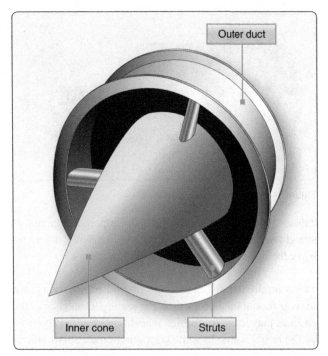

Figure 1-73. *Exhaust collector with welded support struts.*

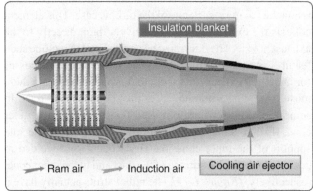

Figure 1-74. *Exhaust system insulation blanket.*

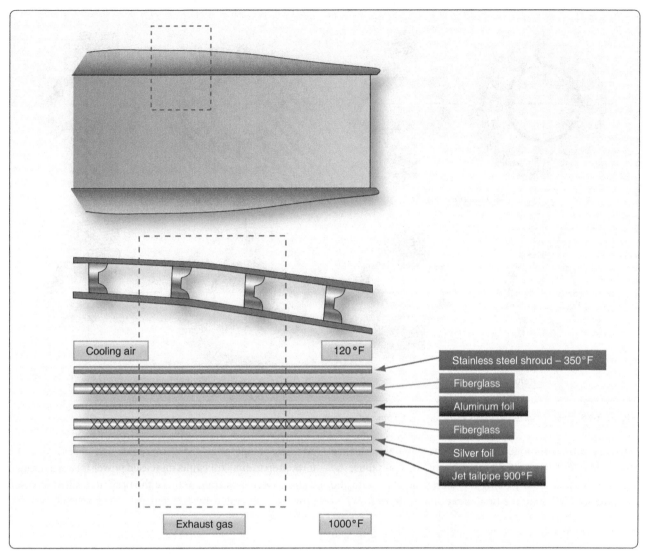

Figure 1-75. *Insulation blanket with the temperatures obtained at the various locations shown.*

- Offer little rotational resistance,

- Facilitate precision alignment of rotating elements,

- Are relatively inexpensive,

- Are easily replaced,

- Withstand high momentary overloads,

- Are simple to cool, lubricate, and maintain,

- Accommodate both radial and axial loads, and

- Are relatively resistant to elevated temperatures.

The main disadvantages are their vulnerability to foreign matter and tendency to fail without appreciable warning. Usually the ball bearings are positioned on the compressor or turbine shaft so that they can absorb any axial (thrust) loads or radial loads. Because the roller bearings present a larger working surface, they are better equipped to support radial loads than thrust loads. Therefore, they are used primarily for this purpose. A typical ball or roller bearing assembly includes a bearing support housing, which must be strongly constructed and supported in order to carry the radial and axial loads of the rapidly rotating rotor. The bearing housing usually contains oil seals to prevent the oil leaking from its normal path of flow. It also delivers the oil to the bearing for its lubrication, usually through spray nozzles. The oil seals may be the labyrinth or thread (helical) type. These seals also may be pressurized to minimize oil leaking along the compressor shaft. The labyrinth seal is usually pressurized, but the helical seal depends solely on reverse threading to stop oil leakage. These two types of seals are very similar, differing only in thread size and the fact that the labyrinth seal is pressurized.

Another type of oil seal used on some of the later engines is the carbon seal. These seals are usually spring loaded and are similar in material and application to the carbon brushes

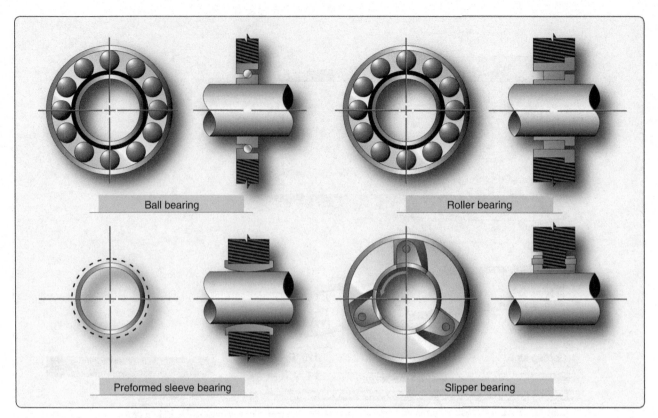

Figure 1-76. *Types of main bearings used for gas turbine rotor support.*

used in electrical motors. Carbon seals rest against a surface provided to create a sealed bearing cavity or void; thus, the oil is prevented from leaking out along the shaft into the compressor airflow or the turbine section. *[Figure 1-77]*

The ball or roller bearing is fitted into the bearing housing and may have a self-aligning feature. If a bearing is self-aligning, it is usually seated in a spherical ring. This allows the shaft a certain amount of radial movement without transmitting stress to the bearing inner race.

The bearing surface is usually provided by a machined journal on the appropriate shaft. The bearing is usually locked in position by a steel snap ring or other suitable locking device. The rotor shaft also provides the matching surface for the oil seals in the bearing housing. These machined surfaces are called lands and fit in rather close to the oil seal.

Turboprop Engines

The turbopropeller (turboprop) engine is a combination of a gas turbine engine, reduction gear box, and a propeller. *[Figure 1-78]* Turboprops are basically gas turbine engines that have a compressor, combustion chamber(s), turbine, and an exhaust nozzle (gas generator), all of which operate in the same manner as any other gas engine. However, the difference is that the turbine in the turboprop engine usually has extra stages to extract energy to drive the propeller. In addition to operating the compressor and accessories, the turboprop turbine transmits increased power forward through a shaft and a gear train to drive the propeller. The increased power is generated by the exhaust gases passing through additional stages of the turbine.

Some engines use a multirotor turbine with coaxial shafts for independent driving of the compressor and propeller. Although there are three turbines utilized in this illustration, as many as five turbine stages have been used for driving the two rotor elements, propeller, and accessories.

The exhaust gases also contribute to engine power output through thrust production, although the amount of energy available for thrust is considerably reduced. Two basic types of turboprop engine are in use: fixed turbine and free turbine. The fixed turbine has a mechanical connection from the gas generator (gas-turbine engine) to the reduction gear box and propeller. The free turbine has only an air link from gas generator to the power turbines. There is no mechanical link from the propeller to the gas turbine engine (gas generator). There are advantages and disadvantages of each system, with the airframe generally dictating the system used.

Since the basic components of normal gas-turbine and turboprop engines differ slightly only in design features, it should be fairly simple to apply acquired knowledge of the

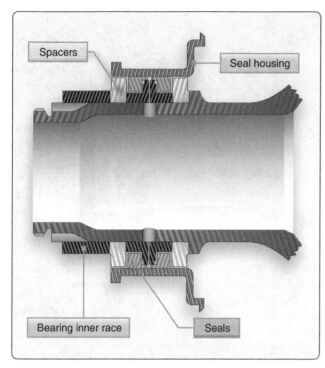

Figure 1-77. *Carbon oil seal.*

basic gas turbine to the turboprop.

The typical turboprop engine can be broken down into assemblies as follows:

1. The power section assembly—contains the usual major components of a gas turbine engine (i.e., compressor, combustion chamber, turbine, and exhaust sections).

2. The reduction gear or gearbox assembly—contains

those sections unique to turboprop configurations.

3. The torquemeter assembly—transmits the torque from the engine to the gearbox of the reduction section.

4. The accessory drive housing assembly—mounted on the bottom of the compressor air inlet housing. It includes the necessary gear trains for driving all power section driven accessories at their proper rpm in relation to engine rpm.

Turboshaft Engines

A gas-turbine engine that delivers power through a shaft to operate something other than a propeller is referred to as a turboshaft engine. *[Figure 1-79]* The output shaft may be coupled directly to the engine turbine, or the shaft may be driven by a turbine of its own (free turbine) located in the exhaust stream. As mentioned with the turboprop, the free turbine rotates independently. This principle is used extensively in current production of turboshaft engines. The turboshaft engine's output is measured in horsepower instead of thrust because the power output is a turning shaft.

Turbofan Engines

The turbofan gas turbine engine is, in principle, the same as a turboprop, except that the propeller is replaced by a duct-enclosed axial-flow fan. *[Figure 1-80]* The fan can be a part of the first-stage compressor blades or can be mounted as a separate set of fan blades. The blades are mounted forward of the compressor.

The general principle of the fan engine is to convert more of the fuel energy into pressure. With more of the energy

Figure 1-78. *PT6 turboprop engine.*

Figure 1-79. *Turboshaft engine.*

converted to pressure, a greater product of pressure times area can be achieved. One of the major advantages is turbofan production of this additional thrust without increasing fuel flow. The end result is fuel economy with the consequent increase in range. Because more of the fuel energy is turned into pressure in the turbofan engine, additional stages must be added in the turbine section to provide the power to drive the fan. This means there is less energy left over and less thrust from the core exhaust gases. Also, in a mixed-exhaust nozzle (where fan air and core air mix in a common nozzle before entering ambient conditions) the exhaust nozzle must be larger in area. The result is that the fan develops most of the thrust. The thrust produced by the fan more than makes up for the decrease in thrust of the core (gas generator) of the engine. Depending on the fan design and bypass ratio, it produces 80 percent of the turbofan engine's total thrust.

Two different exhaust nozzle designs are used with turbofan engines. The air leaving the fan can be ducted overboard by a separate fan nozzle *[Figure 1-43]*, or it can be ducted along the outer case of the basic engine to be discharged through the mixed nozzle (core and fan exhaust together). The fan air is either mixed with the exhaust gases before it is discharged (mixed or common nozzle), or it passes directly to the atmosphere without prior mixing (separate nozzle). Turbofans are the most widely used gas turbine engine for air transport aircraft. The turbofan is a compromise between the good

operating efficiency and high thrust capability of a turboprop and the high speed, high altitude capability of a turbojet.

Turbine Engine Operating Principles

The principle used by a gas turbine engine as it provides force to move an airplane is based on Newton's Third Law. This law states that for every action there is an equal and opposite reaction; therefore, if the engine accelerates a mass of air (action), it applies a force on the aircraft (reaction). The turbofan generates thrust by giving a relatively slower acceleration to a large quantity of air. The old pure turbojet engine achieves thrust by imparting greater acceleration to a smaller quantity of air. This was its main problem with fuel consumption and noise.

The mass of air is accelerated within the engine by the use of a continuous-flow cycle. Ambient air enters the inlet diffuser where it is subjected to changes in temperature, pressure, and velocity due to ram effect. The compressor then increases pressure and temperature of the air mechanically. The air continues at constant pressure to the burner section where its temperature is increased by combustion of fuel. The energy is taken from the hot gas by expanding through a turbine which drives the compressor, and by expanding through an exhaust nozzle designed to discharge the exhaust gas at high velocity to produce thrust.

Figure 1-80. *Turbofan engine.*

The high velocity gases from the engine may be considered continuous, imparting this force against the aircraft in which it is installed, thereby producing thrust. The formula for thrust can be derived from Newton's second law, which states that force is proportional to the product of mass and acceleration. This law is expressed in the following formula:

$$F = M \times A$$

where:

F = force in pounds
M = mass in pounds per second
A = acceleration in feet per second

In the above formula, mass is similar to weight, but it is actually a different quantity. Mass refers to the quantity of matter, while weight refers to the pull of gravity on that quantity of matter. At sea level under standard conditions, 1 pound of mass has a weight of 1 pound. To calculate the acceleration of a given mass, the gravitational constant is used as a unit of comparison. The force of gravity is 32.2 feet per second squared (ft/sec^2). This means that a free falling 1 pound object accelerates at the rate of 32.2 feet per second each second that gravity acts on it. Since the object mass weighs 1 pound, which is also the actual force imparted to it by gravity, it can be assumed that a force of 1 pound accelerates a 1 pound object at the rate of 32.2 ft/sec^2.

Also, a force of 10 pound accelerates a mass of 10 pound at the rate of 32.2 ft/sec^2. This is assuming there is no friction or other resistance to overcome. It is now apparent that the ratio of the force (in pounds) is to the mass (in pounds) as the acceleration in ft/sec^2 is to 32.2. Using M to represent the mass in pounds, the formula may be expressed thus:

$$\frac{F}{M} = \frac{A}{G} \text{ or } F = \frac{MA}{G}$$

where:

F = force
M = mass
A = acceleration
G = gravity

In any formula involving work, the time factor must be considered. It is convenient to have all time factors in equivalent units (i.e., seconds, minutes, or hours). In calculating jet thrust, the term "pounds of air per second" is convenient, since the second is the same unit of time used for the force of gravity.

Thrust

Using the following formula, compute the force necessary to accelerate a mass of 50 pounds by 100 ft/sec^2.

$$F = \frac{MA}{G}$$

$$F = \frac{50 \text{ lb} \times 100 \text{ ft/sec}^2}{32.2 \text{ ft/sec}^2}$$

$$F = \frac{5{,}000 \text{ lb-ft/sec}^2}{32.2 \text{ ft/sec}^2}$$

$$F = 155 \text{ lb}$$

This illustrates that if the velocity mass per second is increased by 100, the resulting thrust is 155 pounds.

Since the turbojet engine accelerates air, the following formula can be used to determine jet thrust:

$$F = \frac{Ms (V_2 - V_1)}{G}$$

where:

F = force in pounds

Ms = mass flow in lb/sec

V_1 = inlet velocity

V_2 = jet velocity (exhaust)

$V_2 - V_1$ = change in velocity; difference between inlet velocity and jet velocity

G = acceleration of gravity or 32.2 ft/sec^2

As an example, to use the formula for changing the velocity of 100 pounds of mass airflow per second from 600 ft/sec to 800 ft/sec, the formula can be applied as follows:

$$F = \frac{100 \text{ lb/sec} (800 \text{ ft/sec} - 600 \text{ ft/sec})}{32.2 \text{ ft/sec}^2}$$

$$F = \frac{20{,}000 \text{ lb/sec}}{32.2 \text{ ft/sec}^2}$$

$$F = 621 \text{ lb}$$

As shown by the formula, if the mass airflow per second and the difference in the velocity of the air from the intake to the exhaust are known, it is easy to compute the force necessary to produce the change in the velocity. Therefore, the thrust of the engine must be equal to the force required to accelerate the air mass through the engine. Then, by using the symbol "Fn" for thrust pounds, the formula becomes:

$$Fn = \frac{Ms (V_2 - V_1)}{G}$$

Thrust of a gas turbine engine can be increased by two methods: increasing the mass flow of air through the engine or increasing the gas velocity. If the velocity of the turbojet engine remains constant with respect to the aircraft, the thrust decreases if the speed of the aircraft is increased. This is because V_1 increases in value. This does not present a serious problem, however, because as the aircraft speed increases, more air enters the engine, and jet velocity increases. The resultant net thrust is almost constant with increased airspeed.

The Brayton cycle is the name given to the thermodynamic cycle of a gas turbine engine to produce thrust. This is a variable volume constant-pressure cycle of events and is commonly called the constant-pressure cycle. A more recent term is "continuous combustion cycle." The four continuous and constant events are intake, compression, expansion (includes power), and exhaust. These cycles are discussed as they apply to a gas-turbine engine. In the intake cycle, air enters at ambient pressure and a constant volume. It leaves the intake at an increased pressure and a decrease in volume. At the compressor section, air is received from the intake at an increased pressure, slightly above ambient, and a slight decrease in volume. Air enters the compressor where it is compressed. It leaves the compressor with a large increase in pressure and decrease in volume, created by the mechanical action of the compressor. The next step, expansion, takes place in the combustion chamber by burning fuel, which expands the air by heating it. The pressure remains relatively constant, but a marked increase in volume takes place. The expanding gases move rearward through the turbine assembly and are converted from velocity energy to mechanical energy by the turbine. The exhaust section, which is a convergent duct, converts the expanding volume and decreasing pressure of the gases to a final high velocity. The force created inside the engine to keep this cycle continuous has an equal and opposite reaction (thrust) to move the aircraft forward.

Bernoulli's principle (whenever a stream of any fluid has its velocity increased at a given point, the pressure of the stream at that point is less than the rest of the stream) is applied to gas turbine engines through the design of convergent and divergent air ducts. The convergent duct increases velocity and decreases pressure. The divergent duct decreases velocity and increases pressure. The convergent principle is usually used for the exhaust nozzle. The divergent principle is used in the compressor and diffuser where the air is slowing and pressurizing.

Gas Turbine Engine Performance

Thermal efficiency is a prime factor in gas turbine performance. It is the ratio of net work produced by the engine to the chemical energy supplied in the form of fuel. The three most important factors affecting the thermal efficiency are turbine inlet temperature, compression ratio, and the component efficiencies of the compressor and turbine. Other factors that affect thermal efficiency are compressor inlet temperature and combustion efficiency. *Figure 1-81* shows the effect that changing compression ratio (compressor pressure ratio) has on thermal efficiency when compressor inlet temperature and the component efficiencies of the compressor and turbine remain constant. The effects that compressor and turbine component efficiencies have on thermal efficiency when turbine and compressor inlet temperatures remain constant are shown in *Figure 1-82*. In actual operation, the turbine engine exhaust temperature varies directly with turbine inlet temperature at a constant compression ratio.

Rpm is a direct measure of compression ratio; therefore, at constant rpm, maximum thermal efficiency can be obtained by maintaining the highest possible exhaust temperature. Since engine life is greatly reduced at high turbine inlet temperatures, the operator should not exceed the exhaust temperatures specified for continuous operation. *Figure 1-83* illustrates the effect of turbine inlet temperature on turbine blade life. In the previous discussion, it was assumed that the state of the air at the inlet to the compressor remains constant. Since this is a practical application of a turbine engine, it becomes necessary to analyze the effect of varying inlet conditions on the thrust or power produced. The three principal variables that affect inlet conditions are the speed of the aircraft, the altitude of the aircraft, and the ambient temperature. To make the analysis simpler, the combination of these three variables can be represented by a single variable called stagnation density.

The power produced by a turbine engine is proportional to the stagnation density at the inlet. The next three illustrations show how changing the density by varying altitude, airspeed, and outside air temperature affects the power level of the engine. *Figure 1-84* shows that the thrust output improves rapidly with a reduction in outside air temperature (OAT) at constant altitude, rpm, and airspeed. This increase occurs partly because the energy required per pound of airflow to drive the compressor varies directly with the temperature, leaving more energy to develop thrust. In addition, the thrust output increases since the air at reduced temperature has an increased density. The increase in density causes the mass flow through the engine to increase. The altitude effect on thrust, as shown in *Figure 1-85*, can also be discussed as a density and temperature effect. In this case, an increase in altitude causes a decrease in pressure and temperature.

Since the temperature lapse rate is lower than the pressure lapse rate as altitude is increased, the density is decreased. Although the decreased temperature increases thrust, the effect of decreased density more than offsets the effect of the colder temperature. The net result of increased altitude is a reduction in the thrust output.

The effect of airspeed on the thrust of a gas-turbine engine is shown in *Figure 1-86*. To explain the airspeed effect, it is necessary to understand first the effect of airspeed on the factors that combine to produce net thrust: specific thrust and engine airflow. Specific thrust is the net thrust in pounds developed per pound of airflow per second. It is the remainder of specific gross thrust minus specific ram drag. As airspeed is increased, ram drag increases rapidly. The exhaust velocity remains relatively constant; thus, the effect of the increase in

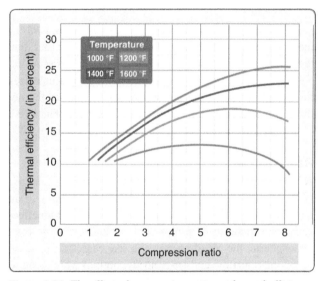

Figure 1-81. *The effect of compression ratio on thermal efficiency.*

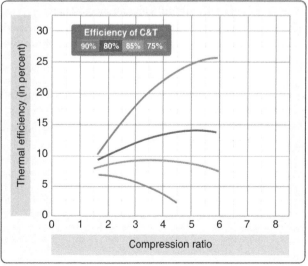

Figure 1-82. *Turbine and compressor efficiency vs. thermal efficiency.*

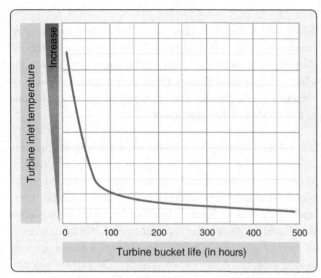

Figure 1-83. *Effect of turbine inlet temperature on turbine bucket life.*

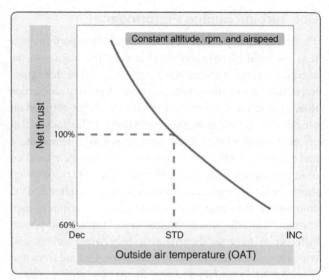

Figure 1-84. *Effect of OAT on thrust output.*

airspeed results in decreased specific thrust. *[Figure 1-86]* In the low-speed range, the specific thrust decreases faster than the airflow increases and causes a decrease in net thrust. As the airspeed increases into the higher range, the airflow increases faster than the specific thrust decreases and causes the net thrust to increase until sonic velocity is reached. The effect of the combination on net thrust is illustrated in *Figure 1-87*.

Ram Recovery

A rise in pressure above existing outside atmospheric pressure at the engine inlet, as a result of the forward velocity of an aircraft, is referred to as ram pressure. Since any ram effect causes an increase in compressor entrance pressure over atmospheric, the resulting pressure rise causes an increase in the mass airflow and gas velocity, both of which tend to increase thrust. Although ram effect increases engine thrust, the thrust being produced by the engine decreases for a given throttle setting as the aircraft gains airspeed. Therefore, two opposing trends occur when an aircraft's speed is increased. What actually takes place is the net result of these two different effects.

An engine's thrust output temporarily decreases as aircraft speed increases from static, but soon ceases to decrease. Moving toward higher speeds, thrust output begins to increase again due to the increased pressure of ram recovery.

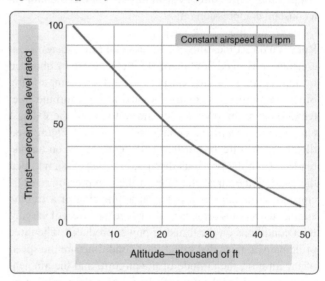

Figure 1-85. *Effect of altitude on thrust output.*

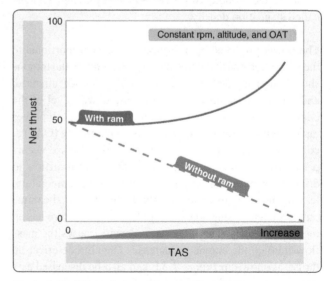

Figure 1-86. *Effect of airspeed on net thrust.*

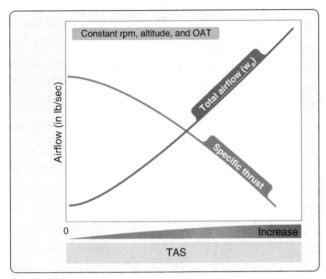

Figure 1-87. *Effect of airspeed on specific thrust and total engine airflow.*

Chapter 2
Engine Fuel & Fuel Metering Systems

Fuel System Requirements

The engine fuel system must supply fuel to the engine's fuel metering device under all conditions of ground and air operation. It must function properly at constantly changing altitudes and in any climate. The most common fuels are AVGAS for reciprocating engines and Jet A for turbine engines. AVGAS is generally either 80 (red) or 100LL (blue) octane. The LL stands for low lead although it contains four times the lead of 80 octane AVGAS. Jet A is a kerosene-based fuel that is clear to straw in color.

Electronic engine controls have allowed great increases in controlling the metered fuel flow to the engine. Engine fuel systems have become very accurate at providing the correct mixture of fuel and air to the engines. Gas turbine fuel controls have also greatly improved the ability to schedule (meter) the fuel correctly during all flight regimes. Improvements in electronics and the use of digital computers have enabled the aircraft and engines to be electronically interfaced together. By the use of electronic sensors and computer logic built in to electronic controls, the engines can be controlled with much more accuracy. Fuel cost and availability have also become factors in providing engines with fuel systems that are efficient and very precise in scheduling fuel flow to the engine. Many engines use an interactive system that senses engine parameters and feeds the information to the onboard computer (electronic engine control). The computer determines the amount of fuel needed and then sends a signal to the metering device. This signal sent to the metering device determines the correct amount of fuel needed by the engine. Electronic controls have become quite common with gas turbines and have increased the capabilities of the fuel system, making it less complicated for the technician and decreasing maintenance problems.

Engine fuel systems can be fairly complicated, yet some are quite simple, such as on small aircraft with a simple gravity-feed fuel system. This system, consisting of a tank to supply fuel to the engine, is often installed in the overhead wing and feeds a small float-type carburetor. On multiengine aircraft, complex systems are necessary so that fuel can be pumped from any combination of tanks to any combination of engines through a crossfeed system. Provisions for transferring fuel from one tank to another may also be included on large aircraft.

Vapor Lock

All fuel systems should be designed so that vapor lock cannot take place. Older gravity-feed systems were more prone to vapor lock. The fuel system should be free of tendency to vapor lock, which can result from changes in ground and in-flight climatic conditions. Normally, the fuel remains in a liquid state until it is discharged into the air stream and then instantly changes to a vapor. Under certain conditions, the fuel may vaporize in the lines, pumps, or other units. The vapor pockets formed by this premature vaporization restrict the fuel flow through units which are designed to handle liquids rather than gases. The resulting partial or complete interruption of the fuel flow is called vapor lock. The three general causes of vapor lock are the lowering of the pressure on the fuel, high fuel temperatures, and excessive fuel turbulence.

At high altitudes, the pressure on the fuel in the tank is low. This lowers the boiling point of the fuel and causes vapor bubbles to form. This vapor trapped in the fuel may cause vapor lock in the fuel system.

Transfer of heat from the engine tends to cause boiling of the fuel in the lines and the pump. This tendency is increased if the fuel in the tank is warm. High fuel temperatures often combine with low pressure to increase vapor formation. This is most apt to occur during a rapid climb on a hot day. As the aircraft climbs, the outside temperature drops, but the fuel does not lose temperature rapidly. If the fuel is warm enough at takeoff, it retains enough heat to boil easily at high altitude. The chief causes of fuel turbulence are sloshing of the fuel in the tanks, the mechanical action of the engine-driven pump, and sharp bends or rises in the fuel lines. Sloshing in the tank tends to mix air with the fuel. As this mixture passes through the lines, the trapped air separates from the fuel and forms vapor pockets at any point where there are abrupt changes in direction or steep rises. Turbulence in the fuel pump often combines with the low pressure at the pump inlet to form a vapor lock at this point.

Vapor lock can become serious enough to block the fuel flow completely and stop the engine. Even small amounts of vapor in the inlet line restrict the flow to the engine-driven pump and reduce its output pressure. To reduce the possibility of vapor lock, fuel lines are kept away from sources of heat; also, sharp bends and steep rises are avoided. In addition, the volatility of the fuel is controlled in manufacture so that it does not vaporize too readily. The major improvement in reducing vapor lock, however, is the incorporation of booster pumps in the fuel system. These booster pumps, which are used widely in most modern aircraft, keep the fuel in the lines to the engine-driven pump under pressure. The pressure on the fuel reduces vapor formation and aids in moving a vapor pocket along. The boost pump also releases vapor from the fuel as it passes through the pump. The vapor moves upward through the fuel in the tank and out the tank vents. To prevent the small amount of vapor that remains in the fuel from upsetting its metering action, vapor eliminators are installed in some fuel systems ahead of the metering device or are built into this unit.

Basic Fuel System

The basic parts of a fuel system include tanks, boost pumps, lines, selector valves, strainers, engine-driven pumps, and pressure gauges. A review of fuel systems in the Aviation Maintenance Technician—General Handbook provides some information concerning these components.

Generally, there are several tanks, even in a simple system, to store the required amount of fuel. The location of these tanks depends on both the fuel system design and the structural design of the aircraft. From each tank, a line leads to the selector valve. This valve is set from the flight deck to select the tank from which fuel is to be delivered to the engine. The boost pump forces fuel through the selector valve to the main line strainer. This filtering unit, located in the lowest part of the system, removes water and dirt from the fuel. During starting, the boost pump forces fuel through a bypass in the engine-driven pump to the metering device. Once the engine-driven pump is rotating at sufficient speed, it takes over and delivers fuel to the metering device at the specified pressure.

The airframe fuel system begins with the fuel tank and ends at the engine fuel system. The engine fuel system usually includes the engine-driven pumps and the fuel metering systems. In aircraft powered with a reciprocating engine, the engine-driven fuel pump and metering system consists of the main components from the point at which the fuel enters the first control unit until the fuel is injected into the intake pipe or cylinder. For example, the engine fuel system of a typical engine has an engine-driven fuel pump, the air-fuel control unit (metering device), the fuel manifold valve, and the fuel discharge nozzles. The fuel metering system on current reciprocating engines meters the fuel at a predetermined ratio to airflow. The airflow to the engine is controlled by the carburetor or air-fuel control unit.

The fuel metering system of the typical gas turbine engine consists of an engine-driven pump, fuel flow transmitter, fuel control with an electronic engine control, a distribution system or manifold, flow divider, and fuel discharge nozzles. On some turboprop engines, a fuel heater and a start control is a part of the engine fuel system. The rate of fuel delivery can be a function of air mass flow, compressor inlet temperature, compressor discharge pressure, compressor revolutions per minute (rpm), exhaust gas temperature, and combustion chamber pressure.

Fuel Metering Devices for Reciprocating Engines

Basic principles of operation are discussed here with no attempt being made to give detailed maintenance instructions. For the specific information needed to inspect or maintain a particular installation or unit, consult the manufacturer's instructions.

The basic requirement of a reciprocating fuel metering system is the same, regardless of the type of system used or the model engine on which the equipment is installed. It must meter fuel proportionately to air to establish the proper air-fuel mixture ratio for the engine at all speeds and altitudes at which the engine may be operated. In the air-fuel mixture curves shown in *Figure 2-1*, note that the basic best power and best economy air-fuel mixture requirements for reciprocating engines are approximately the same. The fuel metering system must atomize and distribute the fuel from the carburetor into the mass airflow. This must be accomplished so that the air-fuel charges going to all cylinders holds equal amounts of fuel. Each one of the engine's cylinders should receive the same quantity of air-fuel mixture and at the same air-fuel ratio.

Due to the drop in atmospheric pressure as altitude is increased, the density of the air also decreases. A normally-aspirated engine has a fixed amount or volume of air that it can draw in during the intake stroke, therefore less air is drawn into the engine as altitude increases. Less air tends to make carburetors run richer at altitude than at ground level, because of the decreased density of the airflow through the carburetor throat for a given volume of air. Thus, it is necessary that a mixture control be provided to lean the mixture and compensate for this natural enrichment. Some aircraft use carburetors in which the mixture control is operated manually. Other aircraft employ carburetors which automatically lean the carburetor mixture at altitude to maintain the proper air-fuel mixture.

The rich mixture requirements for an aircraft engine are established by running a power curve to determine the air-fuel mixture for obtaining maximum usable power. This curve is plotted at 100 rpm intervals from idle speed to

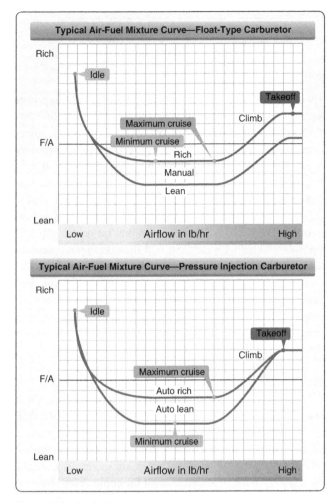

Figure 2-1. *Air-fuel mixture curves.*

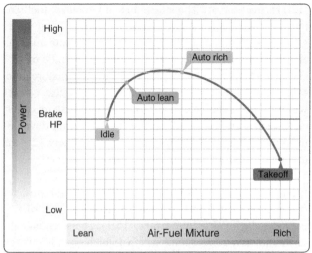

Figure 2-2. *Power versus air-fuel mixture curve.*

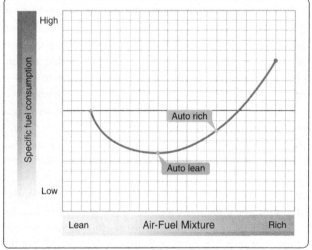

Figure 2-3. *Specific fuel consumption curve.*

takeoff speed. *[Figure 2-2]* Since it is necessary in the power range to add fuel to the basic air-fuel mixture requirements to keep cylinder-head temperatures in a safe range, the fuel mixture must become gradually richer as powers above cruise are used. *[Figure 2-1]* In the power range, the engine runs on a much leaner mixture, as indicated in the curves. However, on the leaner mixture, cylinder-head temperature would exceed the maximum permissible temperatures and detonation would occur.

The best economy setting is established by running a series of curves through the cruise range, as shown in the graph in *Figure 2-3,* the low point (auto-lean) in the curve being the air-fuel mixture where the minimum fuel per horsepower is used. In this range the engine operates normally on slightly leaner mixtures and obviously operates on richer mixtures than the low-point mixture. If a mixture leaner than that specified for the engine is used, the leanest cylinder of the engine is apt to backfire because the slower burning rate of the lean mixture results in a continued burning in the cylinder when the next intake stroke starts.

Air-Fuel Mixtures

Gasoline and other liquid fuels do not burn at all unless they are mixed with air. If the mixture is to burn properly within the engine cylinder, the ratio of air to fuel must be kept within a certain range. It would be more accurate to state that the fuel is burned with the oxygen in the air. Seventy-eight percent of air by volume is nitrogen, which is inert and does not participate in the combustion process, and 21 percent is oxygen. Heat is generated by burning the mixture of gasoline and oxygen. Nitrogen and gaseous byproducts of combustion absorb this heat energy and turn it into power by expansion. The mixture proportion of fuel and air by weight is of extreme importance to engine performance. The characteristics of a given mixture can be measured in terms of flame speed and combustion temperature.

The composition of the air-fuel mixture is described by the mixture ratio. For example, a mixture with a ratio of 12 to 1

(12:1) is made up of 12 pounds of air and 1 pound of fuel. The ratio is expressed in weight because the volume of air varies greatly with temperature and pressure. The mixture ratio can also be expressed as a decimal. Thus, an air-fuel ratio of 12:1 and an air-fuel ratio of 0.083 describe the same mixture ratio. Mixtures of air and gasoline as rich as 8:1 and as lean as 16:1 will burn in an engine cylinder, but beyond these mixtures, either lean or rich blow out could occur. The engine develops maximum power with a mixture of approximately 12 parts of air and 1 part of gasoline by weight.

From a chemist's point of view, the perfect mixture for combustion of fuel and air would be 0.067 pounds of fuel to 1 pound of air (mixture ratio of 15:1). The scientist calls this chemically correct combination a stoichiometric mixture (pronounced stoy-key-o-metric). With this mixture (given sufficient time and turbulence), all the fuel and all the oxygen in the air is completely used in the combustion process. The stoichiometric mixture produces the highest combustion temperatures because the proportion of heat released to a mass of charge (fuel and air) is the greatest. If more fuel is added to the same quantity of air charge than the amount giving a chemically perfect mixture, changes of power and temperature occur. The combustion gas temperature is lowered as the mixture is enriched, and the power increases until the air-fuel ratio is approximately 0.0725. For mixtures from 0.0725 air-fuel ratio to 0.080 air-fuel ratio, the power remains essentially constant even though the combustion temperature continues downward. Mixtures from 0.0725 air-fuel ratio to 0.080 air-fuel ratio are called best power mixtures, since their use results in the greatest power for a given airflow or manifold pressure. In this air-fuel ratio range, there is no increase in the total heat released, but the weight of nitrogen and combustion products is augmented by the vapor formed with the excess fuel. Thus, the working mass of the charge is increased. In addition, the extra fuel in the charge (over the stoichiometric mixture) speeds up the combustion process, which provides a favorable time factor in converting fuel energy into power.

If the air-fuel ratio is enriched above 0.080, there is loss of power and a reduction in temperature. The cooling effects of excess fuel overtake the favorable factor of increased mass. This reduced temperature and slower rate of burning lead to an increasing loss of combustion efficiency. If, with constant airflow, the mixture is leaned below 0.067, air-fuel ratio power and temperature decrease together. This time, the loss of power is not a liability but an asset. The purpose in leaning is to save fuel. Air is free and available in limitless quantities. The object is to obtain the required power with the least fuel flow. A measure of the economical use of fuel is called specific fuel consumption (SFC), which is the fuel weight in pounds per hour per horsepower.

$$SFC = \frac{pounds\ fuel/hour}{horsepower}$$

By using this ratio, the engine's use of fuel at various power settings can be compared. When leaning below 0.067 air-fuel ratio with constant airflow, even though the power diminishes, the cost in fuel to support each horsepower hour (SFC) also is lowered. While the mixture charge is becoming weaker, this loss of strength occurs at a rate lower than that of the reduction of fuel flow. This favorable tendency continues until a mixture strength known as best economy is reached. With this air-fuel ratio, the required hp is developed with the least fuel flow or, to put it another way, the greatest power produced by a given fuel flow. The best economy air-fuel ratio varies somewhat with rpm and other conditions, but for cruise powers on most reciprocating engines, it is sufficiently accurate to define this range of operation as being from 0.060 to 0.065 air-fuel ratios on aircraft where manual leaning is practiced.

Below the best economical mixture strength, power and temperature continue to fall with constant airflow while the SFC increases. As the air-fuel ratio is reduced further, combustion becomes so cool and slow that power for a given manifold pressure gets so low as to be uneconomical. The cooling effect of rich or lean mixtures results from the excess fuel or air over that needed for combustion. Internal cylinder cooling is obtained from unused fuel when air-fuel ratios above 0.067 are used. The same function is performed by excess air when air-fuel ratios below 0.067 are used.

Varying the mixture strength of the charge produces changes in the engine operating condition affecting power, temperature, and spark-timing requirements. The best power air-fuel ratio is desirable when the greatest power from a given airflow is required. The best economy mixture results from obtaining the given power output with the least fuel flow. The air-fuel ratio which gives most efficient operation varies with engine speed and power output.

In the graph showing this variation in air-fuel ratio, note that the mixture is rich at both idling and high-speed operation and is lean through the cruising range. *[Figure 2-1]* At idling speed, some air or exhaust gas is drawn into the cylinder through the exhaust port during valve overlap. The mixture that enters the cylinder through the intake port must be rich enough to compensate for this gas or additional air. At cruising power, lean mixtures save fuel and increase the range of the airplane. An engine running near full power requires a rich mixture to prevent overheating and detonation. Since the engine is operated at full power for only short periods, the

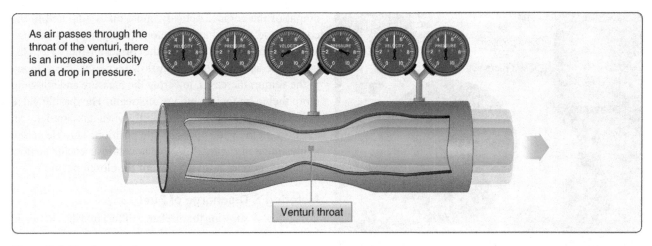

As air passes through the throat of the venturi, there is an increase in velocity and a drop in pressure.

Venturi throat

Figure 2-4. *Simple venturi.*

high fuel consumption is not a serious matter. If an engine is operating on a mixture that is too lean, and adjustments are made to increase the amount of fuel, the power output of the engine increases rapidly at first, then gradually until maximum power is reached. With a further increase in the amount of fuel, the power output drops gradually at first, then more rapidly as the mixture is further enriched.

There are specific instructions concerning mixture ratios for each type of engine under various operating conditions. Failure to follow these instructions results in poor performance and often in damage to the engine. Excessively rich mixtures result in loss of power and waste of fuel. With the engine operating near its maximum output, very lean mixtures cause a loss of power and, under certain conditions, serious overheating. When the engine is operated on a lean mixture, the cylinder head temperature gauge should be watched closely. If the mixture is excessively lean, the engine may backfire through the induction system or stop completely. Backfire results from slow burning of the lean mixture. If the charge is still burning when the intake valve opens, it ignites the fresh mixture and the flame travels back through the combustible mixture in the induction system.

Carburetion Principles

Venturi Principles

The carburetor must measure the airflow through the induction system and use this measurement to regulate the amount of fuel discharged into the airstream. The air measuring unit is the venturi, which makes use of a basic law of physics: as the velocity of a gas or liquid increases, the pressure decreases. As shown in *Figure 2-4,* simple venturi is a passageway or tube in which there is a narrow portion called the throat. As the velocity of the air increases to get through the narrow portion, its pressure drops. Note that the pressure in the throat is lower than that in any other part of the venturi. This pressure drop is proportional to the velocity

and is, therefore, a measure of the airflow. The basic operating principle of most carburetors depends on the differential pressure between the inlet and the venturi throat.

Application of Venturi Principle to Carburetor

The carburetor is mounted on the engine so that air to the cylinders passes through the barrel, the part of the carburetor which contains the venturi. The size and shape of the venturi depends on the requirements of the engine for which the carburetor is designed. A carburetor for a high-powered engine may have one large venturi or several small ones. The air may flow either up or down the venturi, depending on the design of the engine and the carburetor. Those in which the air passes downward are known as downdraft carburetors, and those in which the air passes upward are called updraft carburetors. Some carburetors are made to use a side draft or horizontal air entry into the engine induction system, as shown in *Figure 2-5*.

Air flows through the induction system covered in Chapter 3. When a piston moves toward the crankshaft (down) on the intake stroke, the pressure in the cylinder is lowered.

Figure 2-5. *Side draft horizontal flow carburetor.*

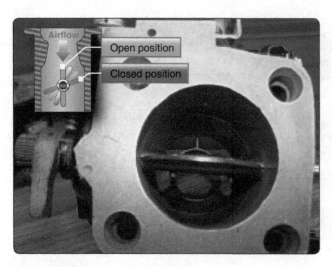

Figure 2-6. *Wide open throttle position.*

Air rushes through the carburetor and intake manifold to the cylinder to replace the air displaced by the piston as it moved down on the intake stroke. Due to this low pressure area caused by the piston moving down, the higher pressure air in the atmosphere flows in to fill the low pressure area. As it does, the airflow must pass through the carburetor venturi. The throttle valve is located between the venturi and the engine. Mechanical linkage connects this valve with the throttle lever in the flight deck. By means of the throttle, airflow to the cylinders is regulated and controls the power output of the engine. Actually, more air is admitted to the engine, and the carburetor automatically supplies enough additional gasoline to maintain the correct air-fuel ratio. This is because as the volume of airflow increases, the velocity in the venturi increases, lowering the pressure and allowing more fuel to be forced into the airstream. The throttle valve obstructs the passage of air very little when it is parallel with the flow, in the wide open throttle position. Throttle action is illustrated in *Figure 2-6*. Note how it restricts the airflow more and more as it rotates toward the closed position.

Metering & Discharge of Fuel

In *Figure 2-7,* showing the discharge of fuel into the airstream, locate the inlet through which fuel enters the carburetor from the engine-driven pump. The float-operated needle valve regulates the flow through the inlet, which maintains the correct level in the fuel float chamber. *[Figures 2-8* and *2-9]* This level must be slightly below the outlet of the discharge nozzle to prevent overflow when the engine is not running.

The discharge nozzle is located in the throat of the venturi at the point where the lowest drop in pressure occurs as air passes through the carburetor to the engine cylinders. There are two different pressures acting on the fuel in the carburetor—a low pressure at the discharge nozzle and a higher (atmospheric) pressure in the float chamber. The higher pressure in the float chamber forces the fuel through

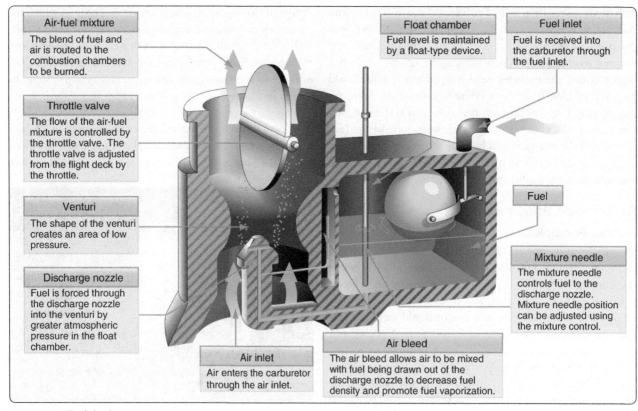

Figure 2-7. *Fuel discharge.*

Figure 2-8. *Needle valve and seat.*

Figure 2-9. *Float chamber discharge nozzle and float.*

the discharge nozzle into the airstream. If the throttle is opened wider to increase the airflow to the engine, there is a greater drop in pressure at the venturi throat. Because of the higher differential pressure, the fuel discharge increases in proportion to the increase in airflow. If the throttle is moved toward the "closed" position, the airflow and fuel flow decrease.

The fuel must pass through the metering jet to reach the discharge nozzle. *[Figure 2-7]* A metering jet is really a certain size hole that the fuel passes through. The size of this jet determines the rate of fuel discharge at each differential pressure. If the jet is replaced with a larger one, the fuel flow increases, resulting in a richer mixture. If a smaller jet is installed, there is a decrease in fuel flow and a leaner mixture.

Carburetor Systems

To provide for engine operation under various loads and at different engine speeds, each carburetor has six systems:

1. Main metering,

2. Idling,

3. Accelerating,

4. Mixture control,

5. Idle cutoff, and

6. Power enrichment or economizer.

Each of these systems has a definite function. It may act alone or with one or more of the others.

The main metering system supplies fuel to the engine at all speeds above idling. The fuel discharged by this system is determined by the drop in pressure in the venturi throat.

A separate system is necessary for idling because the main metering system can be erratic at very low engine speeds. At low speeds the throttle is nearly closed. As a result, the velocity of the air through the venturi is low and there is little drop in pressure. Consequently, the differential pressure is not sufficient to operate the main metering system, and no fuel is discharged from this system. Therefore, most carburetors have an idling system to supply fuel to the engine at low engine speeds.

The accelerating system supplies extra fuel during sudden increases in engine power. When the throttle is opened, the airflow through the carburetor increases to obtain more power from the engine. The main metering system then increases the fuel discharge. During sudden acceleration, however, the increase in airflow is so rapid that there is a slight time lag before the increase in fuel discharge is sufficient to provide the correct mixture ratio with the new airflow. By supplying extra fuel during this period, the accelerating system prevents a temporary leaning out of the mixture and gives smooth acceleration.

The mixture control system determines the ratio of fuel to air in the mixture. By means of a flight deck control, the manual mixture control can select the mixture ratio to suit operating conditions. In addition to these manual controls, many carburetors have automatic mixture controls so that the air-fuel ratio, once it is selected, does not change with variations in air density. This is necessary because as the airplane climbs and the atmospheric pressure decreases, there is a corresponding decrease in the weight of air passing through the induction system. The volume, however, remains constant. Since it is the volume of airflow that determines the pressure drop at the throat of the venturi, the carburetor tends to meter the same amount of fuel to this thin air as to the dense air at sea level. Thus, the natural tendency is for the mixture to become richer as the airplane gains altitude. The automatic mixture control prevents this by decreasing the rate of fuel discharge to compensate for the decrease in air density.

The carburetor has an idle cutoff system so that the fuel can be shut off to stop the engine. This system, incorporated in the manual mixture control, stops the fuel discharge from the carburetor completely when the mixture control lever is set to the "idle cutoff" position. An aircraft engine is stopped by shutting off the fuel rather than by turning off the ignition. If the ignition is turned off with the carburetor still supplying fuel, fresh air-fuel mixture continues to pass through the induction system to the cylinders. As the engine is coasting to a stop and if it is excessively hot, this combustible mixture may be ignited by local hot spots within the combustion chambers. This can cause the engine to continue running or kick backward. Also, the mixture may pass through the cylinders unburned, but be ignited in the hot exhaust manifold. Or, the engine comes to an apparently normal stop, but a combustible mixture remains in the induction passages, the cylinders, and the exhaust system. This is an unsafe condition since the engine may kick over after it has been stopped and seriously injure anyone near the propeller. When the engine is shut down by means of the idle cutoff system, the spark plugs continue to ignite the air-fuel mixture until the fuel discharge from the carburetor ceases. This alone should prevent the engine from coming to a stop with a combustible mixture in the cylinders. Some engine manufacturers suggest that just before the propeller stops turning, the throttle be opened wide so that the pistons can pump fresh air through the induction system, the cylinders, and the exhaust system as an added precaution against accidental kick-over. After the engine has come to a complete stop, the ignition switch is turned to the "off" position.

The power enrichment system automatically increases the richness of the mixture during high power operation. It makes possible the variation in air-fuel ratio necessary to fit different operating conditions. Remember that at cruising speeds, a lean mixture is desirable for economy reasons, while at high power output, the mixture must be rich to obtain maximum power and to aid in cooling the engine cylinders. The power enrichment system automatically brings about the necessary change in the air-fuel ratio. Essentially, it is a valve that is closed at cruising speeds and opened to supply extra fuel to the mixture during high power operation. Although it increases the fuel flow at high power, the power enrichment system is actually a fuel saving device. Without this system, it would be necessary to operate the engine on a rich mixture over the complete power range. The mixture would then be richer than necessary at cruising speed to ensure safe operation at maximum power. The power enrichment system is sometimes called an economizer or a power compensator.

Although the various systems have been discussed separately, the carburetor functions as a unit. The fact that one system is in operation does not necessarily prevent another from functioning. At the same time that the main metering system is discharging fuel in proportion to the airflow, the mixture control system determines whether the resultant mixture is rich or lean. If the throttle is suddenly opened wide, the accelerating and power enrichment systems act to add fuel to that already being discharged by the main metering system.

Carburetor Types

There are two types of carburetor used on aircraft with reciprocating engines—the float-type and the pressure-type. The float-type carburetor, the most common of the two, has several distinct disadvantages. The effect that abrupt maneuvers have on the float action and the fact that its fuel must be discharged at low pressure leads to incomplete vaporization and difficulty in discharging fuel into some types of supercharged systems. The chief disadvantage of the float carburetor, however, is its icing tendency. Since the float carburetor must discharge fuel at a point of low pressure, the discharge nozzle must be located at the venturi throat, and the throttle valve must be on the engine side of the discharge nozzle. This means that the drop in temperature due to fuel vaporization takes place within the venturi. As a result, ice readily forms in the venturi and on the throttle valve.

A pressure-type carburetor discharges fuel into the airstream at a pressure well above atmospheric. This results in better vaporization and permits the discharge of fuel into the airstream on the engine side of the throttle valve. With the discharge nozzle located at this point, the drop in temperature due to fuel vaporization takes place after the air has passed the throttle valve and at a point where engine heat tends to offset it. Thus, the danger of fuel vaporization icing is practically eliminated. The effects of rapid maneuvers and rough air on the pressure-type carburetors are negligible since its fuel chambers remain filled under all operating conditions. Pressure carburetors have been replaced mostly by fuel injection systems and have limited use on modern aircraft engines.

Carburetor Icing

There are three general classifications of carburetor icing:

1. Fuel evaporation ice,

2. Throttle ice, and

3. Impact ice.

Fuel evaporation ice or refrigeration ice is formed because of the decrease in air temperature resulting from the evaporation of fuel after it is introduced into the airstream. As the fuel evaporates, the temperature is lowered in the area where the evaporation takes place. Any moisture in the incoming air can form ice in this area. It frequently occurs in those systems in which fuel is injected into the air upstream from the carburetor throttle, as in the case of float-type carburetors. It occurs less frequently in systems in which the fuel is injected into the air downstream from the carburetor. Refrigeration ice can be formed at carburetor air temperatures as high as 100 °F over a wide range of atmospheric humidity conditions, even at relative humidity well below 100 percent. Generally, fuel evaporation ice tends to accumulate on the fuel distribution nozzle in the carburetor. This type of ice can lower manifold pressure, interfere with fuel flow, and affect mixture distribution.

Throttle ice is formed on the rear side of the throttle, usually when the throttle is in a partially "closed" position. The rush of air across and around the throttle valve causes a low pressure on the rear side; this sets up a pressure differential across the throttle, which has a cooling effect on the air-fuel charge. Moisture freezes in this low pressure area and collects as ice on the low pressure side. Throttle ice tends to accumulate in a restricted passage. The occurrence of a small amount of ice may cause a relatively large reduction in airflow and manifold pressure. A large accumulation of ice may jam the throttles and cause them to become inoperable. Throttle ice seldom occurs at temperatures above 38 °F.

Impact ice is formed either from water present in the atmosphere as snow, sleet, or from liquid water which impinges on surfaces that are at temperatures below 32 °F. Because of inertia effects, impact ice collects on or near a surface that changes the direction of the airflow. This type of ice

may build up on the carburetor elbow, as well as the carburetor screen and metering elements. The most dangerous impact ice is that which collects on the carburetor screen and causes a very rapid reduction of airflow and power. In general, danger from impact ice normally exists only when ice forms on the leading edges of the aircraft structure. Under some conditions, ice may enter the carburetor in a comparatively dry state and will not adhere to the inlet screen or walls or affect engine airflow or manifold pressure. This ice may enter the carburetor and gradually build up internally in the carburetor air metering passages and affect carburetor metering characteristics.

Float-Type Carburetors

A float-type carburetor consists essentially of six subsystems that control the quantity of fuel discharged in relation to the flow of air delivered to the engine cylinders. These systems work together to provide the engine with the correct fuel flow during all engine operating ranges.

The essential subsystems of a float-type carburetor are illustrated in *Figure 2-10*. These systems are:

1. Float chamber mechanism system,
2. Main metering system,
3. Idling system,
4. Mixture control system,

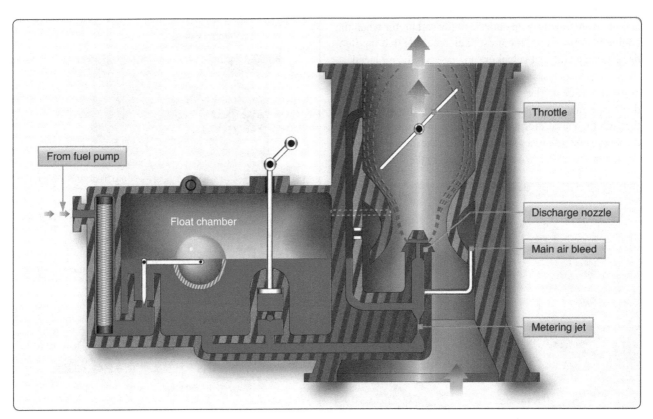

Figure 2-10. *A float-type carburetor.*

5. Accelerating system, and

6. Economizer system.

Float Chamber Mechanism System

A float chamber is provided between the fuel supply and the main metering system of the carburetor. The float chamber, or bowl, serves as a reservoir for fuel in the carburetor. *[Figure 2-11]* This chamber provides a nearly constant level of fuel to the main discharge nozzle which is usually about ⅛" below the holes in the main discharge nozzle. The fuel level must be maintained slightly below the discharge nozzle outlet holes to provide the correct amount of fuel flow and to prevent fuel leakage from the nozzle when the engine is not operating.

The level of fuel in the float chamber is kept nearly constant by means of a float-operated needle valve and a seat. The needle seat is usually made of bronze. The needle valve is constructed of hardened steel, or it may have a synthetic rubber section which fits the seat. With no fuel in the float chamber, the float drops toward the bottom of the chamber and allows the needle valve to open wide. As fuel is admitted from the supply line, the float rises (floats in the fuel) and closes the needle valve when the fuel reaches a predetermined level. When the engine is running, and fuel is being drawn out of the float chamber, the valve assumes an intermediate position so that the valve opening is just sufficient to supply the required amount of fuel and keep the level constant. *[Figure 2-10]* If fuel is found leaking from the discharge nozzle of the carburetor when the engine is not running, the most likely cause is that the float needle valve and seat is leaking and needs to be replaced.

With the fuel at the correct level (float chamber), the discharge rate is controlled accurately by the air velocity through the carburetor venturi where a pressure drop at the discharge nozzle causes fuel to flow into the intake airstream. Atmospheric pressure on top of the fuel in the float chamber

forces the fuel out the discharge nozzle. A vent or small opening in the top of the float chamber allows air to enter or leave the chamber as the level of fuel rises or falls.

Main Metering System

The main metering system supplies fuel to the engine at all speeds above idling and consists of:

1. Venturi,

2. Main metering jet,

3. Main discharge nozzle,

4. Passage leading to the idling system, and

5. Throttle valve.

Since the throttle valve controls the mass airflow through the carburetor venturi, it must be considered a major unit in the main metering system as well as in other carburetor systems. A typical main metering system is illustrated in *Figure 2-12*. The venturi performs three functions:

1. Proportions the air-fuel mixture,

2. Decreases the pressure at the discharge nozzle, and

3. Limits the airflow at full throttle.

The fuel discharge nozzle is located in the carburetor barrel so that its open end is in the throat or narrowest part of the

Figure 2-11. *Float chamber (bowl) with float removed.*

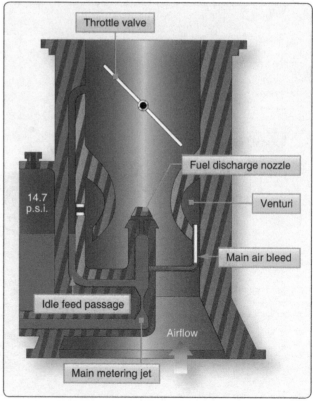

Figure 2-12. *Main metering system.*

venturi. A main metering orifice, or jet, is placed in the fuel passage between the float chamber and the discharge nozzle to limit the fuel flow when the throttle valve is wide open.

When the engine crankshaft is revolved with the carburetor throttle open, the low pressure created in the intake manifold acts on the air passing through the carburetor barrel. Due to the difference in pressure between the atmosphere and the intake manifold, air flows from the air intake through the carburetor barrel into the intake manifold. The volume of airflow depends upon the degree of throttle opening. As the air flows through the venturi, its velocity increases. This velocity increase creates a low pressure area in the venturi throat. The fuel discharge nozzle is exposed to this low pressure. Since the float chamber is vented to atmospheric pressure, a pressure drop across the discharge nozzle is created. It is this pressure difference, or metering force, that causes fuel to flow from the discharge nozzle. The fuel comes out of the nozzle in a fine spray, and the tiny particles of fuel in the spray quickly vaporize in the air.

The metering force (pressure differential) in most carburetors increases as the throttle opening is increased. The fuel must be raised in the discharge nozzle to a level at which it discharges into the airstream. To accomplish this, a pressure differential of 0.5 "Hg is required. When the metering force is considerably reduced at low engine speeds, the fuel delivery from the discharge nozzle decreases if an air bleed (air metering jet) is not incorporated in the carburetor. The decrease in fuel flow in relation to airflow is due to two factors:

1. The fuel tends to adhere to the walls of the discharge nozzle and break off intermittently in large drops instead of forming a fine spray.

2. A part of the metering force is required to raise the fuel level from the float chamber level to the discharge

nozzle outlet.

The basic principle of the air bleed can be explained by simple diagrams, as shown in *Figure 2-13*. In each case, the same degree of suction is applied to a vertical tube placed in the container of liquid. As shown in A, the suction applied on the upper end of the tube is sufficient to lift the liquid a distance of about 1 inch above the surface. If a small hole is made in the side of the tube above the surface of the liquid, as in B, and suction is applied, bubbles of air enter the tube and the liquid is drawn up in a continuous series of small slugs or drops. Thus, air "bleeds" into the tube and partially reduces the forces tending to retard the flow of liquid through the tube. However, the large opening at the bottom of the tube effectively prevents any great amount of suction from being exerted on the air bleed hole or vent. Similarly, an air bleed hole that is too large in proportion to the size of the tube would reduce the suction available to lift the liquid. If the system is modified by placing a metering orifice in the bottom of the tube and air is taken in below the fuel level by means of an air bleed tube, a finely divided mixture of air and liquid is formed in the tube, as shown in C.

In a carburetor, a small air bleed is bled into the fuel nozzle slightly below the fuel level. The open end of the air bleed is in the space behind the venturi wall where the air is relatively motionless and at approximately atmospheric pressure. The low pressure at the tip of the nozzle not only draws fuel from the float chamber but also draws air from behind the venturi. Air bled into the main metering fuel system decreases the fuel density and destroys surface tension. This results in better vaporization and control of fuel discharge, especially at lower engine speeds. The throttle, or butterfly valve, is located in the carburetor barrel near one end of the venturi. It provides a means of controlling engine speed or power output by regulating the airflow to the engine. This valve is

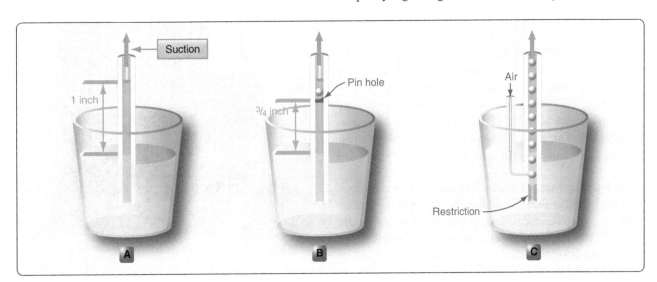

Figure 2-13. *Air bleed principle.*

Figure 2-14. *Throttle action in idle position.*

a disc that can rotate on an axis, so that it can be turned to open or close the carburetor air passage.

Idling System

With the throttle valve closed at idling speeds, air velocity through the venturi is so low that it cannot draw enough fuel from the main discharge nozzle; in fact, the spray of fuel may stop altogether. However, low pressure (piston suction) exists on the engine side of the throttle valve. In order to allow the engine to idle, a fuel passageway is incorporated to discharge fuel from an opening in the low pressure area near the edge of the throttle valve. *[Figure 2-14]* This opening is called the idling jet. With the throttle open enough so that the main discharge nozzle is operating, fuel does not flow out of the idling jet. As soon as the throttle is closed far enough to stop the spray from the main discharge nozzle, fuel flows out the idling jet. A separate air bleed, known as the idle air bleed, is included as part of the idling system. It functions in the same manner as the main air bleed. An idle mixture adjusting device is also incorporated. A typical idling system is illustrated in *Figure 2-15*.

Mixture Control System

As altitude increases, the air becomes less dense. At an altitude of 18,000 feet, the air is only half as dense as it is at sea level. This means that a cubic foot of space contains only half as much air at 18,000 feet as at sea level. An engine cylinder full of air at 18,000 feet contains only half as much oxygen as a cylinder full of air at sea level.

The low pressure area created by the venturi is dependent upon air velocity rather than air density. The action of the venturi draws the same volume of fuel through the discharge nozzle at a high altitude as it does at a low altitude. Therefore, the fuel mixture becomes richer as altitude increases. This can be overcome either by a manual or an automatic mixture control.

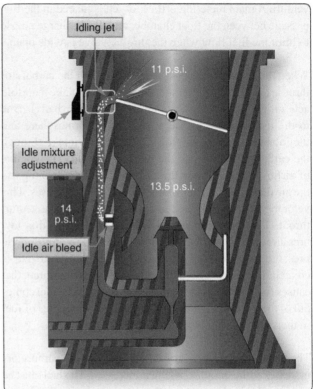

Figure 2-15. *Idling system.*

On float-type carburetors, two types of purely manual or flight deck controllable devices are in general use for controlling air-fuel mixtures, the needle type and the back-suction type. *[Figures 2-16 and 2-17]*

With the needle-type system, manual control is provided by a

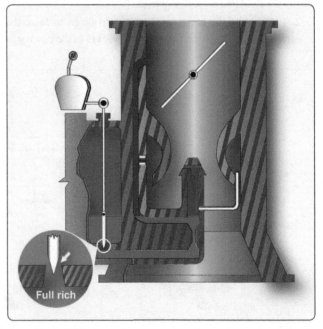

Figure 2-16. *Needle-type mixture control system.*

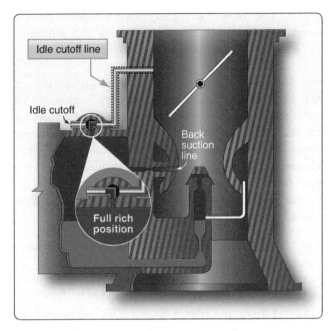

Figure 2-17. *Back-suction-type mixture control system.*

needle valve in the base of the float chamber. *[Figure 2-16]* This can be raised or lowered by adjusting a control in the flight deck. Moving the control to "rich," opens the needle valve wide, which permits the fuel to flow unrestricted to the nozzle. Moving the control to "lean," partially closes the valve and restricts the flow of fuel to the nozzle.

The back-suction-type mixture control system is the most widely used. *[Figure 2-17]* In this system, a certain amount of venturi low pressure acts upon the fuel in the float chamber so that it opposes the low pressure existing at the main discharge nozzle. An atmospheric line, incorporating an adjustable valve, opens into the float chamber. When the valve is completely closed, pressures on the fuel in the float chamber and at the discharge nozzle are almost equal, and fuel flow is reduced to maximum lean. With the valve wide open, pressure on the fuel in the float chamber is greatest and fuel mixture is richest. Adjusting the valve to positions between these two extremes controls the mixture. The quadrant in the flight deck is usually marked "lean" near the back end and "rich" at the forward end. The extreme back position is marked "idle cutoff" and is used when stopping the engine.

On float carburetors equipped with needle-type mixture control, placing the mixture control in idle cutoff seats the needle valve, thus shutting off fuel flow completely. On carburetors equipped with back-suction mixture controls, a separate idle cutoff line, leading to the extreme low pressure on the engine side of the throttle valve, is incorporated. (See the dotted line in *Figure 2-17.*) The mixture control is so linked that when it is placed in the "idle cutoff" position, it opens another passage that leads to piston suction. When placed in other positions, the valve opens a passage leading to

the atmosphere. To stop the engine with such a system, close the throttle and place the mixture in the "idle cutoff" position. Leave the throttle in the closed position until the engine has stopped running and then open the throttle completely.

Accelerating System

When the throttle valve is opened quickly, a large volume of air rushes through the air passage of the carburetor; the amount of fuel that is mixed with the air is less than normal due to the slow response rate of the main metering system. As a result, after a quick opening of the throttle, the air-fuel mixture leans out momentarily. This can cause the engine to accelerate slowly or stumble as it tries to accelerate.

To overcome this tendency, the carburetor is equipped with a small fuel pump called an accelerating pump. A common type of accelerating system used in float carburetors is illustrated in *Figure 2-18*. It consists of a simple piston pump operated through linkage by the throttle control and a passageway opening into the main metering system or the carburetor barrel near the venturi. When the throttle is closed, the piston moves back, and fuel fills the cylinder. If the piston is pushed forward slowly, the fuel seeps past it back into the float chamber; if pushed rapidly, it sprays fuel in the venturi and enriches the mixture. An example of a cutaway accelerator pump is shown in *Figure 2-19*.

Economizer System

For an engine to develop maximum power at full throttle, the fuel mixture must be richer than for cruise. The additional fuel is used for cooling the engine combustion chambers to prevent detonation. An economizer is essentially a valve that is closed at throttle settings below approximately 60–70

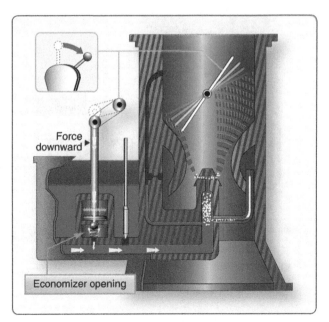

Figure 2-18. *Accelerating system.*

Figure 2-19. *Accelerating pump shown in cutaway.*

percent of rated power. This system, like the accelerating system, is operated by the throttle control.

A typical economizer system consists of a needle valve which begins to open when the throttle valve reaches a predetermined point near the wide-open position. *[Figure 2-20]* As the throttle continues to open, the needle valve is opened further and additional fuel flows through it. This additional fuel supplements the flow from the main metering jet direct to the main discharge nozzle.

A pressure-operated economizer system is shown in *Figure 2-21.* This type has a sealed bellows located in an enclosed compartment. The compartment is vented to engine manifold pressure. When the manifold pressure reaches a certain value, the bellows is compressed and opens a valve in a carburetor fuel passage, supplementing the normal quantity of fuel being discharged through the main nozzle.

Another type of economizer is the back-suction system. *[Figure 2-22]* Fuel economy in cruising is provided by reducing the effective pressure acting on the fuel level in the float compartment. With the throttle valve in cruising position, suction is applied to the float chamber through an economizer hole and back-suction economizer channel and jet. The suction applied to the float chamber opposes the nozzle suction applied by the venturi. Fuel flow is reduced, leaning the mixture for cruising economy.

Another type of mixture control system uses a metering valve that is free to rotate in a stationary metering sleeve. Fuel enters the main and idling systems through a slot cut in the mixture sleeve. Fuel metering is accomplished by the relative position between one edge of the slot in the hollow metering valve and one edge of the slot in the metering sleeve. Moving the mixture control to reduce the size of the slot provides a leaner mixture for altitude compensation.

Pressure Injection Carburetors

Pressure injection carburetors are distinctly different from float-type carburetors as they do not incorporate a vented float chamber or suction pickup from a discharge nozzle located in the venturi tube. Instead, they provide a pressurized fuel system that is closed from the engine fuel pump to the discharge nozzle. The venturi serves only to create pressure differentials for controlling the quantity of fuel to the metering jet in proportion to airflow to the engine.

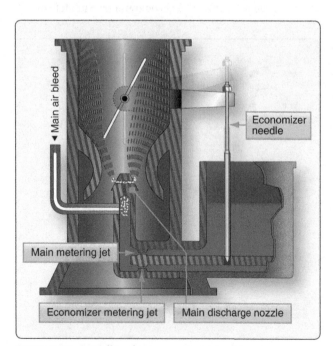

Figure 2-20. *A needle-valve type economizer system.*

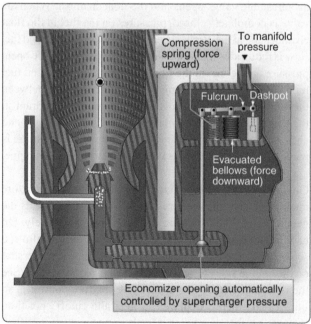

Figure 2-21. *A pressure operated economizer system.*

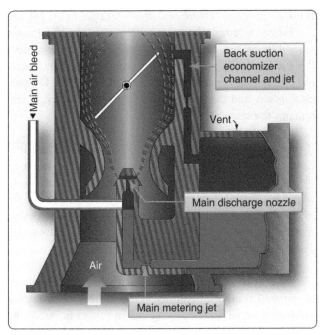

Figure 2-22. *Float-type carburetor.*

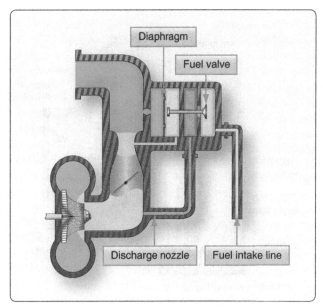

Figure 2-23. *Pressure type carburetor.*

Typical Injection Carburetor

The injection carburetor is a hydromechanical device employing a closed feed system from the fuel pump to the discharge nozzle. It meters fuel through fixed jets according to the mass airflow through the throttle body and discharges it under a positive pressure.

The illustration in *Figure 2-23* represents a pressure-type carburetor simplified so that only the basic parts are shown. Note the two small passages, one leading from the carburetor air inlet to the left side of the flexible diaphragm and the other from the venturi throat to the right side of the diaphragm.

When air passes through the carburetor to the engine, the pressure on the right of the diaphragm is lowered because of the drop in pressure at the venturi throat. As a result, the diaphragm moves to the right, opening the fuel valve. Pressure from the engine-driven pump then forces fuel through the open valve to the discharge nozzle, where it sprays into the airstream. The distance the fuel valve opens is determined by the difference between the two pressures acting on the diaphragm. This difference in pressure is proportional to the airflow through the carburetor. Thus, the volume of airflow determines the rate of fuel discharge.

The pressure injection carburetor is an assembly of the following units:

1. Throttle body,

2. Automatic mixture control,

3. Regulator unit, and

4. Fuel control unit (some are equipped with an adapter).

Throttle Body

The throttle body contains the throttle valves, main venturi, boost venturi, and the impact tubes. All air entering the cylinders must flow through the throttle body; therefore, it is the air control and measuring device. The airflow is measured by volume and by weight so that the proper amount of fuel can be added to meet the engine demands under all conditions. As air flows through the venturi, its velocity is increased, and its pressure is decreased (Bernoulli's principle). This low pressure is vented to the low pressure side of the air diaphragm *[Figure 2-24 chamber B]* in the regulator assembly. The impact tubes sense carburetor inlet air pressure and direct it to the automatic mixture control, which measures the air density. From the automatic mixture control, the air is directed to the high pressure side of the air diaphragm (chamber A). The pressure differential of the two chambers acting upon the air diaphragm is known as the air metering force which opens the fuel poppet valve.

The throttle body controls the airflow with the throttle valves. The throttle valves may be either rectangular or disc shaped, depending on the design of the carburetor. The valves are mounted on a shaft, which is connected by linkage to the idle valve and to the throttle control in the flight deck. A throttle stop limits the travel of the throttle valve and has an adjustment which sets engine idle speed.

Regulator Unit

The regulator is a diaphragm-controlled unit divided into five chambers and contains two regulating diaphragms and a poppet valve assembly. *[Figure 2-24]* Chamber A is

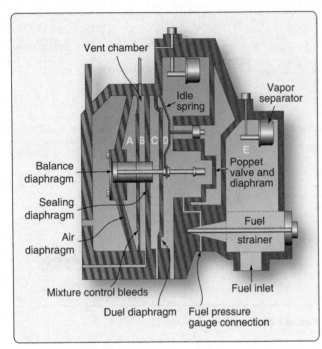

Figure 2-24. *Regulator unit.*

regulated air-inlet pressure from the air intake. Chamber B is boost venturi pressure. Chamber C contains metered fuel pressure controlled by the discharge nozzle or fuel feed valve. Chamber D contains unmetered fuel pressure controlled by the opening of the poppet valve. Chamber E is fuel pump pressure controlled by the fuel pump pressure relief valve. The poppet valve assembly is connected by a stem to the two main control diaphragms. The purpose of the regulator unit is to regulate the fuel pressure to the inlet side of the metering jets in the fuel control unit. This pressure is automatically regulated according to the mass airflow to the engine.

The carburetor fuel strainer (also called the gascolator), located in the inlet to chamber E, is a fine mesh screen through which all the fuel must pass as it enters chamber D. The strainer must be removed and cleaned at scheduled intervals.

Referring to *Figure 2-24*, assume that for a given airflow in lb/hr through the throttle body and venturi, a negative pressure of ¼ psi is established in chamber B. This tends to move the diaphragm assembly and the poppet valve in a direction to open the poppet valve permitting more fuel to enter chamber D. The pressure in chamber C is held constant at 5 psi (10 psi on some installations) by the discharge nozzle or impeller fuel feed valve. Therefore, the diaphragm assembly and poppet valve moves in the open direction until the pressure in chamber D is 5¼ psi. Under these pressures, there is a balanced condition of the diaphragm assembly with a pressure drop of ¼ psi across the jets in the fuel control unit (auto-rich or auto-lean).

If nozzle pressure (chamber C pressure) rises to 5½ psi, the diaphragm assembly balance is upset, and the diaphragm assembly moves to open the poppet valve to establish the necessary 5¾ psi pressure in chamber D. Thus, the ¼ psi differential between chamber C and chamber D is re-established, and the pressure drop across the metering jets remains the same.

If the fuel inlet pressure is increased or decreased, the fuel flow into chamber D tends to increase or decrease with the pressure change causing the chamber D pressure to do likewise. This upsets the balanced condition previously established, and the poppet valve and diaphragm assembly respond by moving to increase or decrease the flow to re-establish the pressure at the ¼ psi differential.

The fuel flow changes when the mixture control plates are moved from auto-lean to auto-rich, thereby selecting a different set of jets or cutting one or two in or out of the system. When the mixture position is altered, the diaphragm and poppet valve assembly repositions to maintain the established pressure differential of ¼ psi between chambers C and D, maintaining the established differential across the jets. Under low power settings (low airflows), the difference in pressure created by the boost venturi is not sufficient to accomplish consistent regulation of the fuel. Therefore, an idle spring, shown in *Figure 2-24*, is incorporated in the regulator. As the poppet valve moves toward the closed position, it contacts the idle spring. The spring holds the poppet valve off its seat far enough to provide more fuel than is needed for idling. This potentially overrich mixture is regulated by the idle valve. At idling speed, the idle valve restricts the fuel flow to the proper amount. At higher speeds, it is withdrawn from the fuel passage and has no metering effect.

Vapor vent systems are provided in these carburetors to eliminate fuel vapor created by the fuel pump, heat in the engine compartment, and the pressure drop across the poppet valve. The vapor vent is located in the fuel inlet (chamber E) or, on some models of carburetors, in both chambers D and E.

The vapor vent system operates in the following way. When air enters the chamber in which the vapor vent is installed, the air rises to the top of the chamber, displacing the fuel and lowering its level. When the fuel level has reached a predetermined position, the float (which floats in the fuel) pulls the vapor vent valve off its seat, permitting the vapor in the chamber to escape through the vapor vent seat, its connecting line, and back to the fuel tank.

If the vapor vent valve sticks in a closed position or the vent line from the vapor vent to the fuel tank becomes clogged, the vapor-eliminating action is stopped. This causes the vapor to build up within the carburetor to the extent that vapor passes through the metering jets with the fuel. With a given size carburetor metering jet, the metering of vapor reduces the quantity of fuel metered. This causes the air-fuel mixture to lean out, usually intermittently.

If the vapor vent valve sticks open or the vapor vent float becomes filled with fuel and sinks, a continuous flow of fuel and vapor occurs through the vent line. It is important to detect this condition, as the fuel flow from the carburetor to the fuel supply tank may cause an overflowing tank with resultant increased fuel consumption.

To check the vent system, disconnect the vapor vent line where it attaches to the carburetor, and turn the fuel booster pump on while observing the vapor vent connection at the carburetor. Move the carburetor mixture control to auto-rich; then return it to idle cutoff. When the fuel booster pump is turned on, there should be an initial ejection of fuel and air followed by a cutoff with not more than a steady drip from the vent connection. Installations with a fixed bleed from the D chamber connected to the vapor vent in the fuel inlet by a short external line should show an initial ejection of fuel and air followed by a continuing small stream of fuel. If there is no flow, the valve is sticking closed; if there is a steady flow, it is sticking open.

Fuel Control Unit

The fuel control unit is attached to the regulator assembly and contains all metering jets and valves. *[Figure 2-25]* The idle and power enrichment valves, together with the mixture control plates, select the jet combinations for the various settings (i.e., auto-rich, auto-lean, and idle cutoff).

The purpose of the fuel control unit is to meter and control the fuel flow to the discharge nozzle. The basic unit consists of three jets and four valves arranged in series, parallel, and series-parallel hookups. *[Figure 2-25]* These jets and valves receive fuel under pressure from the regulator unit and then meter the fuel as it flows to the discharge nozzle. The manual mixture control valve controls the fuel flow. By using proper size jets and regulating the pressure differential across the jets, the right amount of fuel is delivered to the discharge nozzle, giving the desired air-fuel ratio in the various power settings. It should be remembered that the inlet pressure to the jets is regulated by the regulator unit and the outlet pressure is controlled by the discharge nozzle.

The jets in the basic fuel control unit are the auto-lean jet, the auto-rich jet, and power enrichment jet. The basic fuel flow is the fuel required to run the engine with a lean mixture and is metered by the auto-lean jet. The auto-rich jet adds enough fuel to the basic flow to give a slightly richer mixture than

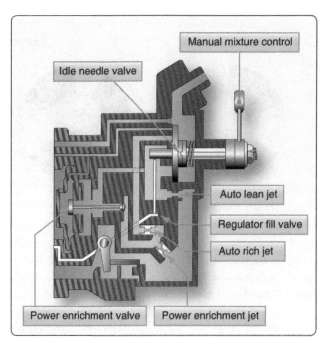

Figure 2-25. *Fuel control unit.*

best power mixture when the manual mixture control is in the auto-rich position.

The four valves in the basic fuel control unit are:

1. Idle needle valve,
2. Power enrichment valve,
3. Regulator fill valve, and
4. Manual mixture control.

The functions of these valves are:

1. The idle needle valve meters the fuel in the idle range only. It is a round, contoured needle valve, or a cylinder valve placed in series with all other metering devices of the basic fuel control unit. The idle needle valve is connected by linkage to the throttle shaft so that it restricts the fuel flowing at low power settings (idle range).

2. The manual mixture control is a rotary disc valve consisting of a round stationary disc with ports leading from the auto-lean jet, the auto-rich jet, and two smaller ventholes. Another rotating part, resembling a cloverleaf, is held against the stationary disc by spring tension and rotated over the ports in that disc by the manual mixture control lever. All ports and vents are closed in the idle cutoff position. In the auto-lean position, the ports from the auto-lean jet and the two ventholes are open. The port from the auto-rich jet remains closed in this position. In the auto-rich position, all ports are open. The valve plate positions

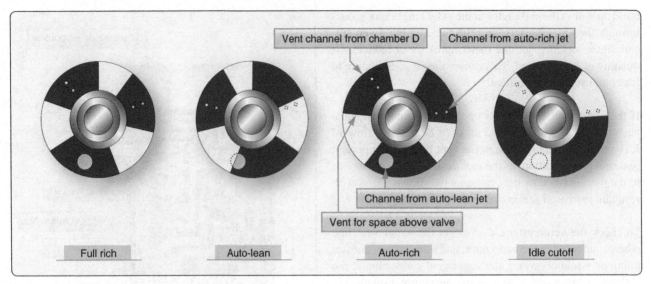

Figure 2-26. *Manual mixture control valve plate positions.*

are illustrated in *Figure 2-26*. The three positions of the manual mixture control lever make it possible to select a lean mixture a rich mixture, or to stop fuel flow entirely. The idle cutoff position is used for starting or stopping the engine. During starting, fuel is supplied by the primer.

3. The regulator fill valve is a small poppet-type valve located in a fuel passage which supplies chamber C of the regulator unit with metered fuel pressure. In idle cutoff, the flat portion of the cam lines up with the valve stem, and a spring closes the valve. This provides a means of shutting off the fuel flow to chamber C and thus provides for a positive idle cutoff.

4. The power enrichment valve is another poppet-type valve. It is in parallel with the auto-lean and auto-rich jets, but it is in series with the power enrichment jet. This valve starts to open at the beginning of the power range. It is opened by the unmetered fuel pressure overcoming metered fuel pressure and spring tension. The power enrichment valve continues to open wider during the power range until the combined flow through the valve and the auto-rich jet exceeds that of the power enrichment jet. At this point the power enrichment jet takes over the metering and meters fuel throughout the power range.

5. Carburetors equipped for water injection are modified by the addition of a derichment valve and a derichment jet. The derichment valve and derichment jet are in series with each other and parallel with the power enrichment jet.

The carburetor controls fuel flow by varying two basic factors. The fuel control unit, acting as a pressure-reducing valve, determines the metering pressure in response to the metering forces. The regulator unit, in effect, varies the size of the orifice through which the metering pressure forces the fuel. It is a basic law of hydraulics that the amount of fluid that passes through an orifice varies with the size of the orifice and the pressure drop across it. The internal automatic devices and mixture control act together to determine the effective size of the metering passage through which the fuel passes. The internal devices, fixed jets, and variable power enrichment valve are not subject to direct external control.

Automatic Mixture Control (AMC)

The automatic mixture control unit consists of a bellows assembly, calibrated needle, and seat. *[Figure 2-27]* The purpose of the automatic mixture control is to compensate for changes in air density due to temperature and altitude changes.

The automatic mixture control contains a metallic bellows, which is sealed at 28 "Hg absolute pressure. The bellows responds to changes in pressure and temperature. In the illustration, the automatic mixture control is located at the carburetor air inlet. As the density of the air changes, the expansion and contraction of the bellows moves the tapered needle in the atmospheric line. At sea level, the bellows is contracted, and the needle is not in the atmospheric passage. As the aircraft climbs and the atmospheric pressure decreases, the bellows expands, inserting the tapered needle farther and farther into the atmospheric passage and restricting the flow of air to chamber A of the regulator unit. *[Figure 2-24]* At the same time, air leaks slowly from chamber A to chamber B through the small bleed (often referred to as the back-suction bleed or mixture control bleed). The rate at which air leaks through this bleed is about the same at high altitude as it is at sea level. As the tapered needle restricts the flow of air into chamber A, the pressure on the left side of the air diaphragm decreases. As

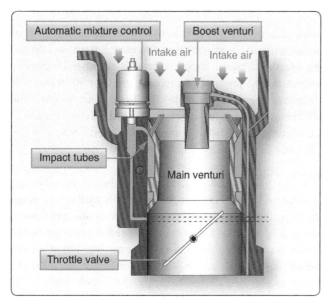

Figure 2-27. *Automatic mixture control and throttle body.*

The automatic mixture control can be removed and cleaned if the lead seal at the point of adjustment is not disturbed.

Stromberg PS Carburetor

The PS series carburetor is a low-pressure, single-barrel, injection-type carburetor. The carburetor consists basically of the air section, the fuel section, and the discharge nozzle mounted together to form a complete fuel metering system. This carburetor is similar to the pressure-injection carburetor; therefore, its operating principles are the same.

In this type carburetor, metering is accomplished on a mass airflow basis. *[Figure 2-28]* Air flowing through the main venturi creates suction at the throat of the venturi, which is transmitted to the B chamber in the main regulating part of the carburetor and to the vent side of the fuel discharge nozzle diaphragm. The incoming air pressure is transmitted to a chamber A of the regulating part of the carburetor and to the main discharge bleed in the main fuel discharge jet. The discharge nozzle consists of a spring-loaded diaphragm connected to the discharge nozzle valve, which controls the

a result, the poppet valve moves toward its seat, reducing the fuel flow to compensate for the decrease in air density.

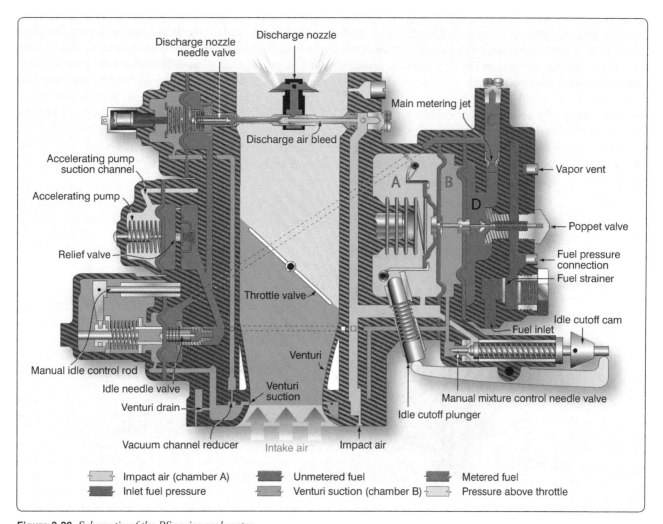

Figure 2-28. *Schematic of the PS series carburetor.*

flow of fuel injected into the main discharge jet. Here, it is mixed with air to accomplish distribution and atomization into the airstream entering the engine.

In the PS series carburetor, as in the pressure-injection carburetor, the regulator spring has a fixed tension, which tends to hold the poppet valve open during idling speeds or until the D chamber pressure equals approximately 4 psi. The discharge nozzle spring has a variable adjustment which, when tailored to maintain 4 psi, results in a balanced pressure condition of 4 psi in chamber C of the discharge nozzle assembly and 4 psi in chamber D. This produces a zero drop across the main jets at zero fuel flow.

At a given airflow, if the suction created by the venturi is equivalent to ¼ pound, the pressure decrease is transmitted to chamber B and to the vent side of the discharge nozzle. Since the area of the air diaphragm between chambers A and B is twice as great as that between chambers B and D, the ¼ pound decrease in pressure in chamber B moves the diaphragm assembly to the right to open the poppet valve. Meanwhile, the decreased pressure on the vent side of the discharge nozzle assembly causes a lowering of the total pressure from 4 pounds to 3¾ pounds. The greater pressure of the metered fuel (4¼ pounds) results in a differential across the metering head of ¼ pound (for the ¼ pound pressure differential created by the venturi).

The same ratio of pressure drop across the jet to venturi suction applies throughout the range. Any increase or decrease in fuel inlet pressure tends to upset the balance in the various chambers in the manner already described. When this occurs, the main fuel regulator diaphragm assembly repositions to restore the balance.

The mixture control, whether operated manually or automatically, compensates for enrichment at altitude by bleeding impact air pressure into chamber B, thereby increasing the pressure (decreasing the suction) in chamber B. Increasing the pressure in chamber B tends to move the diaphragm and poppet valve more toward the closed position, restricting fuel flow to correspond proportionately to the decrease in air density at altitude.

The idle valve and economizer jet can be combined in one assembly. The unit is controlled manually by the movement of the valve assembly. At low airflow positions, the tapered section of the valve becomes the predominant jet in the system, controlling the fuel flow for the idle range. As the valve moves to the cruise position, a straight section on the valve establishes a fixed orifice effect which controls the cruise mixture. When the valve is pulled full-open by the throttle valve, the jet is pulled completely out of the seat, and the seat side becomes the controlling jet. This jet is calibrated for takeoff power mixtures.

An airflow-controlled power enrichment valve can also be used with this carburetor. It consists of a spring-loaded, diaphragm-operated metering valve. Refer to *Figure 2-29* for a schematic view of an airflow power enrichment valve. One side of the diaphragm is exposed to unmetered fuel pressure and the other side to venturi suction plus spring tension. When the pressure differential across the diaphragm establishes a force strong enough to compress the spring, the valve opens and supplies an additional amount of fuel to the metered fuel circuit in addition to the fuel supplied by the main metering jet.

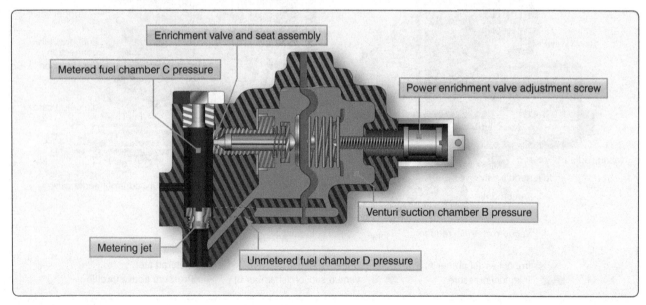

Figure 2-29. *Airflow power enrichment valve.*

Accelerating Pump

The accelerating pump of the Stromberg PS carburetor is a spring-loaded diaphragm assembly located in the metered fuel channel with the opposite side of the diaphragm vented to the engine side of the throttle valve. With this arrangement, opening the throttle results in a rapid decrease in suction. This decrease in suction permits the spring to extend and move the accelerating pump diaphragm. The diaphragm and spring action displace the fuel in the accelerating pump and force it out the discharge nozzle.

Vapor is eliminated from the top of the main fuel chamber D through a bleed hole, then through a vent line back to the main fuel tank in the aircraft.

Manual Mixture Control

A manual mixture control provides a means of correcting for enrichment at altitude. It consists of a needle valve and seat that form an adjustable bleed between chamber A and chamber B. The valve can be adjusted to bleed off the venturi suction to maintain the correct air-fuel ratio as the aircraft gains altitude.

When the mixture control lever is moved to the idle cutoff position, a cam on the linkage actuates a rocker arm which moves the idle cutoff plunger inward against the release lever in chamber A. The lever compresses the regulator diaphragm spring to relieve all tension on the diaphragm between chambers A and B. This permits fuel pressure plus poppet valve spring force to close the poppet valve, stopping the fuel flow. Placing the mixture control lever in idle cutoff also positions the mixture control needle valve off its seat and allows metering suction within the carburetor to bleed off.

Fuel-Injection Systems

The fuel-injection system has many advantages over a conventional carburetor system. There is less danger of induction system icing, since the drop in temperature due to fuel vaporization takes place in or near the cylinder. Acceleration is also improved because of the positive action of the injection system. In addition, fuel injection improves fuel distribution. This reduces the overheating of individual cylinders often caused by variation in mixture due to uneven distribution. The fuel-injection system also gives better fuel economy than a system in which the mixture to most cylinders must be richer than necessary so that the cylinder with the leanest mixture operates properly.

Fuel-injection systems vary in their details of construction, arrangement, and operation. The Bendix and Continental fuel-injection systems are discussed in this section. They are described to provide an understanding of the operating principles involved. For the specific details of any one system, consult the manufacturer's instructions for the equipment involved.

Bendix/Precision Fuel-Injection System

The Bendix inline stem-type regulator injection system (RSA) series consists of an injector, flow divider, and fuel discharge nozzle. It is a continuous-flow system which measures engine air consumption and uses airflow forces to control fuel flow to the engine. The fuel distribution system to the individual cylinders is obtained by the use of a fuel flow divider and air bleed nozzles.

Fuel Injector

The fuel injector assembly consists of:

1. An airflow section,

2. A regulator section, and

3. A fuel metering section. Some fuel injectors are equipped with an automatic mixture control unit.

Airflow Section

The airflow consumption of the engine is measured by sensing impact pressure and venturi throat pressure in the throttle body. These pressures are vented to the two sides of an air diaphragm. A cutaway view of the airflow measuring section is shown in *Figure 2-30*. Movement of the throttle valve causes a change in engine air consumption. This results in a change in the air velocity in the venturi. When airflow through the engine increases, the pressure on the left of the diaphragm is lowered due to the drop in pressure at the venturi throat. *[Figure 2-31]* As a result, the diaphragm moves to the left, opening the ball valve. Contributing to this force is the impact pressure that is picked up by the impact tubes. *[Figure 2-32]* This pressure differential is referred to as the "air metering force." This force is accomplished by channeling the impact and venturi suction pressures to opposite sides of a diaphragm. The difference between these

Figure 2-30. *Cutaway view of airflow measuring section.*

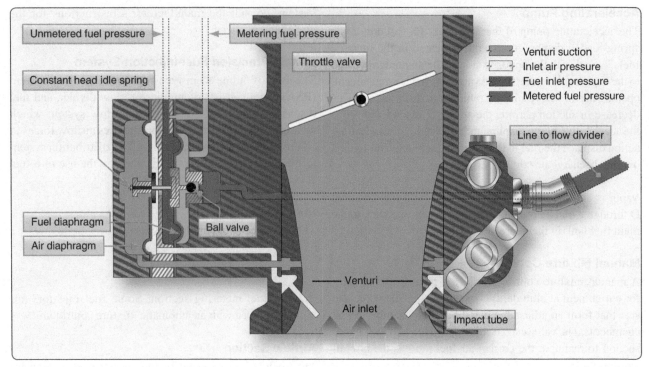

Figure 2-31. *Airflow section of a fuel injector.*

two pressures becomes a usable force that is equal to the area of the diaphragm times the pressure difference.

Regulator Section

The regulator section consists of a fuel diaphragm that opposes the air metering force. Fuel inlet pressure is applied to one side of the fuel diaphragm and metered fuel pressure is applied to the other side. The differential pressure across the fuel diaphragm is called the fuel metering force. The fuel pressure shown on the ball side of the fuel diaphragm is the pressure after the fuel has passed through the fuel strainer and the manual mixture control rotary plate and is referred to as metered fuel pressure. Fuel inlet pressure is applied to the

opposite side of the fuel diaphragm. The ball valve attached to the fuel diaphragm controls the orifice opening and fuel flow through the forces placed on it. *[Figure 2-33]*

The distance the ball valve opens is determined by the difference between the pressures acting on the diaphragms. This difference in pressure is proportional to the airflow through the injector. Thus, the volume of airflow determines the rate of fuel flow.

Under low power settings, the difference in pressure created by the venturi is insufficient to accomplish consistent regulation of the fuel. A constant-head idle spring is

Figure 2-32. *Impact tubes for inlet air pressure.*

Figure 2-33. *Fuel diaphragm with ball valve attached.*

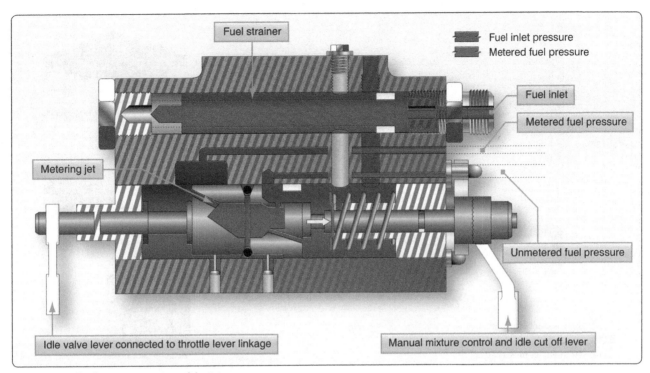

Figure 2-34. *Fuel metering section of the injector.*

incorporated to provide a constant fuel differential pressure. This allows an adequate final flow in the idle range.

Fuel Metering Section

The fuel metering section is attached to the air metering section and contains an inlet fuel strainer, a manual mixture control valve, an idle valve, and the main metering jet. *[Figure 2-34]* The idle valve is connected to the throttle valve by means of an external adjustable link. In some injector models, a power enrichment jet is also located in this section. The purpose of the fuel metering section is to meter and control the fuel flow to the flow divider. *[Figure 2-35]* The manual mixture control valve produces full rich condition when the lever is against the rich stop, and a progressively leaner mixture as the lever is moved toward idle cutoff. Both idle speed and idle mixture may be adjusted externally to meet individual engine requirements.

Flow Divider

The metered fuel is delivered from the fuel control unit to a pressurized flow divider. This unit keeps metered fuel under pressure, divides fuel to the various cylinders at all engine speeds, and shuts off the individual nozzle lines when the control is placed in idle cutoff.

Referring to the diagram in *Figure 2-36*, metered fuel pressure enters the flow divider through a channel that permits fuel to pass through the inside diameter of the flow divider needle. At idle speed, the fuel pressure from the regulator must build up to overcome the spring force applied to the diaphragm and

valve assembly. This moves the valve upward until fuel can pass out through the annulus of the valve to the fuel nozzle. *[Figure 2-37]* Since the regulator meters and delivers a fixed amount of fuel to the flow divider, the valve opens only as far as necessary to pass this amount to the nozzles. At idle, the opening required is very small; the fuel for the individual cylinders is divided at idle by the flow divider.

As fuel flow through the regulator is increased above idle requirements, fuel pressure builds up in the nozzle lines. This pressure fully opens the flow divider valve, and fuel distribution to the engine becomes a function of the discharge nozzles.

Figure 2-35. *Fuel inlet and metering.*

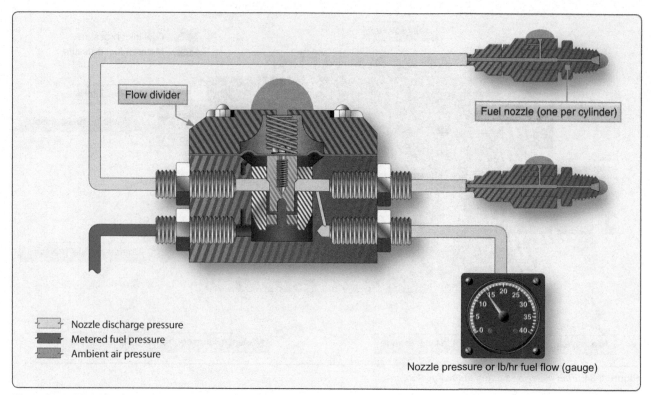

Figure 2-36. *Flow divider.*

Figure 2-37. *Flow divider cutaway.*

Figure 2-38. *Fuel nozzle assembly.*

A fuel pressure gauge, calibrated in pounds per hour fuel flow, can be used as a fuel flow meter with the Bendix RSA injection system. This gauge is connected to the flow divider and senses the pressure being applied to the discharge nozzle. This pressure is in direct proportion to the fuel flow and indicates the engine power output and fuel consumption.

Fuel Discharge Nozzles

The fuel discharge nozzles are of the air bleed configuration. There is one nozzle for each cylinder located in the cylinder head. *[Figure 2-38]* The nozzle outlet is directed into the intake port. Each nozzle incorporates a calibrated jet. The jet size is determined by the available fuel inlet pressure and the maximum fuel flow required by the engine. The fuel is discharged through this jet into an ambient air pressure chamber within the nozzle assembly. Before entering the individual intake valve chambers, the fuel is mixed with air to aid in atomizing the fuel. Fuel pressure, before the individual nozzles, is in direct proportion to fuel flow; therefore, a simple pressure gauge can be calibrated in fuel flow in gallons per hour and be employed as a flow meter. Engines modified with turbosuperchargers must use shrouded nozzles. By the use of an air manifold, these nozzles are vented to the injector air inlet pressure.

Continental/TCM Fuel-Injection System

The Continental fuel-injection system injects fuel into the intake valve port in each cylinder head. *[Figure 2-39]* The system consists of a fuel injector pump, a control unit, a fuel manifold, and a fuel discharge nozzle. It is a continuous-flow type, which controls fuel flow to match engine airflow. The continuous-flow system permits the use of a rotary vane pump which does not require timing to the engine.

Fuel-Injection Pump

The fuel pump is a positive-displacement, rotary-vane type with a splined shaft for connection to the accessory drive system of the engine. *[Figure 2-40]* A spring-loaded, diaphragm-type relief valve is provided. The relief valve diaphragm chamber is vented to atmospheric pressure. A sectional view of a fuel-injection pump is shown in *Figure 2-41*.

Fuel enters at the swirl well of the vapor separator. Here, vapor is separated by a swirling motion so that only liquid fuel is delivered to the pump. The vapor is drawn from the top center of the swirl well by a small pressure jet of fuel and is directed into the vapor return line. This line carries the vapor back to the fuel tank.

Ignoring the effect of altitude or ambient air conditions, the use of a positive-displacement, engine-driven pump means that changes in engine speed affect total pump flow proportionally. Since the pump provides greater capacity than is required by the engine, a recirculation path is required. By arranging a calibrated orifice and relief valve in this path, the pump delivery pressure is also maintained in proportion to engine speed. These provisions assure proper pump pressure and fuel delivery for all engine operating speeds.

A check valve is provided so that boost pump pressure to the system can bypass the engine-driven pump for starting. This feature also suppresses vapor formation under high ambient temperatures of the fuel and permits use of the auxiliary pump as a source of fuel pressure in the event of engine-driven pump failure.

Air-Fuel Control Unit

The function of the air-fuel control assembly is to control engine air intake and to set the metered fuel pressure for proper air-fuel ratio. The air throttle is mounted at the manifold inlet and its butterfly valve, positioned by the throttle control in the aircraft, controls the flow of air to the engine. *[Figure 2-42]*

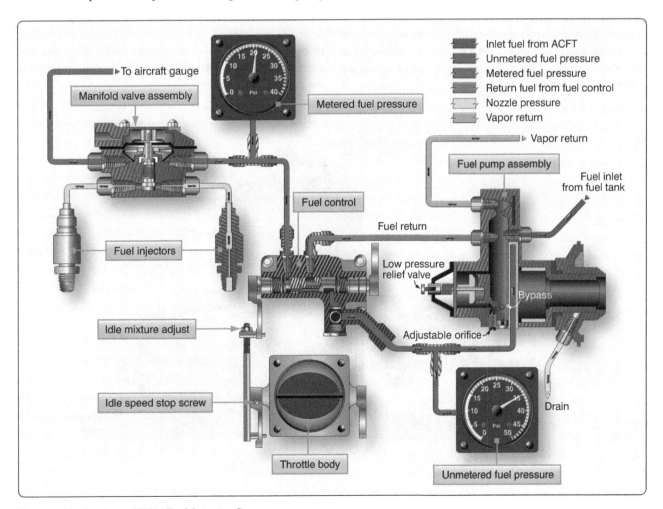

Figure 2-39. *Continental/TCM Fuel-Injection System.*

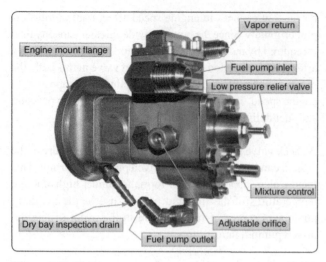

Figure 2-40. *Fuel pump.*

The air throttle assembly is an aluminum casting which contains the shaft and butterfly-valve assembly. The casting bore size is tailored to the engine size, and no venturi or other restriction is used.

Fuel Control Assembly

The fuel control body is made of bronze for best bearing action with the stainless steel valves. Its central bore contains a metering valve at one end and a mixture control valve at the other end. Each stainless steel rotary valve includes a groove which forms a fuel chamber.

Fuel enters the control unit through a strainer and passes to the metering valve. *[Figure 2-43]* This rotary valve has a cam-shaped edge on the outer part of the end face. The position of the cam at the fuel delivery port controls the fuel passed to the manifold valve and the nozzles. The fuel return port connects to the return passage of the center metering plug. The alignment of the mixture control valve with this passage determines the amount of fuel returned to the fuel pump.

By connecting the metering valve to the air throttle, the fuel flow is properly proportioned to airflow for the correct air-fuel ratio. A control level is mounted on the mixture control valve shaft and connected to the flight deck mixture control.

Fuel Manifold Valve

The fuel manifold valve contains a fuel inlet, a diaphragm chamber, and outlet ports for the lines to the individual nozzles. *[Figure 2-44]* The spring-loaded diaphragm operates a valve in the central bore of the body. Fuel pressure provides the force for moving the diaphragm. The diaphragm is enclosed by a cover that retains the diaphragm loading spring. When the valve is down against the lapped seat in the body, the fuel lines to the cylinders are closed off. The valve is drilled for passage of fuel from the diaphragm chamber to its base, and a ball valve is installed within the valve. All incoming fuel must pass through a fine screen installed in the diaphragm chamber.

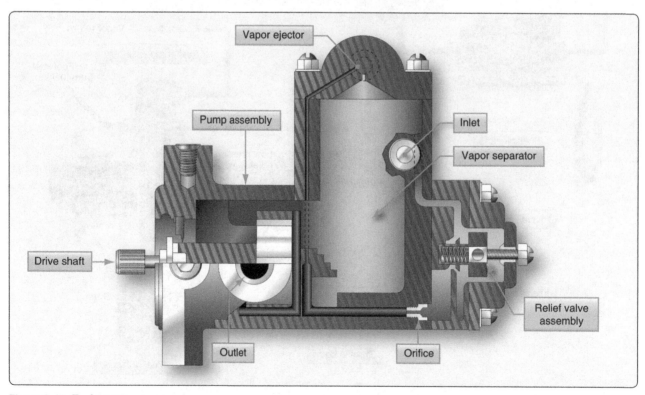

Figure 2-41. *Fuel injection pump.*

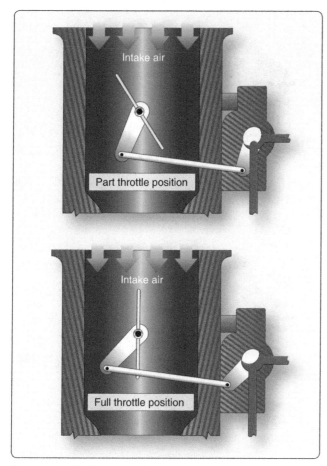

Figure 2-42. *Fuel air control unit.*

Fuel Discharge Nozzle

The fuel discharge nozzle is located in the cylinder head with its outlet directed into the intake port. The nozzle body contains a drilled central passage with a counterbore at each end. *[Figure 2-45]* The lower end is used as a chamber for air-fuel mixing before the spray leaves the nozzle. The upper bore contains a removable orifice for calibrating the nozzles. Nozzles are calibrated in several ranges, and all nozzles furnished for one engine are of the same range and are identified by a letter stamped on the hex of the nozzle body.

Drilled radial holes connect the upper counterbore with the outside of the nozzle body. These holes enter the counterbore above the orifice and draw air through a cylindrical screen fitted over the nozzle body. A shield is press-fitted on the nozzle body and extends over the greater part of the filter screen, leaving an opening near the bottom. This provides both mechanical protection and an abrupt change in the direction of airflow which keeps dirt and foreign material out of the nozzle interior.

Carburetor Maintenance

Carburetor Removal

The removal procedures vary with both the type of carburetor concerned and the type of engine on which it is used. Always refer to the applicable manufacturer's technical instructions for a particular installation. Generally, the procedures are much the same, regardless of the type of carburetor concerned.

Before removing a carburetor, make sure the fuel shutoff (or selector) valve is closed. Disconnect the throttle and mixture control linkages, and lockwire the throttle valve in the closed position. Disconnect the fuel inlet line and all vapor return, gauge, and primer lines. If the same carburetor is to be re-installed, do not alter the rigging of the throttle and mixture controls. Remove the airscoop or airscoop adapter. Remove

From the fuel-injection control valve, fuel is delivered to the fuel manifold valve, which provides a central point for dividing fuel flow to the individual cylinders. In the fuel manifold valve, a diaphragm raises or lowers a plunger valve to open or close the individual cylinder fuel supply ports simultaneously.

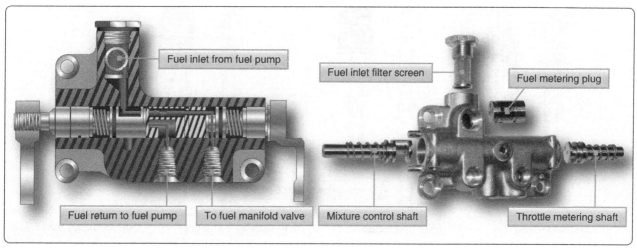

Figure 2-43. *Dual fuel control assembly.*

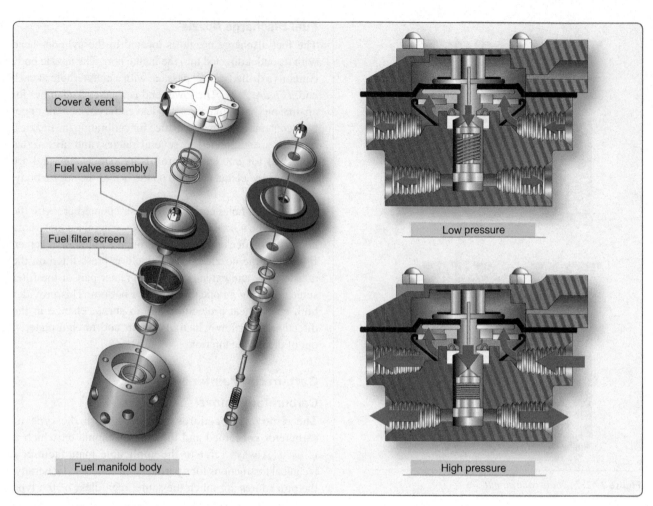

Figure 2-44. *Fuel manifold valve assembly.*

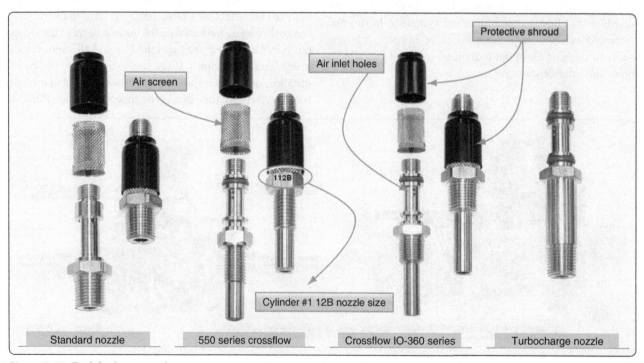

Figure 2-45. *Fuel discharge nozzles.*

the air screens and gaskets from the carburetor. Remove the nuts and washers securing the carburetor to the engine. When removing a downdraft carburetor, use extreme care to ensure that nothing is dropped into the engine. Remove the carburetor. Immediately install a protective cover on the carburetor mounting flange of the engine to prevent small parts or foreign material from falling into the engine. When there is danger of foreign material entering open fuel lines during removal or installation of the carburetor, plug them using the proper cover fittings.

Installation of Carburetor

Check the carburetor for proper lockwiring before installation on an engine. Be sure that all shipping plugs have been removed from the carburetor openings.

Remove the protective cover from the carburetor mounting flange on the engine. Place the carburetor mounting flange gasket in position. On some engines, bleed passages are incorporated in the mounting pad. The gasket must be installed so that the bleed hole in the gasket is aligned with the passage in the mounting flange.

Inspect the induction passages for the presence of any foreign material before installing the carburetor. As soon as the carburetor is placed in position on the engine, close and lockwire the throttle valves in the closed position until the remainder of the installation is completed. Place the carburetor deck screen, when feasible, in position to further eliminate the possibility of foreign objects entering the induction system.

When installing a carburetor that uses diaphragms for controlling fuel flow, connect the fuel lines and fill the carburetor with fuel. To do this, turn on the fuel boost pump and move the mixture control from the idle cutoff position to rich position. Continue the flow until oil-free fuel flows from the drain valve. This indicates that the preservative oil has been flushed from the carburetor. Turn off the fuel flow, plug the fuel inlet and vapor vent outlet, and then allow the carburetor, filled with fuel, to stand for a minimum of 8 hours. This is necessary in order to soak the diaphragms and render them pliable to the same degree as when the unit was originally calibrated. Tighten the carburetor mounting bolts to the value specified in the table of torque limits in the applicable maintenance manual. Tighten and safety any other nuts and bolts incidental to the installation of the carburetor before connecting the throttle and mixture-control levers. After the carburetor has been bolted to the engine, check the throttle and mixture-control lever on the unit for freedom of movement before connecting the control cables or linkage. Check the vapor vent lines or return lines from the carburetor to the aircraft fuel tank for restriction.

Rigging Carburetor Controls

Connect and adjust carburetor or fuel metering equipment throttle controls so that full movement of the throttle is obtained from corresponding full movement of the control in the flight deck. In addition, check and adjust the throttle control linkages so that springback on the throttle quadrant in the aircraft is equal in both the full-open and full-closed positions. Correct any excess play or looseness of control linkage or cables. Controls should be checked so that they go stop-to-stop on the carburetor. Check for complete and full travel of each control.

When installing carburetors or fuel metering equipment incorporating manual-type mixture controls that do not have marked positions, adjust the mixture control mechanism to provide an equal amount of springback at both the rich and lean ends of the control quadrant in the flight deck when the mixture control on the carburetor or fuel metering equipment is moved through the full range. Where mixture controls with detents are used, rig the control mechanism so that the designated positions on the control quadrant in the aircraft agree with the corresponding positions on the carburetor or fuel metering equipment. Controls should move freely and smoothly without binding throughout their total travel. In all cases, check the controls for proper positioning in both the advance and retard positions. Correct excess play or looseness of control linkage or cables. Safety all controls properly to eliminate the possibility of loosening from vibration during operation.

Adjusting Idle Mixtures

Excessively rich or lean idle mixtures result in incomplete combustion within the engine cylinder, with resultant formation of carbon deposits on the spark plugs and subsequent spark plug fouling. In addition, excessively rich or lean idle mixtures make it necessary to taxi at high idle speeds with resultant fast taxi speeds and excessive brake wear. Each engine must have the carburetor idle mixture tailored for the particular engine and installation if best operation is to be obtained.

Engines that are properly adjusted, insofar as valve operation, cylinder compression, ignition, and carburetor idle mixture are concerned, idle at the prescribed rpm for indefinite periods without loading up, overheating, or spark plug fouling. If an engine does not respond to idle mixture adjustment with the resultant stable idling characteristics previously outlined, some other phase of engine operation is not correct. In such cases, determine and correct the cause of the difficulty. A general guide to check and adjust the idle mixture and speed on many types of reciprocating engine is discussed in the following paragraphs. Always refer to the appropriate manual for specific information.

Before checking the idle mixture on any engine, warm up the engine until oil and cylinder head temperatures are normal. Keep the propeller control in the increase rpm setting throughout the entire process of warming up the engine. Always make idle mixture adjustments with cylinder head temperatures at normal values. The idle mixture adjustment is made on the idle mixture fuel control valve. *[Figure 2-46]* It should not be confused with the adjustment of the idle speed stop. The importance of idle mixture adjustment cannot be overstressed. Optimum engine operation at low speeds can be obtained only when proper air-fuel mixtures are delivered to every cylinder of the engine. Excessively rich idle mixtures and the resultant incomplete combustion are responsible for more spark plug fouling than any other single cause. Excessively lean idle mixtures result in faulty acceleration. Furthermore, the idle mixture adjustment affects the air-fuel mixture and engine operation well up into the cruise range.

On an engine with a conventional carburetor, the idle mixture is checked by manually leaning the mixture with the flight deck mixture control. Move the carburetor mixture control slowly and smoothly toward the idle cutoff position. On installations that do not use a manifold pressure gauge, it is necessary to observe the tachometer for an indication of a rpm change. With most installations, the idle mixture should be adjusted to provide an rpm rise prior to decreasing as the engine ceases to fire. This rpm increase varies from 10 to 50 rpm, depending on the installation. Following the momentary increase in rpm, the engine speed starts to drop. Immediately move the mixture control back to rich to prevent the engine from stopping completely.

On RSA fuel-injection engines, the optimum idle setting is one that is rich enough to provide a satisfactory acceleration under all conditions and lean enough to prevent spark plug

Figure 2-46. *Idle mixture adjustment for carburetor.*

fouling or rough operation. A rise of 25–50 rpm as the mixture control is moved to the idle cutoff position usually satisfies both of these conditions. The actual idle mixture adjustment is made by the lengthening or shortening of the linkage between the throttle lever and the idle lever. *[Figure 2-47]*

If the check of the idle mixture reveals it to be too lean or too rich, increase or decrease the idle fuel flow as required. Then, repeat the check. Continue checking and adjusting the idle mixture until it checks out properly. During this process, it may be desirable to move the idle speed stop completely out of the way and to hold the engine speed at the desired rpm by means of the throttle. This eliminates the need for frequent readjustments of the idle stop as the idle mixture is improved and the idle speed picks up. After each adjustment, clear the engine by briefly running it at higher rpm. This prevents fouling of the plugs which might otherwise be caused by incorrect idle mixture. After adjusting the idle mixture, recheck it several times to determine definitively that the mixture is correct and remains constant on repeated changes from high power back to idle. Correct any inconsistency in engine idling before releasing the aircraft for service.

Setting the idle mixture on the continental TCM fuel injection system consists of a conventional spring loaded screw located in the air throttle lever. *[Figure 2-48]* The fuel pump pressure is part of the basic calibration and requires servicing to make sure the pump pressure are set correctly before making idle adjustments. The idle mixture adjustment is the locknut at the metering valve end of the linkage between the metering valve and the air throttle levers. Tightening the nut to shorten the linkage provides a richer mixture. A leaner mixture is obtained by backing off the nut to lengthen the linkage. Adjust to obtain a slight and momentary gain in idle speed as the mixture control is slowly moved toward idle cut off. If the idle mixture is set too lean, the idle speed drops with no gain in speed.

Idle Speed Adjustment

After adjusting the idle mixture, reset the idle stop to the idle rpm specified in the aircraft maintenance manual. The engine must be warmed up thoroughly and checked for ignition system malfunctioning. Throughout any carburetor adjustment procedure, periodically run the engine up to approximately half of normal rated speed to clear the engine.

Some carburetors are equipped with an eccentric screw to adjust idle rpm. Others use a spring-loaded screw to limit the throttle valve closing. In either case, adjust the screw as required to increase or decrease rpm with the throttle retarded against the stop. Open the throttle to clear the engine; close the throttle and allow the rpm to stabilize. Repeat this operation until the desired idling speed is obtained.

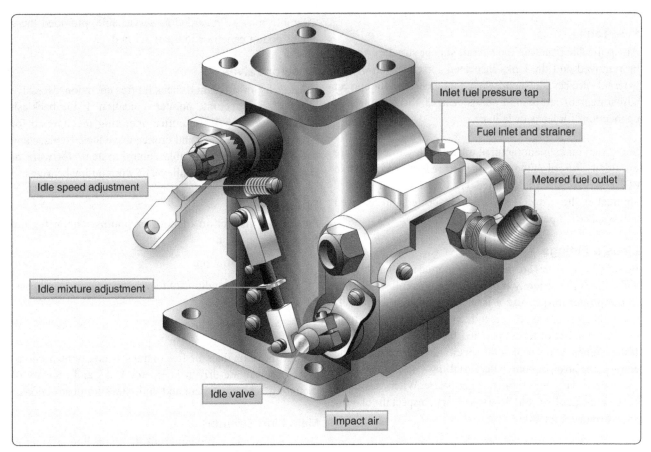

Figure 2-47. *Bendix adjustment of idle mixture linkage.*

Fuel System Inspection & Maintenance

The inspection of a fuel system installation consists basically of an examination of the system for conformity to design requirements together with functional tests to prove correct operation. Since there are considerable variations in the fuel systems used on different aircraft, no attempt has been made

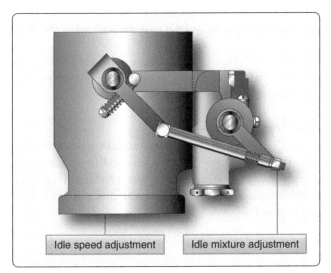

Figure 2-48. *TCM adjustment points.*

to describe any particular system in detail. It is important that the manufacturer's instructions for the aircraft concerned be followed when performing inspection or maintenance functions.

Complete System

Inspect the entire system for wear, damage, or leaks. Make sure that all units are securely attached and properly safetied. The drain plugs or valves in the fuel system should be opened to check for the presence of sediment or water. The filter and sump should also be checked for sediment, water, or slime. The filters or screens, including those provided for flow meters and auxiliary pumps, must be clean and free from corrosion. The controls should be checked for freedom of movement, security of locking, and freedom from damage due to chafing. The fuel vents should be checked for correct positioning and freedom from obstruction; otherwise, fuel flow or pressure fueling may be affected. Filler neck drains should be checked for freedom from obstruction.

If booster pumps are installed, the system should be checked for leaks by operating the pumps. During this check, the ammeter or load meter should be read and the readings of all the pumps, where applicable, should be approximately the same.

Fuel Tanks

All applicable panels in the aircraft skin or structure should be removed and the tanks inspected for corrosion on the external surfaces, for security of attachment, and for correct adjustment of straps and slings. Check the fittings and connections for leaks or failures.

Some fuel tanks manufactured of light alloy materials are provided with inhibitor cartridges to reduce the corrosive effects of combined leaded fuel and water. Where applicable, the cartridge should be inspected and renewed at the specified periods.

Lines & Fittings

Be sure that the lines are properly supported and that the nuts and clamps are securely tightened. To tighten hose clamps to the proper torque, use a hose-clamp torque wrench. If this wrench is not available, tighten the clamp finger-tight plus the number of turns specified for the hose and clamp. If the clamps do not seal at the specified torque, replace the clamps, the hose, or both. After installing a new hose, check the clamps daily and tighten if necessary. When this daily check indicates that cold flow has ceased, inspect the clamps at less frequent intervals.

Replace the hose if the plys have separated, if there is excessive cold flow, or if the hose is hard and inflexible. Permanent impressions from the clamp and cracks in the tube or cover stock indicate excessive cold flow. Replace any hose that has collapsed at the bends or as a result of misaligned fittings or lines. Some hoses tend to flare at the ends beyond the clamps. This is not an unsatisfactory condition unless leakage is present.

Blisters may form on the outer synthetic rubber cover of the hose. These blisters do not necessarily affect the serviceability of the hose. When a blister is discovered on a hose, remove the hose from the aircraft and puncture the blister with a pin. The blister should then collapse. If fluid (oil, fuel, or hydraulic) emerges from the pinhole in the blister, reject the hose. If only air emerges, then test the hose pressure at 1½ times the working pressure. If no fluid leakage occurs, the hose can be regarded as serviceable.

Puncturing the outer cover of the hose may permit the entry of corrosive elements, such as water, which could attack the wire braiding and ultimately result in failure. For this reason, puncturing the outer covering of hoses exposed to the elements should be avoided.

The external surface of hose may develop fine cracks, usually short in length, which are caused by surface aging. The hose assembly may be regarded as serviceable, provided these cracks do not penetrate to the first braid.

Selector Valves

Rotate selector valves and check for free operation, excessive backlash, and accurate pointer indication. If the backlash is excessive, check the entire operating mechanism for worn joints, loose pins, and broken drive lugs. Replace any defective parts. Inspect cable control systems for worn or frayed cables, damaged pulleys, or worn pulley bearings.

Pumps

During an inspection of booster pumps, check for the following conditions:

1. Proper operation;

2. Leaks and condition of fuel and electrical connections; and

3. Wear of motor brushes.

Be sure the drain lines are free of traps, bends, or restrictions. Check the engine-driven pump for leaks and security of mounting. Check the vent and drain lines for obstructions.

Main Line Strainers

Drain water and sediment from the main line strainer at each preflight inspection. Remove and clean the screen at the periods specified in the airplane maintenance manual. Examine the sediment removed from the housing. Particles of rubber are often early warnings of hose deterioration. Check for leaks and damaged gaskets.

Fuel Quantity Gauges

If a sight gauge is used, be sure that the glass is clear and that there are no leaks at the connections. Check the lines leading to it for leaks and security of attachment. Check the mechanical gauges for free movement of the float arm and for proper synchronization of the pointer with the position of the float.

On the electrical and electronic gauges, be sure that both the indicator and the tank units are securely mounted and that their electrical connections are tight.

Fuel Pressure Gauge

Check the pointer for zero tolerance and excessive oscillation. Check the cover glass for looseness and for proper range markings. Check the lines and connections for leaks. Be sure that there is no obstruction in the vent. Replace the instrument if it is defective.

Pressure Warning Signal

Inspect the entire installation for security of mounting and condition of the electrical, fuel, and air connections.

Check the lamp by pressing the test switch to see that it lights. Check the operation by turning the battery switch on, building up pressure with the booster pump, and observing the pressure at which the light goes out. If necessary, adjust the contact mechanism.

Water Injection Systems for Reciprocating Engines

These systems have very limited use in modern aircraft engines. Water injection was used mostly on large radial engines. The water injection system enabled more power to be obtained from the engine at takeoff than is possible without water injection. The carburetor (operating at high power settings) delivers more fuel to the engine than it actually needs. A leaner mixture would produce more power; however, the additional fuel is necessary to prevent overheating and detonation. With the injection of the antidetonant fluid, the mixture can be leaned out to that which produces maximum power, and the vaporization of the water-alcohol mixture then provides the cooling formerly supplied by the excess fuel.

Turbine Engine Fuel System—General Requirements

The fuel system is one of the more complex aspects of the gas turbine engine. It must be possible to increase or decrease the power at will to obtain the thrust required for any operating condition. In turbine-powered aircraft, this control is provided by varying the flow of fuel to the combustion chambers. However, some turboprop aircraft also use variable-pitch propellers; thus, the selection of thrust is shared by two controllable variables, fuel flow and propeller blade angle.

The quantity of fuel supplied must be adjusted automatically to correct for changes in ambient temperature or pressure. If the quantity of fuel becomes excessive in relation to mass airflow through the engine, the limiting temperature of the turbine blades can be exceeded, or it will produce compressor stall and a condition referred to as rich blowout. Rich blowout occurs when the amount of oxygen in the air supply is insufficient to support combustion and when the mixture is cooled below the combustion temperature by the excess fuel. The other extreme, lean flameout, occurs if the fuel quantity is reduced proportionally below the air quantity. The engine must operate through acceleration and deceleration without any fuel-control-related problems.

The fuel system must deliver fuel to the combustion chambers not only in the right quantity, but also in the right condition for satisfactory combustion. The fuel nozzles form part of the fuel system and atomize or vaporize the fuel so that it ignites and burns efficiently. The fuel system must also supply fuel so that the engine can be easily started on the ground and in the air. This means that the fuel must be injected into the combustion chambers in a combustible condition during engine starting, and that combustion must be sustained while the engine is accelerating to its normal idling speed. Another critical condition to which the fuel system must respond occurs during a rapid acceleration. When the engine is accelerated, energy must be furnished to the turbine in excess of that necessary to maintain a constant rpm. However, if the fuel flow increases too rapidly, an over rich mixture can be produced, with the possibility of a rich blowout or compressor stall.

Turbofan, turbojet, turboshaft, and turboprop engines are equipped with a fuel control unit which automatically satisfies the requirements of the engine. Although the basic requirements apply generally to all gas turbine engines, the way in which individual fuel controls meet these needs cannot be conveniently generalized.

Turbine Fuel Controls

Gas turbine engine fuel controls can be divided into three basic groups:

1. Hydromechanical,

2. Hydromechanical/electronic, and

3. Full Authority Digital Engine (or Electronics) Control (FADEC).

The hydromechanical/electronic fuel control is a hybrid of the two types of fuel control but can function solely as a hydromechanical control. In the dual mode, inputs and outputs are electronic, and fuel flow is set by servo motors. The third type, FADEC, uses electronic sensors for its inputs and controls fuel flow with electronic outputs. The FADEC-type control gives the electronic controller (computer) complete control. The computing section of the FADEC system depends completely on sensor inputs to the electronic engine control (EEC) to meter the fuel flow. The fuel metering device meters the fuel using only outputs from the EEC. Most turbine fuel controls are quickly going to the FADEC type of control. This electronically controlled fuel control is very accurate in scheduling fuel by sensing many of the engine parameters.

Regardless of the type, all fuel controls accomplish essentially the same function. That function is to schedule the fuel flow to match the power required by the pilot. Some sense more engine variables than others. The fuel control can sense many different inputs, such as power lever position, engine rpm for each spool, compressor inlet pressure and temperature, burner pressure, compressor discharge pressure, and many more parameters as needed by the specific engine. These variables affect the amount of thrust that an engine produces for a given fuel flow. By sensing these parameters, the fuel control has a clear picture of what is happening in the engine and can adjust fuel flow as needed. Each type of turbine engine has its own specific needs for fuel delivery and control.

Hydromechanical Fuel Control

Hydromechanical fuel controls were used and are still used on many engines, but their use is becoming limited giving way to electronic based controls. Fuel controls have two sections, computing and metering, to provide the correct fuel flow for the engine. A pure hydromechanical fuel control has no electronic interface assisting in computing or metering the fuel flow. It also is generally driven by the gas generator gear train of the engine to sense engine speed. Other mechanical engine parameters that are sensed are compressor discharge pressure, burner pressure, exhaust temperature, and inlet air temperature and pressure. Once the computing section determines the correct amount of fuel flow, the metering section through cams and servo valves delivers the fuel to the engine fuel system. Actual operating procedures for a hydromechanical fuel control is very complicated and still the fuel metering is not as accurate as with an electronic type of interface or control. Electronic controls can receive more inputs with greater accuracy than hydromechanical controls. Early electronic controls used a hydromechanical control with an electronic system added on the system to fine tune the metering of the fuel. This arrangement also used the hydromechanical system as a backup if the electronic system failed. *[Figure 2-49]*

Hydromechanical/Electronic Fuel Control

The addition of the electronic control to the basic hydromechanical fuel control was the next step in the development of turbine engine fuel controls. Generally, this type of system used a remotely located EEC to adjust the fuel flow. A description of a typical system is explained in the following information. The basic function of the engine fuel system is to pressurize the fuel, meter fuel flow, and deliver atomized fuel to the combustion section of the engine. Fuel flow is controlled by a hydromechanical fuel control assembly, which contains a fuel shutoff section and a fuel metering section.

This fuel control unit is sometimes mounted on the vane fuel pump assembly. It provides the power lever connection and the fuel shutoff function. The unit provides mechanical overspeed protection for the gas generator spool during normal (automatic mode) engine operation. In automatic mode, the EEC is in control of metering the fuel. In manual mode, the hydromechanical control takes over.

During normal engine operation, a remotely mounted electronic fuel control unit (EFCU) (same as an EEC) performs the functions of thrust setting, speed governing and acceleration, and deceleration limiting through EFCU outputs

to the fuel control assembly in response to power lever inputs. In the event of electrical or EFCU failure, or at the option of the pilot, the fuel control assembly functions in manual mode to allow engine operation at reduced power under control of the hydromechanical portion of the controller only.

The total engine fuel and control system consists of the following components and provides the functions as indicated:

1. The vane fuel pump assembly is a fixed displacement fuel pump that provides high pressure fuel to the engine fuel control system. *[Figure 2-50]*

2. The filter bypass valve in the fuel pump allows fuel to bypass the fuel filter when the pressure drop across the fuel filter is excessive. An integral differential pressure indicator visually flags an excessive differential pressure condition before bypassing occurs, by extending a pin from the fuel filter bowl. Fuel pump discharge flow in excess of that required by the fuel control assembly is returned from the control to the pump interstage.

3. The hydromechanical fuel control assembly provides the fuel metering function of the EFCU.

   Fuel is supplied to the fuel control through a 200-micron inlet filter screen and is metered to the engine by the servo-operated metering valve. It is a fuel flow/compressor discharge pressure (Wf/P3) ratio device that positions the metering valve in response to engine compressor discharge pressure (P3). Fuel pressure differential across the servo valve is maintained by the servo-operated bypass valve in response to commands from the EFCU. *[Figure 2-49]* The manual mode solenoid valve is energized in the automatic mode. The automatic mode restricts operation of the mechanical speed governor. It is restricted to a single overspeed governor setting above the speed range controlled electronically. Deenergizing the manual mode valve enables the mechanical speed governor to function as an all speed governor in response to power lever angle (PLA). The fuel control system includes a low power sensitive torque motor which may be activated to increase or decrease fuel flow in the automatic mode (EFCU mode). The torque motor provides an interface to an electronic control unit that senses various engine and ambient parameters and activates the torque motor to meter fuel flow accordingly. This torque motor provides electromechanical conversion of an electrical signal from the EFCU. The torque motor current is zero in the manual mode, which establishes a fixed Wf/P3 ratio.

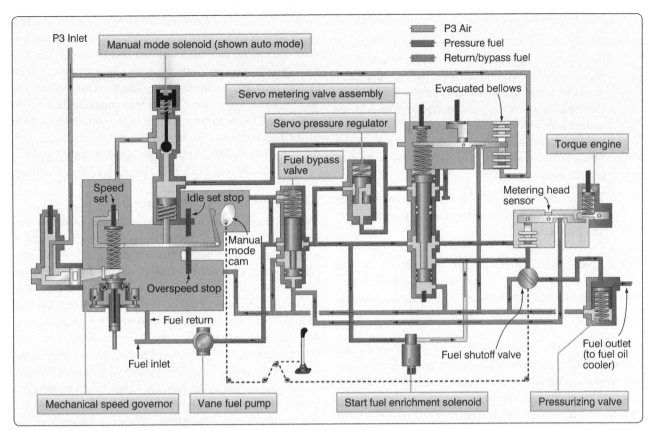

Figure 2-49. *Fuel control assembly schematic hydromechanical/electronic.*

Figure 2-50. *Fuel pump and filter.*

This fixed Wf/P3 ratio is such that the engine operates surge free and is capable of producing a minimum of 90 percent thrust up to 30,000 feet for this example system. All speed governing of the high-pressure spool (gas generator) is achieved by the flyweight governor. The flyweight governor modulates a pneumatic servo, consistent with the speed set point as determined by the power lever angle (PLA) setting. The pneumatic

servo accomplishes Wf/P3 ratio modulation to govern the gas generator speed by bleeding down the P3 acting on the metering valve servo. The P3 limiter valve bleeds down the P3 pressure acting in the metering valve servo when engine structural limits are encountered in either control mode. The start fuel enrichment solenoid valve provides additional fuel flow in parallel with the metering valve when required for engine cold starting or altitude restarts. The valve is energized by the EFCU when enrichment is required. It is always deenergized in the manual mode to prevent high altitude sub-idle operation. Located downstream of the metering valve are the manual shutoff and pressurizing valves. The shutoff valve is a rotary unit connected to the power lever. It allows the pilot to direct fuel to the engine manually. The pressurizing valve acts as a discharge restrictor to the hydromechanical control. It functions to maintain minimum operating pressures throughout the control. The pressurizing valve also provides a positive leak-tight fuel shutoff to the engine fuel nozzles when the manual valve is closed.

4. The flow divider and drain valve assembly proportions fuel to the engine primary and secondary fuel nozzles. It drains the nozzles and manifolds at engine shutdown. It also incorporates an integral solenoid for

modifying the fuel flow for cold-starting conditions.

During an engine start, the flow divider directs all flow through the primary nozzles. After start, as the engine fuel demand increases, the flow divider valve opens to allow the secondary nozzles to function. During all steady-state engine operation, both primary and secondary nozzles are flowing fuel. A 74-micron, self-bypassing screen is located under the fuel inlet fitting and provides last chance filtration of the fuel prior to the fuel nozzles.

5. The fuel manifold assembly is a matched set consisting of both primary and secondary manifolds and the fuel nozzle assemblies.

Twelve fuel nozzles direct primary and secondary fuel through the nozzles causing the fuel to swirl and form a finely atomized spray. The manifold assembly provides fuel routing and atomizing to ensure proper combustion.

The EEC system consists of the hydromechanical fuel control, EFCU, and aircraft mounted power lever angle potentiometer. Aircraft-generated control signals include inlet pressure, airstream differential pressure, and inlet temperature plus pilot selection of either manual or auto mode for the EFCU operation. Engine-generated control signals include fan spool speed, gas generator spool speed, inner turbine temperature, fan discharge temperature, and compressor discharge pressure. Aircraft- and engine-generated control signals are directed to the EFCU where these signals are interpreted. The PLA potentiometer is mounted in the throttle quadrant. The PLA potentiometer transmits an electrical signal to the EFCU, which represents engine thrust demand in relation to throttle position. If the EFCU determines a power change is required, it commands the torque motor to modulate differential pressure at the head sensor. This change in differential pressure causes the metering valve to move, varying fuel flow to the engine as required. The EFCU receives electrical signals which represent engine operating variables. It also receives a pilot-initiated signal (by power-lever position) representing engine thrust demand. The EFCU computes electrical output signals for use by the engine fuel control for scheduling engine operation within predetermined limits. The EFCU is programmed to recognize predetermined engine operating limits and to compute output signals such that these operating limits are not exceeded. The EFCU is remotely located and airframe mounted. An interface between the EFCU and aircraft/engine is provided through the branched wiring harness assembly. *[Figure 2-51]*

FADEC Fuel Control Systems

A full authority digital electronic control (FADEC) has been developed to control fuel flow on most new turbine engine models. A true FADEC system has no hydromechanical fuel control backup system. The system uses electronic sensors that feed engine parameter information into the EEC. The EEC gathers the needed information to determine the amount of fuel flow and transmits it to a fuel metering valve. The fuel metering valve simply reacts to the commands from the EEC. The EEC is a computer that is the computing section of the fuel delivery system and the metering valve meters the fuel flow. FADEC systems are used on many types of turbine engines from APUs to the largest propulsion engines.

FADEC for an Auxiliary Power Unit

An APU engine uses the aircraft fuel system to supply fuel to the fuel control. An electric boost pump may be used to supply fuel under pressure to the control. The fuel usually passes through an aircraft shutoff valve that is tied to the fire detecting/extinguishing system. An aircraft furnished inline fuel filter may also be used. Fuel entering the fuel control unit first passes through a 10-micron filter. If the filter becomes contaminated, the resulting pressure drop opens the filter bypass valve and unfiltered fuel then is supplied to the APU. Shown in *Figure 2-52* is a pump with an inlet pressure access plug so that a fuel pressure gauge might be installed for troubleshooting purposes. Fuel then enters a positive displacement, gear-type pump. Upon discharge from the pump, the fuel passes through a 70-micron screen. The screen is installed at this point to filter any wear debris that might be discharged from the pump element. From the screen, fuel branches to the metering valve, differential pressure valve, and the ultimate relief valve. Also shown at this point is a pump discharge pressure access plug, another point where a pressure gauge might be installed.

The differential pressure valve maintains a constant pressure drop across the metering valve by bypassing fuel to the pump inlet so that metered flow is proportional to metering valve area. The metering valve area is modulated by the torque motor, which receives variable current from the engine control unit (ECU). The ultimate relief valve opens to bypass excess fuel back to the pump inlet whenever system pressure exceeds a predetermined pressure. This occurs during each shutdown since all flow is stopped by the shutoff valve and the differential pressure valve, is unable to bypass full pump capacity. Fuel flows from the metering valve out of the fuel control unit (FCU), through the solenoid shutoff valve and on to the atomizer. Initial flow is through the primary nozzle tip only. The flow divider opens at higher pressure and adds flow through the secondary path.

FADEC Fuel Control Propulsion Engine

Many large high-bypass turbofan engines use the FADEC type of fuel control system. The EEC is the primary component of the FADEC engine fuel control system. The

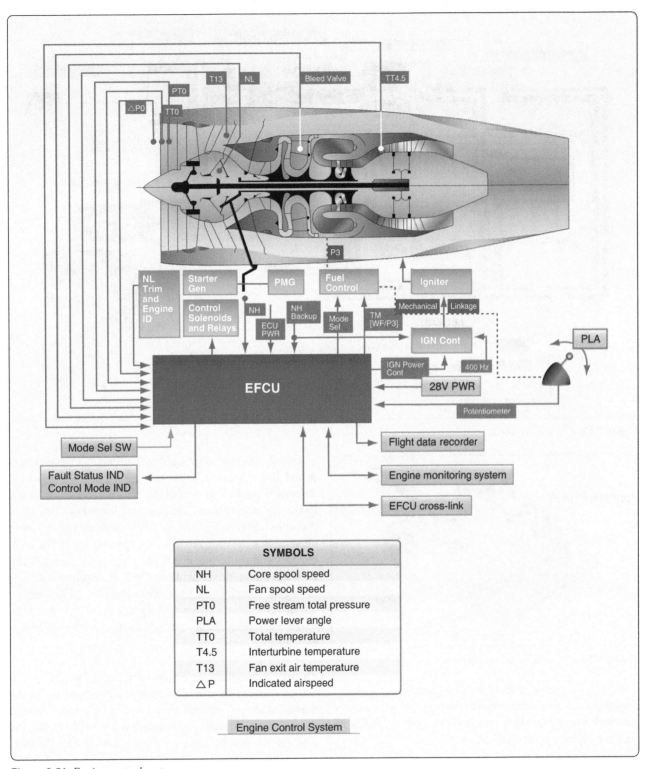

Figure 2-51. *Engine control system.*

EEC is a computer that controls the operation of the engine. The EEC housing contains two electronic channels (two separate computers) that are physically separated internally and is naturally cooled by convection. The EEC is generally placed in an area of the engine nacelle that is cool during engine operation. It attaches to the lower-left fan case with shock mounts. *[Figure 2-53]*

The EEC computer uses data it receives from many engine sensors and airplane systems to control the engine operation. It receives electronic signals from the flight deck to set engine power or thrust. The throttle lever angle resolver supplies the

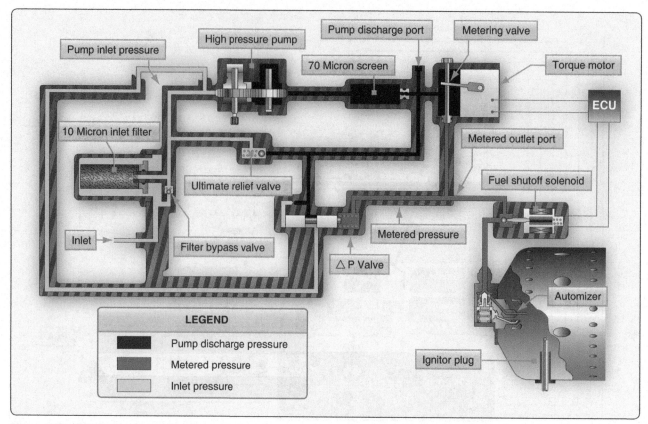

Figure 2-52. *APU fuel system schematic.*

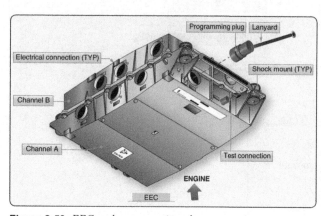

Figure 2-53. *EEC and programming plug.*

EEC with a signal in proportion to the thrust lever position. The EEC controls most engine components and receives feedback from them. Many components supply the EEC with data for engine operation.

Power for the EEC comes from the aircraft electrical system or the permanent magnet alternator (PMA). When the engine is running, the PMA supplies power to the EEC directly. The EEC is a two channel computer that controls every aspect of engine operation. Each channel, which is an independent computer, can completely control the operation of the engine. The processor does all of the control calculations and supplies all the data for the control signals for the torque motors and

solenoids. The cross-talk logic compares data from channels A and B and uses the cross-talk logic to find which EEC channel is the best to control the output driver for a torque motor or solenoid bank. The primary channel controls all of the output drivers. If the cross-talk logic finds that the other channel is better for control of a specific bank, the EEC changes control of that one bank to the other channel. The EEC has output driver banks that supply the control signals to engine components. Each channel of the EEC supplies the driver banks with control signals. The EEC has both volatile and nonvolatile memory to store performance and maintenance data.

The EEC can control the engine thrust in two modes, which can be selected by use of a mode selection switch. In the normal mode, engine thrust is set with engine pressure ratio (EPR); in the alternate mode, thrust is set by N1. When the fuel control switch is moved from run to cutoff, the EEC resets. During this reset, all fault data is recorded in the nonvolatile memory. The EEC controls the metering valve in the fuel metering unit to supply fuel flow for combustion. *[Figure 2-54]* The fuel metering unit is mounted on the front face of the gearbox and is attached to the front of the fuel pump. *[Figure 2-55]* The EEC also sends a signal to the minimum pressure and shutoff valve in the fuel metering unit to start or stop fuel flow. The EEC receives position feedback for several engine components by using

Figure 2-54. *Fuel metering unit.*

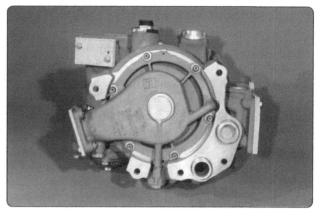

Figure 2-55. *Fuel pump.*

rotary differential transformer, linear variable differential transformer, and thermocouples. These sensors feed engine parameter information from several systems back to the EEC. The fuel control run cutoff switch controls the high pressure fuel shut off valve that allows or cuts off fuel flow. The fuel temperature sensor thermocouple attaches to the fuel outlet line on the rear of the fuel/oil cooler and sends this information to the EEC. The EEC uses a torque motor driver to control the position of the metering valve in the fuel metering unit. The EEC uses solenoid drivers to control the other functions of the fuel metering unit (FMU). The EEC also controls several other subsystems of the engine, as shown in *Figure 2-56*, through torque motors and solenoids, such as fuel and air oil coolers, bleed valves, variable stator vanes, turbine cooling air valves, and the turbine case cooling system.

Each channel of the EEC has seven electrical connections, three on each side and one on the bottom. Both channels share the inputs of the two connections on the top of the EEC. These are the programming plug and test connector. The programming plug selects the proper software in the EEC for the thrust rating of the engine. The plug attaches to the engine fan case with a lanyard. When removing the EEC, the plug remains with the engine. Each channel of the EEC has three pneumatic connections on the bottom of the EEC. Transducers inside the EEC supply the related and opposite EEC channel with a signal in proportion to the pressure. The pressures that are read by the EEC are ambient pressure, burner pressure, low pressure compressor (LPC) exit pressure, and fan inlet pressure. Each channel has its own wire color that connects the EEC to its sensors. Channel A wiring is blue and channel B sensor signals are green. The non-EEC circuit wire is gray while the thermocouple signals are yellow. This color coding helps simplify which sensors are used with each channel.

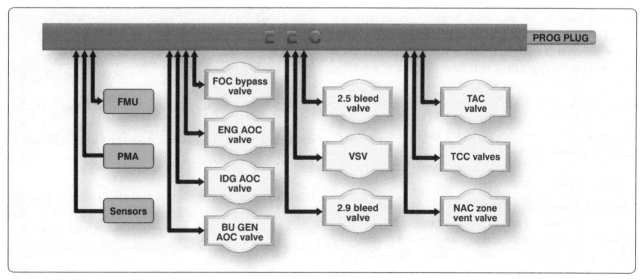

Figure 2-56. *Systems controlled by EEC.*

Fuel System Operation

The fuel pump receives fuel from the airplane fuel system. The low pressure boost stage of the pump pressurizes the fuel and sends it to the fuel/oil cooler (FOC). The fuel flows from the FOC, through the fuel pump filter element, and then to the high pressure main stage of the pump. The high pressure main stage increases the fuel pressure and sends it to the fuel metering unit (FMU). It also supplies servo fuel to the servo fuel heater and engine components. Fuel for combustion (metered fuel) goes through the fuel flow transmitter to the distribution valve. *[Figure 2-57]* The fuel distribution valve supplies metered fuel to the fuel supply manifolds. *[Figure 2-58]* The fuel injectors get the metered fuel from the fuel supply manifolds and spray the fuel into the engine for combustion. *[Figure 2-59]* The fuel pump housing contains a disposable fuel filter element. The fuel filter differential pressure switch supplies a signal to the EEC that indicates an almost clogged filter condition. Unfiltered fuel can then bypass the filter element if the element becomes clogged.

Water Injection System

On warm days, thrust is reduced because of the decrease in air density. This can be compensated for by injecting water at the compressor inlet or diffuser case. This lowers the air temperature and increases air density. A microswitch in the fuel control is actuated by the control shaft when the power lever is moved toward the maximum power position.

A water injection speed reset servo resets the speed adjustment to a higher value during water injection. Without this adjustment, the fuel control would decrease rpm so that no additional thrust would be realized during water injection. The servo is a shuttle valve that is acted upon by water pressure during water injection. Movement of the servo displaces a lever on the cam-operated lever linkage to the speed governor speeder spring, increasing the force of the speeder spring and increasing the set speed. Because the resulting rpm is usually higher while water is flowing, increased thrust during water injection is ensured. If the water injection system is not armed in the flight deck or if there is no water available, nothing happens when the water injection switch in the fuel control unit is actuated. When water is available, a portion of it is directed to the water injection speed re-set servo. Water injection systems are not normally used on high-bypass turbofan engines.

Fuel Control Maintenance

The field repair of the turbine engine fuel control is very limited. The only repairs permitted in the field are the replacement of the control and adjustments afterwards. These adjustments are limited to the idle rpm and the maximum speed adjustment, commonly called trimming the engine. Both adjustments are made in the normal range of operation.

Figure 2-57. *Fuel flow transmitter.*

Figure 2-58. *Fuel distribution valve.*

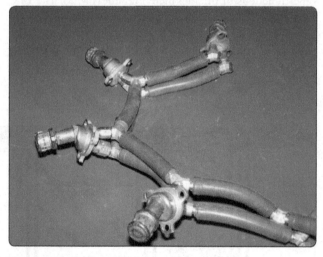

Figure 2-59. *Fuel manifolds.*

During engine trimming, the fuel control is checked for idle rpm, maximum rpm, acceleration, and deceleration. The procedures used to check the fuel control vary depending on the aircraft and engine installation.

The engine is trimmed in accordance with the procedures in

the maintenance or overhaul manual for a particular engine. In general, the procedure consists of obtaining the ambient air temperature and the field barometric pressure (not sea level) immediately preceding the trimming of the engine. Care must be taken to obtain a true temperature reading comparable to that of the air that enters the engine. Using these readings, the desired turbine discharge pressure or EPR (engine pressure ratio) reading is computed from charts published in the maintenance manual.

The engine is operated at full throttle (or at the part power control trim stop) for a sufficient period of time to ensure that it has completely stabilized. Five minutes is the usual recommended stabilization period. A check should be made to ensure that the compressor air-bleed valves have fully closed and that all accessory drive air bleed for which the trim curve has not been corrected (such as a cabin air-conditioning unit) has been turned off. When the engine has stabilized, a comparison is made of the observed and the computed turbine discharge pressure Pt7 (or EPR) to determine the approximate amount of trimming required. If a trim is necessary, the engine fuel control is then adjusted to obtain the target turbine discharge pressure Pt7 or EPR on the gauge. Immediately following the fuel control adjustment, the tachometer reading is observed and recorded. Fuel flow and exhaust gas temperature readings should also be taken.

On Pratt and Whitney engines, using a dual-spool compressor, the observed N2 tachometer reading is next corrected for speed bias by means of temperature/rpm curve. The observed tachometer reading is divided by the percent trim speed obtained from the curve. The result is the new engine trim speed in percent, corrected to standard day (59 °F or 15 °C) temperature. The new trim speed in rpm may be calculated when the rpm at which the tachometer reads 100 percent is known. This value may be obtained from the appropriate engine manual. If all these procedures have been performed satisfactorily, the engine has been properly trimmed.

Engine trimming should always be carried out under precisely controlled conditions with the aircraft headed into the wind. Precise control is necessary to ensure maintenance of a minimum thrust level upon which the aircraft performance is based. In addition, precise control of engine trimming contributes to better engine life in terms of both maximum time between overhaul and minimum out-of-commission time due to engine maintenance requirements. Engines should never be trimmed if icing conditions exist.

Most electronic control fuel control systems do not require trimming or mechanical adjustments. Changes to the EEC in the FADEC system is normally accomplished through software changes or changing the EEC.

Engine Fuel System Components

Main Fuel Pumps (Engine Driven)

Main fuel pumps deliver a continuous supply of fuel at the proper pressure and at all times during operation of the aircraft engine. The engine-driven fuel pump must be capable of delivering the maximum needed flow at appropriate pressure to obtain satisfactory nozzle spray and accurate fuel regulation.

These engine driven fuel pumps may be divided into two distinct system categories:

1. Nonconstant displacement and

2. Nonpositive displacement.

Their use depends on where in the engine fuel system they are used. A nonpositive-displacement pump produces a continuous flow. However, because it does not provide a positive internal seal against slippage, its output varies considerably as pressure varies. Centrifugal and propeller pumps are examples of nonpositive-displacement pumps. If the output port of a nonpositive-displacement pump was blocked off, the pressure would rise and output would decrease to zero. Although the pumping element would continue moving, flow would stop because of slippage inside the pump. In a positive displacement pump, slippage is negligible compared to the pump's volumetric output flow. If the output port were plugged, pressure would increase instantaneously to the point that the pump pressure relief valve opens. Generally, a nonpositive-displacement is used at the inlet of the engine-driven pump to provide positive flow to the second stage of the pump. The output of a centrifugal pump can be varied as needed and is sometimes referred to as a boost stage of the engine-driven pump.

The second or main stage of the engine-driven fuel pump for turbine engines is generally a positive displacement type of pump. The term "positive displacement" means that the gear supplies a fixed quantity of fuel to the engine for every revolution of the pump gears. Gear-type pumps have approximately straight line flow characteristics, whereas fuel requirements fluctuate with flight or ambient air conditions. Hence, a pump of adequate capacity at all engine operating conditions has excess capacity over most of the range of operation. This is the characteristic that requires the use of a pressure relief valve for bypassing excess fuel back to the inlet. A typical two-stage turbine engine driven pump is illustrated in *Figure 2-60*. The impeller, which is driven at a greater speed than the high pressure elements, increases the fuel pressure depending upon engine speed.

The fuel is discharged from the boost element (impeller) to the two high-pressure gear elements. A relief valve is

Figure 2-60. *Dual element fuel pump.*

incorporated in the discharge port of the pump. This valve opens at a predetermined pressure and is capable of bypassing the total fuel flow. This allows fuel in excess of that required for engine operation at the time to be recirculated. The bypass fuel is routed to the inlet side of the second stage pump. Fuel flows from the pump to the fuel metering unit or fuel control. The fuel control is often attached to the fuel pump. The fuel pump is also lubricated by the fuel passing through the pump, and it should never be turned without fuel flow supplied to the inlet of the pump. As the engine coasts down at shutdown, the fuel pump should be provided with fuel until it comes to a stop.

Fuel Heater

Gas turbine engine fuel systems are very susceptible to the formation of ice in the fuel filters. When the fuel in the aircraft fuel tanks cools to 32 °F or below, residual water in the fuel tends to freeze, forming ice crystals. When these ice crystals in the fuel become trapped in the filter, they block fuel flow to the engine, which causes a very serious problem. To prevent this problem, the fuel is kept at a temperature above freezing. Warmer fuel also can improve combustion, so some means of regulating the fuel temperature is needed.

The method of regulating fuel temperature is to use a fuel heater which operates as a heat exchanger to warm the fuel. The heater can use engine bleed air or engine lubricating oil as a source of heat. The bleed air type is called an air-to-liquid exchanger and the oil type is known as a liquid-to-liquid heat exchanger. The function of a fuel heater is to protect the engine fuel system from ice formation. However, should ice form in the filter, the heater can also be used to thaw ice on the fuel screen to allow fuel to flow freely again. On most installations, the fuel filter is fitted with a pressure-drop warning switch, which illuminates a warning light on the flight deck instrument panel. If ice begins to collect on the filter surface, the pressure across the filter slowly decreases. When the pressure reaches a predetermined value, the warning light alerts the flight deck personnel.

Fuel deicing systems are designed to be used intermittently. The control of the system may be manual, by a switch in the flight deck, or automatic, using a thermostatic sensing element in the fuel heater to open or close the air or oil shutoff valve. A fuel heater system is shown in *Figure 2-61*. In a FADEC system, the computer controls the fuel temperature by sensing the fuel temperature and heating it as needed.

Fuel Filters

A low-pressure filter is installed between the supply tanks and the engine fuel system to protect the engine-driven fuel pump and various control devices. An additional high-pressure fuel filter is installed between the fuel pump and the fuel control to protect the fuel control from contaminants that could come from the low pressure pump.

The three most common types of filters in use are the micron filter, the wafer screen filter, and the plain screen mesh filter. The individual use of each of these filters is dictated by the filtering treatment required at a particular location. The micron filter has the greatest filtering action of any present-day filter type and, as the name implies, is rated in microns. *[Figure 2-62]* (A micron is one thousandth of 1 millimeter.) The porous cellulose material frequently used in construction of the filter cartridges is capable of removing foreign matter measuring from 10–25 microns. The minute openings make this type of filter susceptible to clogging; therefore, a bypass valve is a necessary safety factor.

Since the micron filter does such a thorough job of removing foreign matter, it is especially valuable between the fuel tank and engine. The cellulose material also absorbs water, preventing it from passing through the pumps. If water does seep through the filter, which happens occasionally when filter elements become saturated with water, the water can and does quickly damage the working elements of the fuel pump and control units, since these elements depend solely on the fuel for their lubrication. To reduce water damage to pumps and control units, periodic servicing and replacement of filter elements is imperative. Daily draining of fuel tank sumps and low-pressure filters eliminates much filter trouble and undue maintenance of pumps and fuel control units.

The most widely used fuel filters are the 200-mesh and the 35-mesh micron filters. They are used in fuel pumps, fuel controls, and between the fuel pump and fuel control where removal of micronic particles is needed. These filters, usually made of fine-mesh steel wire, are a series of layers of wire.

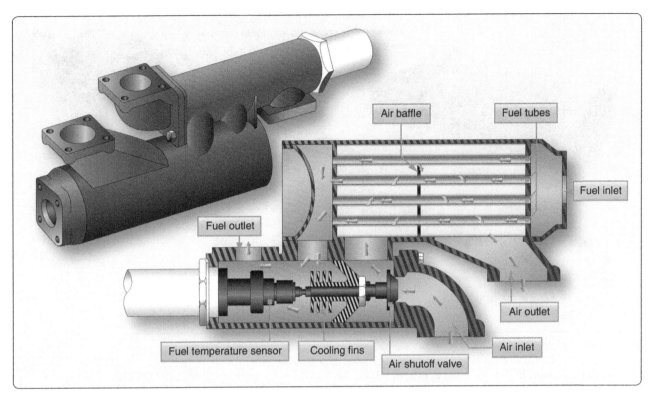

Figure 2-61. *Fuel heater.*

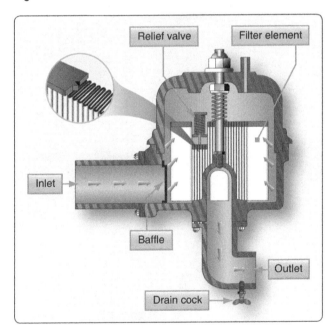

Figure 2-62. *Aircraft fuel filter.*

The wafer screen type of filter has a replaceable element, which is made of layers of screen discs of bronze, brass, steel, or similar material. *[Figure 2-63]* This type of filter is capable of removing micronic particles. It also has the strength to withstand high pressure.

Fuel Spray Nozzles & Fuel Manifolds

Although fuel spray nozzles are an integral part of the fuel system, their design is closely related to the type of combustion chamber in which they are installed. The fuel nozzles inject fuel into the combustion area in a highly atomized, precisely patterned spray so that burning is completed evenly, in the shortest possible time, and in the smallest possible space. It is very important that the fuel be evenly distributed and well centered in the flame area within the liners. This is to preclude the formation of any hot spots or hot streaking in the combustion chambers and to prevent the flame burning through the liner.

Fuel nozzle types vary considerably between engines, although for the most part fuel is sprayed into the combustion area under pressure through small orifices in the nozzles. The two types of fuel nozzles generally used are the simplex and the duplex configurations. The duplex nozzle usually requires a dual manifold and a pressurizing valve or flow divider for dividing primary and secondary (main) fuel flow, but the simplex nozzle requires only a single manifold for proper fuel delivery.

The fuel nozzles can be constructed to be installed in various ways. The two methods used quite frequently are:

1. External mounting wherein a mounting pad is provided for attachment of the nozzles to the case or the inlet air elbow, with the nozzle near the dome; or

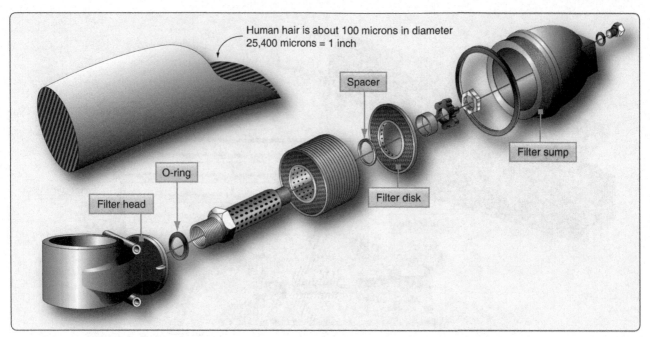

Human hair is about 100 microns in diameter
25,400 microns = 1 inch

Spacer

Filter sump

O-ring

Filter disk

Filter head

Figure 2-63. *Wafer screen filter.*

2. Internal mounting at the liner dome, in which the chamber cover must be removed for replacement or maintenance of the nozzle.

The nozzles used in a specific engine should be matched so that they flow equal amounts of fuel. Even fuel distribution is important to efficient combustion in the burner section. The fuel nozzle must present a fine spray with the correct pattern and optimum atomization.

Simplex Fuel Nozzle

The simplex fuel nozzle was the first nozzle type used in turbine engines and was replaced in most installations with the duplex nozzle, which gave better atomization at starting and idling speeds. The simplex nozzle is still being used in several installations. *[Figure 2-64]* Each of the simplex nozzles consists of a nozzle tip, an insert, and a strainer made up of fine-mesh screen and a support.

Duplex Fuel Nozzle

The duplex fuel nozzle is widely used in present day gas turbine engines and produces two different spray patterns. As mentioned previously, its use requires a flow divider, but at the same time it offers a desirable spray pattern for combustion over a wide range of operating pressures. *[Figure 2-65]* A nozzle typical of this type is illustrated in *Figure 2-66.*

Airblast Nozzles

Airblast nozzles are used to provide improved mixing of the fuel and airflow to provide an optimum spray for combustion. As can be seen in *Figure 2-64*, swirl vanes are used to mix the

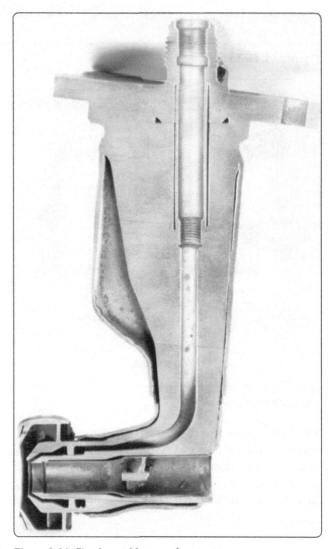

Figure 2-64. *Simplex airblast nozzle cutaway.*

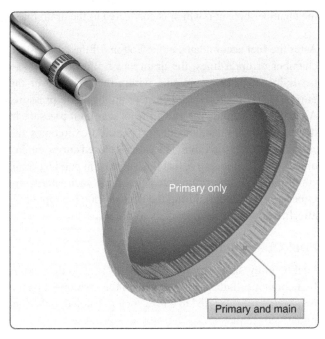

Figure 2-65. *Duplex nozzle spray pattern.*

air and fuel at the nozzle opening. By using a proportion of the primary combustion airflow in the fuel spray, locally rich fuel concentrations can be reduced. This type of fuel nozzle can be either simplex or duplex, depending upon the engine. This nozzle type can operate at lower working pressures than other nozzles which allows for lighter pumps. This airblast nozzle also helps in reducing the tendency of the nozzle to carbon up which can disturb the flow pattern.

Flow Divider

A flow divider creates primary and secondary fuel supplies that are discharged through separate manifolds, providing two separate fuel flows. *[Figure 2-67]* Metered fuel from the fuel control enters the inlet of the flow divider and passes through an orifice and then on to the primary nozzles. A passage in the flow divider directs fuel flow from both sides of the orifice to a chamber. This chamber contains a differential pressure bellows, a viscosity compensated restrictor (VCR), and a surge dampener. During engine start, fuel pressure is applied to the inlet port and across the VCR, surge dampener, and on to the primary side of the nozzles. Fuel is also applied under pressure to the outside of the flow divider bellows and through the surge dampener to the inside of the flow divider bellows. This unequal pressure causes the flow divider valve to remain closed. When fuel flow increases, the differential pressure on the bellows also increases. At a predetermined pressure, the bellows compresses, allowing the flow divider valve to open. This action starts fuel flow to the secondary manifold, which increases the fuel flow to the engine. This fuel flows out of the secondary opening in the nozzles.

Fuel Pressurizing & Dump Valves

The fuel pressurizing valve is usually required on engines incorporating duplex fuel nozzles to divide the flow into primary and secondary manifolds. As the fuel required for starting and altitude idling flows, it passes through the primary line. As the fuel flow increases, the valve begins to open the main line until at maximum flow the secondary line is passing approximately 90 percent of the fuel.

Fuel pressurizing valves usually trap fuel forward of the

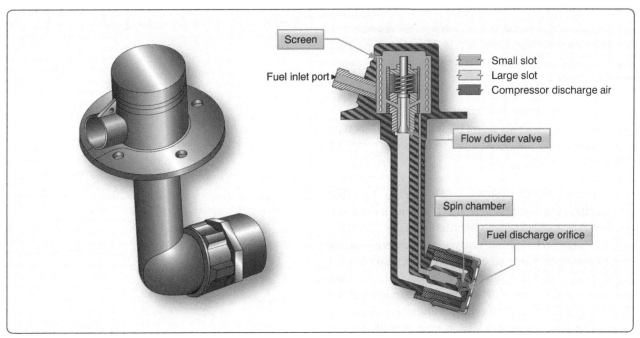

Figure 2-66. *Duplex fuel nozzle.*

Figure 2-67. *Flow divider.*

manifold, giving a positive cutoff. This cutoff prevents fuel from dribbling into the manifold and through the fuel nozzles, limiting afterfires and carbonization of the fuel nozzles. Carbonization occurs because combustion chamber temperatures are lowered, and the fuel is not completely burned.

A flow divider performs essentially the same function as a pressurizing valve. It is used, as the name implies, to divide flow to the duplex fuel nozzles. It is not unusual for units performing identical functions to have different nomenclature between engine manufacturers.

Combustion Drain Valves

The drain valves are units used for draining fuel from the various components of the engine where accumulated fuel is most likely to present operating problems. The possibility of combustion chamber accumulation with the resultant fire hazard is one problem. A residual problem is leaving gum deposits, after evaporation, in such places as fuel manifolds and fuel nozzles.

In some instances, the fuel manifolds are drained by an individual unit known as a drip or dump valve. This type of valve may operate by pressure differential, or it may be solenoid operated.

The combustion chamber drain valve drains fuel that accumulates in the combustion chamber after each shutdown and fuel that may have accumulated during a false start. If the combustion chambers are the can type, fuel drains by gravity down through the flame tubes or interconnector tubes until it gathers in the lower chambers, which are fitted with drain lines to the drain valve. If the combustion chamber is of the basket or annular type, the fuel merely drains through the air holes in the liner and accumulates in a trap in the bottom of the chamber housing, which is connected to the drain line.

After the fuel accumulates in the bottom of the combustion chamber or drain lines, the drain valve allows the fuel to be drained whenever pressure within the manifold or the burner(s) has been reduced to near atmospheric pressure. A small spring holds the valve off its seat until pressure in the combustion chamber during operation overcomes the spring and closes the valve. The valve is closed during engine operation. It is imperative that this valve be in good working condition to drain accumulated fuel after each shutdown. Otherwise, a hot start during the next starting attempt or an afterfire after shutdown is likely to occur.

Fuel Quantity Indicating Units

Fuel quantity units vary from one installation to the next. A fuel counter or indicator, mounted on the instrument panel, is electrically connected to a flow meter installed in the fuel line to the engine.

The fuel counter, or totalizer, is used to keep record of fuel use. When the aircraft is serviced with fuel, the counter is manually set to the total number of pounds of fuel in all tanks. As fuel passes through the measuring element of the flow meter, it sends electrical impulses to the fuel counter. These impulses actuate the fuel counter mechanism so that the number of pounds passing to the engine is subtracted from the original reading. Thus, the fuel counter continually shows the total quantity of fuel, in pounds, remaining in the aircraft. However, there are certain conditions that cause the fuel counter indication to be inaccurate. Any jettisoned fuel is indicated on the fuel counter as fuel still available for use. Any fuel that leaks from a tank or a fuel line upstream of the flow meter is not counted.

Chapter 3
Induction & Exhaust Systems

Reciprocating Engine Induction Systems

The basic induction system of an aircraft reciprocating engine consists of an air scoop used to collect the inlet air and ducting that transfers the air to the inlet filter. The air filter is generally housed in the carburetor heat box or other housing close by that is attached to the carburetor or fuel injection controller. The engine used in light aircraft is usually equipped with either a carburetor or a fuel-injection system. After air passes through the fuel metering device, an intake manifold with long curved pipes or passages is used to send the air-fuel mixture to the cylinders. An induction air scoop is shown in *Figure 3-1*. The air scoop is located on the engine cowling to allow maximum airflow into the engine's induction system. The air filter, shown in *Figure 3-2*, prevents dirt and other foreign matter from entering the engine. Filtered air enters the fuel metering device (carburetor/fuel injector) where the throttle plate controls the amount of air flowing to the engine. The air coming out of the throttle is referred to as manifold pressure. This pressure is measured in inches of mercury ("Hg) and controls engine power output.

Induction systems can consist of several different arrangements. Two that are used are the updraft and downdraft induction systems. An updraft induction system consists of

Figure 3-1. *Inlet scoop in engine cowling.*

two runners and a balance tube with intake pipes for each cylinder to deliver induction air to each cylinder's intake port. *[Figure 3-3]* The balance tube is used to reduce pressure imbalances between the two side induction runners. With carbureted engines, it is important to maintain a constant and even pressure in the induction system so that each cylinder receives equal amounts of fuel. On fuel-injected engines, the fuel is injected at the intake port just before the intake valve. It is important with this system to keep the pressure consistent at each intake port.

A downdraft balanced induction system provides optimum airflow to each of the individual cylinders throughout a wide operational range. *[Figure 3-4]* Better matched air-fuel ratios provide a much smoother and more efficient engine operation. Air from the induction manifold flows into the intake ports where it is mixed with fuel from the fuel nozzles and then enters the cylinders as a combustible mixture as the intake valve opens.

Basic Carburetor Induction System

Figure 3-2 is a diagram of an induction system used in an engine equipped with a carburetor. In this induction system, carburetor normal flow air is admitted at the lower front nose cowling below the propeller spinner and is passed through an air filter into air ducts leading to the carburetor. A carburetor heat air valve is located below the carburetor for selecting an alternate warm air source (carburetor heat) to prevent carburetor icing. *[Figure 3-5]* Carburetor icing occurs when the temperature is lowered in the throat of the carburetor and enough moisture is present to freeze and block the flow of air to the engine. The carburetor heat valve admits air from the outside air scoop for normal operation, and it admits warm air from the engine compartment for operation during icing conditions. The carburetor heat is operated by a push-pull control in the flight deck. When the carburetor heat air door is closed, warm ducted air from around the exhaust is directed into the carburetor. This raises the intake air temperature. An alternate air door can be opened by engine suction if the normal route of airflow should be blocked by something. The valve is spring loaded closed and is sucked open by the engine if needed.

The carburetor air filter, shown in *Figure 3-6*, is installed in the air scoop in front of the carburetor air duct. Its purpose is

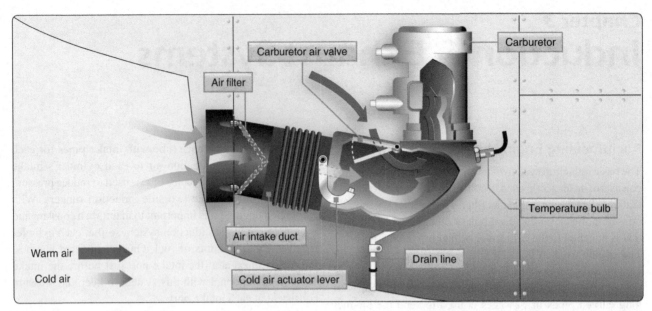

Figure 3-2. *Nonsupercharged induction system using a carburetor.*

Carburetor air valve

Air filter

Carburetor

Temperature bulb

Warm air

Cold air

Air intake duct

Drain line

Cold air actuator lever

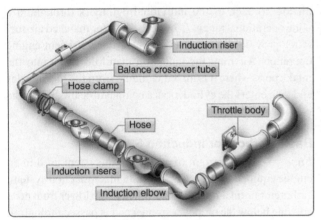

Induction riser

Balance crossover tube

Hose clamp

Throttle body

Hose

Induction risers

Induction elbow

Figure 3-3. *Updraft induction system.*

Figure 3-5. *Location of a carburetor heat air valve.*

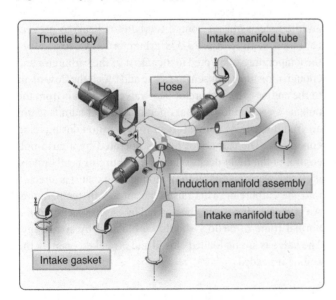

Throttle body

Intake manifold tube

Hose

Induction manifold assembly

Intake manifold tube

Intake gasket

Figure 3-4. *Downdraft balanced induction system.*

to stop dust and other foreign matter from entering the engine through the carburetor. The screen consists of an aluminum alloy frame and a deeply crimped screen, arranged to present maximum screen area to the airstream. There are several types of air filters in use including paper, foam, and other types of filters. Most air filters require servicing at regular intervals and the specific instructions for the type of filter must be followed. *[Figure 3-6]*

The carburetor air ducts consist of a fixed duct riveted to the nose cowling and a flexible duct between the fixed duct and the carburetor air valve housing. The carburetor air ducts normally provide a passage for outside air to the carburetor. Applying carburetor heat to an operating engine decreases the density of the air, which leans the air-fuel mixture. Air enters the system through the ram-air intake. The intake opening is located in the slipstream so the air is forced into the induction system giving a ram effect to the incoming airflow. The air passes through the air ducts to the carburetor. The carburetor meters the fuel in proportion to the air and mixes the air with

Figure 3-6. *Location of air filter.*

the correct amount of fuel. The throttle plate of the carburetor can be controlled from the flight deck to regulate the flow of air (manifold pressure), and in this way, power output of the engine can be controlled.

Although many newer aircraft are not so-equipped, some engines are equipped with carburetor air temperature indicating systems which shows the temperature of the air at the carburetor inlet. If the bulb is located at the engine side of the carburetor, the system measures the temperature of the air-fuel mixture.

Induction System Icing

A short discussion concerning the formation and location of induction system ice is helpful, even though a technician is not normally concerned with operations that occur when the aircraft is in flight. *[Figure 3-7]* Technicians should know something about induction system icing because of its effect on engine performance and troubleshooting. Even when an inspection shows that everything is in proper working order and the engine performs perfectly on the ground, induction system ice can cause an engine to act erratically and lose power in the air. Many engine troubles commonly attributed to other sources are actually caused by induction system icing.

Induction system icing is an operating hazard because it can cut off the flow of the air-fuel charge or vary the air-fuel ratio. Ice can form in the induction system while an aircraft is flying in clouds, fog, rain, sleet, snow, or even clear air that has high moisture content (high humidity). Induction system icing is generally classified in three types:

* Impact ice,
* Fuel evaporation ice, and
* Throttle ice.

Induction system ice can be prevented or eliminated by raising the temperature of the air that passes through the system, using a carburetor heat system located upstream near the induction system inlet and well ahead of the dangerous icing

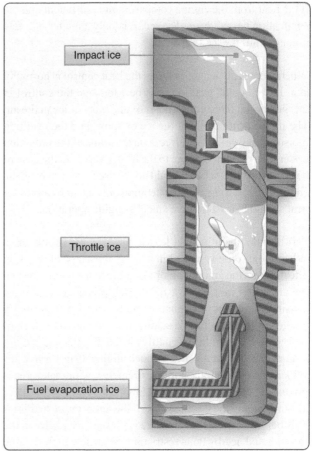

Figure 3-7. *Location of a carburetor heat air valve.*

zones. This air is collected by a duct surrounding the exhaust manifold. Heat is usually obtained through a control valve that opens the induction system to the warm air circulating in the engine compartment and around the exhaust manifold.

Improper or careless use of carburetor heat can be just as dangerous as the most advanced stage of induction system ice. Increasing the temperature of the air causes it to expand and decrease in density. This action reduces the weight of the charge delivered to the cylinder and causes a noticeable loss in power because of decreased volumetric efficiency. If icing is not present when carburetor heat or induction system anti-icing is applied and the throttle setting does not change, the mixture will become richer. In addition, high intake air temperature may cause detonation and engine failure, especially during takeoff and high power operation. Therefore, during all phases of engine operation, the carburetor temperature must afford the greatest protection against icing and detonation.

When there is danger of induction system icing, the flight deck carburetor heat control is moved to the hot position. Throttle ice or any ice that restricts airflow or reduces manifold pressure can best be removed by using full carburetor heat.

If the heat from the engine compartment is sufficient and the application has not been delayed, it is only a matter of a few minutes until the ice is cleared.

When there is no danger of icing, the heat control is normally kept in the "cold" position. It is best to leave the control in this position if there are particles of dry snow or ice in the air. The use of heat may melt the ice or snow, and the resulting moisture may collect and freeze on the walls of the induction system. To prevent damage to the heater valves in the case of backfire, carburetor heat should not be used while starting the engine. Also, during ground operation only enough carburetor heat should be used to give smooth engine operation.

Part-throttle operation can lead to icing in the throttle area. When the throttle is placed in a partly closed position, it, in effect, limits the amount of air available to the engine. When the aircraft is in a glide, a fixed-pitch propeller windmills, causing the engine to consume more air than it normally would at this same throttle setting, thus adding to the lack of air behind the throttle. The partly closed throttle, under these circumstances, establishes a much higher than normal air velocity past the throttle, and an extremely low-pressure area is produced. The low-pressure area lowers the temperature of the air surrounding the throttle valve. If the temperature in this air falls below freezing and moisture is present, ice forms on the throttles and nearby units restricting the airflow to the engine causing it to quit. Throttle ice may be minimized on engines equipped with controllable-pitch propellers by the use of a higher than normal brake mean effective pressure (BMEP) at this low power. The high BMEP decreases the icing tendency because a large throttle opening at low engine revolutions per minute (rpm) partially removes the temperature-reducing obstruction that part-throttle operation offers.

Induction System Filtering

Dust and dirt can be a serious source of trouble to an aircraft engine. Dust consists of small particles of hard, abrasive material that can be carried by the air and drawn into the engine cylinders. It can also collect on the fuel-metering elements of the carburetor, upsetting the proper relation between airflow and fuel flow at all engine power settings. It acts on the cylinder walls by grinding down these surfaces and the piston rings. Then, it contaminates the oil and is carried through the engine, causing further wear on the bearings and gears. In extreme cases, an accumulation may clog an oil passage and cause oil starvation. Although dust conditions are most critical at ground level, continued operation under such conditions without engine protection results in extreme engine wear and can produce excessive oil consumption. When operation in a dusty atmosphere is necessary, the engine can be protected by an alternate induction system air inlet which incorporates a dust filter. This type of air filter system normally consists of a filter element, a door, and an electrically operated actuator. When the filter system is operating, air is drawn through a louvered access panel that does not face directly into the airstream. With this entrance location, considerable dust is removed as the air is forced to turn and enter the duct. Since the dust particles are solid, they tend to continue in a straight line, and most of them are separated at this point. Those that are drawn into the louvers are easily removed by the filter.

In flight, with air filters operating, consideration must be given to possible icing conditions which may occur from actual surface icing or from freezing of the filter element after it becomes rain soaked. Some installations have a spring-loaded filter door which automatically opens when the filter is excessively restricted. This prevents the airflow from being cut off when the filter is clogged with ice or dirt. Other systems use an ice guard in the filtered-air entrance.

The ice guard consists of a coarse-mesh screen located a short distance from the filtered-air entrance. In this location, the screen is directly in the path of incoming air so that the air must pass through or around the screen. When ice forms on the screen, the air, which has lost its heavy moisture particles, passes around the iced screen and into the filter element. The efficiency of any filter system depends upon proper maintenance and servicing. Periodic removal and cleaning of the filter element is essential to satisfactory engine protection.

Induction System Inspection & Maintenance

The induction system should be checked for cracks and leaks during all regularly scheduled engine inspections. The units of the system should be checked for security of mounting. The system should be kept clean at all times, since pieces of rags or paper can restrict the airflow if allowed to enter the air intakes or ducts. Loose bolts and nuts can cause serious damage if they pass into the engine.

On systems equipped with a carburetor air filter, the filter should be checked regularly. If it is dirty or does not have the proper oil film, the filter element should be removed and cleaned. After it has dried, it is usually immersed in a mixture of oil and rust-preventive compound. The excess fluid should be allowed to drain off before the filter element is reinstalled. Paper-type filters should be inspected and replaced as needed. The efficiency of any filter system depends upon proper maintenance and servicing. Periodic removal and cleaning of the filter element is essential to satisfactory engine protection. If the induction system air filter becomes excessively dirty, it will cause a loss of power or the engine will not start.

Extinguishing Engine Fires

In all cases, a fireguard should stand by with a CO_2 fire

Probable Cause	Isolation Procedure	Correction
1 Engine fails to start		
a Induction system obstructed	Inspect air scoop and air ducts	Remove obstructions
b Air leaks	Inspect carburetor mounting and intake pipes	Tighten carburetor and repair or replace intake pipe
2 Engine runs rough		
a Loose air ducts	Inspect air ducts	Tighten air ducts
b Leaking intake pipes	Inspect intake pipe packing nuts	Tighten nuts
c Engine valves sticking	Remove rocker arm cover and check valve action	Lubricate and free sticking valves
d Bent or worn valve push rods	Inspect push rods	Replace worn or damaged push rods
3 Low power		
a Restricted intake duct	Examine intake duct	Remove restrictions
b Broken door in carburetor air valve	Inspect air valve	Replace air valve
c Dirty air filter	Inspect air filter	Clean air filter
4 Engine idles improperly		
a Shrunken intake packing	Inspect packing for proper fit	Replace packing
b Hole in intake pipe	Inspect intake pipe	Replace defective intake pipes
c Loose carburetor mounting	Inspect mount bolts	Tighten mount bolts

Figure 3-8. *Common problems for troubleshooting induction systems.*

extinguisher while the aircraft engine is being started. This is a necessary precaution against fire during the starting procedure. The fireguard must be familiar with the induction system of the engine so that in case of fire, they can direct the CO_2 into the air intake of the engine to extinguish it. A fire could also occur in the exhaust system of the engine from liquid fuel being ignited in the cylinder and expelled during the normal rotation of the engine.

If an engine fire develops during the starting procedure, continue cranking to start the engine and blow out the fire. If the engine does not start and the fire continues to burn, discontinue the start attempt. The fireguard then extinguishes the fire using the available equipment. The fireguard must observe all safety practices at all times while standing by during the starting procedure.

Induction System Troubleshooting
Figure 3-8 provides a general guide to the most common induction system troubles.

Supercharged Induction Systems
Since aircraft operate at altitudes where the air pressure is lower, it is useful to provide a system for compressing the air-fuel mixture. Some systems are used to normalize the air pressure entering the engine. These systems are used to regain the air pressure lost by the increase in altitude. This type of system is not a ground boost system and it is not used to ever boost the manifold pressure above 30 inches of mercury. A true supercharged engine, called ground boosted engines, can

boost the manifold pressure above 30 inches of mercury. In other words, a true supercharger boosts the manifold pressure above ambient pressure.

Since many engines installed in light aircraft do not use any type of compressor or supercharging device, induction systems for reciprocating engines can be broadly classified as supercharged or nonsupercharged. *[Figure 3-9]* Supercharging systems used in reciprocating engine induction systems are normally classified as either internally driven or externally driven (turbosupercharged). Internally driven superchargers compress the air-fuel mixture after it leaves the carburetor, while externally driven superchargers (turbochargers) compress the air before it is mixed with the

Figure 3-9. *An example of a naturally aspirated reciprocating engine.*

metered fuel from the carburetor.

Internally Driven Superchargers

Internally-driven superchargers were used almost exclusively in high horsepower radial reciprocating engines and are engine driven through a mechanical connection. Although their use is very limited, some are still used in cargo carriers and spray planes. Except for the construction and arrangement of the various types of superchargers, all induction systems with internally driven superchargers were very similar. Aircraft engines require the same air temperature control to produce good combustion in the engine cylinders. For example, the charge must be warm enough to ensure complete fuel vaporization and, thus, even distribution. At the same time, it must not be so hot that it reduces volumetric efficiency or causes detonation. All reciprocating engines must guard against intake air that is too hot. As with any type of supercharging (compressing intake air), the air gains heat as it is compressed. Sometimes this air requires cooling before it is routed to the engine's intake ports. With these requirements, most induction systems that use internally driven superchargers must include pressure and temperature-sensing devices and the necessary units required to warm or cool the air.

The simple internally driven supercharger induction system is used to explain the location of units and the path of the air and air-fuel mixture. [*Figure 3-10*] Air enters the system through the ram air intake. The intake opening is located so that the air is forced into the induction system, giving a ram effect caused by the aircraft moving through the air. The air passes through ducts to the carburetor. The carburetor meters the fuel in proportion to the air and mixes the air with the correct

amount of fuel. The carburetor can be controlled from the flight deck to regulate the flow of air. In this way, the power output of the engine can be controlled. The manifold pressure gauge measures the pressure of the air-fuel mixture before it enters the cylinders. It is an indication of the performance that can be expected of the engine. The carburetor air temperature indicator measures either the temperature of the inlet air or of the air-fuel mixture. Either the air inlet or the mixture temperature indicator serves as a guide so that the temperature of the incoming charge may be kept within safe limits. If the temperature of the incoming air at the entrance to the carburetor scoop is 100 °F, there is approximately a 50 °F drop in temperature because of the partial vaporization of the fuel at the carburetor discharge nozzle. Partial vaporization takes place and the air temperature falls due to absorption of the heat by vaporization. The final vaporization takes place as the mixture enters the cylinders where higher temperatures exist. The fuel, as atomized into the airstream that flows in the induction system, is in a globular form. The problem, then, becomes one of uniformly breaking up and distributing the fuel, remaining in globular form to the various cylinders. On engines equipped with a large number of cylinders, the uniform distribution of the mixture becomes a greater problem, especially at high engine speeds when full advantage is taken of large air capacity.

One method used mainly on radial reciprocating engines of improving fuel distribution is shown in *Figure 3-11*. This device is known as a distribution impeller. The impeller is attached directly to the end of the rear shank of the crankshaft by bolts or studs. Since the impeller is attached to the end of the crankshaft and operates at the same speed, it does not materially boost or increase the pressure on the mixture

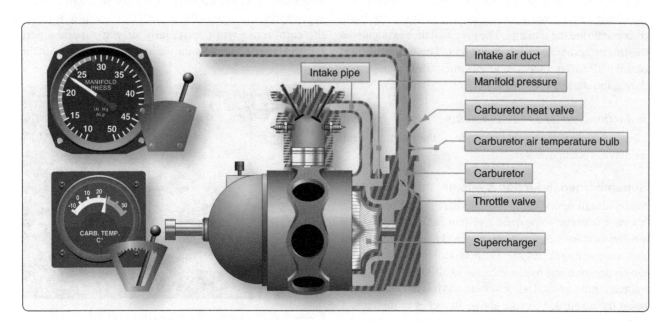

Figure 3-10. *Internally-driven supercharger induction system.*

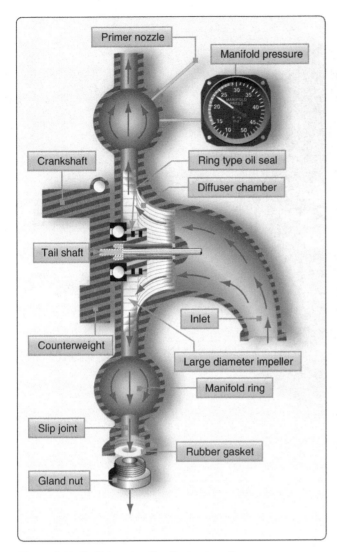

Figure 3-11. *Radial engine distribution impeller.*

flowing into the cylinders. But, the fuel remaining in the globular form is broken up into finer particles as it strikes the impeller, thereby coming in contact with more air. This creates a more homogeneous mixture with a consequent improvement in distribution to the various cylinders, especially on acceleration of the engine or when low temperatures prevail.

To obtain greater pressure of the air-fuel mixture within the cylinders, the diffuser or blower section contains a high speed impeller. Unlike the distribution impeller, which is connected directly to the crankshaft, the supercharger, or blower impeller, is driven through a gear train from the crankshaft.

Turbosuperchargers

Externally driven superchargers (turbosuperchargers) are designed to deliver compressed air to the inlet of the carburetor or air-fuel control unit of an engine. Externally driven superchargers derive their power from the energy of engine exhaust gases directed against a turbine that drives

an impeller that compresses the incoming air. For this reason, they are commonly called turbosuperchargers or turbochargers. To be a true supercharger, it must boost the manifold pressure above 30 "Hg.

The typical turbosupercharger, shown in *Figure 3-12,* is composed of three main parts:

1. Compressor assembly,
2. Turbine wheel assembly, and
3. A full floating shaft bearing assembly.

Detail examples of a turbosupercharger are shown in *Figure 3-13.* In addition to the major assemblies, there is a baffle between the compressor casing and the exhaust-gas turbine that directs cooling air to the pump and bearing casing, and also shields the compressor from the heat radiated by the turbine. In installations where cooling air is limited, the baffle is replaced by a regular cooling shroud that receives its air directly from the induction system.

The compressor assembly is made up of an impeller, a diffuser, and a casing. The air for the induction system enters through a circular opening in the center of the compressor casing, where it is picked up by the blades of the impeller, which gives it high velocity as it travels outward toward the diffuser. The diffuser vanes direct the airflow as it leaves the impeller and also converts the high velocity of the air to high-pressure.

Motive power for the impeller is furnished through the impeller's attachment to the turbine wheel shaft of the exhaust-gas turbine. This complete assembly is referred to as the rotor. (The rotor revolves on the oil feed bearings.) The exhaust gas turbine assembly consists of the turbocharger and waste gate valve. *[Figure 3-14]* The turbine wheel, driven by exhaust gases, drives the impeller. The turbo housing collects and directs the exhaust gases onto the turbine wheel, and the waste gate regulates the amount of exhaust gases directed to the turbine. The waste gate controls the volume of the exhaust gas that is directed onto the turbine and thereby regulates the speed of the rotor (turbine and impeller). *[Figure 3-15]*

If the waste gate is completely closed, all the exhaust gases are "backed up" and forced through the turbine wheel. If the waste gate is partially closed, a corresponding amount of exhaust gas is directed to the turbine. The exhaust gasses, thus directed, strike the turbine blades, arranged radially around the outer edge of the turbine, and cause the rotor (turbine and impeller) to rotate. The gases, having exhausted most of their energy, are then exhausted overboard. When the waste gate is fully open, nearly all of the exhaust gases pass overboard providing little or no boost.

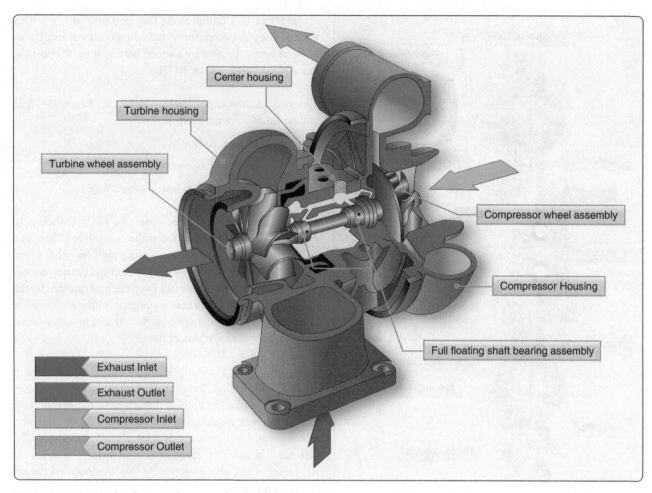

Figure 3-12. *A typical turbosupercharger and its main parts.*

Center housing

Turbine housing

Turbine wheel assembly

Compressor wheel assembly

Compressor Housing

Full floating shaft bearing assembly

Exhaust Inlet

Exhaust Outlet

Compressor Inlet

Compressor Outlet

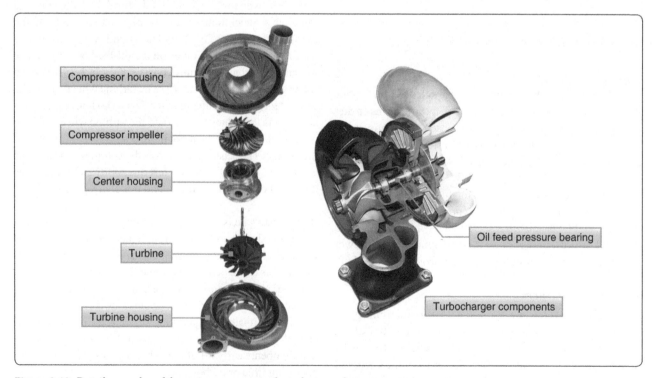

Figure 3-13. *Detail examples of the main components of a turbosupercharger.*

Compressor housing

Compressor impeller

Center housing

Turbine

Turbine housing

Oil feed pressure bearing

Turbocharger components

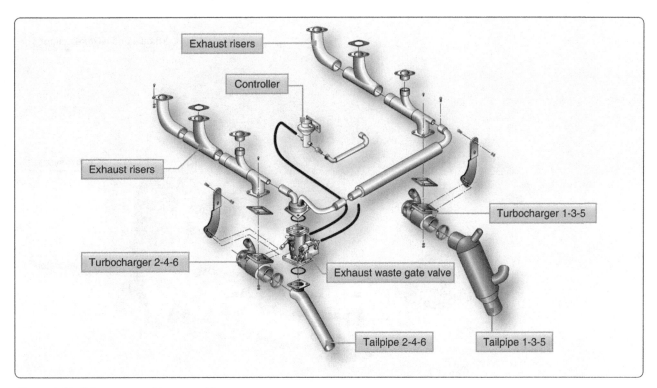

Figure 3-14. *Exhaust gas turbine assembly.*

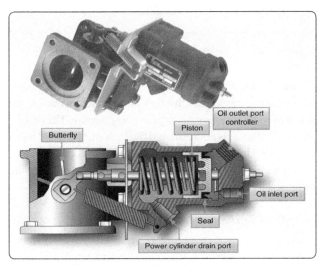

Figure 3-15. *Waste gate control of exhaust.*

Normalizer Turbocharger

Some engines used in light aircraft are equipped with an externally driven normalizing system. These systems are powered by the energy of exhaust gases and are usually referred to as "normalizing turbocharger" systems. These systems were not designed to be used as a true supercharger (boost manifold pressure over 30 "Hg). They compensate for the power lost due to the pressure drop resulting from increased altitude. On many small aircraft engines, the turbocharger (normalizing) system is designed to be operated only above a certain altitude, 5,000 feet for example, since maximum power without normalizing is

available below that altitude. The location of the air induction and exhaust systems of a typical normalizing turbocharger system for a small aircraft is shown in *Figure 3-16*.

Ground-Boosted Turbosupercharger System

Some ground-boosted (sea level) turbosupercharged systems are designed to operate from sea level up to their critical altitude. These engines, sometimes referred to as sea level-boosted engines, can develop more power at sea level than an engine without turbosupercharging. As was mentioned earlier, an engine must be boosted above 30 "Hg to truly be supercharged. This type of turbocharger accomplishes this by increasing the manifold pressure above 30 "Hg to around 40 "Hg.

The turbosupercharger air induction system consists of a filtered ram-air intake located on the side of the nacelle. *[Figure 3-17]* An alternate air door within the nacelle permits compressor suction automatically to admit alternate air (heated engine compartment air) if the induction air filter becomes clogged. In many cases, the alternate air door can be operated manually in the event of filter clogging.

Almost all turbocharger systems use engine oil as the control fluid for controlling the amount of boost (extra manifold pressure) provided to the engine. The waste-gate actuator and controllers use pressurized engine oil for their power supply. The turbocharger is controlled by the waste gate and waste gate actuator. The waste gate actuator, which is physically connected to the waste gate by mechanical linkage, controls

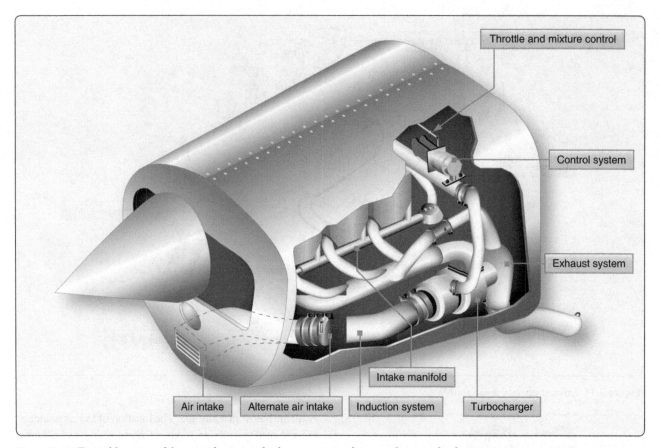

Figure 3-16. *Typical location of the air induction and exhaust systems of a normalizing turbocharger system.*

the position of the waste gate butterfly valve. The waste gate bypasses the engine exhaust gases around the turbocharger turbine inlet. By controlling the amount of exhaust gases that pass through the turbine of the turbocharger, the speed of the compressor and the amount of intake boost (upper deck pressure) is controlled. Engine oil is also used to cool and lubricate the bearings that support the compressor and turbine in the turbocharger. Turbocharger lubricating oil is engine oil supplied through the engine oil system. An oil supply hose from the rear of the oil cooler directs oil to the turbocharger center housings and bearings. Oil hoses return oil from the turbochargers to the oil scavenge pump located on the rear of the engine. The one-way check valve in the oil supply line prevents oil from draining into the turbocharger while the engine is not operating. Piston ring-like oil seals are used on the compressor wheel shaft to prevent the lubricating oil from entering the turbine and compressor housings from the center housing.

The position of the waste gate is controlled by adjusting the oil pressure in the waste gate actuator. Several different types of controllers are used to provide the correct pressure in the waste gate actuator. This is done either by restricting the oil flow or by allowing the oil to return to the engine. The more the oil is restricted, the more pressure is in the waste gate

actuator and the more closed the waste gate is. This causes the exhaust gases to pass through the turbine, increasing the speed of the compressor raising the inlet pressure. The reverse happens if the oil is not restricted by the controllers and boost is reduced. The pressure from the outlet of the compressor of the turbocharger to the throttle is referred to as deck pressure or upper deck pressure.

A Typical Turbosupercharger System

Figure 3-18 is a schematic of a sea level booster turbosupercharger system. This system used widely is automatically regulated by three components:

- Exhaust bypass valve assembly,
- Density controller, and
- Differential pressure controller.

By regulating the waste gate position and the "fully open" and "closed" positions, a constant power output can be maintained. When the waste gate is fully open, all the exhaust gases are directed overboard to the atmosphere, and no air is compressed and delivered to the engine air inlet. Conversely, when the waste gate is fully closed, a maximum volume of exhaust gases flows into the turbocharger turbine, and maximum supercharging is accomplished. Between these

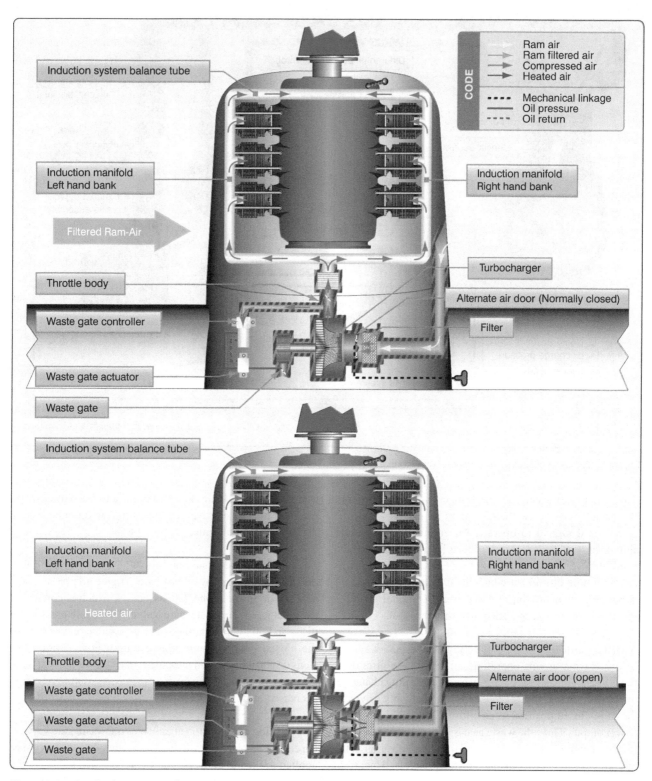

CODE

→	Ram air
→	Ram filtered air
→	Compressed air
→	Heated air
- - - -	Mechanical linkage
——	Oil pressure
– – –	Oil return

Induction system balance tube

Induction manifold Left hand bank

Induction manifold Right hand bank

Filtered Ram-Air

Throttle body

Turbocharger

Alternate air door (Normally closed)

Waste gate controller

Filter

Waste gate actuator

Waste gate

Induction system balance tube

Induction manifold Left hand bank

Induction manifold Right hand bank

Heated air

Throttle body

Turbocharger

Waste gate controller

Alternate air door (open)

Waste gate actuator

Filter

Waste gate

Figure 3-17. *A turbocharger air induction system.*

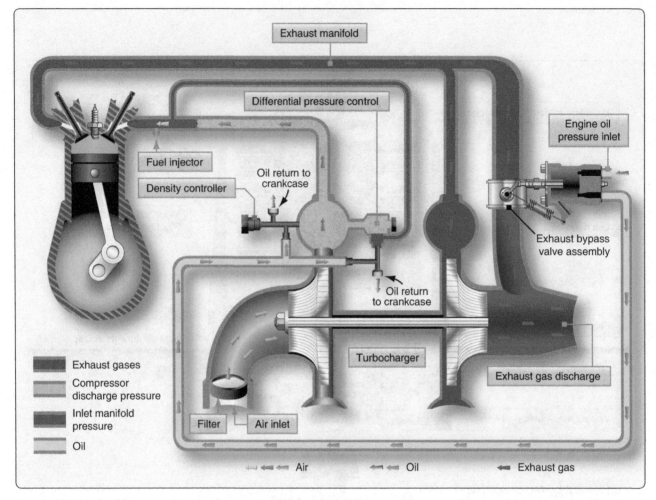

Figure 3-18. *Sea level booster turbosupercharger system.*

two extremes of waste gate position, constant power output can be achieved below the maximum altitude at which the system is designed to operate. An engine with a critical altitude of 16,000 feet cannot produce 100 percent of its rated manifold pressure above 16,000 feet. Critical altitude means the maximum altitude at which, in standard atmosphere, it is possible to maintain, at a specified rotational speed, a specified power or a specified manifold pressure.

A critical altitude exists for every possible power setting below the maximum operating ceiling. If the aircraft is flown above this altitude without a corresponding change in the power setting, the waste gate is automatically driven to the fully closed position in an effort to maintain a constant power output. Thus, the waste gate is almost fully open at sea level and continues to move toward the closed position as the aircraft climbs, in order to maintain the preselected manifold pressure setting. When the waste gate is fully closed (leaving only a small clearance to prevent sticking), the manifold pressure begins to drop if the aircraft continues to climb. If a higher power setting cannot be selected, the turbocharger's critical altitude has been reached. Beyond this altitude, the

power output continues to decrease. If a turbocharger waste gate will not close fully, then the aircraft will not be able to reach its critical altitude.

The position of the waste gate valve, which determines power output, is controlled by oil pressure. Engine oil pressure acts on a piston in the waste gate assembly, which is connected by linkage to the waste gate valve. When oil pressure is increased on the piston, the waste gate valve moves toward the closed position, and engine output power increases. Conversely, when the oil pressure is decreased, the waste gate valve moves toward the open position, and output power is decreased as described earlier.

The position of the piston attached to the waste gate valve is dependent on bleed oil, which controls the engine oil pressure applied to the top of the piston. Oil is returned to the engine crankcase through two control devices, the density controller and the differential pressure controller. These two controllers, acting independently, determine how much oil is bled back to the crankcase and establishes the oil pressure on the piston.

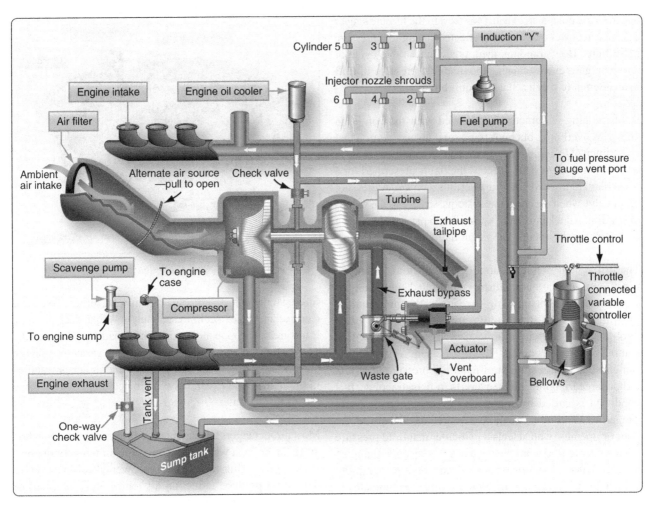

Figure 3-19. *Components of a turbocharger system engine.*

The density controller is designed to limit the manifold pressure below the turbocharger's critical altitude and regulates bleed oil only at the full throttle position. The pressure- and temperature-sensing bellows of the density controller react to pressure and temperature changes between the fuel injector inlet and the turbocharger compressor. The bellows, filled with dry nitrogen, maintain a constant density by allowing the pressure to increase as the temperature increases. Movement of the bellows repositions the bleed valve, causing a change in the quantity of bleed oil, which changes the oil pressure on top of the waste gate piston. *[Figure 3-18]*

The differential pressure controller functions during all positions of the waste gate valve other than the fully open position, which is controlled by the density controller. One side of the diaphragm in the differential pressure controller senses air pressure upstream from the throttle; the other side samples pressure on the cylinder side of the throttle valve. *[Figure 3-18]* At the "wide open" throttle position when the density controller controls the waste gate, the pressure across the differential pressure controller diaphragm is at a minimum and the controller spring holds the bleed valve closed. At "part

throttle" position, the air differential is increased, opening the bleed valve to bleed oil to the engine crankcase and reposition the waste gate piston. Thus, the two controllers operate independently to control turbocharger operation at all positions of the throttle. Without the overriding function of the differential pressure controller during part-throttle operation, the density controller would position the waste gate valve for maximum power. The differential pressure controller reduces injector entrance pressure and continually repositions the valve over the whole operating range of the engine.

The differential pressure controller reduces the unstable condition known as "bootstrapping" during part-throttle operation. Bootstrapping is an indication of unregulated power change that results in the continual drift of manifold pressure. This condition can be illustrated by considering the operation of a system when the waste gate is fully closed. During this time, the differential pressure controller is not modulating the waste gate valve position. Any slight change in power caused by a change in temperature or rpm fluctuation is magnified and results in manifold pressure change since the slight change causes a change in the amount of exhaust gas flowing to the turbine. Any change in exhaust gas flow to the turbine causes a change in power output and is reflected in manifold pressure indications. Bootstrapping, then, is an undesirable cycle of turbocharging events causing the manifold pressure to drift in an attempt to reach a state of equilibrium.

Bootstrapping is sometimes confused with the condition known as overboost, but bootstrapping is not a condition that is detrimental to engine life. An overboost condition is one in which manifold pressure exceeds the limits prescribed for a particular engine and can cause serious damage. A pressure relief valve when used in some systems, set slightly in excess of maximum deck pressure, is provided to prevent damaging over boost in the event of a system malfunction.

The differential pressure controller is essential to smooth functioning of the automatically controlled turbocharger, since it reduces bootstrapping by reducing the time required to bring a system into equilibrium. There is still extra throttle sensitivity with a turbocharged engine than with a naturally aspirated engine. Rapid movement of the throttle can cause a certain amount of manifold pressure drift in a turbocharged engine. Less severe than bootstrapping, this condition is called overshoot. While overshoot is not a dangerous condition, it can be a source of concern to the pilot or operator who selects a particular manifold pressure setting only to find it has changed in a few seconds and must be reset. Since the automatic controls cannot respond rapidly enough to abrupt changes in throttle settings to eliminate the inertia of turbocharger speed changes, overshoot must be controlled by the operator. This can best be accomplished by slowly making changes in throttle setting, accompanied by a few seconds' wait for the system to reach a new equilibrium. Such a procedure is effective with turbocharged engines, regardless of the degree of throttle sensitivity.

Turbocharger Controllers & System Descriptions

Turbocharger system engines contain many of the same components mentioned with the previous systems. [Figure 3-19] Some systems use special lines and fittings that are connected to the upper-deck pressure for air reference to the fuel injection system and in some cases for pressurizing the magnetos. Basic system operation is similar to other turbocharger systems with the main differences being in the controllers. The controller monitors deck pressure by sensing the output of the compressor. The controller controls the oil flow through the waste gate actuator, which opens or closes the exhaust bypass valve. If a malfunction occurs causing the waste gate or controller to not be stable, it will cause the engine to surge due to an erratic manifold pressure. When deck pressure is insufficient, the controller restricts oil flow thereby increasing oil pressure at the waste gate actuator. This pressure acts on the piston to close off the waste gate valve,

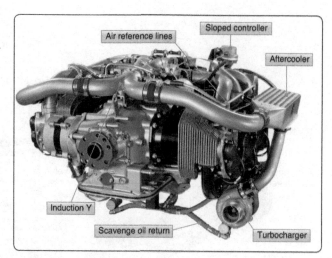

Figure 3-20. *An aftercooler installation.*

forcing more exhaust gas pulses to turn the turbine faster and cause an increase in compressor output. When deck pressure is too great, the opposite occurs. The exhaust waste gate fully opens and bypasses some of the exhaust gases to decrease exhaust flow across the turbine. An aftercooler is installed in the induction air path between the compressor stage and the air throttle inlet. *[Figure 3-20]*

Most turbochargers are capable of compressing the induction air to the point at which it can raise the air temperature by a factor of five. This means that full power takeoff on a 100 °F day could produce induction air temperatures exiting the compressor at up to 500 °F. This would exceed the allowable throttle air inlet temperature on all reciprocating engine models. Typically, the maximum air throttle inlet temperature ranges from a low 230 °F to a high of 300 °F. Exceeding these maximums can place the combustion chambers closer to detonation. The function of the aftercooler is to cool the compressed air, which decreases the likelihood of detonation and increases the charge air density, which improves the turbocharger performance for that engine design. On engine start, the controller senses insufficient compressor discharge pressure (deck pressure) and restricts the flow of oil from the waste gate actuator to the engine. This causes the waste gate butterfly valve to close. As the throttle is advanced, exhaust gas flows across the turbine increases, thereby increasing turbine/compressor shaft speed and compressor discharge pressure. The controller senses the difference between upper deck and manifold pressure. If either deck pressure or throttle differential pressure rises, the controller poppet valve opens, relieving oil pressure to the waste gate actuator. This decreases turbocharger compressor discharge pressure (deck pressure).

Variable Absolute Pressure Controller (VAPC)

The VAPC contains an oil control valve similar to the other

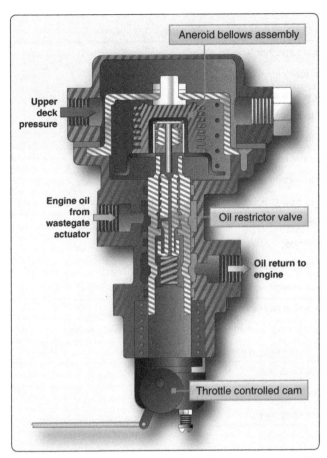

Figure 3-21. *A diagram of a variable absolute pressure controller (VAPC).*

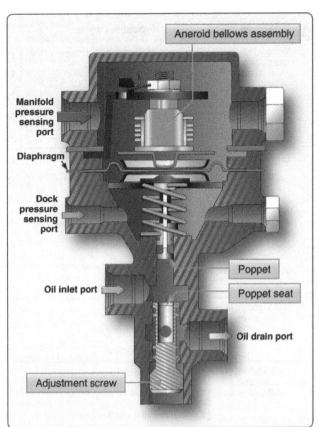

Figure 3-22. *A diagram of a sloped controller used to maintain the rated compressor discharge pressure at wide-open throttle.*

controllers that were discussed. *[Figure 3-21]* The oil restrictor is actuated by an aneroid bellows that is referenced to upper deck pressure. A cam connected to the throttle mechanism applies pressure to the restrictor valve and aneroid. As the throttle is opened to greater values, the cam applies a greater pressure to the aneroid. This increases the amount of upper deck pressure necessary to compress the aneroid and thereby open the oil restrictor valve. This means that the scheduled absolute value of upper deck pressure that is required to overcome the aneroid is variable by throttle position. As the throttle is opened wide, the manifold pressure and upper deck pressure requirements greatly increase.

Sloped Controller

The sloped controller is designed to maintain the rated compressor discharge pressure at wide-open throttle and to reduce this pressure at part throttle settings. *[Figure 3-22]* A diaphragm, coupled with a spring-supported bellows for absolute pressure reference, is exposed to deck pressure and intake manifold pressure through ports located before and after the throttle, respectively. This arrangement constantly monitors deck pressure and the pressure differential between the deck and manifold pressure due to a partially closed throttle. If either deck pressure or throttle differential pressure

rises, the controller poppet opens and decreases turbocharger discharge (deck) pressure. The sloped controller is more sensitive to the throttle differential pressure than to deck pressure, thereby accomplishing deck pressure reduction as the throttle is closed.

Absolute Pressure Controller

One device used to control the speed and output of the turbocharger, but controls the system only at maximum output, is the absolute pressure controller. The absolute pressure controller contains an aneroid bellows that is referenced to upper deck pressure. It operates the waste gate, which diverts, more or less, exhaust gas over the turbine. As an absolute pressure setting is reached, it bypasses oil, and relieves the pressure on the waste gate actuator. This allows the absolute pressure controller to control the maximum turbocharger compressor discharge pressure. The turbocharger is completely automatic, requiring no pilot action up to the critical altitude.

Turbocharger System Troubleshooting

Figure 3-23 includes some of the most common turbocharger system malfunctions together with their cause and repair. These troubleshooting procedures are presented as a guide only and should not be substituted for applicable

Trouble	Probable Cause	Remedy
Aircraft fails to reach critical altitude	Damaged compressor or turbine wheel	Replace turbocharger
	Exhaust system leaks	Repair leaks
	Faulty turbocharger bearings	Replace turbocharge
	Wastegate will not close fully	Refer to wastegate in the trouble column
	Malfunctioning controller	Refer to differential controller in the trouble column
Engine surges	Bootstrapping	Ensure engine is operated in proper range
	Wastegate malfunction	Refer to wastegate in the trouble column
	Controller malfunction	Refer to differential controller in the trouble column
Wastegate will not close fully	Wastegate bypass valve bearing tight	Replace bypass valve
	Oil inlet orifice blocked	Clean orifice
	Differential controller malfunction	Refer to controller in the trouble column
	Broken wastegate linkage	Replace linkage and adjust waste gate for proper opening and closing
Wastegate will not open	Oil outlet obstructed	Clean and reconnect oil return line
	Broken wastegate linkage	Replace linkage and adjust waste gate opening and closing
	Controller malfunction	Refer to controller in the trouble column
Differential controller malfunctions	Seals leaking	Replace controller
	Diaphragm broken	Replace controller
	Controller valve stuck	Replace controller
Density controller malfunctions	Seals leaking	Replace controller
	Bellows damaged	Replace controller
	Valve stuck	Replace controller

Figure 3-23. *Common issues when troubleshooting turbocharger systems.*

manufacturer's instructions or troubleshooting procedures.

Turbine Engine Inlet Systems

The engine inlet of a turbine engine is designed to provide a relatively distortion-free flow of air, in the required quantity, to the inlet of the compressor. *[Figure 3-24]* Many engines use inlet guide vanes (IGV) to help straighten the airflow and direct it into the first stages of the compressor. A uniform and steady airflow is necessary to avoid compressor stall (airflow tends to stop or reverse direction of flow) and excessive internal engine temperatures in the turbine section. Normally, the air-inlet duct is considered an airframe part and not a part of the engine. However, the duct is very important to the engine's overall performance and the engine's ability to produce an optimum amount of thrust.

A gas turbine engine consumes considerable more airflow than a reciprocating engine. The air entrance passage is correspondingly larger. Furthermore, it is more critical in determining engine and aircraft performance, especially at high airspeeds. Inefficiencies of the inlet duct result in successively magnified losses through other components of the engine. The inlet varies according to the type of turbine engine. Small turboprop and turboshaft engines have a lower airflow than large turbofan engines which require a completely different type of inlet. Many turboprop, auxiliary power units, and turboshaft engines use screens that cover the inlet to prevent foreign object damage (FOD).

Figure 3-24. *An example of a turbine engine inlet.*

As aircraft speed increases, thrust tends to decrease somewhat; as the aircraft speed reaches a certain point, ram recovery compensates for the losses caused by the increases in speed. The inlet must be able to recover as much of the total pressure of the free airstream as possible. As air molecules are trapped and begin to be compressed in the inlet, much of the pressure loss is recovered. This added pressure at the inlet of the engine increases the pressure and airflow to the engine. This is known as "ram recovery" or "total pressure recovery." The inlet duct must uniformly deliver air to the compressor inlet with as little turbulence and pressure variation as possible. The engine inlet duct must also hold the drag effect on the aircraft to a minimum.

Air pressure drop in the engine inlet is caused by the friction of the air along both sides of the duct and by the bends in the duct system. Smooth flow depends upon keeping the amount of turbulence to a minimum as the air enters the duct. On engines with low flow rates, turning the airflow allows the engine nacelle to be smaller and have less drag. On turbofan engines, the duct must have a sufficiently straight section to ensure smooth, even airflow because of the high airflows. The choice of configuration of the entrance to the duct is dictated by the location of the engine within the aircraft and the airspeed, altitude, and attitude at which the aircraft is designed to operate. To accomplish this, inlet ducts are designed to function as diffusers with a divergent shape, decreasing the velocity and increasing the static pressure of the air passing through them.

Divided-Entrance Duct

The requirements of high-speed, single- or twin-engine military aircraft, in which the pilot sits low in the fuselage and close to the nose, render it difficult to employ the older type single-entrance duct, which is not used on modern aircraft. Some form of a divided duct, which takes air from either side of the fuselage, has become fairly widely used. This divided duct can be either a wing-root inlet or a scoop at each side of the fuselage. *[Figure 3-25]* Either type of duct presents more problems to the aircraft designer than a single-entrance duct because of the difficulty of obtaining sufficient airscoop area without imposing prohibitive amounts of drag. Internally, the problem is the same as that encountered with the single-entrance duct: to construct a duct of reasonable length with as few bends as possible. Scoops at the sides of the fuselage are often used. These side scoops are placed as far forward as possible to permit a gradual bend toward the compressor inlet, making the airflow characteristics approach those of a single-entrance duct. A series of turning vanes is sometimes placed in the side-scoop inlet to assist in straightening the incoming airflow and to prevent turbulence.

Figure 3-25. *An example of a divided-entrance duct.*

Variable-Geometry Duct

The main function of an inlet duct is to furnish the proper amount of air to the engine inlet. In a typical military aircraft using a turbojet or low bypass turbofan engine, the maximum airflow requirements are such that the Mach number of the airflow directly ahead of the face of the engine is less than Mach 1. Airflow through the engine must be less than Mach 1 at all times. Therefore, under all flight conditions, the velocity of the airflow as it enters the air-inlet duct must be reduced through the duct before the airflow is ready to enter the compressor. To accomplish this, inlet ducts are designed to function as diffusers, decreasing the velocity and increasing the static pressure of the air passing through them. *[Figure 3-26]*

As with military supersonic aircraft, a diffuser progressively decreases in area in the downstream direction. Therefore, a supersonic inlet duct follows this general configuration until the velocity of the incoming air is reduced to Mach 1. The

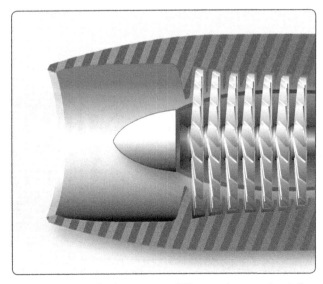

Figure 3-26. *An inlet duct acts as a diffuser to decrease the airflow velocity and to increase the static pressure of air.*

aft section of the duct then increases in area, since this part must act as a subsonic diffuser. *[Figure 3-27]* In practice, inlet ducts for supersonic aircraft follows this general design only as much as practical, depending upon the design features of the aircraft. For very high speed aircraft, the inside area of configuration of the duct is changed by a mechanical device as the speed of the aircraft increases or decreases. A duct of this type is usually known as a variable-geometry inlet duct. Military aircraft use the three methods described above to diffuse the inlet air and slow the inlet airflow at supersonic flight speeds. One is to vary the area, or geometry, of the inlet duct either by using a movable restriction, such as a ramp or wedge, inside the duct. Another system is some sort of a variable airflow bypass arrangement, which extracts part of the inlet airflow from the duct ahead of the engine. In some cases, a combination of both systems is used.

The third method is the use of a shock wave in the airstream. A shock wave is a thin region of discontinuity in a flow of air or gas, during which the speed, pressure, density, and temperature of the air or gas undergo a sudden change. Stronger shock waves produce larger changes in the properties of the air or gas. A shock wave is willfully set up in the supersonic flow of the air entering the duct, by means of some restriction or small obstruction which automatically protrudes into the duct at high flight Mach numbers. The shock wave results in diffusion of the airflow, which, in turn, decreases the velocity of the airflow. In at least one aircraft installation, both the shock method and the variable-geometry method of causing diffusion are used in combination. The same device that changes the area of the duct also sets up a shock wave that further reduces the speed of the incoming air within the duct. The amount of change in duct area and the magnitude of the shock are varied automatically with the airspeed of the aircraft.

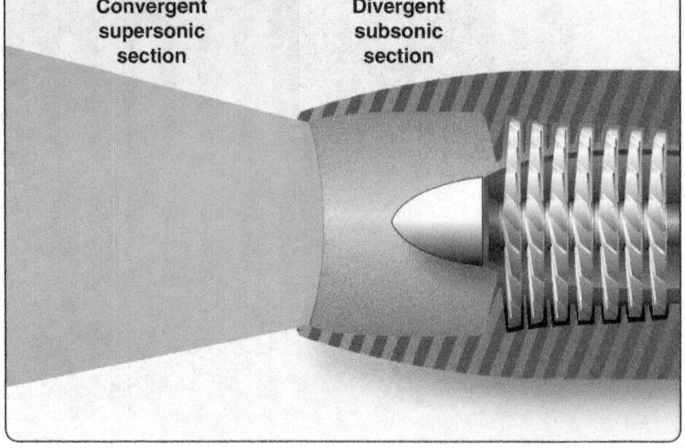

Compressor Inlet Screens

To prevent the engine from readily ingesting any items that can be drawn in the intake, a compressor inlet screen is sometimes placed across the engine air inlet at some location along the inlet duct. Engines that incorporate inlet screens, such as turboprops *[Figure 3-28]* and APUs *[Figure 3-29]* are not as vulnerable to FOD. The advantages and disadvantages of a screen vary. If the engine is readily subjected to internal damage, as would be the case for an engine having an axial compressor fitted with aluminum compressor blades, an inlet screen is almost a necessity. Screens, however, add appreciably to inlet duct pressure loss and are very susceptible to icing. Failure due to fatigue is also a problem. A failed screen can sometimes cause more damage than no screen at all. In some instances, inlet screens are made retractable and may be withdrawn from the airstream after takeoff or whenever icing conditions prevail. Such screens are subject to mechanical failure and add both weight and bulk to the installation. In large turbofan engines having steel or titanium compressor (fan) blades, which do not damage

Figure 3-28. *An example of a turboprop engine that incorporates inlet screens.*

Figure 3-29. *An example of an inlet screen on an APU.*

Figure 3-27. *The aft section of an inlet duct acting as a subsonic diffuser.*

easily, the disadvantages of compressor screens outweigh the advantages, so they are not generally used.

Bellmouth Compressor Inlets

A bellmouth inlet is usually installed on an engine undergoing testing in a test cell. For this reason, the bellmouth inlet duct is most often used on helicopter airframes. *[Figure 3-30]* It is generally equipped with probes that, with the use of instruments, can measure intake temperature and pressure (total and static). *[Figure 3-31]* During testing, it is important that the outside static air is allowed to flow into the engine with as little resistance as possible. The bellmouth is attached to the movable part of the test stand and moves with the engine. The thrust stand is made up of two components, one nonmoving and one moving. This is so the moving component can push against a load cell and measure thrust during the testing of the engine. The bellmouth is designed with the single objective of obtaining very high aerodynamic efficiency. Essentially, the inlet is a bell-shaped funnel having carefully rounded shoulders which offer practically no air resistance. *[Figure 3-30]* Duct loss is so slight that it

is considered zero. The engine can, therefore, be operated without the complications resulting from losses common to an installed aircraft inlet duct. Engine performance data, such as rated thrust and thrust specific fuel consumption, are obtained while using a bellmouth inlet. Usually, the inlets are fitted with protective screening. In this case, the efficiency lost as the air passes through the screen must be taken into account when very accurate engine data are necessary.

Turboprop & Turboshaft Compressor Inlets

The air inlet on a turboprop is more of a problem than some other gas turbine engines because the propeller drive shaft, the hub, and the spinner must be considered in addition to other inlet design factors. The ducted arrangement is generally considered the best inlet design of the turboprop engine as far as airflow and aerodynamic characteristics are concerned. *[Figure 3-32]* The inlet for many types of turboprops are anti-iced by using electrical elements in the lip opening of the intake. Ducting either part of the engine or nacelle directs the airflow to the intake of the engine. Deflector doors are sometimes used to deflect ice or dirt away from the intake. *[Figure 3-33]* The air then passes through a screen and into the engine on some models. A conical spinner, which does not allow ice to build up on the surface, is sometimes used with turboprop and turbofan engines. In either event, the arrangement of the spinner and the inlet duct plays an important function in the operation and performance of the engine.

Turbofan Engine Inlet Sections

High-bypass turbofan engines are usually constructed with the fan at the forward end of the compressor. A typical turbofan intake section is shown in *Figure 3-34*. Sometimes, the inlet cowl is bolted to the front of the engine and provides the airflow path into the engine. In dual compressor (dual spool) engines, the fan is integral with the relatively slow-turning, low-pressure compressor, which allows the fan

Figure 3-30. *A bellmouth inlet used during system tests.*

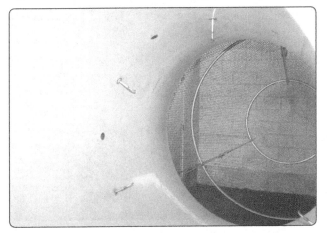

Figure 3-31. *Probes within a bellmouth inlet used to measure intake temperature and pressure.*

Figure 3-32. *An example of a ducted arrangement on a turboprop engine.*

Figure 3-33. *Deflector doors used to deflect ice or dirt away from the intake.*

Figure 3-35. *Rubber stripping inside a turbofan engine inlet allows for friction for short periods of time during changes in the flightpath.*

Figure 3-34. *A typical turbofan intake section.*

blades to rotate at low tip speed for best fan efficiency. The fan permits the use of a conventional air inlet duct, resulting in low inlet duct loss. The fan reduces engine damage from ingested foreign material because much of any material that may be ingested is thrown radially outward and passes through the fan discharge rather than through the core of the engine. Warm bleed air is drawn from the engine and circulated on the inside of the inlet lip for anti-icing. The fan hub or spinner is either heated by warm air or is conical as mentioned earlier. Inside the inlet by the fan blade tips is an abraidable rub strip that allows the fan blades to rub for short times due to flightpath changes. *[Figure 3-35]* Also, inside the inlet are sound-reducing materials to lower the noise generated by the fan.

The fan on high-bypass engines consists of one stage of rotating blades and stationary vanes that can range in diameter from less than 84 inches to more than 112 inches. *[Figure 3-36]* The fan blades are either hollow titanium or composite materials. The air accelerated by the outer part of

the fan blades forms a secondary airstream, which is ducted overboard without passing through the main engine. This secondary air (fan flow) produces 80 percent of the thrust in high-bypass engines. The air that passes through the inner part of the fan blades becomes the primary airstream (core flow) through the engine itself. *[Figure 3-36]*

The air from the fan exhaust, which is ducted overboard, may be discharged in either of two ways:

1. To the outside air through short ducts (dual exhaust nozzles) directly behind the fan. *[Figure 3-37]*

2. Ducted fan, which uses closed ducts all the way to the rear of the engine, where it is exhausted to the outside air through a mixed exhaust nozzle. This type engine is called a ducted fan and the core airflow and fan airflow mix in a common exhaust nozzle.

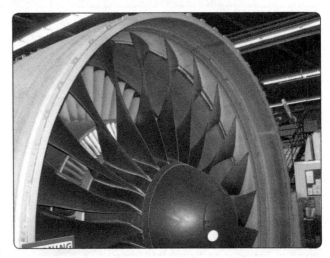

Figure 3-36. *The air that passes through the inner part of the fan blades becomes the primary airstream.*

Figure 3-37. *Air from the fan exhaust can be discharged overboard through short ducts directly behind the fan.*

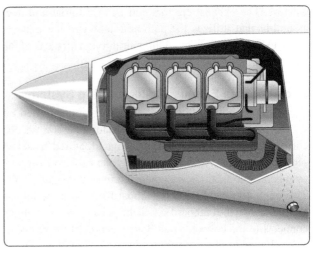

Figure 3-38. *Location of a typical collector exhaust system.*

Reciprocating Engine Exhaust Systems

The reciprocating engine exhaust system is fundamentally a scavenging system that collects and disposes of the high temperature, noxious gases being discharged by the engine. Its main function is to dispose of the gases with complete safety to the airframe and the occupants of the aircraft. The exhaust system can perform many useful functions, but its first duty is to provide protection against the potentially destructive action of the exhaust gases. Modern exhaust systems, though comparatively light, adequately resist high temperatures, corrosion, and vibration to provide long, trouble-free operation with minimum maintenance.

There are two general types of exhaust systems in use on reciprocating aircraft engines: the short stack (open) system and the collector system. The short stack system is generally used on nonsupercharged engines and low-powered engines where noise level is not too objectionable. The collector system is used on most large nonsupercharged engines and on all turbosupercharged engines and installations on which it would improve nacelle streamlining or provide easier maintenance in the nacelle area. On turbosupercharged engines, the exhaust gases must be collected to drive the turbine compressor of the supercharger. Such systems have individual exhaust headers that empty into a common collector ring with only one outlet. From this outlet, the hot exhaust gas is routed via a tailpipe to the turbosupercharger that drives the turbine. Although the collector system raises the back pressure of the exhaust system, the gain in horsepower from turbosupercharging more than offsets the loss in horsepower that results from increased back pressure. The short stack system is relatively simple, and its removal and installation consists essentially of removing and installing the hold-down nuts and clamps. Short stack systems have limited use on most modern aircraft.

In *Figure 3-38,* the location of typical collector exhaust

system components of a horizontally opposed engine is shown in a side view. The exhaust system in this installation consists of a down-stack from each cylinder, an exhaust collector tube on each side of the engine, and an exhaust ejector assembly protruding aft and down from each side of the firewall. The down-stacks are connected to the cylinders with high temperature locknuts and secured to the exhaust collector tube by ring clamps. A cabin heater exhaust shroud is installed around each collector tube. *[Figure 3-39]*

The collector tubes terminate at the exhaust ejector openings at the firewall and are tapered to deliver the exhaust gases at the proper velocity to induce airflow through the exhaust

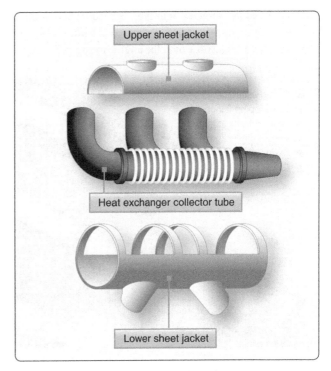

Upper sheet jacket

Heat exchanger collector tube

Lower sheet jacket

Figure 3-39. *A cabin heater exhaust shroud.*

ejectors. The exhaust ejectors consist of a throat-and-duct assembly that utilizes the pumping action of the exhaust gases to induce a flow of cooling air through all parts of the engine compartment (augmenter tube action).

Radial Engine Exhaust Collector Ring System

Figure 3-40 shows the exhaust collector ring installed on a 14-cylinder radial engine. The collector ring is a welded corrosion-resistant steel assembly manufactured in seven sections, with each section collecting the exhaust from two cylinders. The sections are graduated in size. *[Figure 3-41]* The small sections are on the inboard side, and the largest sections are on the outboard side at the point where the tailpipe connects to the collector ring. Each section of the collector ring is bolted to a bracket on the blower section of the engine and is partly supported by a sleeve connection between the collector ring ports and the short stack on the engine exhaust ports. The exhaust tailpipe is joined to the collector ring by a telescoping expansion joint, which allows enough slack for the removal of segments of the collector ring without removing the tailpipe. The exhaust tailpipe is a welded, corrosion-resistant steel assembly consisting of the exhaust tailpipe and, on some aircraft, a muff-type heat exchanger.

Manifold & Augmentor Exhaust Assembly

Some radial engines are equipped with a combination exhaust manifold and augmentor assembly. On a typical 18-cylinder engine, two exhaust assemblies and two augmentor assemblies are used. Each manifold assembly collects exhaust gases from nine cylinders and discharges the gases into the forward end of the augmentor assembly. The exhaust gases are directed into the augmentor bellmouths. The augmentors are designed to produce a venturi effect to draw an increased airflow over the engine to augment engine

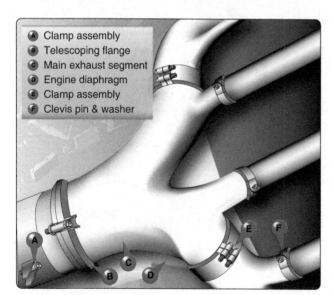

A   Clamp assembly
B   Telescoping flange
C   Main exhaust segment
D   Engine diaphragm
E   Clamp assembly
F   Clevis pin & washer

Figure 3-40. *Elements of an exhaust collector ring installed on a radial engine.*

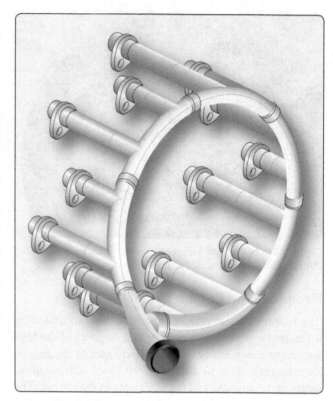

Figure 3-41. *A radial engine exhaust collector ring is graduated in size from the inboard side to the outboard side.*

cooling. An augmentor vane is located in each tailpipe. When the vane is fully closed, the cross-sectional area of the tailpipe is reduced by approximately 45 percent. The augmentor vanes are operated by an electrical actuator, and indicators adjacent to the augmentor vane switches in the flight deck show vane positions. The vanes may be moved toward the "closed" position to decrease the velocity of flow through the augmentor to raise the engine temperature. This system is only used with older aircraft that generally use radial engines.

Reciprocating Engine Exhaust System Maintenance Practices

Any exhaust system failure should be regarded as a severe hazard. Depending on the location and type of failure, an exhaust system failure can result in carbon monoxide poisoning of crew and passengers, partial or complete loss of engine power, or an aircraft fire. Cracks in components, leaking gaskets, or complete failure can cause serious problems in flight. Often, these failures can be detected before complete failure. Black soot around an exhaust gasket shows the gasket has failed. The exhaust system should be inspected very thoroughly.

Exhaust System Inspection

While the type and location of exhaust system components vary somewhat with the type of aircraft, the inspection requirements for most reciprocating engine exhaust systems are very similar. The following paragraphs include a discussion of the most common exhaust system inspection items and procedures for all reciprocating engines. *Figure 3-42* shows the primary inspection areas of three types of exhaust systems.

When performing maintenance on exhaust systems, never use galvanized or zinc-plated tools on the exhaust system. Exhaust system parts should never be marked with a lead pencil. The lead, zinc, or galvanized mark is absorbed by the metal of the exhaust system when heated, creating a distinct change in its molecular structure. This change softens the metal in the area of the mark, causing cracks and eventual failure.

After the installation of a complete exhaust system and all pieces of engine cowl are installed and secured, the engine should be operated to allow the exhaust system to heat up to normal operating temperatures. The engine is then shut down and the cowling removed to expose the exhaust system. Each clamped connection and each exhaust port connection should be inspected for evidence of exhaust gas leakage.

An exhaust leak is indicated by a flat gray or a sooty black streak on the pipes in the area of the leak. An exhaust leak is usually the result of poor alignment of two mated exhaust system members. When a leaking exhaust connection is discovered, the clamps should be loosened, and the leaking units repositioned to ensure a gas-tight fit.

After repositioning, the system nuts should be retightened enough to eliminate any looseness without exceeding the specified torque. If tightening to the specified torque does not eliminate looseness, the bolts and nuts should be replaced since they have probably stretched. After tightening to the specified torque, all nuts should be safetied. With the cowling removed, all necessary cleaning operations can be performed. Some exhaust units are manufactured with a plain sandblast finish. Others may have a ceramic-coated finish. Ceramic-coated stacks should be cleaned by degreasing only. They should never be cleaned with sandblast or alkali cleaners.

During the inspection of an exhaust system, close attention should be given to all external surfaces of the exhaust system for cracks, dents, or missing parts. This also applies to welds, clamps, supports, support attachment lugs, bracing, slip joints, stack flanges, gaskets, and flexible couplings. Each bend should be examined, as well as areas adjacent to welds. Any dented areas or low spots in the system should be inspected for thinning and pitting due to internal erosion by combustion products or accumulated moisture. An ice pick or similar pointed instrument is useful in probing suspected areas.

The system should be disassembled as necessary to inspect internal baffles or diffusers. If a component of the exhaust system is inaccessible for a thorough visual inspection or is hidden by nonremovable parts, it should be removed and checked for possible leaks. This can often be accomplished best by plugging the openings of the component, applying a suitable internal pressure (approximately 2 psi), and submerging it in water. Any leaks cause bubbles that can readily be detected. The procedures required for an installation inspection are also performed during most regular inspections. Daily inspection of the exhaust system usually consists of checking the exposed exhaust system for cracks, scaling, excessive leakage, and loose clamps.

Muffler & Heat Exchanger Failures

Approximately half of all muffler and heat exchanger failures can be traced to cracks or ruptures in the heat exchanger surfaces used for cabin and carburetor heat sources. Failures in the heat exchanger surface (usually in the outer wall) allow

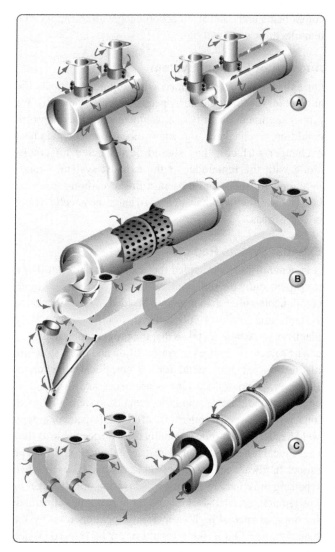

Figure 3-42. *Primary inspection areas of three types of exhaust systems.*

exhaust gases to escape directly into the cabin heat system. These failures, in most cases, are caused by thermal and vibration fatigue cracking in areas of stress concentration. Failure of the spot-welds, which attach the heat transfer pins, can result in exhaust gas leakage. In addition to a carbon monoxide hazard, failure of heat exchanger surfaces can permit exhaust gases to be drawn into the engine induction system, causing engine overheating and power loss.

Exhaust Manifold & Stack Failures

Exhaust manifold and stack failures are usually fatigue failures at welded or clamped points (e.g., stack-to-flange, stack-to-manifold, and crossover pipe or muffler connections). Although these failures are primarily fire hazards, they also present carbon monoxide problems. Exhaust gases can enter the cabin via defective or inadequate seals at firewall openings, wing strut fittings, doors, and wing root openings.

Internal Muffler Failures

Internal failures (baffles, diffusers, etc.) can cause partial or complete engine power loss by restricting the flow of the exhaust gases. If pieces of the internal baffling breaks loose and partially or totally blocks the flow of exhaust gases, engine failure can occur. *[Figure 3-43]* As opposed to other failures, erosion and carburization caused by the extreme thermal conditions are the primary causes of internal failures. Engine backfiring and combustion of unburned fuel within the exhaust system are probable contributing factors. In addition, local hot-spot areas caused by uneven exhaust gas flow can result in burning, bulging, or rupture of the outer muffler wall.

Figure 3-43. *An example of internal muffler failure. Muffler failure can be caused by erosion and carbonization, which in turn can lead to breakage blocking exhaust flow.*

Exhaust Systems with Turbocharger

When a turbocharger or a turbosupercharger system is included, the engine exhaust system operates under greatly increased pressure and temperature conditions. Extra precautions should be taken in exhaust system care and maintenance. During high-pressure altitude operation, the exhaust system pressure is maintained at or near sea level values. Due to the pressure differential, any leaks in the system allow the exhaust gases to escape with torch-like intensity that can severely damage adjacent structures. A common cause of malfunction is coke deposits (carbon buildup) in the waste gate unit causing erratic system operation. Excessive deposit buildups may cause the waste gate valve to stick in the "closed" position, causing an overboost condition. Coke deposit buildup in the turbo itself causes a gradual loss of power in flight and low manifold pressure reading prior to takeoff. Experience has shown that periodic de-coking, or removal of carbon deposits, is necessary to maintain peak efficiency. Clean, repair, overhaul, and adjust the system components and controls in accordance with the applicable manufacturer's instructions.

Augmentor Exhaust System

On exhaust systems equipped with augmentor tubes, the augmentor tubes should be inspected at regular intervals for proper alignment, security of attachment, and general overall condition. Even where augmentor tubes do not contain heat exchanger surfaces, they should be inspected for cracks along with the remainder of the exhaust system. Cracks in augmentor tubes can present a fire or carbon monoxide hazard by allowing exhaust gases to enter the nacelle, wing, or cabin areas.

Exhaust System Repairs

It is generally recommended that exhaust stacks, mufflers, tailpipes, etc., be replaced with new or reconditioned components rather than repaired. Welded repairs to exhaust systems are complicated by the difficulty of accurately identifying the base metal so that the proper repair materials can be selected. Changes in composition and grain structure of the original base metal further complicate the repair. However, when welded repairs are necessary, the original contours should be retained; the exhaust system alignment must not be warped or otherwise affected. Repairs or sloppy weld beads that protrude internally are not acceptable as they cause local hot spots and may restrict exhaust gas flow. The proper hardware and clamps should always be used when repairing or replacing exhaust system components. Steel or low temperature, self-locking nuts should not be substituted for brass or special high temperature locknuts used by the manufacturer. Old gaskets should never be re-used. When disassembly is necessary, gaskets should be replaced with new ones of the same type provided by the manufacturer.

Turbine Engine Exhaust Nozzles

Turbine engines have several different types of exhaust nozzles depending upon the type of engine. Turboshaft engines in helicopters can have an exhaust nozzle that forms a divergent duct. This type of nozzle would not provide any thrust, all engine power going to rotate the rotors, improving helicopter hovering abilities. Turbofan engines tend to fall into either ducted fan of unducted fan engines. Ducted fan engines take the fan airflow and direct it through closed ducts along the engine. Then, it flows into a common exhaust nozzle. The core exhaust flow and the fan flow mix and flow from the engine through this mixed nozzle. The unducted fan has two nozzles, one for the fan airflow and one for the core airflow. These both flow to ambient air separate from each other and have separate nozzles. *[Figure 3-44]*

The unducted engine or the separate nozzle engine handles high amounts of airflow. The fan air which creates most of the thrust (80–85 percent total thrust) must be directed through the fan blades and exit vanes with little turbulence as possible. *[Figure 3-45]* The core airflow needs to be straightened as it comes from the turbine. Through the use of a converging nozzle, the exhaust gases increase in velocity before they are discharged from the exhaust nozzle. Increasing the velocity of the gases increases their momentum and increases the thrust produced (20–15 percent total thrust). Most of the energy of the gases have been absorbed to drive the fan through the low-pressure turbine stages.

Turboprop exhaust nozzles provide small amounts of thrust (10–15 percent) but are mainly used to discharge the exhaust gases from the aircraft. Most of the energy has been transferred to the propeller. On some turboprop aircraft, an exhaust duct is often referred to as a tailpipe, although the duct itself is

Figure 3-45. *Fan air is directed through the fan blades and exit vanes.*

essentially a simple, stainless steel, conical or cylindrical pipe. The assembly also includes an engine tail cone and the struts inside the duct. The tail cone and the struts add strength to the duct, impart an axial direction to the gas flow, and smooth the gas flow. In a typical installation, the tailpipe assembly is mounted in the nacelle and attached at its forward end to the firewall. The forward section of the tailpipe is funnel shaped and surrounds but does not contact the turbine exhaust section. This arrangement forms an annular gap that serves as an air ejector for the air surrounding the engine hot section. As the high-velocity exhaust gases enter the tailpipe, a low-pressure effect is produced which causes the air around the engine hot section to flow through the annular gap into the tailpipe. The rear section of the tailpipe is secured to the airframe by two support arms, one on each side of the tailpipe. The support arms are attached to the upper surface of the wing in such

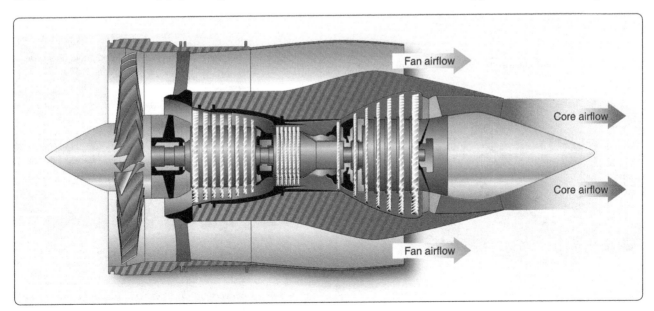

Figure 3-44. *Path of both core exhaust flow and fan flow from the engine to separate nozzles.*

a way that allow movement fore and aft to compensate for expansion. The tailpipe assembly is wrapped in an insulating blanket to shield the surrounding area from the high heat produced by the exhaust gases. Such blankets may be made of a stainless steel laminated sheet on the outside and fiberglass on the inside. This is used when the engine exhaust is located some distance from the edge of the wing or aircraft structure.

Immediately aft of the turbine outlet, and usually just forward of the flange to which the exhaust duct is attached, the engine is instrumented for turbine discharge pressure. One or more pressure probes are inserted into the exhaust duct to provide adequate sampling of the exhaust gases. In large engines, it is not practical to measure the internal temperature at the turbine inlet, so the engine is often also instrumented for exhaust gas temperature at the turbine outlet.

Convergent Exhaust Nozzle

As the exhaust gases exit the rear of the engine, they flow into the exhaust nozzle. *[Figure 3-46]* The very first part of the exhaust nozzle and the exhaust plug form a divergent duct to reduce turbulence in the airflow, then the exhaust gases flow into the convergent component of the exhaust nozzle where the flow is restricted by a smaller outlet opening. Since this forms a convergent duct, the gas velocity is increased providing increased thrust. The restriction of the opening of the outlet of the exhaust nozzle is limited by two factors. If the nozzle opening is too big, thrust is being wasted. If it is too little, the flow is choked in the other components of the engine. In other words, the exhaust nozzle acts as an orifice, the size of which determines the density and velocity of the gases as they emerge from the engine. This is critical to thrust performance. Adjusting the area of the exhaust nozzle changes both the engine performance and the exhaust gas temperature. When the velocity of the exhaust gases at the

nozzle opening becomes Mach 1, the flow passes only at this speed—it does not increase or decrease. Sufficient flow to maintain Mach 1 at the nozzle opening and have extra flow (flow that is being restricted by the opening) creates what is called a choked nozzle. The extra flow builds up pressure in the nozzle, which is sometimes called pressure thrust. A differential in pressure exists between the inside of the nozzle and the ambient air. By multiplying this difference in pressure times the area of the nozzle opening, pressure thrust can be calculated. Many engines cannot develop pressure thrust because most of the energy is used to drive turbines that turn propellers, large fans, or helicopter rotors.

Convergent-Divergent Exhaust Nozzle

Whenever the engine pressure ratio is high enough to produce exhaust gas velocities which might exceed Mach 1 at the engine exhaust nozzle, more thrust can be gained by using a convergent-divergent type of nozzle. *[Figure 3-47]* The advantage of a convergent-divergent nozzle is greatest at high Mach numbers because of the resulting higher pressure ratio across the engine exhaust nozzle.

To ensure that a constant weight or volume of a gas flows past any given point after sonic velocity is reached, the rear part of a supersonic exhaust duct is enlarged to accommodate the additional weight or volume of a gas that flows at supersonic rates. If this is not done, the nozzle does not operate efficiently. This is the divergent section of the exhaust duct.

When a divergent duct is used in combination with a conventional exhaust duct, it is called a convergent-divergent exhaust duct. In the convergent-divergent, or C-D nozzle, the convergent section is designed to handle the gases while they remain subsonic, and to deliver the gases to the throat of the nozzle just as they attain sonic velocity. The divergent section

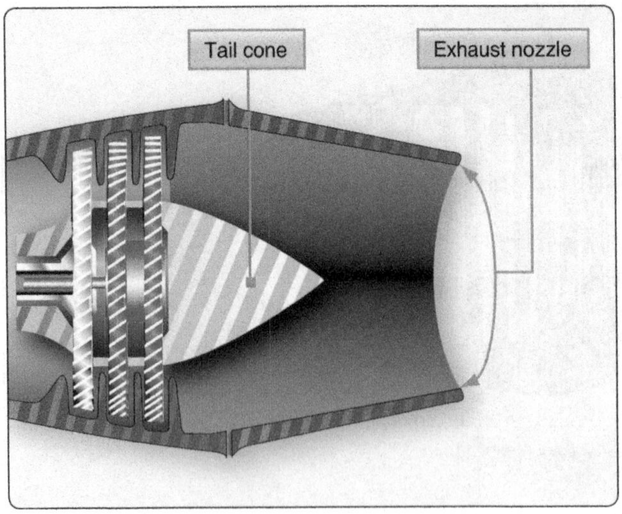

Figure 3-46. *Exhaust gases exit the rear of the engine through the exhaust nozzle.*

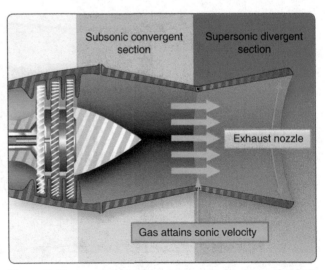

Figure 3-47. *A convergent-divergent nozzle can be used to help produce more thrust when exhaust gas velocities are greater than Mach 1.*

handles the gases, further increasing their velocity, after they emerge from the throat and become supersonic. As the gas flows from the throat of the nozzle, it becomes supersonic (Mach 1 and above) and then passes into the divergent section of the nozzle. Since it is supersonic, it continues to increase in velocity. This type of nozzle is generally used on very high speed aerospace vehicles.

Thrust Reversers

As aircraft have increased in gross weights with higher landing airspeeds, the problem of stopping an aircraft after landing has greatly increased. In many instances, the aircraft brakes can no longer be relied upon solely to slow the aircraft within a reasonable distance, immediately after touchdown. Most thrust reverser systems can be divided into two categories: mechanical-blockage and aerodynamic-blockage. Mechanical blockage is accomplished by placing a removable obstruction in the exhaust gas stream, usually somewhat to the rear of the nozzle. The engine exhaust gases are mechanically blocked and diverted at a suitable angle in the reverse direction by an inverted cone, half-sphere, or clam shell. *[Figure 3-48]* This is placed in position to reverse the flow of exhaust gases. This type is generally used with ducted turbofan engines, where the fan and core flow mix in a common nozzle before exiting the engine. The clamshell-type or mechanical-blockage reverser operates to form a barrier in the path of escaping exhaust gases, which nullifies and reverses the forward thrust of the engine. The reverser system must be able to withstand high temperatures, be mechanically strong, relatively light in weight, reliable, and "fail-safe." When not in use, it must be streamlined into the configuration of the engine nacelle. When the reverser is not in use, the clamshell doors retract and nest neatly around the engine exhaust duct, usually forming the rear section of the engine nacelle.

In the aerodynamic blockage type of thrust reverser, used mainly with unducted turbofan engines, only fan air is used to slow the aircraft. A modern aerodynamic thrust reverser system consists of a translating cowl, blocker doors, and cascade vanes that redirect the fan airflow to slow the aircraft. *[Figure 3-49]* If the thrust levers are at idle position and the aircraft has weight on the wheels, moving the thrust levers aft activates the translating cowl to open, closing the blocker doors. This action stops the fan airflow from going aft and redirects it through the cascade vanes, which direct the airflow forward to slow the aircraft. Since the fan can produce approximately 80 percent of the engine's thrust, the fan is the best source for reverse thrust. By returning the thrust levers (power levers) to the idle position, the blocker doors open and the translating cowl closes.

A thrust reverser must not have any adverse effect on engine operation either deployed or stowed. Generally, there is

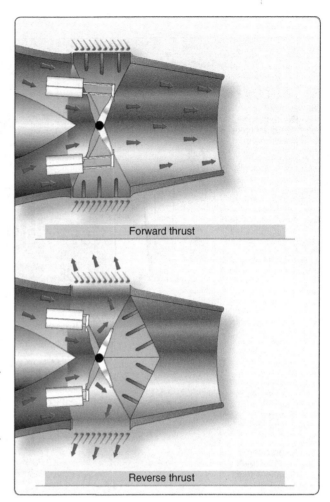

Figure 3-48. *Engine exhaust gases are blocked and diverted in a reserve direction during thrust reversal.*

an indication in the flight deck with regard to the status of the reverser system. The thrust reverser system consists of several components that move either the clam shell doors or the blocker door and translating cowl. Actuating power is generally pneumatic or hydraulic and uses gearboxes, flexdrives, screwjacks, control valves, and air or hydraulic motors to deploy or stow the thrust reverser systems. The systems are locked in the stowed position until commanded to deploy by the flight deck. Since there are several moving parts, maintenance and inspection requirements are very important. While performing any type of maintenance, the reverser system must be mechanically locked out from deploying while personnel are in the area of the reverser system.

Afterburning/Thrust Augmentation

The terms afterburning and thrust augmentation generally pertain to military engine applications. The terms are used to describe the same system. Normally, this is used to increase the thrust of the engine up to double the original thrust. The required additions to the exhaust nozzle for this system are a flame stabilizer, fuel manifold, flame holder, igniter, and a

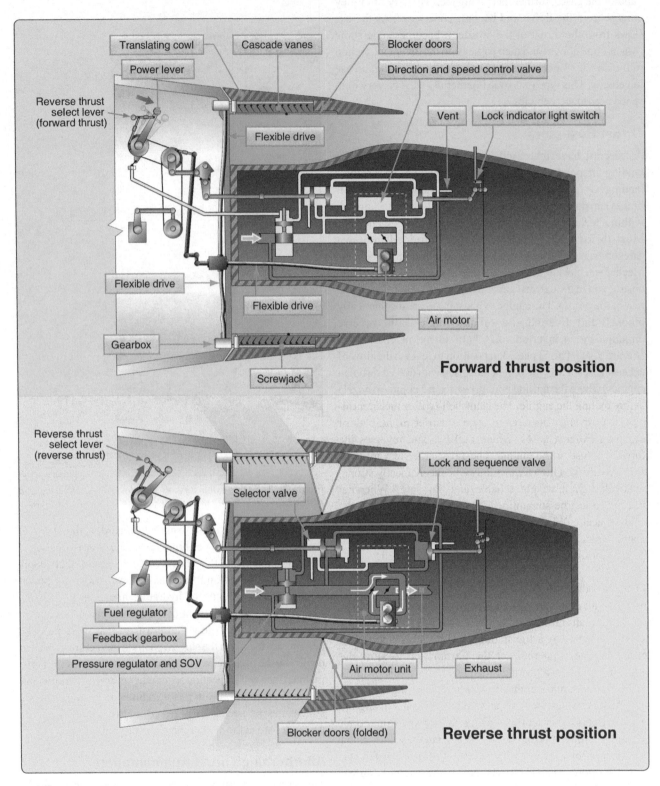

Figure 3-49. *Components of a thrust reverser system.*

variable area exhaust nozzle. *[Figure 3-50]* After the engine has reached full power under normal operation, the power lever can be advanced to activate the afterburner. This allows more fuel to flow into the exhaust nozzle where it is ignited and burned. As energy and mass are added to the gas flow, the exhaust nozzle must open wider to allow greater flow. As the power lever is moved back out of the afterburner, the exhaust nozzle closes down again. Some low-bypass turbofan engines used in military aircraft use bypass (fan air) to flow into the exhaust nozzle. Just as in a ducted fan, this air is used in the afterburner. It contains more oxygen and assists combustion in the afterburner. Since fuel is being burned in the exhaust nozzle, the heat buildup around the nozzle is a problem. A special type of liner is used around the nozzle to allow cooler air to circulate around the nozzle. This operates somewhat like a single burner can combustion chamber. Operation in the afterburner mode is somewhat limited by high fuel consumption, which can be almost double normal consumption.

Thrust Vectoring

Thrust vectoring is the ability of an aircraft's main engines to direct thrust other than parallel to the vehicle's longitudinal axis, allowing the exhaust nozzle to move or change position to direct the thrust in varied directions. Vertical takeoff aircraft use thrust vectoring as takeoff thrust and then change direction to propel the aircraft in horizontal flight. Military aircraft use thrust vectoring for maneuvering in flight to change direction. Thrust vectoring is generally accomplished by relocating the direction of the exhaust nozzle to direct the thrust to move the aircraft in the desired path. At the rear of a gas turbine engine, a nozzle directs the flow of hot exhaust gases out of the engine and afterburner. Usually, the nozzle points straight out of the engine. The pilot can move, or vector, the vectoring nozzle up and down by 20°. This makes the aircraft much more maneuverable in flight. *[Figure 3-51]*

Figure 3-50. *An example of a variable area exhaust nozzle used to increase or decrease exhaust flow during afterburn.*

Figure 3-51. *A pilot can direct thrust via the vectoring nozzle 20° up or down to increase flight maneuverability.*

Engine Noise Suppression

Aircraft powered by gas turbine engines sometimes require noise suppression for the engine exhaust gases when operating from airports located in or near highly populated areas. Several types of noise suppressors are used. A common type of noise suppressor is an integral, airborne part of the aircraft engine installation or engine exhaust nozzle. Engine noise comes from several sources on the engine, the fan, or compressor and the air discharge from the core of the engine. There are three sources of noise involved in the operation of a gas turbine engine. The engine air intake and vibration from engine housing are sources of some noise, but the noise generated does not compare in magnitude with that produced by the engine exhaust. *[Figure 3-52]* The noise produced by the engine exhaust is caused by the high degree of turbulence of a high-velocity jet stream moving through a relatively quiet atmosphere. For a distance of a few nozzle diameters downstream behind the engine, the velocity of the jet stream is high, and there is little mixing of the atmosphere with the jet stream. In this region, the turbulence within the high speed jet stream is very fine grain turbulence and produces relatively high-frequency noise. This noise is caused by violent, turbulent mixing of the exhaust gases with the atmosphere and is influenced by the shearing action caused by the relative speeds between the velocity and the atmosphere.

Farther downstream, as the velocity of the jet stream slows down, the jet stream mixes with the atmosphere and turbulence of a coarser type begins. Compared with noise from other portions of the jet stream, noise from this portion has a much lower frequency. As the energy of the jet stream finally is dissipated in large turbulent swirls, a greater portion of the energy is converted into noise. The noise generated as the exhaust gases dissipate is at a frequency near the low end of the audible range. The lower the frequency of the noise,

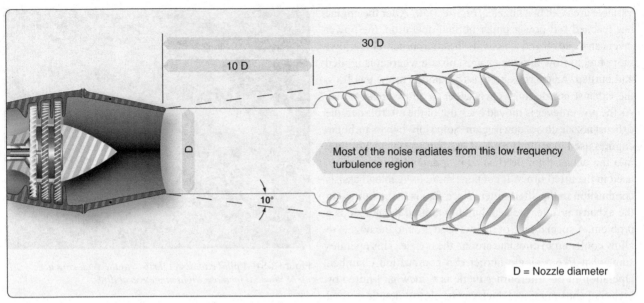

Most of the noise radiates from this low frequency turbulence region

D = Nozzle diameter

Figure 3-52. *Engine noise from engine exhaust is created by the turbulence of a high velocity jet stream moving through the relatively quiet atmosphere.*

the greater the distance the noise travels. This means that the low-frequency noises reach an individual on the ground in greater volume than the high-frequency noises, and hence are more objectionable. High-frequency noise is weakened more rapidly than low-frequency noise, both by distance and the interference of buildings, terrain, and atmospheric disturbances. A deep-voiced, low-frequency foghorn, for example, may be heard much farther than a shrill, high-frequency whistle, even though both may have the same overall volume (decibels) at their source.

Noise levels vary with engine thrust and are proportional to the amount of work done by the engine on the air that passes through it. An engine having relatively low airflow but high thrust due to high turbine discharge (exhaust gas) temperature, pressure, and/or afterburning produces a gas stream of high velocity and, therefore, high noise levels. A larger engine, handling more air, is quieter at the same thrust and large engines operating at partial thrust are less noisy than smaller engines operating at full thrust. Thus, the noise level can be reduced considerably by operating the engine at lower power settings. Compared with a turbojet, a turbofan version of the same engine is quieter during takeoff. The noise level produced by a fan-type engine is lower, principally because the exhaust gas velocities ejected at the engine tailpipe are slower than those for a turbojet of comparative size.

Fan engines require a larger turbine to provide additional power to drive the fan. The large turbine, which usually has an additional turbine stage, reduces the velocity of the gas and, therefore, reduces the noise produced because exhaust gas noise is proportional to exhaust gas velocity. The exhaust from the fan is at a relatively low velocity and, therefore, does

not create a noise problem. Because of the characteristic of low-frequency noise to linger at a relatively high volume, effective noise reduction for a turbojet aircraft must be achieved by revising the noise pattern or by changing the frequency of the noise emitted by the jet nozzle.

The noise suppressors in current use are either of the corrugated-perimeter type, or the multi-tube type. *[Figure 3-53]* Both types of suppressors break up the single, main jet exhaust stream into a number of smaller jet streams. This increases the total perimeter of the nozzle area and reduces the size of the air stream eddies created as the gases are discharged into the open air. Although the total noise-energy remains unchanged, the frequency is raised considerably. The size of the air stream eddies scales down at a linear rate with the size of the exhaust stream. This has two effects: 1) the change in frequency may put some of the noise above the audibility range of the human ear, and 2) high frequencies

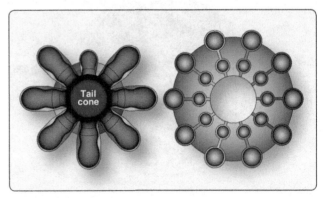

Figure 3-53. *Noise suppressors currently in use are corrugated-perimeter type, or multi-tube type.*

within the audible range, while perhaps more annoying, are more highly attenuated by atmospheric absorption than are low frequencies. Thus, the falloff in intensity is greater and the noise level is less at any given distance from the aircraft. In the engine nacelle, the area between the engine and the cowl has acoustic linings surrounding the engine. This noise-absorbing lining material converts acoustic energy into heat. These linings normally consist of a porous skin supported by a honeycomb backing and provide a separation between the face sheet and the engine duct. For optimum suppression, the acoustic properties of the skin and the liner are carefully matched.

Turbine Engine Emissions

Engineers are introducing new combustion technology that has dramatically reduced emissions from gas turbine engines. Lowering exhaust emissions from gas turbine, especially oxides of nitrogen (NO_X), continue to require improvement. Most of the research has centered around the combustion section of the engine. New technology with unique combustor design has greatly reduced emissions. One manufacturer has a design called the Twin Annular, Pre-mixing Swirler (TAPS) combustor. Most advanced designs rely on a method of pre-mixing the air-fuel before it enters the combustion burner area. In the TAPS design, air from the high-pressure compressor is directed into the combustor through two high-energy swirlers adjacent to the fuel nozzles. This swirl creates a more thorough and leaner mix of fuel and air, which burns at lower temperatures than in previous gas turbine engine designs. Most of the NO_X is formed by the reaction of oxygen and nitrogen at high temperatures. The NO_X levels are higher if the burning air-fuel mixture stays at high temperatures for a longer time. Newly designed combustors also produce lower levels of carbon monoxide and unburned hydrocarbons. The increases in gas turbine engine component efficiencies have resulted in fewer emissions from gas turbine engines.

Chapter 4
Engine Ignition & Electrical Systems

Reciprocating Engine Ignition Systems

The basic requirements for reciprocating engine ignition systems are similar, regardless of the type of engine. All ignition systems must deliver a high-tension spark across the electrodes of each spark plug in each cylinder of the engine in the correct firing order. At a predetermined number of degrees ahead of the top dead center position of the piston, as measured by crankshaft travel in degrees of rotation, the spark occurs in the cylinder. The potential output voltage of the system must be adequate to arc the gap in the spark plug electrodes under all operating conditions. The spark plug is threaded into the cylinder head with the electrodes exposed to the combustion area of the engine's cylinder.

Ignition systems can be divided into two classifications: magneto-ignition systems or electronic Full Authority Digital Engine Control (FADEC) systems for reciprocating engines. Ignition systems can also be subclassified as either single or dual magneto-ignition systems. The single magneto-ignition system, usually consisting of one magneto and the necessary wiring, was used with another single magneto on the same engine. Dual magnetos generally use one rotating magnet that feeds two complete magnetos in one magneto housing. An example of each type is shown in *Figure 4-1*.

Aircraft magneto-ignition systems can be classified as either high-tension or low-tension. The low-tension magneto system, covered in a later section of this chapter, generates a low-voltage that is distributed to a transformer coil near each spark plug. This system eliminates some problems inherent in the high-tension system that was containing the high-voltage until it passed through the spark plug. The materials that were used for ignition leads could not withstand the high-voltage

and were prone to leak to ground before the spark would get to the cylinder. As new materials evolved and shielding was developed, the problems with high-tension magnetos were overcome. The high-tension magneto system is still the most widely used aircraft ignition system.

Some very old antique aircraft used a battery-ignition system. In this system, the source of energy is a battery or generator, rather than a magneto. This system was similar to that used in most automobiles at the time. *Figure 4-2* shows a simplified schematic of a battery-ignition system.

Magneto-Ignition System Operating Principles

The magneto, a special type of engine-driven alternating current (AC) generator, uses a permanent magnet as a source of energy. By the use of a permanent magnet (basic magnetic field), coil of wire (concentrated lengths of conductor), and relative movement of the magnetic field, current is generated in the wire. At first, the magneto generates electrical power by the engine rotating the permanent magnet and inducing a current to flow in the coil windings. As current flows through the coil windings, it generates its own magnetic field that surrounds the coil windings. At the correct time, this current flow is stopped and the magnetic field collapses across a second set of windings in the coil and a high-voltage is generated. This is the voltage used to arc across the spark plug gap. In both cases, the three basic things needed to generate electrical power are present to develop the high-voltage that forces a spark to jump across the spark plug gap

Figure 4-1. *Single and dual magnetos.*

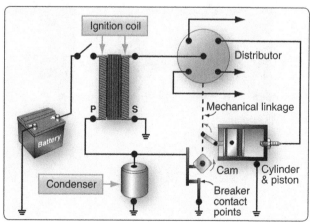

Figure 4-2. *Battery-ignition system.*

in each cylinder. Magneto operation is timed to the engine so that a spark occurs only when the piston is on the proper stroke at a specified number of crankshaft degrees before the top dead center piston position.

High-Tension Magneto System Theory of Operation

The high-tension magneto system can be divided, for purposes of discussion, into three distinct circuits: magnetic, primary electrical, and secondary electrical circuits.

Magnetic Circuit

The magnetic circuit consists of a permanent multi-pole rotating magnet, a soft iron core, and pole shoes. *[Figure 4-3]* The magnet is geared to the aircraft engine and rotates in the gap between two pole shoes to furnish the magnetic lines of force (flux) necessary to produce an electrical voltage. The poles of the magnet are arranged in alternate polarity so that the flux can pass out of the north pole through the coil core and back to the south pole of the magnet. When the magnet is in the position shown in *Figure 4-3A*, the number of magnetic lines of force through the coil core is maximum because two magnetically opposite poles are perfectly aligned with the pole shoes.

This position of the rotating magnet is called the full register position and produces a maximum number of magnetic lines of force, flux flow clockwise through the magnetic circuit and from left to right through the coil core. When the magnet is moved away from the full register position, the amount of flux passing through the coil core begins to decrease. This occurs because the magnet's poles are moving away from the pole shoes, allowing some lines of flux to take a shorter path through the ends of the pole shoes.

As the magnet moves farther from the full register position, more lines of flux are short circuited through the pole shoe ends. Finally, at the neutral position 45° from the full register position, all flux lines are short circuited, and no flux flows through the coil core. *[Figure 4-3B]* As the magnet moves from full register to the neutral position, the number of flux lines through the coil core decreases in the same manner as the gradual collapse of flux in the magnetic field of an ordinary electromagnet.

The neutral position of the magnet is where one of the poles of the magnet is centered between the pole shoes of the magnetic circuit. As the magnet is moved clockwise from this position, the lines of flux that had been short circuited through the pole shoe ends begin to flow through the coil core again. But this time, the flux lines flow through the coil core in the opposite direction. *[Figure 4-3C]* The flux flow reverses as the magnet moves out of the neutral position because the north pole of the rotating permanent magnet is opposite the right pole shoe instead of the left. *[Figure 4-3A]*

When the magnet is again moved a total of 90°, another full register position is reached with a maximum flux flow in the opposite direction. The 90° of magnet travel is shown in *Figure 4-4,* where a curve shows how the flux density in the coil core, without a primary coil around the core, changes as the magnet is rotated.

Figure 4-4 shows that as the magnet moves from the full register position 0°, flux flow decreases and reaches a zero value as it moves into the neutral position 45°. While the magnet moves through the neutral position, flux flow reverses and begins to increase as indicated by the curve below the

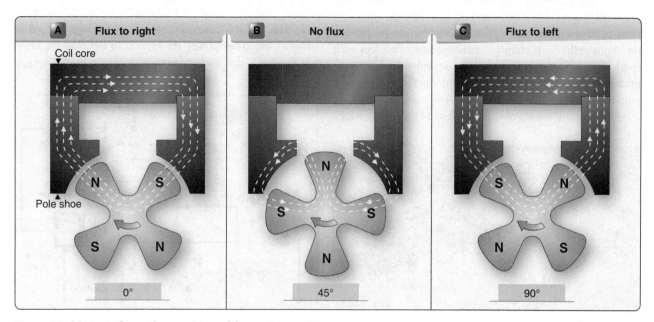

Figure 4-3. *Magnetic flux at three positions of the rotating magnet.*

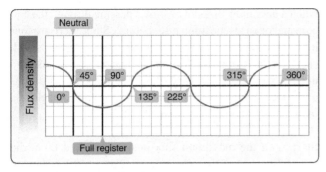

Figure 4-4. *Change in flux density as magnet rotates.*

horizontal line. At the 90° position, another position of maximum flux is reached. Thus, for one revolution 360° of the four pole magnet, there are four positions of maximum flux, four positions of zero flux, and four flux reversals.

This discussion of the magnetic circuit demonstrates how the coil core is affected by the rotating magnet. It is subjected to an increasing and decreasing magnetic field and a change in polarity each 90° of magnet travel.

When a coil of wire as part of the magneto's primary electrical circuit is wound around the coil core, it is also affected by the varying magnetic field.

Primary Electrical Circuit

The primary electrical circuit consists of a set of breaker contact points, a condenser, and an insulated coil. *[Figure 4-5]* The coil is made up of a few turns of heavy copper wire, one end is grounded to the coil core and the other end to the ungrounded side of the breaker points. *[Figure 4-5]* The primary circuit is complete only when the ungrounded breaker point contacts the grounded breaker point. The third unit in the circuit, the condenser (capacitor), is wired in parallel with the breaker points. The condenser prevents arcing at the points when the circuit is opened and hastens the collapse of the magnetic field about the primary coil.

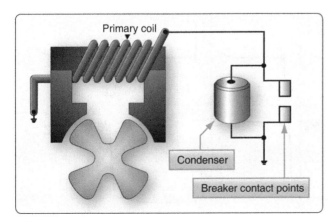

Figure 4-5. *Primary electrical circuit of a high-tension magneto.*

The primary breaker closes at approximately full register position. When the breaker points are closed, the primary electrical circuit is completed and the rotating magnet induces current flow in the primary circuit. This current flow generates its own magnetic field, which is in such a direction that it opposes any change in the magnetic flux of the permanent magnet's circuit.

While the induced current is flowing in the primary circuit, it opposes any decrease in the magnetic flux in the core. This is in accordance with Lenz's Law that states: "An induced current always flows in such a direction that its magnetism opposes the motion or the change that induced it." (For a review of Lenz's Law, refer to the Aviation Maintenance Technician—General Handbook, FAA-H-8083-30). Thus, the current flowing in the primary circuit holds the flux in the core at a high value in one direction until the rotating magnet has time to rotate through the neutral position to a point a few degrees beyond neutral. This position is called the E-gap position (E stands for efficiency).

There are three basic events required to fire a spark plug when its piston is in the prescribed position:

1. the magneto must be in the E-gap position,

2. the breaker contact points must be open, and

3. the distributor must be aligned correctly.

With the magnetic rotor in E-gap position and the primary coil holding the magnetic field of the magnetic circuit in the opposite polarity, a very high rate of flux change can be obtained by opening the primary breaker points. Opening the breaker points stops the flow of current in the primary circuit and allows the magnetic rotor to quickly reverse the field through the coil core. This sudden flux reversal produces a high rate of flux change in the core, that cuts across the secondary coil of the magneto (wound over and insulated from the primary coil), inducing the pulse of high-voltage electricity in the secondary needed to fire a spark plug. As the rotor continues to rotate to approximately full register position, the primary breaker points close again and the cycle is repeated to fire the next spark plug in firing order. The sequence of events can now be reviewed in greater detail to explain how the state of extreme magnetic stress occurs.

With the breaker points, cam, and condenser connected in the circuit as shown in *Figure 4-6*, the action that takes place as the magnetic rotor turns is depicted by the graph curve in *Figure 4-7*. At the top (A) of *Figure 4-7*, the original static flux curve of the magnets is shown. Shown below the static flux curve is the sequence of opening and closing the magneto breaker points. Note that opening and closing the breaker points is timed by the breaker cam. The points close

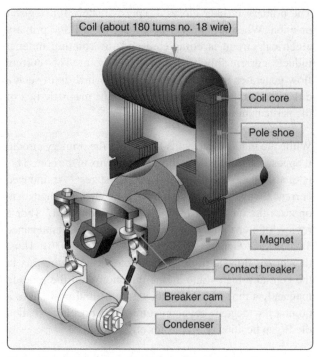

Figure 4-6. *Components of a high-tension magneto circuit.*

when a maximum amount of flux is passing through the coil core and open at a position after neutral. Since there are four lobes on this particular cam (there are some magnetos with cams that have only two lobes), the breaker points close and open in the same relation to each of the four neutral positions of the rotor magnet. Also, the point opening and point closing intervals are approximately equal.

Starting at the maximum flux position marked 0° at the top of *Figure 4-7*, the sequence of events in the following paragraphs occurs.

As the magnet rotor is turned toward the neutral position, the amount of flux through the core starts to decrease. *[Figure 4-7D]* This change in flux linkages induces a current in the primary winding. *[Figure 4-7C]* This induced current creates a magnetic field of its own that opposes the change of flux linkages inducing the current. Without current flowing in the primary coil, the flux in the coil core decreases to zero as the magnet rotor turns to neutral and starts to increase in the opposite direction (dotted static flux curve in *Figure 4-7D*). But, the electromagnetic action of the primary current prevents the flux from changing and temporarily holds the field instead of allowing it to change (resultant flux line in *Figure 4-7D*).

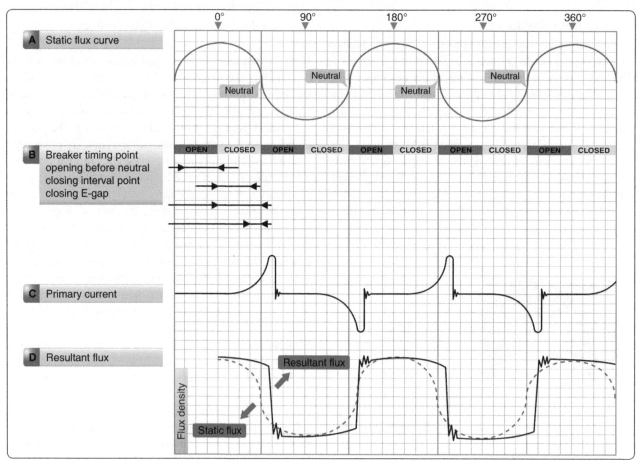

Figure 4-7. *Magneto flux curves.*

Figure 4-8. *Pivotless type breaker assembly and cam.*

flow is accomplished through a pair of breaker contact points made of an alloy that resists pitting and burning.

Most breaker points used in aircraft ignition systems are of the pivotless type in which one of the breaker points is movable and the other stationary. *[Figure 4-8]* The movable breaker point attached to the leaf spring is insulated from the magneto housing and is connected to the primary coil. *[Figure 4-8]* The stationary breaker point is grounded to the magneto housing to complete the primary circuit when the points are closed and can be adjusted so that the points can open at the proper time.

Another part of the breaker assembly is the cam follower, which is spring-loaded against the cam by the metal leaf spring. The cam follower is a Micarta block or similar material that rides the cam and moves upward to force the movable breaker contact away from the stationary breaker contact each time a lobe of the cam passes beneath the follower. A felt oiler pad is located on the underside of the metal spring leaf to lubricate and prevent corrosion of the cam.

As a result of the holding process, there is a very high stress in the magnetic circuit by the time the magnet rotor has reached the position where the breaker points are about to open. The breaker points, when opened, function with the condenser to interrupt the flow of current in the primary coil, causing an extremely rapid change in flux linkages. The high-voltage in the secondary winding discharges across the gap in the spark plug to ignite the air-fuel mixture in the engine cylinder. Each spark actually consists of one peak discharge, after which a series of small oscillations takes place.

They continue to occur until the voltage becomes too low to maintain the discharge. Current flows in the secondary winding during the time that it takes for the spark to completely discharge. The energy or stress in the magnetic circuit is completely dissipated by the time the contacts close for the production of the next spark. Breaker assemblies, used in high-tension magneto-ignition systems, automatically open and close the primary circuit at the proper time in relation to piston position in the cylinder to which an ignition spark is being furnished. The interruption of the primary current

The breaker-actuating cam may be directly driven by the magneto rotor shaft or through a gear train from the rotor shaft. Most large radial engines use a compensated cam that is designed to operate with a specific engine and has one lobe for each cylinder to be fired by the magneto. The cam lobes are machine ground at unequal intervals to compensate for the elliptical path of the articulated connecting rods. This path causes the pistons top dead center position to vary from cylinder to cylinder with regard to crankshaft rotation. A compensated 14-lobe cam, together with a two-, four-, and eight-lobe uncompensated cam, is shown in *Figure 4-9.*

The unequal spacing of the compensated cam lobes, although it provides the same relative piston position for ignition to occur, causes a slight variation of the E-gap position of the rotating magnet and thus a slight variation in the high-voltage impulses generated by the magneto. Since the spacing between each lobe is tailored to a particular cylinder of a particular engine, compensated cams are marked to show

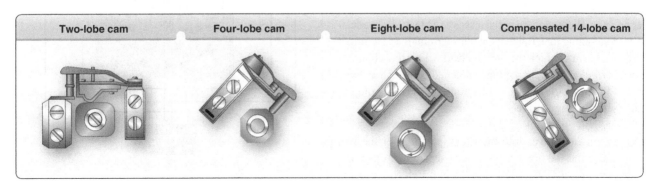

Figure 4-9. *Typical breaker assemblies.*

the series of the engine, the location of the master rods, the lobe used for magneto timing, the direction of cam rotation, and the E-gap specification in degrees past neutral of magnet rotation. In addition to these markings, a step is cut across the face of the cam, that, when aligned with scribed marks on the magneto housing, places the rotating magnet in the E-gap position for the timing cylinder. Since the breaker points should begin to open when the rotating magnet moves into the E-gap position, alignment of the step on the cam with marks in the housing provides a quick and easy method of establishing the exact E-gap position to check and adjust the breaker points.

Secondary Electrical Circuit

The secondary circuit contains the secondary windings of the coil, distributor rotor, distributor cap, ignition lead, and spark plug. The secondary coil is made up of a winding containing approximately 13,000 turns of fine, insulated wire; one end of which is electrically grounded to the primary coil or to the coil core and the other end connected to the distributor rotor. The primary and secondary coils are encased in a non-conducting material. The whole assembly is then fastened to the pole shoes with screws and clamps.

When the primary circuit is closed, the current flow through the primary coil produces magnetic lines of force that cut across the secondary windings, inducing an electromotive force. When the primary circuit current flow is stopped, the magnetic field surrounding the primary windings collapses, causing the secondary windings to be cut by the lines of force. The strength of the voltage induced in the secondary windings, when all other factors are constant, is determined by the number of turns of wire. Since most high-tension magnetos have many thousands of turns of wire in the secondary coil windings, a very high-voltage, often as high as 20,000 volts, is generated in the secondary circuit. The high-voltage induced in the secondary coil is directed to the distributor, which consists of two parts: revolving and stationary. The revolving part is called a distributor rotor and the stationary part is called a distributor block. The rotating part, which may take the shape of a disc, drum, or finger, is made of a non-conducting material with an embedded conductor. The stationary part consists of a block also made of non-conducting material that contains terminals and terminal receptacles into which the ignition lead wiring that connects the distributor to the spark plug is attached. This high-voltage is used to jump the air gap of electrodes of the spark plug in the cylinder to ignite the air-fuel mixture.

As the magnet moves into the E-gap position for the No. 1 cylinder and the breaker points just separate or open, the distributor rotor aligns itself with the No. 1 electrode in the distributor block. The secondary voltage induced as the

breaker points open enters the rotor where it arcs a small air gap to the No. 1 electrode in the block.

Since the distributor rotates at one-half crankshaft speed on all four-stroke cycle engines, the distributor block has as many electrodes as there are engine cylinders, or as many electrodes as cylinders served by the magneto. The electrodes are located circumferentially around the distributor block so that, as the rotor turns, a circuit is completed to a different cylinder and spark plug each time there is alignment between the rotor finger and an electrode in the distributor block. The electrodes of the distributor block are numbered consecutively in the direction of distributor rotor travel. *[Figure 4-10]*

The distributor numbers represent the magneto sparking order rather than the engine cylinder numbers. The distributor electrode marked "1" is connected to the spark plug in the No. 1 cylinder; distributor electrode marked "2" to the second

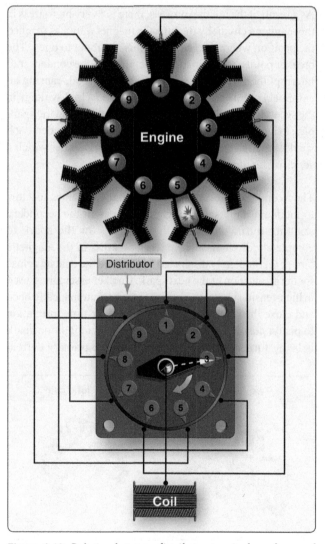

Figure 4-10. *Relation between distributor terminal numbers and cylinder numbers.*

cylinder to be fired; distributor electrode marked "3" to the third cylinder to be fired, and so forth.

In *Figure 4-10*, the distributor rotor finger is aligned with the distributor electrode marked "3," which fires the No. 5 cylinder of a nine-cylinder radial engine. Since the firing order of a nine-cylinder radial engine is 1-3-5-7-9-2-4-6-8, the third electrode in the magneto sparking order serves the No. 5 cylinder.

Magneto & Distributor Venting

Since magneto and distributor assemblies are subjected to sudden changes in temperature, the problems of condensation and moisture are considered in the design of these units. Moisture in any form is a good conductor of electricity. If absorbed by the nonconducting material in the magneto, such as distributor blocks, distributor fingers, and coil cases, it can create a stray electrical conducting path. The high-voltage current that normally arcs across the air gaps of the distributor can flash across a wet insulating surface to ground, or the high-voltage current can be misdirected to some spark plug other than the one that should be fired. This condition is called flashover and usually results in cylinder misfiring. This can cause a serious engine condition called pre-ignition, which can damage the engine. For this reason, coils, condensers, distributors, and distributor rotors are waxed so that moisture on such units stand in separate beads and do not form a complete circuit for flashover.

Flashover can lead to carbon tracking, which appears as a fine pencil-like line on the unit across which flashover occurs. The carbon trail results from the electric spark burning dirt particles that contain hydrocarbon materials. The water in the hydrocarbon material is evaporated during flashover, leaving carbon to form a conducting path for current. When moisture is no longer present, the spark continues to follow the carbon track to the ground. This prevents the spark from getting to the spark plug, so the cylinder does not fire.

Magnetos cannot be hermetically sealed to prevent moisture from entering a unit, because the magneto is subject to pressure and temperature changes in altitude. Thus, adequate drains and proper ventilation reduce the tendency of flashover and carbon tracking. Good magneto circulation also ensures that corrosive gases produced by normal arcing across the distributor air gap, such as ozone, are carried away. In some installations, pressurization of the internal components of the magnetos and other various parts of the ignition system is essential to maintain a higher absolute pressure inside the magneto and to eliminate flashover due to high altitude flight. This type of magneto is used with turbocharged engines that operate at higher altitudes. Flashover becomes more likely at high altitudes because of the lower air pressure,

which makes it easier for the electricity to jump air gaps. By pressurizing the interior of the magneto, the normal air pressure is maintained and the electricity or the spark is held within the proper areas of the magneto even though the ambient pressure is very low.

Even in a pressurized magneto, the air is allowed to flow through and out of the magneto housing. By providing more air and allowing small amounts of air to bleed out for ventilation, the magneto remains pressurized. Regardless of the method of venting employed, the vent bleeds or valves must be kept free of obstructions. Further, the air circulating through the components of the ignition system must be free of oil since even minute amounts of oil on ignition parts result in flashover and carbon tracking.

Ignition Harness

The ignition lead directs the electrical energy from the magneto to the spark plug. The ignition harness contains an insulated wire for each cylinder that the magneto serves in the engine. *[Figure 4-11]* One end of each wire is connected to the magneto distributor block and the other end is connected to the proper spark plug. The ignition harness leads serve a dual purpose. It provides the conductor path for the high-tension voltage to the spark plug. It also serves as a shield for stray magnetic fields that surround the wires as they momentarily carry high-voltage current. By conducting these magnetic lines of force to the ground, the ignition harness cuts down electrical interference with the aircraft radio and other electrically sensitive equipment.

A magneto is a high frequency radiation emanating (radio wave) device during its operation. The wave oscillations produced in the magneto are uncontrolled and cover a wide range of frequencies and must be shielded. If the magneto and ignition leads were not shielded, they would form antennas and pick up the random frequencies from the ignition system.

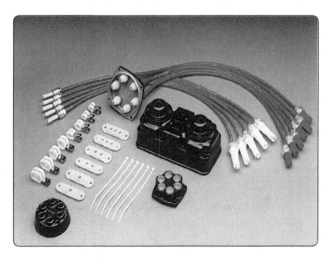

Figure 4-11. *A high-tension ignition harness.*

The lead shielding is a metal mesh braid that surrounds the entire length of the lead. The lead shielding prevents the radiation of the energy into the surrounding area.

Capacitance is the ability to store an electrostatic charge between two conducting plates separated by a dielectric. Lead insulation is called a dielectric, meaning it can store electrical energy as an electrostatic charge. An example of electrostatic energy storage in a dielectric is the static electricity stored in a plastic hair comb. When shielding is placed around the ignition lead, capacitance increases by bringing the two plates closer together. Electrically, the ignition lead acts as a capacitor and has the ability to absorb and store electrical energy. The magneto must produce enough energy to charge the capacitance caused by the ignition lead and have enough energy left over to fire the plug.

Ignition lead capacitance increases the electrical energy required to provide a spark across the plug gap. More magneto primary current is needed to fire the plug with the shielded lead. This capacitance energy is discharged as fire across the plug gap after each firing of the plug. Reversing the polarity during servicing by rotating the plugs to new locations, the plug wear is equalized across the electrodes. The very center of the ignition lead is the high-voltage carrier surrounded by a silicone insulator material that is surrounded by a metal mesh, or shielding, covered with a thin silicone rubber coating that prevents damage by engine heat, vibration, or weather.

A sectional view of the typical ignition lead is shown in *Figure 4-12*. Ignition leads must be routed and clamped correctly to avoid hot spots on the exhaust and vibration points as the leads are routed from the magneto to the individual cylinders. Ignition leads are normally of the all-weather type and are hard connected at the magneto distributor and affixed to the spark plug by threads. The shielded ignition lead spark plug terminal is available in all-weather ¾ inch diameter and ⅝ inch diameter barrel ignition lead nut. *[Figure 4-13]* The ⅝ – 24 plug takes a ¾ wrench on the lead nut and the ¾ – 20 plug takes a ⅞ wrench on the lead nut. The ¾ inch all-weather design utilizes a terminal seal that results in greater terminal well insulation. This

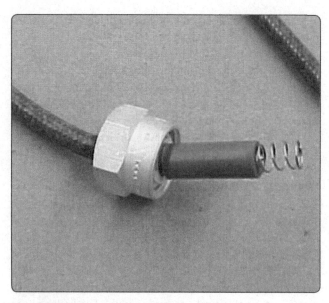

Figure 4-13. *Ignition lead spark plug end.*

is recommended because the lead end of the spark plug is completely sealed from moisture.

An older radial engine type of ignition harness is a manifold formed to fit around the crankcase of the engine with flexible extensions terminating at each spark plug. A typical high-tension ignition harness is shown in *Figure 4-14*. Many older single-row radial engine aircraft ignition systems employ a dual-magneto system, in which the right magneto supplies the electric spark for the front plugs in each cylinder, and the left magneto fires the rear plugs.

Ignition Switches

All units in an aircraft ignition system are controlled by an ignition switch. The type of switch used varies with the number of engines on the aircraft and the type of magnetos used. All switches, however, turn the system off and on in much the same manner. The ignition switch is different in at least one respect from all other types of switches: when the ignition switch is in the off position, a circuit is completed through the switch to ground. In other electrical switches, the off position normally breaks or opens the circuit.

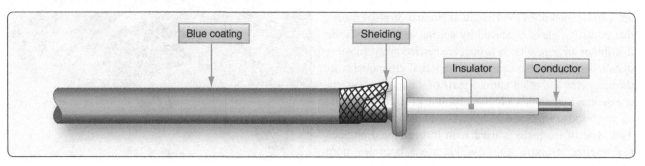

Figure 4-12. *Ignition lead.*

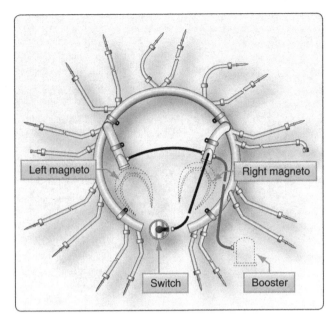

Figure 4-14. *Accessory-mounted nine cylinder engine ignition harness.*

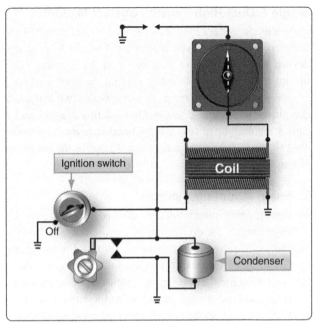

Figure 4-15. *Typical ignition switch in off position.*

The ignition switch has one terminal connected to the primary electrical circuit between the coil and the breaker contact points. The other terminal of the switch is connected to the aircraft ground structure. As shown in *Figure 4-15*, two ways to complete the primary circuit are:

1. Through the closed breaker points to ground and

2. Through the closed ignition switch to ground.

Figure 4-15 shows that the primary current is not interrupted when the breaker contacts open since there is still a path to ground through the closed, or off, ignition switch. Since primary current is not stopped when the contact points open, there can be no sudden collapse of the primary coil flux field and no high-voltage induced in the secondary coil to fire the spark plug.

As the magnet rotates past the electrical gap (E-gap) position, a gradual breakdown of the primary flux field occurs. But that breakdown occurs so slowly that the induced voltage is too low to fire the spark plug. Thus, when the ignition switch is in the off position with the switch closed, the contact points are as completely short-circuited as if they were removed from the circuit, and the magneto is inoperative.

When the ignition switch is placed in the on position, switch open, the interruption of primary current and the rapid collapse of the primary coil flux field is once again controlled or triggered by the opening of the breaker contact points. *[Figure 4-16]* When the ignition switch is in the on position, the switch has absolutely no effect on the primary circuit.

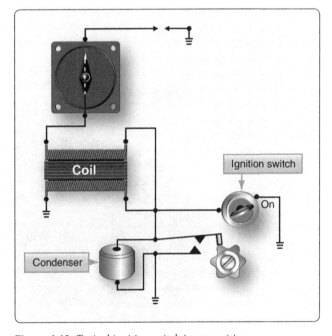

Figure 4-16. *Typical ignition switch in on position.*

The ignition/starter switch, or magneto switch, controls the magnetos on or off and can also connect the starter solenoid for turning the starter. When a starting vibrator, a box that emits pulsating direct current (DC), is used on the engine, the ignition/starter switch is used to control the vibrator and retard points. This system is explained in detail later in this chapter. Some ignition starter switches have a push to prime feature during the starting cycle. This system allows additional fuel to spray into the intake port of the cylinder during the starting cycle.

Single & Dual High-Tension System Magnetos

High-tension system magnetos used on aircraft engines are either single or dual type magnetos. The single magneto design incorporates the distributor in the housing with the magneto breaker assembly, rotating magnet, and coil. *[Figure 4-17]* The dual magneto incorporates two magnetos contained in a single housing. One rotating magnet and a cam are common to two sets of breaker points and coils. Two separate distributor units are mounted in the magneto. *[Figure 4-18]*

Magneto Mounting Systems

Flange-mounted magnetos are attached to the engine by a flange around the driven end of the rotating shaft of the magneto. *[Figure 4-19]* Elongated slots in the mounting flange permit adjustment through a limited range to aid in timing the magneto to the engine. Some magnetos mount by the flange and use clamps on each side to secure the magneto to the engine. This design also allows for timing adjustments. Base mounted magnetos are only used on very old or antique aircraft engines.

Figure 4-18. *A dual magneto with two distributors.*

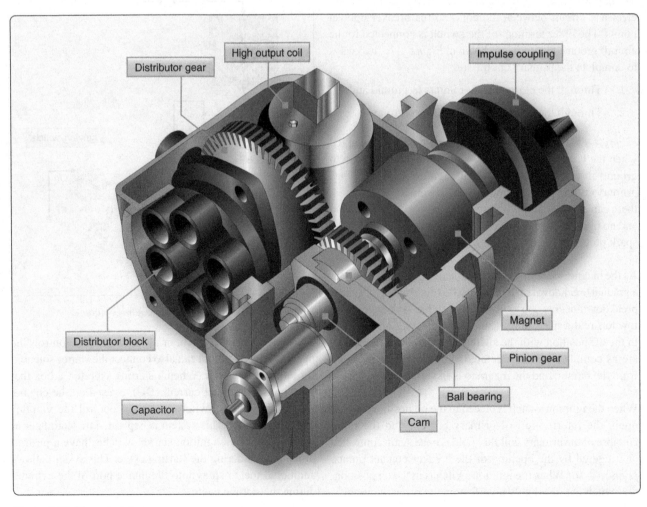

Figure 4-17. *Magneto cutaway.*

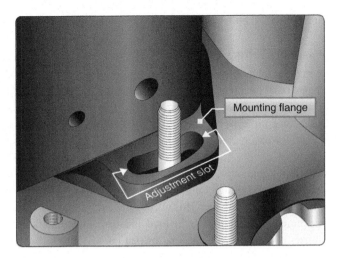

Figure 4-19. *Magneto mounting flange.*

High- & Low-Tension Magneto Systems

High-tension ignition systems have undergone many refinements and improvements in design. This includes new electronic systems that control more than just providing ignition to the cylinders. High-tension voltage presents certain problems with carrying the high-voltage from the magneto internally and externally to the spark plugs. In early years, it was difficult to provide insulators that could contain the high-voltage, especially at high altitudes when the air pressures were reduced. Another requirement of high-tension systems was that all weather and radio-equipped aircraft have

ignition wires enclosed in shielding to prevent radio noise due to high-voltages. Many aircraft were turbosupercharged and operated at increased high altitudes. The low pressure at these altitudes would allow the high-voltage to leak out even more. To meet these problems, low-tension ignition systems were developed.

Electronically, the low-tension system is different from the high-tension system. In the low-tension system, low-voltage is generated in the magneto and flows to the primary winding of a transformer coil located near the spark plug. There, the voltage is increased to high by transformer action and conducted to the spark plug by very short high-tension leads. *[Figure 4-20]*

The low-tension system virtually eliminates flashover in both the distributor and the harness because the air gaps within the distributor have been eliminated by the use of a brush-type distributor, and high-voltage is present only in short leads between the transformer and spark plug.

Although a certain amount of electrical leakage is characteristic of all ignition systems, it is more pronounced on radio-shielded installations because the metal conduit is at ground potential and close to the ignition wires throughout their entire length. In low-tension systems, however, this leakage is reduced considerably because the current throughout most of the system is transmitted at a low-voltage potential. Although

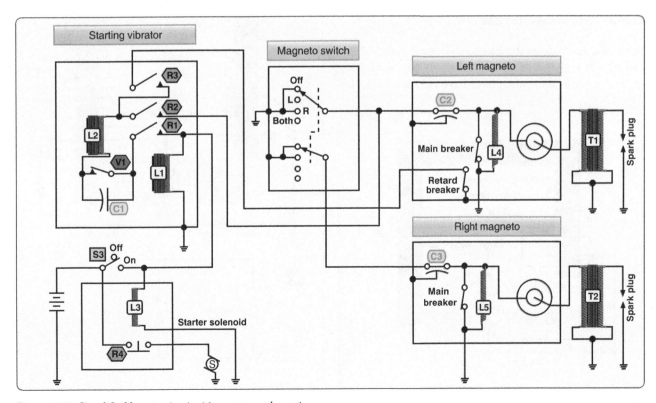

Figure 4-20. *Simplified low-tension ignition system schematic.*

the leads between the transformer coils and the spark plugs of a low-tension ignition system are short, they are high-tension high-voltage conductor, and are subject to the same failures that occur in high-tension systems. Low-tension ignition systems have limited use in modern aircraft because of the excellent materials and shielding available to construct high-tension ignition leads and the added cost of a coil for each spark plug with the low-tension system.

Types of DC Generators

There are three types of DC generators: series wound, parallel (shunt) wound, and series-parallel (compound) wound. The appropriate generator is determined by the connections to the armature and field circuits with respect to the external circuit. The external circuit is the electrical load powered by the generator. In general, the external circuit is used for charging the aircraft battery and supplying power to all electrical equipment being used by the aircraft. As their names imply, windings in series have characteristics different from windings in parallel.

Series Wound DC Generators

The series generator contains a field winding connected in series with the external circuit. *[Figure 4-21]* Series generators have very poor voltage regulation under changing load, since the greater the current is through the field coils to the external circuit, the greater the induced EMFs and the greater the output voltage is. When the aircraft electrical load is increased, the voltage increases; when the load is decreased, the voltage decreases.

Since the series wound generator has such poor voltage and current regulation, it is never employed as an airplane generator. Generators in airplanes have field windings, that

are connected either in shunt or in compound formats.

Parallel (Shunt) Wound DC Generators

A generator having a field winding connected in parallel with the external circuit is called a shunt generator. *[Figure 4-22]* It should be noted that, in electrical terms, shunt means parallel. Therefore, this type of generator could be called either a shunt generator or a parallel generator.

In a shunt generator, any increase in load causes a decrease in the output voltage, and any decrease in load causes an increase output voltage. This occurs since the field winding is connected in parallel to the load and armature, and all the current flowing in the external circuit passes only through the armature winding (not the field).

As shown in *Figure 4-22A*, the output voltage of a shunt generator can be controlled by means of a rheostat inserted in series with the field windings. As the resistance of the field circuit is increased, the field current is reduced; consequently, the generated voltage is also reduced. As the field resistance is decreased, the field current increases and the generator output increases. In the actual aircraft, the field rheostat would be replaced with an automatic control device, such

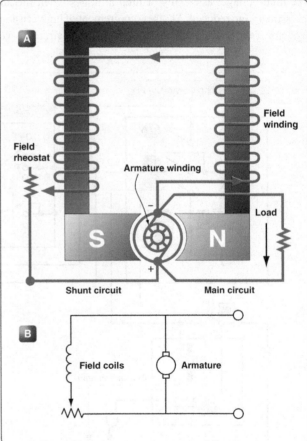

Figure 4-22. *Shunt wound generator.*

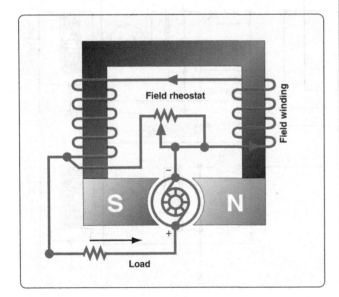

Figure 4-21. *Diagram of a series wound generator.*

as a voltage regulator.

Compound Wound DC Generators

A compound wound generator employs two field windings one in series and another in parallel with the load. *[Figure 4-23]*

This arrangement takes advantage of both the series and parallel characteristics described earlier. The output of a compound wound generator is relatively constant, even with changes in the load.

Generator Ratings

A DC generator is typically rated for its voltage and power output. Each generator is designed to operate at a specified voltage, approximately 14 or 28 volts. It should be noted that aircraft electrical systems are designed to operate at one of these two voltage values. The aircraft's voltage depends on which battery is selected for that aircraft. Batteries are either 12 or 24 volts when fully charged. The generator selected must have a voltage output slightly higher than the battery voltage. Hence, the 14- or 28-volt rating is required for aircraft DC generators.

The power output of any generator is given as the maximum number of amperes the generator can safely supply. Generator rating and performance data are stamped on the nameplate

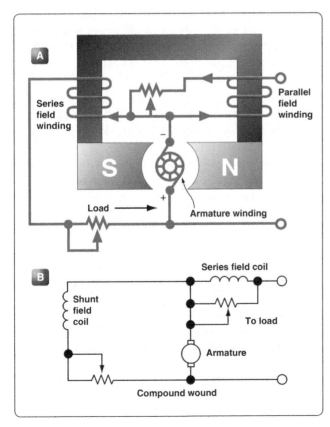

Figure 4-23. *Compound wound generator.*

attached to the generator. When replacing a generator, it is important to choose one of the proper ratings.

The rotation of generators is termed either clockwise or counterclockwise, as viewed from the driven end. The direction of rotation may also be stamped on the data plate. It is important that a generator with the correct rotation be used; otherwise, the polarity of the output voltage is reversed. The speed of an aircraft engine varies from idle rpm to takeoff rpm; however, during the major portion of a flight, it is at a constant cruising speed. The generator drive is usually geared to turn the generator between $1\frac{1}{8}$ and $1\frac{1}{2}$ times the engine crankshaft speed. Most aircraft generators have a speed at which they begin to produce their normal voltage. Called the "coming in" speed, it is usually about 1,500 rpm.

DC Generator Maintenance

The following information about the inspection and maintenance of DC generator systems is general in nature because of the large number of differing aircraft generator systems. These procedures are for familiarization only. Always follow the applicable manufacturer's instructions for a given generator system. In general, the inspection of the generator installed in the aircraft should include the following items:

1. Security of generator mounting.
2. Condition of electrical connections.
3. Dirt and oil in the generator. If oil is present, check engine oil seals. Blow out any dirt with compressed air.
4. Condition of generator brushes.
5. Generator operation.
6. Voltage regulator operation.

Sparking of brushes quickly reduces the effective brush area in contact with the commutator bars. The degree of such sparking should be determined. Excessive wear warrants a detailed inspection and possible replacement of various components. *[Figure 4-24]*

Manufacturers usually recommend the following procedures to seat brushes that do not make good contact with slip rings or commutators. Lift the brush sufficiently to permit the insertion of a strip of extra-fine 000 (triple aught) grit, or finer, sandpaper under the brush, rough side towards the carbon brush. *[Figure 4-25]*

Pull the sandpaper in the direction of armature rotation, being careful to keep the ends of the sandpaper as close to the slip ring or commutator surface as possible in order to avoid rounding the edges of the brush. When pulling the sandpaper back to the starting point, raise the brush so it does not ride

Figure 4-24. *Wear areas of commutator and brushes.*

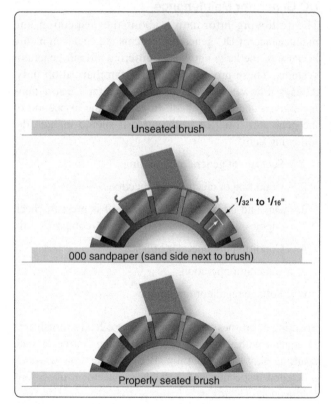

Unseated brush

1/32" to 1/16"

000 sandpaper (sand side next to brush)

Properly seated brush

Figure 4-25. *Seating brushes with sandpaper.*

on the sandpaper. Sand the brush only in the direction of rotation. Carbon dust resulting from brush sanding should be thoroughly cleaned from all parts of the generators after a sanding operation.

After the generator has run for a short period, brushes should be inspected to make sure that pieces of sand have not become embedded in the brush. Under no circumstances should emery cloth or similar abrasives be used for seating brushes (or smoothing commutators), since they contain conductive materials that cause arcing between brushes and commutator

bars. It is important that the brush spring pressure be correct. Excessive pressure causes rapid wear of brushes. Too little pressure, however, allows bouncing of the brushes, resulting in burned and pitted surfaces. The pressure recommended by the manufacturer should be checked by the use of a spring scale graduated in ounces. Brush spring tension on some generators can be adjusted. A spring scale is used to measure the pressure that a brush exerts on the commutator.

Flexible low-resistance pigtails are provided on most heavy current carrying brushes, and their connections should be securely made and checked at frequent intervals. The pigtails should never be permitted to alter or restrict the free motion of the brush. The purpose of the pigtail is to conduct the current from the armature, through the brushes, to the external circuit of the generator.

FADEC System Description

A FADEC is a solid-state digital electronic ignition and electronic sequential port fuel injection system with only one moving part that consists of the opening and closing of the fuel injector. FADEC continuously monitors and controls ignition, timing, and fuel mixture/delivery/injection, and spark ignition as an integrated control system. FADEC monitors engine operating conditions (crankshaft speed, top dead center position, the induction manifold pressure, and the induction air temperature) and then automatically adjusts the fuel-to-air ratio mixture and ignition timing accordingly for any given power setting to attain optimum engine performance. As a result, engines equipped with FADEC require neither magnetos nor manual mixture control.

This microprocessor-based system controls ignition timing for engine starting and varies timing with respect to engine speed and manifold pressure. *[Figure 4-26]*

PowerLink provides control in both specified operating conditions and fault conditions. The system is designed to prevent adverse changes in power or thrust. In the event of loss of primary aircraft-supplied power, the engine controls continue to operate using a secondary power source (SPS). As a control device, the system performs self-diagnostics to determine overall system status and conveys this information to the pilot by various indicators on the health status annunciator (HSA) panel. PowerLink is able to withstand storage temperature extremes and operate at the same capacity as a non-FADEC-equipped engine in extreme heat, cold, and high humidity environments.

Low-Voltage Harness

The low-voltage harness connects all essential components of the FADEC System. *[Figure 4-26]* This harness acts as a signal transfer bus interconnecting the electronic control

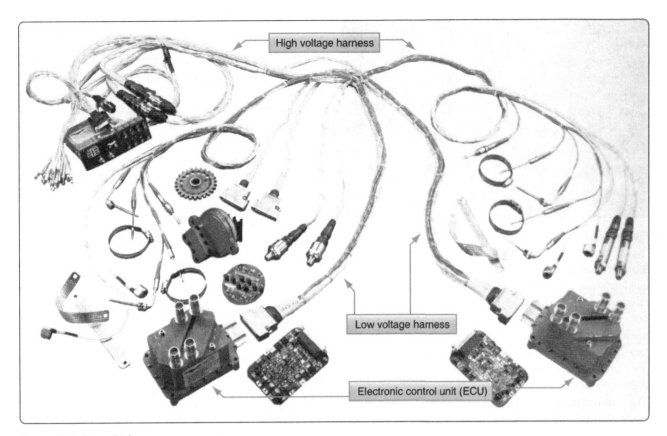

Figure 4-26. *PowerLink system components.*

units (ECUs) with aircraft power sources, the ignition switch, speed sensor assembly (SSA), temperature and pressure sensors. The fuel injector coils and all sensors, except the SSA and fuel pressure and manifold pressure sensors, are hardwired to the low-voltage harness. This harness transmits sensor inputs to the ECUs through a 50-pin connector. The harness connects to the engine-mounted pressure sensors via cannon plug connectors. The 25-jpin connectors connect the harness to the speed sensor signal conditioning unit. The low-voltage harness attaches to the cabin harness by a firewall-mounted data port through the same cabin harness/bulkhead connector assembly. The bulkhead connectors also supply the aircraft electrical power required to run the system.

The ECU is at the heart of the system, providing both ignition and fuel injection control to operate the engine with the maximum efficiency realizable. Each ECU contains two microprocessors, referred to as a computer, that control two cylinders. Each computer controls its own assigned cylinder and is capable of providing redundant control for the other computer's cylinder.

The computer constantly monitors the engine speed and timing pulses developed from the camshaft gear as they are detected by the SSA. Knowing the exact engine speed and the timing sequence of the engine, the computers monitor the manifold air pressure and manifold air temperature to calculate air density and determine the mass air flow into the cylinder during the intake stroke. The computers calculate the percentage of engine power based on engine revolutions per minute (rpm) and manifold air pressure.

From this information, the computer can then determine the fuel required for the combustion cycle for either best power or best economy mode of operation. The computer precisely times the injection event, and the duration of the injector should be on time for the correct fuel-to-air ratio. Then, the computer sets the spark ignition event and ignition timing, again based on percentage of power calculation. Exhaust gas temperature is measured after the burn to verify that the fuel-to-air ratio calculations were correct for that combustion event. This process is repeated by each computer for its own assigned cylinder on every combustion/power cycle.

The computers can also vary the amount of fuel to control the fuel-to-air ratio for each individual cylinder to control both cylinder head temperature (CHT) and exhaust gas temperature (EGT).

Electronic Control Unit (ECU)
An ECU is assigned to a pair of engine cylinders. *[Figure 4-27]* The ECUs control the fuel mixture and spark timing for their respective engine cylinders; ECU 1 controls opposing cylinders 1 and 2, ECU 2 controls cylinders 3 and 4, and ECU

Figure 4-27. *Electronic control unit.*

3 controls cylinders 5 and 6. Each ECU is divided into upper and lower portions. The lower portion contains an electronic circuit board, while the upper portion houses the ignition coils. Each electronic control board contains two independent microprocessor controllers that serve as control channels. During engine operation, one control channel is assigned to operate a single engine cylinder. Therefore, one ECU can control two engine cylinders, one control channel per cylinder. The control channels are independent, and there are no shared electronic components within one ECU. They also operate on independent and separate power supplies. However, if one control channel fails, the other control channel in the pair within the same ECU is capable of operating both its assigned cylinder and the other opposing engine cylinder as backup control for fuel injection and ignition timing. Each control channel on the ECU monitors the current operating conditions and operates its cylinder to attain engine operation within specified parameters. The following transmit inputs to the control channels across the low-voltage harness:

1. Speed sensor that monitors engine speed and crank position,

2. Fuel pressure sensors,

3. Manifold pressure sensors,

4. Manifold air temperature (MAT) sensors,

5. CHT sensors, and

6. EGT sensors.

All critical sensors are dually redundant with one sensor from each type of pair connected to control channels in different ECUs. Synthetic software default values are also used in the unlikely event that both sensors of a redundant pair fail. The control channel continuously monitors changes in engine speed, manifold pressure, manifold temperature, and fuel pressure based on sensor input relative to operating conditions to determine how much fuel to inject into the intake port of the cylinder.

PowerLink Ignition System

The ignition system consists of the high-voltage coils atop the ECU, the high-voltage harness, and spark plugs. Since there are two spark plugs per cylinder on all engines, a six-cylinder engine has 12 leads and 12 spark plugs. One end of each lead on the high-voltage harness attaches to a spark plug, and the other end of the lead wire attaches to the spark plug towers on each ECU. The spark tower pair is connected to opposite ends of one of the ECU's coil packs. Two coil packs are located in the upper portion of the ECU. Each coil pack generates a high-voltage pulse for two spark plug towers. One tower fires a positive polarity pulse and the other of the same coil fires a negative polarity pulse. Each ECU controls the ignition spark for two engine cylinders. The control channel within each ECU commands one of the two coil packs to control the ignition spark for the engine cylinders. *[Figure 4-28]* The high-voltage harness carries energy from the ECU spark towers to the spark plugs on the engine.

For both spark plugs in a given cylinder to fire on the compression stroke, both control channels must fire their coil packs. Each coil pack has a spark plug from each of the two cylinders controlled by that ECU unit.

The ignition spark is timed to the engine's crankshaft position. The timing is variable throughout the engine's operating range and is dependent upon the engine load conditions. The spark energy is also varied with respect to the engine load.

Note: Engine ignition timing is established by the ECUs and cannot be manually adjusted.

Engine Indicating & Crew Alerting System (EICAS)

An engine indicating and crew alerting system (EICAS) performs many of the same functions as an ECAM system. The objective is still to monitor the aircraft systems for the pilot. All EICAS display engine, as well as airframe, parameters. Traditional gauges are not utilized, other than a standby combination engine gauge in case of total system failure.

EICAS is also a two-monitor, two-computer system with a display select panel. Both monitors receive information from the same computer. The second computer serves as a standby. Digital and analog inputs from the engine and airframe systems are continuously monitored. Caution and warning lights, as well as aural tones, are incorporated. *[Figure 4-29]*

EICAS provides full time primary engine parameters (EPR, N1, EGT) on the top, primary monitor. Advisories and warnings are also shown there. Secondary engine parameters

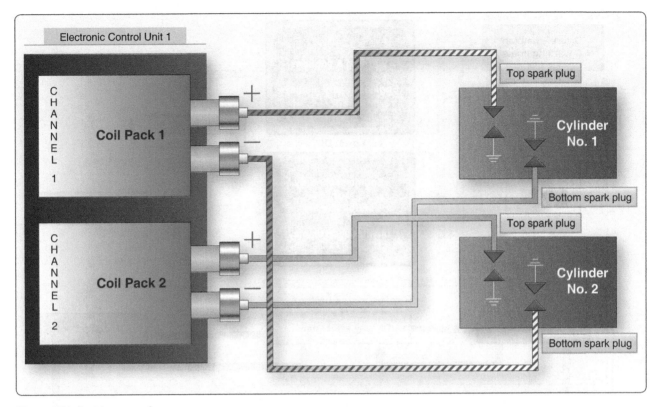

Figure 4-28. *Ignition control.*

and nonengine system status are displayed on the bottom screen. The lower screen is also used for maintenance diagnosis when the aircraft is on the ground. Color coding is used, as well as message prioritizing.

The display select panel allows the pilot to choose which computer is actively supplying information. It also controls the display of secondary engine information and system status displays on the lower monitor. EICAS has a unique feature that automatically records the parameters of a failure event to be regarded afterwards by maintenance personnel. Pilots that suspect a problem may be occurring during flight can press the event record button on the display select panel. This also records the parameters for that flight period to be studied later by maintenance. Hydraulic, electrical, environmental, performance, and APU data are examples of what may be recorded.

EICAS uses BITE for systems and components. A maintenance panel is included for technicians. From this panel, when the aircraft is on the ground, push-button switches display information pertinent to various systems for analysis. *[Figure 4-30]*

Auxiliary Ignition Units

During engine starting, the output of a magneto is low because the cranking speed of the engine is low. This is understandable when the factors that determine the amount of voltage induced in a circuit are considered.

To increase the value of an induced voltage, the strength of the magnetic field must be increased by using a stronger magnet, by increasing the number of turns in the coil, or by increasing the rate of relative motion between the magnet and the conductor.

Since the strength of the rotating magnet and the number of turns in the coil are constant factors in magneto ignition systems, the voltage produced depends upon the speed at which the rotating magnet is turned. When the engine is being cranked for starting, the magnet is rotated at about 80 rpm. Since the value of the induced voltage is so low, a spark may not jump the spark plug gap. To facilitate engine starting, an auxiliary device is connected to the magneto to provide a high ignition voltage.

Ordinarily, such auxiliary ignition units are energized by the battery and connected to the left magneto. Reciprocating engine starting systems normally include one of the following types of auxiliary starting systems: booster coil (older style), starting vibrator (sometimes called shower of sparks), impulse coupling, or electronic ignition systems.

During the starting cycle, the engine is turning very slowly compared to normal speed. The ignition must be retarded or moved back to prevent kickback of the piston trying to

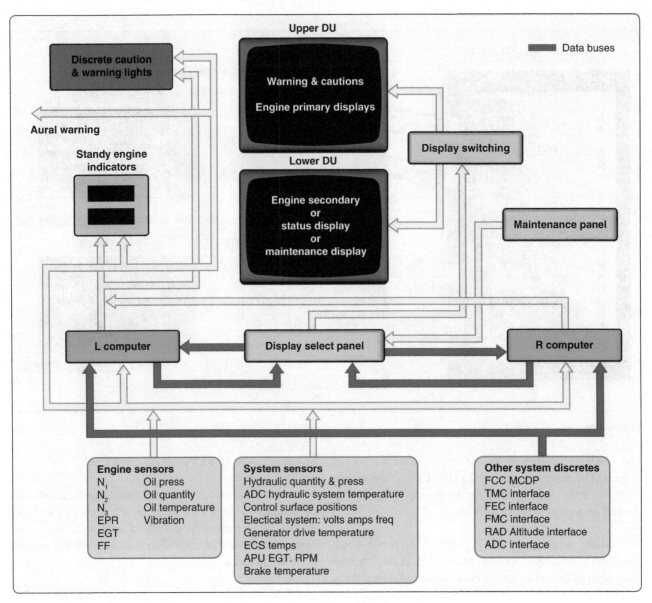

Figure 4-29. *Schematic of an engine indicating and crew alerting system (EICAS).*

rotate opposite normal rotation. Each starting system has a method of retarding the spark during starting of the engine.

Booster Coil

Used mainly with older radial engine ignition systems, the booster coil assembly consists of two coils wound on a soft iron core, a set of contact points, and a condenser. *[Figure 4-31]* The booster coil is separate from the magneto and can generate a series of sparks on its own. During the start cycle, these sparks are routed to the trailing finger on the distributor rotor and then to the appropriate cylinder ignition lead. The primary winding has one end grounded at the internal grounding strip and its other end connected to the moving contact point. The stationary contact is fitted with a terminal to which battery voltage is applied when the magneto switch is placed in the start position, or automatically applied when the starter is engaged. The secondary winding, which

contains several times as many turns as the primary coil, has one end grounded at the internal grounding strip and the other terminated at a high-tension terminal. The high-tension terminal is connected to an electrode in the distributor by an ignition cable.

Since the regular distributor terminal is grounded through the primary or secondary coil of a high-tension magneto, the high-voltage furnished by the booster coil must be distributed by a separate circuit in the distributor rotor. This is accomplished by using two electrodes in one distributor rotor. The main electrode, or finger, carries the magneto output voltage; the auxiliary electrode or trailing finger, distributes only the output of the booster coil. The auxiliary electrode is always located so that it trails the main electrode, thus retarding the spark during the starting period.

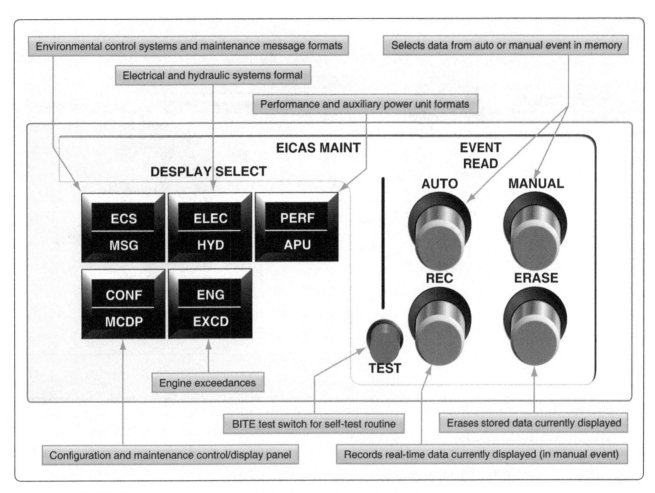

Environmental control systems and maintenance message formats

Electrical and hydraulic systems formal

Performance and auxiliary power unit formats

Selects data from auto or manual event in memory

EICAS MAINT

DESPLAY SELECT

ECS
MSG

ELEC
HYD

PERF
APU

CONF
MCDP

ENG
EXCD

EVENT READ

AUTO

MANUAL

REC

ERASE

TEST

Engine exceedances

BITE test switch for self-test routine

Erases stored data currently displayed

Configuration and maintenance control/display panel

Records real-time data currently displayed (in manual event)

Figure 4-30. *The EICAS maintenance control panel is for the exclusive use of technicians.*

Figure 4-32 illustrates, in schematic form, the booster coil components shown in *Figure 4-31*. In operation, battery voltage is applied to the positive (+) terminal of the booster coil through the start switch. This causes current to flow through the closed contact points to the primary coil and ground. *[Figure 4-32]* Current flow through the primary coil sets up a magnetic field about the coil that magnetizes the coil core. As the core is magnetized, it attracts the movable contact point, which is normally held against the stationary contact point by a spring.

As the movable contact point is pulled toward the iron core, the primary circuit is broken, collapsing the magnetic field that extended about the coil core. Since the coil core acts as an electromagnet only when current flows in the primary coil, it loses its magnetism as soon as the primary coil circuit is broken. This permits the action of the spring to close the contact points and again complete the primary coil circuit. This remagnetizes the coil core, and again attracts the movable contact point, which again opens the primary coil circuit. This action causes the movable contact point to vibrate rapidly, as long as the start switch is held in the closed, or on, position. The result of this action is a

continuously expanding and collapsing magnetic field that links the secondary coil of the booster coil. With several times as many turns in the secondary as in the primary, the induced voltage that results from lines of force linking the secondary is high enough to furnish ignition for the engine.

The condenser, which is connected across the contact points, has an important function in this circuit. *[Figure 4-32]* As current flow in the primary coil is interrupted by the opening of the contact points, the high self-induced voltage that accompanies each collapse of the primary magnetic field surges into the condenser. Without a condenser, an arc would jump across the points with each collapse of the magnetic field. This would burn and pit the contact points and greatly reduce the voltage output of the booster coil. The booster coil generates a pulsating DC in the primary winding that induces a high-voltage spark in the secondary windings of the booster coil.

Impulse Coupling

Many opposed reciprocating engines are equipped with an impulse coupling as the auxiliary starting system. An impulse coupling gives one of the magnetos attached to the

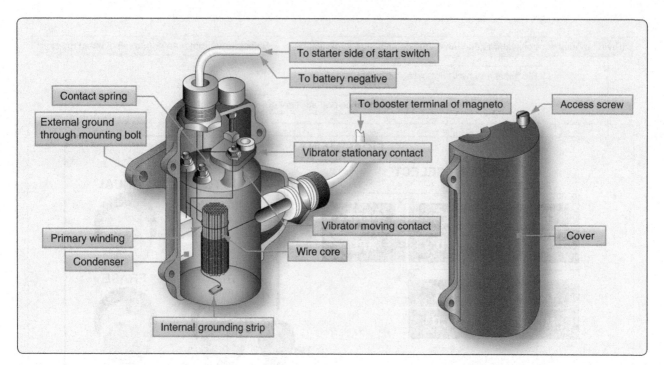

Figure 4-31. *Booster coil.*

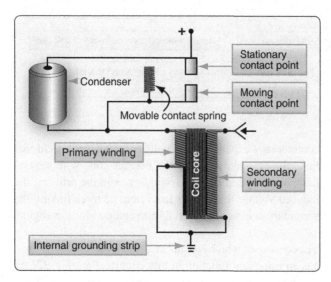

Figure 4-32. *Booster coil schematic.*

engine, generally the left, a brief acceleration, that produces an intense spark for starting. This device consists of a cam and flyweight assembly, spring, and a body assembly. *[Figure 4-33]* The assembled impulse coupling is shown installed on a typical magneto in *Figure 4-34*.

The magneto is flexibly connected through the impulse coupling by means of the spring so that at low speed the magneto is temporarily held. *[Figure 4-35]* The flyweight, because of slow rotation, catches on a stud or stop pins, and the magneto spring is wound as the engine continues to turn. The engine continues to rotate until the piston of the

cylinder to be fired reaches approximately a top dead center position. At this point, the magneto flyweight contacts the body of the impulse coupling and is released. The spring kicks back to its original position, resulting in a quick twist of the rotating magnet of the magneto. *[Figure 4-36]* This, being equivalent to high-speed magneto rotation, produces a spark that jumps the gap at the spark plug electrodes. The impulse coupling has two functions: rotating the magneto fast enough to produce a good spark and retarding the timing of the spark during the start cycle. After the engine is started and the magneto reaches a speed at which it furnishes sufficient current, the flyweights in the impulse coupling fly outward due to centrifugal force or rapid rotation. This action prevents the two flyweight coupling members from contact with the stop pin. That makes it a solid unit, returning the magneto to a normal timing position relative to the engine. The presence of an impulse coupling is identified by a sharp clicking noise as the crankshaft is turned at starter cranking speed past top center on each cylinder.

A problem that can arise from impulse couplings is that the flyweights can become magnetized and not engage the stop pins. Congealed oil or sludge on the flyweights during cold weather may produce the same results. This prevents the flyweight weights from engaging the stop pins, which results in no starting spark being produced. Wear can cause problems with impulse couplings. They should be inspected and any maintenance should be performed as set forth by the manufacturer. Another disadvantage of the impulse coupling is that it can produce only one spark for each firing cycle of the cylinder. This is a disadvantage, especially during

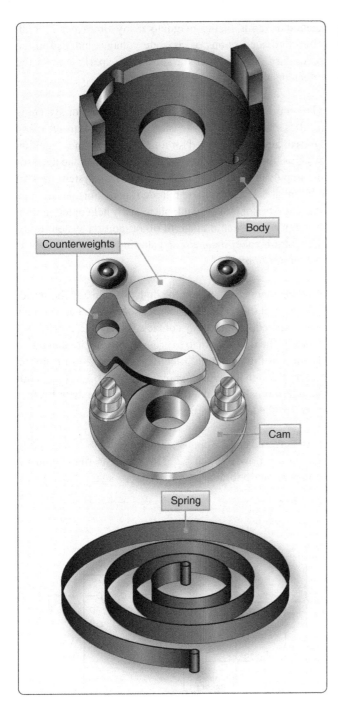

Figure 4-33. *Parts of an impulse coupling.*

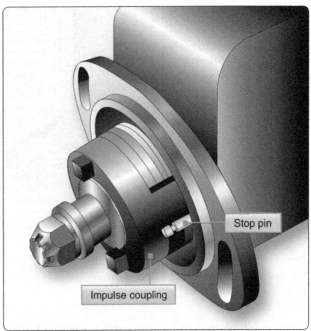

Figure 4-34. *Impulse coupling on a magneto.*

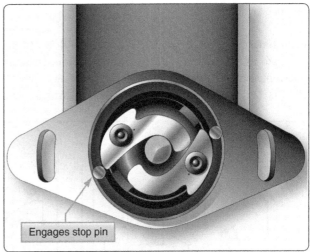

Figure 4-35. *Flyweights engage stop pins.*

adverse starting conditions. Even with these disadvantages, the impulse coupling is still in wide use.

High-Tension Retard Breaker Vibrator

To provide for more spark power during the starting cycle, the shower of sparks system was developed, which provides several sparks at the spark plug electrodes during starting. The starting vibrator, or shower of sparks, consists essentially of an electrically operated vibrator, a condenser, and a relay. *[Figure 4-37]* These units are mounted on a base plate and

enclosed in a metal case.

The starting vibrator, unlike the booster coil, does not produce the high ignition voltage within itself. The function of this starting vibrator is to change the DC of the battery into a pulsating DC and deliver it to the primary coil of the magneto. Closing the ignition switch energizes the starter solenoid and causes the engine to rotate. At the same time, current also flows through the vibrator coil and its contact points. Current flow in the vibrator coil sets up a magnetic field that attracts and opens the vibrator points. When the vibrator points open, current flow in the coil stops, and the magnetic field that attracted the movable vibrator contact point disappears. This allows the vibrator points to close and

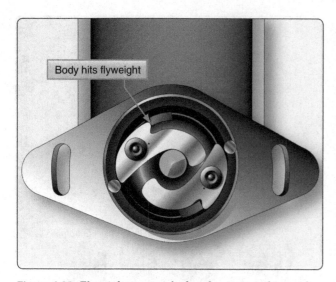

Figure 4-36. *Flyweight contacts body, releasing impulse coupling to spin.*

secondary coils of the magneto many times per second. The rapid successions of separate voltages induced in the secondary coil produces a shower of sparks across the selected spark plug air gap.

The retard breaker magneto and starting vibrator system is used as part of the high-tension starting system on many types of aircraft. Designed for four- and six-cylinder ignition systems, the retard breaker magneto eliminates the need for the impulse coupling in light aircraft. This system uses an additional breaker to obtain retarded sparks for starting. The starting vibrator is also adaptable to many helicopter ignition systems. A schematic diagram of an ignition system using the retard breaker magneto and starting vibrator concept is shown in *Figure 4-37*.

With the magneto switch in the both position and the starter switch S1 in the on position, starter solenoid L3 and coil L1 are energized, closing relay contacts R4, R1, R2, and R3. R3 connects the right magneto to ground, keeping it inoperative during starting operation. Electrical current flows from the battery through R1, vibrator points V1, coil L2, through both the retard breaker points, through R2, and the main breaker points of the left magneto to ground.

The energized coil L2 opens vibrator points V1, interrupting the current flow through L2. The magnetic field about L2 collapses, and vibrator points V1 close again. Once more,

again permits battery current to flow in the vibrator coil. This completes a cycle of operation. The cycle, however, occurs many times per second, so rapidly that the vibrator points produce an audible buzz.

Each time the vibrator points close, current flows to the magneto as a pulsating DC. Since this current is being interrupted many times per second, the resulting magnetic field is building and collapsing across the primary and

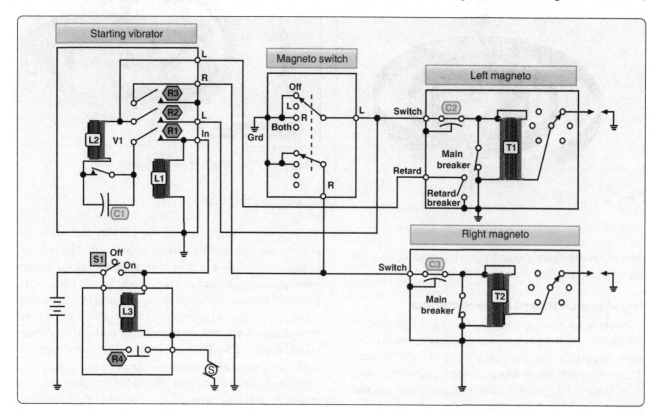

Figure 4-37. *High-tension retard breaker magneto and starting vibrator circuit.*

current flows through L2, and again V1 vibrator points open. This process is repeated continuously, and the interrupted battery current flows to ground through the main and retard breaker points of the left magneto.

Since relay R4 is closed, the starter is energized and the engine crankshaft is rotated. When the engine reaches its normal advance firing position, the main breaker points of the left magneto begin to open. The interrupted surges of current from the vibrator can still find a path to ground through the retard breaker points, which do not open until the retarded firing position of the engine is reached. At this point in crankshaft travel, the retard points open. Since the main breaker points are still open, the magneto primary coil is no longer shorted, and current produces a magnetic field around T1.

Each time the vibrator points V1 open, current flow through V1 is interrupted. The collapsing field about T1 cuts through the magneto coil secondary and induces a high-voltage surge of energy used to fire the spark plug. Since the V1 points are opening and closing rapidly and continuously, a shower of sparks is furnished to the cylinders when both the main and retard breaker points are open.

After the engine begins to accelerate, the manual starter switch is released, causing L1 and L3 to become deenergized. This action causes both the vibrator and retard breaker circuits to become inoperative. It also opens relay contact R3, which removes the ground from the right magneto. Both magnetos now fire at the normal running advanced degree position of crankshaft rotation before top dead center piston position.

Low-Tension Retard Breaker Vibrator

This system, which is in limited use, is designed for light aircraft reciprocating engines. A typical system consists of a retard breaker magneto, a single breaker magneto, a starting vibrator, transformer coils, and a starter and ignition switch. *[Figure 4-38]*

To operate the system, place the starter switch S3 in the on position. This energizes starter solenoid L3 and coil L1, closing relay contacts R1, R2, R3, and R4. With the magneto switch in the L position, current flows through R1, the vibrator points, L2, R2, and through the main breaker points to ground. Current also flows through R3 and the retard breaker points to ground. Current through L2 builds up a magnetic field that opens the vibrator points. Then, the current stops flowing through L2, reclosing the points. These surges of current flow through both the retard and main breaker points to ground.

Since the starter switch is closed, the engine crankshaft is turning. When it has turned to the normal advance or running ignition position, the main breaker points of the magneto

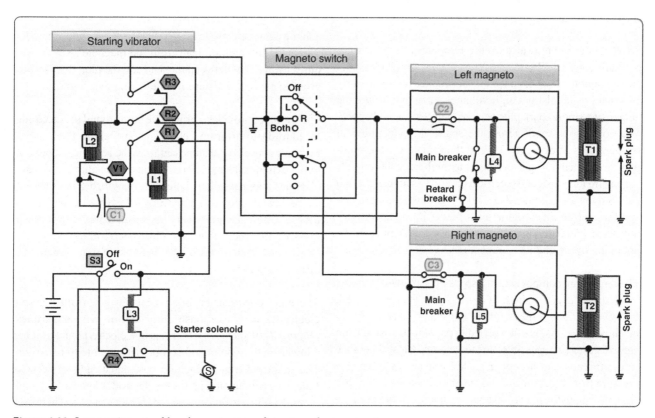

Figure 4-38. *Low-tension retard breaker magneto and starting vibrator circuit.*

open. However, current still flows to ground through the closed retard breaker points. As the engine continues to turn, the retard ignition position is reached, and the retard breaker points open. Since the main breaker points are still open, current must flow to ground through coil L4, producing a magnetic field around the coil L4.

As the engine continues to turn, the vibrator breaker points open, collapsing the L4 magnetic field through T1 primary, inducing a high-voltage in the secondary of T1 to fire the spark plug.

When the engine fires, the starter switch is released, de-energizing L1 and L3. This opens the vibrator circuit and retard breaker points circuit. The ignition switch is then turned to the both position, permitting the right magneto to operate in time with the left magneto.

Spark Plugs

The function of the spark plug in an ignition system is to conduct a short impulse of high-voltage current through the wall of the combustion chamber. Inside the combustion chamber, it provides an air gap across which the impulse can produce an electric spark to ignite the air-fuel charge. While the aircraft spark plug is simple in construction and operation, it can be the cause of malfunctions in aircraft engines. Despite this fact, spark plugs provide a great deal of trouble-free operation when properly maintained and when correct engine operating procedures are practiced.

Spark plugs operate at extreme temperatures, electrical pressures, and very high cylinder pressures. A cylinder of an engine operating at 2,100 rpm must produce approximately 17 separate and distinct high-voltage sparks that bridge the air gap of a single spark plug each second. This would appear as a continuous spark across the spark plug electrodes at temperatures of over 3,000 °F. At the same time, the spark plug is subjected to gas pressures as high as 2,000 pounds per square inch (psi) and electrical pressure as high as 20,000 volts. Given the extremes that spark plugs must operate under, and the fact that the engine loses power if one spark does not occur correctly, proper function of a spark plug in the operation of the engine is imperative.

The three main components of a spark plug are the electrode, insulator, and outer shell. *[Figure 4-39]* The outer shell, threaded to fit into the cylinder, is usually made of finely machined steel and is often plated to prevent corrosion from engine gases and possible thread seizure. Close-tolerance screw threads and a copper gasket prevent cylinder gas pressure from escaping around the plug. Pressure that might escape through the plug is retained by inner seals between the outer metal shell and the insulator, and between the insulator

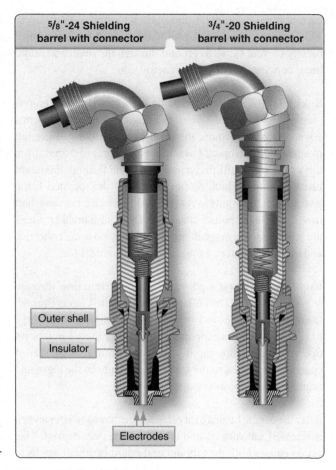

Figure 4-39. *Spark plug cutaway.*

and the center electrode assembly. The other end is threaded to receive the ignition lead from the magneto. All-weather plugs form a seal between the lead and the plug that is water proof to prevent moisture from entering this connection.

The insulator provides a protective core around the electrode. In addition to affording electrical insulation, the ceramic insulator core also transfers heat from the ceramic tip, or nose, to the cylinder. The insulator is made from aluminum oxide ceramic having excellent dielectric strength, high mechanical strength, and thermal conductivity. The types of spark plugs used in different engines vary in respect to heat range, reach, massive electrode, fine wire electrode (Iridium/platinum), or other characteristics of the installation requirements for different engines.

The electrodes can be of several designs from massive electrodes or Nickel-base alloy to fine wire electrodes. *[Figure 4-39 and 4-40]* The massive electrode material has a lower melting point and is more susceptible to corrosion. The main differences include cost and length of service. Fine wire iridium and platinum electrodes have a very high melting point and are considered precious metals. Therefore, the cost of this type of spark plug is higher, but they have

Figure 4-40. *Fine wire electrodes.*

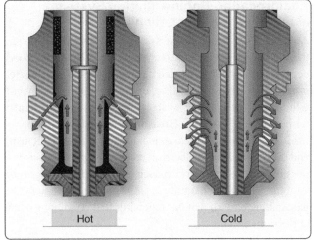

Figure 4-41. *Hot and cold spark plugs.*

a longer service life with increased performance. Fine wire spark plugs are more effective than massive electrode plugs because the size shields its own spark from some of the fuel air mixture. Less than efficient combustion occurs due to uneven ignition. The iridium electrode allows for a larger spark gap, which creates a more intense spark that increases performance. The spark gap of any electrode is vulnerable to erosion and the melting point of the electrode material.

The heat range of a spark plug is a measure of its ability to transfer the heat of combustion to the cylinder head. The plug must operate hot enough to burn off carbon deposits, which can cause fouling, a condition where the plug no longer produces a spark across the electrodes, yet remain cool enough to prevent a pre-ignition condition. Spark plug pre-ignition is caused by plug electrodes glowing red hot as a glow plug, setting off the air-fuel mixture before the normal firing position. The length of the nose core is the principal factor in establishing the plug's heat range. *[Figure 4-41]* Hot plugs have a long insulator nose that creates a long heat transfer path; cold plugs have a relatively short insulator to provide a rapid transfer of heat to the cylinder head. *[Figure 4-41]*

If an engine were operated at only one speed, spark plug design would be greatly simplified. Because flight demands impose different loads on the engine, spark plugs must be designed to operate as hot as possible at slow speeds and light loads, and as cool as possible at cruise and takeoff power.

The choice of spark plugs to be used in a specific aircraft engine is determined by the engine manufacturer after extensive tests. When an engine is certificated to use hot

or cold spark plugs, the plug used is determined by the compression ratio, the degree of supercharging, and how the engine is to be operated. High-compression engines tend to use colder range plugs while low-compression engines tend to use hot range plugs.

A spark plug with the proper reach ensures that the electrode end inside the cylinder is in the best position to achieve ignition. The spark plug reach is the length of the threaded portion that is inserted in the spark plug bushing of the cylinder. *[Figure 4-42]* Spark plug seizure and/or improper combustion within the cylinder can occur if a plug with the wrong reach is used. In extreme cases, if the reach is too long, the plug may contact a piston or valve and damage the engine. If the plug threads are too long, they extend into the combustion chamber and carbon adheres to the threads making it almost impossible to remove the plug. This can also be a source of pre-ignition. Heat of combustion can make some of the carbon a source for ignition, which can ignite the air-fuel mixture prematurely. It is very important to select the approved spark plugs for the engine.

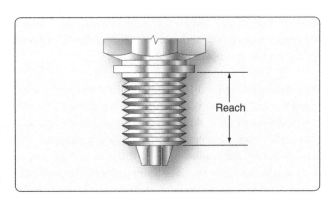

Figure 4-42. *Spark plug reach.*

Reciprocating Engine Ignition System Maintenance & Inspection

An aircraft's ignition system is the result of careful design and thorough testing. The ignition system usually provides good, dependable service, provided it is maintained and inspected properly. However, difficulties can occur with normal wear, which affects ignition system performance, especially with magneto systems. Breakdown and deterioration of insulating materials, breaker point wear, corrosion, bearing and oil seal wear, and electrical connection problems are all possible defects that can be associated with magneto-ignition systems. The ignition timing requires precise adjustment and painstaking care so that the following four conditions occur at the same instant:

1. The piston in the No. 1 cylinder must be in a position a prescribed number of degrees before top dead center on the compression stroke.

2. The rotating magnet of the magneto must be in the E-gap position.

3. The breaker points must be just opening on the No. 1 cam lobe.

4. The distributor finger must be aligned with the electrode serving the No. 1 cylinder.

If one of these conditions is out of synchronization with any of the others, the ignition system is out of time. If the spark is out of time, it is not delivered to the cylinder at the correct time and engine performance decreases.

When ignition in the cylinder occurs before the optimum crankshaft position is reached, the timing is said to be early. If ignition occurs too early, the piston rising in the cylinder is opposed by the full force of combustion. This condition results in a loss of engine power, overheating, and possible detonation and pre-ignition.

If ignition occurs at a time after the optimum crankshaft position is reached, the ignition timing is said to be late. If it occurs too late, not enough time is allowed to consume the air-fuel charge, and combustion is incomplete. As a result, the engine loses power and requires a greater throttle opening to carry a given propeller load.

Moisture forming on different parts of the ignition system causes more common irregularities. Moisture can enter ignition system units through cracks or loose covers, or it can result from condensation. Breathing, a situation that occurs during the readjustment of the system from low to high atmospheric pressure, can result in drawing in moisture-laden air. Ordinarily, the heat of the engine is sufficient to evaporate this moisture, but occasionally the moist air condenses as the engine cools. The result is an appreciable moisture accumulation which causes the insulation materials to lose electrical resistance. A slight amount of moisture contamination may cause reduction in magneto output by short-circuiting to ground a part of the high-voltage current intended for the spark plug. If the moisture accumulation is appreciable, the entire magneto output may be dissipated to ground by way of flashover and carbon tracking. Moisture accumulation during flight is extremely rare because the high operating temperature of the system is effective in preventing condensation. Difficulties from moisture accumulation are probably more evident during starting and ground operation.

Spark plugs are often diagnosed as being faulty when the real malfunction exists in a different system. Malfunctioning of the carburetor, poor fuel distribution, too much valve overlap, leaking primer system, or poor idle speed and mixture settings show symptoms that are the same as those for faulty ignition. Unfortunately, many of these conditions can be temporarily improved by a spark plug change, but the trouble recurs in a short time because the real cause of the malfunction has not been eliminated. A thorough understanding of the various engine systems, along with meticulous inspection and good maintenance methods, can substantially reduce such errors.

Magneto-Ignition Timing Devices

Built-In Engine Timing Reference Marks

Most reciprocating engines have timing reference marks built into the engine. The timing reference marks vary by manufacturer. *[Figure 4-43]* When the starter gear hub is installed correctly, the timing marks are marked on it that line up with the mark on the starter. On an engine that has no starter gear hub, the timing mark is normally on the propeller flange edge. *[Figure 4-44]* The top center (TC) mark stamped on the edge aligns with the crankcase split line below the crankshaft when the No. 1 piston is at top dead center. Other flange marks indicate degrees before top center.

Some engines have degree markings on the propeller reduction drive gear. To time these engines, the plug provided on the exterior of the reduction gear housing must be removed to view the timing marks. On other engines, the timing marks are on a crankshaft flange and can be viewed by removing a plug from the crankcase. In every case, the engine manufacturer's instructions give the location of built-in timing reference marks.

In using built-in timing marks to position the crankshaft, be sure to sight straight across the stationary pointer or mark on the nose section, the propeller shaft, crankshaft flange, or bell gear. *[Figure 4-45]* Sighting at an angle results in an error in positioning the crankshaft. Normally, the No. 1 cylinder is used to time or check the timing of the magnetos. When installing magnetos, the timing marks must be lined up and the No. 1 cylinder must be on the compression stroke.

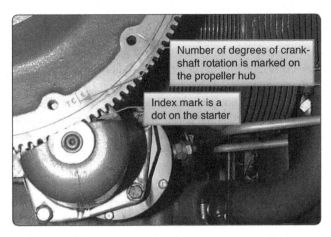

Figure 4-43. *Lycoming timing marks.*

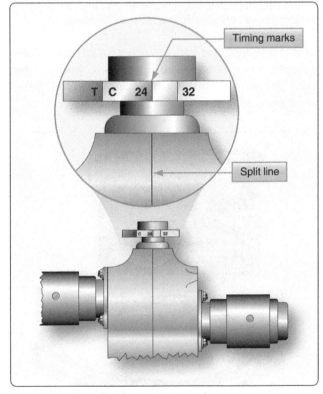

Figure 4-44. *Propeller flange timing marks.*

The amount of gear backlash in any system of gears varies between installations because there is clearance between the gear teeth. Always take timing when reading, or stop movement of the engine for timing set up, in the direction of rotation. Another unfavorable aspect in the use of timing marks on the reduction gear is the small error that exists when sighting down the reference mark to the timing mark inside the housing on the reduction gear. This can occur because there is depth between the two reference marks.

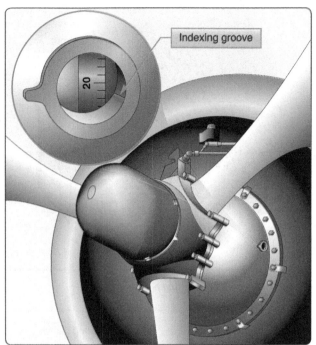

Figure 4-45. *Typical built-in timing mark on propeller reduction gear.*

Timing Discs

Most timing disc devices are mounted to the crankshaft flange and use a timing plate. *[Figure 4-46]* The markings vary according to the specifications of the engine. This plate is temporarily installed on the crankshaft flange with a scale numbered in crankshaft degrees and the pointer attached to the timing disc.

Piston Position Indicators

Any given piston position, whether it is to be used for ignition, valve, or injection pump timing, is referenced to a piston position called top dead center. This piston position is not to be confused with a piston position called top center. A piston in top center has little value from a timing standpoint because the corresponding crankshaft position may vary from 1° to 5° for this piston position. This is illustrated in *Figure 4-47,* which is exaggerated to emphasize the no-travel zone of the piston. Notice that the piston does not move while the crankshaft describes the small arc from position A to position B. This no-travel zone occurs between the time the crankshaft and connecting rod stop pushing the piston upward, and continues until the crankshaft has swung the lower end of the connecting rod into a position where the crankshaft can start pulling the piston downward. Top dead center is a piston and crankshaft position from which all other piston and crankshaft locations are referenced. When a piston is in the top dead center position of the crankshaft, it is also in the center of the no-travel zone. The piston is in a position where a straight line can be drawn through the center of the crankshaft journal, the crankpin, and the piston pin. This is

Figure 4-46. *A timing plate and pointer.*

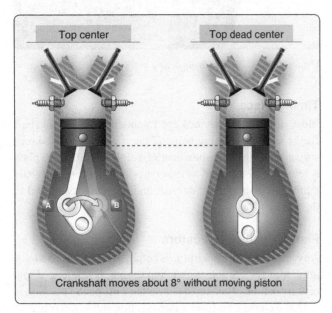

Figure 4-47. *Difference between top center and top dead center.*

shown on the right of *Figure 4-47*. With such an alignment, a force applied to the piston could not move the crankshaft.

Timing Lights

Timing lights are used to help determine the exact instant that the magneto points open. There are two general types of timing lights in common use. Both have two lights and three external wire connections. Although both have internal circuits that are somewhat different, their function is very much the same. *[Figures 4-48* and *4-49]*

Three wires plug into the light box. *[Figure 4-49]* There are two lights on the front face of the unit, one green and one red, and a switch to turn the unit on and off. To use the timing light, the center lead, which is black, marked "ground lead"

is connected to the case of the magneto being tested. The other leads are connected to the primary leads of the breaker point assembly of the magnetos being timed. The color of the lead corresponds to the color of the light on the timing light.

With the leads connected in this manner, it can be easily determined whether the points are open or closed by turning on the switch and observing the two lights. If the points are closed, most of the current flows through the breaker points and not through the transformers, and the lights do not come on. If the points are open, the current flows through the transformer and the lights glow. Some models of timing lights operate in the reverse manner (i.e., the light goes out when the points open). Each of the two lights is operated separately by the set of breaker points to which it is connected. This makes it possible to observe the time, or point in reference to magneto rotor rotation, that each set of points opens.

Most timing lights use batteries that must be replaced after long use. Attempts to use a timing light with weak batteries may result in erroneous readings because of low current flow in the circuits.

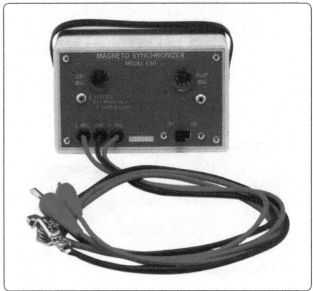

Figure 4-48. *E50 Magneto Synchronizer.*

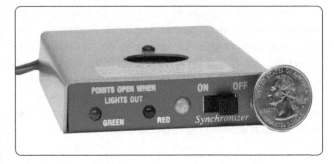

Figure 4-49. *Timing light.*

Checking the Internal Timing of a Magneto

When replacing or preparing a magneto for installation, the first concern is with the internal timing of the magneto. For each magneto model, the manufacturer determines how many degrees beyond the neutral position a pole of the rotor magnet should be to obtain the strongest spark at the instant of breaker point separation. This angular displacement from the neutral position, known as the E-gap angle, varies with different magneto models. On one model, a step is cut on the end of the breaker cam to check internal timing of the magneto. When a straightedge is laid along this step and it coincides with the timing marks on the rim of the breaker housing, the magneto rotor is then in the E-gap position, and the breaker contact points should just begin to open.

Another method for checking E-gap is to align a timing mark with a pointed chamfered tooth. *[Figure 4-50]* The breaker points should be just starting to open when these marks line up.

In a third method, the E-gap is correct when a timing pin is in place and red marks visible through a vent hole in the side of the magneto case are aligned. *[Figure 4-51]* The contact points should be just opening when the rotor is in the position just described.

Bench timing the magneto, or setting the E-gap, involves positioning the magneto rotor at the E-gap position and setting the breaker points to open when the timing lines or marks provided for that purpose are perfectly aligned.

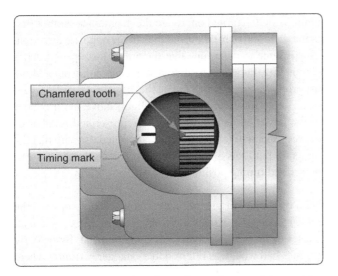

Figure 4-50. *Timing marks indicate the number one firing position of a magneto.*

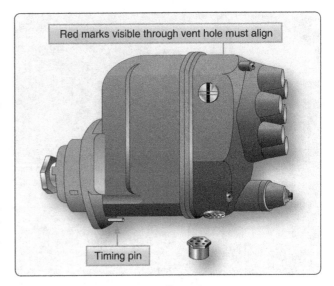

Figure 4-51. *Checking magneto E-gap.*

High-Tension Magneto E-Gap Setting (Bench Timing)

The following steps are taken to check and adjust the timing of the breaker points for the S-200 magneto, which does not have timing marks in the breaker compartment:

1. Remove the timing inspection plug from the top of the magneto. Turn the rotating magnet in its normal direction of rotation until the painted, chamfered tooth on the distributor gear is approximately in the center of the inspection window. Then, turn the magnet back a few degrees until it is in its neutral position. Because of its magnetism, the rotating magnet holds itself in the neutral position.

2. Install the timing kit and place the pointer in the zero position. *[Figure 4-52]*

3. Connect a suitable timing light across the main breaker points and turn the magnet in its normal direction of rotation 10° as indicated by the pointer. This is the E-gap position. The main breaker points should be adjusted to open at this point.

4. Turn the rotating magnet until the cam follower is at the highpoint on the cam lobe, and measure the clearance between the breaker points. This clearance must be 0.018 inch ± 0.006 inch [0.46 millimeter (mm) ± 0.15 mm]. If the breaker point clearance is not within these limits, the points must be adjusted for correct setting. It is then necessary to recheck and readjust the timing for breaker opening. If the breaker points cannot be adjusted to open at the correct time, they should be replaced.

Figure 4-52. *Installing timing kit.*

Timing the High-tension Magneto to the Engine

When replacing magnetos on aircraft engines, two factors must be considered:

1. The internal timing of the magneto, including breaker point adjustment, which must be correct to obtain maximum potential voltage from the magneto.

2. The engine crankshaft position where the spark occurs. The engine is usually timed by using the No. 1 cylinder on the compression stroke.

The magneto must be timed by first adjusting or checking the internal timing with the magneto off the engine. This is done by checking and adjusting the ignition points to open at the E-gap position. The chamfered tooth should line up (reference timing mark for the magneto) in the middle of the timing window. The magneto is set to fire the No. 1 cylinder. Remove the most accessible spark plug from the No. 1 cylinder. Pull the propeller through in the direction of rotation until the No. 1 piston is coming up on the compression stroke. This can be determined by holding a thumb over the spark plug hole until the compression air is felt. Set the engine crankshaft at the prescribed number of degrees ahead of true top dead center as specified in the applicable manufacturer's instruction, usually using the timing marks on the engine. With the engine set at a prescribed number of degrees ahead of true top dead center on the compression stroke and with final movement of the engine stopped in the direction of normal rotation, the magneto can be installed on the engine. *[Figure 4-53]*

While holding the magneto drive in the firing position for the No. 1 cylinder as indicated by the alignment of the reference marks for the magneto, install the magneto drive into the

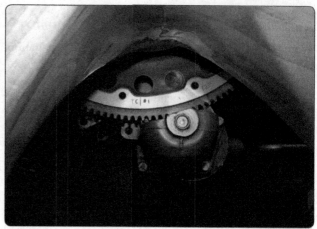

Figure 4-53. *Timing marks aligned.*

engine drive. It should be installed in the middle of its slotted flange to allow for fine timing of the magneto to the engine. Attach a timing light to both magnetos. With the engine still in the firing position, the magnetos should be timed by moving them in the flange slots until the breaker points in the magneto just open. If the slots in the mounting flange of the magneto do not permit sufficient movement to effect breaker point opening for the No. 1 cylinder, move the magneto out of position far enough to permit turning the magneto drive shaft. Then, install the magneto in position again and repeat the previous check for point opening.

Install the magneto attaching nuts on the studs and tighten slightly. The nuts must not be tight enough to prevent the movement of the magneto assembly when the magneto mounting flange is tapped with a mallet. Reconnect the timing light to the magneto and breaker points. With the light and ignition switch turned on, rotate the magneto assembly first in the direction of rotation and then in the opposite direction. This is done to determine that the points just opened. After completing this adjustment, tighten the mounting nuts. Move the propeller one blade opposite the direction of rotation and then, while observing the timing light, move the propeller in the direction of rotation until the prescribed number of degrees ahead of top dead center is reached. Be sure that the lights for both sets of points come on points open, within the prescribed timing position.

Both right and left sets of breaker points should open at the same instant, proper magneto-to-engine timing exists, and all phases of magneto operation are synchronized. Some early engines had what was referred to as staggered timing where one magneto would fire at a different number of degrees before top dead center on the compression stroke. In this case, each magneto had to be timed separately. If staggered ignition timing is used, the spark plug nearest the exhaust valve will fire first.

In the following example, a timing light is used for timing the magneto to the engine. The timing light is designed in such a way that one of two lights come on when the points open. The timing light incorporates two lights. When connecting the timing light to the magneto, the leads should be connected so that the light on the right side of the box represents the breaker points on the right magneto, and the light on the left side represents the left magneto breaker points. The black lead or ground lead must be attached to the engine or an effective ground. When using the timing light to check a magneto in a complete ignition system installed on the aircraft, the ignition switch for the engine must be turned to both. Otherwise, the lights do not indicate breaker point opening.

Performing Ignition System Checks

The ignition system has checks performed on it during the aircraft engine run-up, which is the engine check before each flight. The magneto check, as it is usually referred to, is performed during the engine run-up check list.

One other check is accomplished prior to engine shutdown. The ignition system check is used to check the individual magnetos, harnesses, and spark plugs. After reaching the engine rpm specified for the ignition system check, allow the rpm to stabilize. Place the ignition switch in the right position and note the rpm drop on the tachometer. Return the switch to the both position. Allow the switch to remain in the both position for a few seconds so that the rpm stabilizes again. Place the ignition switch in the left position and again note the rpm drop. Return the ignition switch to the both position. Note the amount of total rpm drop that occurs for each magneto position. The magneto drop should be even for both magnetos and is generally in the area of a 25–75 rpm drop for each magneto. Always refer to the aircraft operating manual for specific information. This rpm drop is because operating on one magneto combustion is not as efficient as it is with two magnetos providing sparks in the cylinder.

Remember, this tests not only the magnetos but also the ignition leads and spark plugs. If either magneto has excessive rpm drop while operating by itself, the ignition system needs to be checked for problems. If only one magneto has a high magneto drop, the problem can be isolated and corrected by operating on that magneto. This ignition system check is usually performed at the beginning of the engine run-up because rpm drops not within the prescribed limits affect later checks.

Ignition Switch Check

The ignition switch check is performed to see that all magneto ground leads are electrically grounded. The ignition switch check is usually made at 700 rpm. On those aircraft engine installations that do not idle at this low rpm, set the engine speed to the lowest possible to perform this check. When the speed to perform this check is obtained, momentarily turn the ignition switch to the off position. The engine should completely quit firing. After a drop of 200–300 rpm is observed, return the switch to the both position as rapidly as possible. Do this quickly to eliminate the possibility of afterfire and backfire when the ignition switch is returned to both.

If the ignition switch is not returned quickly enough, the engine rpm drops off completely and the engine stops. In this case, leave the ignition switch in the off position and place the mixture control in the idle-cutoff position to avoid overloading the cylinders and exhaust system with raw fuel. When the engine has completely stopped, allow it to remain inoperative for a short time before restarting.

If the engine does not cease firing in the off position, the magneto ground lead, more commonly referred to as the P lead, is open, and the trouble must be corrected. This means that one or more of the magnetos are not being shut off even when the ignition switch is in the off position. Turning the propeller of this engine can result in personnel injury or death. If the propeller is turned in this condition, the engine can start with personnel in the propeller arch.

Maintenance & Inspection of Ignition Leads

Inspection of ignition leads should include both a visual and an electrical test. During the visual test, the lead cover should be inspected for cracks or other damage, abrasions, mutilated braid, or other physical damage. Inspect leads for overheating if routed close to exhaust stacks. Disconnect the harness coupling nuts from the top of the spark plugs and remove the leads from the spark plug lead well. Inspect the contact springs and compression springs for any damage or distortion and the sleeves for cracks or carbon tracking. The coupling nut that connects to the spark plug should be inspected for damaged threads or other defects.

Each lead should be checked for continuity using a high-tension lead tester by connecting the black lead to the contact spring and the red lead to the eyelet of the same lead in the cover. The continuity lamp on the tester should illuminate when tested. The insulation resistance test of each lead is accomplished using the high-tension lead tester by attaching the red, or high-voltage, lead to the spring of the harness lead. Then, attach the black lead to the ferrule of the same lead. Depress the press-to-test push button switch on the lead tester. Observe that the indictor lamp flashes and gap fires simultaneously as long as the press-to-test switch is held in the depressed position.

If the indicator lamp flashes and the gap fails to fire, the lead under test is defective and must be replaced. The indicator lamp flashes to show that a high-voltage impulse was sent out. If it fails to pass through the tester, then the electrical pulse leaked through the wire showing it to be defective.

When defective leads are revealed by an ignition harness test, continue the test to determine whether the leads or distributor block are defective. If the difficulty is in an individual ignition lead, determine whether the electrical leak is at the spark plug elbow or elsewhere. Remove the elbow, pull the ignition lead out of the manifold a slight amount, and repeat the harness test on the defective lead. If this stops the leakage, cut away the defective portion of the lead and reinstall the elbow assembly, integral seal, and terminal (sometimes referred to as cigarette). *[Figure 4-54]* When installing the leads, you should avoid bends because weak points may develop in the insulation through which high-tension current can leak.

If the lead is too short to repair in the manner described, or the electrical leak is inside the harness, replace the defective lead. Single ignition lead replacement procedures are as follows:

1. Disassemble the magneto or distributor so that the distributor block is accessible.

2. Loosen the piercing screw in the distributor block for the lead to be replaced, and remove the lead from the distributor block.

3. Remove approximately 1 inch of insulation from the distributor block end of the defective lead and approximately 1 inch of insulation from the end of the replacement cable. Splice this end to the end of the lead to be replaced and solder the splice.

4. Remove the elbow adapter from the spark plug end of the defective lead, then pull the old lead out and put the new lead into the harness. While pulling the leads through the harness, have someone push the replacement lead into the ignition manifold at the distributor end to reduce the force required to pull the lead through the ignition manifold.

5. When the replacement lead has been pulled completely through the manifold, force the ignition lead up into the manifold from the distributor block end to provide extra length for future repairs, which may be necessary because of chafing at the spark plug elbow.

6. Remove approximately ⅜ inch of insulation from the distributor block end. Bend the ends of the wire back and prepare the ends of the cable for installation into the distributor block well. Insert the lead in the distributor and tighten the piercing screw.

7. Remove approximately ¼ inch of insulation from the spark plug end of the lead and install the elbow, integral seal, and cigarette. *[Figure 4-54]*

8. Install a marker on the distributor end of the cable to identify its cylinder number. If a new marker is not available, use the marker removed from the defective cable.

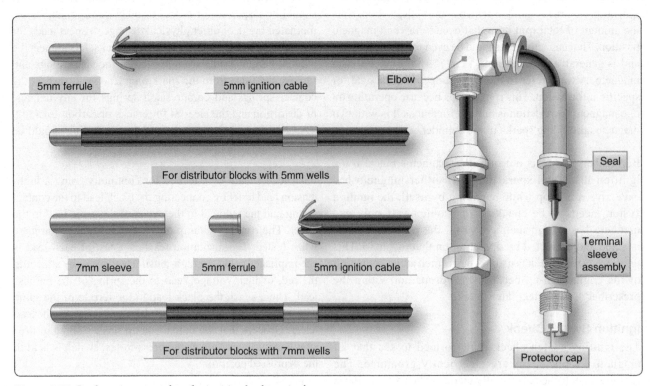

Figure 4-54. *Replacement procedure for ignition lead terminals.*

Replacement of Ignition Harness

Replace a complete ignition harness only when the shielding of the manifold is damaged or when the number of defective leads makes it more practical to replace the harness than to replace the individual leads. Replace a cast-filled harness only when leakage in the cast-filled portion is indicated. Before replacing any harness to correct engine malfunctioning, make extensive ignition harness tests. Typical procedures for installing an ignition harness are detailed in the following paragraphs.

Install the ignition harness on the engine. Tighten and safety the hold down nuts and bolts and install and tighten the individual lead brackets according to instructions. The ignition harness is then ready for connection of the individual leads to the distributor block. A band is attached to each lead at the distributor end of the harness to identify the cylinder for the lead. However, each lead should be checked individually with a continuity or timing light prior to connecting it.

Check for continuity by grounding the lead at the cylinder and then checking at the distributor block end to establish that the lead grounded is as designated on the band for the lead. After checking all leads for proper identification, cut them to the proper length for installation into the distributor block. Before cutting the leads, however, force them back into the manifold as far as possible to provide surplus wire in the ignition manifold. This extra wire may be needed at a later date in the event that chafing of a lead at the spark plug elbow necessitates cutting a short section of wire from the spark plug end of the harness. After cutting each lead to length, remove approximately ⅜ inch of insulation from the end and prepare the lead for insertion into the distributor block. Before installing the lead, back out the set screw in the distributor block far enough to permit slipping the end of the wire into the hole without force. Insert the lead into the block and tighten the set screw. Connect the wires in firing order (the first cylinder to fire No. 1 location on the block, the second in the firing order to No. 2 location, etc).

After connecting each lead, check continuity between the lead and its distributor block electrode with continuity light or timing light. To perform one test lead, touch the other test lead to the proper distributor block electrode. If the light does not indicate a complete circuit, the set screw is not making contact with the ignition wire or the lead is connected to the wrong block location. Correct any faulty connections before installing the distributor block.

Checking Ignition Induction Vibrator Systems

To check the induction vibrator, ensure that the manual mixture control is in idle cutoff, the fuel shutoff valve and booster pump for that engine are in the off position, and the

battery switch is on. Since the induction vibrator buzzes whether the ignition switch is on or off, leave the switch off during the check. If the engine is equipped with an inertia or combination starter, make the check by closing the engage mesh switch; if the engine is equipped with a direct-cranking starter, see that the propeller is clear and close the start switch. An assistant stationed close to the induction vibrator should listen for an audible buzzing sound. If the unit buzzes when the starter is engaged or cranked, the induction vibrator is operating properly.

Spark Plug Inspection & Maintenance

Spark plug operation can often be a major source of engine malfunctions because of lead, oil, graphite, carbon fouling, and spark plug gap erosion. Most of these failures, which usually accompany normal spark plug operation, can be minimized by good operational and maintenance practices. A spark plug is considered fouled if it has stopped allowing the spark to bridge the gap either completely or intermittently.

Carbon Fouling of Spark Plugs

Carbon fouling from fuel is associated with mixtures that are too rich to burn or mixtures that are so lean they cause intermittent firing. [Figure 4-55] Each time a spark plug does not fire, raw fuel and oil collect on the nonfiring electrodes and nose insulator. These difficulties are almost invariably associated with an improper idle mixture adjustment, a leaking primer, or carburetor malfunctions that cause too rich a mixture in the idle range. A rich air-fuel mixture is detected by soot or black smoke coming from the exhaust and by an increase in rpm when the idling air-fuel mixture is leaned to best power. The soot that forms as a result of overly rich idle air-fuel mixtures settles on the inside of the combustion chamber because the heat of the engine and the turbulence

Figure 4-55. *Carbon fouled spark plug.*

in the combustion chamber are slight. At higher engine speeds and powers, however, the soot is swept out and does not condense out of the charge in the combustion chamber.

Oil Fouling of Spark Plugs

Even though the idling air-fuel mixture is correct, there is a tendency for oil to be drawn into the cylinder past the piston rings, valve guides, and impeller shaft oil seal rings. At low engine speeds, the oil combines with the soot in the cylinder to form a solid that is capable of shorting out the spark plug. Spark plugs that are wet or covered with lubricating oil are usually grounded out during the engine start. In some cases, these plugs may clear up and operate properly after a short period of engine operation.

Engine oil that has been in service for any length of time holds in suspension minute carbon particles that are capable of conducting an electric current. Thus, a spark plug will not arc the gap between the electrodes when the plug is full of oil. Instead, the high-voltage impulse flows through the oil from one electrode to the other without a spark as though a wire conductor were placed between the two electrodes. Combustion in the affected cylinder does not occur until, at a higher rpm, increased airflow has carried away the excess oil. Then, when intermittent firing starts, combustion assists in emitting the remaining oil. In a few seconds, the engine is running clean with white fumes of evaporating and burning oil coming from the exhaust.

Lead Fouling of Spark Plugs

Lead fouling of aviation spark plugs is a condition likely to occur in any engine using leaded fuels. Lead is added to aviation fuel to improve its anti-knock qualities. The lead, however, has the undesirable effect of forming lead oxide during combustion. This lead oxide forms as a solid with varying degrees of hardness and consistency. Lead deposits on combustion chamber surfaces are good electrical conductors at high temperatures and cause misfiring. At low temperatures, the same deposits may be good insulators. In either case, lead formations on aircraft spark plugs prevent their normal operation. *[Figure 4-56]* To minimize the formation of lead deposits, ethylene dibromide is added to the fuel as a scavenging agent that combines with the lead during combustion.

Lead fouling may occur at any power setting, but perhaps the power setting most conducive to lead fouling is cruising with lean mixtures. At this power, the cylinder head temperature is relatively low and there is more oxygen than needed to consume all the fuel in the air-fuel mixture. Oxygen, when hot, is very active and aggressive. When all the fuel has been consumed, some of the excess oxygen unites with some of the lead and some of the scavenger agent to form

Figure 4-56. *Lead fouled spark plug.*

oxygen compounds of lead or bromine or both. Some of these undesirable lead compounds solidify and build up in layers as they contact the relatively cool cylinder walls and spark plugs. Although lead fouling may occur at any power setting, experience indicates that the lead buildup is generally confined to a specific combustion temperature range. Combustion temperatures outside this specific range minimize the lead fouling tendency.

If lead fouling is detected before the spark plugs become completely fouled, the lead can usually be eliminated or reduced by either a sharp rise or a sharp decrease in combustion temperature. This imposes a thermal shock on cylinder parts, causing them to expand or contract. Since there is a different rate of expansion between deposits and metal parts on which they form, the deposits chip off or are loosened and then scavenged from the combustion chamber by the exhaust or are burned in the combustion process.

Several methods of producing thermal shock to cylinder parts are used. The method used depends on the accessory equipment installed on the engine. A sharp rise in combustion temperatures can be obtained on all engines by operating them at full takeoff power for approximately 1 minute. When using this method to eliminate fouling, the propeller control must be placed in low pitch, or high rpm, and the throttle advanced slowly to produce takeoff rpm and manifold pressure. Slow movement of the throttle control provides reasonable freedom from backfiring in the affected cylinders during the application of power.

Another method of producing thermal shock is the use of excessively rich air-fuel mixtures. This method suddenly cools the combustion chamber because the excess fuel does not contribute to combustion; instead, it absorbs heat from the combustion area. Some carburetor installations use two-position manual mixture controls that provide a lean mixture setting for cruising economy and a richer mixture setting for all powers above cruising. Neither manual mixture control setting in this type of configuration is capable of producing an excessively rich air-fuel mixture. Even when the engine is operated in auto-rich at powers where an auto-lean setting would be entirely satisfactory, the mixture is not rich enough.

Graphite Fouling of Spark Plugs

As a result of careless and excessive application of thread lubricant, called antiseize compound, to the spark plug, the lubricant flows over the electrodes and causes shorting. Shorting occurs because graphite is a good electrical conductor. The elimination of service difficulties caused by graphite is up to the aircraft technician. Use care when applying the lubricant to make certain that smeared fingers, shop towels, or brushes do not contact the electrodes or any part of the ignition system except the spark plug threads. Never apply to the first set of threads.

Gap Erosion of Spark Plugs

Erosion of the electrodes takes place in all aircraft spark plugs as the spark jumps the air gap between the electrodes. [Figure 4-57]

The spark carries with it a portion of the electrode, part of which is deposited on the other electrode. The remainder is blown off in the combustion chamber. As the airgap is enlarged by erosion, the resistance that the spark must

Figure 4-57. *Spark plug gap erosion.*

overcome in jumping the air gap also increases. This means that the magneto must produce a higher voltage to overcome the higher resistance. With higher voltages in the ignition system, a greater tendency exists for the spark to discharge at some weak insulation point in the ignition system. Since the resistance of an air gap also increases as the pressure in the engine cylinder increases, a double danger exists at takeoff and during sudden acceleration with enlarged airgaps. Insulation breakdown, premature flashover, and carbon tracking result in misfiring of the spark plug and go hand in hand with excessive spark plug gap. Wide gap settings also raise the coming in speed of a magneto and therefore cause hard starting.

Spark plug manufacturers have partially overcome the problem of gap erosion by using a hermetically sealed resistor in the center electrode of spark plugs. This added resistance in the high-tension circuit reduces the peak current at the instant of firing. This reduced current flow helps prevent metal disintegration in the electrodes. Also, due to the high erosion rate of steel or any of its known alloys, spark plug manufacturers are using tungsten or an alloy of nickel for their massive electrode plugs and iridium/platinum plating for their fine wire electrode plugs.

Spark Plug Removal

Spark plugs should be removed for inspection and servicing at the intervals recommended by the manufacturer. Since the rate of gap erosion varies with different operating conditions, engine models, and type of spark plug, engine malfunction traceable to faulty spark plugs may occur before the regular servicing interval is reached. Normally, in such cases, only the faulty plugs are replaced.

Since spark plugs can be easily damaged, careful handling of the used and replacement plugs during installation and removal of spark plugs from an engine cannot be overemphasized. To prevent damage, spark plugs should always be handled individually and new and reconditioned plugs should be stored in separate cartons. A common method of storage is illustrated in *Figure 4-58*. This is a drilled tray, which prevents the plugs from bumping against one another and damaging the fragile insulators and threads. Additionally, the tray helps identify where the spark plug came from and its proper rotation. If a plug is dropped on the floor or other hard surface, it should not be installed in an engine, since the shock of impact usually causes small, invisible cracks in the insulators. A dropped spark plug should be discarded.

Before a spark plug can be removed, the ignition harness lead must be disconnected. Using the special spark plug coupling elbow wrench, loosen and remove the spark plug to elbow coupling nut from the spark plug. Take care to pull the lead

Figure 4-58. *Spark plug tray.*

straight out and in line with the centerline of the plug barrel. If a side load is applied, damage to the barrel insulator and the ceramic lead terminal may result. *[Figure 4-59]* If the lead cannot be removed easily in this manner, the neoprene collar may be stuck to the shielding barrel. Break loose the neoprene collar by twisting the collar as though it were a nut being unscrewed from a bolt.

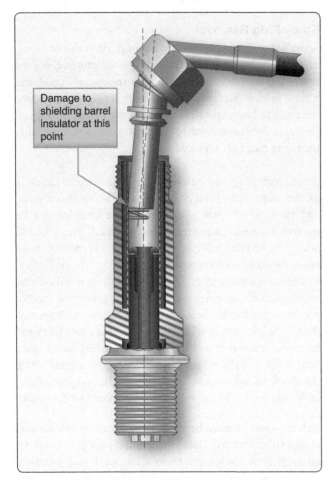

Damage to shielding barrel insulator at this point

Figure 4-59. *Improper lead removal technique.*

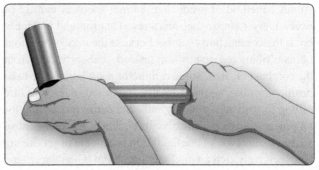

Figure 4-60. *Proper spark plug removal technique.*

After the lead has been disconnected, select the proper size deep socket for spark plug removal. Apply steady pressure with one hand on the hinge handle, holding the socket in alignment with the other hand. Failure to hold the socket in correct alignment causes the socket to tilt to one side and damage the spark plug. *[Figure 4-60]*

In the course of engine operation, carbon and other products of combustion are deposited across the spark plug and cylinder, and some carbon may even penetrate the lower threads of the shell. As a result, a high torque is generally required to break the spark plug loose. This factor imposes a shearing load on the shell section of the plug. After removing the plugs, they should be placed in a spark plug tray. *[Figure 4-58]*

Spark Plug Reconditioning Service

A visual inspection should be the first step in servicing spark plugs. The threads on the shielding barrel and on the shell that screws into the cylinder should be inspected for damaged or nicked threads. Inspect the lead shielding barrel for corrosion, nicks, and cracks. The firing end should be checked for insulator cracks, chips, and excessive electrode wear. The shell hex or wrench hex should be checked to see if it is rounded off or mutilated. If the spark plug passes the visual check, then it should be degreased using petroleum solvent. Take care to keep solvent out of the shielding barrel.

Never soak the plugs in solvent. After drying the firing end of the plugs, remove the lead compound deposits using a vibrator cleaner. *[Figure 4-61]* The firing end can now be cleaned by using an abrasive blaster. This is usually done using a spark plug cleaner tester. *[Figure 4-62]* As the firing end is subjected to the abrasive blast, the plug should be rotated so all the area of the firing end is cleaned. After the abrasive blast, the firing end gets a thorough air blast to remove the abrasive material. The shielding barrel insulators may be cleaned with a cotton cloth or felt swab saturated with solvent, wood alcohol, or other approved cleaner. The firing end should be inspected using a light and a magnifying glass. If the plug passes the firing end visual and cleaning checks, then the spark gap should be set using a round thickness

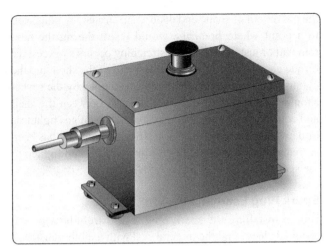

Figure 4-61. *Spark plug vibrator cleaner.*

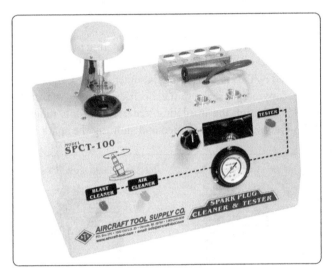

Figure 4-62. *Spark plug cleaner tester.*

gauge. The spark plug should be tested by using a tester as shown in *Figure 4-62*, which passes a high-voltage through the spark plug and fire the gap. As this test takes place, the firing end of the plug is subjected to air pressure to simulate the pressure in the engine's cylinder. If the firing pattern is good, the plug should be returned to its holder ready for installation in the engine.

Inspection Prior to Installation

Before installing new or reconditioned spark plugs in the engine cylinders, clean the spark plug bushings or Heli-Coil inserts.

Brass or stainless steel spark plug bushings are usually cleaned with a spark plug bushing cleanout tap. Before inserting the cleanout tap in the spark plug hole, fill the flutes of the tap, or channels between threads, with clean grease to prevent hard carbon or other material removed by the tap from dropping into the inside of the cylinder. Align the tap with the

bushing threads by sight where possible, and start the tap by hand until there is no possibility of it being cross-threaded in the bushing. To start the tap on installations where the spark plug hole is located deeper than can be reached by a clenched hand, it may be necessary to use a short length of hose slipped over the square end of the tap to act as an extension. When screwing the tap into the bushing, be sure that the full tap cutting thread reaches the bottom thread of the bushing. This removes carbon deposits from the bushing threads without removing bushing metal, unless the pitch diameter of the threads has contracted as the result of shrinkage or some other unusual condition. Replace the cylinder if, during the thread-cleaning process, the bushing is found to be loose, loosened in the cylinder, or the threads are cross-threaded or otherwise seriously damaged.

Spark plug Heli-Coil inserts are cleaned with a round wire brush, preferably one having a diameter slightly larger than the diameter of the spark plug hole. A brush considerably larger than the hole may cause removal of material from the Heli-Coil proper or from the cylinder head surrounding the insert. Also, the brush should not disintegrate with use, allowing wire bristles to fall into the cylinder. Clean the insert by carefully rotating the wire brush with a power tool. When using the power brush, be careful that no material is removed from the spark plug gasket seating surface, since this may cause a change in the spark plug's heat range, combustion leakage, and eventual cylinder damage. Never clean the Heli-Coil inserts with a cleaning tap, since permanent damage to the insert results. If a Heli-Coil insert is damaged as a result of normal operation or while cleaning it, replace it according to the applicable manufacturer's instructions.

Using a lint-free rag and cleaning solvent, wipe the spark plug gasket seating surface of the cylinder to eliminate the possibility of dirt or grease being accidentally deposited on the spark plug electrodes at the time of installation.

Before the new or reconditioned plugs are installed, they must be inspected for each of the following conditions:

1. Ensure that the plug is of the approved type, as indicated by the applicable manufacturer's instructions.

2. Check for evidence of rust-preventive compound on the spark plug exterior and core insulator and on the inside of the shielding barrel. Rust-preventive compound accumulations are removed by washing the plug with a brush and cleaning solvent. It must then be dried with a dry air blast.

3. Check both ends of the plug for nicked or cracked threads and any indication of cracks in the nose insulator.

4. Inspect the inside of the shielding barrel for cracks in the barrel insulator, and the center electrode contact for rust and foreign material that might cause poor electrical contact.

5. Install a new spark plug gasket. When the thermocouple gasket is used, do not use an additional gasket.

The gap setting should be checked with a round wire-thickness gauge. *[Figure 4-63]* A flat-type gauge gives an incorrect clearance indication because the massive ground electrodes are contoured to the shape of the round center electrode. When using the wire thickness gauge, insert the gauge in each gap parallel to the centerline of the center electrode. If the gauge is tilted slightly, the indication is incorrect. Do not install a plug that does not have an air gap within the specified clearance range.

Spark Plug Installation

Prior to spark plug installation, carefully coat the first two or three threads from the electrode end of the shell with a graphite base antiseize compound. Prior to application, stir the antiseize compound to ensure thorough mixing. When applying the antiseize compound to the threads, be extremely careful that none of the compound gets on the ground, center electrodes, or on the nose of the plug, where it can spread to the ground or center electrode during installation. This precaution is mentioned because the graphite in the compound is an excellent electrical conductor and could cause permanent fouling.

To install a spark plug, start it into the cylinder without using a wrench of any kind, and turn it until the spark plug is seated on the gasket. If you can screw the plug into the cylinder with comparative ease using your fingers, this indicates good, clean threads. In this case, only a small amount of additional tightening torque is needed to compress the gasket to form a gastight seal. If a high torque is needed to install the plug, dirty or damaged threads on either the plug or plug bushing are indicated. The use of excessive torque might compress

the gasket out of shape and distort and stretch the plug shell to a point where breakage would result during the next removal or installation. Shell stretching occurs as excessive torque continues to screw the lower end of the shell into the cylinder after the upper end has been stopped by the gasket shoulder. As the shell stretches, the seal between the shell and core insulator is opened, creating a loss of gas tightness or damage to the core insulator. After a spark plug has been seated with the fingers, use a torque wrench and tighten to the specified torque. *[Figure 4-64]*

Spark Plug Lead Installation

Before installing the spark plug lead, carefully wipe the terminal sleeve and the integral seal with a cloth moistened with acetone or an approved solvent. After the plug lead is cleaned, inspect it for cracks and scratches. If the terminal sleeve is damaged or heavily stained, replace it.

Application of a light coating of an insulating material to the outer surface of the terminal sleeve, as well as filling the space occupied by the contact spring, is sometimes recommended. By occupying the space in the electrical contact area of the shielding barrel, the insulating material prevents moisture from entering the contact area and shorting the spark plug.

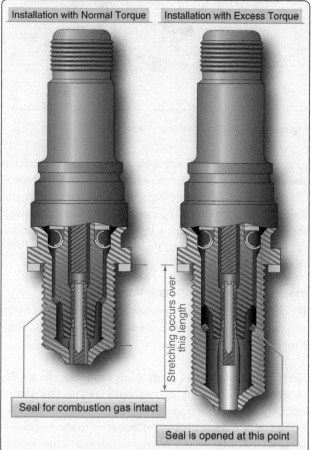

Figure 4-64. *Effect of excessive torque in installing a spark plug.*

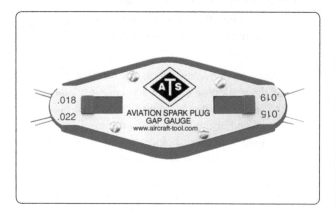

Figure 4-63. *Wire gap gauge.*

Some manufacturers recommend the use of such insulating compounds only when moisture in the ignition system becomes a problem, and others have discontinued the use of such materials.

After inspection of the spark plug lead, slip the lead into the shielding barrel of the plug with care. Then, tighten the spark plug coupling elbow nut with the proper tool. Most manufacturers' instructions specify the use of a tool designed to help prevent an overtorque condition. After the coupling nut is tightened, avoid checking for tightness by twisting the body of the elbow.

After all plugs have been installed, torqued, and the leads properly installed, start the engine and perform a complete ignition system operational check.

Breaker Point Inspection

Inspection of the magneto consists essentially of a periodic breaker point and dielectric inspection. After the magneto has been inspected for security of mounting, remove the magneto cover, or breaker cover, and check the cam for proper lubrication. Under normal conditions, there is usually ample oil in the felt oiler pad of the cam follower to keep the cam lubricated between overhaul periods. However, during the regular routine inspection, examine the felt pad on the cam follower to be sure it contains sufficient oil for cam lubrication. Make this check by pressing the thumbnail against the oiler pad. If oil appears on the thumbnail, the pad contains sufficient oil for cam lubrication. If there is no evidence of oil on the fingernail, apply one drop of a light aircraft engine oil to the bottom felt pad and one drop to the upper felt pad of the follower assembly. *[Figure 4-65]*

After application, allow at least 15 minutes for the felt to absorb the oil. At the end of 15 minutes, blot off any excess oil with a clean, lint-free cloth. During this operation, or any time the magneto cover is off, use extreme care to keep the breaker compartment free of oil, grease, or engine cleaning solvents, since each of these have an adhesiveness that collects dirt and grime that could foul an otherwise good set of breaker contact points.

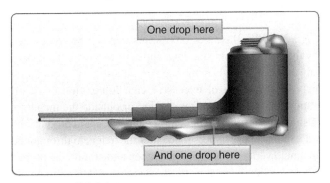

Figure 4-65. *Felt lubricator.*

After the felt oiler pad has been inspected, serviced, and found to be satisfactory, visually inspect the breaker contacts for any condition that may interfere with proper operation of the magneto. If the inspection reveals an oily or gummy substance on the sides of the contacts, swab the contacts with a flexible wiper, such as a pipe cleaner dipped in acetone or other approved solvent. By forming a hook on the end of the wiper, ready access can be gained to the back side of the contacts.

To clean the contact mating surfaces, force open the breaker points enough to admit a small swab. Whether spreading the points for purposes of cleaning or checking the surfaces for condition, always apply the opening force at the outer end of the mainspring and never spread the contacts more than ¹⁄₁₆ inch. If the contacts are spread wider than recommended, the mainspring, the spring carrying the movable contact point, is likely to take a permanent set. If the mainspring takes a permanent set, the movable contact point loses some of its closing tension and the points then either bounce or float, preventing the normal induction buildup of the magneto.

A swab can be made by wrapping a piece of linen tape or a small piece of lint-free cloth over one of the leaves of a clearance gauge and dipping the swab in an approved solvent. Pass the swab between the carefully separated contact surfaces until the surfaces are clean. During this entire operation, take care that drops of solvent do not fall on lubricated parts, such as the cam, follower block, or felt oiler pad.

To inspect the breaker contact surfaces, it is necessary to know what a normal operating set of contacts looks like, what surface condition is considered as permissible wear, and what surface condition is cause for dressing or replacement. The probable cause of an abnormal surface condition can be determined from the contact appearance. The normal contact surface has a dull gray, sandblasted, almost rough appearance over the area where electrical contact is made. *[Figure 4-66]* This gray, sandblasted appearance indicates that the points have worn in and have mated to each other and are providing the best possible electrical contact. This does not imply that this is the only acceptable contact surface condition. Slight, smooth-surfaced irregularities, without deep pits or high peaks, such as shown in *Figure 4-67*, are considered normal wear and are not cause for replacement.

However, when wear advances to a point where the slight, smooth irregularities develop into well-defined peaks extending noticeably above the surrounding surface, the breaker contacts must be replaced. *[Figure 4-68]*

Unfortunately, when a peak forms on one contact, the mating

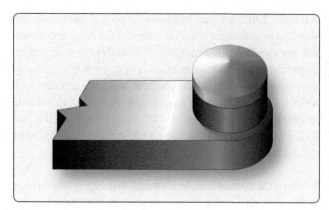

Figure 4-66. *Normal contact surface.*

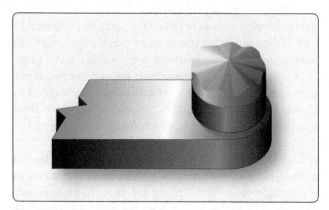

Figure 4-67. *Points with normal irregularities.*

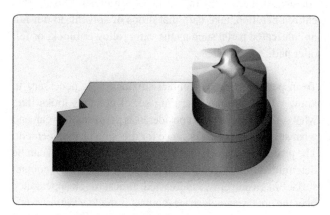

Figure 4-68. *Points with well-defined peaks.*

contact has a corresponding pit or hole. This pit is more troublesome than the peak because it penetrates the platinum pad of the contact surface. It is sometimes difficult to judge whether a contact surface is pitted deeply enough to require replacement because this depends on how much of the original platinum is left on the contact surface. The danger arises from the possibility that the platinum pad may already be thin as a result of long service life and previous dressings. At overhaul facilities, a gauge is used to measure the remaining thickness of the pad, and no difficulty in determining the condition of the pad exists. But at line maintenance activities, this gauge is generally unavailable.

Therefore, if the peak is quite high or the pit quite deep, remove and replace them with a new assembly. A comparison between *Figures 4-67* and *4-68* will help to draw the line between minor irregularities and well-defined peaks.

Some examples of possible breaker contact surface conditions are illustrated in *Figure 4-69*. Item A illustrates an example of erosion or wear called frosting. This condition results from an open-circuited condenser and is easily recognized by the coarse, crystalline surface and the black "sooty" appearance of the sides of the points. The lack of effective condenser action results in an arc of intense heat being formed each time the points open. This, together with the oxygen in the air, rapidly oxidizes and erodes the platinum surface of the points, producing the coarse, crystalline, or frosted appearance. Properly operating points have a fine-grained, frosted, or silvery appearance and should not be confused with the coarse-grained and sooty point caused by faulty condenser action.

Figure 4-69B and *C* illustrate badly pitted points. In the early stage, these points are identified by a fairly even contact edge and minute pits or pocks in or near the center of the contact surface with an overall smoky appearance. In more advanced stages, the pit may develop into a large, jagged crater, and eventually the entire contact surface takes on a burned, black, and crumpled appearance. Pitted points, as a general rule, are caused by dirt and impurities on the contact surfaces. If points are excessively pitted, a new breaker assembly must be installed.

Figure 4-69E illustrates a built-up point that can be recognized by the mound of metal that has been transferred from one point to another. Buildup, like the other conditions mentioned, results primarily from the transfer of contact material by means of the arc as the points separate. But, unlike the others, there is no burning or oxidation in the process because of the closeness of the pit of one point and the buildup of the other. This condition may result from excessive breaker point spring tension that retards the opening of the points or causes a slow, lazy break. It can also be caused by a poor primary condenser or a loose connection at the primary coil. If excessive buildup has occurred, a new breaker assembly must be installed.

Figure 4-69F illustrates oily points that can be recognized by their smoked and smudged appearance and by the lack of any of the previously mentioned irregularities. This condition may be the result of excessive cam lubrication or of oil vapors that may come from within or outside the magneto. A smoking or fuming engine, for example, could produce the oil vapors. These vapors then enter the magneto through the magneto ventilator and pass between and around the points. These conductive vapors produce arcing and burning on the

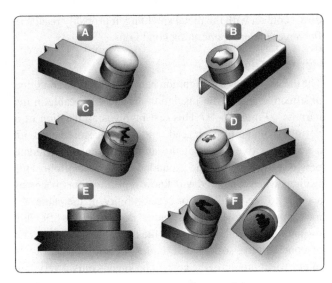

Figure 4-69. *Examples of contact surface conditions.*

contact surfaces. The vapors also adhere to the other surfaces of the breaker assembly and form the sooty deposit. If so, install new breaker assembly.

Dielectric Inspection

Another phase of magneto inspection is the dielectric inspection. This inspection is a visual check for cleanliness and cracks. If inspection reveals that the coil cases, condensers, distributor rotor, or blocks are oily or dirty or have any trace of carbon tracking, they require cleaning and possibly waxing to restore their dielectric qualities.

Clean all accessible condensers and coil cases that contain condensers by wiping them with a lint-free cloth moistened with acetone. Many parts of this type have a protective coating. This protective coating is not affected by acetone, but it may be damaged by scraping or by the use of other cleaning fluids. Never use unapproved cleaning solvents or improper cleaning methods. Also, when cleaning condensers or parts that contain condensers, do not dip, submerge, or saturate the parts in any solution because the solution used may seep inside the condenser and short out the plates.

Coil cases, distributor blocks, distributor rotors, and other dielectric parts of the ignition system are treated with a wax coating when they are new and again at overhaul. The waxing of dielectrics aids their resistance to moisture absorption, carbon tracking, and acid deposits. When these parts become dirty or oily, some of the original protection is lost, and carbon tracking may result.

If any hairline carbon tracks or acid deposits are present on the surface of the dielectric, immerse the part in approved cleaning solvent and scrub it vigorously with a stiff bristle brush. When the carbon track or acid deposits have been removed, wipe the part with a clean, dry cloth to remove all traces of the solvent used for cleaning. Then, coat the part with a special ignition-treating wax. After wax treating the part, remove excess wax deposits and reinstall the part in the magneto.

Ignition Harness Maintenance

Although the ignition harness is simple, it is a vital link between the magneto and spark plug. Because the harness is mounted on the engine and exposed to the atmosphere, it is vulnerable to heat, moisture, and the effects of changing altitude. These factors, plus aging insulation and normal gap erosion, work against efficient engine operation. The insulation may break down on a wire inside the harness and allow the high-voltage to leak through the insulation to the harness shielding instead of going to the spark plug. Open circuits may result from broken wires or poor connections. A bare wire may be in physical contact with the shielding, or two wires may be shorted together.

Any serious defect in an individual lead prevents the high-tension impulse from reaching the spark plug to which the lead is connected. As a result, this plug will not fire. When only one spark plug is firing in a cylinder, the charge is not consumed as quickly as it would be if both plugs were firing. This factor causes the peak pressure of combustion to occur later on in the power stroke. If the peak pressure in the cylinder occurs later, a loss of power in that cylinder results. However, the power loss from a single cylinder becomes a minor factor when the effects of a longer burning time is considered. A longer burning time overheats the affected cylinder, causing detonation, possible pre-ignition, and perhaps permanent damage to the cylinder.

High-Tension Ignition Harness Faults

Perhaps the most common and most difficult high-tension ignition system faults to detect are high-voltage leaks. This is leakage from the core conductor through insulation to the ground of the shielded manifold. A certain small amount of leakage exists even in brand new ignition cable during normal operation. Various factors combine to produce first a high rate of leakage and then complete breakdown. Of these factors, moisture in any form is probably the worst.

Under high-voltage stress, an arc forms and burns a path across the insulator where the moisture exists. If there is gasoline, oil, or grease present, it breaks down and forms carbon. The burned path is called a carbon track, since it is actually a path of carbon particles. With some types of insulation, it may be possible to remove the carbon track and restore the insulator to its former useful condition. This is generally true of porcelain, ceramics, and some of the plastics because these materials are not hydrocarbons and any carbon track forming on them is the result of a dirt film that can be wiped away.

Differences in location and amount of leakage produce different indications of malfunction during engine operation. Indications are generally misfiring or crossfiring. The indication may be intermittent, changing with manifold pressure or with climate conditions. An increase in manifold pressure increases the compression pressure and the resistance of the air across the air gap of the spark plugs. An increase in the resistance at the air gap opposes the spark discharge and produces a tendency for the spark to discharge at some weak point in the insulation. A weak spot in the harness may be aggravated by moisture collecting in the harness manifold. With moisture present, continued engine operation causes the intermittent faults to become permanent carbon tracks. Thus, the first indication of ignition harness unserviceability may be engine misfiring or roughness caused by partial leakage of the ignition voltage.

Figure 4-70 demonstrates four faults that may occur. Fault A shows a short from one cable conductor to another. This fault usually causes misfiring, since the spark is short circuited to a plug in a cylinder where the cylinder pressure is low. Fault B illustrates a cable with a portion of its insulation scuffed away. Although the insulation is not completely broken down, more than normal leakage exists, and the spark plug to which this cable is connected may be lost during takeoff when the manifold pressure is quite high. Fault C is the result of condensation collecting in the lowest portion of the ignition manifold. This condensation may completely evaporate during engine operation, but the carbon track that is formed by the initial flashover remains to allow continued flashover whenever high manifold pressure exists. Fault D may be caused by a flaw in the insulation or the result of a weak spot in the insulation that is aggravated by the presence of moisture. However, since the carbon track is in direct contact with the metal shielding, it probably results in flashover under all operating conditions.

Harness Testing

The electrical test of the ignition harness checks the condition or effectiveness of the insulation around each cable in the harness. *[Figure 4-71]* This test involves application of a definite voltage to each lead, and then measurement with a very sensitive meter of the amount of current leakage between the lead and the grounded harness manifold. This reading, when compared with known specifications, becomes a guide to the condition or serviceability of the cable. As mentioned earlier, there is a gradual deterioration of flexible insulating material. When new, the insulation has a low rate of conductivity; so low that, under several thousand volts of electrical pressure, the current leakage is only a very few millionths of an ampere. Natural aging causes an extremely slow, but certain, change in the resistance of insulating material, allowing an ever-increasing rate of current leakage. The procedures for testing ignition harness and leads were discussed earlier in this chapter.

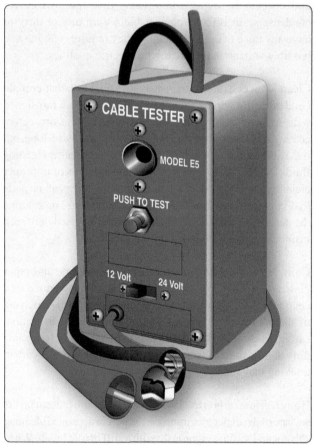

Figure 4-71. *Harness tester.*

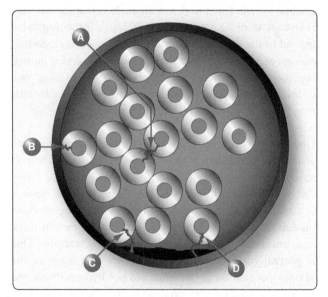

Figure 4-70. *Cross section of an ignition harness.*

Turbine Engine Ignition Systems

Since turbine ignition systems are operated mostly for a brief period during the engine-starting cycle, they are, as a rule, more trouble-free than the typical reciprocating engine ignition system. The turbine engine ignition system does not need to be timed to spark during an exact point in the operational cycle. It is used to ignite the fuel in the combustor and then it is switched off. Other modes of turbine ignition system operation, such as continuous ignition that is used at a lower voltage and energy level, are used for certain flight conditions.

Continuous ignition is used in case the engine were to flame out. This ignition could relight the fuel and keep the engine from stopping. Examples of critical flight modes that use continuous ignition are takeoff, landing, and some abnormal and emergency situations.

Most gas turbine engines are equipped with a high-energy, capacitor-type ignition system and are air cooled by fan airflow. Fan air is ducted to the exciter box, and then flows around the igniter lead and surrounds the igniter before flowing back into the nacelle area. Cooling is important when continuous ignition is used for some extended period of time. Gas turbine engines may be equipped with an electronic-type ignition system, which is a variation of the simpler capacitor-type system.

The typical turbine engine is equipped with a capacitor-type, or capacitor discharge, ignition system consisting of two identical independent ignition units operating from a common low-voltage (DC) electrical power source: the aircraft battery, 115AC, or its permanent magnet generator. The generator is turned directly by the engine through the accessory gear box and produces power any time the engine is turning. The fuel in turbine engines can be ignited readily in ideal atmospheric conditions, but since they often operate in the low temperatures of high altitudes, it is imperative that the system be capable of supplying a high heat intensity spark. Thus, a high-voltage is supplied to arc across a wide igniter spark gap, providing the ignition system with a high degree of reliability under widely varying conditions of altitude, atmospheric pressure, temperature, fuel vaporization, and input voltage.

A typical ignition system includes two exciter units, two transformers, two intermediate ignition leads, and two high-tension leads. Thus, as a safety factor, the ignition system is actually a dual system designed to fire two igniter plugs. *[Figure 4-72]*

Figure 4-73 is a functional schematic diagram of a typical older style capacitor-type turbine ignition system. A 24-volt

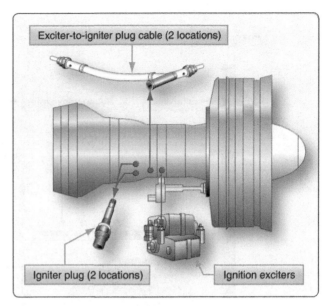

Figure 4-72. *Turbine ignition system components.*

DC input voltage is supplied to the input receptacle of the exciter unit. Before the electrical energy reaches the exciter unit, it passes through a filter that prevents noise voltage from being induced into the aircraft electrical system. The low-voltage input power operates a DC motor that drives one multilobe cam and one single-lobe cam. At the same time, input power is supplied to a set of breaker points that are actuated by the multilobe cam.

From the breaker points, a rapidly interrupted current is delivered to an auto transformer. When the breaker closes, the flow of current through the primary winding of the transformer establishes a magnetic field. When the breaker opens, the flow of current stops, and the collapse of the field induces a voltage in the secondary of the transformer. This voltage causes a pulse of current to flow into the storage capacitor through the rectifier, which limits the flow to a single direction. With repeated pulses, the storage capacitor assumes a charge, up to a maximum of approximately 4 joules. (Note: 1 joule per second equals 1 watt.) The storage capacitor is connected to the spark igniter through the triggering transformer and a contactor, normally open.

When the charge on the capacitor has built up, the contactor is closed by the mechanical action of the single-lobe cam. A portion of the charge flows through the primary of the triggering transformer and the capacitor connected with it. This current induces a high-voltage in the secondary, which ionizes the gap at the spark igniter.

When the spark igniter is made conductive, the storage capacitor discharges the remainder of its accumulated energy along with the charge from the capacitor in series with the primary of the triggering transformer. The spark rate at the

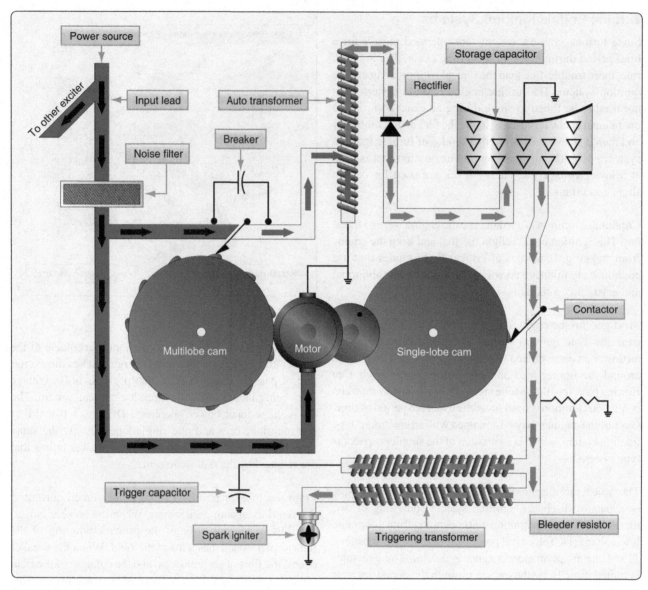

Figure 4-73. *Capacitor type ignition system schematic.*

spark igniter varies in proportion to the voltage of the DC power supply that affects the rpm of the motor. However, since both cams are geared to the same shaft, the storage capacitor always accumulates its store of energy from the same number of pulses before discharge. The employment of the high-frequency triggering transformer, with a low-reactance secondary winding, holds the time duration of the discharge to a minimum. This concentration of maximum energy in minimum time achieves an optimum spark for ignition purposes, capable of blasting carbon deposits and vaporizing globules of fuel.

All high-voltage in the triggering circuits is completely isolated from the primary circuits. The complete exciter is hermetically sealed, protecting all components from adverse operating conditions, eliminating the possibility of flashover at altitude due to pressure change. This also ensures shielding

against leakage of high-frequency voltage interfering with the radio reception of the aircraft.

Capacitor Discharge Exciter Unit

This capacity-type system provides ignition for turbine engines. Like other turbine ignition systems, it is required only for starting the engine; once combustion has begun, the flame is continuous. *[Figure 4-74]*

The energy is stored in capacitors. Each discharge circuit incorporates two storage capacitors; both are located in the exciter unit. The voltage across these capacitors is stepped up by transformer units. At the instant of igniter plug firing, the resistance of the gap is lowered sufficiently to permit the larger capacitor to discharge across the gap. The discharge of the second capacitor is of low-voltage, but of very high

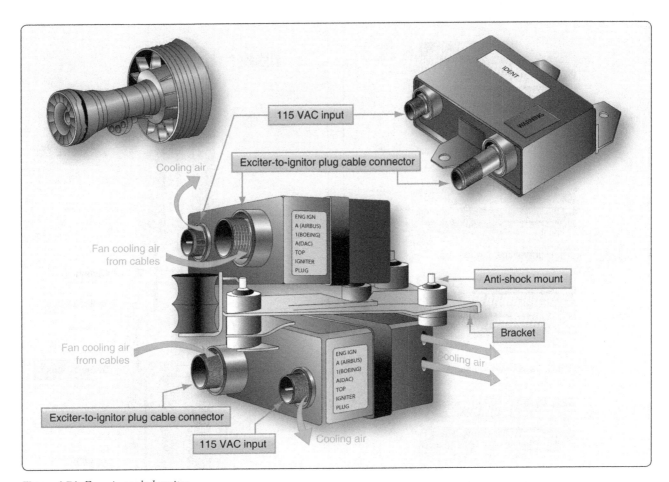

Figure 4-74. *Fan air-cooled exciter.*

energy. The result is a spark of great heat intensity, capable of not only igniting abnormal fuel mixtures but also burning away any foreign deposits on the plug electrodes.

The exciter is a dual unit that produces sparks at each of the two igniter plugs. A continuous series of sparks is produced until the engine starts. The power is then cut off, and the plugs do not fire while the engine is operating other than on continuous ignition for certain flight conditions. This is why the exciters are air cooled to prevent overheating during long use of continuous ignition.

Igniter Plugs

The igniter plug of a turbine engine ignition system differs considerably from the spark plug of a reciprocating engine ignition system. *[Figure 4-75]* Its electrode must be capable of withstanding a current of much higher energy than the electrode of a conventional spark plug. This high energy current can quickly cause electrode erosion, but the short periods of operation minimize this aspect of igniter maintenance. The electrode gap of the typical igniter plug is designed much larger than that of a spark plug since the operating pressures are much lower and the spark can arc more easily than in a spark plug. Finally, electrode fouling,

common to the spark plug, is minimized by the heat of the high-intensity spark.

Figure 4-76 is a cutaway illustration of a typical annular-gap igniter plug, sometimes referred to as a long reach igniter because it projects slightly into the combustion chamber liner to produce a more effective spark.

Another type of igniter plug, the constrained-gap plug, is used in some types of turbine engines. *[Figure 4-77]* It operates at a much cooler temperature because it does not project into the combustion-chamber liner. This is possible because the spark does not remain close to the plug, but arcs beyond the face of the combustion chamber liner.

Turbine Ignition System Inspection & Maintenance

Maintenance of the typical turbine engine ignition system consists primarily of inspection, test, troubleshooting, removal, and installation.

Inspection

Inspection of the ignition system normally includes the following:

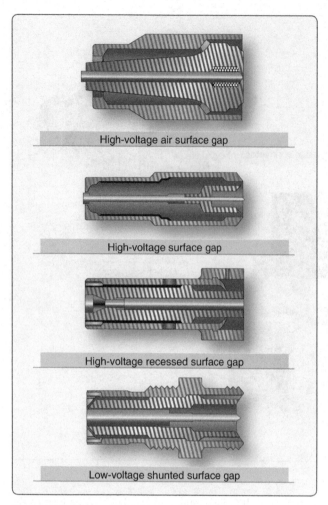

Figure 4-75. *Igniter plugs.*

- Ignition lead terminal inspection; ceramic terminal should be free of arcing, carbon tracking and cracks.

- The grommet seal should be free of flashover and carbon tracking. *[Figure 4-78]*

- The wire insulation should remain flexible with no evidence of arcing through the insulation.

- Inspect the complete system for security of component mounting, shorts or high-voltage arcing, and loose connections.

Check System Operation

The igniter can be checked by listening for a snapping noise as the engine begins to turn, driven by the starter. Though the following procedure is not common practice and should only be used when the maintenance manual suggests it as an alternative method, the igniter can also be checked by removing it and activating the start cycle, noting the spark across the igniter.

Caution: The high energy level and voltage associated with turbine ignition systems can cause injury or death to personnel coming into contact with the activated system.

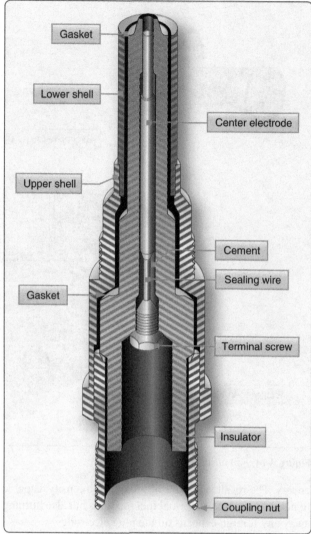

Figure 4-76. *Typical annular gap igniter plug.*

Repair

Tighten and secure as required and replace faulty components and wiring. Secure, tighten, and safety as required.

Removal, Maintenance, & Installation of Ignition System Components

The following instructions are typical procedures suggested by many gas turbine manufacturers. These instructions are applicable to the engine ignition components. Always consult the applicable manufacturer's instructions before performing any ignition system maintenance.

Ignition System Leads

1. Remove clamps securing ignition leads to engine.

2. Remove safety wire and disconnect electrical connectors from exciter units.

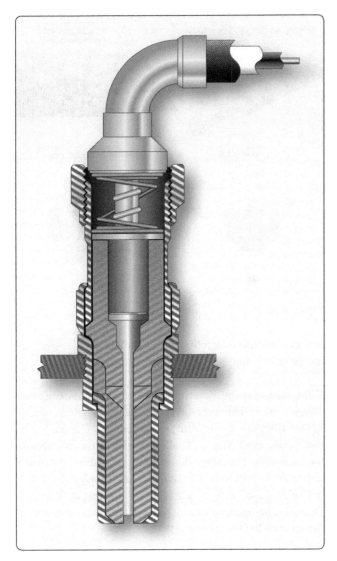

Figure 4-77. *Constrained gap igniter plug.*

3. Remove safety wire and disconnect lead from igniter plug.

4. Discharge any electrical charge stored in the system by grounding and remove ignition leads from engine.

5. Clean leads with approved dry cleaning solvent.

6. Inspect connectors for damaged threads, corrosion, cracked insulators, and bent or broken connector pins.

7. Inspect leads for worn or burned areas, deep cuts, fraying, and general deterioration.

8. Perform continuity check of ignition leads.

9. Reinstall leads, reversing the removal procedure.

Igniter Plugs

1. Disconnect ignition leads from igniter plugs. A good procedure to perform before disconnecting the ignition lead is to disconnect the low-voltage primary lead

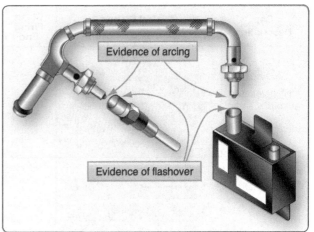

Figure 4-78. *Flashover inspection.*

from the ignition exciter unit and wait at least one minute to permit the stored energy to dissipate before disconnecting the high-voltage cable from the igniter.

2. Remove igniter plugs from mounts.

3. Inspect igniter plug gap surface material. Before inspection, remove residue from the shell exterior using a dry cloth. Do not remove any deposits or residue from the firing end of the low-voltage igniters. High-voltage igniters can have the firing end cleaned to aid in inspection. *[Figure 4-79]*

4. Inspect for fretting of igniter plug shank.

5. Replace an igniter plug whose surface is granular, chipped, or otherwise damaged.

6. Replace dirty or carbonized igniter plugs.

7. Install igniter plugs in mounting pads.

8. Check for proper clearance between chamber liner and igniter plug.

9. Tighten igniter plugs to manufacturer's specified torque.

10. Safety wire igniter plugs.

Powerplant Electrical Systems

The satisfactory performance of any modern aircraft depends to a great degree on the continuing reliability of electrical systems and subsystems. Improperly or carelessly installed or maintained wiring can be a source of both immediate and potential danger. The continued proper performance of electrical systems depends upon the knowledge and technique of the mechanic who installs, inspects, and maintains the electrical wire and cable of the electrical systems.

The procedures and practices outlined in this section are general recommendations and are not intended to replace the manufacturer's instructions in approved practices.

Gap Description	Typical Firing End Configuration	Clean Firing End
High-voltage air surface gap		Yes
High-voltage surface gap		Yes
High-voltage recessed surface gap		Yes
Low-voltage shunted surface gap		No

Figure 4-79. *Firing end cleaning.*

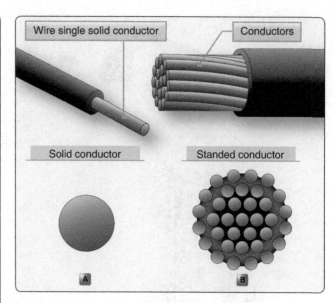

Figure 4-80. *Two types of aircraft wire.*

and the smallest is number 40. Larger and smaller sizes are manufactured but are not commonly used.

Wire size may be determined by using a wire gauge. *[Figure 4-82]* This type of gauge measures wires ranging in size from number 0 (zero) to number 36. The wire to be measured is inserted in the smallest slot that just accommodates the bare wire. The gauge number corresponding to that slot indicates the wire size. The slot has parallel sides and should not be confused with the semicircular opening at the end of the slot. The opening simply permits the free movement of the wire all the way through the slot.

Gauge numbers are useful in comparing the diameter of wires, but not all types of wire or cable can be accurately measured with a gauge. Large wires are usually stranded to increase their flexibility. In such cases, the total area can be determined by multiplying the area of one strand, usually computed in circular mils (commonly used as a reference to wire size) when diameter or gauge number is known by the number of strands in the wire or cable.

Factors Affecting the Selection of Wire Size

Several factors must be considered in selecting the size of wire for transmitting and distributing electric power. One factor is the allowable power loss (PR loss) in the line. This loss represents electrical energy converted into heat. The use of large conductors reduces the resistance and therefore the PR loss. However, large conductors are more expensive initially than small ones; they are heavier and require more substantial supports.

A second factor is the permissible voltage drop (IR drop) in the line. If the source maintains a constant voltage at

For the purpose of this discussion, a wire is described as a single solid conductor, or a stranded conductor, covered with an insulating material. *[Figure 4-80]* The term "cable," as used in aircraft electrical installations, includes the following:

1. Multiconductor cable—two or more separately insulated conductors in the same jacket.

2. Twisted pair—two or more separately insulated conductors twisted together.

3. Shielded cable—one or more insulated conductors, covered with a metallic braided shield.

4. Radio frequency cable—a single, insulated center conductor with a metallic braided outer conductor. The concentricity of the center conductor and the outer conductor is carefully controlled during manufacture to ensure that they are coaxial.

Wire Size

Wire is manufactured in sizes according to a standard known as the American wire gauge (AWG). The wire diameters become smaller as the gauge numbers become larger. The largest wire size shown in *Figure 4-81* is number 0000,

Cross Section			Ohms per 1,000 ft		
Gauge Number	Diameter (mils)	Circular (mils)	Square inches	25 °C (77 °F)	65 °C (149 °F)
0000	460.0	212,000.0	0.166	0.0500	0.0577
000	410.0	168,000.0	0.132	0.0630	0.0727
00	365.0	133,000.0	0.105	0.0795	0.0917
0	325.0	106,000.0	0.0829	0.100	0.166
1	289.0	83,700.0	0.0657	0.126	0.146
2	258.0	66,400.0	0.0521	0.159	0.184
3	229.0	52,600.0	0.0413	0.201	0.232
4	204.0	41,700.0	0.0328	0.253	0.292
5	182.0	33,100.0	0.0260	0.319	0.369
6	162.0	26,300.0	0.0206	0.403	0.465
7	144.0	20,800.0	0.0164	0.508	0.586
8	128.0	16,500.0	0.0130	0.641	0.739
9	114.0	13,100.0	0.0103	0.808	0.932
10	102.0	10,400.0	0.00815	1.02	1.18
11	91.0	8,230.0	0.00647	1.28	1.48
12	81.0	6,530.0	0.00513	1.62	1.87
13	72.0	5,180.0	0.00407	2.04	2.36
14	64.0	4,110.0	0.00323	2.58	2.97
15	57.0	3,260.0	0.00256	3.25	3.75
16	51.0	2,580.0	0.00203	4.09	4.73
17	45.0	2,050.0	0.00161	5.16	5.96
18	40.0	1,620.0	0.00128	6.51	7.51
19	36.0	1,290.0	0.00101	8.21	9.48
20	32.0	1,020.0	0.000802	10.40	11.90
21	28.5	810.0	0.000636	13.10	15.10
22	25.3	642.0	0.000505	16.50	19.00
23	22.6	509.0	0.000400	20.80	24.00
24	20.1	404.0	0.000317	26.20	30.20
25	17.9	320.0	0.000252	33.00	38.10
26	15.9	254.0	0.000200	41.60	48.00
27	14.2	202.0	0.000158	52.50	60.60
28	12.6	160.0	0.000126	66.20	76.40
29	11.3	127.0	0.0000995	83.40	96.30
30	10.0	101.0	0.0000789	105.00	121.00
31	8.9	79.7	0.0000626	133.00	153.00
32	8.0	63.2	0.0000496	167.00	193.00
33	7.1	50.1	0.0000394	211.00	243.00
34	6.3	39.8	0.0000312	266.00	307.00
35	5.6	31.5	0.0000248	335.00	387.00
36	5.0	25.0	0.0000196	423.00	488.00
37	4.5	19.8	0.0000156	533.00	616.00
38	4.0	15.7	0.0000123	673.00	776.00
39	3.5	12.5	0.0000098	848.00	979.00
40	3.1	9.9	0.0000078	1.070.00	1,230.00

Figure 4-81. *American wire gauge for standard annealed solid copper wire.*

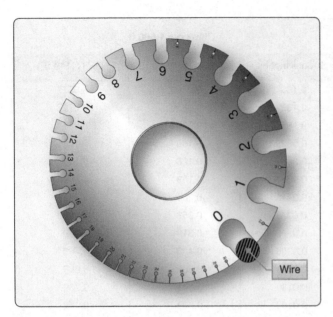

Figure 4-82. *Wire gauge.*

the input to the line, any variation in the load on the line causes a variation in line current and a consequent variation in the IR drop in the line. A wide variation in the IR drop in the line causes poor voltage regulation at the load. The obvious remedy is to reduce either current or resistance. A reduction in load current lowers the amount of power being transmitted, whereas a reduction in line resistance increases the size and weight of conductors required. A compromise is generally reached whereby the voltage variation at the load is within tolerable limits and the weight of line conductors is not excessive.

A third factor is the current carrying ability of the conductor. When current is drawn through the conductor, heat is generated. The temperature of the wire rises until the heat radiated, or otherwise dissipated, is equal to the heat generated by the passage of current through the line. If the conductor is insulated, the heat generated in the conductor is not so readily removed as it would be if the conductor were not insulated. Thus, to protect the insulation from too much heat, the current through the conductor must be maintained below a certain value.

When electrical conductors are installed in locations where the ambient temperature is relatively high, the heat generated by external sources constitutes an appreciable part of the total conductor heating. Allowance must be made for the influence of external heating on the allowable conductor current, and each case has its own specific limitations. The maximum allowable operating temperature of insulated conductors varies with the type of conductor insulation being used.

Tables are available that list the safe current ratings for various sizes and types of conductors covered with various types of insulation. The chart in *Figure 4-83* shows the current carrying capacity and resistance of copper wire continuous duty wire in bundles at various temperature ratings.

Factors Affecting Selection of Conductor Material

Although silver is the best conductor, its cost limits its use to special circuits where a substance with high conductivity is needed. The two most generally used conductors are copper and aluminum. Each has characteristics that make its use advantageous under certain circumstances; also, each has certain disadvantages.

Copper has a higher conductivity; it is more ductile, can be drawn out, has relatively high tensile strength, and can be easily soldered. It is more expensive and heavier than aluminum.

Although aluminum has only about 60 percent of the conductivity of copper, it is used extensively. Its light weight makes possible long spans, and its relatively large diameter for a given conductivity reduces corona, the discharge of electricity from the wire when it has a high potential. The discharge is greater when smaller diameter wire is used than when larger diameter wire is used. Some bus bars are made of aluminum which has a greater radiating surface than copper for the same conductance. The characteristics of copper and aluminum are compared in *Figure 4-84*.

Voltage Drop in Aircraft Wire & Cable

The voltage drop in the main power cables from the aircraft generation source or the battery to the bus should not exceed 2 percent of the regulated voltage when the generator is carrying rated current or the battery is being discharged at a 5-minute rate. The 5-minute rate in this case means that the battery should last a minimum of 5 minutes in an emergency, with all battery operated equipment running. *Figure 4-85* shows the recommended maximum voltage drop in the load circuits between the bus and the utilization equipment.

The resistance of the current return path through the aircraft structure is always considered negligible. However, this is based on the assumption that adequate bonding of the structure or a special electric current return path has been provided that is capable of carrying the required electric current with a negligible voltage drop. A resistance measurement of 0.005 ohms from ground point of the generator or battery to ground terminal of any electrical device is considered satisfactory.

Another satisfactory method of determining circuit resistance is to check the voltage drop across the circuit. If the voltage drop does not exceed the limit established by the aircraft or product manufacturer, the resistance value for the circuit is considered satisfactory. When using the voltage drop method of checking a circuit, the input voltage must be maintained at a constant value.

Wire Size	Continuous Duty Current (Amps)-Wires in Bundles, Groups, Harnesses, or Conduits (See Note #1)			Max. Resistance ohms/1,000 ft@20 °C Tin Plated Conductor (See Note #2)	Nominal Conductor Area (circ.mils)
	Wire Conductor Temperature Rating				
	105 °C	150 °C	200 °C		
24	2.5	4	5	28.40	475
22	3	5	6	16.20	755
20	4	7	9	9.88	1,216
18	6	9	12	6.23	1,900
16	7	11	14	4.81	2,426
14	10	14	18	3.06	3,831
12	13	19	25	2.02	5,874
10	17	26	32	1.26	9,354
8	38	57	71	0.70	16,983
6	50	76	97	0.44	26,818
4	68	103	133	0.28	42,615
2	95	141	179	0.18	66,500
1	113	166	210	0.15	81,700
0	128	192	243	0.12	104,500
00	147	222	285	0.09	133,000
000	172	262	335	0.07	166,500
0000	204	310	395	0.06	210,900

Note 1:   Rating is for 70 °C ambient, 33 or more wires in the bundle for sizes 24 through 10, and 9 wires for size 8 and larger, with no more than 20 percent of harness current carrying capacity being used, at an operating altitude of 60,000 feet.

Note 2:   For resistance of silver or nickel-plated conductors, see wire specifications.

Figure 4-83. *Current-carrying capacity and resistance of copper wire.*

Characteristic	Copper	Aluminum
Tensile strength (lb/in²)	55,000	25,000
Tensile strength for same conductivity (lb)	55,000	40,000
Weight for same conductivity (lb)	100	48
Cross section for same conductivity (CM)	100	160
Specific resistance (Ω/mil ft)	10.6	17

Figure 4-84. *Characteristics of copper and aluminum.*

Nominal System Voltage	Allowable Voltage Drop	
	Continuous Operation	Intermittent Operation
14	0.5	1
28	1	2
115	4	8
200	7	14

Figure 4-85. *Recommended voltage drop in load circuits.*

The graph in *Figure 4-86* applies to copper conductors carrying direct current. To select the correct size of conductor, two major requirements must be met. First, the size must be sufficient to prevent an excessive voltage drop while carrying the required current over the required distance. Second, the size must be sufficient to prevent overheating of the cable while carrying the required current. The graphs in *Figures 4-86* and *4-87* can simplify these determinations. To use this graph to select the proper size of conductor, the following must be known:

1. The conductor length in feet;

2. The number of amperes of current to be carried;

3. The amount of voltage drop permitted;

4. Whether the current to be carried is intermittent or continuous;

5. The estimated or measured temperature of the conductor;

6. Whether the wire to be installed is in a conduit or in a bundle; and

7. Whether it is a single conductor in free air.

Suppose that you want to install a 50-foot conductor from the aircraft bus to the equipment in a 28-volt system. For this length, a 1-volt drop is permissible for continuous operation with a conductor temperature of 20 °C or less. By referring to the chart in *Figure 4-86*, the maximum number of feet a conductor may be run carrying a specified current with a 1-volt drop can be determined. In this example, the number 50 is selected.

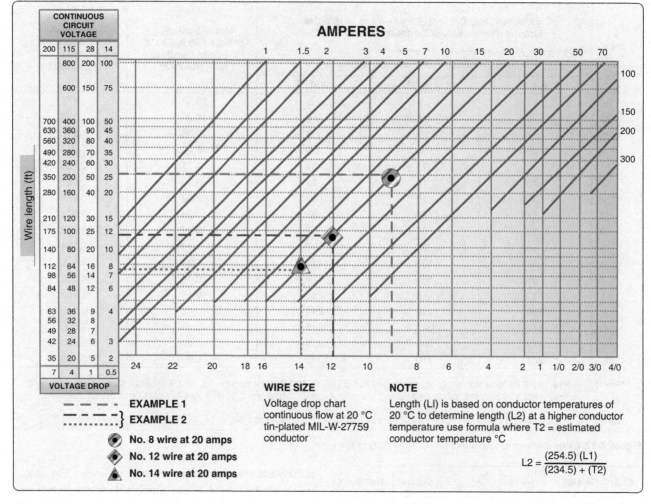

Figure 4-86. *Conductor graph—continuous flow.*

Assuming the current required by the equipment is 20 amperes, the line indicating the value of 20 amperes should be selected from the diagonal lines. Follow this diagonal line downward until it intersects the horizontal line number 50. From this point, drop straight down to the bottom of the graph to find that a conductor between size No. 8 and No. 10 is required to prevent a greater drop than 1 volt. Since the indicated value is between two numbers, the larger size, No. 8, should be selected. This is the smallest size that should be used to avoid an excessive voltage drop.

If the installation is for equipment having only an intermittent (maximum 2 minutes) requirement for power, the graph in *Figure 4-87* is used in the same manner.

Conductor Insulation

Two fundamental properties of insulation materials (e.g., rubber, glass, asbestos, and plastic) are insulation resistance and dielectric strength. These are entirely different and distinct properties.

Insulation resistance is the resistance to current leakage through and over the surface of insulation materials. Insulation resistance can be measured with a megger without damaging the insulation. This serves as a useful guide in determining the general condition of insulation. However, the data obtained in this manner may not give a true picture of the condition of the insulation. Clean, dry insulation having cracks or other faults may show a high value of insulation resistance but would not be suitable for use.

Dielectric strength is the ability of the insulator to withstand potential difference and is usually expressed in terms of the voltage at which the insulation fails due to electrostatic stress. Maximum dielectric strength values can be measured by raising the voltage of a test sample until the insulation breaks down.

Because of the expense of insulation, its stiffening effect, and the great variety of physical and electrical conditions under which the conductors are operated, only the necessary minimum insulation is applied for any particular type of cable

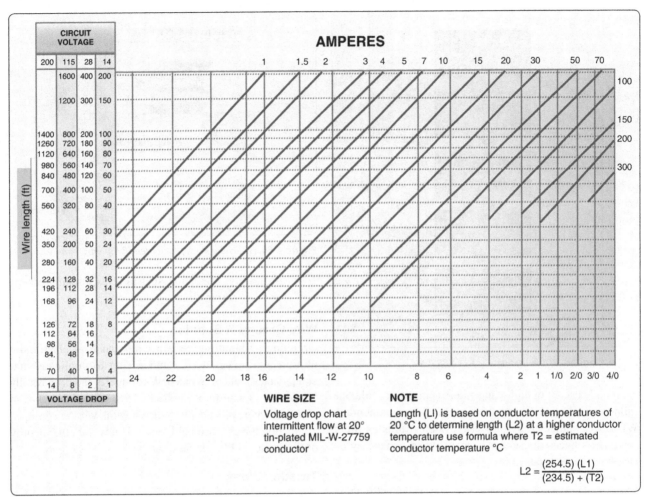

Figure 4-87. *Conductor graph—intermittent flow.*

designed to do a specific job.

The type of conductor insulation material varies with the type of installation. Rubber, silk, and paper insulation are no longer used extensively in aircraft systems. More common today are such materials as vinyl, cotton, nylon, Teflon, and Rockbestos.

Identifying Wire & Cable

To aid in testing and repair operations, many maintenance activities mark wire or cable with a combination of letters and numbers that identify the wire, the circuit it belongs to, the gauge number, and other information necessary to relate the wire or cable to a wiring diagram. Such markings are the identification code.

There is no standard procedure for marking and identifying wiring; each manufacturer normally develops its own identification code. *Figure 4-88* illustrates one identification system and shows the usual spacing in marking a wire. Some system components, especially plugs and jacks, are identified by a letter or group of letters and numbers added to the basic identification number. These letters and numbers

may indicate the location of the component in the system. Interconnected cables are also marked in some systems to indicate location, proper termination, and use. In any system, the marking should be legible, and the stamping color should contrast with the color of the wire insulation. For example, use black stamping with light-colored backgrounds, or white stamping on dark-colored backgrounds.

Most manufacturers mark the wires at intervals of not more than 15 inches lengthwise and within 3 inches of each junction or terminating point. *[Figure 4-89]*

Coaxial cable and wires at terminal blocks and junction boxes are often identified by marking or stamping a wiring sleeve rather than the wire itself. For general purpose wiring, flexible vinyl sleeving, either clear or white opaque, is

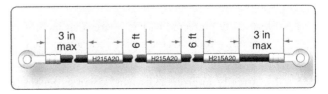

Figure 4-88. *Spacing of printed identification marks.*

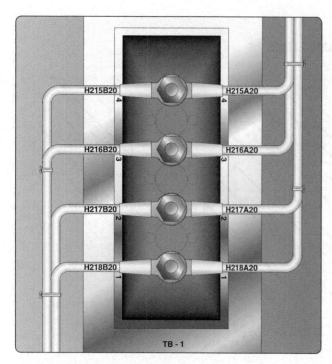

Figure 4-89. *Wire identification at a terminal block.*

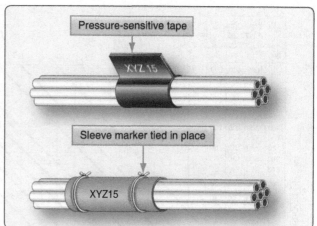

Figure 4-90. *Alternate methods of identifying wire bundles.*

commonly used. For high-temperature applications, silicone rubber or silicone fiberglass sleeving is recommended. Where resistance to synthetic hydraulic fluids or other solvents is necessary, either clear or white opaque nylon sleeving can be used.

While the preferred method is to stamp the identification marking directly on the wire or on sleeving, other methods are often employed. One method uses a marked sleeve tied in place. The other uses a pressure-sensitive tape. *[Figure 4-90]*

Electrical Wiring Installation

The following recommended procedures for installing aircraft electrical wiring are typical of those used on most types of aircraft. For purposes of this discussion, the following definitions are applicable:

1. Open wiring—any wire, wire group, or wire bundle not enclosed in conduit.

2. Wire group—two or more wires in the same location, tied together to identity the group.

3. Wire bundle—two or more wire groups tied together because they are going in the same direction at the point where the tie is located. The bundle facilitates maintenance.

4. Electrically protected wiring—wires that include in the circuit protections against overloading, such as fuses, circuit breakers, or other limiting devices.

5. Electrically unprotected wiring—wires, generally from generators to main bus distribution points, that

do not have protection, such as fuses, circuit breakers, or other current-limiting devices.

Wire Groups & Bundles

Grouping or bundling certain wires, such as electrically unprotected power wiring and wiring to duplicate vital equipment, should be avoided. Wire bundles should generally be limited in size to a bundle of 75 wires, or 2 inches in diameter where practicable. When several wires are grouped at junction boxes, terminal blocks, panels, etc., the identity of the group within a bundle can be retained. *[Figure 4-91]*

Twisting Wires

When specified on the engineering drawing, parallel wires must be twisted. The most common examples are:

1. Wiring in the vicinity of magnetic compass or flux valve,

2. Three-phase distribution wiring, and

3. Certain other wires (usually radio wiring).

Twist the wires so that they lie snugly against each other, making approximately the number of twists per foot as listed in *Figure 4-92.* Always check wire insulation for damage after twisting. If the insulation is torn or frayed, replace the wire.

Spliced Connections in Wire Bundles

Spliced connections in wire groups or bundles should be located so that they can be easily inspected. Splices should also be staggered so that the bundle does not become excessively enlarged. *[Figure 4-93]* All noninsulated splices should be covered with plastic, securely tied at both ends.

Slack in Wiring Bundles

Single wires or wire bundles should not be installed with excessive slack. Slack between supports should normally not exceed ½ inch. This is the maximum it should be possible

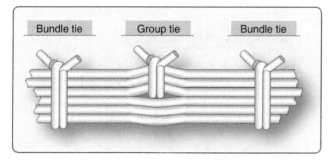

Figure 4-91. *Group and bundle ties.*

	#22	#20	#18	#16	#14	#12	#10	#8	#6	#4
2 Wires	10	10	9	8	7 ½	7	6 ½	6	5	4
3 Wires	10	10	8 ½	7	6 ½	6	5 ½	5	4	3

Figure 4-92. *Recommended number of twists per foot.*

to deflect the wire with normal hand force. However, this may be exceeded if the wire bundle is thin and the clamps are far apart. But the slack should never be so great that the wire bundle can abrade against any surface it touches. *[Figure 4-94]* A sufficient amount of slack should be allowed near each end of a bundle to:

1. Permit easy maintenance;

2. Allow replacement of terminals;

3. Relieve mechanical strain on the wires, wire junctions, or supports;

4. Permit free movement of shock and vibration-mounted equipment; and

5. Permit shifting of equipment for purposes of maintenance.

Bend Radii

Bends in wire groups or bundles should not be less than ten times the outside diameter of the wire group or bundle. However, at terminal strips, where wire is suitably supported at each end of the bend, a minimum radius of three times the outside diameter of the wire, or wire bundle, is usually acceptable. There are exceptions to these guidelines in the case of certain types of cable; for example, coaxial cable should never be bent to a smaller radius than six times the outside diameter.

Routing & Installation

All wiring should be installed so that it is mechanically and electrically sound and neat in appearance. Whenever practicable, wires and bundles should be routed parallel with, or at right angles to, the stringers or ribs of the area involved. An exception to this general rule is the coaxial cables, which are routed as directly as possible.

The wiring must be adequately supported throughout its length. A sufficient number of supports must be provided to prevent undue vibration of the unsupported lengths. All wires and wire groups should be routed and installed to protect them from:

1. Chafing or abrasion;

2. High temperature;

3. Being used as handholds, or as support for personal belongings and equipment;

4. Damage by personnel moving within the aircraft;

5. Damage from cargo stowage or shifting;

6. Damage from battery acid fumes, spray, or spillage;

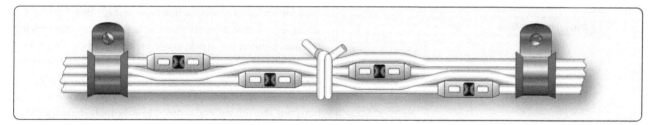

Figure 4-93. *Staggered splices in wire bundle.*

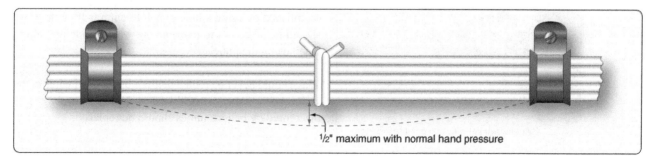

½" maximum with normal hand pressure

Figure 4-94. *Slack in wire bundle between supports.*

and

7. Damage from solvents and fluids.

Protection Against Chafing

Wires and wire groups should be installed so that they are protected against chafing or abrasion in those locations where contact with sharp surfaces or other wires would damage the insulation. Damage to the insulation can cause short circuits, malfunctions, or inadvertent operation of equipment. Cable clamps should be used to support wire bundles at each hole through a bulkhead. *[Figure 4-95]* If wires come closer than ¼ inch to the edge of the hole, a suitable grommet is used in the hole. *[Figure 4-96]*

Sometimes, it is necessary to cut nylon or rubber grommets to facilitate installation. In these instances, after insertion, the grommet can be secured in place with general purpose cement. The slot should be at the top of the hole, and the cut should be made at an angle of 45° to the axis of the wire bundle hole.

Protection Against High Temperature

To prevent insulation deterioration, wires should be kept separate from high-temperature equipment, such as resistors, exhaust stacks, heating ducts. The amount of separation is usually specified by engineering drawings. Some wires must be run through hot areas. These wires must be insulated with high-temperature rated material, such as asbestos, fiberglass, or Teflon. Additional protection is also often required in the form of conduits. A low-temperature insulated wire should never be used to replace a high-temperature insulated wire.

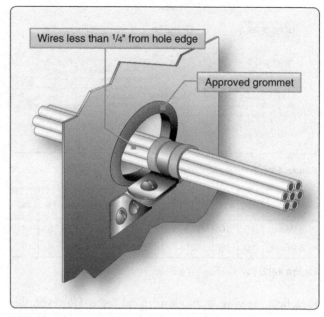

Figure 4-96. *Cable clamp and grommet at bulkhead hole.*

Many coaxial cables have soft plastic insulation, such as polyethylene, that is especially subject to deformation and deterioration at elevated temperatures. All high-temperature areas should be avoided when installing these cables.

Additional abrasion protection should be given to asbestos wires enclosed in conduit. Either conduit with a high temperature rubber liner should be used or asbestos wires can be enclosed individually in high-temperature plastic tubes before being installed in the conduit.

Protection Against Solvents & Fluids

Avoid installing wires in areas where they are subjected to damage from fluids. Wires should not be placed in the lowest four inches of the aircraft fuselage, except those that must terminate in that area. If there is a possibility that wiring without a protective nylon outer jacket may be soaked with fluids, plastic tubing should be used to protect it. This tubing should extend past the exposure area in both directions and should be tied at each end. If the wire has a low point between the tubing ends, provide a ⅛-inch drainage hole. *[Figure 4-97]* This hole should be punched into the tubing after the installation is complete and the low point definitely established by using a hole punch to cut a half circle. Care should be taken not to damage any wires inside the tubing when using the punch.

Wire should never be routed below a battery. All wires in the vicinity of a battery should be inspected frequently. Wires discolored by battery fumes should be replaced.

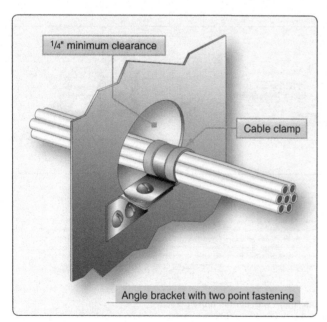

Figure 4-95. *Cable clamp at bulkhead hole.*

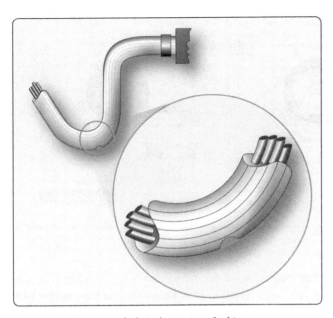

Figure 4-97. *Drainage hole in low point of tubing.*

Protection of Wires in Wheel Well Area

Wires located in wheel wells are subject to many additional hazards, such as exposure to fluids, pinching, and severe flexing in service. All wire bundles should be protected by sleeves of flexible tubing securely held at each end. There should be no relative movement at points where flexible tubing is secured. These wires and the insulating tubing should be inspected carefully at very frequent intervals, and wires or tubing should be replaced at the first sign of wear. There should be no strain on attachments when parts are fully extended, but slack should not be excessive.

Routing Precautions

When wiring must be routed parallel to combustible fluid or oxygen lines for short distances, as much separation as possible should be maintained. The wires should be on a level with, or above, the plumbing lines. Clamps should be spaced so that if a wire is broken at a clamp, it will not contact the line. Where a 6-inch separation is not possible, both the wire bundle and the plumbing line can be clamped to the same structure to prevent any relative motion. If the separation is less than 2 inches but more than ½ inch, two cable clamps back to back can be used to maintain a rigid separation only and not for support of the bundle. *[Figure 4-98]* No wire should be routed so that it is located nearer than ½ inch to a plumbing line, nor should a wire or wire bundle be supported from a plumbing line that carries flammable fluids or oxygen. Wiring should be routed to maintain a minimum clearance of at least 3 inches from control cables. If this cannot be accomplished, mechanical guards should be installed to prevent contact between wiring and control cables.

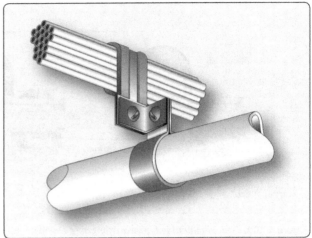

Figure 4-98. *Separation of wires from plumbing lines.*

Installation of Cable Clamps

Cable clamps should be installed with regard to the proper mounting angle. *[Figure 4-99]* The mounting screw should be above the wire bundle. It is also desirable that the back of the cable clamp rest against a structural member where practicable. *Figure 4-100* shows some typical mounting hardware used in installing cable clamps. Be sure that wires are not pinched in cable clamps. Where possible, mount them directly to structural members. *[Figure 4-101]*

Clamps can be used with rubber cushions to secure wire bundles to tubular structures. *[Figure 4-102]* Such clamps must fit tightly but should not be deformed when locked in place.

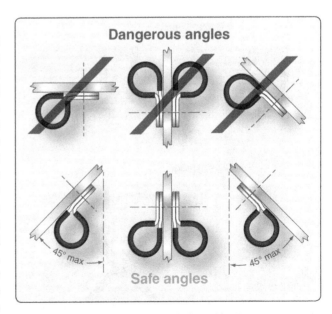

Figure 4-99. *Proper mounting angle for cable clamps.*

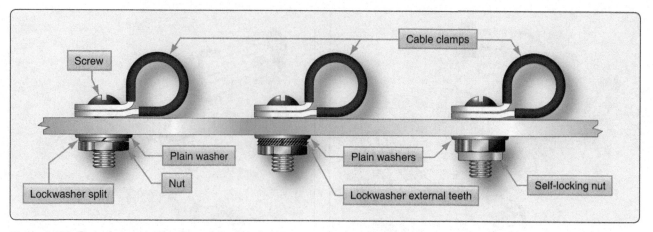

Figure 4-100. *Typical mounting hardware for cable clamps.*

Lacing & Tying Wire Bundles

Wire groups and bundles are laced or tied with cord to provide ease of installation, maintenance, and inspection. This section describes and illustrates recommended procedures for lacing and tying wires with knots that hold tightly under all conditions. For the purposes of this discussion, the following terms are defined:

1. Tying is the securing together of a group or bundle of wires by individual pieces of cord tied around the group or bundle at regular intervals.

2. Lacing is the securing together of a group or bundle of wires by a continuous piece of cord forming loops at regular intervals around the group or bundle.

The material used for lacing and tying is either cotton or nylon cord. Nylon cord is moisture- and fungus-resistant, but cotton cord must be waxed before using to give it these necessary protective characteristics.

Single-Cord Lacing

Figure 4-103 shows the steps in lacing a wire bundle with a single cord. The lacing procedure is started at the thick end of the wire group or bundle with a knot consisting of a clove hitch with an extra loop. The lacing is then continued at regular intervals with half hitches along the wire group or bundle and at each point where a wire or wire group branches off. The half hitches should be spaced so that the bundle is neat and secure.

Figure 4-101. *Mounting cable clamps to structure.*

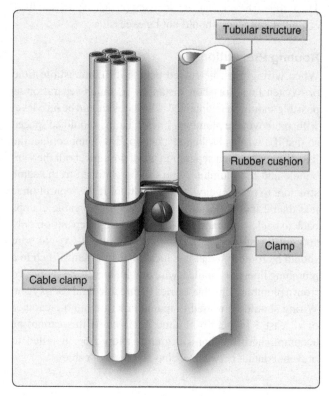

Figure 4-102. *Installing cable clamps to tubular structure.*

The lacing is ended by tying a knot consisting of a clove hitch with an extra loop. After the knot is tied, the free ends of the lacing cord should be trimmed to approximately ⅜ inch.

Double-Cord Lacing

Figure 4-104 illustrates the procedure for double-cord lacing. The lacing is started at the thick end of the wire group or bundle with a bowline-on-a-bight knot. *[Figure 4-104A]* At regular intervals along the wire group or bundle, and at each point where a wire branches off, the lacing is continued using half hitches, with both cords held firmly together. The half hitches should be spaced so that the group or bundle is neat and secure. The lacing is ended with a knot consisting of a half hitch, continuing one of the cords clockwise and the other counterclockwise and then tying the cord ends with a square knot. The free ends of the lacing cord should be trimmed to approximately ⅜ inch.

Lacing Branch-Offs

Figure 4-105 illustrates a recommended procedure for lacing a wire group that branches off the main wire bundle. The branch-off lacing is started with a knot located on the main bundle just past the branch-off point. Continue the lacing along the branched-off wire group using regularly spaced half hitches. If a double cord is used, both cords should be held snugly together. The half hitches should be spaced to lace the bundle neatly and securely. End the lacing with the regular terminal knot used in single- or double-cord lacing, as applicable, and trim the free ends of the lacing cord neatly.

Tying

All wire groups or bundles should be tied where supports are more than 12 inches apart. Ties are made using waxed cotton cord, nylon cord, or fiberglass cord. Some manufacturers permit the use of pressure-sensitive vinyl electrical tape. When permitted, the tape should be wrapped three turns around the bundle and the ends heat sealed to prevent unwinding of the tape. *Figure 4-106* illustrates a recommended procedure for tying a wire group or bundle. The tie is started by wrapping the cord around the wire group to tie a clove-hitch knot. Then, a square knot with an extra loop is tied and the free ends of the cord trimmed.

Temporary ties are sometimes used in making up and installing wire groups and bundles. Colored cord is normally used to make temporary ties, since they are removed when the installation is complete.

Whether lacing or tying, bundles should be secured tightly enough to prevent slipping, but not so tightly that the cord cuts into or deforms the insulation. This applies especially to coaxial cable, which has a soft dielectric insulation between the inner and outer conductor. Coaxial cables have been damaged by the use of lacing materials or by methods of lacing or tying wire bundles that cause a concentrated force on the cable insulation. Elastic lacing materials, small-diameter lacing cord, and excessive tightening deform the interconductor insulation and result in short circuits or impedance changes. Flat nylon braided waxed lacing tape

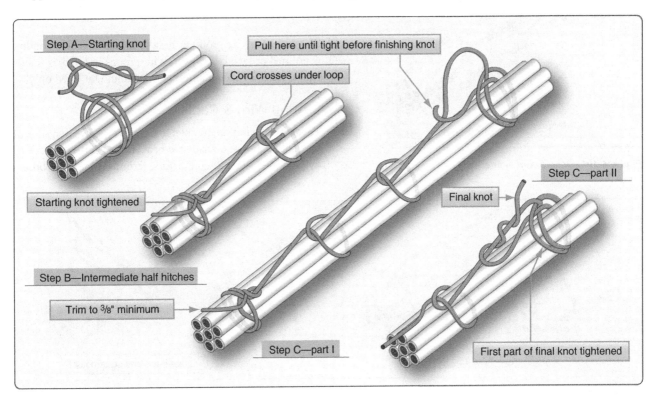

Figure 4-103. *Single cord lacing.*

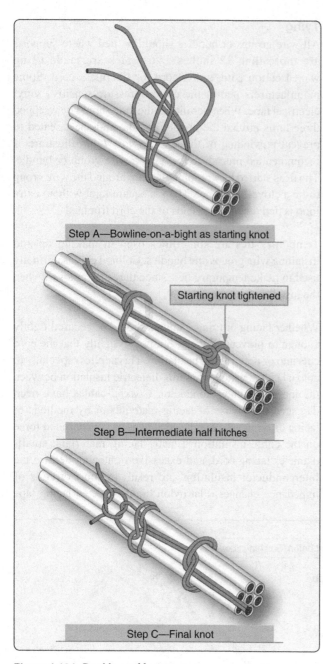

Step A—Bowline-on-a-bight as starting knot

Starting knot tightened

Step B—Intermediate half hitches

Step C—Final knot

Figure 4-104. *Double cord lacing.*

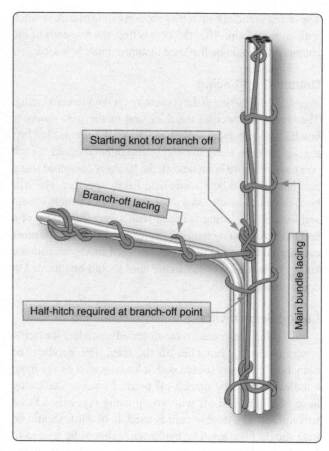

Starting knot for branch off

Branch-off lacing

Half-hitch required at branch-off point

Main bundle lacing

Figure 4-105. *Lacing a branch off.*

should be used for lacing or tying any wire bundles containing coaxial cables.

The part of a wire group or bundle located inside a conduit is not tied or laced; however, wire groups or bundles inside enclosures, such as junction boxes, should be laced only.

Cutting Wire & Cable

To make installation, maintenance, and repair easier, runs of wire and cable in aircraft are broken at specified locations by junctions, such as connectors, terminal blocks, or buses.

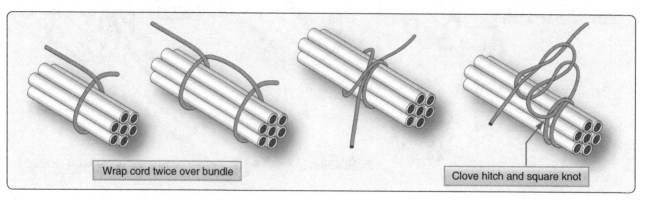

Wrap cord twice over bundle

Clove hitch and square knot

Figure 4-106. *Tying a wire group of bundle.*

Before assembly to these junctions, wires and cables must be cut to length.

All wires and cables should be cut to the lengths specified on drawings and wiring diagrams. The cut should be made clean and square, and the wire or cable should not be deformed. If necessary, large diameter wire should be reshaped after cutting. Good cuts can be made only if the blades of cutting tools are sharp and free from nicks. A dull blade deforms and extrudes wire ends.

Stripping Wire & Cable

Nearly all wire and cable used as electrical conductors are covered with some type of insulation. In order to make electrical connections with the wire, a part of this insulation must be removed to expose the bare conductor. Copper wire can be stripped in a number of ways depending on the size and insulation. *Figure 4-107* lists some types of stripping tools recommended for various wire sizes and types of insulation. Aluminum wire must be stripped using extreme care, since individual strands break very easily after being nicked.

The following general precautions are recommended when stripping any type of wire:

1. When using any type of wire stripper, hold the wire so that it is perpendicular to the cutting blades.

2. Adjust automatic stripping tools carefully; follow the manufacturer's instructions to avoid nicking, cutting, or otherwise damaging strands. This is especially important for aluminum wires and for copper wires smaller than No. 10. Examine stripped wires for damage. Cut off and restrip, if length is sufficient, or reject and replace any wires with more than the allowable number of nicked or broken strands listed in the manufacturer's instructions.

3. Make sure insulation is clean cut with no frayed or ragged edges. Trim, if necessary.

4. Make sure all insulation is removed from stripped area. Some types of wires are supplied with a transparent layer of insulation between the conductor and the primary insulation. If this is present, remove it.

5. When using hand wire strippers to remove lengths of

insulation longer than ¾ inch, it is easier to accomplish in two or more operations.

6. Retwist copper strands by hand or with pliers, if necessary, to restore natural lay and tightness of strands.

A pair of hand wire strippers is shown in *Figure 4-108*. This tool is commonly used to strip most types of wire. The following general procedures describe the steps for stripping wire with a hand stripper. *[Figure 4-109]*

1. Insert wire into exact center of correct cutting slot for wire size to be stripped. Each slot is marked with wire size.

2. Close handles together as far as they will go.

3. Release handles allowing wire holder to return to the open position.

4. Remove stripped wire.

Solderless Terminals & Splices

Splicing of electrical cable should be kept to a minimum and avoided entirely in locations subject to extreme vibrations. Individual wires in a group or bundle can usually be spliced if the completed splice is located where it can be inspected periodically. The splices should be staggered so that the bundle does not become excessively enlarged. Many types of aircraft splice connectors are available for splicing individual wires. Self-insulated splice connectors are usually preferred; however, a noninsulated splice connector can be used if the splice is covered with plastic sleeving secured at both ends. Solder splices may be used, but they are particularly brittle and not recommended.

Electric wires are terminated with solderless terminal lugs to permit easy and efficient connection to and disconnection from terminal blocks, bus bars, or other electrical equipment. Solderless splices join electric wires to form permanent continuous runs. Solderless terminal lugs and splices are made of copper or aluminum and are preinsulated or uninsulated, depending on the desired application.

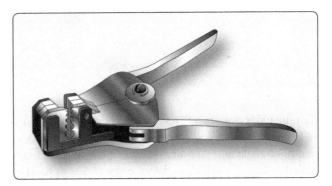

Figure 4-108. *Light duty hand wire strippers.*

Stripper	Wire Size	Insulations
Hot blade	#26–#4	All except asbestos
Rotary, electric	#26–#4	All
Bench	#20–#6	All
Hand pliers	#26–#8	All
Knife	#2–#0000	All

Figure 4-107. *Wire strippers for copper wire.*

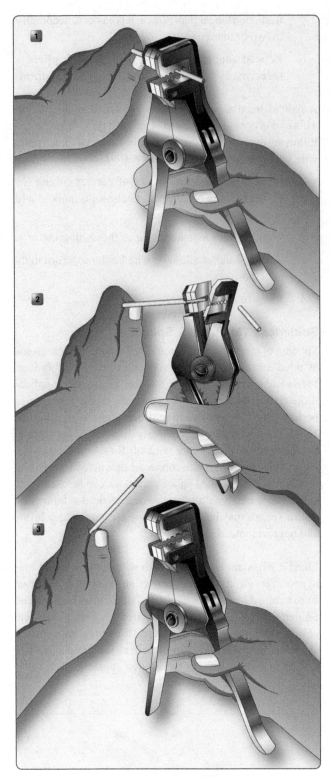

Figure 4-109. *Stripping wire with hand strippers.*

Terminal lugs are generally available in three types for use in different space conditions. These are the flag, straight, and right-angle lugs. Terminal lugs are crimped, sometimes called staked or swaged, to the wires by means of hand or power crimping tools.

Copper Wire Terminals

Copper wires are terminated with solderless, preinsulated straight copper terminal lugs. The insulation is part of the terminal lug and extends beyond its barrel so that it covers a portion of the wire insulation, making the use of an insulation sleeve unnecessary. *[Figure 4-110]*

In addition, preinsulated terminal lugs contain an insulation grip (a metal reinforcing sleeve) beneath the insulation for extra gripping strength on the wire insulation. Preinsulated terminals accommodate more than one size of wire; the insulation is usually color coded to identify the wire sizes that can be terminated with each of the terminal lug sizes.

Crimping Tools

Hand, portable power, and stationary power tools are available for crimping terminal lugs. These tools crimp the barrel of the terminal lug to the conductor and simultaneously crimp the insulation grip to the wire insulation.

Hand crimping tools all have a self-locking ratchet that prevents opening the tool until the crimp is complete. Some hand crimping tools are equipped with a nest of various size inserts to fit different size terminal lugs. Others are used on one terminal lug size only. All types of hand crimping tools are checked by gauges for proper adjustment of crimping jaws.

Figure 4-111 shows a terminal lug inserted into a hand tool. The following general guidelines outline the crimping procedure:

1. Strip the wire insulation to proper length.

2. Insert the terminal lug, tongue first, into the hand tool barrel crimping jaws until the terminal lug barrel butts flush against the tool stop.

3. Insert the stripped wire into the terminal lug barrel until the wire insulation butts flush against the end of the barrel.

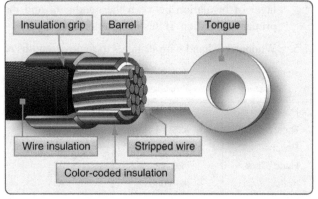

Figure 4-110. *Preinsulated terminal lug.*

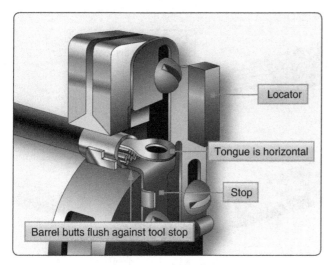

Figure 4-111. *Inserting terminal lug into hand tool.*

4. Squeeze the tool handles until the ratchet releases.

5. Remove the completed assembly and examine it for proper crimp.

Some types of uninsulated terminal lugs are insulated after assembly to a wire by means of pieces of transparent flexible tubing called sleeves. The sleeve provides electrical and mechanical protection at the connection. When the size of the sleeves used is such that it fits tightly over the terminal lug, the sleeves need not be tied; otherwise, it should be tied with lacing cord *[Figure 4-112]*

Aluminum Wire Terminals

Aluminum wire is being used increasingly in aircraft systems because of its weight advantage over copper. However, bending aluminum causes "work hardening" of the metal, making it brittle. This results in failure or breakage of strands much sooner than in a similar case with copper wire. Aluminum also forms a high-resistant oxide

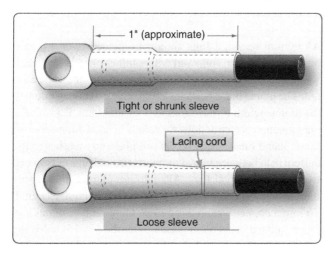

Figure 4-112. *Insulating sleeves.*

film immediately upon exposure to air. To compensate for these disadvantages, it is important to use the most reliable installation procedures. Only aluminum terminal lugs are used to terminate aluminum wires.

All aluminum terminals incorporate an inspection hole that permits checking the depth of wire insertion. *[Figure 4-113]* The barrel of aluminum terminal lugs is filled with a petrolatum-zinc dust compound. This compound removes the oxide film from the aluminum by a grinding process during the crimping operation. The compound also minimizes later oxidation of the completed connection by excluding moisture and air. The compound is retained inside the terminal lug barrel by a plastic or foil seal at the end of the barrel.

Splicing Copper Wires Using Preinsulated Wires

Preinsulated permanent copper splices join small wires of sizes 22 through 10. Each splice size can be used for more than one wire size. Splices are usually color coded in the same manner as preinsulated, small copper terminal lugs. Some splices are insulated with white plastic. Splices are also used to reduce wire sizes *[Figure 4-114]*

Crimping tools are used to accomplish this type of splice. The crimping procedures are the same as those used for terminal lugs, except that the crimping operation must be done twice, one for each end of the splice.

Emergency Splicing Repairs

Broken wires can be repaired by means of crimped splices, by using terminal lugs from which the tongue has been cut off, or by soldering together and potting broken strands. These repairs are applicable to copper wire. Damaged aluminum wire must not be temporarily spliced. These repairs are for temporary emergency use only and should be replaced as soon as possible with permanent repairs. Since some manufacturers prohibit splicing, the applicable manufacturer's instructions should always be consulted.

Splicing with Solder & Potting Compound

When neither a permanent splice nor a terminal lug is available, a broken wire can be repaired as follows *[Figure 4-115]*:

1. Install a piece of plastic sleeving about 3 inches long and of the proper diameter to fit loosely over the insulation on one piece of the broken wire.

2. Strip approximately 1½ inches from each broken end of the wire.

3. Lay the stripped ends side by side and twist one wire around the other with approximately four turns.

4. Twist the free end of the second wire around the first

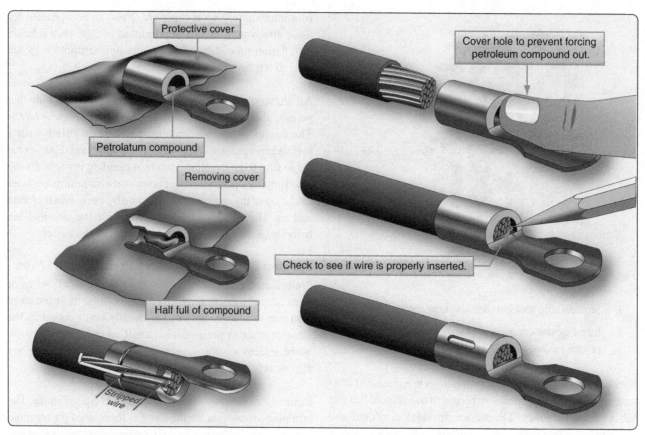

Figure 4-113. *Inserting aluminum wire into aluminum terminal lugs.*

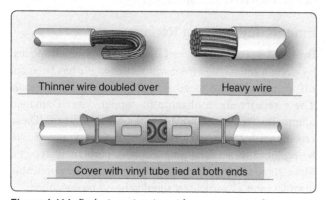

Figure 4-114. *Reducing wire size with a permanent splice.*

wire with approximately four turns. Solder the wire turns together using ⁶⁰⁄₄₀ tin-lead resin-core solder.

5. When solder is cool, draw the sleeve over the soldered wires and tie at one end. If potting compound is available, fill the sleeve with potting material and tie securely.

6. Allow the potting compound to set without touching for 4 hours. Full cure and electrical characteristics are achieved in 24 hours.

Connecting Terminal Lugs to Terminal Blocks

Terminal lugs should be installed on terminal blocks in such a manner that they are locked against movement in the direction of loosening. *[Figure 4-116]*

Terminal blocks are normally supplied with studs secured in place by a plain washer, an external tooth lockwasher, and a nut. In connecting terminals, a recommended practice is to place copper terminal jugs directly on top of the nut, followed with a plain washer and elastic stop nut, or with a plain washer, split steel lockwasher, and plain nut.

Aluminum terminal lugs should be placed over a plated brass plain washer, followed with another plated brass plain washer, split steel lockwasher, and plain nut or elastic stop nut. The plated brass washer should have a diameter equal to the tongue width of the aluminum terminal lug. Consult the manufacturer's instructions for recommended dimensions of these plated brass washers. Do not place any washer in the current path between two aluminum terminal lugs or between two copper terminal lugs. Also, do not place a lockwasher directly against the tongue or pad of the aluminum terminal. To join a copper terminal lug to an aluminum terminal lug, place a plated brass plain washer over the nut that holds the stud in place; follow with the aluminum terminal lug, a plated

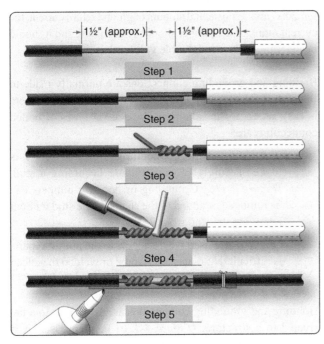

Figure 4-115. *Repairing broken wire by soldering and potting.*

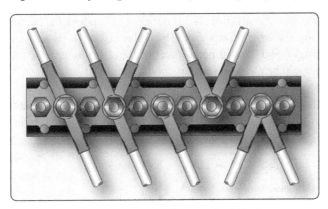

Figure 4-116. *Connecting terminals to terminal block.*

brass plain washer, the copper terminal lug, plain washer, split steel lockwasher and plain nut or self-locking, all metal nut. As a general rule, use a torque wrench to tighten nuts to ensure sufficient contact pressure. Manufacturer's instructions provide installation torques for all types of terminals.

Bonding & Grounding

Bonding is the electrical connecting of two or more conducting objects not otherwise connected adequately. Grounding is the electrical connecting of a conducting object to the primary structure for return of current. Primary structure is the main frame, fuselage, or wing structure of the aircraft. Bonding and grounding connections are made in aircraft electrical systems to:

1. Protect aircraft and personnel against hazards from lightning discharge,

2. Provide current return paths,

3. Prevent development of radio-frequency potentials,

4. Protect personnel from shock hazard,

5. Provide stability of radio transmission and reception, and

6. Prevent accumulation of static charge.

General Bonding & Grounding Procedures

The following general procedures and precautions are recommended when making bonding or grounding connections.

1. Bond or ground parts to the primary aircraft structure, where practicable.

2. Make bonding or grounding connections in such a manner that no part of the aircraft structure is weakened.

3. Bond parts individually, if possible.

4. Install bonding or grounding connections against smooth, clean surfaces.

5. Install bonding or grounding connections so that vibration, expansion or contraction, or relative movement in normal service does not break or loosen the connection.

6. Install bonding and grounding connections in protected areas whenever possible.

Bonding jumpers should be kept as short as practicable, and installed so that the resistance of each connection does not exceed 0.003 ohm. The jumper should not interfere with the operation of movable aircraft elements, such as surface controls; normal movement of these elements should not result in damage to the bonding jumper.

To be sure a low resistance connection has been made, nonconducting finishes, such as paint and anodizing films, should be removed from the surface to be contacted by the bonding terminal.

Electrolytic action can rapidly corrode a bonding connection if suitable precautions are not observed. Aluminum alloy jumpers are recommended for most cases; however, copper jumpers can be used to bond together parts made of stainless steel, cadmium-plated steel, copper, brass, or bronze. Where contact between dissimilar metals cannot be avoided, the choice of jumper and hardware should be such that corrosion is minimized, and the part most likely to corrode is the jumper or associated hardware. Parts A and B of *Figure 4-117* illustrate some proper hardware combinations for making bonding connections. At locations where finishes are removed, a protective finish should be applied to the completed connection to prevent corrosion.

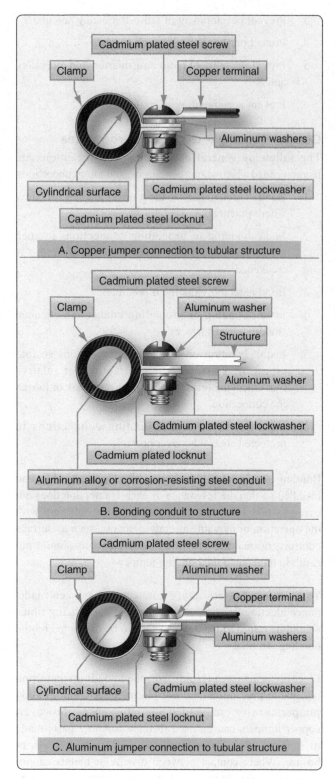

Figure 4-117. *Hardware combinations used in making bonding connections.*

The use of solder to attach bonding jumpers should be avoided. Tubular members should be bonded by means of clamps to which the jumper is attached. The proper choice of clamp material minimizes the probability of corrosion. When bonding jumpers carry a substantial amount of ground return current, the current rating of the jumper should be adequate, and it should be determined that a negligible voltage drop is produced.

Bonding and grounding connections are normally made to flat surfaces by means of through-bolts or screws where there is easy access for installation. The general types of bolted connections are:

1. In making a stud connection, a bolt or screw is locked securely to the structure becoming a stud. *[Figure 4-118]* Grounding or bonding jumpers can be removed or added to the shank of the stud without removing the stud from the structure.

2. Nutplates are used where access to the nut for repairs is difficult. Nutplates are riveted or welded to a clean area of the structure. *[Figure 4-119]*

Bonding and grounding connections are also made to a tab riveted to a structure. *[Figure 4-120]* In such cases, it is important to clean the bonding or grounding surface and make the connection as though the connection were being made to the structure. If it is necessary to remove the tab for any reason, the rivets should be replaced with rivets one size larger, and the mating surfaces of the structure and the tab should be clean and free of anodic film.

Bonding or grounding connections can be made to aluminum alloy, magnesium, or corrosion-resistant steel tubular

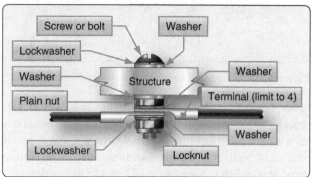

Figure 4-118. *Stud bonding or grounding to a flat surface.*

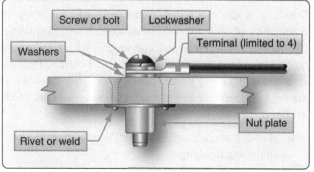

Figure 4-119. *Nut plate bonding or grounding to flat surface.*

4-66

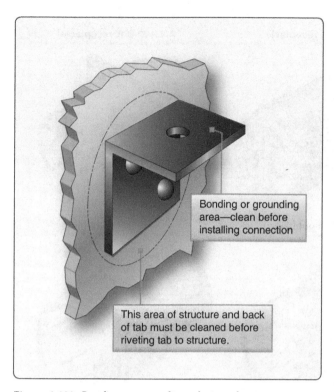

Figure 4-120. *Bonding or grounding tab riveted to structure.*

structure. *Figure 4-121* shows the arrangement of hardware for bonding with an aluminum jumper. Because of the ease with which aluminum is deformed, it is necessary to distribute screw and nut pressure by means of plain washers.

Hardware used to make bonding or grounding connections should be selected on the basis of mechanical strength, current to be carried, and ease of installation. If connection is made by aluminum or copper jumpers to the structure of a dissimilar material, a washer of suitable material should be installed between the dissimilar metals so that any corrosion occurs on the washer.

Hardware material and finish should be selected on the basis

of the material of the structure to which attachment is made and on the material of the jumper and terminal specified for the bonding or grounding connection. Either a screw or bolt of the proper size for the specified jumper terminal should be used. When repairing or replacing existing bonding or grounding connections, the same type of hardware used in the original connection should always be used.

Connectors

Connectors (plugs and receptacles) facilitate maintenance when frequent disconnection is required. Since the cable is soldered to the connector inserts, the joints should be individually installed and the cable bundle firmly supported to avoid damage by vibration. Connectors have been particularly vulnerable to corrosion in the past, due to condensation within the shell. Special connectors with waterproof features have been developed that may replace nonwaterproof plugs in areas where moisture causes a problem. A connector of the same basic type and design should be used when replacing a connector. Connectors that are susceptible to corrosion difficulties may be treated with a chemically inert waterproof jelly. When replacing connector assemblies, the socket-type insert should be used on the half that is " live" or "hot" after the connector is disconnected to prevent unintentional grounding.

Types of Connectors

Connectors are identified by Air Force-Navy (AN) numbers and are divided into classes with the manufacturer's variations in each class. The manufacturer's variations are differences in appearance and in the method of meeting a specification. Some commonly used connectors are shown in *Figure 4-122*. There are five basic classes of AN connectors used in most aircraft. Each class of connector has slightly different construction characteristics. Classes A, B, C, and D are made of aluminum, and class K is made of steel.

1. Class A—solid, one-piece back shell general-purpose connector.

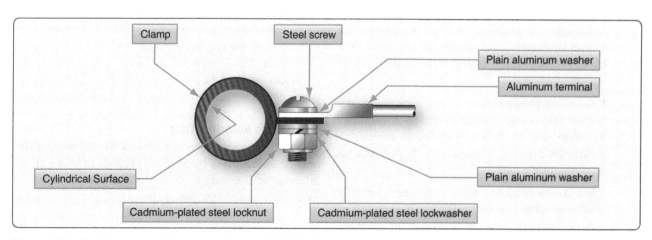

Figure 4-121. *Bonding or grounding connections to a cylindrical structure.*

Figure 4-122. *AN connectors.*

2. Class B—connector back shell separates into two parts lengthwise. Used primarily where it is important that the soldered connectors are readily accessible. The back shell is held together by a threaded ring or by screws.

3. Class C—a pressurized connector with inserts that are not removable. Similar to a class A connector in appearance, but the inside sealing arrangement is sometimes different. It is used on walls or bulkheads of pressurized equipment.

4. Class D—moisture and vibration resistant connector that has a sealing grommet in the back shell. Wires are threaded through tight fitting holes in the grommet, sealing against moisture.

5. Class K—a fireproof connector used in areas where it is vital that the electric current is not interrupted, even though the connector may be exposed to continuous open flame. Wires are crimped to the pin or socket contacts and the shells are made of steel. This class of connector is normally longer than other connectors.

Connector Identification

Code letters and numbers are marked on the coupling ring or shell to identify a connector. This code provides all the information necessary to obtain the correct replacement for a defective or damaged part. *[Figure 4-123]*

Many special-purpose connectors have been designed for

use in aircraft applications. These include subminiature and rectangular shell connectors, and connectors with short body shells, or of split-shell construction.

Installation of Connectors

The following procedures outline one recommended method of assembling connectors to receptacles:

1. Locate the proper position of the plug in relation to the receptacle by aligning the key of one part with the groove or keyway of the other part.

2. Start the plug into the receptacle with a slight forward pressure and engage the threads of the coupling ring and receptacle.

3. Alternately push in the plug and tighten the coupling ring until the plug is completely seated.

4. Use connector pliers to tighten coupling rings one-sixteenth to one-eighth turn beyond finger tight if space around the connector is too small to obtain a good finger grip.

5. Never use force to mate connectors to receptacles.

Do not hammer a plug into its receptacle and never use a torque wrench or pliers to lock coupling rings.

A connector is generally disassembled from a receptacle in the following manner:

1. Use connector pliers to loosen coupling rings that are too tight to be loosened by hand.

2. Alternately pull on the plug body and unscrew the coupling ring until the connector is separated.

3. Protect disconnected plugs and receptacles with caps or plastic bags to keep debris from entering and causing faults.

4. Do not use excessive force and do not pull on attached wires.

Conduit

Conduit is used in aircraft installations for the mechanical protection of wires and cables. It is available in metallic and nonmetallic materials and in both rigid and flexible form.

When selecting conduit size for a specific cable bundle application, it is common practice to allow for ease in maintenance and possible future circuit expansion by specifying the conduit inner diameter about 25 percent larger than the maximum diameter of the conductor bundle. The nominal diameter of a rigid metallic conduit is the outside diameter. Therefore, to obtain the inside diameter, subtract twice the tube wall thickness.

From the abrasion standpoint, the conductor is vulnerable at the ends of the conduit. Suitable fittings are affixed to conduit ends in such a manner that a smooth surface comes in contact with the conductor within the conduit. When fittings are not used, the conduit end should be flared to prevent wire insulation damage. The conduit is supported by clamps along the conduit run.

Many of the common conduit installation problems can be avoided by proper attention to the following details:

1. Do not locate conduit where it can be used as a handhold or footstep.

2. Provide drain holes at the lowest point in a conduit run. Drilling burrs should be carefully removed from the drain holes.

3. Support the conduit to prevent chafing against the structure and to avoid stressing its end fittings.

Damaged conduit sections should be repaired to prevent injury to the wires or wire bundle. The minimum acceptable tube bend radii for rigid conduit as prescribed by the manufacturer's instructions should be carefully followed. Kinked or wrinkled bends in a rigid conduit are normally not considered acceptable.

Flexible aluminum conduit is widely available in two types: bare flexible and rubber-covered conduit. Flexible brass conduit is normally used instead of flexible aluminum where it is necessary to minimize radio interference. Flexible conduit may be used where it is impractical to use rigid conduit, such as areas that have motion between conduit ends or where complex bends are necessary. Transparent adhesive tape is recommended when cutting flexible conduit with a hacksaw to minimize fraying of the braid.

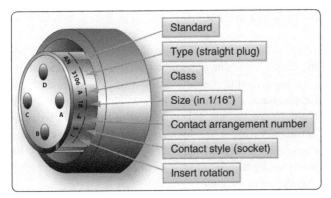

Figure 4-123. *AN connector markings.*

Electrical Equipment Installation

This section provides general procedures and safety precautions for installation of commonly used aircraft electrical equipment and components. Electrical load limits, acceptable means of controlling or monitoring electrical loads, and circuit protection devices are subjects with which mechanics must be familiar to properly install and maintain aircraft electrical systems.

Electrical Load Limits

When installing additional electrical equipment that consumes electrical power in an aircraft, the total electrical load must be safely controlled or managed within the rated limits of the affected components of the aircraft's power-supply system. Regulation of the field current strength is used to control DC generator voltage.

Before any aircraft electrical load is increased, the associated wires, cables, and circuit-protection devices, such as fuses or circuit breakers, should be checked to determine that the new electrical load—previous maximum load plus added load—does not exceed the rated limits of the existing wires, cables, or protection devices.

The generator or alternator output ratings prescribed by the manufacturer should be compared with the electrical loads that can be imposed on the affected generator or alternator by installed equipment. When the comparison shows that the probable total connected electrical load can exceed the output load limits of the generator(s) or alternator(s), the load should be reduced so that an overload cannot occur. When a storage battery is part of the electrical power system, ensure that the battery is continuously charged in flight, except when short intermittent loads are connected, such as a radio transmitter, a landing gear motor, or other similar devices that may place short-time demand loads on the battery.

Controlling or Monitoring the Electrical Load

Placards are recommended to inform crewmembers of an aircraft about the combinations of loads that can safely be connected to the power source.

In installations where the ammeter is in the battery lead and the regulator system limits the maximum current that the generator or alternator can deliver, a voltmeter can be installed on the system bus. As long as the ammeter does not read discharge, except for short intermittent loads such as operating the gear and flaps, and the voltmeter remains at system voltage, the generator or alternator is not overloaded.

The ammeter can be redlined at 100 percent of the generator or alternator rating in installations with the ammeter in the generator or alternator lead, and the regulator system does

not limit the maximum current that the generator or alternator can deliver. If the ammeter reading is never allowed to exceed the red line, except for short, intermittent loads, the generator or alternator is not overloaded.

Where the use of placards or monitoring devices is not practical or desired, and where assurance is needed that the battery in a typical small aircraft generator or battery power source is charged in flight, the total continuous connected electrical load may be held to approximately 80 percent of the total rated generator output capacity. When more than one generator is used in parallel, the total rated output is the combined output of the installed generators.

Means must be provided for quickly coping with the sudden overloads that can be caused by generator or engine failure if two or more generators are operated in parallel and the total connected system load can exceed the rated output of one generator. A quick load-reduction system can be employed or a specified procedure where the total load is reduced to a quantity that is within the rated capacity of the remaining operable generator or generators.

Electrical loads should be connected to inverters, alternators, or similar aircraft electrical power sources in such a manner that the rated limits of the power source are not exceeded, unless some type of effective monitoring means is provided to keep the load within prescribed limits.

Circuit Protection Devices

Conductors should be protected with circuit breakers or fuses located as close as possible to the electrical power source bus. Normally, the manufacturer of the electrical equipment specifies the fuse or circuit breaker to be used when installing the equipment.

The circuit breaker or fuse should open the circuit before the conductor emits smoke. To accomplish this, the time/current characteristic of the protection device must fall below that of the associated conductor. Circuit protector characteristics should be matched to obtain the maximum utilization of the connected equipment.

Figure 4-124 shows an example of the table used in selecting the circuit breaker and fuse protection for copper conductors. This limited table is applicable to a specific set of ambient temperatures and wire bundle sizes and is presented as a typical example only. It is important to consult such guides before selecting a conductor for a specific purpose. For example, a wire run individually in the open air may be protected by the circuit breaker of the next higher rating to that shown in the table.

Wire AN Gauge Copper	Circuit Breaker Amperage	Fuse Amperage
22	5.0	5
20	7.5	5
18	10.0	10
16	15.0	10
14	20.0	15
12	30.0	20
10	40.0	30
8	50.0	50
6	80.0	70
4	100.0	70
2	125.0	100
1		150
0		150

Figure 4-124. *Wire and circuit protector table.*

All resettable circuit breakers should open the circuit in which they are installed, regardless of the position of the operating control when an overload or circuit fault exists. Such circuit breakers are referred to as trip-free. Automatic reset circuit breakers automatically reset themselves periodically. They should not be used as circuit protection devices in aircraft.

Switches

A specifically designed switch should be used in all circuits in which a switch malfunction would be hazardous. Such switches are of rugged construction and have sufficient contact capacity to break, make, and carry continuously the connected load current. Snap-action design is generally preferred to obtain rapid opening and closing of contacts regardless of the speed of the operating toggle or plunger, thereby minimizing contact arcing.

The nominal current rating of the conventional aircraft switch is usually stamped on the switch housing. This rating represents the continuous current rating with the contacts closed. Switches should be derated from their nominal current rating for the following types of circuits:

1. High rush-in circuits—circuits containing incandescent lamps can draw an initial current that is 15 times greater than the continuous current. Contact burning or welding may occur when the switch is closed.

2. Inductive circuits—magnetic energy stored in solenoid coils or relays is released and appears as an arc as the control switch is opened.

3. Motors—direct current motors draw several times their rated current during starting, and magnetic energy stored in their armature and field coils is released when the control switch is opened.

Figure 4-125 is typical of those tables available for selecting the proper nominal switch rating when the continuous load current is known. This selection is essentially a derating to obtain reasonable switch efficiency and service life.

Hazardous errors in switch operation can be avoided by logical and consistent installation. Two position on-off switches should be mounted so that the on position is reached by an upward or forward movement of the toggle. When the switch controls movable aircraft elements, such as landing gear or flaps, the toggle should move in the same direction as the desired motion. Inadvertent operation of a switch can be prevented by mounting a suitable guard over the switch.

Relays

Relays are used as switching devices in which a weight reduction can be achieved or electrical controls can be simplified. A relay is an electrically operated switch and is therefore subject to dropout under low system voltage conditions. The previous discussion of switch ratings is generally applicable to relay contact ratings.

Nominal System Voltage	Type of Load	Derating Factor
24 VDC	Lamp	8
24 VDC	Inductive (Relay-Solenoid)	4
24 VDC	Resistive (Heater)	2
24 VDC	Motor	3
12 VDC	Lamp	5
12 VDC	Inductive (Relay-Solenoid)	2
12 VDC	Resistive (Heater)	1
12 VDC	Motor	2

Figure 4-125. *Switch derating factors.*

Chapter 5
Engine Starting Systems

Introduction

Most aircraft engines, reciprocating or turbine, require help during the starting process. Hence, this device is termed the starter. A starter is an electromechanical mechanism capable of developing large amounts of mechanical energy that can be applied to an engine, causing it to rotate. Reciprocating engines need only to be turned through at a relatively slow speed until the engine starts and turns on its own. Once the reciprocating engine has fired and started, the starter is disengaged and has no further function until the next start. In the case of a turbine engine, the starter must turn the engine up to a speed that provides enough airflow through the engine for fuel to be ignited. Then, the starter must continue to help the engine accelerate to a self-sustaining speed. Turbine engine starters have a critical role in starting of the engine.

If the starter turns the turbine engine up to a self-sustaining speed, the engine start process will be successful. There are only a few types or methods used to turn the engine. Almost all reciprocating engines use a form of electric motor geared to the engine. Modern turbine engines use electric motors, starter/generators (electric motor and a generator in the same housing), and air turbine starters. Air turbine starters are driven by compressed air through a turbine wheel that is mechanically connected through reduction gears to one of the engine's compressors, generally the highest pressure compressor.

Reciprocating Engine Starting Systems

In the early stages of aircraft development, relatively low powered reciprocating engines were started by pulling the propeller through a part of a revolution by hand. Difficulty was often experienced in cold weather starting when lubricating oil temperatures were near the congealing point. In addition, the magneto systems delivered a weak starting spark at the very low cranking speeds. This was often compensated for by providing a hot spark using such ignition system devices as the booster coil, induction vibrator, or impulse coupling.

Some small, low-powered aircraft which use hand-cranking of the propeller, or propping, for starting are still being operated. For general instructions on starting this type of aircraft, refer to the Aviation Maintenance Technician—General Handbook, Chapter 1, Safety, Ground Operations, and Servicing. Throughout the development of the aircraft reciprocating engine from the earliest use of starting systems to the present, a number of different starter systems have been used. Most reciprocating engine starters are the direct cranking electric type. A few older model aircraft are still equipped with inertia starters. Thus, only a brief description of these starting systems is included in this section.

Inertia Starters

There are three general types of inertia starters: hand, electric, and combination hand and electric. The operation of all types of inertia starters depends on the kinetic energy stored in a rapidly rotating flywheel for cranking ability. Kinetic energy is energy possessed by a body by virtue of its state of motion, which may be movement along a line or spinning action.

In the inertia starter, energy is stored slowly during an energizing process by a manual hand crank or electrically

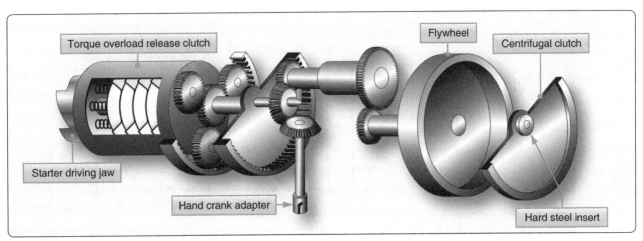

Figure 5-1. *Combination hand and electric inertia starter.*

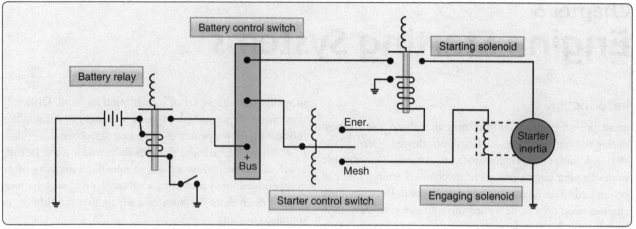

Figure 5-2. *Electric inertia starting circuit.*

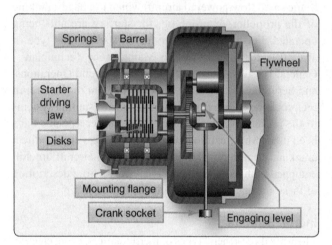

Figure 5-3. *Torque overload release clutch.*

with a small motor. The flywheel and movable gears of a combination hand electric inertia starter are shown in *Figure 5-1*. The electrical circuit for an electric inertia starter is shown in *Figure 5-2*. During the energizing of the starter, all movable parts within it, including the flywheel, are set in motion. After the starter has been fully energized, it is engaged to the crankshaft of the engine by a cable

pulled manually or by a meshing solenoid that is energized electrically. When the starter is engaged, or meshed, flywheel energy is transferred to the engine through sets of reduction gears and a torque overload release clutch. *[Figure 5-3]*

Direct Cranking Electric Starter

The most widely used starting system on all types of reciprocating engines utilizes the direct cranking electric starter. This type of starter provides instant and continual cranking when energized. The direct cranking electric starter consists basically of an electric motor, reduction gears, and an automatic engaging and disengaging mechanism that is operated through an adjustable torque overload release clutch. A typical circuit for a direct cranking electric starter is shown in *Figure 5-4*. The engine is cranked directly when the starter solenoid is closed. As shown in *Figure 5-4*, the main cables leading from the starter to the battery are heavy duty to carry the high current flow, which may be in a range from as high as 350 amperes to 100 amperes (amps), depending on the starting torque required. The use of solenoids and heavy wiring with a remote control switch reduces overall cable weight and total circuit voltage drop.

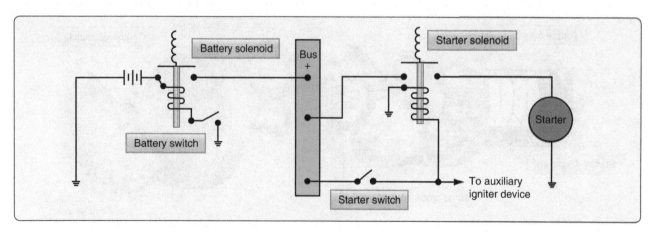

Figure 5-4. *Typical starting circuit using a direct cranking electric starter.*

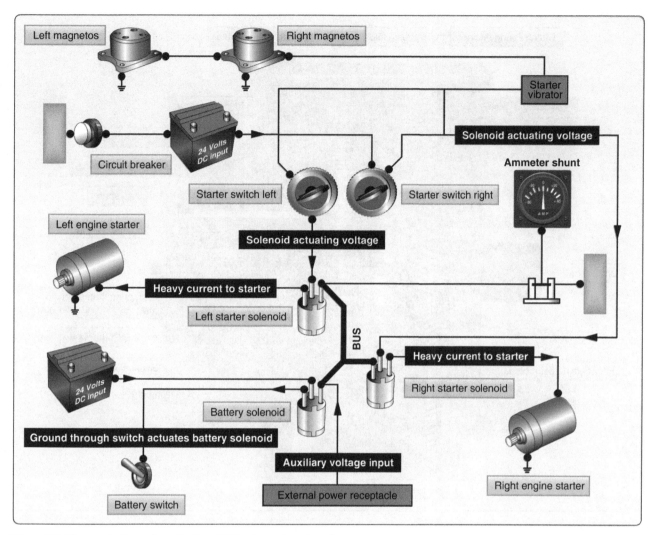

Figure 5-5. *Engine starting schematic for a light twin-engine aircraft.*

The typical starter motor is a 12- or 24-volt, series-wound motor that develops high starting torque. The torque of the motor is transmitted through reduction gears to the overload release clutch. Typically, this action actuates a helically-splined shaft moving the starter jaw outward to engage the engine cranking jaw before the starter jaw begins to rotate. After the engine reaches a predetermined speed, the starter automatically disengages. The schematic in *Figure 5-5* provides a pictorial arrangement of an entire starting system for a light twin-engine aircraft.

Direct Cranking Electric Starting System for Large Reciprocating Engines

In a typical high horsepower reciprocating engine starting system, the direct cranking electric starter consists of two basic components: a motor assembly and a gear section. The gear section is bolted to the drive end of the motor to form a complete unit.

The motor assembly consists of the armature and motor

pinion assembly, the end bell assembly, and the motor housing assembly. The motor housing also acts as the magnetic yoke for the field structure.

The starter motor is a nonreversible, series interpole motor. Its speed varies directly with the applied voltage and inversely with the load. The starter gear section consists of an external housing with an integral mounting flange, planetary gear reduction, a sun and integral gear assembly, a torque-limiting clutch, and a jaw and cone assembly. *[Figure 5-6]* When the starter circuit is closed, the torque developed in the starter motor is transmitted to the starter jaw through the reduction gear train and clutch. The starter gear train converts the high speed low torque of the motor to the low speed high torque required to crank the engine. In the gear section, the motor pinion engages the gear on the intermediate countershaft. *[Figure 5-6]* The pinion of the countershaft engages the internal gear. The internal gear is an integral part of the sun gear assembly and is rigidly attached to the sun gear shaft. The sun gear drives three planet gears that are part of the

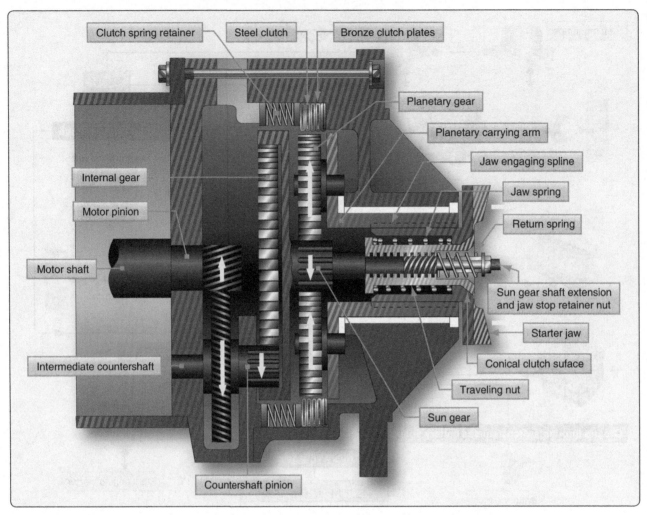

Figure 5-6. *Starter gear section.*

planetary gear assembly. The individual planet gear shafts are supported by the planetary carrying arm, a barrel-like part shown in *Figure 5-6.*

The carrying arm transmits torque from the planet gears to the starter jaw as follows:

1. The cylindrical portion of the carrying arm is splined longitudinally around the inner surface.

2. Mating splines are cut on the exterior surface of the cylindrical part of the starter jaw.

3. The jaw slides fore and aft inside the carrying arm to engage and disengage with the engine.

The three planet gears also engage the surrounding internal teeth on the six steel clutch plates. *[Figure 5-6]* These plates are interleaved with externally splined bronze clutch plates that engage the sides of the housing, preventing them from turning. The proper pressure is maintained upon the clutch pack by a clutch spring retainer assembly. A cylindrical traveling nut inside the starter jaw extends and retracts the jaw. Spiral jaw-engaging splines around the inner wall of the nut mate with similar splines cut on an extension of the sun gear shaft. *[Figure 5-6]*

Being splined in this fashion, rotation of the shaft forces the nut out and the nut carries the jaw with it. A jaw spring around the traveling nut carries the jaw with the nut and tends to keep a conical clutch surface around the inner wall of the jaw head seated against a similar surface around the underside of the nut head. A return spring is installed on the sun gear shaft extension between a shoulder, formed by the splines around the inner wall of the traveling nut, and a jaw stop retaining nut on the end of the shaft. Because the conical clutch surfaces of the traveling nut and the starter jaw are engaged by jaw spring pressure, the two parts tend to rotate at the same speed. However, the sun gear shaft extension turns six times faster than the jaw. The spiral splines on it are cut left hand, and the sun gear shaft extension, turning to the right in relation to the jaw, forces the traveling nut and the jaw out from the starter its full travel (about 5/16 inches) in approximately 12° of rotation of the jaw.

The jaw moves out until it is stopped either by engagement with the engine or by the jaw stop retaining nut. The travel nut continues to move slightly beyond the limit of jaw travel, just enough to relieve some of the spring pressure on the conical clutch surfaces. As long as the starter continues to rotate, there is just enough pressure on the conical clutch surfaces to provide torque on the spiral splines that balance most of the pressure of the jaw spring. If the engine fails to start, the starter jaw does not retract since the starter mechanism provides no retracting force. However, when the engine fires and the engine jaw overruns the starter jaw, the sloping ramps of the jaw teeth force the starter jaw into the starter against the jaw spring pressure. This disengages the conical clutch surfaces entirely, and the jaw spring pressure forces the traveling nut to slide in along the spiral splines until the conical clutch surfaces are again in contact.

When the starter and engine are both running, there is an engaging force keeping the jaws in contact that continue until the starter is de-energized. However, the rapidly moving engine jaw teeth, striking the slowly moving starter jaw teeth, hold the starter jaw disengaged. As soon as the starter comes to rest, the engaging force is removed, and the small return spring throws the starter jaw into its fully retracted position where it remains until the next start. When the starter jaw first engages the engine jaw, the motor armature has had time to reach considerable speed because of its high starting torque. The sudden engagement of the moving starter jaw with the stationary engine jaw would develop forces sufficiently high enough to severely damage the engine or the starter were it not for the plates in the clutch pack that slip when the engine torque exceeds the clutch-slipping torque.

In normal direct cranking action, the internal steel gear clutch plates are held stationary by the friction of the bronze plates with which they are interleaved. When the torque imposed by the engine exceeds the clutch setting, however, the internal gear clutch plates rotate against the clutch friction, allowing the planet gears to rotate while the planetary carrying arm and the jaw remain stationary. When the engine reaches the speed that the starter is trying to achieve, the torque drops off to a value less than the clutch setting, the internal gear clutch plates are again held stationary, and the jaw rotates at the speed that the motor is attempting to drive it. The starter control switches are shown schematically in *Figure 5-7*.

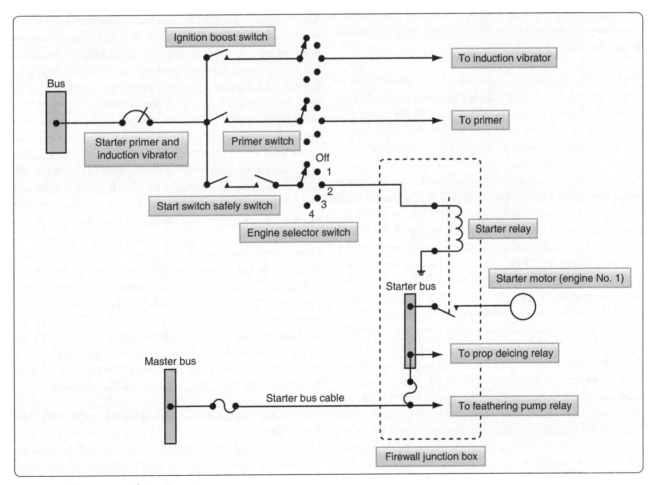

Figure 5-7. *Starter control circuit.*

The engine selector switch must be positioned and the starter switch and the safety switch—wired in series—must be closed before the starter can be energized. Current is supplied to the starter control circuit through a circuit breaker labeled "Starter, Primer, and Induction Vibrator." *[Figure 5-7]* When the engine selector switch is in position for the engine start, closing the starter energizes the starter relay located in the engine nacelle area. Energizing the starter relay completes the power circuit to the starter motor. The current necessary for this heavy load is taken directly from the master bus through the starter bus cable.

All starting systems have operating time limits because of the high energy used during cranking or rotation of the engine. These limits are referred to as starter limits and must be observed or overheating and damage of the starter occurs. After energizing the starter for 1 minute, it should be allowed to cool for at least 1 minute. After a second or subsequent cranking period of 1 minute, it should cool for 5 minutes.

Direct Cranking Electric Starting System for Small Aircraft

Most small, reciprocating engine aircraft employ a direct cranking electric starting system. Some of these systems are automatically engaged starting systems, while others are manually engaged.

Manually engaged starting systems used on many older, small aircraft employ a manually operated overrunning clutch drive pinion to transmit power from an electric starter motor to a crankshaft starter drive gear. *[Figure 5-8]* A knob or handle on the instrument panel is connected by a flexible control to a lever on the starter. This lever shifts the starter drive pinion into the engaged position and closes the starter switch contacts when the starter knob or handle is pulled. The starter lever is attached to a return spring that returns the lever and the flexible control to the off position. When the engine starts, the overrunning action of the clutch protects the starter drive pinion until the shift lever can be released to disengage the pinion. For the typical unit, there is a specified length of travel for the starter gear pinion. *[Figure 5-8]* It is important that the starter lever move the starter pinion gear this proper distance before the adjustable lever stud contacts the starter switch.

The automatic, or remote solenoid engaged, starting systems employ an electric starter mounted on an engine adapter. A starter solenoid is activated by either a push button or turning the ignition key on the instrument panel. When the solenoid is activated, its contacts close, and electrical energy energizes the starter motor. Initial rotation of the starter motor engages the starter through an overrunning clutch in the starter adapter, which incorporates worm reduction gears.

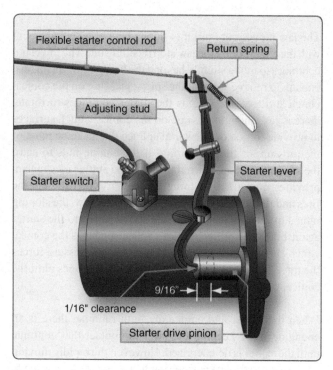

Figure 5-8. *Starter level controls and adjustment.*

Some engines incorporate an automatic starting system that employs an electric starter motor mounted on a right angle drive adapter. As the starter motor is electrically energized, the adapter worm shaft and gear engage the starter shaft gear by means of a spring and clutch assembly. The shaft gear, in turn, rotates the crankshaft. When the engine begins to turn on its own power, the clutch spring disengages from the shaft gear. The starter adapter uses a worm drive gear shaft and worm gear to transfer torque from the starter motor to the clutch assembly. *[Figure 5-9]* As the worm gear rotates the worm wheel and clutch spring, the clutch spring is tightened around the drum of the starter shaft gear. As the shaft gear turns, torque is transmitted directly to the crankshaft gear.

Other engines use a starter that drives a ring gear mounted to the propeller hub. *[Figure 5-10]* It uses an electric motor and a drive gear that engages as the motor is energized and spins the gear, which moves out and engages the ring gear on the propeller hub cranking the engine for start. *[Figure 5-11]* As the engine starts, the starter drive gear is spun back by the engine turning, which disengages the drive gear. *[Figure 5-12]* The starter motors on small aircraft also have operational limits with cool down times that should be observed.

Reciprocating Engine Starting System Maintenance Practices

Most starting system maintenance practices include replacing the starter motor brushes and brush springs, cleaning dirty commutators, and turning down burned or out-of-round

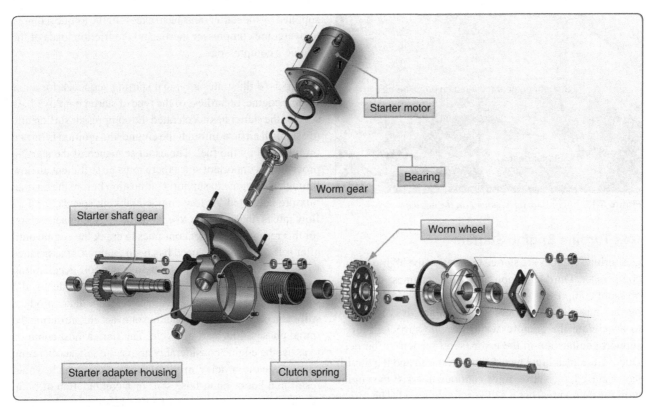

Figure 5-9. *Starter adapter.*

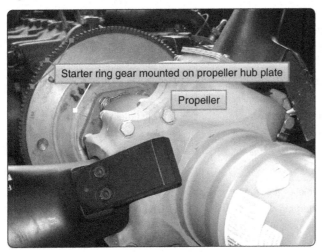

Figure 5-10. *Starter ring gear mounted on the propeller hub.*

Figure 5-11. *Starter drive gear mounting holes and electrical connector.*

starter commutators. As a rule, starter brushes should be replaced when worn down to approximately one-half the original length. Brush spring tension should be sufficient to give brushes a good firm contact with the commutator. Brush leads should be unbroken and lead terminal screws tight.

A glazed or dirty starter commutator can be cleaned by holding a strip of double-0 sandpaper or a brush seating stone against the commutator as it is turned. The sandpaper or stone should be moved back and forth across the commutator to avoid wearing a groove. Emery paper or carborundum should never be used for this purpose because of their possible shorting action.

Roughness, out-of-roundness, or high-mica conditions are reasons for turning down the commutator. In the case of a high-mica condition, the mica should be undercut after the turning operation is accomplished. Refer to FAA-H-8083-30, Aviation Maintenance Technician—General for a review of high-mica commutators in motors.

The drive gear should be checked for wear along with the ring gear. The electrical connections should be checked for looseness and corrosion. Also, check the security of the mounting of the housing of the starter.

Troubleshooting Small Aircraft Starting Systems

The troubleshooting procedures listed in *Figure 5-13* are typical of those used to isolate malfunctions in small aircraft starting systems.

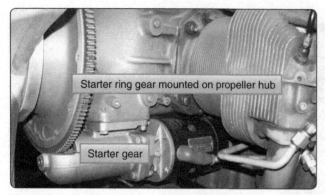
Figure 5-12. *Engine starter mounted on the engine.*

Gas Turbine Engine Starters

Gas turbine engines are started by rotating the high-pressure compressor. On dual-spool, axial flow engines, the high-pressure compressor and N1 turbine system is only rotated by the starter. To start a gas turbine engine, it is necessary to accelerate the compressor to provide sufficient air to support combustion in the combustion section, or burners. Once ignition and fuel have been introduced and the lite-off has occurred, the starter must continue to assist the engine until the engine reaches a self-sustaining speed. The torque supplied by the starter must be in excess of the torque required to overcome compressor inertia and the friction loads of the engine's compressor.

Figure 5-14 illustrates a typical starting sequence for a gas turbine engine, regardless of the type of starter employed. As soon as the starter has accelerated the compressor sufficiently to establish airflow through the engine, the ignition is turned on followed by the fuel. The exact sequence of the starting procedure is important since there must be sufficient airflow through the engine to support combustion before the air-fuel mixture is ignited. At low engine cranking speeds, the fuel flow rate is not sufficient to enable the engine to accelerate; for this reason, the starter continues to crank the engine until after self-accelerating speed has been attained. If assistance from the starter were cut off below the self-accelerating speed, the engine would either fail to accelerate to idle speed or might even decelerate because it could not produce sufficient energy to sustain rotation or to accelerate during the initial phase of the starting cycle. The starter must continue to assist the engine considerably above the self-accelerating speed to avoid a delay in the starting cycle, which would result in a hot or hung false start or a combination of both. At the proper points in the sequence, the starter and ignition

Small Aircraft Troubleshooting Procedures			
	Probable Cause	**Isolation Procedure**	**Remedy**
Starter will not operate	• Defective master switch or circuit. • Defective starter switch or switch circuit. • Starter lever does not activate switch. • Defective starter.	• Check master circuit. • Check switch circuit continuity. • Check starter lever adjustment. • Check through items above. If another cause is not apparent, starter is defective.	• Repair circuit. • Replace switch or wires. • Adjust starter lever in accordance with manufacturer's instructions. • Remove and repair or replace starter.
Starter motor runs but does not turn crankshaft	• Starter lever adjusted to activate switch without engaging pinion with crankshaft gear. • Defective overrunning clutch or drive. • Damaged starter pinion gear or crankshaft gear.	• Check starter lever adjustment. • Remove starter and check starter drive and overrunning clutch. • Remove and check pinion gear and crankshaft gear.	• Adjust starter lever in accordance with manufacturer's instructions. • Replace defective parts. • Replace defective parts.
Starter drags	• Low battery. • Starter switch or relay contacts burned or dirty. • Defective starter.	• Check battery. • Check contacts. • Check starter brushes, brush spring tension for solder thrown on brush cover.	• Charge or replace battery. • Replace with serviceable unit. • Repair or replace starter.
Starter excessively noisy	• Dirty, worn commutator. • Worn starter pinion. • Worn or broken teeth on crankshaft gears.	• Clean and check visually. • Remove and examine pinion. • Remove starter and turn over engine by hand to examine crankshaft gear.	• Turn down commutator. • Replace starter drive. • Replace crankshaft gear.

Figure 5-13. *Small aircraft troubleshooting procedures.*

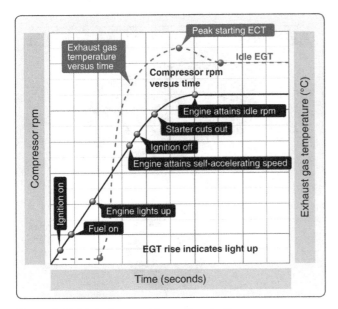

Figure 5-14. *Typical gas turbine engine starting sequence.*

are automatically cut off. The basic types of starters that are in current use for gas turbine engines are direct current (DC) electric motor, starter/generators, and the air turbine type of starters.

Many types of turbine starters have included several different methods for turning the engine for starting. Several methods have been used but most of these have given way to electric or air turbine starters. An air impingement starting system, which is sometimes used on small engines, consists of jets of compressed air piped to the inside of the compressor or turbine case so that the jet air blast is directed onto the compressor or turbine rotor blades, causing them to rotate.

A typical cartridge/pneumatic turbine engine starter may be operated as an ordinary air turbine starter from a ground-operated air supply or an engine cross-bleed source. It may also be operated as a cartridge starter. *[Figure 5-15]* To accomplish a cartridge start, a cartridge is first placed in the breech cap. The breech is then closed on the breech chamber by means of the breech handle and then rotated a partial turn to engage the lugs between the two breech sections. The cartridge is ignited by applying voltage through the connector at the end of the breech handle. Upon ignition, the cartridge begins to generate gas. The gas is forced out of the breech to the hot gas nozzles that are directed toward the buckets on the turbine rotor, and rotation is produced via the overboard exhaust collector. Before reaching the nozzle, the hot gas passes an outlet leading to the relief valve. This valve directs hot gas to the turbine, bypassing the hot gas nozzle, as the pressure rises above the preset maximum. Thus, the pressure of the gas within the hot gas circuit is maintained at the optimum level.

The air-fuel combustion starter was used to start gas turbine

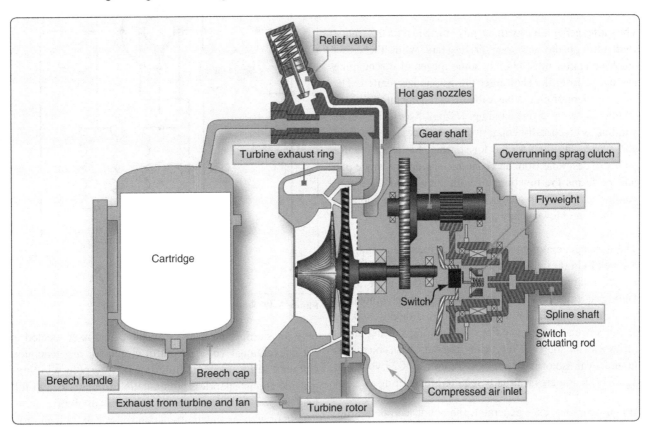

Figure 5-15. *Cartridge/pneumatic starter schematic.*

engines by using the combustion energy of jet A fuel and compressed air. The starter consists of a turbine-driven power unit and auxiliary fuel, air, and ignition systems. Operation of this type starter is, in most installations, fully automatic; actuation of a single switch causes the starter to fire and accelerate the engine from rest to starter cutoff speed.

Hydraulic pumps and motors have also been used for some smaller engines. Many of these systems are not often used on modern commercial aircraft because of the high power demands required to turn the large turbofan engines during the starting cycle on transport aircraft.

Electric Starting Systems & Starter Generator Starting System

Electric starting systems for gas turbine aircraft are of two general types: direct cranking electrical systems and starter generator systems. Direct cranking electric starting systems are used mostly on small turbine engines, such as Auxiliary Power Units (APUs), and some small turboshaft engines. Many gas turbine aircraft are equipped with starter generator systems. Starter generator starting systems are also similar to direct cranking electrical systems except that after functioning as a starter, they contain a second series of windings that allow it to switch to a generator after the engine has reached a self-sustaining speed. This saves weight and space on the engine.

The starter generator is permanently engaged with the engine shaft through the necessary drive gears, while the direct cranking starter must employ some means of disengaging the starter from the shaft after the engine has started. The starter generator unit is basically a shunt generator with an additional heavy series winding. *[Figure 5-16]* This series winding is electrically connected to produce a strong field and a resulting high torque for starting. Starter generator units are desirable from an economic standpoint, since one unit performs the functions of both starter and generator. Additionally, the total weight of starting system components is reduced and fewer spare parts are required.

The starter generator internal circuit has four field windings: a series field (C field), a shunt field, a compensating field, and an interpole or commutating winding. *[Figure 5-17]* During starting, the C field, compensating, and commutating windings are used. The unit is similar to a direct cranking starter since all of the windings used during starting are in series with the source. While acting as a starter, the unit makes no practical use of its shunt field. A source of 24 volts and 1,500 peak amperes is usually required for starting.

When operating as a generator, the shunt, compensating, and commutating windings are used. The C field is used

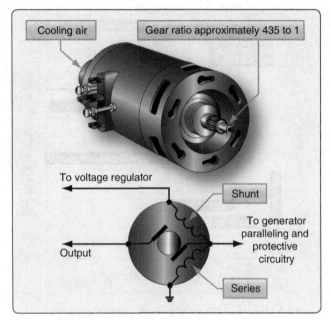

Figure 5-16. *Typical starter generator.*

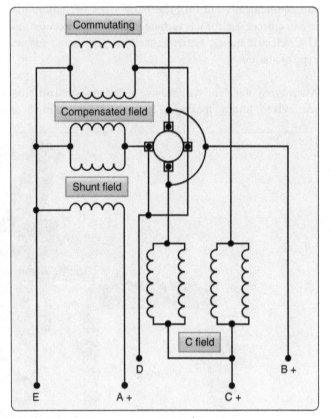

Figure 5-17. *Starter generator internal circuit.*

only for starting purposes. The shunt field is connected in the conventional voltage control circuit for the generator. Compensating and commutating or interpole windings provide almost sparkless commutation from no load to full load. *Figure 5-18* illustrates the external circuit of a starter generator with an undercurrent controller. This unit controls the starter generator when it is used as a starter. Its purpose is

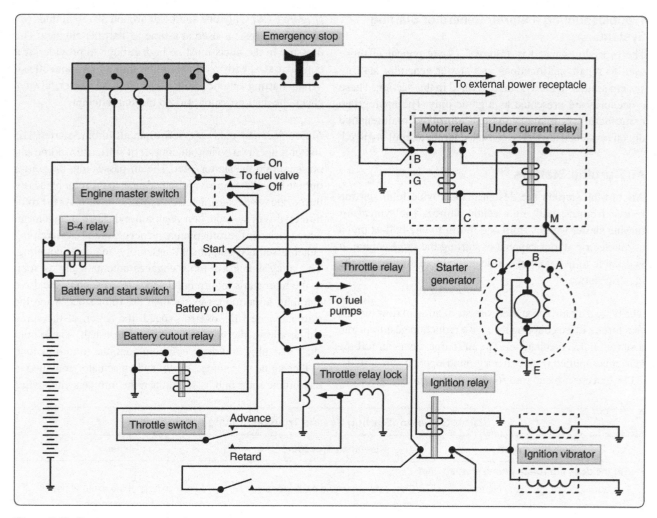

Figure 5-18. *Starter generator circuit.*

to assure positive action of the starter and to keep it operating until the engine is rotating fast enough to sustain combustion. The control block of the undercurrent controller contains two relays. One is the motor relay that controls the input to the starter; the other, the undercurrent relay, controls the operation of the motor relay.

The sequence of operation for the starting system is discussed in the following paragraphs. *[Figure 5-18]* To start an engine equipped with an undercurrent relay, it is first necessary to close the engine master switch. This completes the circuit from the aircraft's bus to the start switch, to the fuel valves, and to the throttle relay. Energizing the throttle relay starts the fuel pumps and completing the fuel valve circuit gives the necessary fuel pressure for starting the engine. As the battery and start switch is turned on, three relays close: the motor relay, ignition relay, and battery cutout relay. The motor relay closes the circuit from the power source to the starter motor; the ignition relay closes the circuit to the ignition units; the battery cutout relay disconnects the battery. Opening the battery circuit is necessary because the heavy drain of the

starter motor would damage the battery. Closing the motor relay allows a very high current to flow to the motor. Since this current flows through the coil of the undercurrent relay, it closes. Closing the undercurrent relay completes a circuit from the positive bus to the motor relay coil, ignition relay coil, and battery cutout relay coil. The start switch is allowed to return to its normal off position, and all units continue to operate.

As the motor builds up speed, the current draw of the motor begins to decrease. As it decreases to less than 200 amps, the undercurrent relay opens. This action breaks the circuit from the positive bus to the coils of the motor, ignition, and battery cutout relays. The de-energizing of these relay coils halts the start operation.

After these procedures are completed, the engine should be operating efficiently, and ignition should be self-sustaining. If, however, the engine fails to reach sufficient speed to halt the starter operation, the stop switch may be used to break the circuit from the positive bus to the main contacts of the undercurrent relay.

Troubleshooting a Starter Generator Starting System

The procedures listed in *Figure 5-19* are typical of those used to repair malfunctions in a starter generator starting system similar to the system described in this section. These procedures are presented as a guide only. The appropriate manufacturer's instructions and approved maintenance directives should always be consulted for the aircraft involved.

Air Turbine Starters

Air turbine starters are designed to provide high starting torque from a small, lightweight source. The typical air turbine starter weighs from one-fourth to one-half as much as an electric starter capable of starting the same engine. It is capable of developing considerable more torque than the electric starter.

The typical air turbine starter consists of an axial flow turbine that turns a drive coupling through a reduction gear train and a starter clutch mechanism. The air to operate an air turbine starter is supplied from either a ground-operated air cart, the APU, or a cross-bleed start from an engine already operating.

[Figure 5-20] Only one source of around 30–50 pounds per square inch (psi) is used at a time to start the engines. The pressure in the ducts must be high enough to provide for a complete start with a normal limit minimum of about 30 psi. When starting engines with an air turbine starter, always check the duct pressure prior to the start attempt.

Figure 5-21 is a cutaway view of an air turbine starter. The starter is operated by introducing air of sufficient volume and pressure into the starter inlet. The air passes into the starter turbine housing where it is directed against the rotor blades by the nozzle vanes causing the turbine rotor to turn. As the rotor turns, it drives the reduction gear train and clutch arrangement, which includes the rotor pinion, planet gears and carrier, sprag clutch assembly, output shaft assembly, and drive coupling. The sprag clutch assembly engages automatically as soon as the rotor starts to turn but disengages as soon as the drive coupling turns more rapidly than the rotor side. When the starter reaches this overrun speed, the action of the sprag clutch allows the gear train to coast to a halt. The output shaft assembly and drive coupling continue to turn as long as the engine is running. A rotor switch actuator, mounted in the turbine rotor hub, is set to open the turbine switch when

Starter Generator Starting System Troubleshooting Procedures		
Probable Cause	**Isolation Procedure**	**Remedy**
Engine does not rotate during start attempt		
• Low supply voltage to the starter. • Power switch is defective. • Ignition switch in throttle quadrant. • Start-lockout relay is defective. • Battery series relay is defective. • Starter relay is defective. • Defective starter. • Start lock-in relay defective. • Starter drive shaft in component drive gearbox is sheared.	• Check voltage of the battery or external power source. • Check switch for continuity. • Check switch for continuity. • Check position of generator control switch. • With start circuit energized, check for 48 volts DC across series relay coil. • With start circuit energized, check for 48 volts DC across starter relay coil. • With start circuit energized, check for proper voltage at the starter. • With start circuit energized, check for 28 volts DC across the relay coil. • Listen for sounds of starter rotation during an attempted start. If the starter rotates but the engine does not, the drive shaft is sheared.	• Adjust voltage of the external power source or charge batteries. • Replace switch. • Replace switch. • Place switch in OFF position. • Replace relay if no voltage is present. • Replace relay if no voltage is present. • Replace the starter if voltage is present. • Replace relay if voltage is not present. • Replace the engine.
Engine starts but does not accelerate to idle		
• Insufficient starter voltage.	• Check starter terminal voltage.	• Use larger capacity ground power unit or charge batteries.
Engine fails to start when throttle is placed in idle		
• Defective ignition system.	• Turn on system and listen for spark-igniter operation.	• Clean or replace spark igniters, or replace exciters or leads to igniters.

Figure 5-19. *Starter generator starting system troubleshooting procedures.*

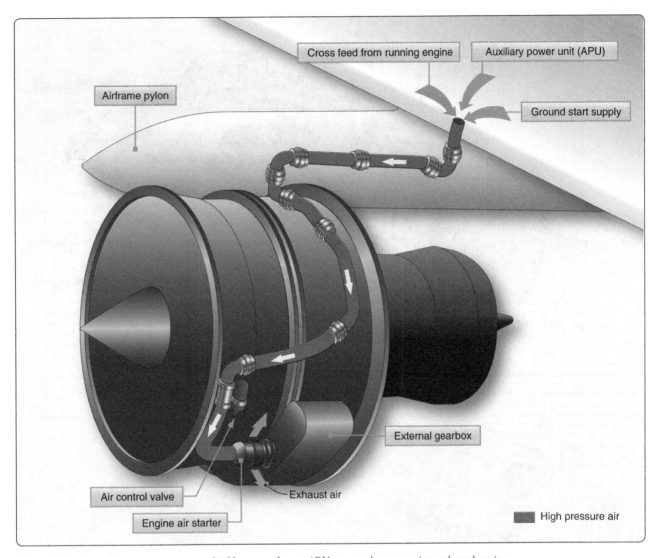

Figure 5-20. *Air turbine starters are supplied by ground cart, APU, or another operating onboard engine.*

the starter reaches cutout speed. Opening the turbine switch interrupts an electrical signal to the start valve. This closes the valve and shuts off the air supply to the starter.

The turbine housing contains the turbine rotor, the rotor switch actuator, and the nozzle components that direct the inlet air against the rotor blades. The turbine housing incorporates a turbine rotor containment ring designed to dissipate the energy of blade fragments and direct their discharge at low energy through the exhaust duct in the event of rotor failure due to excessive turbine overspeed.

The transmission housing contains the reduction gears, the clutch components, and the drive coupling. The transmission housing also provides a reservoir for the lubricating oil. *[Figure 5-22]* Normal maintenance for air turbine starters includes checking the oil level, inspecting the magnetic chip detector for metal particles, and checking for leaks. Oil can be added to the transmission housing sump through a port in

the starter. This port is closed by a vent plug containing a ball valve that allows the sump to be vented to the atmosphere during normal flight. The housing also incorporates a sight gauge that is used to check the oil quantity. A magnetic drain plug in the transmission drain opening attracts any ferrous particles that may be in the oil. The starter uses turbine oil, the same as the engine, but this oil does not circulate through the engine.

The ring gear housing, which is internal, contains the rotor assembly. The switch housing contains the turbine switch and bracket assembly. To facilitate starter installation and removal, a mounting adapter is bolted to the mounting pad on the engine. Quick-detach clamps join the starter to the mounting adapter and inlet duct. *[Figure 5-22]* Thus, the starter is easily removed for maintenance or overhaul by disconnecting the electrical line, loosening the clamps, and carefully disengaging the drive coupling from the engine starter drive as the starter is withdrawn.

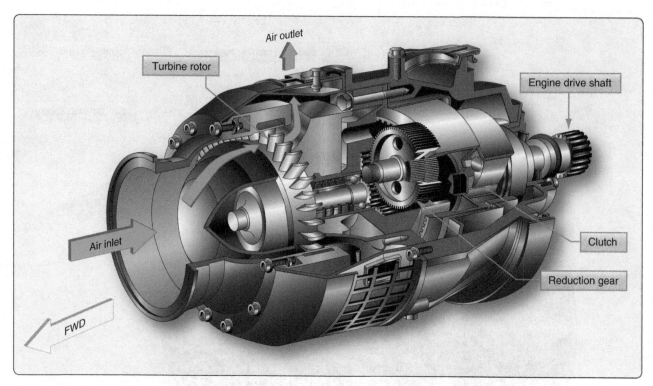

Figure 5-21. *Cutaway view of an air turbine starter.*

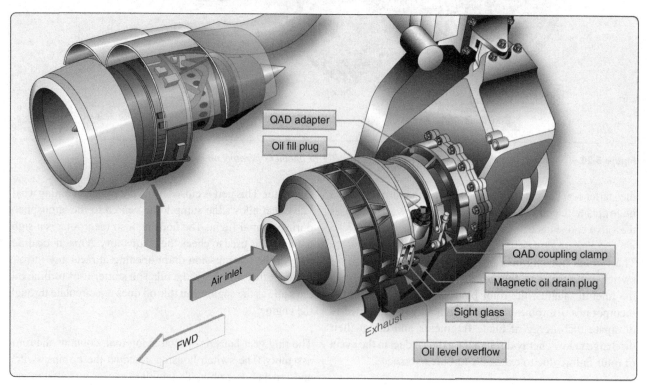

Figure 5-22. *Air turbine starter.*

The air path is directed through a combination pressure-regulating and shutoff valve, or bleed valve, that controls all duct pressure flowing to the starter inlet ducting. This valve regulates the pressure of the starter operating air and shuts off the air supply to the engine when selected off. Downstream from the bleed valve is the start valve, which is used to control air flow into the starter. *[Figure 5-23]*

The pressure-regulating and shutoff valve consists of two subassemblies: pressure-regulating valve and pressure-

Figure 5-23. *Regulating and shutoff bleed valve.*

regulating valve control. *[Figure 5-24]* The regulating valve assembly consists of a valve housing containing a butterfly-type valve. *[Figure 5-24]* The shaft of the butterfly valve is connected through a cam arrangement to a servo piston. When the piston is actuated, its motion on the cam causes rotation of the butterfly valve. The slope of the cam track is designed to provide small initial travel and high initial torque when the starter is actuated. The cam track slope also provides more stable action by increasing the opening time of the valve.

The control assembly is mounted on the regulating valve housing and consists of a control housing in which a solenoid is used to stop the action of the control crank in the off position. *[Figure 5-24]* The control crank links a pilot valve that meters pressure to the servo piston, with the bellows connected by an air line to the pressure-sensing port on the starter.

Turning on the starter switch energizes the regulating valve solenoid. The solenoid retracts and allows the control crank to rotate to the open position. The control crank is rotated by the control rod spring moving the control rod against the closed end of the bellows. Since the regulating valve is closed and downstream pressure is negligible, the bellows can be fully extended by the bellows spring.

As the control crank rotates to the open position, it causes the pilot valve rod to open the pilot valve, allowing upstream air, which is supplied to the pilot valve through a suitable filter and a restriction in the housing, to flow into the servo piston chamber. The drain side of the pilot valve, which bleeds the servo chamber to the atmosphere, is now closed by the pilot valve rod and the servo piston moves inboard. *[Figure 5-24]* This linear motion of the servo piston is translated to rotary motion of the valve shaft by the rotating cam, thus opening the regulating valve. As the valve opens, downstream pressure increases. This pressure is bled back to the bellows through the pressure-sensing line and compresses the bellows. This action moves the control rod, thereby turning the control crank, and moving the pilot valve rod gradually away from the servo chamber to vent to the atmosphere. *[Figure 5-24]* When downstream (regulated) pressure reaches a preset value, the amount of air flowing into the servo through the restriction equals the amount of air being bled to the atmosphere through the servo bleed; the system is in a state of equilibrium.

When the bleed valve and the start valve are open, the regulated air passing through the inlet housing of the starter

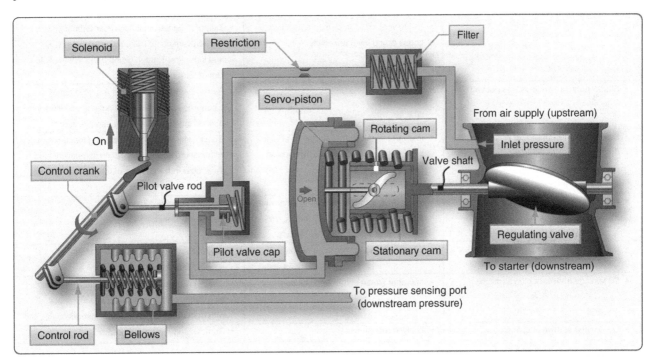

Figure 5-24. *Pressure-regulating and shutoff valve in on position.*

impinges on the turbine causing it to turn. As the turbine turns, the gear train is activated and the inboard clutch gear, which is threaded onto a helical screw, moves forward as it rotates; its jaw teeth engage those of the outboard clutch gear to drive the output shaft of the starter. The clutch is an overrunning type to facilitate positive engagement and minimize chatter. When starter cut-out speed is reached, the start valve is closed. When the air to the starter is terminated, the outboard clutch gear, driven by the engine, begins to turn faster than the inboard clutch gear; the inboard clutch gear, actuated by the return spring, disengages the outboard clutch gear allowing the rotor to coast to a halt. The outboard clutch shaft continues to turn with the engine.

Air Turbine Starter Troubleshooting Guide

The troubleshooting procedures listed in *Figure 5-25* are applicable to air turbine starting systems equipped with a combination pressure-regulating and shutoff valve. These procedures should be used as a guide only and are not intended to replace the manufacturer's instructions.

Air Turbine Starter System Troubleshooting Procedures		
Trouble	Probable Cause	Remedy
• Starter does not operate (no rotation).	• No air supply. • Electrical open in cutout switch. • Sheared starter drive coupling. • Internal starter discrepancy.	• Check air supply. • Check switch continuity. If no continuity, remove starter and adjust or replace switch. • Remove starter and replace drive coupling. • Remove and replace starter.
• Starter will not accelerate to normal cutoff speed.	• Low starter air supply. • Starter cutout switch set improperly. • Valve pressure regulated too low. • Internal starter malfunction.	• Check air source pressure. • Adjust rotor switch actuator. • Replace valve. • Remove and replace starter.
• Starter will not cut off.	• Low air supply. • Rotor switch actuator set too high. • Starter cutout switch shorted.	• Check air supply. • Adjust switch actuator assembly. • Replace switch and bracket assembly.
• External oil leakage.	• Oil level too high. • Loose vent, oil filler, or magnetic plugs. • Loose clamp band assembly.	• Drain oil and re-service properly. • Tighten magnetic plug to proper torque. • Tighten vent and oil filler plugs as necessary and lock wire. Tighten clamp band assembly to higher torque.
• Starter runs, but engine does not turn over.	• Sheared drive coupling.	• Remove starter and replace the drive coupling. If couplings persist in breaking in unusually short periods of time, remove and replace starter.
• Starter inlet will not line up with supply ducting.	• Improper installation of starter on engine, or improper indexing of turbine housing on starter.	• Check installation and/or indexing for conformance with manufacturer's installation instructions and the proper index position of the turbine housing specified for the aircraft.
• Metallic particles on magnetic drain plug.	• Small fuzzy particles indicate normal wear. • Particles coarser than fuzzy (chips, slivers, etc.) indicate internal difficulty.	• No remedial action required. • Remove and replace starter.
• Broken nozzle vanes.	• Large foreign particles in air supply.	• Remove and replace starter and check air supply filter.
• Oil leakage from vent plug assembly.	• Improper starter installation position.	• Check installed position for levelness of oil plugs and correct as required in accordance with manufacturer's installation instructions.
• Oil leakage at drive coupling.	• Leaking rear seal assembly.	• Remove and replace starter.

Figure 5-25. *Air turbine starter system troubleshooting procedures.*

Chapter 6
Lubrication & Cooling Systems

Principles of Engine Lubrication

The primary purpose of a lubricant is to reduce friction between moving parts. Because liquid lubricants or oils can be circulated readily, they are used universally in aircraft engines. In theory, fluid lubrication is based on the actual separation of the surfaces so that no metal-to-metal contact occurs. As long as the oil film remains unbroken, metallic friction is replaced by the internal fluid friction of the lubricant. Under ideal conditions, friction and wear are held to a minimum. Oil is generally pumped throughout the engine to all areas that require lubrication. Overcoming the friction of the moving parts of the engine consumes energy and creates unwanted heat. The reduction of friction during engine operation increases the overall potential power output. Engines are subjected to several types of friction.

Types of Friction

Friction may be defined as the rubbing of one object or surface against another. One surface sliding over another surface causes sliding friction, as found in the use of plain bearings. The surfaces are not completely flat or smooth and have microscopic defects that cause friction between the two moving surfaces. [Figure 6-1] Rolling friction is created when a roller or sphere rolls over another surface, such as with ball or roller bearings, also referred to as antifriction bearings. The amount of friction created by rolling friction is less than that created by sliding friction and this bearing uses an outer race and an inner race with balls, or steel spheres, rolling between the moving parts or races. Another type of friction is wiping friction, which occurs between gear teeth. With this type of friction, pressure can vary widely and loads applied to the gears can be extreme, so the lubricant must be able to withstand the loads.

Functions of Engine Oil

In addition to reducing friction, the oil film acts as a cushion between metal parts. [Figure 6-2] This cushioning effect is particularly important for such parts as reciprocating engine crankshafts and connecting rods, which are subject to shock-loading. As the piston is pushed down on the power stroke, it applies loads between the connecting rod bearing and the crankshaft journal. The load-bearing qualities of the oil must prevent the oil film from being squeezed out, causing metal-to-metal contact in the bearing. Also, as oil circulates through the engine, it absorbs heat from the pistons and cylinder walls. In reciprocating engines, these components are especially dependent on the oil for cooling.

Oil cooling can account for up to 50 percent of the total engine cooling and is an excellent medium to transfer the heat from the engine to the oil cooler. The oil also aids in forming a seal between the piston and the cylinder wall to prevent leakage of the gases from the combustion chamber.

Oils clean the engine by reducing abrasive wear by picking up foreign particles and carrying them to a filter where they are removed. The dispersant, an additive, in the oil holds the particles in suspension and allows the filter to trap them as the oil passes through the filter. The oil also prevents corrosion on the interior of the engine by leaving a coating of oil on parts when the engine is shut down. This is one of the reasons why the engine should not be shut down for long periods of time. The coating of oil preventing corrosion will not last on

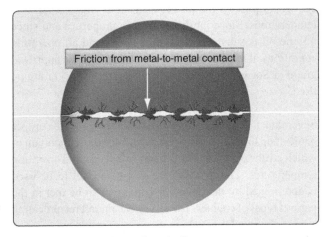

Figure 6-1. *Two moving surfaces in direct contact create excessive friction.*

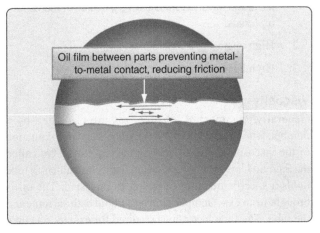

Figure 6-2. *Oil film acts as a cushion between two moving surfaces.*

the parts, allowing them to rust or corrode.

The engine's oil is the life blood of the engine and it is very important for the engine to perform its function and to extend the length between overhauls.

Requirements & Characteristics of Reciprocating Engine Lubricants

While there are several important properties that satisfactory reciprocating engine oil must possess, its viscosity is most important in engine operation. The resistance of an oil to flow is known as its viscosity. Oil that flows slowly is viscous or has a high viscosity; if it flows freely, it has a low viscosity. Unfortunately, the viscosity of oil is affected by temperature. It was not uncommon for earlier grades of oil to become practically solid in cold weather, increasing drag and making circulation almost impossible. Other oils may become so thin at high temperatures that the oil film is broken, causing a low load carrying ability, resulting in rapid wear of the moving parts.

The oil selected for aircraft engine lubrication must be light enough to circulate freely at cold temperatures, yet heavy enough to provide the proper oil film at engine operating temperatures. Since lubricants vary in properties and since no one oil is satisfactory for all engines and all operating conditions, it is extremely important that only the approved grade or Society of Automotive Engineers (SAE) rating be used.

Several factors must be considered in determining the proper grade of oil to use in a particular engine, the most important of which are the operating load, rotational speeds, and operating temperatures. The grade of the lubricating oil to be used is determined by the operating conditions to be met in the various types of engines. The oil used in aircraft reciprocating engines has a relatively high viscosity required by:

1. Large engine operating clearances due to the relatively large size of the moving parts, the different materials used, and the different rates of expansion of the various materials;

2. High operating temperatures; and

3. High bearing pressures.

Viscosity

Generally, commercial aviation oils are classified by a number, (such as 80, 100, 140, etc.) that is an approximation of the viscosity as measured by a testing instrument called the Saybolt Universal Viscosimeter. In this instrument, a tube holds a specific quantity of the oil to be tested. The oil is brought to an exact temperature by a liquid bath surrounding the tube. The time in seconds required for exactly 60 cubic centimeters of oil to flow through an accurately calibrated

orifice is recorded as a measure of the oil's viscosity. If actual Saybolt values were used to designate the viscosity of oil, there would probably be several hundred grades of oil.

To simplify the selection of oils, they are often classified under an SAE system that divides all oils into seven groups (SAE 10 to 70) according to viscosity at either 130 °F or 210 °F. SAE ratings are purely arbitrary and bear no direct relationship to the Saybolt or other ratings.

The letter W occasionally is included in the SAE number giving a designation, such as SAE 20W. This W indicates that the oil, in addition to meeting the viscosity requirements at the testing temperature specifications, is satisfactory oil for winter use in cold climates. This should not be confused with the W used in front of the grade or weight number that indicates the oil is of the ashless dispersant type.

Although the SAE scale has eliminated some confusion in the designation of lubricating oils, it must not be assumed that this specification covers all the important viscosity requirements. An SAE number indicates only the viscosity grade or relative viscosity; it does not indicate quality or other essential characteristics. It is well known that there are good oils and inferior oils that have the same viscosities at a given temperature and, therefore, are subject to classification in the same grade.

The SAE letters on an oil container are not an endorsement or recommendation of the oil by the SAE. Although each grade of oil is rated by an SAE number, depending on its specific use, it may be rated with a commercial aviation grade number or an Army and Navy specification number. The correlation between these grade numbering systems is shown in *Figure 6-3.*

Viscosity Index

The viscosity index is a number that indicates the effect of temperature changes on the viscosity with the oil. When oil has a low viscosity index, it signifies a relatively large change of viscosity of increased temperature. The oil becomes thin at high temperatures and thick at low temperatures. Oils with a high viscosity index have small changes in viscosity over a wide temperature range.

Commercial Aviation No.	Commercial SAE No.	Army and Navy Specification No.
65	30	1065
80	40	1080
100	50	1100
120	60	1120
140	70	

Figure 6-3. *Grade designations for aviation oils.*

The best oil for most purposes is one that maintains a constant viscosity throughout temperature changes. Oil having a high viscosity index resists excessive thickening when the engine is subjected to cold temperatures. This allows for rapid cranking speeds during starting and prompt oil circulation during initial startup. This oil resists excessive thinning when the engine is at operating temperature and provides full lubrication and bearing load protection.

Flash Point & Fire Point

Flash point and fire point are determined by laboratory tests that show the temperature at which a liquid begins to give off ignitable vapors, flash, and the temperature at which there are sufficient vapors to support a fire. These points are established for engine oils to determine that they can withstand the high temperatures encountered in an engine.

Cloud Point & Pour Point

Cloud point and pour point also help to indicate suitability. The cloud point of oil is the temperature at which its wax content, normally held in solution, begins to solidify and separate into tiny crystals, causing the oil to appear cloudy or hazy. The pour point of oil is the lowest temperature at which it flows or can be poured.

Specific Gravity

Specific gravity is a comparison of the weight of the substance to the weight of an equal volume of distilled water at a specified temperature. As an example, water weighs approximately 8 pounds to the gallon; oil with a specific gravity of 0.9 would weigh 7.2 pounds to the gallon.

In the early years, the performance of aircraft piston engines was such that they could be lubricated satisfactorily by means of straight mineral oils, blended from specially selected petroleum base stocks. Oil grades 65, 80, 100, and 120 are straight mineral oils blended from selected high-viscosity index base oils. These oils do not contain any additives except for very small amounts of pour point depressant, which helps improve fluidity at very low temperatures, and an antioxidant. This type of oil is used during the break-in period of a new aviation piston engine or those recently overhauled.

Demand for oils with higher degrees of thermal and oxidation stability necessitated fortifying them with the addition of small quantities of nonpetroleum materials. The first additives incorporated in straight mineral piston engine oils were based on the metallic salts of barium and calcium. In most engines, the performance of these oils with respect to oxidation and thermal stability was excellent, but the combustion chambers of the majority of engines could not tolerate the presence of the ash deposits derived from these metal-containing additives. To overcome the disadvantages of harmful combustion chamber deposits, a nonmetallic (i.e., non-ash forming, polymeric) additive was developed that was incorporated in blends of selected mineral oil base stocks. W oils are of the ashless type and are still in use. The ashless dispersant grades contain additives, one of which has a viscosity stabilizing effect that removes the tendency of the oil to thin out at high oil temperatures and thicken at low oil temperatures.

The additives in these oils extend operating temperature range and improve cold engine starting and lubrication of the engine during the critical warm-up period permitting flight through wider ranges of climatic changes without the necessity of changing oil.

Semi-synthetic multigrade SAE W15 W50 oil for piston engines has been in use for some time. Oils W80, W100, and W120 are ashless dispersant oils specifically developed for aviation piston engines. They combine nonmetallic additives with selected high viscosity index base oils to give exceptional stability, dispersancy, and antifoaming performance. Dispersancy is the ability of the oil to hold particles in suspension until they can either be trapped by the filter or drained at the next oil change. The dispersancy additive is not a detergent and does not clean previously formed deposits from the interior of the engine.

Some multigrade oil is a blend of synthetic and mineral-based oil semisynthetic, plus a highly effective additive package, that is added due to concern that fully synthetic oil may not have the solvency to handle the lead deposits that result from the use of leaded fuel. As multigrade oil, it offers the flexibility to lubricate effectively over a wider range of temperatures than monograde oils. Compared to monograde oil, multigrade oil provides better cold-start protection and a stronger lubricant film (higher viscosity) at typical operating temperatures. The combination of nonmetallic, antiwear additives and selected high viscosity index mineral and synthetic base oils give exceptional stability, dispersancy, and antifoaming performance. Startup can contribute up to 80 percent of normal engine wear due to lack of lubrication during the start-up cycle. The more easily the oil flows to the engine's components at start up, the less wear occurs.

The ashless dispersant grades are recommended for aircraft engines subjected to wide variations of ambient temperature, particularly the turbocharged series engines that require oil to activate the various turbo controllers. At temperatures below 20 °F, preheating of the engine and oil supply tank is normally required regardless of the type of oil used.

Premium, semisynthetic multigrade ashless dispersant oil is a special blend of a high-quality mineral oil and synthetic hydrocarbons with an advanced additive package that has been specifically formulated for multigrade applications. The ashless antiwear additive provides exceptional wear protection for wearing surfaces.

Many aircraft manufacturers add approved preservative lubricating oil to protect new engines from rust and corrosion at the time the aircraft leaves the factory. This preservative oil should be removed at end of the first 25 hours of operation. When adding oil during the period when preservative oil is in the engine, use only aviation grade straight mineral oil or ashless dispersant oil, as required, of the viscosity desired.

If ashless dispersant oil is used in a new engine, or a newly overhauled engine, high oil consumption might possibly be experienced. The additives in some of these ashless dispersant oils may retard the break in of the piston rings and cylinder walls. This condition can be avoided by the use of mineral oil until normal oil consumption is obtained, then change to the ashless dispersant oil. Mineral oil should also be used following the replacement of one or more cylinders or until the oil consumption has stabilized.

In all cases, refer to the manufacturers' information when oil type or time in service is being considered.

Reciprocating Engine Lubrication Systems

Aircraft reciprocating engine pressure lubrication systems can be divided into two basic classifications: wet sump and dry sump. The main difference is that the wet sump system stores oil in a reservoir inside the engine. After the oil is circulated through the engine, it is returned to this crankcase-based reservoir. A dry sump engine pumps the oil from the engine's crankcase to an external tank that stores the oil. The dry sump system uses a scavenge pump, some external tubing, and an external tank to store the oil.

Other than this difference, the systems use similar types of components. Because the dry sump system contains all the components of the wet sump system, the dry sump system is explained as an example system.

Combination Splash & Pressure Lubrication

The lubricating oil is distributed to the various moving parts of a typical internal combustion engine by one of the three following methods: pressure, splash, or a combination of pressure and splash.

The pressure lubrication system is the principal method of lubricating aircraft engines. Splash lubrication may be used in addition to pressure lubrication on aircraft engines, but it is never used by itself; aircraft-engine lubrication systems are always either the pressure type or the combination pressure

and splash type, usually the latter.

The advantages of pressure lubrication are:

1. Positive introduction of oil to the bearings.
2. Cooling effect caused by the large quantities of oil that can be pumped, or circulated, through a bearing.
3. Satisfactory lubrication in various attitudes of flight.

Lubrication System Requirements

The lubrication system of the engine must be designed and constructed so that it functions properly within all flight attitudes and atmospheric conditions that the aircraft is expected to operate. In wet sump engines, this requirement must be met when only half of the maximum lubricant supply is in the engine. The lubrication system of the engine must be designed and constructed to allow installing a means of cooling the lubricant. The crankcase must also be vented to the atmosphere to preclude leakage of oil from excessive pressure.

Dry Sump Oil Systems

Many reciprocating and turbine aircraft engines have pressure dry sump lubrication systems. The oil supply in this type of system is carried in a tank. A pressure pump circulates the oil through the engine. Scavenger pumps then return it to the tank as quickly as it accumulates in the engine sumps. The need for a separate supply tank is apparent when considering the complications that would result if large quantities of oil were carried in the engine crankcase. On multiengine aircraft, each engine is supplied with oil from its own complete and independent system.

Although the arrangement of the oil systems in different aircraft varies widely and the units of which they are composed differ in construction details, the functions of all such systems are the same. A study of one system clarifies the general operation and maintenance requirements of other systems.

The principal units in a typical reciprocating engine dry sump oil system include an oil supply tank, an engine-driven pressure oil pump, a scavenge pump, an oil cooler with an oil cooler control valve, oil tank vent, necessary tubing, and pressure and temperature indicators. *[Figure 6-4]*

Oil Tanks

Oil tanks are generally associated with a dry sump lubrication system, while a wet sump system uses the crankcase of the engine to store the oil. Oil tanks are usually constructed of aluminum alloy and must withstand any vibration, inertia, and fluid loads expected in operation.

Each oil tank used with a reciprocating engine must have expansion space of not less than the greater of 10 percent

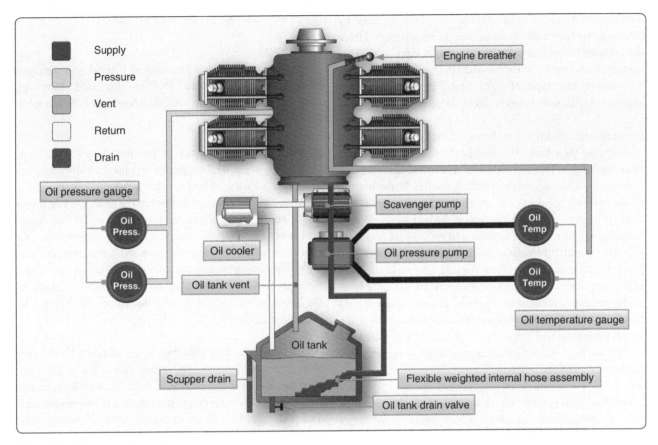

Figure 6-4. *Oil system schematic.*

of the tank capacity or 0.5 gallons. Each filler cap of an oil tank that is used with an engine must provide an oil-tight seal. The oil tank usually is placed close to the engine and high enough above the oil pump inlet to ensure gravity feed.

Oil tank capacity varies with the different types of aircraft, but it is usually sufficient to ensure an adequate supply of oil for the total fuel supply. The tank filler neck is positioned to provide sufficient room for oil expansion and for foam to collect.

The filler cap or cover is marked with the word OIL. A drain in the filler cap well disposes of any overflow caused by the filling operation. Oil tank vent lines are provided to ensure proper tank ventilation in all attitudes of flight. These lines are usually connected to the engine crankcase to prevent the loss of oil through the vents. This indirectly vents the tanks to the atmosphere through the crankcase breather.

Early large radial engines had many gallons of oil in their tank. To help with engine warm up, some oil tanks had a built-in hopper or temperature accelerating well. *[Figure 6-5]* This well extended from the oil return fitting on top of the oil tank to the outlet fitting in the sump in the bottom of the tank. In some systems, the hopper tank is open to the main oil supply at the lower end. Other systems have flapper-type valves

that separate the main oil supply from the oil in the hopper. The opening at the bottom of the hopper in one type and the flapper valve-controlled openings in the other allow oil from the main tank to enter the hopper and replace the oil consumed by the engine. Whenever the hopper tank includes

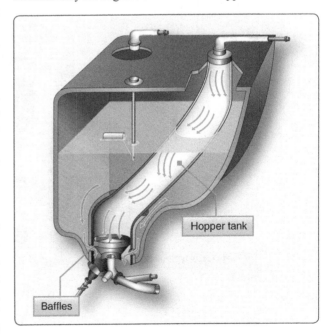

Figure 6-5. *Oil tank with hopper.*

the flapper controlled openings, the valves are operated by differential oil pressure. By separating the circulating oil from the surrounding oil in the tank, less oil is circulated. This hastens the warming of the oil when the engine was started. Very few of these types of tanks are still in use and most are associated with radial engine installations.

Generally, the return line in the top of the tank is positioned to discharge the returned oil against the wall of the tank in a swirling motion. This method considerably reduces foaming that occurs when oil mixes with air. Baffles in the bottom of the oil tank break up this swirling action to prevent air from being drawn into the inlet line of the oil pressure pump. Foaming oil increases in volume and reduces its ability to provide proper lubrication. In the case of oil-controlled propellers, the main outlet from the tank may be in the form of a standpipe so that there is always a reserve supply of oil for propeller feathering in case of engine failure. An oil tank sump, attached to the undersurface of the tank, acts as a trap for moisture and sediment. *[Figure 6-4]* The water and sludge can be drained by manually opening the drain valve in the bottom of the sump.

Most aircraft oil systems are equipped with the dipstick-type quantity gauge, often called a bayonet gauge. Some larger aircraft systems also have an oil quantity indicating system that shows the quantity of oil during flight. One type system consists essentially of an arm and float mechanism that rides the level of the oil and actuates an electric transmitter on top of the tank. The transmitter is connected to a flight deck gauge that indicates the quantity of oil.

Oil Pump

Oil entering the engine is pressurized, filtered, and regulated by units within the engine. They are discussed along with the external oil system to provide a concept of the complete oil system.

As oil enters the engine, it is pressurized by a gear-type pump. *[Figure 6-6]* This pump is a positive displacement pump that consists of two meshed gears that revolve inside the housing. The clearance between the teeth and housing is small. The pump inlet is located on the left and the discharge port is connected to the engine's system pressure line. One gear is attached to a splined drive shaft that extends from the pump housing to an accessory drive shaft on the engine. Seals are used to prevent leakage around the drive shaft. As the lower gear is rotated counterclockwise, the driven idler gear turns clockwise.

As oil enters the gear chamber, it is picked up by the gear teeth, trapped between them and the sides of the gear chamber, then it is carried around the outside of the gears and discharged from the pressure port into the oil screen passage. The pressurized oil flows to the oil filter, where any solid particles suspended in the oil are separated from it, preventing possible damage to moving parts of the engine.

Oil under pressure then opens the oil filter check valve mounted in the top of the filter. This valve is used mostly

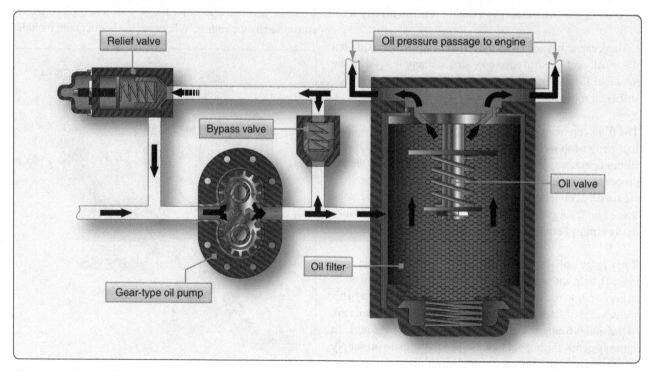

Figure 6-6. *Engine oil pump and associated units.*

with dry sump radial engines and is closed by a light spring loading of 1 to 3 pounds per square inch (psi) when the engine is not operating to prevent gravity-fed oil from entering the engine and settling in the lower cylinders or sump area of the engine. If oil were allowed to gradually seep by the rings of the piston and fill the combustion chamber, it could cause a liquid lock. This could happen if the valves on the cylinder were both closed, and the engine was cranked for start. Damage could occur to the engine.

The oil filter bypass valve, located between the pressure side of the oil pump and the oil filter, permits unfiltered oil to bypass the filter and enter the engine if the oil filter is clogged or during cold weather if congealed oil is blocking the filter during engine start. The spring loading on the bypass valve allows the valve to open before the oil pressure collapses the filter; in the case of cold, congealed oil, it provides a low-resistance path around the filter. Dirty oil in an engine is better than no lubrication.

Oil Filters

The oil filter used on an aircraft engine is usually one of four types: screen, Cuno, canister, or spin-on. A screen-type filter with its double-walled construction provides a large filtering area in a compact unit. *[Figure 6-6]* As oil passes through the fine-mesh screen, dirt, sediment, and other foreign matter are removed and settle to the bottom of the housing. At regular intervals, the cover is removed, and the screen and housing cleaned with a solvent. Oil screen filters are used mostly as suction filters on the inlet of the oil pump.

The Cuno oil filter has a cartridge made of discs and spacers. A cleaner blade fits between each pair of discs. The cleaner blades are stationary, but the discs rotate when the shaft is turned. Oil from the pump enters the cartridge well that surrounds the cartridge and passes through the spaces between the closely spaced discs of the cartridge, then through the hollow center, and on to the engine. Any foreign particles in the oil are deposited on the outer surface of the cartridge. When the cartridge is rotated, the cleaner blades comb the foreign matter from the discs. The cartridge of the manually operated Cuno filter is turned by an external handle. Automatic Cuno filters have a hydraulic motor built into the filter head. This motor, operated by engine oil pressure, rotates the cartridge whenever the engine is running. There is a manual turning nut on the automatic Cuno filter for rotating the cartridge manually during inspections. This filter is not often used on modern aircraft.

A canister housing filter has a replaceable filter element that is replaced with the rest of the components other than seals and gaskets being reused. *[Figure 6-7]* The filter element is designed with a corrugated, strong steel center

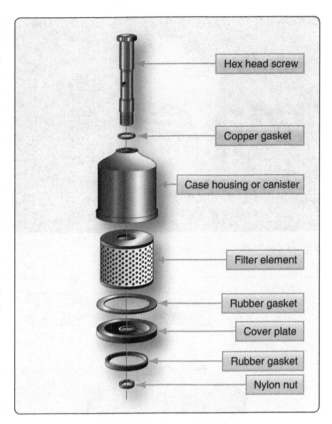

Figure 6-7. *Housing filter element type oil filter.*

tube supporting each convoluted pleat of the filter media, resulting in a higher collapse pressure rating. The filter provides excellent filtration, because the oil flows through many layers of locked-in-fibers.

Full flow spin-on filters are the most widely used oil filters for reciprocating engines. *[Figure 6-8]* Full flow means that all the oil is normally passed through the filter. In a full flow system, the filter is positioned between the oil pump and the engine bearings, which filters the oil of any contaminants before they pass through the engine bearing surfaces. The filter also contains an antidrain back valve and a pressure relief valve, all sealed in a disposable housing. The relief valve is used in case the filter becomes clogged. It would open to allow the oil to bypass, preventing the engine components from oil starvation. A cutaway of the micronic filter element shows the resin-impregnated cellulosic full-pleat media that is used to trap harmful particles, keeping them from entering the engine. *[Figure 6-9]*

Oil Pressure Regulating Valve

An oil pressure regulating valve limits oil pressure to a predetermined value, depending on the installation. *[Figure 6-6]* This valve is sometimes referred to as a relief valve, but its real function is to regulate the oil pressure at a preset pressure level. The oil pressure must be sufficiently high to ensure adequate lubrication of the engine and its

Figure 6-8. *Full flow spin-on filter.*

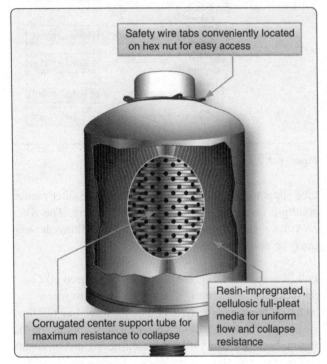

Safety wire tabs conveniently located on hex nut for easy access

Resin-impregnated, cellulosic full-pleat media for uniform flow and collapse resistance

Corrugated center support tube for maximum resistance to collapse

Figure 6-9. *Cutaway view of a filter.*

accessories at high speeds and powers. This pressure helps ensure that the oil film between the crankshaft journal and bearing is maintained. However, the pressure must not be too high, as leakage and damage to the oil system may result. The oil pressure is generally adjusted by loosening the locknut and turning the adjusting screw. *[Figure 6-10]* On most aircraft engines, turning the screw clockwise increases the tension of the spring that holds the relief valve on its seat and increases the oil pressure; turning the adjusting screw counterclockwise decreases the spring tension and lowers the pressure. Some engines use washers under the spring that are either removed or added to adjust the regulating valve and pressure. The oil pressure should be adjusted only after the engine's oil is at operating temperature and the correct

viscosity is verified. The exact procedure for adjusting the oil pressure and the factors that vary an oil pressure setting are included in applicable manufacturer's instructions.

Oil Pressure Gauge

Usually, the oil pressure gauge indicates the pressure that oil enters the engine from the pump. This gauge warns of possible engine failure caused by an exhausted oil supply, failure of the oil pump, burned-out bearings, ruptured oil lines, or other causes that may be indicated by a loss of oil pressure.

One type of oil pressure gauge uses a Bourdon-tube mechanism that measures the difference between oil pressure and cabin, or atmospheric, pressure. This gauge is constructed similarly to other Bourdon-type gauges, except that it has a small restriction built into the instrument case, or into the nipple connection leading to the Bourdon tube. This restriction prevents the surging action of the oil pump from damaging the gauge or causing the pointer to oscillate too violently with each pressure pulsation. The oil pressure gauge has a scale ranging from 0–200 psi, or from 0–300 psi. Operation range markings are placed on the cover glass, or the face of the gauge, to indicate the safe range of oil pressure for a given installation.

A dual-type oil pressure gauge is available for use on

Figure 6-10. *Oil pressure adjustment screw.*

multiengine aircraft. The dual indicator contains two Bourdon tubes, housed in a standard instrument case; one tube being used for each engine. The connections extend from the back of the case to each engine. There is one common movement assembly, but the moving parts function independently. In some installations, the line leading from the engine to the pressure gauge is filled with light oil. Since the viscosity of this oil does not vary much with changes in temperature, the gauge responds better to changes in oil pressure. In time, engine oil mixes with some of the light oil in the line to the transmitter; during cold weather, the thicker mixture causes sluggish instrument readings. To correct this condition, the gauge line must be disconnected, drained, and refilled with light oil.

The current trend is toward electrical transmitters and indicators for oil and fuel pressure-indicating systems in all aircraft. In this type of indicating system, the oil pressure being measured is applied to the inlet port of the electrical transmitter where it is conducted to a diaphragm assembly by a capillary tube. The motion produced by the diaphragm's expansion and contraction is amplified through a lever and gear arrangement. The gear varies the electrical value of the indicating circuit, which in turn, is reflected on the indicator in the flight deck. This type of indicating system replaces long fluid-filled tubing lines with an almost weightless piece of wire.

Oil Temperature Indicator

In dry-sump lubricating systems, the oil temperature bulb may be anywhere in the oil inlet line between the supply tank and the engine. Oil systems for wet-sump engines have the temperature bulb located where it senses oil temperature after the oil passes through the oil cooler. In either system, the bulb is located so that it measures the temperature of the oil before it enters the engine's hot sections. An oil temperature gauge in the flight deck is connected to the oil temperature bulb by electrical leads. The oil temperature is indicated on the gauge. Any malfunction of the oil cooling system appears as an abnormal reading.

Oil Cooler

The cooler, either cylindrical or elliptical shaped, consists of a core enclosed in a double-walled shell. The core is built of copper or aluminum tubes with the tube ends formed to a hexagonal shape and joined together in the honeycomb effect. *[Figure 6-11]* The ends of the copper tubes of the core are soldered, whereas aluminum tubes are brazed or mechanically joined. The tubes touch only at the ends so that a space exists between them along most of their lengths. This allows oil to flow through the spaces between the tubes while the cooling air passes through the tubes.

The space between the inner and outer shells is known as

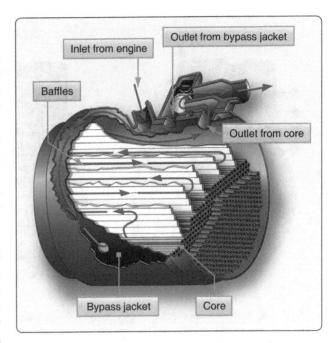

Figure 6-11. *Oil cooler.*

the annular or bypass jacket. Two paths are open to the flow of oil through a cooler. From the inlet, it can flow halfway around the bypass jacket, enter the core from the bottom, and then pass through the spaces between the tubes and out to the oil tank. This is the path the oil follows when it is hot enough to require cooling. As the oil flows through the core, it is guided by baffles that force the oil to travel back and forth several times before it reaches the core outlet. The oil can also pass from the inlet completely around the bypass jacket to the outlet without passing through the core. Oil follows this bypass route when the oil is cold or when the core is blocked with thick, congealed oil.

Oil Cooler Flow Control Valve

As discussed previously, the viscosity of the oil varies with its temperature. Since the viscosity affects its lubricating properties, the temperature at which the oil enters an engine must be held within close limits. Generally, the oil leaving an engine must be cooled before it is recirculated. Obviously, the amount of cooling must be controlled if the oil is to return to the engine at the correct temperature. The oil cooler flow control valve determines which of the two possible paths the oil takes through the oil cooler. *[Figure 6-12]*

There are two openings in a flow control valve that fit over the corresponding outlets at the top of the cooler. When the oil is cold, a bellows within the flow control contracts and lifts a valve from its seat. Under this condition, oil entering the cooler has a choice of two outlets and two paths. Following the path of least resistance, the oil flows around the jacket and out past the thermostatic valve to the tank. This allows the oil to warm up quickly and, at the same time, heats the

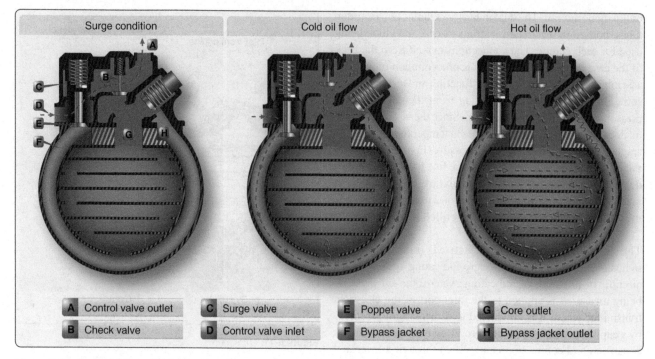

| A | Control valve outlet | C | Surge valve | E | Poppet valve | G | Core outlet |
| B | Check valve | D | Control valve inlet | F | Bypass jacket | H | Bypass jacket outlet |

Figure 6-12. *Control valve with surge protection.*

oil in the core. As the oil warms up and reaches its operating temperature, the bellows of the thermostat expand and closes the outlet from the bypass jacket. The oil cooler flow control valve, located on the oil cooler, must now flow oil through the core of the oil cooler. No matter which path it takes through the cooler, the oil always flows over the bellows of the thermostatic valve. As the name implies, this unit regulates the temperature by either cooling the oil or passing it on to the tank without cooling, depending on the temperature at which it leaves the engine.

Surge Protection Valves

When oil in the system is congealed, the scavenger pump may build up a very high pressure in the oil return line. To prevent this high pressure from bursting the oil cooler or blowing off the hose connections, some aircraft have surge protection valves in the engine lubrication systems. One type of surge valve is incorporated in the oil cooler flow control valve; another type is a separate unit in the oil return line. *[Figure 6-12]*

The surge protection valve incorporated in a flow control valve is the more common type. Although this flow control valve differs from the one just described, it is essentially the same except for the surge protection feature. The high-pressure operation condition is shown in *Figure 6-12*, in which the high oil pressure at the control valve inlet has forced the surge valve (C) upward. Note how this movement has opened the surge valve and, at the same time, seated the poppet valve (E). The closed poppet valve prevents oil

from entering the cooler proper; therefore, the scavenge oil passes directly to the tank through outlet (A) without passing through either the cooler bypass jacket or the core. When the pressure drops to a safe value, the spring forces the surge and poppet valves downward, closing the surge valve (C) and opening the poppet valve (E). Oil then passes from the control valve inlet (D), through the open poppet valve, and into the bypass jacket (F). The thermostatic valve, according to oil temperature, determines oil flow either through the bypass jacket to port (H) or through the core to port (G). The check valve (B) opens to allow the oil to reach the tank return line.

Airflow Controls

By regulating the airflow through the cooler, the temperature of the oil can be controlled to fit various operating conditions. For example, the oil reaches operating temperature more quickly if the airflow is cut off during engine warm-up. There are two methods in general use: shutters installed on the rear of the oil cooler, and a flap on the air-exit duct. In some cases, the oil cooler air-exit flap is opened manually and closed by a linkage attached to a flight deck lever. More often, the flap is opened and closed by an electric motor.

One of the most widely used automatic oil temperature control devices is the floating control thermostat that provides manual and automatic control of the oil inlet temperatures. With this type of control, the oil cooler air-exit door is opened and closed automatically by an electrically operated actuator. Automatic operation of the actuator is determined by electrical impulses received from a controlling thermostat inserted in the oil pipe leading from the oil cooler to the oil supply tank. The actuator may be operated manually by an oil cooler air-exit door control switch. Placing this switch in the "open" or "closed" position produces a corresponding movement of the cooler door. Placing the switch in the "auto" position puts the actuator under the automatic control of the floating control thermostat. *[Figure 6-13]* The thermostat shown in *Figure 6-13* is adjusted to maintain a normal oil temperature so that it does not vary more than approximately 5° to 8 °C, depending on the installation.

During operation, the temperature of the engine oil flowing over the bimetal element causes it to wind or unwind slightly. *[Figure 6-13B]* This movement rotates the shaft (A) and the grounded center contact arm (C). As the grounded contact arm is rotated, it is moved toward either the open or closed floating contact arm (G). The two floating contact arms are oscillated by the cam (F), which is continuously rotated by an electric motor (D) through a gear train (E). When the grounded center contact arm is positioned by the bimetal element so that it touches one of the floating contact arms, an electric circuit to the oil cooler exit-flap actuator motor is completed, causing the actuator to operate and position the oil cooler air-exit flap. Newer systems use electronic control systems, but the function or the overall operation is basically the same regarding control of the oil temperature through control of the air flow through the cooler.

In some lubrication systems, dual oil coolers are used. If the typical oil system previously described is adapted to two oil coolers, the system is modified to include a flow divider, two identical coolers and flow regulators, dual air-exit doors, a two-door actuating mechanism, and a Y-fitting. *[Figure 6-14]*

Oil is returned from the engine through a single tube to the flow divider (E), where the return oil flow is divided equally into two tubes (C), one for each cooler. The coolers and regulators have the same construction and operations as the cooler and flow regulator just described. Oil from the coolers is routed through two tubes (D) to a Y-fitting, where the floating control thermostat (A) samples oil temperature and positions the two oil cooler air-exit doors through the use of a two-door actuating mechanism. From the Y-fitting, the lubricating oil is returned to the tank where it completes its circuit.

Dry Sump Lubrication System Operation

The following lubrication system is typical of those on small, single-engine aircraft. The oil system and components are those used to lubricate a 225 horsepower (hp) six-cylinder, horizontally opposed, air-cooled engine. In a typical dry sump pressure-lubrication system, a mechanical pump supplies oil under pressure to the bearings throughout the engine. *[Figure 6-4]* The oil flows into the inlet or suction side of the oil pump through a suction screen and a line connected to the external tank at a point higher than the bottom of the oil sump. This prevents sediment that falls into the sump from being drawn into the pump. The tank outlet is higher than the pump inlet, so gravity can assist the flow into the pump. The engine-driven, positive-displacement, gear-type pump forces the oil into the full flow filter. *[Figure 6-6]* The oil either passes through the filter under normal conditions or, if the filter were to become clogged, the filter bypass valve would open as mentioned earlier. In the bypass position, the oil would not be filtered. As seen in *Figure 6-6*, the regulating (relief) valve senses when system pressure is reached and opens enough to bypass oil to the inlet side of the oil pump. Then, the oil flows into a manifold that distributes the oil

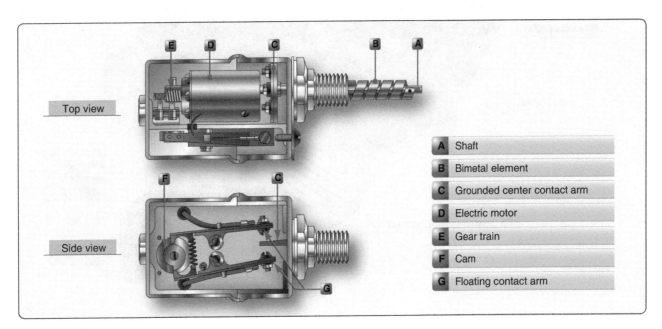

Figure 6-13. *Floating control thermostat.*

A	Shaft
B	Bimetal element
C	Grounded center contact arm
D	Electric motor
E	Gear train
F	Cam
G	Floating contact arm

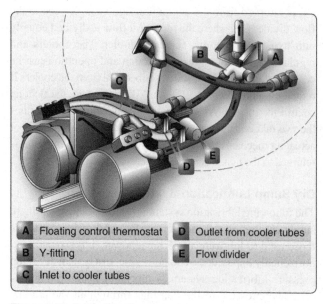

A	Floating control thermostat	D	Outlet from cooler tubes
B	Y-fitting	E	Flow divider
C	Inlet to cooler tubes		

Figure 6-14. *Dual oil cooler system.*

through drilled passages to the crankshaft bearings and other bearings throughout the engine. Oil flows from the main bearings through holes drilled in the crankshaft to the lower connecting rod bearings. *[Figure 6-15]*

Oil reaches a hollow camshaft (in an inline or opposed engine), or a cam plate or cam drum (in a radial engine), through a connection with the end bearing or the main oil manifold; it then flows out to the various camshaft, cam drum, or cam plate bearings and the cams.

The engine cylinder surfaces receive oil sprayed from the crankshaft and also from the crankpin bearings. Since oil seeps slowly through the small crankpin clearances before it is sprayed on the cylinder walls, considerable time is required for enough oil to reach the cylinder walls, especially on a cold day when the oil flow is more sluggish. This is one of the chief reasons for using modern multiviscosity oils that flow well at low temperatures.

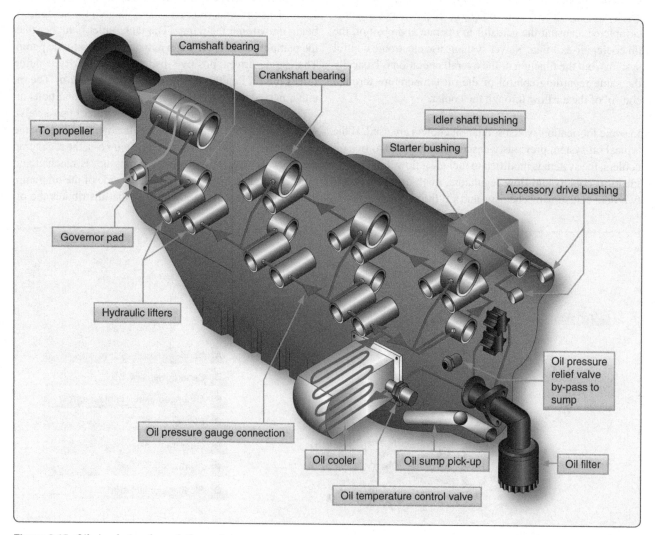

Figure 6-15. *Oil circulation through the engine.*

When the circulating oil has performed its function of lubricating and cooling the moving parts of the engine, it drains into the sumps in the lowest parts of the engine. Oil collected in these sumps is picked up by gear or gerotor-type scavenger pumps as quickly as it accumulates. These pumps have a greater capacity than the pressure pump. This is needed because the volume of the oil has generally increased due to foaming (mixing with air). On dry sump engines, this oil leaves the engine, passes through the oil cooler, and returns to the supply tank.

A thermostat attached to the oil cooler controls oil temperature by allowing part of the oil to flow through the cooler and part to flow directly into the oil supply tank. This arrangement allows hot engine oil with a temperature still below 65 °C (150 °F) to mix with the cold uncirculated oil in the tank. This raises the complete engine oil supply to operating temperature in a shorter period of time.

Wet-Sump Lubrication System Operation

A simple form of a wet-sump system is shown in *Figure 6-16*. The system consists of a sump or pan in which the oil supply is contained. The oil supply is limited by the sump (oil pan) capacity. The level (quantity) of oil is indicated or measured by a vertical rod that protrudes into the oil from an elevated hole on top of the crankcase. In the bottom of the sump (oil pan) is a screen strainer having a suitable mesh, or series of openings, to strain undesirable particles from the oil and yet pass sufficient quantity to the inlet or (suction) side of the oil pressure pump. *Figure 6-17* shows a typical oil sump that has the intake tube running through it. This preheats the air-fuel mixture before it enters the cylinders.

The rotation of the pump, which is driven by the engine, causes the oil to pass around the outside of the gears. *[Figure 6-6]* This develops a pressure in the crankshaft oiling system (drilled passage holes). The variation in the

Figure 6-17. *Wet-sump system's sump with intake tube running through it.*

speed of the pump from idling to full-throttle operating range of the engine and the fluctuation of oil viscosity because of temperature changes are compensated by the tension on the relief valve spring. The pump is designed to create a greater pressure than required to compensate for wear of the bearings or thinning out of oil. The parts oiled by pressure throw a lubricating spray into the cylinder and piston assemblies. After lubricating the various units it sprays, the oil drains back into the sump and the cycle is repeated. The system is not readily adaptable to inverted flying since the entire oil supply floods the engine.

Lubrication System Maintenance Practices
Oil Tank

The oil tank, constructed of welded aluminum, is serviced (filled) through a filler neck located on the tank and equipped with a spring-loaded locking cap. Inside the tank, a weighted, flexible rubber oil hose is mounted so that it is repositioned automatically to ensure oil pickup during all maneuvers. A dipstick guard is welded inside the tank for the protection of the flexible oil hose assembly. During normal flight, the oil tank is vented to the engine crankcase by a flexible line at the top of the tank. The location of the oil system components in relation to each other and to the engine is shown in *Figure 6-18*.

Repair of an oil tank usually requires that the tank be removed. The removal and installation procedures normally remain the same regardless of whether the engine is removed or not. First, the oil must be drained. Most light aircraft provide an oil drain similar to that shown in *Figure 6-19*. On some aircraft, the normal ground attitude of the aircraft may prevent the oil tank from draining completely. If the amount of undrained oil is excessive, the aft portion of the tank can be raised slightly after the tank straps have been loosened to complete the drainage.

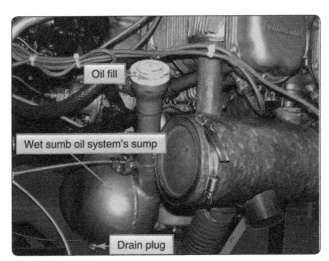

Figure 6-16. *Basic wet-sump oil system.*

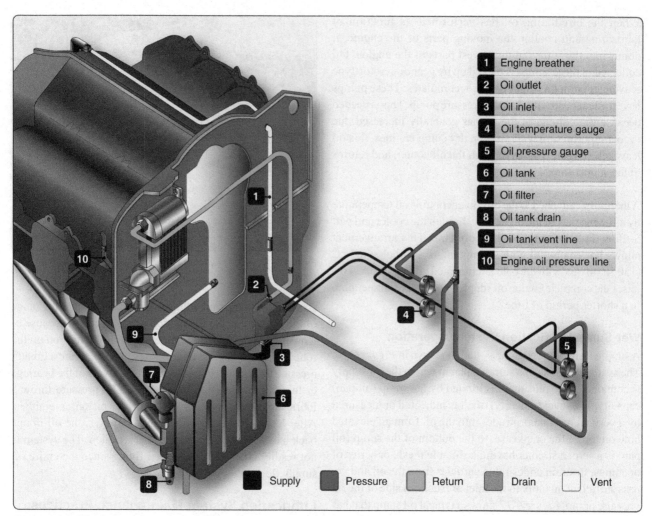

1	Engine breather
2	Oil outlet
3	Oil inlet
4	Oil temperature gauge
5	Oil pressure gauge
6	Oil tank
7	Oil filter
8	Oil tank drain
9	Oil tank vent line
10	Engine oil pressure line

■ Supply ■ Pressure ■ Return ■ Drain □ Vent

Figure 6-18. *Oil system perspective.*

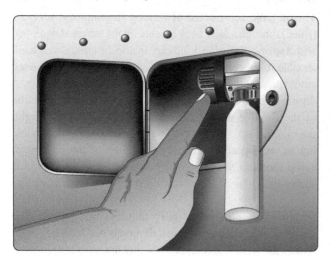

Figure 6-19. *Oil tank drain.*

After disconnecting the oil inlet and vent lines, the scupper drain hose and bonding wire can be removed. *[Figure 6-20]* The securing straps fitted around the tank can now be removed. *[Figure 6-21]* Any safety wire securing the clamp must be removed before the clamp can be loosened and the

strap disconnected. The tank can now be lifted out of the aircraft. The tank is reinstalled by reversing the sequence used in the tank removal. After installation, the oil tank should be filled to capacity. *[Figure 6-22]*

After the oil tank has been filled, the engine should be run for at least two minutes. Then, the oil level should be checked and, if necessary, sufficient oil should be added to bring the oil up to the proper level on the dipstick. *[Figure 6-23]*

Oil Cooler

The oil cooler used with this aircraft's opposed-type engine is the honeycomb type. *[Figure 6-24]* With the engine operating and an oil temperature below 65 °C (150 °F), oil cooler bypass valve opens allowing oil to bypass the core. This valve begins to close when the oil temperature reaches approximately 65 °C (150 °F). When the oil temperature reaches 85 °C (185 °F), ±2 °C, the valve is closed completely, diverting all oil flow through the cooler core.

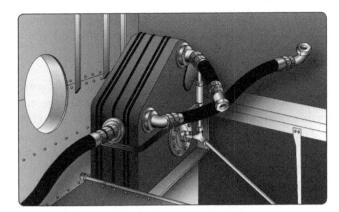

Figure 6-20. *Disconnect oil lines.*

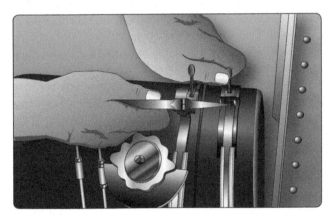

Figure 6-21. *Removal of securing straps.*

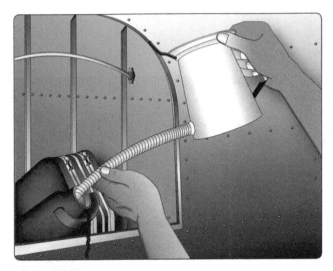

Figure 6-22. *Filling an oil tank.*

Oil Temperature Bulbs

Most oil temperature bulbs are mounted in the pressure oil screen housing. They relay an indication of engine oil inlet temperature to the oil temperature indicators mounted on the instrument panel. Temperature bulbs can be replaced by removing the safety wire and disconnecting the wire leads from the temperature bulbs, then removing the temperature bulbs using the proper wrench. *[Figure 6-25]*

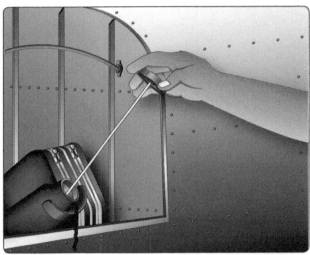

Figure 6-23. *Checking oil level with dipstick.*

Figure 6-24. *Oil cooler.*

Pressure & Scavenge Oil Screens

Sludge accumulates on the pressure and scavenges oil screens during engine operation. *[Figure 6-26]* These screens must be removed, inspected, and cleaned at the intervals specified by the manufacturer.

Typical removal procedures include removing the safety devices and loosening the oil screen housing or cover plate. A suitable container should be provided to collect the oil that drains from the filter housing or cavity. The container must be clean so that the oil collected in it can be examined for foreign particles. Any contamination already present in the container gives a false indication of the engine condition. This could result in a premature engine removal.

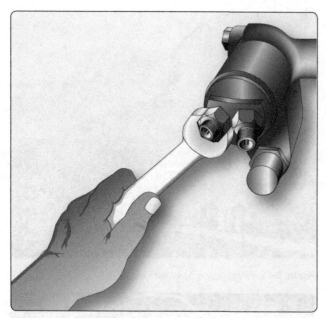

Figure 6-25. *Removing oil temperature bulb.*

After the screens are removed, they should be inspected for contamination and for the presence of metal particles that may indicate possible engine internal wear, damage, or in extreme cases, engine failure. The screen must be cleaned prior to reinstalling in the engine. In some cases, it is necessary to disassemble the filter for inspection and cleaning. The manufacturer's procedures should be followed when disassembling and reassembling an oil screen assembly. When reinstalling a filter or screen, use new O-rings and gaskets and tighten the filter housing or cover retaining nuts to the torque value specified in the applicable maintenance manual. Filters should be safetied as required.

Oil Pressure Relief Valve

An oil pressure regulating (relief) valve limits oil pressure to the value specified by the engine manufacturer. Oil pressure settings can vary from around 35 psi minimum to around 90 psi maximum, depending on the installation. The oil pressure must be high enough to ensure adequate lubrication of the engine and accessories at high speeds and power settings. On the other hand, the pressure must not be too high, since leakage and damage to the oil system may result. Before any attempt is made to adjust the oil pressure, the engine must be at the correct operating temperature and a check should be made to assure that the correct viscosity oil is being used in the engine. One example of adjusting the oil pressure is done by removing a cover nut, loosening a locknut, and turning the adjusting screw. *[Figure 6-27]* Turn the adjusting screw clockwise to increase the pressure, or counterclockwise to decrease the pressure. Make the pressure adjustments while the engine is idling and tighten the adjustment screw lock-nut after each adjustment. Check the oil pressure reading while the engine is running at the rpm specified in the manufacturer's maintenance manual. This may be from around 1,900 rpm to 2,300 rpm. The oil pressure reading should be between the limits prescribed by the manufacturer at all throttle settings.

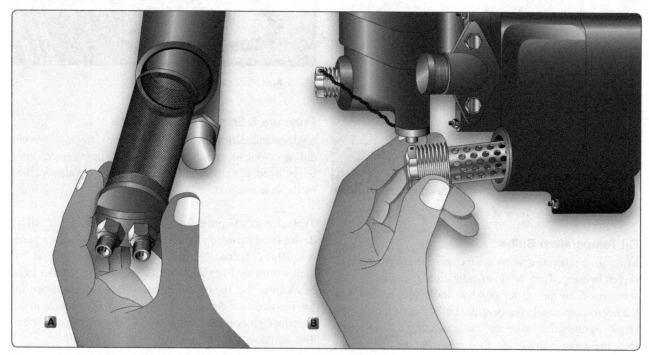

Figure 6-26. *Oil pressure screen (A) and scavenge oil screen assembly (B).*

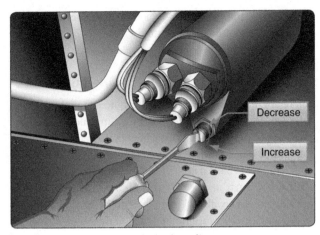

Figure 6-27. *Oil pressure relief valve adjustment.*

Recommendations for Changing Oil

Draining Oil

Oil, in service, is constantly exposed to many harmful substances that reduce its ability to protect moving parts. The main contaminants are:

- Gasoline,
- Moisture,
- Acids,
- Dirt,
- Carbon, and
- Metallic particles.

Because of the accumulation of these harmful substances, common practice is to drain the entire lubrication system at regular intervals and refill with new oil. The time between oil changes varies with each make and model aircraft and engine combination.

In engines that have been operating on straight mineral oil for several hundred hours, a change to ashless dispersant oil should be made with a degree of caution as the cleaning action of some ashless dispersant oils tends to loosen sludge deposits and cause plugged oil passages. When an engine has been operating on straight mineral oil, and is known to be in excessively dirty condition, the switch to ashless dispersant oil should be deferred until after the engine is overhauled.

When changing from straight mineral oil to ashless dispersant oil, the following precautionary steps should be taken:

1. Do not add ashless dispersant oil to straight mineral oil. Drain the straight mineral oil from the engine and fill with ashless dispersant oil.

2. Do not operate the engine longer than 5 hours before the first oil change.

3. Check all oil filters and screens for evidence of sludge or plugging. Change oil every 10 hours if sludge conditions are evident. Repeat 10-hour checks until clean screen is noted, then change oil at recommended time intervals.

4. All turbocharged engines must be broken in and operated with ashless dispersant oil.

Oil & Filter Change & Screen Cleaning

One manufacturer recommends that for new, remanufactured; or newly overhauled engines and for engines with any newly installed cylinders, the oil should be changed after the first replacement/screen cleaning at 25 hours. The oil should be changed, filter replaced, or pressure screen cleaned, and oil sump suction screen cleaned and inspected. A typical interval for oil change is 25 hours, along with a pressure screen cleaning and oil sump suction screen check for all engines employing a pressure screen system. Typical 50-hour interval oil changes generally include the oil filter replacement and suction screen check for all engines using full-flow filtration systems. A time maximum of 4 months between servicing is also recommended for oil system service.

Oil Filter Removal Canister Type Housing

Remove the filter housing from the engine by removing the safety wire and loosening the hex head screw and housing by turning counterclockwise and removing the filter from the engine. *[Figure 6-7]* Remove the nylon nut that holds the cover plate on the engine side of the filter. Remove the cover plate, hex head screw from the housing. To remove the spin-on type of filter, cut the safety wire and use the wrench pad on the rear of the filter to turn the filter counterclockwise, and remove filter. Inspect the filter element as described in the following paragraph. Discard old gaskets and replace with new replacement kit gaskets.

Oil Filter/Screen Content Inspection

Check for premature or excessive engine component wear that is indicated by the presence of metal particles, shavings, or flakes in the oil filter element or screens. The oil filter can be inspected by opening the filter paper element. Check the condition of the oil from the filter for signs of metal contamination. Then, remove the paper element from the filter and carefully unfold the paper element; examine the material trapped in the filter. If the engine employs a pressure screen system, check the screen for metal particles. After draining the oil, remove the suction screen from the oil sump and check for metal particles. *[Figure 6-28]* If examination of the used oil filter or pressure screen and the oil sump suction screen indicates abnormal metal content, additional service may be required to determine the source and possible need for corrective maintenance. To inspect the spin on filter the can must be cut open to remove the filter element for inspection.

Figure 6-28. *Oil sump screen.*

Using the special filter cutting tool, slightly tighten the cutter blade against filter and rotate 360° until the mounting plate separates from the can. *[Figure 6-29]* Using a clean plastic bucket containing varsol, move the filter to remove contaminants. Use a clean magnet and check for any ferrous metal particles in the filter or varsol solution. Then, take the remaining varsol and pour it through a clean filter or shop towel. Using a bright light, inspect for any nonferrous metals.

Assembly of & Installation of Oil Filters

After cleaning the parts, installation of the canister or filter element type filter is accomplished by lightly oiling the new rubber gaskets and installing a new copper gasket on the hex head screw. Assemble the hex head screw into the filter case using the new copper gasket. Install the filter element and place the cover over the case, then manually thread on the nylon nut by hand. Install the housing on the engine by turning it clockwise, then torque and safety it. Spin-on filters generally have installation instructions on the filter. Place a coating of engine oil on the rubber gasket, install the filter, torque and safety it. Always follow the manufacturer's current instructions to perform any maintenance.

Troubleshooting Oil Systems

The outline of malfunctions and their remedies listed in *Figure 6-30* can expedite troubleshooting of the lubrication system. The purpose of this section is to present typical troubles. It is not intended to imply that any of the troubles are exactly as they may be in a particular aircraft.

Requirements for Turbine Engine Lubricants

There are many requirements for turbine engine lubricating oils. Due to the absence of reciprocating motion and the presence of ball and roller bearings (antifriction bearings), the turbine engine uses a less viscous lubricant. Gas turbine engine oil must have a high viscosity for good load-carrying ability but must also be of sufficiently low viscosity to provide good flowability. It must also be of low volatility to prevent loss by evaporation at the high altitudes at which the engines operate. In addition, the oil should not foam and should be essentially nondestructive to natural or synthetic rubber seals in the lubricating system. Also, with high-speed antifriction bearings, the formation of carbons or varnishes must be held to a minimum. Synthetic oil for turbine engines are usually supplied in sealed one-quart cans.

The many requirements for lubricating oils are met in the synthetic oils developed specifically for turbine engines. Synthetic oil has two principal advantages over petroleum oil. It has a lower tendency to deposit lacquer and coke (solids left after solvents have been evaporated) because it does not evaporate the solvents from the oil at high temperature. Oil grades used in some turbine engines normally contain thermal and oxidation preventives, load-carrying additives, and substances that lower the pour point in addition to synthetic chemical-base materials. MIL-L-7808, which is a military specification for turbine oil, is a type I turbine oil. Turbine synthetic oil has a viscosity of around 5 to 5.5 centistokes at 210° F that is approved against the military specification MIL-PRF-23699F. This oil is referred to as type II turbine oil. Most turbine oils meet this type II specification and are made with the following characteristics:

1. Vapor phase deposits—carbon deposits formed from oil mist and vapor contact with hot engine surfaces.

2. Load-carrying ability—provides for heavy loads on

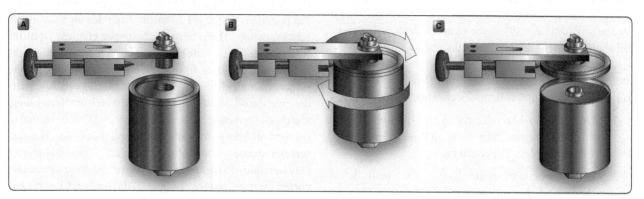
Figure 6-29. *Cutting open a spin-on type oil filter using a special filter cutter.*

Trouble	Isolation Procedure	Remedy
1 Excessive oil consumption		
Oil line leakage.	Check external lines for evidence of oil leakage.	Replace or repair defective lines.
Accessory seal leakage.	Check for leak at accessories immediately after engine operation.	Replace accessory and/or defective accessory oil seal.
Low grade of oil.		Fill tank with proper grade oil.
Failing or failed bearing.	Check sump and oil pressure pump screen for material particles.	Replace engine if metal particles are found.
2 High or low indicated oil pressure		
Defective pressure gauge.	Check indicator.	Replace indicator if defective.
Improper operation of oil pressure.	Erratic pressure indications either excessively high or low.	Remove, clean, and inspect relief valve accessory oil seal.
Inadequate oil supply.	Check oil quantity.	Fill oil tank.
Diluted or contaminated oil.		Drain engine and tank; refill tank.
Clogged oil screen.		Remove and clean oil screen.
Oil viscosity incorrect.	Make sure correct oil is being used.	Drain engine and tank; refill tank.
Oil pump pressure relief valve adjustment incorrect.	Check pressure relief valve adjustment.	Make correct adjustment on oil pump pressure relief valve.
3 High or low indicated oil temperature		
Defective temperature gauge.	Check indicator.	Replace indicator if defective.
Inadequate oil supply.	Check oil quantity.	Fill oil tank.
Diluted or contaminated oil.		Drain engine and tank; refill tank.
Obstruction in oil tank.	Check tank.	Drain oil and remove obstruction.
Clogged oil screen.		Remove and clean oil screens.
Obstruction in oil cooler passages.	Check cooler for blocked or deformed passages.	Replace oil cooler if defective.
4 Oil foaming		
Diluted or contaminated oil.		Drain engine and tank; refill tank.
Oil level in tank too high.	Check oil quantity.	Drain excess oil from tank.

Figure 6-30. *Oil system troubleshooting procedures.*

the bearing systems of turbine engines.

3. Cleanliness—minimum formation of sludge deposits during severe operation.

4. Bulk stability—resistance to physical or chemical change resulting from oxidation. Permits long periods of service operation without significant increase in viscosity or total acidity, the main indicators of oxidation.

5. Compatibility—most turbine oil is compatible with other oils that meet the same military specification. But, most engine manufacturers do not recommend the indiscriminate mixing of approved oil brands and

this is not a generally accepted practice.

6. Seal Wear—essential for the life of engines with carbon seals that lubricant properties prevent wear of the carbon at the carbon seal face.

Turbine Oil Health & Safety Precautions

Under normal conditions, the use of turbine oil presents a low health risk for humans. Although each person reacts somewhat differently to exposure, contact with liquids, vapors, and mist of turbine oil should be minimized. Information on established limits on exposure to turbine oil can generally be found in the material safety data sheets (MSDS). Prolonged breathing of hydrocarbon vapor concentrations in excess of the prescribed limits may result in lightheadedness, dizziness, and nausea. If turbine oil is ingested, call a doctor immediately; identify the product and how much was ingested. Because of the risk of ingestion, petroleum products should never be siphoned by mouth.

Prolonged or repeated contact of turbine oil with the skin can cause irritation and dermatitis. In case of skin contact, wash the skin thoroughly with soap and warm water. Promptly remove oil-soaked clothing and wash. If turbine oil contacts the eyes, flush the eyes with fresh water until the irritation subsides. Protective clothing, gloves, and eye protection should be used when handling turbine oil.

During operation, it is possible for the oil to be subjected to very high temperatures that can break it down and produce a product of unknown toxicity. If this happens, all precautions to avoid explosives should be taken. It can also have a tendency to blister, discolor, or remove paint whenever it is spilled. Painted surfaces should be wiped clean with a petroleum solvent after spillage.

Spectrometric Oil Analysis Program

The Spectrometric Oil Analysis Program allows an oil sample to be analyzed and searched for the presence of minute metallic elements. Due to oil circulation throughout an aircraft engine, every lubricant that is in service contains microscopic particles of metallic elements called wear metals. As the engine operates over time, the oil picks up very small particles that stay suspended in the oil. Oil analysis programs identify and measure these particles in parts per million (PPM) by weight. The analyzed elements are grouped into categories, such as wear metals and additives, and their measurement in PPM provides data that expert analysts can use as one of many tools to determine the engine's condition. An increase in PPM of certain materials can be a sign of component wear or impending failure of the engine. When you take a sample, note and record the amount of wear metals. If the amount of wear metals increases beyond a normal rate, then the operator can be notified quickly so repair or a recommendation of a specific maintenance procedure or inspection can be ordered.

Oil analysis increases safety by identifying an engine problem before engine failure. It also saves money by finding engine problems before they become large problems or complete engine failure. This procedure can be used for both turbine and reciprocating engines.

Typical Wear Metals & Additives

The following examples of wear metals are associated with areas of the engine that could be lead to their source. Identifying the metal can help identify the engine components that are wearing or failing.

- Iron—wear from rings, shafts, gears, valve train, cylinder walls, and pistons in some engines.
- Chromium—primary sources are chromed parts (such as rings, liners, etc.) and some coolant additives.
- Nickel—secondary indicator of wear from certain types of bearings, shafts, valves, and valve guides.
- Aluminum—indicates wear of pistons, rod bearings, and certain types of bushings.
- Lead—mostly from tetraethyl lead contamination.
- Copper—wear from bearings, rocker arm bushings, wrist pin bushings, thrust washers, and other bronze or brass parts, and oil additive or antiseize compound.
- Tin—wear from bearings.
- Silver—wear of bearings that contain silver and, in some instances, a secondary indicator of oil cooler problems.
- Titanium—alloy in high-quality steel for gears and bearings.
- Molybdenum—gear or ring wear and used as an additive in some oils.
- Phosphorous—antirust agents, spark plugs, and combustion chamber deposits.

Turbine Engine Lubrication Systems

Both wet- and dry-sump lubrication systems are used in gas turbine engines. Wet-sump engines store the lubricating oil in the engine proper, while dry-sump engines utilize an external tank mounted on the engine or somewhere in the aircraft structure near the engine, similar to reciprocating piston engines mentioned earlier.

Turbine engine's oil systems can also be classified as a pressure relief system that maintains a somewhat constant pressure: the full flow type of system, in which the pressure varies with engine speed, and the total loss system, used in engines that are for short duration operation (target drones, missiles, etc.). The most widely used system is the pressure relief system with the full flow used mostly on large fan-type engines. One of the main functions of the oil system in turbine engines is cooling the bearings by carrying the heat away from the bearing by circulating oil around the bearing.

The exhaust turbine bearing is the most critical lubricating point in a gas turbine engine because of the high temperature normally present. In some engines, air cooling is used in addition to oil cooling the bearing, which supports the turbine. Air cooling, referred to as secondary air flow, is cooling air provided by bleed air from the early stages of the compressor.

This internal air flow has many uses on the inside of the engine. It is used to cool turbine disc, vanes, and blades. Also, some turbine wheels may have bleed air flowing over the turbine disc, which reduces heat radiation to the bearing surface. Bearing cavities sometimes use compressor air to aid in cooling the turbine bearing. This bleed air, as it is called, is usually bled off a compressor stage at a point where air has enough pressure but has not yet become too warm (as the air is compressed, it becomes heated).

The use of cooling air substantially reduces the quantity of oil necessary to provide adequate cooling of the bearings. Since cooling is a major function of the oil in turbine engines, the lubricating oil for bearing cooling normally requires an oil cooler. When an oil cooler is required, usually a greater quantity of oil is necessary to provide for circulation between the cooler and engine. To ensure proper temperature, oil is routed through either air-cooled and/or fuel-cooled oil coolers. This system is used to also heat (regulate) the fuel to prevent ice in the fuel.

Turbine Lubrication System Components

The following component descriptions include most found in the various turbine lubrication systems. However, since engine oil systems vary somewhat according to engine model and manufacturer, not all of these components are necessarily found in any one system.

Oil Tank

Although the dry-sump systems use an oil tank that contains most of the oil supply, a small sump is usually included on the engine to hold a small supply of oil. It usually contains the oil pump, the scavenge and pressure inlet strainers, scavenge return connection, pressure outlet ports, an oil filter, and mounting bosses for the oil pressure gauge and temperature bulb connections.

A view of a typical oil tank is shown in *Figure 6-31*. It is designed to furnish a constant supply of oil to the engine during any aircraft attitude. This is done by a swivel outlet assembly mounted inside the tank, a horizontal baffle mounted in the center of the tank, two flapper check valves mounted on the baffle, and a positive vent system.

The swivel outlet fitting is controlled by a weighted end that is free to swing below the baffle. The flapper valves in the baffle are normally open; they close only when the oil in the bottom of the tank tends to rush to the top of the tank during decelerations. This traps the oil in the bottom of the tank where it is picked up by the swivel fitting. A sump drain is located in the bottom of the tank. The vent system inside the tank is so arranged that the airspace is vented at all times even though oil may be forced to the top of the tank by deceleration of the aircraft.

All oil tanks are provided with expansion space. This allows

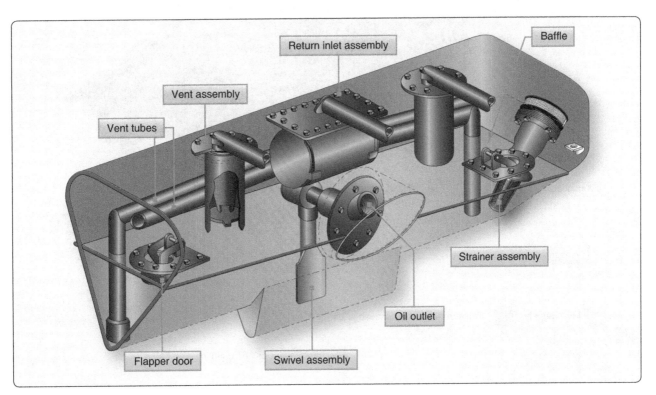

Figure 6-31. *Oil tank.*

expansion of the oil after heat is absorbed from the bearings and gears and after the oil foams as a result of circulating through the system. Some tanks also incorporate a deaerator tray for separating air from the oil returned to the top of the tank. Usually these deaerators are the can type in which oil enters at a tangent. The air released is carried out through the vent system in the top of the tank. In most oil tanks, a pressure buildup is desired within the tank to ensure a positive flow of oil to the oil pump inlet. This pressure buildup is made possible by running the vent line through an adjustable check relief valve. The check relief valve is usually set to relieve at about 4 psi, keeping positive pressure on the oil pump inlet. If the air temperature is abnormally low, the oil may be changed to a lighter grade. Some engines may provide for the installation of an immersion-type oil heater.

Oil Pump

The oil pump is designed to supply oil under pressure to the parts of the engine that require lubrication, then circulate the oil through coolers as needed, and return the oil to the oil tank. Many oil pumps consist of not only a pressure supply element, but also scavenge elements, such as in a dry-sump system. However, there are some oil pumps that serve a single function; that is, they either supply or scavenge the oil. These pump elements can be located separate from each other and driven by different shafts from the engine. The numbers of pumping elements (two gears that pump oil), pressure and scavenge, depend largely on the type and model of the engine. Several scavenge oil pump elements can be used to accommodate the larger capacity of oil and air mix. The scavenge elements have a greater pumping capacity than the pressure element to prevent oil from collecting in the bearing sumps of the engine.

The pumps may be one of several types, each type having certain advantages and limitations. The two most common oil pumps are the gear and gerotor, with the gear-type being the most commonly used. Each of these pumps has several possible configurations.

The gear-type oil pump has only two elements: one for pressure oil and one for scavenging. *[Figure 6-32]* However, some types of pumps may have several elements: one or more

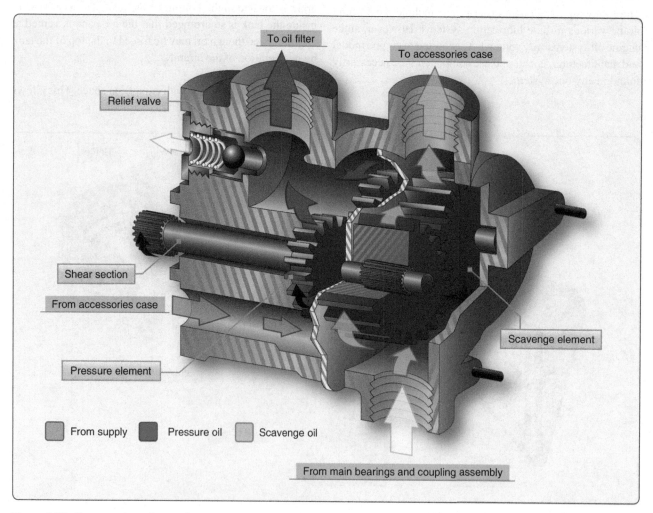

Figure 6-32. *Cutaway view of gear oil pump.*

elements for pressure and two or more for scavenging. The clearances between the gear teeth and the sides of the pump wall and plate are critical to maintain the correct output of the pump.

A regulating (relief) valve in the discharge side of the pump limits the output pressure of the pump by bypassing oil to the pump inlet when the outlet pressure exceeds a predetermined limit. [Figure 6-32] The regulating valve can be adjusted, if needed, to bring the oil pressure within limits. Also shown is the shaft shear section that causes the shaft to shear if the pump gears should seize up and not turn.

The gerotor pump, like the gear pump, usually contains a single element for oil pressure and several elements for scavenging oil. Each of the elements, pressure and scavenge, is almost identical in shape; however, the capacity of the elements can be controlled by varying the size of the gerotor elements. For example, the pressure element may have a pumping capacity of 3.1 gallons per minute (gpm) as compared to 4.25 gpm capacity for the scavenge elements. Consequently, the pressure element is smaller since the elements are all driven by a common shaft. The pressure is determined by engine rpm with a minimum pressure at idling speed and maximum pressure at intermediate and maximum engine speeds.

A typical set of gerotor pumping elements is shown in Figure 6-33. Each set of gerotors is separated by a steel plate, making each set an individual pumping unit consisting of an inner and an outer element. The small star-shaped inner element has external lobes that fit within and are matched with the outer element that has internal lobes. The small element fits on and is keyed to the pump shaft and acts as a drive for the outer free-turning element. The outer element fits within a steel plate having an eccentric bore. In one engine model, the oil pump has four elements: one for oil feed and three for scavenge. In some other models, pumps have six elements: one for feed and five for scavenge. In each case, the oil flows as long as the engine shaft is turning.

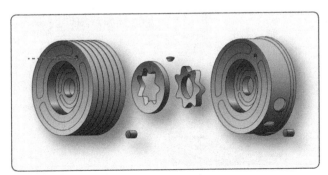

Figure 6-33. *Typical gerotor pumping elements.*

Turbine Oil Filters

Filters are an important part of the lubrication system because they remove foreign particles that may be in the oil. This is particularly important in gas turbines as very high engine speeds are attained; the antifriction types of ball and roller bearings would become damaged quite rapidly if lubricated with contaminated oil. Also, there are usually numerous drilled or core passages leading to various points of lubrication. Since these passages are usually rather small, they are easily clogged.

There are several types and locations of filters used for filtering the turbine lubricating oil. The filtering elements come in a variety of configurations and mesh sizes. Mesh sizes are measured in microns, which is a linear measurement equal to one millionth of a meter (a very small opening).

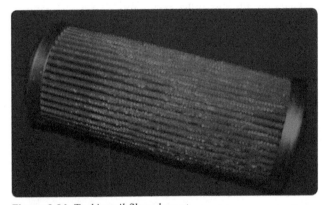

Figure 6-34. *Turbine oil filter element.*

Figure 6-35. *Turbine oil filter paper element.*

A main oil strainer filter element is shown in *Figure 6-34*. The filtering element interior is made of varying materials including paper and metal mesh. *[Figure 6-35]* Oil normally flows through the filter element from the outside into the filter body. One type of oil filter uses a replaceable laminated paper element, while others use a very fine stainless steel metal mesh of about 25–35 microns.

Most filters are located close to the pressure pump and consist of a filter body or housing, filter element, a bypass valve, and a check valve. The filter bypass valve prevents the oil flow from being stopped if the filter element becomes clogged. The bypass valve opens whenever a certain pressure is reached. If this occurs, the filtering action is lost, allowing unfiltered oil to be pumped to the bearings. However, this prevents the bearings from receiving no oil at all. In the bypass mode, many engines have a mechanical indicator that pops out to indicate the filter is in the bypass mode. This indication is visual and can only be seen by inspecting the engine directly. An antidrain check valve is incorporated into the assembly to prevent the oil in the tank from draining down into the engine sumps when the engine is not operating. This check valve is normally spring loaded closed with 4 to 6 psi needed to open it.

The filters generally discussed are used as main oil filters; that is, they strain the oil as it leaves the pump before being piped to the various points of lubrication. In addition to the main oil filters, there are also secondary filters located throughout the system for various purposes. For instance, there may be a finger screen filter that is sometimes used for straining scavenged oil. These screens tend to be large mesh screens that trap larger contaminants. Also, there are fine-mesh screens called last chance filters for straining the oil just before it passes from the spray nozzles onto the bearing surfaces. *[Figure 6-36]* These filters are located at each bearing and help screen out contaminants that could plug the oil spray nozzle.

Oil Pressure Regulating Valve

Most turbine engine oil systems are pressure regulating type systems that keep the pressure fairly constant. An oil pressure regulating valve is included in the oil system on the pressure side of the pressure pump. A regulating valve system controls the systems pressure to a limited pressure within the system. It is more of a regulating valve than a relief valve because it keeps the pressure in the system within certain limits other than only opening when the absolute maximum pressure of the system is exceeded.

The regulating valve *Figure 6-37* has a valve held against a seat by a spring. By adjusting the tension (increase) on the spring, you change the pressure at which the valve opens, and you also increase the system pressure. A screw pressing on the spring adjusts the tension on the valve and the system pressure.

Oil Pressure Relief Valve

Some large turbofan oil systems do not have a regulating valve. The system pressure varies with engine rpm and pump speed. There is a wide range of pressure in this system. A relief valve is used to relieve pressure only if it exceeds the maximum limit for the system. *[Figure 6-38]* This true relief valve system is preset to relieve pressure and bypass the oil back to the inlet side of the oil pump whenever the pressure exceeds the maximum preset system limit. This relief valve is especially important when oil coolers are incorporated in the system since the coolers are easily ruptured because of their thin-wall construction. Under normal operation, it should never open.

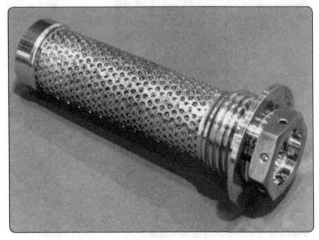

Figure 6-36. *Last-chance filter before spray nozzle.*

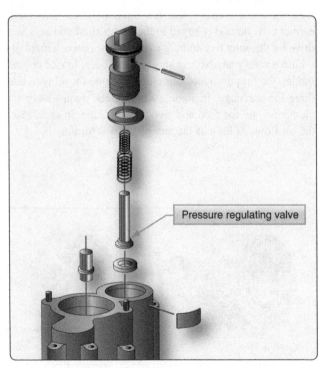

Pressure regulating valve

Figure 6-37. *Pressure regulating valve.*

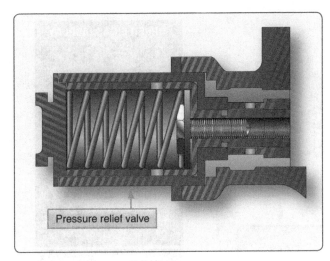

Figure 6-38. *Pressure relief valve.*

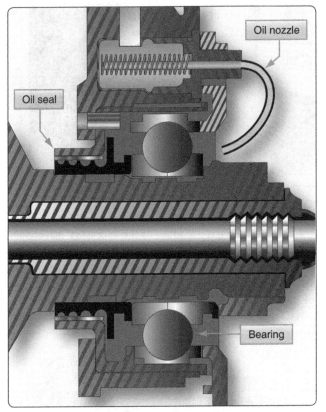

Figure 6-39. *Oil nozzles spray lubricate on bearings.*

Oil Jets

Oil jets (or nozzles) are located in the pressure lines adjacent to, or within, the bearing compartments and rotor shaft couplings. *[Figure 6-39]* The oil from these nozzles is delivered in the form of an atomized spray. Some engines use an air-oil mist spray that is produced by tapping high-pressure bleed air from the compressor to the oil nozzle outlet. This method is considered adequate for ball and roller bearings; however, the solid oil spray method is considered the better of the two methods.

The oil jets are easily clogged because of the small orifice in their tips; consequently, the oil must be free of any foreign particles. If the last-chance filters in the oil jets should become clogged, bearing failure usually results since nozzles are not accessible for cleaning except during engine maintenance. To prevent damage from clogged oil jets, main oil filters are checked frequently for contamination.

Lubrication System Instrumentation

Gauge connection provisions are incorporated in the oil system for oil pressure, oil quantity, low oil pressure, oil filter differential pressure switch, and oil temperature. The oil pressure gauge measures the pressure of the lubricant as it leaves the pump and enters the pressure system. The oil pressure transmitter connection is located in the pressure line between the pump and the various points of lubrication. An electronic sensor is placed to send a signal to the Full Authority Digital Engine Control (FADEC) control unit and through the Engine Indicating and Crew Alerting System (EICAS) computers, and on to the displays in the flight deck. *[Figure 6-40]* The tank quantity transmitter information is sent to the EICAS computers. The low oil pressure switch alerts the crew if the oil pressure falls below a certain pressure during engine operation. The differential oil pressure switch alerts the flight crew of an impending oil filter bypass because of a clogged filter. A message is sent to the display in the upper EICAS display in the flight deck as can be seen in *Figure 6-40*. Oil temperature can be sensed at one or more points in the engine's oil flow path. The signal is sent to the FADEC/EICAS computer and is displayed on the lower EICAS display.

Lubrication System Breather Systems (Vents)

Breather subsystems are used to remove excess air from the bearing cavities and return the air to the oil tank where it is separated from any oil mixed in the vapor of air and oil by the deaerator. Then, the air is vented overboard and back to the atmosphere. All engine bearing compartments, oil tanks, and accessory cases are vented together so the pressure in the system remains the same.

The vent in an oil tank keeps the pressure within the tank from rising above or falling below that of the outside atmosphere. However, the vent may be routed through a check relief valve that is preset to maintain a slight (approximately 4 psi) pressure on the oil to assure a positive flow to the oil pump inlet.

In the accessory case, the vent (or breather) is a screen-protected opening that allows accumulated air pressure within the accessory case to escape to the atmosphere. The scavenged oil carries air into the accessory case and this air

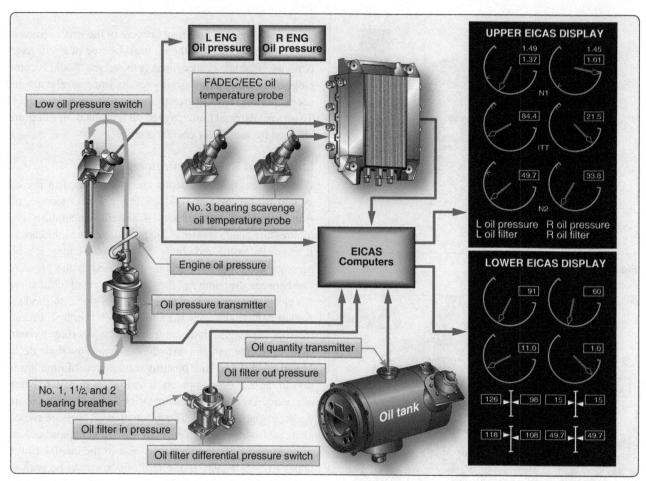

Figure 6-40. *Oil indicating system.*

must be vented. Otherwise, the pressure buildup within the accessory case would stop the flow of oil draining from the bearing, forcing this oil past the bearing oil seals and into the compressor housing. If in enough quantity, oil leakage could cause burning and seal and bearing malfunction. The screened breathers are usually located in the front center of the accessory case to prevent oil leakage through the breather when the aircraft is in unusual flight attitudes. Some breathers may have a baffle to prevent oil leakage during flight maneuvers. A vent that leads directly to the bearing compartment may be used in some engines. This vent equalizes pressure around the bearing surface so that the lower pressure at the first compressor stage does not cause oil to be forced past the bearing rear oil seal into the compressor.

Lubrication System Check Valve

Check valves are sometimes installed in the oil supply lines of dry-sump oil systems to prevent reservoir oil from seeping (by gravity) through the oil pump elements and high-pressure lines into the engine after shutdown. Check valves, by stopping flow in an opposite direction, prevent accumulations of undue amounts of oil in the accessory gearbox, compressor rear housing, and combustion chamber. Such accumulations could cause excessive loading of the accessory drive gears during starts, contamination of the cabin pressurization air, or internal oil fires. The check valves are usually the spring-loaded ball-and-socket type constructed for free flow of pressure oil. The pressure required to open these valves

Figure 6-41. *Typical thermostatic bypass valve.*

Figure 6-42. *Air oil cooler.*

varies, but the valves generally require from 2 to 5 psi to permit oil to flow to the bearings.

Lubrication System Thermostatic Bypass Valves

Thermostatic bypass valves are included in oil systems using an oil cooler. Although these valves may be called different names, their purpose is always to maintain proper oil temperature by varying the proportion of the total oil flow passing through the oil cooler. A cutaway view of a typical thermostatic bypass valve is shown in *Figure 6-41*. This valve consists of a valve body, having two inlet ports and one outlet port, and a spring-loaded thermostatic element valve. The valve is spring loaded because the pressure drop through the oil cooler could become too great due to denting or clogging of the cooler tubing. In such a case, the valve

opens, bypassing the oil around the cooler.

Air-Oil Coolers

Two basic types of oil coolers in general use are the air-cooled and the fuel-cooled. Air-oil coolers are used in the lubricating systems of some turbine engines to reduce the temperature of the oil to a degree suitable for recirculation through the system. The air-cooled oil cooler is normally installed at the forward end of the engine. It is similar in construction and operation to the air-cooled cooler used on reciprocating engines. An air-oil cooler is usually included in a dry-sump oil system. *[Figure 6-42]* This cooler may be air-cooled or fuel-cooled and many engines use both. Dry-sump lubrication systems require coolers for several reasons. First, air cooling of bearings by using compressor bleed-air is not sufficient to cool the turbine bearing cavities because of the heat present in area of the turbine bearings. Second, the large turbofan engines normally require a greater number of bearings, which means that more heat is transferred to the oil. Consequently, the oil coolers are the only means of dissipating the oil heat.

Fuel-Oil Coolers

The fuel-cooled oil cooler acts as a fuel-oil heat exchanger in that the fuel cools the hot oil and the oil heats the fuel for combustion. *[Figure 6-43]* Fuel flowing to the engine must pass through the heat exchanger; however, there is a thermostatic valve that controls the oil flow, and the oil may bypass the cooler if no cooling is needed. The fuel-oil heat

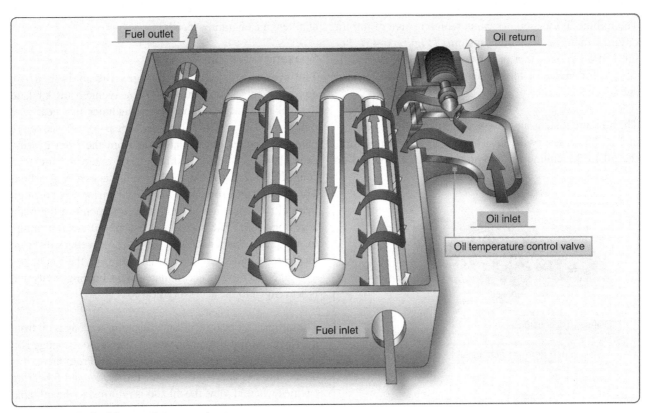

Figure 6-43. *Fuel-oil heat exchanger cooler.*

exchanger consists of a series of joined tubes with an inlet and outlet port. The oil enters the inlet port, moves around the fuel tubes, and goes out the oil outlet port.

Deoiler

The deoiler removes the oil from the breather air. The breather air goes into an impeller that turns in the deoiler housing. Centrifugal force drives the oil towards the outer wall of the impeller. Then, the oil drains from the deoiler into a sump or oil tank. Because the air is much lighter than the oil, it goes through the center of the impeller and is vented overboard.

Magnetic Chip Detectors

Magnetic chip detectors are used in the oil system to detect and catch ferrous (magnetic) particles present in the oil. *[Figure 6-44]* Scavenge oil generally flows past chip detectors so any magnetic particles are attracted and stick to the chip detector. Chip detectors are placed in several locations but generally are in the scavenge lines for each scavenge pump, oil tank, and in the oil sumps. Some engines have several detectors to one detector. During maintenance, the chip detectors are removed from the engine and inspected for metal; if none is found, the detector is cleaned, replaced, and safety wired. If metal is found on a chip detector, an investigation should be made to find the source of the metal on the chip.

Typical Dry-Sump Pressure Regulated Turbine Lubrication System

The turbine lubrication system is representative of turbine engines using a dry-sump system. *[Figure 6-45]* The lubrication system is a pressure regulated, high-pressure design. It consists of the pressure, scavenge, and breather subsystems.

The pressure system supplies oil to the main engine bearings and to the accessory drives. The scavenger system returns the oil to the engine oil tank that is usually mounted on

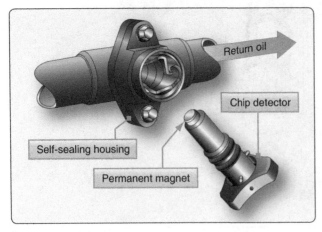

Figure 6-44. *Chip detector.*

the compressor case. It is connected to the inlet side of the pressure oil pump and completes the oil flow cycle. A breather system connecting the individual bearing compartments and the oil tank with the breather pressurizing valve completes the engine lubrication system. In a turbine pressure relief dry-sump lubrication system, the oil supply is carried in a tank mounted on the engine. With this type of system, a larger oil supply can be carried, and the temperature of the oil can be readily controlled.

Pressure System

The oil pressure branch of the engine lubrication system is pressurized by a gear-type pressure pump located in the oil pump and accessory drive housing. *[Figure 6-45]* The pressure pump receives engine oil at its lower (inlet) side and discharges pressurized oil to an oil filter located on the housing. From the oil filter, which is equipped with a bypass valve for operation in case the filter clogs, the pressurized oil is transmitted to a cored passage running through to the pressure regulating (relief) valve that maintains system pressure. The pressure regulating (relief) valve is located downstream of the pump. It is adjusted to maintain a proper pressure to the oil metering jets in the engine. The pressure regulating (relief) valve is usually easily accessible for adjustment. Then, the oil flows through the fuel-oil cooler and on to the bearing cavities through last-chance filters and out spray nozzles to the bearings. Pressurized oil distributed to the engine main bearings is sprayed on the bearings through fixed orifice nozzles providing a relatively constant oil flow at all engine operating speeds.

Scavenge System

The scavenge system scavenges the main bearing compartments and circulates the scavenged oil back to the tank. The scavenge oil system includes five gear-type pumps. *[Figure 6-45]* The No.1 bearing oil scavenge pump scavenges accumulated oil from the front bearing case. It directs the oil through an external line to a central collecting point in the main accessory gearbox. The oil return from No. 2 and 3 bearings is through internal passages to a central collecting point in the main accessory case. The accessory gearbox oil suction pump, located in the main accessory gearbox, scavenges oil from the gearbox housing to the oil tank. Oil from the No. 4, No. 4½ and No. 5 bearing accumulates in the bearing cavity and is scavenged to the accessory gearbox.

The turbine rear bearing oil suction pump scavenges oil from the No. 6 bearing compartment and directs the scavenged oil through a passage in the turbine case strut. From there, it is directed to the bearing cavity for the 4, 4½, and 5 bearing cavities where it joins the oil and is returned to the oil tank. The scavenge oil passes through the deaerator as it enters the

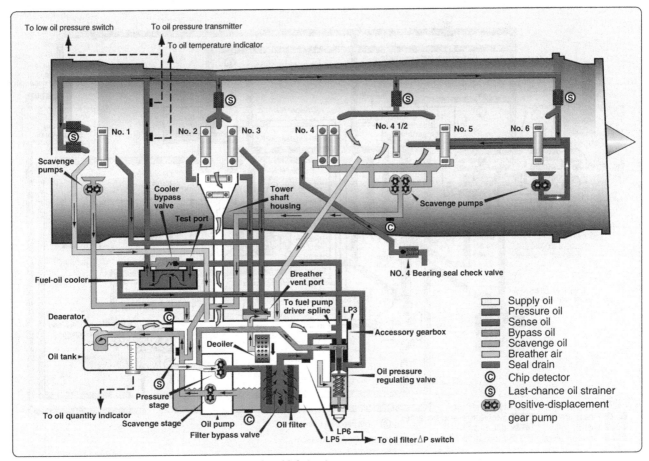

Figure 6-45. *Typical turbine dry-sump pressure regulated lubrication system.*

oil tank, which separates the air mixed in the return oil. The oil stays in the tank while the air flows into the accessory gearbox and enters the deoiler.

Breather Pressurizing System

The breather pressurizing system ensures a proper oil spray pattern from the main bearing oil jets and furnishes a pressure head to the scavenge system. Breather tubes in the compressor inlet case, the oil tank, the diffuser case, and the turbine exhaust case are connected to external tubing at the top of the engine. By means of this tubing, the vapor-laden atmospheres of the various bearing compartments and the oil tank are brought together in the deoiler in the accessory gearbox. The deoiler separates out the oil from the air-oil mist and vents the air back to the atmosphere.

Typical Dry-Sump Variable Pressure Lubrication System

The dry-sump variable-pressure lubrication system uses the same basic subsystems that the regulated systems use (pressure Scavenge breather). *[Figure 6-46]* The main difference is that the pressure in this system is not regulated by a regulating bypass valve. Most large turbofan engine pressure systems are variable-pressure systems in which the

pump outlet pressure (oil pressure) depends on the engine rpm. In other words, the pump output pressure is proportional to the engine speed. Since the resistance to flow in the system does not vary much during operation and the pump has only the variable of turning faster or slower, the pressure is a function of engine speed. As an example, oil pressure can vary widely in this type of system, from 100 psi to over 260 psi, with the relief valve opening at about 540 psi.

Pressure Subsystem

The oil flows from the oil tank down to the pressure stage of the oil pump. A slight pressure in the tank assures that the flow of oil into the pressure pump is continuous. After being pressurized, it moves on to the oil filter where it is filtered. If the filter is clogged, the bypass valve sends the oil around the filter. There is no regulating valve but there is a relief valve to prevent the system pressure from exceeding the maximum limits. This valve is usually set to open well above the system's operating pressure. The oil flows from the filter housing to the engine air-oil cooler. The oil either bypasses the cooler (cold) or passes through the cooler (hot) and then on to the fuel-oil cooler. Through the use of the coolers, the fuel temperature is adjusted to meet the requirements needed for the engine. Some of the oil passes through the classified

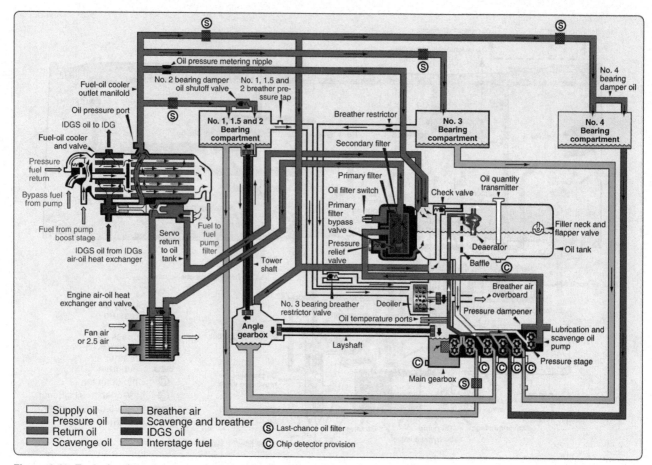

Figure 6-46. *Typical turbine dry-sump variable pressure lubrication system.*

oil pressure trim orifice that helps adjust oil pressure at low speeds. The oil now flows through the last-chance oil filters (strainers) that remove particles from the oil if the oil filter has been bypassed. The engine oil passes through the nozzles to lubricate the bearings, gearboxes, seals, and accessory drive splines. After performing its functions of lubricating, cleaning, and cooling the bearings, the oil needs to be returned to the old tank by the scavenge system.

Scavenger Subsystem

The scavenger oil pump has several stages that pull oil from the bearing compartments and gearboxes and sends the oil to the tank. At the tank, the oil enters the deaerator, which separates the air from the scavenge oil. The oil returns to the tank and the air is vented through a check valve overboard. Each stage of the scavenge pump has a magnetic chip detector that can be removed for inspection.

Breather Subsystem

The purpose of the breather system is to remove air from the bearing compartments, separate breather air from oil, and vent the air overboard. The breather air from the bearing compartments is drawn to the gearbox by the deoiler. The deoiler is turned at high speed and causes the oil to

separate from the air. The air is then vented with air from the deaerator overboard. By referring to *Figure 6-46*, notice that the deaerator is in the oil tank and the deoiler is in the main gearbox.

Turbine Engine Wet-Sump Lubrication System

In some engines, the lubrication system is the wet-sump type. There are relatively few engines using a wet-sump type of oil system. The components of a wet-sump system are similar to those of a dry-sump system. The major difference between the two systems is the location of the oil reservoir. The reservoir for the wet-sump oil system may be the accessory gear case or it may be a sump mounted on the bottom of the accessory case. Regardless of configuration, reservoirs for wet-sump systems are an integral part of the engine and contain the bulk of the engine oil supply. *[Figure 6-47]*

Included in the wet-sump reservoir are the following components:

1. A sight gauge indicates the oil level in the sump.

2. A vent or breather equalizes pressure within the accessory casing.

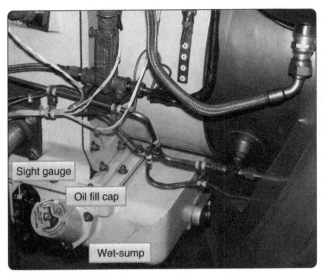

Figure 6-47. *Typical turbine wet-sump system.*

3. A magnetic drain plug may be provided to drain the oil and also to trap any ferrous metal particles in the oil. This plug should always be examined closely during inspections. The presence of metal particles may indicate gear or bearing failure.

4. Provision may also be made for a temperature bulb and an oil pressure fitting.

This system is typical of all engines using a wet-sump lubrication system. The bearing and drive gears in the accessory drive casing are lubricated by a splash system. The oil for the remaining points of lubrication leaves the pump under pressure and passes through a filter to jet nozzles that direct the oil into the rotor bearings and couplings. The oil is returned to the reservoir (sump) by gravity. Oil from the compressor bearing and the accessories drive coupling shaft drains directly into the reservoir. Turbine oil drains into a sump where the oil was originally pumped.

Turbine Engine Oil System Maintenance

Maintenance of gas turbine lubrication systems consists mainly of adjusting, removing, cleaning, and replacing various components. Oil filter maintenance and oil change intervals for turbine engines vary widely from model to model, depending on the severity of the oil temperature conditions imposed by the specific airframe installation and engine configuration. The applicable manufacturer's instructions should be followed. The oil filter should be removed at every regular inspection. It should be disassembled, cleaned, and any worn or damaged filter elements replaced. The following steps illustrate typical oil filter removal cleaning and replacement procedures:

1. Provide a suitable container for collecting the drained oil, if needed.

Figure 6-48. *Oil filter housing.*

2. Remove the filter housing and withdraw the filter assembly. *[Figure 6-48]* Discard the old seals.

3. Immerse the screen or filter in an approved carbon remover at room temperature for a few minutes. Rinse them in a degreaser fluid or cleaning solvent. Then, blow them dry with an air jet.

4. Then, install the filter in the filter housing assembly. Place a new seal and tighten it to the torque prescribed in the manufacturer's instructions.

5. Secure with lock wire.

To adjust the oil pressure, first remove the adjusting screw acorn cap on the oil pressure relief valve. Then, loosen the locknut and turn the adjusting screw clockwise to increase, or counterclockwise to decrease, the oil pressure. In a typical turbojet lubrication system, the adjusting screw is adjusted to provide an oil pressure of 45, ±5 psi, at approximately 75 percent of normal rated thrust. The adjustment should be made while the engine is idling; it may be necessary to perform several adjustments before the desired pressure is obtained. When the proper pressure setting is achieved, tighten the adjusting screw locknut, and install the acorn cap with a new gasket, then tighten and secure with lock wire.

Checking or servicing aircraft engine oil is an important maintenance function. Before servicing any aircraft engine, consult the specific aircraft maintenance manual to determine the proper type of servicing equipment and procedures. In general, aircraft engine oil is checked with a dipstick or a sight gauge. There are markings on the stick or around the sight gauge to determine the correct level. Turbine engines must be checked just after shutdown.

Maintenance of scavenge and breather systems at regular inspections includes checks for oil leaks and security of mounting of system components. Also, check chip detectors for particles of ferrous material and clean last-chance filters; install and safety.

Engine Cooling Systems

Excessive heat is always undesirable in both reciprocating and turbine aircraft engines. If means were not available for its control or elimination, major damage or complete engine failure would occur. Although the vast majority of reciprocating engines are air cooled, some diesel liquid-cooled engines are being made available for light aircraft. *[Figure 6-49]* In a liquid-cooled engine, around the cylinder are water jackets, in which liquid coolant is circulated and the coolant takes away the excess heat. The excess heat is then dissipated by a heat exchanger or radiator using air flow. Turbine engines use secondary airflow to cool the inside components and many of the exterior components.

Reciprocating Engine Cooling Systems

An internal-combustion engine is a heat machine that converts chemical energy in the fuel into mechanical energy at the crankshaft. It does not do this without some loss of energy, however, and even the most efficient aircraft engines may waste 60 to 70 percent of the original energy in the fuel. Unless most of this waste heat is rapidly removed, the cylinders may become hot enough to cause complete engine failure. Excessive heat is undesirable in any internal-combustion engine for three principal reasons:

1. It affects the behavior of the combustion of the air-fuel charge.

2. It weakens and shortens the life of engine parts.

3. It impairs lubrication.

If the temperature inside the engine cylinder is too great, the air-fuel mixture is preheated, and combustion occurs before the desired time. Since premature combustion causes detonation, knocking, and other undesirable conditions, there must be a way to eliminate heat before it causes damage.

One gallon of aviation gasoline has enough heat value to boil 75 gallons of water; thus, it is easy to see that an engine that burns 4 gallons of fuel per minute releases a tremendous amount of heat. About one-fourth of the heat released is changed into useful power. The remainder of the heat must be dissipated so that it is not destructive to the engine. In a typical aircraft powerplant, half of the heat goes out with the exhaust and the other is absorbed by the engine. Circulating oil picks up part of this soaked-in heat and transfers it to the airstream through the oil cooler. The engine cooling system takes care of the rest. Cooling is a matter of transferring the excess heat from the cylinders to the air, but there is more to such a job than just placing the cylinders in the airstream. A cylinder on a large engine is roughly the size of a gallon jug. Its outer surface, however, is increased by the use of cooling fins so that it presents a barrel-sized exterior to the cooling air. Such an arrangement increases the heat transfer by convection. If too much of the cooling fin area is broken off, the cylinder cannot cool properly, and a hotspot develops. Therefore, cylinders are normally replaced if a specified number of square inches of fins are missing.

Cowling and baffles are designed to force air over the cylinder cooling fins. *[Figure 6-50]* The baffles direct the air close around the cylinders and prevent it from forming hot pools of stagnant air while the main streams rush by unused. Blast tubes are built into the baffles to direct jets of cooling air onto the rear spark plug elbows of each cylinder to prevent overheating of ignition leads. Blast tubes also provide cooling to engine accessories such as alternators, generators, and starters.

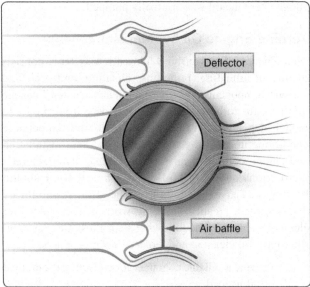

Figure 6-50. *Cylinder baffle and deflector system.*

Figure 6-49. *Diesel liquid cooled aircraft engine.*

An engine can have an operating temperature that is too low. For the same reasons that an engine is warmed up before takeoff, it is kept warm during flight. Fuel evaporation and distribution and oil circulation depend on an engine being kept at its optimum operating temperature. The aircraft engine has temperature controls that regulate air circulation over the engine. Unless some controls are provided, the engine could overheat on takeoff and get too cold in high altitude, high-speed and low-power letdowns.

The most common means of controlling cooling is the use of cowl flaps. *[Figure 6-51]* These flaps are opened and closed by electric motor-driven jackscrews, by hydraulic actuators, or manually in some light aircraft. When extended for increased cooling, the cowl flaps produce drag and sacrifice streamlining for the added cooling. On takeoff, the cowl flaps are opened only enough to keep the engine below the red-line temperature. Heating above the normal range is allowed so that drag is as low as possible. During ground operations, the cowl flaps should be opened wide since drag does not matter and cooling needs to be set at maximum. Cowl flaps are used mostly with older aircraft and radial engine installations.

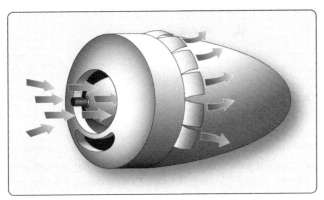

Figure 6-51. *Regulating the cooling airflow.*

Some aircraft use augmentors to provide additional cooling airflow. *[Figure 6-52]* Each nacelle has two pairs of tubes running from the engine compartment to the rear of the nacelle. The exhaust collectors feed exhaust gas into the inner augmentor tubes. The exhaust gas mixes with air that has passed over the engine and heats it to form a high-temperature, low-pressure, jet-like exhaust. This low-pressure area in the augmentors draws additional cooling air over the engine. Air entering the outer shells of the augmentors is heated through contact with the augmentor tubes but is not contaminated with exhaust gases. The heated air from the shell goes to the cabin heating, defrosting, and anti-icing system.

Augmentors use exhaust gas velocity to cause airflow over the engine so that cooling is not entirely dependent on the prop wash. Vanes installed in the augmentors control the volume of air. These vanes are usually left in the trail position to permit maximum flow. They can be closed to increase the heat for cabin or anti-icing use or to prevent the engine from cooling too much during descent from altitude. In addition to augmentors, some aircraft have residual heat doors or nacelle flaps that are used mainly to let the retained heat escape after engine shutdown. The nacelle flaps can be opened for more cooling than that provided by the augmentors. A modified form of the previously described augmentor cooling system is used on some light aircraft. *[Figure 6-53]* Augmentor systems are not used much on modern aircraft.

As shown in *Figure 6-53*, the engine is pressure cooled by air taken in through two openings in the nose cowling, one on each side of the propeller spinner. A pressure chamber is sealed off on the top side of the engine with baffles properly directing the flow of cooling air to all parts of the engine compartment. Warm air is drawn from the lower part of the engine compartment by the pumping action of the exhaust

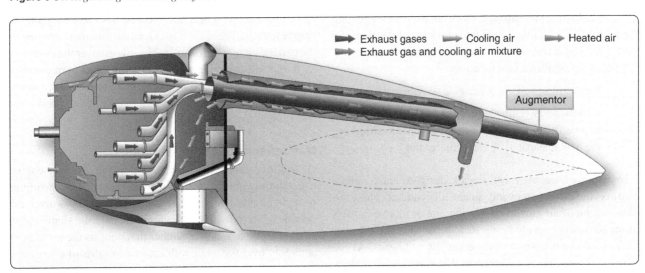

Exhaust gases ⇒ Cooling air ⇒ Heated air ⇒
Exhaust gas and cooling air mixture ⇒

Augmentor

Figure 6-52. *Augmentor.*

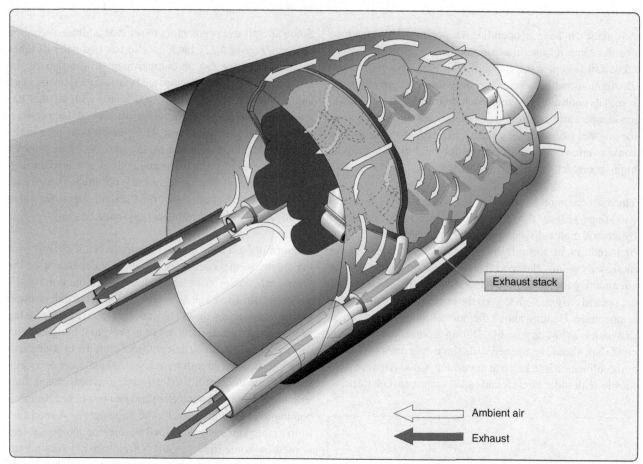

Figure 6-53. *Engine cooling and exhaust system.*

gases through the exhaust ejectors. This type of cooling system eliminates the use of controllable cowl flaps and assures adequate engine cooling at all operating speeds.

Reciprocating Engine Cooling System Maintenance

The engine cooling system of most reciprocating engines usually consists of the engine cowling, cylinder baffles, cylinder fins, and some use a type of cowl flaps. In addition to these major units, there are also some temperature-indicating systems, such as cylinder head temperature, oil temperature, and exhaust gas temperature.

The cowling performs two functions:

1. It streamlines the bulky engine to reduce drag.

2. It forms an envelope around the engine that forces air to pass around and between the cylinders, absorbing the heat dissipated by the cylinder fins.

The cylinder bases are metal shields, designed and arranged to direct the flow of air evenly around all cylinders. This even distribution of air aids in preventing one or more cylinders from being excessively hotter than the others. The cylinder fins radiate heat from the cylinder walls and heads. As the air passes over the fins, it absorbs this heat, carries it away

from the cylinder, and is exhausted overboard through the bottom rear of the cowl.

The controllable cowl flaps provide a means of decreasing or increasing the exit area at the rear of the engine cowling. *[Figure 6-54]* Closing the cowl flaps decreases the exit area, which effectively decreases the amount of air that can circulate over the cylinder fins. The decreased airflow cannot carry away as much heat; therefore, it has a tendency for the engine temperature to increase. Opening the cowl flaps makes the exit area larger. The flow of cooling air over the cylinders increases, absorbing more heat and the engine temperature tends to decrease. Good inspection and maintenance in the care of the engine cooling system aids in overall efficient and economical engine operation.

Maintenance of Engine Cowling

Of the total ram airflow approaching the airborne engine nacelle, only about 15 to 30 percent enters the cowling to provide engine cooling. The remaining air flows over the outside of the cowling. Therefore, the external shape of the cowl must be faired in a manner that permits the air to flow smoothly over the cowl with a minimum loss of energy.

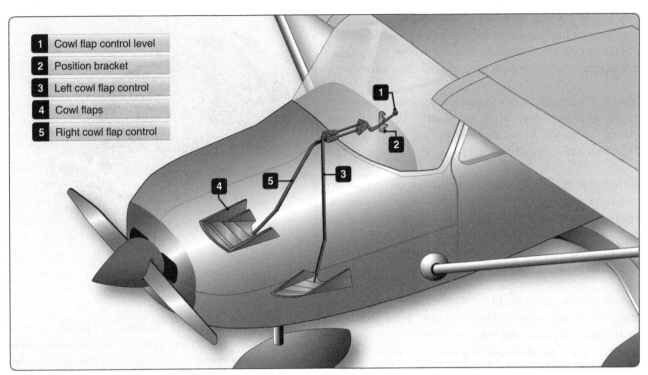

1	Cowl flap control level
2	Position bracket
3	Left cowl flap control
4	Cowl flaps
5	Right cowl flap control

Figure 6-54. *Small aircraft cowl flaps.*

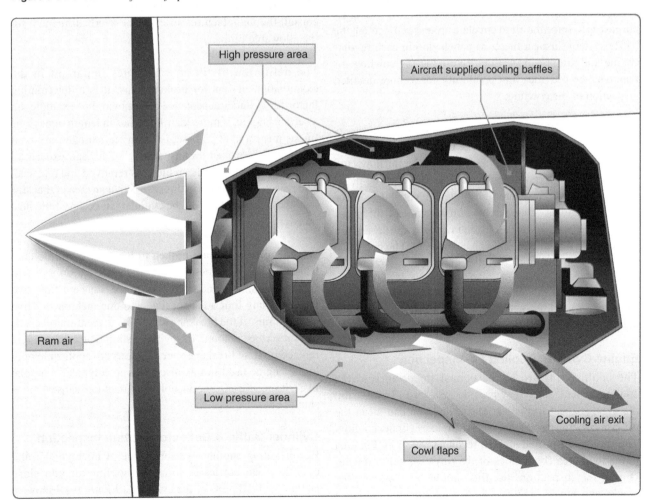

Figure 6-55. *Differential air cooling.*

The engine cowling discussed in this section is typical of that used on many radial or horizontally opposed engines. All cooling systems function in the same manner, with minor engineering changes designed for specific installations.

The cowl is manufactured in removable sections, the number varies with the aircraft make and model. The installation shown in *Figure 6-55* contains two sections that are locked together when installed.

The cowl panels, made from sheet aluminum or composite material, have a smooth external surface to permit undisturbed airflow over the cowl. The internal construction is designed to give strength to the panel and, in addition, to provide receptacles for the toggle latches, cowl support, and engine air seal.

An air seal is constructed of rubber material, bolted to a metal rib riveted to the cowl panel. *[Figure 6-55]* This seal, as the name implies, seals the air in the engine section, preventing the air from escaping along the inner surface of the panel without circulating around the cylinders. The engine air seal must be used on engines that have a complete cylinder baffling system that covers the cylinder heads. Its purpose is to force the air to circulate around and through the baffle system. Inspect the cowl panels during each regular engine and aircraft inspection. Removing the cowling for maintenance provides an opportunity for a more detailed inspection of the cowling.

Inspect the cowling panels for scratches, dents, and tears in the panels. This type of damage causes weakness of the panel structure, increases drag by disrupting airflow, and contributes to the starting of corrosion. The cowling panel latches should be inspected for pulled rivets and loose or damaged handles. The internal construction of the panel should be examined to see that the reinforcing ribs are not cracked and that the air seal is not damaged. The cowl flap hinges, if equipped, and cowl flap hinge bondings should be checked for security of mounting and for breaks or cracks. These inspections are visual checks and should be performed frequently to ensure that the cowling is serviceable and is contributing to efficient engine cooling.

Engine Cylinder Cooling Fin Inspection

The cooling fins are of the utmost importance to the cooling system, since they provide a means of transferring the cylinder heat to the air. Their condition can mean the difference between adequate or inadequate cylinder cooling. The fins are inspected at each regular inspection. Fin area is the total area (both sides of the fin) exposed to the air. During the inspection, the fins should be examined for cracks and breaks. *[Figure 6-56]* Small cracks are not a

Figure 6-56. *A cylinder head and fins.*

reason for cylinder removal. These cracks can be filled or even sometimes stop-drilled to prevent any further cracking. Rough or sharp corners on fins can be smoothed out by filing, and this action eliminates a possible source of new cracks. However, before reprofiling cylinder cooling fins, consult the manufacturer's service or overhaul manual for the allowable limits.

The definition of fin area becomes important in the examination of fins for broken areas. It is a determining factor for cylinder acceptance or removal. For example, on a certain engine, if more than 12 inches in length of any one fin, as measured at its base, is completely broken off, or if the total fins broken on any one cylinder head exceed 83 square inches of area, the cylinder is removed and replaced. The reason for removal in this case is that an area of that size would cause a hot spot on the cylinder; since very little heat transfer could occur.

Where adjacent fins are broken in the same area, the total length of breakage permissible is six inches on any two adjacent fins, four inches on any three adjacent fins, two inches on any four adjacent fins, and one inch on any five adjacent fins. If the breakage length in adjacent fins exceeds this prescribed amount, the cylinder should be removed and replaced. These breakage specifications are applicable only to the engine used in this discussion as a typical example. In each specific case, applicable manufacturer's instructions should be consulted.

Cylinder Baffle & Deflector System Inspection

Reciprocating engines use some type of intercylinder and cylinder head baffles to force the cooling air into close contact with all parts of the cylinders. *Figure 6-50* shows a baffle and deflector system around a cylinder. The air baffle

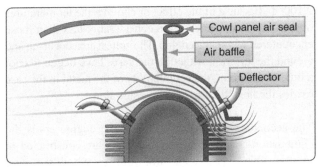

Figure 6-57. *Cylinder head baffle and deflector system.*

blocks the flow of air and forces it to circulate between the cylinder and the deflectors. *Figure 6-57* illustrates a baffle and deflector arrangement designed to cool the cylinder head. The air baffle prevents the air from passing away from the cylinder head and forces it to go between the head and deflector. Although the resistance offered by baffles to the passage of the cooling air demands that an appreciable pressure differential be maintained across the engine to obtain the necessary airflow, the volume of cooling air required is greatly reduced by employing properly designed and located cylinder deflectors.

As shown in *Figure 6-55*, the airflow approaches the nacelle and piles up at the top of the engine, creating a high pressure in the top of the cylinders. This piling up of the air reduces the air velocity. The outlet at the bottom rear of the cowling produces a low-pressure area. As the air nears the cowl exit, it is speeded up again and merges smoothly with the airstream. The pressure differential between the top and the bottom of the engine forces the air past the cylinders through the passages formed by the deflectors. The baffles and deflectors normally are inspected during the regular engine inspection, but they should be checked whenever the cowling is removed for any purpose. Checks should be made for cracks, dents, or loose hold down studs. Cracks or dents, if severe enough, would necessitate repair or removal and replacement of these units. However, a crack that has just started can be stop-drilled, and dents can be straightened, permitting further service from these baffles and deflectors.

Cylinder Temperature Indicating Systems

This system usually consists of an indicator, electrical wiring, and a thermocouple. The wiring is between the instrument and the nacelle firewall. At the firewall, one end of the thermocouple leads connects to the electrical wiring, and the other end of the thermocouple leads connects to the cylinder. The thermocouple consists of two dissimilar metals, generally constantan and iron, connected by wiring to an indicating system. If the temperature of the junction is different from the temperature where the dissimilar metals are connected to wires, a voltage is produced. This voltage sends a current through wires to the indicator, a current-measuring instrument

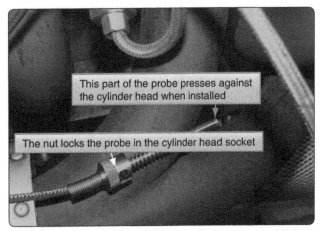

Figure 6-58. *Bayonet type CHT probe.*

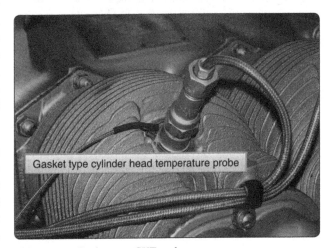

Figure 6-59. *Gasket type CHT probe.*

graduated in degrees.

The thermocouple end that connects to the cylinder is either the bayonet or gasket type. To install the bayonet type, the knurled nut is pushed down and turned clockwise until it is snug. *[Figure 6-58]* In removing this type, the nut is pushed down and turned counterclockwise until released. The gasket type fits under the spark plug and replaces the normal spark plug gasket. *[Figure 6-59]* When installing a thermocouple lead, remember not to cut off the lead because it is too long, but coil and tie up the excess length. The thermocouple is designed to produce a given amount of resistance. If the length of the lead is reduced, an incorrect temperature reading results. The bayonet or gasket of the thermocouple is inserted or installed on the hottest cylinder of the engine, as determined in the block test. When the thermocouple is installed and the wiring connected to the instrument, the indicated reading is the cylinder temperature. Prior to operating the engine, provided it is at ambient temperature, the cylinder head temperature indicator indicates the free outside air temperature; that is one test for determining that the instrument is working correctly. The cover glass of the cylinder head temperature indicator should be checked regularly to see that it has not slipped or cracked. The cover glass should be checked for indications of missing or damaged decals that indicate temperature limitations. If the thermocouple leads were excessive in length and had to be coiled and tied down, the tie should be inspected for security or chafing of the wire. The bayonet or gasket should be inspected for cleanness and security of mounting. When operating the engine, all of the electrical connections should be checked if the cylinder head temperature pointer fluctuates.

Exhaust Gas Temperature Indicating Systems

The exhaust gas temperature indicator consists of a thermocouple placed in the exhaust stream just after the cylinder port. *[Figure 6-60]* It is then connected to the instrument in the instrument panel. This allows for the adjustment of the mixture, which has a large effect on engine temperature. By using this instrument to set the mixture, the engine temperature can be controlled and monitored.

Turbine Engine Cooling

The intense heat generated when fuel and air are burned necessitates that some means of cooling be provided for all internal combustion engines. Reciprocating engines are cooled either by passing air over fins attached to the cylinders or by passing a liquid coolant through jackets that surround the cylinders. The cooling problem is made easier because combustion occurs only during every fourth stroke of a four-stroke-cycle engine.

The burning process in a gas turbine engine is continuous, and nearly all of the cooling air must be passed through the inside of the engine. If only enough air were admitted to the engine to provide an ideal air-fuel ratio of 15:1, internal temperatures would increase to more than 4,000 °F. In practice, a large amount of air in excess of the ideal ratio is admitted to the engine. The large surplus of air cools the hot sections of the engine to acceptable temperatures ranging from 1,500° to

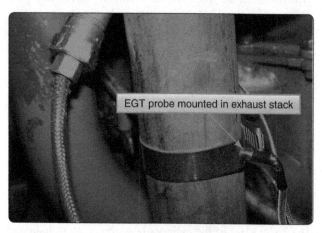

Figure 6-60. *EGT probe in exhaust stack.*

EGT probe mounted in exhaust stack

2,100 °F. Because of the effect of cooling, the temperatures of the outside of the case are considerably less than those encountered within the engine. The hottest area occurs in and around the turbines. Although the gases have begun to cool a little at this point, the conductivity of the metal in the case carries the heat directly to the outside skin.

The secondary air passing through the engine cools the combustion-chamber liners. The liners are constructed to induce a thin, fast-moving film of air over both the inner and outer surfaces of the liner. Can-annular-type burners frequently are provided with a center tube to lead cooling air into the center of the burner to promote high combustion-efficiency and rapid dilution of the hot combustion gases while minimizing pressure losses. In all types of gas turbines, large amounts of relatively cool air join and mix with the burned gases aft of the burners to cool the hot gases just before they enter the turbines.

Cooling-air inlets are frequently provided around the exterior of the engine to permit the entrance of air to cool the turbine case, the bearings, and the turbine nozzle. Internal air is bled from the engine compressor section and is vented to the bearings and other parts of the engine. Air vented into or from the engine is ejected into the exhaust stream. When located on the side of the engine, the case is cooled by outside air flowing around it. The engine exterior and the engine nacelle are cooled by passing fan air around the engine and the nacelle. The engine compartment frequently is divided into two sections. The forward section is referred to as the cold section and the aft section (turbine) is referred to as the hot section. Case drains drain potential leaks overboard to prevent fluids from building up in the nacelle.

Accessory Zone Cooling

Turbine powerplants can be divided into primary zones that are isolated from each other by fireproof bulkheads and seals. The zones are the fan case compartment, intermediate compressor case compartment, and the core engine compartment. *[Figure 6-61]* Calibrated airflows are supplied to the zones to keep the temperatures around the engine at levels that are acceptable. The airflow provides for proper ventilation to prevent a buildup of any harmful vapors. Zone 1, for example, is around the fan case that contains the accessory case and the electronic engine control (EEC). This area is vented by using ram air through an inlet in the nose cowl and is exhausted through a louvered vent in the right fan cowling.

If the pressure exceeds a certain limit, a pressure relief door opens and relieves the pressure. Zone 2 is cooled by fan air from the upper part of the fan duct and is exhausted at the lower end back into the fan air stream. This area has both

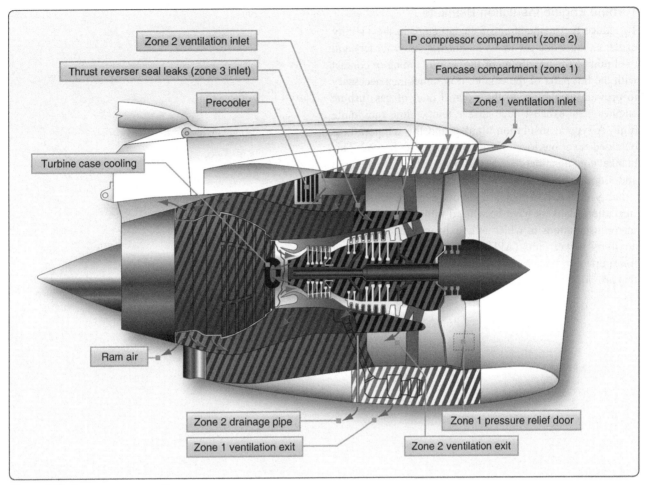

Figure 6-61. *Accessory zone cooling.*

fuel and oil lines, so removing any unwanted vapors would be important.

Zone 3 is the area around the high-pressure compressor to the turbine cases. This zone also contains fuel and oil lines and other accessories. Air enters from the exhaust of the pre-cooler and other areas and is exhausted from the zone through the aft edge of the thrust reverser inner wall and the turbine exhaust sleeve.

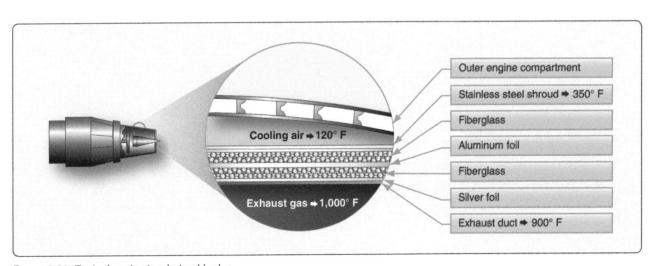

Figure 6-62. *Typical engine insulation blanket.*

Turbine Engine Insulation Blankets

To reduce the temperature of the structure in the vicinity of the exhaust duct or thrust augmentor (afterburner) and to eliminate the possibility of fuel or oil coming in contact with the hot parts of the engine, it is sometimes necessary to provide insulation on the exhaust duct of gas turbine engines. The exhaust duct surface temperature runs quite high. A typical insulation blanket and the temperatures obtained at various locations are shown in *Figure 6-62*. This blanket contains fiberglass as the low conductance material and aluminum foil as the radiation shield. The blanket is suitably covered so that it does not become oil soaked. Insulation blankets have been used rather extensively on many installations in which long exhaust is needed. Some auxiliary power units (APU) mounted in the tail cone of transport aircraft have air that surrounds the exhaust tail pipe that provides cooling and protects the surrounding structure.

Chapter 7
Propellers

General

The propeller, the unit that must absorb the power output of the engine, has passed through many stages of development. Although most propellers are two-bladed, great increases in power output have resulted in the development of four- and six-bladed propellers of large diameters. However, all propeller-driven aircraft are limited by the revolutions per minute (rpm) at which propellers can be turned.

There are several forces acting on the propeller as it turns; a major one is centrifugal force. This force at high rpm tends to pull the blades out of the hub, so blade weight is very important to the design of a propeller. Excessive blade tip speed (rotating the propeller too fast) may result not only in poor blade efficiency, but also in fluttering and vibration. Since the propeller speed is limited, the aircraft speed of a propeller driven aircraft is also limited—to approximately 400 miles per hour (mph). As aircraft speeds increased, turbofan engines were used for higher speed aircraft. Propeller-driven aircraft have several advantages and are widely used for applications in turboprops and reciprocating engine installations. Takeoff and landing can be shorter and less expensive. New blade materials and manufacturing techniques have increased the efficiency of propellers. Many smaller aircraft will continue to use propellers well into the future.

The basic nomenclature of the parts of a propeller is shown in *Figure 7-1* for a simple fixed-pitch, two-bladed wood propeller. The aerodynamic cross-section of a blade in *Figure 7-2* includes terminology to describe certain areas shown.

Many different types of propeller systems have been developed for specific aircraft installation, speed, and mission. Propeller development has encouraged many

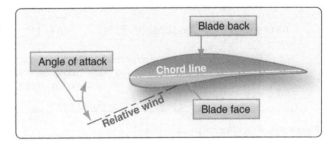

Figure 7-2. *Cross-sectional area of a propeller blade airfoil.*

changes as propulsion systems have evolved. The first propellers were fabric-covered sticks made to force air in a rearward direction. Propellers started as simple two-bladed wood propellers and have advanced to the complex propulsion systems of turboprop aircraft that involve more than just the propeller. As an outgrowth of operating large, more complex propellers, a variable-pitch, constant-speed feathering and reversing propeller system was developed. This system allows the engine rpm to be varied only slightly during different flight conditions and, therefore, increases flying efficiency. A basic constant-speed system consists of a flyweight-equipped governor unit that controls the pitch angle of the blades so that the engine speed remains constant. The governor can be regulated by controls in the flight deck so that any desired blade angle setting and engine operating speed can be obtained. A low-pitch, high-rpm setting, for example, can be utilized for takeoff. Then, after the aircraft is airborne, a higher pitch and lower rpm setting can be used. *Figure 7-3* shows normal propeller movement with the positions of low pitch, high pitch, feather (used if the engine quits to reduce drag), and zero pitch into negative pitch, or reverse pitch.

Basic Propeller Principles

The aircraft propeller consists of two or more blades and a central hub to which the blades are attached. Each blade of

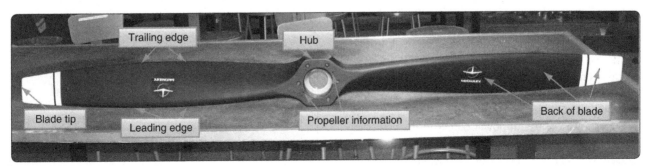

Figure 7-1. *Basic nomenclature of propellers.*

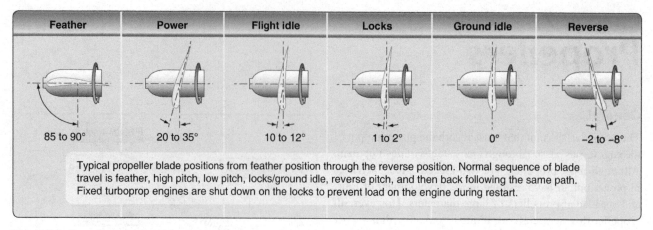

Feather	Power	Flight idle	Locks	Ground idle	Reverse
85 to 90°	20 to 35°	10 to 12°	1 to 2°	0°	–2 to –8°

Typical propeller blade positions from feather position through the reverse position. Normal sequence of blade travel is feather, high pitch, low pitch, locks/ground idle, reverse pitch, and then back following the same path. Fixed turboprop engines are shut down on the locks to prevent load on the engine during restart.

Figure 7-3. *Propeller range positions.*

an aircraft propeller is essentially a rotating wing. As a result of their construction, the propeller blades produce forces that create thrust to pull or push the aircraft through the air. The power needed to rotate the propeller blades is furnished by the engine. The propeller is mounted on a shaft, which may be an extension of the crankshaft on low-horsepower engines; on high-horsepower engines, it is mounted on a propeller shaft that is geared to the engine crankshaft. In either case, the engine rotates the airfoils of the blades through the air at high speeds, and the propeller transforms the rotary power of the engine into thrust.

Propeller Aerodynamic Process

An aircraft moving through the air creates a drag force opposing its forward motion. If an aircraft is to fly on a level path, there must be a force applied to it that is equal to the drag but acting forward. This force is called thrust. The work done by thrust is equal to the thrust times the distance it moves the aircraft.

$$\text{Work} = \text{Thrust} \times \text{Distance}$$

The power expended by thrust is equal to the thrust times the velocity at which it moves the aircraft.

$$\text{Power} = \text{Thrust} \times \text{Velocity}$$

If the power is measured in horsepower units, the power expended by the thrust is termed thrust horsepower.

The engine supplies brake horsepower through a rotating shaft, and the propeller converts it into thrust horsepower. In this conversion, some power is wasted. For maximum efficiency, the propeller must be designed to keep this waste as small as possible. Since the efficiency of any machine is the ratio of the useful power output to the power input, propeller efficiency is the ratio of thrust horsepower to brake horsepower. The usual symbol for propeller efficiency is the

Greek letter η (eta). Propeller efficiency varies from 50 percent to 87 percent, depending on how much the propeller slips.

Pitch is not the same as blade angle, but because pitch is largely determined by blade angle, the two terms are often used interchangeably. An increase or decrease in one is usually associated with an increase or decrease in the other. Propeller slip is the difference between the geometric pitch of the propeller and its effective pitch. *[Figure 7-4]* Geometric pitch is the distance a propeller should advance in one revolution with no slippage; effective pitch is the distance it actually advances. Thus, geometric or theoretical pitch is based on no slippage. Actual, or effective, pitch recognizes propeller slippage in the air. The relationship can be shown as:

$$\text{Geometric pitch} - \text{Effective pitch} = \text{slip}$$

Geometric pitch is usually expressed in pitch inches and calculated by using the following formula:

$$\text{GP} = 2 \times \pi \times \text{R} \times \text{tangent of blade angle at 75 percent station}$$

R = Radius at the 75 percent blade station

$$\pi = 3.14$$

Although blade angle and propeller pitch are closely related, blade angle is the angle between the face or chord of a blade section and the plane in which the propeller rotates.

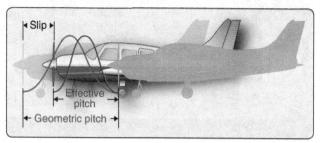

Figure 7-4. *Effective pitch and geometric pitch.*

[*Figure 7-5*] Blade angle, usually measured in degrees, is the angle between the chordline of the blade and the plane of rotation. The chordline of the propeller blade is determined in about the same manner as the chordline of an airfoil. In fact, a propeller blade can be considered as being composed of an infinite number of thin blade elements, each of which is a miniature airfoil section whose chord is the width of the propeller blade at that section. Because most propellers have a flat blade face, the chordline is often drawn along the face of the propeller blade.

The typical propeller blade can be described as a twisted airfoil of irregular planform. Two views of a propeller blade are shown in *Figure 7-6*. For purposes of analysis, a blade can be divided into segments that are located by station numbers in inches from the center of the blade hub. The cross-sections of each 6-inch blade segment are shown as airfoils in the right side of *Figure 7-6*. Also identified in *Figure 7-6* are the blade shank and the blade butt. The blade shank is the thick, rounded portion of the propeller blade near the hub and is designed to give strength to the blade. The blade butt, also called the blade base or root, is the end of the blade that fits in the propeller hub. The blade tip is that part of the propeller blade farthest from the hub, generally defined as the last 6 inches of the blade.

A cross-section of a typical propeller blade is shown in *Figure 7-7*. This section or blade element is an airfoil comparable to a cross-section of an aircraft wing. The blade back is the cambered or curved side of the blade, similar to the upper surface of an aircraft wing. The blade face is the flat side of the propeller blade. The chordline is an imaginary line drawn through the blade from the leading edge to the trailing edge. The leading edge is the thick edge of the blade that meets the air as the propeller rotates.

A rotating propeller is acted upon by centrifugal twisting, aerodynamic twisting, torque bending, and thrust bending forces. The principal forces acting on a rotating propeller

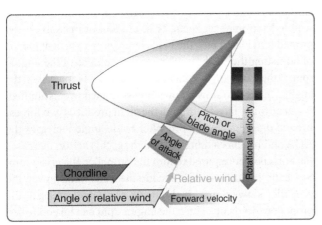

Figure 7-5. *Propeller aerodynamic factors.*

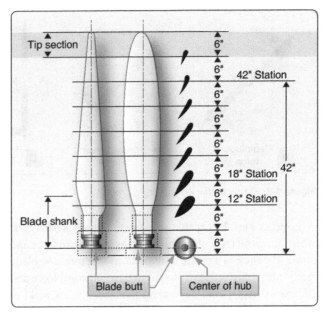

Figure 7-6. *Typical propeller blade elements.*

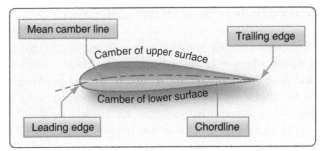

Figure 7-7. *Cross-section of a propeller blade.*

are illustrated in *Figure 7-8*.

Centrifugal force is a physical force that tends to throw the rotating propeller blades away from the hub. [*Figure 7-8A*] This is the most dominant force on the propeller. Torque bending force, in the form of air resistance, tends to bend the propeller blades in the direction opposite that of rotation. [*Figure 7-8B*] Thrust bending force is the thrust load that tends to bend propeller blades forward as the aircraft is pulled through the air. [*Figure 7-8C*] Aerodynamic twisting force tends to turn the blades to a high blade angle. [*Figure 7-8D*] Centrifugal twisting force, being greater than the aerodynamic twisting force, tends to force the blades toward a low blade angle.

At least two of these forces acting on the propellers blades are used to move the blades on a controllable pitch propeller. Centrifugal twisting force is sometimes used to move the blades to the low pitch position, while aerodynamic twisting force is used to move the blades into high pitch. These forces can be the primary or secondary forces that move the blades to the new pitch position.

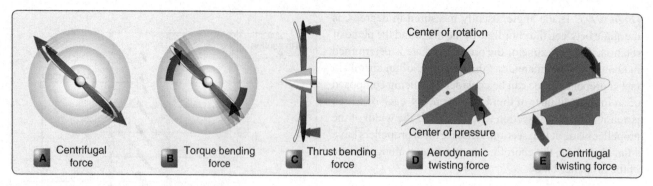

Figure 7-8. *Forces acting on a rotating propeller.*

A propeller must be capable of withstanding severe stresses, which are greater near the hub, caused by centrifugal force and thrust. The stresses increase in proportion to the rpm. The blade face is also subjected to tension from the centrifugal force and additional tension from the bending. For these reasons, nicks or scratches on the blade may cause very serious consequences. These could lead to cracks and failure of the blade and are addressed in the repair section later in this chapter.

A propeller must also be rigid enough to prevent fluttering, a type of vibration in which the ends of the blade twist back and forth at high frequency around an axis perpendicular to the engine crankshaft. Fluttering is accompanied by a distinctive noise, often mistaken for exhaust noise. The constant vibration tends to weaken the blade and eventually causes failure.

Aerodynamic Factors

To understand the action of a propeller, consider first its motion, which is both rotational and forward. Thus, as shown by the vectors of propeller forces in *Figure 7-9,* a section of a propeller blade moves downward and forward. As far as the forces are concerned, the result is the same as if the blade were stationary and the air coming at it from a direction opposite its path. The angle at which this air (relative wind)

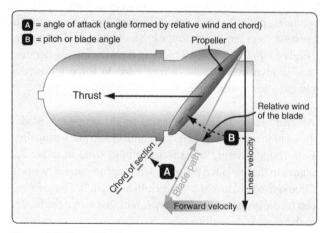

Figure 7-9. *Propeller forces.*

strikes the propeller blade is called angle of attack (AOA). The air deflection produced by this angle causes the dynamic pressure at the engine side of the propeller blade to be greater than atmospheric pressure, creating thrust.

The shape of the blade also creates thrust because it is shaped like a wing. As the air flows past the propeller, the pressure on one side is less than that on the other. As in a wing, this difference in pressure produces a reaction force in the direction of the lesser pressure. The area above a wing has less pressure, and the force (lift) is upward. The area of decreased pressure is in front of a propeller which is mounted in a vertical instead of a horizontal position, and the force (thrust) is in a forward direction. Aerodynamically, thrust is the result of the propeller shape and the AOA of the blade.

Another way to consider thrust is in terms of the mass of air handled. In these terms, thrust is equal to the mass of air handled multiplied by the slipstream velocity minus the velocity of the aircraft. Thus, the power expended in producing thrust depends on the mass of air moved per second. On the average, thrust constitutes approximately 80 percent of the torque (total horsepower absorbed by the propeller). The other 20 percent is lost in friction and slippage. For any speed of rotation, the horsepower absorbed by the propeller balances the horsepower delivered by the engine. For any single revolution of the propeller, the amount of air displaced (moved) depends on the blade angle, which determines the quantity or amount of mass of air the propeller moves. Thus, the blade angle is an excellent means of adjusting the load on the propeller to control the engine rpm. If the blade angle is increased, more load is placed on the engine, tending to slow it down unless more power is applied. As an airfoil is moved through the air, it produces two forces: lift and drag. Increasing propeller blade angle increases the AOA and produces more lift and drag; this action increases the horsepower required to turn the propeller at a given rpm. Since the engine is still producing the same horsepower, the propeller slows down. If the blade angle is decreased, the propeller speeds up. Thus, the engine rpm can be controlled by increasing or decreasing the blade angle.

The blade angle is also an excellent method of adjusting the AOA of the propeller. On constant-speed propellers, the blade angle must be adjusted to provide the most efficient AOA at all engine and aircraft speeds. Lift versus drag curves, which are drawn for propellers as well as wings, indicate that the most efficient AOA is a small one varying from 2° to 4° positive. The actual blade angle necessary to maintain this small AOA varies with the forward speed of the aircraft. This is due to a change in the relative wind direction, which varies with aircraft speed.

Fixed-pitch and ground-adjustable propellers are designed for best efficiency at one rotation and forward speed. In other words, they are designed to fit a given aircraft and engine combination. A propeller may be used that provides the maximum propeller efficiency for takeoff, climb, cruising, or high speeds. Any change in these conditions results in lowering the efficiency of both the propeller and the engine.

A constant-speed propeller, however, keeps the blade angle adjusted for maximum efficiency for most conditions encountered in flight. During takeoff, when maximum power and thrust are required, the constant-speed propeller is at a low propeller blade angle or pitch. The low blade angle keeps the AOA small and efficient with respect to the relative wind. At the same time, it allows the propeller to handle a smaller mass of air per revolution. This light load allows the engine to turn at high rpm and to convert the maximum amount of fuel into heat energy in a given time. The high rpm also creates maximum thrust. Although the mass of air handled per revolution is small, the engine rpm is high, the slipstream velocity (air coming off the propeller) is high, and, with the low aircraft speed, the thrust is maximum.

After liftoff, as the speed of the aircraft increases, the constant-speed propeller changes to a higher angle (or pitch). Again, the higher blade angle keeps the AOA small and efficient with respect to the relative wind. The higher blade angle increases the mass of air handled per revolution. This decreases the engine rpm, reducing fuel consumption and engine wear, and keeps thrust at a maximum.

For climb after takeoff, the power output of the engine is reduced to climb power by decreasing the manifold pressure and increasing the blade angle to lower the rpm. Thus, the torque (horsepower absorbed by the propeller) is reduced to match the reduced power of the engine. The AOA is again kept small by the increase in blade angle. The greater mass of air handled per second, in this case, is more than offset by the lower slipstream velocity and the increase in airspeed.

At cruising altitude, when the aircraft is in level flight and less power is required than is used in takeoff or climb, engine power is again reduced by lowering the manifold pressure and increasing the blade angle to decrease the rpm. Again, this reduces torque to match the reduced engine power; for, although the mass of air handled per revolution is greater, it is more than offset by a decrease in slipstream velocity and an increase in airspeed. The AOA is still small because the blade angle has been increased with an increase in airspeed. Pitch distribution is the twist in the blade from the shank to the blade tip, due to the variation in speeds that each section of the blade is traveling. The tip of the blade is traveling much faster than the inner portion of the blade.

Propeller Controls & Instruments

Fixed pitch propellers have no controls and require no adjustments in flight. The constant-speed propeller has a propeller control in the center pedestal between the throttle and the mixture control. *[Figure 7-10]* The two positions for the control are increase rpm (full forward) and decrease rpm (pulled aft). This control is directly connected to the propeller governor and, by moving the control, adjusts the tension on the governor speeder spring. This control can also be used to feather the propeller in some aircraft by moving the control to the full decrease rpm position. The two main instruments used with the constant-speed propeller are the engine tachometer and the manifold pressure gauge. Rotations per minute (rpm) is controlled by the propeller control and the manifold pressure is adjusted by the throttle.

Propeller Location
Tractor Propeller

Tractor propellers are those mounted on the upstream end of a drive shaft in front of the supporting structure. Most aircraft are equipped with this type of propeller. The tractor type of propeller comes in all types of propellers. A major advantage of the tractor propeller is that lower stresses are induced in the propeller as it rotates in relatively undisturbed air.

Figure 7-10. *Turboprop propeller controls.*

Pusher Propellers

Pusher propellers are those mounted on the downstream end of a drive shaft behind the supporting structure. Pusher propellers are constructed as fixed- or variable-pitch propellers. Seaplanes and amphibious aircraft have used a greater percentage of pusher propellers than other kinds of aircraft. On land aircraft, where propeller-to-ground clearance usually is less than propeller-to-water clearance of watercraft, pusher propellers are subject to more damage than tractor propellers. Rocks, gravel, and small objects dislodged by the wheels are quite often thrown or drawn into a pusher propeller. Similarly, aircraft with pusher propellers are apt to encounter propeller damage from water spray thrown up by the hull during landing or takeoff airspeed. Consequently, the pusher propeller is mounted above and behind the wings to prevent such damage.

Types of Propellers

There are various types or classes of propellers, the simplest of which are the fixed-pitch and ground-adjustable propellers. The complexity of propeller systems increases from these simpler forms to controllable-pitch and complex constant-speed systems (automatic systems). Various characteristics of several propeller types are discussed in the following paragraphs, but no attempt is made to cover all types of propellers.

Fixed-Pitch Propeller

As the name implies, a fixed-pitch propeller has the blade pitch, or blade angle, built into the propeller. *[Figure 7-11]* The blade angle cannot be changed after the propeller is built. Generally, this type of propeller is one piece and is constructed of wood or aluminum alloy.

Fixed-pitch propellers are designed for best efficiency at one rotational and forward speed. They are designed to fit a set of conditions of both aircraft and engine speeds and any change in these conditions reduces the efficiency of both the propeller and the engine. The fixed-pitch propeller is used on aircraft of low power, speed, range, or altitude. Many single-engine aircraft use fixed-pitch propellers and the advantages of these are less expense and their simple operation. This type of propeller does not require any control inputs from the pilot in flight.

Test Club Propeller

A test club is used to test and break in reciprocating engines. *[Figure 7-12]* They are made to provide the correct amount of load on the engine during the test break-in period. The multi-blade design also provides extra cooling air flow during testing.

Figure 7-11. *Fixed-pitch propeller.*

Ground-Adjustable Propeller

The ground-adjustable propeller operates as a fixed-pitch propeller. The pitch, or blade angle, can be changed only when the propeller is not turning. This is done by loosening the clamping mechanism that holds the blades in place. After the clamping mechanism has been tightened, the pitch of the blades cannot be changed in flight to meet variable flight requirements. The ground-adjustable propeller is not often used on present-day aircraft.

Controllable-Pitch Propeller

The controllable-pitch propeller permits a change of blade pitch, or angle, while the propeller is rotating. This allows

Figure 7-12. *Test club.*

the propeller to assume a blade angle that gives the best performance for particular flight conditions. The number of pitch positions may be limited, as with a two-position controllable propeller, or the pitch may be adjusted to any angle between the minimum and maximum pitch settings of a given propeller. The use of controllable-pitch propellers also makes it possible to attain the desired engine rpm for a particular flight condition.

This type of propeller is not to be confused with a constant-speed propeller. With the controllable-pitch type, the blade angle can be changed in flight, but the pilot must change the propeller blade angle directly. The blade angle will not change again until the pilot changes it. The use of a governor is the next step in the evolution of propeller development, making way for constant-speed propellers with governor systems. An example of a two-position propeller is a Hamilton Standard flyweight two-position propeller. These types of propeller are not in wide use today.

Constant-Speed Propellers

The propeller has a natural tendency to slow down as the aircraft climbs and to speed up as the aircraft dives because the load on the engine varies. To provide an efficient propeller, the speed is kept as constant as possible. By using propeller governors to increase or decrease propeller pitch, the engine speed is held constant. When the aircraft goes into a climb, the blade angle of the propeller decreases just enough to prevent the engine speed from decreasing. The engine can maintain its power output if the throttle setting is not changed. When the aircraft goes into a dive, the blade angle increases sufficiently to prevent overspeeding and, with the same throttle setting, the power output remains unchanged. If the throttle setting is changed instead of changing the speed of the aircraft by climbing or diving, the blade angle increases or decreases as required to maintain a constant engine rpm. The power

output (not the rpm) changes in accordance with changes in the throttle setting. The governor-controlled, constant-speed propeller changes the blade angle automatically, keeping engine rpm constant.

One type of pitch-changing mechanism is operated by oil pressure (hydraulically) and uses a piston-and-cylinder arrangement. The piston may move in the cylinder, or the cylinder may move over a stationary piston. The linear motion of the piston is converted by several different types of mechanical linkage into the rotary motion necessary to change the blade angle. The mechanical connection may be through gears, the pitch-changing mechanism that turns the butt of each blade. Each blade is mounted with a bearing that allows the blade to rotate to change pitch. *[Figure 7-13]*

In most cases, the oil pressure for operating the different types of hydraulic pitch-changing mechanisms comes directly from the engine lubricating system. When the engine lubricating system is used, the engine oil pressure is usually boosted by a pump that is integral with the governor to operate the propeller. The higher oil pressure (approximately 300 pounds per square inch (psi)) provides a quicker blade-angle change. The governors direct the pressurized oil for operation of the hydraulic pitch-changing mechanisms.

The governors used to control hydraulic pitch-changing mechanisms are geared to the engine crankshaft and are sensitive to changes in rpm. When rpm increases above the value for which a governor is set, the governor causes the propeller pitch-changing mechanism to turn the blades to a higher angle. This angle increases the load on the engine, and rpm decreases. When rpm decreases below the value for which a governor is set, the governor causes the pitch-changing mechanism to turn the blades to a lower angle; the load on the engine is decreased, and rpm increases. Thus, a propeller governor tends to keep engine rpm constant.

Blade bearing support area for each blade

Figure 7-13. *Blade bearing areas in hub.*

In constant-speed propeller systems, the control system adjusts pitch through the use of a governor, without attention by the pilot, to maintain a specific preset engine rpm within the set range of the propeller. For example, if engine speed increases, an overspeed condition occurs and the propeller needs to slow down. The controls automatically increase the blade angle until desired rpm has been reestablished. A good constant-speed control system responds to such small variations of rpm that for all practical purposes, a constant rpm is maintained.

Each constant-speed propeller has an opposing force that operates against the oil pressure from the governor. Flyweights mounted to the blades move the blades in the high pitch direction as the propeller turns. *[Figure 7-13]* Other forces used to move the blades toward the high pitch direction include air pressure (contained in the front dome), springs, and aerodynamic twisting moment.

Feathering Propellers

Feathering propellers must be used on multi-engine aircraft to reduce propeller drag to a minimum under one or more engine failure conditions. A feathering propeller is a constant-speed propeller used on multi-engine aircraft that has a mechanism to change the pitch to an angle of approximately 90°. A propeller is usually feathered when the engine fails to develop power to turn the propeller. By rotating the propeller blade angle parallel to the line of flight, the drag on the aircraft is greatly reduced. With the blades parallel to the airstream, the propeller stops turning and minimum windmilling, if any, occurs. The blades are held in feather by aerodynamic forces.

Almost all small feathering propellers use oil pressure to take the propeller to low pitch and blade flyweights, springs, and compressed air to take the blades to high pitch. Since the blades would go to the feather position during shutdown, latches lock the propeller in the low pitch position as the propeller slows down at shutdown. *[Figure 7-14]* These can be internal or external and are contained within the propeller hub. In flight, the latches are prevented from stopping the blades from feathering because they are held off their seat by centrifugal force. Latches are needed to prevent excess load on the engine at start up. If the blade were in the feathered position during engine start, the engine would be placed under an undue load during a time when the engine is already subject to wear.

Reverse-Pitch Propellers

Additional refinements, such as reverse-pitch propellers (mainly used on turbo props), are included in some propellers to improve their operational characteristics. Almost all reverse-pitch propellers are of the feathering type. A reverse-pitch propeller is a controllable propeller in which the blade angles can be changed to a negative value during operation.

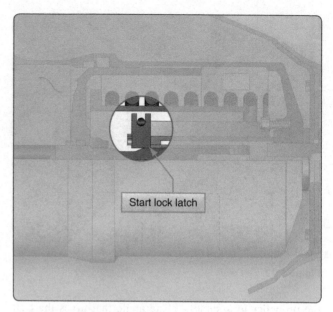

Figure 7-14. *Feathering latches.*

The purpose of the reversible pitch feature is to produce a negative blade angle that produces thrust opposite the normal forward direction. Normally, when the landing gear is in contact with the runway after landing, the propellers blades can be moved to negative pitch (reversed), which creates thrust opposite of the aircraft direction and slows the aircraft. As the propeller blades move into negative pitch, engine power is applied to increase the negative thrust. This aerodynamically brakes the aircraft and reduces ground roll after landing. Reversing the propellers also reduces aircraft speed quickly on the runway just after touchdown and minimizes aircraft brake wear.

Propeller Governor

A governor is an engine rpm-sensing device and high-pressure oil pump. In a constant-speed propeller system, the governor responds to a change in engine rpm by directing oil under pressure to the propeller hydraulic cylinder or by releasing oil from the hydraulic cylinder. The change in oil volume in the hydraulic cylinder changes the blade angle and maintains the propeller system rpm. The governor is set for a specific rpm via the flight deck propeller control, which compresses or releases the governor speeder spring.

A propeller governor is used to sense propeller and engine speed and normally provides oil to the propeller for low pitch position. *[Figure 7-15]* There are a couple of nonfeathering propellers that operate opposite to this; they are discussed later in this chapter. Fundamental forces, some already discussed, are used to control blade angle variations required for constant-speed propeller operation. These forces are:

1. Centrifugal twisting moment—a component of the

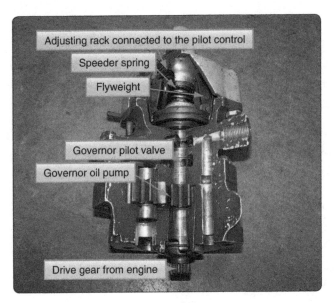

Figure 7-15. *Parts of a governor.*

centrifugal force acting on a rotating blade that tends at all times to move the blade into low pitch.

2. Propeller-governor oil on the propeller piston side—balances the propeller blade flyweights, which moves the blades toward high pitch.

3. Propeller blade flyweights—always move the blades toward high pitch.

4. Air pressure against the propeller piston—pushes toward high pitch.

5. Large springs—push in the direction of high pitch and feather.

6. Centrifugal twisting force—moves the blades toward low pitch.

7. Aerodynamic twisting force—moves the blades toward high pitch.

All of the forces listed are not equal in strength. The most powerful force is the governor oil pressure acting on the propeller piston. This piston is connected mechanically to the blades; as the piston moves, the blades are rotated in proportion. By removing the oil pressure from the governor, the other forces can force the oil from the piston chamber and move the propeller blades in the other direction.

Governor Mechanism

The engine-driven single-acting propeller governor (constant-speed control) receives oil from the lubricating system and boosts its pressure to that required to operate the pitch-changing mechanism. *[Figure 7-16]* It consists of a gear pump to increase the pressure of the engine oil, a pilot valve controlled by flyweights in the governor to control the flow of oil through the governor to and away from the propeller, and a relief valve system that regulates the operating oil pressures in the governor. A spring called the speeder spring opposes the governor flyweight's ability to fly outward when turning. The tension on this spring can be adjusted by the propeller control on the control quadrant. The tension of the speeder spring sets the maximum rpm of the engine in the governor mode. As the engine and propeller rpm is increased at the maximum set point (maximum speed) of the governor, the governor flyweights overcome the tension of the speeder spring and move outward. This action moves the pilot valve in the governor to release oil from the propeller piston and allows the blade flyweights to increase blade pitch, which increases the load on the engine, slowing it down or maintaining the set speed.

In addition to boosting the engine oil pressure to produce one of the fundamental control forces, the governor maintains the required balance between control forces by metering to, or draining from, the propeller piston the exact quantity of oil necessary to maintain the proper blade angle for constant-speed operation. The position of the pilot valve, with respect to the propeller-governor metering port, regulates the quantity of oil that flows through this port to or from the propeller.

A speeder spring above the rack opposes the action of the governor flyweights, which sense propeller speed. If the flyweights turn faster than the tension on the speeder spring, they fly out; this is an overspeed condition. To slow the engine propeller combination down, the blade angle (pitch) must be increased. Oil is allowed to flow away from the propeller piston and the flyweights increase the pitch or blade angle slowing the propeller until it reaches an on-speed condition where the force on the governor flyweights and the tension on the speeder spring are balanced. This balance of forces can be disturbed by the aircraft changing attitude (climb or dive) or the pilot changing the tension on the speeder spring with the propeller control on the instrument panel (i.e., if the pilot selects a different rpm).

Underspeed Condition

When the engine is operating below the rpm set by the pilot using the flight deck control, the governor is operating in an underspeed condition. *[Figure 7-17]* In this condition, the flyweights tilt inward because there is not enough centrifugal force on the flyweights to overcome the force of the speeder spring. The pilot valve, forced down by the speeder spring, meters oil flow to decrease propeller pitch and raise engine rpm. If the nose of the aircraft is raised or the blades are moved to a higher blade angle, this increases the load on the engine and the propeller tries to slow down. To maintain a constant speed, the governor senses the decrease in speed and increases

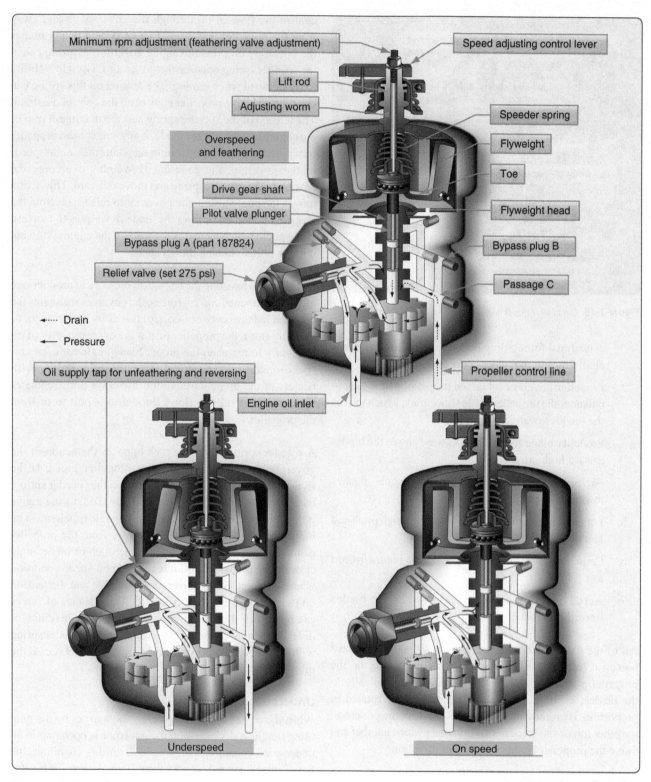

Figure 7-16. *Typical governor.*

oil flow to the propeller, moving the blades to a lower pitch and allowing them to maintain the same speed. When the engine speed starts to drop below the rpm for which the governor is set, the resulting decrease in centrifugal force exerted by the flyweights permits the speeder spring to lower the pilot valve (flyweights inward), thereby opening the propeller-governor metering port. The oil then flows through the valve port and into the propeller piston causing the blades to move to a lower pitch (a decrease in load).

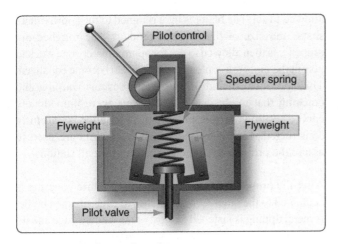

Figure 7-17. *Underspeed condition.*

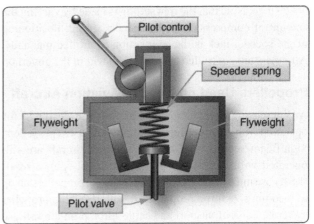

Figure 7-18. *Overspeed condition.*

Overspeed Condition

When the engine is operating above the rpm set by the pilot using the flight deck control, the governor is operating in an overspeed condition. *[Figure 7-18]* In an overspeed condition, the centrifugal force acting on the flyweights is greater than the speeder spring force. The flyweights tilt outward and raise the pilot valve. The pilot valve then meters oil flow to increase propeller pitch and lower engine rpm. When the engine speed increases above the rpm for which the governor is set, note that the flyweights move outward against the force of the speeder spring, raising the pilot valve. This opens the propeller-governor metering port, allowing governor oil flow from the propeller piston allowing flyweights on the blades to increase pitch and slow the engine.

On-Speed Condition

When the engine is operating at the rpm set by the pilot using the flight deck control, the governor is operating on speed. *[Figure 7-19]* In an on-speed condition, the centrifugal force acting on the flyweights is balanced by the speeder spring, and the pilot valve is neither directing oil to nor from the propeller hydraulic cylinder. In the on-speed condition, the forces of the governor flyweights and the tension on the speeder spring are equal; the propeller blades are not moving or changing pitch. If something happens to unbalance these forces, such as if the aircraft dives or climbs, or the pilot selects a new rpm range through the propeller control (changes tension on the speeder spring), then these forces are unequal and an underspeed or overspeed condition would result. A change in rpm comes about in the governing mode by pilot selection of a new position of the propeller control, which changes the tension of the governor speeder spring or by the aircraft changing attitude. The governor, as a speed-sensing device, causes the propeller to maintain a set rpm regardless of the aircraft attitude. The speeder spring propeller governing range is limited to about 200 rpm. Beyond this rpm, the governor cannot maintain the correct rpm.

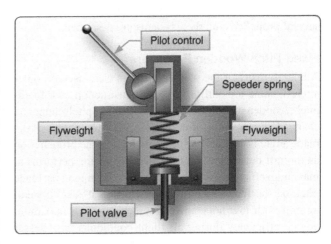

Figure 7-19. *On-speed condition.*

Governor System Operation

If the engine speed drops below the rpm for which the governor is set, the rotational force on the engine-driven governor flyweights becomes less. *[Figure 7-17]* This allows the speeder spring to move the pilot valve downward. With the pilot valve in the downward position, oil from the gear type pump flows through a passage to the propeller and moves the cylinder outward. This in turn decreases the blade angle and permits the engine to return to the on-speed setting. If the engine speed increases above the rpm for which the governor is set, the flyweights move against the force of the speeder spring and raise the pilot valve. This permits the oil in the propeller to drain out through the governor drive shaft. As the oil leaves the propeller, the centrifugal force acting on the flyweights turns the blades to a higher angle, which decreases the engine rpm. When the engine is exactly at the rpm set by the governor, the centrifugal reaction of the flyweights balances the force of the speeder spring, positioning the pilot valve so that oil is neither supplied to nor drained from the propeller. With this condition, propeller blade angle does not change. Note that the rpm setting is made by varying the amount of compression in the speeder spring. Positioning of the speeder rack is the only action controlled manually. All others are controlled automatically within the governor.

Propellers Used on General Aviation Aircraft

An increasing number of light aircraft are designed for operation with governor-regulated, constant-speed propellers. Significant segments of general aviation aircraft are still operated with fixed-pitch propellers. Light-sport aircraft (LSA) use multiblade fixed-pitch composite propellers on up to medium size turbo prop aircraft with reversing propeller systems. Larger transport and cargo turbo prop aircraft use propeller systems with dual or double-acting governors and differential oil pressure to change pitch. Some types of propeller systems are beyond the scope of this text, but several propellers and their systems are described.

Fixed-Pitch Wooden Propellers

Although many of the wood propellers were used on older aircraft, some are still in use. The construction of a fixed-pitch, wooden propeller is such that its blade pitch cannot be changed after manufacture. *[Figure 7-20]* The choice of the blade angle is decided by the normal use of the propeller on an aircraft during level flight when the engine performs at maximum efficiency. The impossibility of changing the blade pitch on the fixed-pitch propeller restricts its use to small aircraft with low horsepower engines in which maximum engine efficiency during all flight conditions is of lesser importance than in larger aircraft. The wooden, fixed-pitch propeller is well suited for such small aircraft because of its light weight, rigidity, economy of production, simplicity of construction, and ease of replacement.

A wooden propeller is not constructed from a solid block but is built up of a number of separate layers of carefully selected and well-seasoned hardwoods. Many woods, such as mahogany, cherry, black walnut, and oak, are used to some extent, but birch is the most widely used. Five to nine separate layers are used, each about ¾ inch thick. The several layers are glued together with a waterproof, resinous glue and

allowed to set. The blank is then roughed to the approximate shape and size of the finished product. The roughed-out propeller is then allowed to dry for approximately one week to permit the moisture content of the layers to become equalized. This additional period of seasoning prevents warping and cracking that might occur if the blank were immediately carved. Following this period, the propeller is carefully constructed. Templates and bench protractors are used to assure the proper contour and blade angle at all stations.

After the propeller blades are finished, a fabric covering is cemented to the outer 12 or 15 inches of each finished blade. A metal tipping is fastened to most of the leading edge and tip of each blade to protect the propeller from damage caused by flying particles in the air during landing, taxiing, or takeoff. *[Figure 7-21]* Metal tipping may be of terneplate, Monel metal, or brass. Stainless steel has been used to some extent. It is secured to the leading edge of the blade by countersunk wood screws and rivets. The heads of the screws are soldered to the tipping to prevent loosening, and the solder is filed

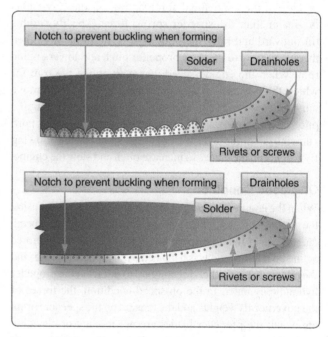

Figure 7-21. *Installation of metal sheath and tipping.*

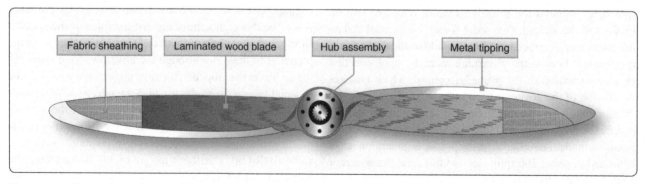

Figure 7-20. *Fix-pitch wooden propeller assembly.*

to make a smooth surface. Since moisture condenses on the tipping between the metal and the wood, the tipping is provided with small holes near the blade tip to allow this moisture to drain away or be thrown out by centrifugal force. It is important that these drain holes be kept open at all times. Since wood is subject to swelling, shrinking, and warping because of changes of moisture content, a protective coating is applied to the finished propeller to prevent a rapid change of moisture content. The finish most commonly used is a number of coats of water-repellent, clear varnish. After these processes are completed, the propeller is mounted on a spindle and very carefully balanced.

Several types of hubs are used to mount wooden propellers on the engine crankshaft. The propeller may have a forged steel hub that fits a splined crankshaft; it may be connected to a tapered crankshaft by a tapered, forged steel hub; or it may be bolted to a steel flange forged on the crankshaft. In any case, several attaching parts are required to mount the propeller on the shaft properly.

Hubs fitting a tapered shaft are usually held in place by a retaining nut that screws onto the end of the shaft. On one model, a locknut is used to safety the retaining nut and to provide a puller for removing the propeller from the shaft. This nut screws into the hub and against the retaining nut. The locknut and the retaining nut are safetied together with lock-wire or a cotter pin.

Front and rear cones may be used to seat the propeller properly on a splined shaft. The rear cone is a one-piece bronze cone that fits around the shaft and against the thrust nut (or spacer) and seats in the rear-cone seat of the hub. The front cone is a two-piece, split-type steel cone that has a groove around its inner circumference so that it can be fitted over a flange of the propeller retaining nut. Then, the retaining nut is threaded into place and the front cone seats in the front cone hub. A snap ring is fitted into a groove in the hub in front of the front cone so that when the retaining nut is unscrewed from the propeller shaft, the front cone acts against the snap ring and pulls the propeller from the shaft.

One type of hub incorporates a bronze bushing instead of a front cone. When this type of hub is used, it may be necessary to use a puller to start the propeller from the shaft. A rear-cone spacer is sometimes provided with the splined-shaft propeller assembly to prevent the propeller from interfering with the engine cowling. The wide flange on the rear face of some types of hubs eliminates the use of a rear-cone spacer.

One type of hub assembly for the fixed-pitch, wooden propeller is a steel fitting inserted in the propeller to mount it on the propeller shaft. It has two main parts: the faceplate and the flange plate. *[Figure 7-22]* The faceplate is a steel disc that forms the forward face of the hub. The flange plate is a steel flange with an internal bore splined to receive the propeller shaft. The end of the flange plate opposite the flange disc is externally splined to receive the faceplate; the faceplate bore has splines to match these external splines. Both faceplate and flange plates have a corresponding series of holes drilled on the disc surface concentric with the hub center. The bore of the flange plate has a 15° cone seat on the rear end and a 30° cone seat on the forward end to center the hub accurately on the propeller shaft.

Metal Fixed-Pitch Propellers

Metal fixed-pitch propellers are similar in general appearance to a wooden propeller, except that the sections are usually thinner. The metal fixed-pitch propeller is widely used on many models of light aircraft and LSA. Many of the earliest metal propellers were manufactured in one piece of forged Duralumin. Compared to wooden propellers, they were lighter in weight because of elimination of blade-clamping devices, offered a lower maintenance cost because they were made in one piece, provided more efficient cooling because of the effective pitch nearer the hub, and, because there was no joint between the blades and the hub, the propeller pitch could be changed, within limits, by twisting the blade slightly by a propeller repair station.

Propellers of this type are now manufactured as one-piece anodized aluminum alloy. They are identified by stamping the propeller hub with the serial number, model number, Federal Aviation Administration (FAA) type certificate number, production certificate number, and the number of times the propeller has been reconditioned. The complete model number of the propeller is a combination of the basic model number and suffix numbers to indicate the propeller diameter and pitch. An explanation of a complete model number, using the McCauley 1B90/CM propeller, is provided in *Figure 7-23.*

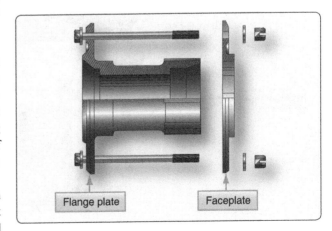

Figure 7-22. *Hub assembly.*

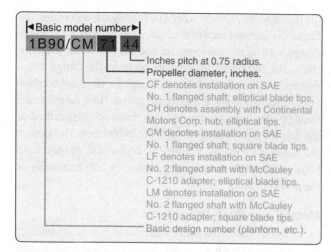

◄Basic model number►
1B90/CM 71 44

- Inches pitch at 0.75 radius.
- Propeller diameter, inches.
- CF denotes installation on SAE No. 1 flanged shaft; elliptical blade tips.
- CH denotes assembly with Continental Motors Corp. hub; elliptical tips.
- CM denotes installation on SAE No. 1 flanged shaft; square blade tips.
- LF denotes installation on SAE No. 2 flanged shaft with McCauley C-1210 adapter; elliptical blade tips.
- LM denotes installation on SAE No. 2 flanged shaft with McCauley C-1210 adapter; square blade tips.
- Basic design number (planform, etc.).

Figure 7-23. *Complete propeller model numbers.*

Constant-Speed Propellers

Hartzell Constant-Speed, Nonfeathering

Hartzell propellers can be divided by Aluminum hub (compact) and steel hub. Hartzell compact aluminum propellers represent new concepts in basic design. They combine low weight and simplicity in design and rugged construction. In order to achieve these ends, the hub is made as compact as possible, utilizing aluminum alloy forgings for most of the parts. The hub shell is made in two halves, bolted together along the plane of rotation. This hub shell carries

the pitch change mechanism and blade roots internally. The hydraulic cylinder, which provides power for changing the pitch, is mounted at the front of the hub. The propeller can be installed only on engines with flanged mounting provisions.

One model of nonfeathering aluminum hub constant-speed propeller utilizes oil pressure from a governor to move the blades into high pitch (reduced rpm). The centrifugal twisting moment of the blades tends to move them into low pitch (high rpm) in the absence of governor oil pressure. This is an exception to most of the aluminum hub models and feathering models. Most of the Hartzell propeller aluminum and steel hub models use centrifugal force acting on blade flyweights to increase blade pitch and governor oil pressure for low pitch. Many types of light aircraft use governor-regulated, constant-speed propellers in two-bladed and up to six-bladed versions. These propellers may be the nonfeathering type, or they may be capable of feathering and reversing. The steel hub contains a central "spider," that supports aluminum blades with a tube extending inside the blade roots. Blade clamps connect the blade shanks with blade retention bearings. A hydraulic cylinder is mounted on the rotational axis connected to the blade clamps for pitch actuation. *[Figure 7-24]*

The basic hub and blade retention is common to all models described. The blades are mounted on the hub spider for angular adjustment. The centrifugal force of the blades, amounting to as much as 25 tons, is transmitted to the hub

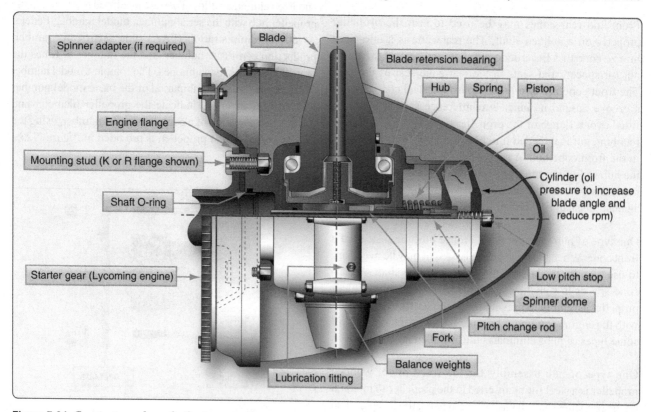

Figure 7-24. *Constant speed non-feathering propeller.*

spider through blade clamps and then through ball bearings. The propeller thrust and engine torque is transmitted from the blades to the hub spider through a bushing inside the blade shank. In order to control the pitch of the blades, a hydraulic piston-cylinder element is mounted on the front of the hub spider. The piston is attached to the blade clamps by means of a sliding rod and fork system for nonfeathering models and a link system for the feathering models. The piston is actuated in the forward direction by means of oil pressure supplied by a governor, which overcomes the opposing force created by the flyweights. Hartzell and McCauley propellers for light aircraft are similar in operation. The manufacturer's specifications and instructions must be consulted for information on specific models.

Constant-Speed Feathering Propeller

The feathering propeller utilizes a single oil supply from a governing device to hydraulically actuate a change in blade angle. *[Figure 7-25]* This propeller has five blades and is used primarily on Pratt & Whitney turbine engines. A two piece aluminum hub retains each propeller blade on a thrust bearing. A cylinder is attached to the hub and contains a feathering spring and piston. The hydraulically actuated piston transmits linear motion through a pitch change rod and fork to each blade to result in blade angle change.

While the propeller is operating, the following forces are constantly present: 1) spring force, 2) flyweight force, 3)

centrifugal twisting moment of each blade, and 4) blade aerodynamic twisting forces. The spring and flyweight forces attempt to rotate the blades to higher blade angle, while the centrifugal twisting moment of each blade is generally toward lower blade angle. Blade aerodynamic twisting force is usually very small in relation to the other forces and can attempt to increase or decrease blade angle. The summation of the propeller forces is toward higher pitch (low rpm) and is opposed by a variable force toward lower pitch (high rpm).

The variable force is oil under pressure from a governor with an internal pump that is mounted on and driven by the engine. The oil from the governor is supplied to the propeller and hydraulic piston through a hollow engine shaft. Increasing the volume of oil within the piston and cylinder decreases the blade angle and increases propeller rpm. If governor-supplied oil is lost during operation, the propeller increases pitch and feather. Feathering occurs because the summation of internal propeller forces causes the oil to drain out of the propeller until the feather stop position is reached. Normal in-flight feathering is accomplished when the pilot retards the propeller condition lever past the feather detent. This permits control oil to drain from the propeller and return to the engine sump. Engine shutdown is normally accomplished during the feathering process. Normal in-flight unfeathering is accomplished when the pilot positions the propeller condition lever into the normal flight (governing) range and restarts the engine. As engine speed increases, the governor supplies oil

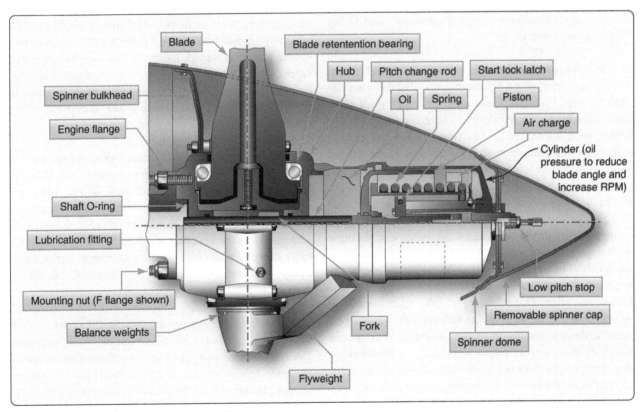

Figure 7-25. *Constant-speed feathering propeller.*

to the propeller and the blade angle decreases. Decreasing the volume of oil increases blade angle and decreases propeller rpm. By changing blade angle, the governor can vary the load on the engine and maintain constant engine rpm (within limits), independent of where the power lever is set. The governor uses engine speed sensing mechanisms that permit it to supply or drain oil as necessary to maintain constant engine speed (rpm). Most of the steel hub Hartzell propellers and many of the aluminum hub are full feathering. These feathering propellers operate similarly to the nonfeathering ones except the feathering spring assists the flyweights to increase the pitch. This propeller is normally placed in the full high pitch position before the engine is shut down to prevent exposure and corrosion of the pitch changing mechanism.

Feathering is accomplished by releasing the governor oil pressure, allowing the flyweights and feathering spring to feather the blades. This is done by pulling the condition lever (pitch control) back to the limit of its travel, which opens up a port in the governor allowing the oil from the propeller to drain back into the engine. Feathering occurs because the summation of internal propeller forces causes the oil to drain out of the propeller until the feather stop position is reached. The time necessary to feather depends upon the size of the oil passage from the propeller to the engine, and the force exerted by the spring and flyweights. The larger the passage is through the governor and the heavier the spring, the quicker the feathering action is. An elapsed time for feathering of between 3 and 10 seconds is usual with this system. Engine shutdown is normally accomplished during the feathering process.

In order to prevent the feathering spring and flyweights from feathering the propeller when the engine is shut down and the engine stopped, automatically removable high-pitch stops were incorporated in the design. These consist of spring-loaded latches fastened to the stationary hub that engage high-pitch stop plates bolted to the movable blade clamps. When the propeller is in rotation at speeds over 600–800 rpm, centrifugal force acts to disengage the latches from the high-pitch stop plates so that the propeller pitch may be increased to the feathering position. At lower rpm, or when the engine is stopped, the latch springs engage the latches with the high-pitch stops, preventing the pitch from increasing further due to the action of the feathering spring. As mentioned earlier, the engine load would be excessive, especially on fixed-turbine turboprop engines. One safety feature inherent in this method of feathering is that the propeller feathers if the governor oil pressure drops to zero for any reason. As the governor obtains its supply of oil from the engine lubricating system, it follows that if the engine runs out of oil or if oil pressure fails due to breakage of a part of the engine, the propeller feathers automatically. This action may save the engine from further

damage in case the pilot is not aware of trouble.

Unfeathering

Unfeathering can be accomplished by any of several methods, as follows:

1. Start the engine, so the governor can pump oil back into the propeller to reduce pitch. In most light twins, this procedure is considered adequate since the feathering of the propeller would happen infrequently. Vibration can occur when the engine starts and the propeller starts to come out of feather.

2. Provide an accumulator connected to the governor with a valve to trap an air-oil charge when the propeller is feathered but released to the propeller when the rpm control is returned to normal position. This system is used with training aircraft because it unfeathers the propeller in a very short time and starts the engine windmilling.

3. Provide an unfeathering pump that provides pressure to force the propeller back to low pitch quickly using engine oil.

Normal in-flight unfeathering is accomplished when the pilot positions the propeller condition lever into the normal flight (governing) range. [Figure 7-26] This causes the governor to disconnect the propeller oil supply from drain and reconnects it to the governed oil supply line from the governor. At that point, there is no oil available from the engine oil pump to the governor; therefore, no governed oil is available from the governor for controlling the propeller blade angle and rpm. As the engine is started, its speed increases, the governor supplies oil to the propeller, and the blade angle decreases. As soon as the engine is operating, the governor starts to unfeather the blades. Soon, windmilling takes place, which speeds up the process of unfeathering.

In general, restarting and unfeathering of propellers can be classified as reciprocating engine restart unfeathering, turboprop engine restart unfeathering, and accumulator unfeathering. When reciprocating unfeathering is used, the engine takes a little longer to start turning enough to provide oil pressure to the governor and then to the propeller. This delay can cause vibration as the propeller is unfeathered. Many aircraft can use an accumulator to provide stored pressure to unfeather the propeller much quicker.

Special unfeathering systems are available for certain aircraft where restarting the engine is difficult or for training purposes. The system consists of an oil accumulator connected to the governor through a valve. [Figure 7-26] The air or nitrogen pressure in one side of the accumulator pushes a piston to force oil from the other side of the accumulator through

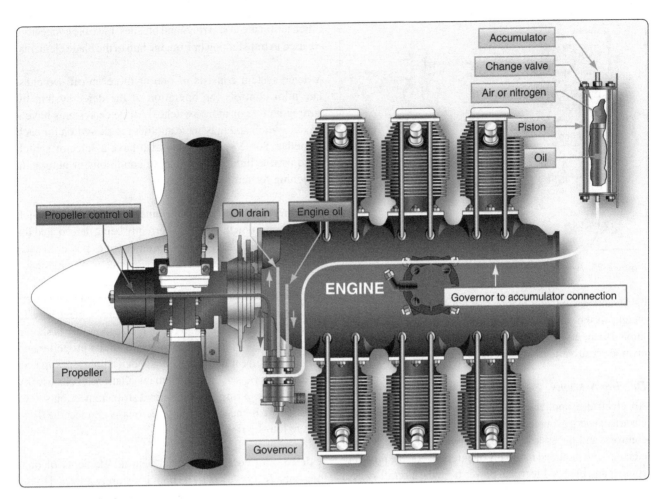

Figure 7-26. *Unfeathering system.*

the governor to the propeller piston to move the propeller blades from feather to a lower blade angle. The propeller then begins to windmill and permits the engine to start. When the unfeathering pump is used, it is an additional pump that, once the propeller control is in the correct position, the (full increase rpm) pump is actuated and the oil pressure from the pump unfeathers the propeller.

Propeller Auxiliary Systems

Ice Control Systems

Ice formation on a propeller blade, in effect, produces a distorted blade airfoil section that causes a loss in propeller efficiency. Generally, ice collects asymmetrically on a propeller blade and produces propeller unbalance and destructive vibration and increases the weight of the blades.

Anti-Icing Systems

A typical fluid system includes a tank to hold a supply of anti-icing fluid. *[Figure 7-27]* This fluid is forced to each propeller by a pump. The control system permits variation in the pumping rate so that the quantity of fluid delivered to a propeller can be varied, depending on the severity of icing.

Fluid is transferred from a stationary nozzle on the engine nose case into a circular U-shaped channel (slinger ring) mounted on the rear of the propeller assembly. The fluid under pressure of centrifugal force is transferred through nozzles to each blade shank.

Because airflow around a blade shank tends to disperse anti-icing fluids to areas where ice does not collect in large quantities, feed shoes, or boots, are installed on the blade leading edge. These feed shoes are a narrow strip of rubber extending from the blade shank to a blade station that is approximately 75 percent of the propeller radius. The feed shoes are molded with several parallel open channels in which fluid flows from the blade shank toward the blade tip by centrifugal force. The fluid flows laterally from the channels over the leading edge of the blade.

Isopropyl alcohol is used in some anti-icing systems because of its availability and low cost. Phosphate compounds are comparable to isopropyl alcohol in anti-icing performance and have the advantage of reduced flammability. However, phosphate compounds are comparatively expensive and, consequently, are not widely used. This system has

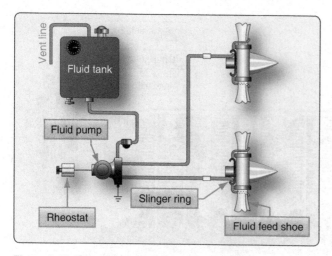

Figure 7-27. *Typical propeller fluid anti-icing system.*

which terminate in slip rings and brushes. Flexible connectors are used to transfer power from the hub to the blade elements.

A deice system consists of one or more on-off switches. The pilot controls the operation of the deice system by turning on one or more switches. All deice systems have a master switch and may have another toggle switch for each propeller. Some systems may also have a selector switch to adjust for light or heavy icing conditions or automatic switching for icing conditions.

The timer or cycling unit determines the sequence of which blades (or portion thereof) are currently being deiced, and for what length of time. The cycling unit applies power to each deice boot, or boot segment, in a sequence or all on order.

A brush block, which is normally mounted on the engine just behind the propeller, is used to transfer electricity to the slip ring. A slip ring and brush block assembly is shown in *Figure 7-29.* The slip ring rotates with the propeller and provides a current path to the blade deice boots. A slip ring wire harness is used on some hub installations to electrically connect the slip ring to the terminal strip connection screw. A deice wire harness is used to electrically connect the deice boot to the slip ring assembly.

A deice boot contains internal heating elements or dual elements. *[Figure 7-30]* The boot is securely attached to the leading edge of each blade with adhesive.

disadvantages in that it requires several components that add weight to the aircraft, and the time of anti-ice available is limited to the amount of fluid on board. This system is not used on modern aircraft, giving way to the electric deicing systems.

Deicing Systems

An electric propeller-icing control system consists of an electrical energy source, a resistance heating element, system controls, and necessary wiring. *[Figure 7-28]* The heating elements are mounted internally or externally on the propeller spinner and blades. Electrical power from the aircraft system is transferred to the propeller hub through electrical leads,

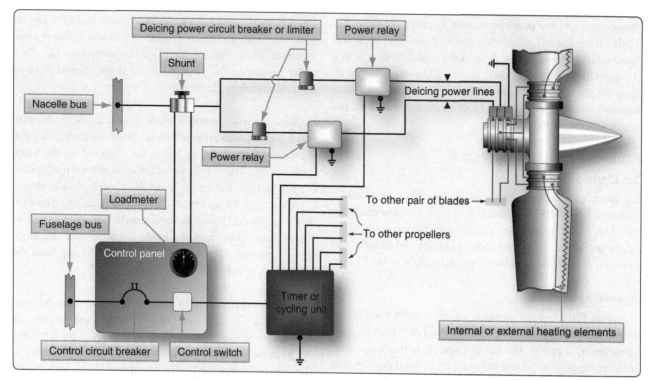

Figure 7-28. *Typical electrical deicing system.*

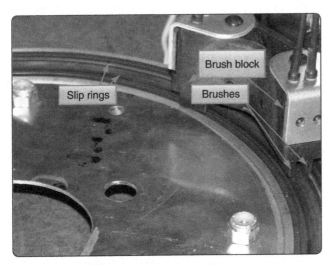

Figure 7-29. *Deicing brush block and slip ring assembly.*

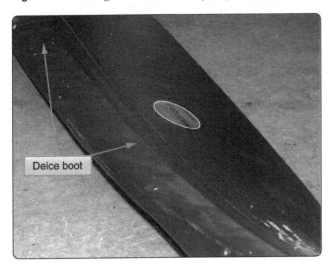

Figure 7-30. *Electric deice boot.*

Icing control is accomplished by converting electrical energy to heat energy in the heating element. Balanced ice removal from all blades must be obtained as nearly as possible if excessive vibration is to be avoided. To obtain balanced ice removal, variation of heating current in the blade elements is controlled so that similar heating effects are obtained in opposite blades.

Electric deicing systems are usually designed for intermittent application of power to the heating elements to remove ice after formation but before excessive accumulation. Proper control of heating intervals aids in preventing runback, since heat is applied just long enough to melt the ice face in contact with the blade. If heat supplied to an icing surface is more than that required for melting just the inner ice face, but insufficient to evaporate all the water formed, water will run back over the unheated surface and freeze. Runback of this nature causes ice formation on uncontrolled icing areas of the blade or surface.

Cycling timers are used to energize the heating element circuits for periods of 15 to 30 seconds, with a complete cycle time of 2 minutes. A cycling timer is an electric motor driven contactor that controls power contactors in separate sections of the circuit. Controls for propeller electrical deicing systems include on-off switches, ammeters or loadmeters to indicate current in the circuits, and protective devices, such as current limiters or circuit breakers. The ammeters or loadmeters permit monitoring of individual circuit currents and reflect operation of the timer. To prevent element overheating, the propeller deicing system is used only when the propellers are rotating and for short test periods of time during the takeoff check list or system inspection.

Propeller Synchronization & Synchrophasing

Most multi-engine aircraft are equipped with propeller synchronization systems. Synchronization systems provide a means of controlling and synchronizing engine rpm. Synchronization reduces vibration and eliminates the unpleasant beat produced by unsynchronized propeller operation. The synchrophasing system is designed to maintain a preset angular relationship between the designated master propeller and the slave propellers.

A typical synchrophasing system is an electronic system. *[Figure 7-31]* It functions to match the rpm of both engines and establish a blade phase relationship between the left and right propellers to reduce cabin noise. The system is controlled by a two-position switch located forward of the throttle quadrant. Turning the control switch on supplies direct current (DC) power to the electronic control box. Input signals representing propeller rpm are received from magnetic pickup on each propeller. The computed input signals are corrected to a command signal and sent to a rpm trimming coil located on the propeller governor of the slow engine. Its rpm is adjusted to that of the other propeller.

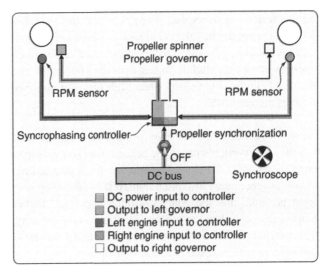

Figure 7-31. *Synchrophasing system.*

Autofeathering System

An autofeather system is used normally only during takeoff, approach, and landing. It is used to feather the propeller automatically if power is lost from either engine. The system uses a solenoid valve to dump oil pressure from the propeller cylinder (this allows the prop to feather) if two torque switches sense low torque from the engine. This system has a test-off-arm switch that is used to arm the system.

The autofeather system automatically energizes the holding coil (pulling in the feather button) when engine power loss results in a propeller thrust drop to a preset value. This system is switch-armed for use during takeoff and can function only when the power lever is near or in the "takeoff" position.

The NTS device mechanically moves the NTS plunger, which actuates a linkage in the propeller control when a predetermined negative torque value is sensed (when the propeller drives the engine). This plunger, working through control linkage, shifts the feather valve plunger, sending the blades toward feather.

As the blade angle increases, negative torque decreases until the NTS signal is removed, closing the feather valve. If the predetermined negative torque value is again exceeded, the NTS plunger again causes the feather valve plunger to shift. The normal effect of the NTS is a cycling of rpm slightly below the rpm at which the negative torque was sensed.

Unfeathering is initiated by pulling the feather button to the "unfeather" position. This action supplies voltage to the auxiliary motor to drive the auxiliary pump. Because the propeller governor is in an underspeed position with the propeller feathered, the blades will move in a decreased pitch direction under auxiliary pump pressure.

The pitch lock operates in the event of a loss of propeller oil pressure or an overspeed. The ratchets of the assembly become engaged when the oil pressure, which keeps them apart, is dissipated through a flyweight-actuated valve, which operates at an rpm slightly higher than the 100% rpm The ratchets become disengaged when high pressure and rpm settings are restored.

At the "flight idle" power lever position, the control beta follow-up low-pitch stop on the beta set cam (on the alpha shaft) is set about 2' below the flight low-pitch stop setting, acting as a secondary low-pitch stop. At the "takeoff" power lever position, this secondary low-pitch stop sets a higher blade angle stop than the mechanical flight low-pitch stop. This provides for control of overspeed after rapid power lever advance, as well as a secondary low-pitch stop.

Propeller Inspection & Maintenance

Propellers must be inspected regularly. The exact time interval for particular propeller inspections is usually specified by the propeller manufacturer. The regular daily inspection of propellers varies little from one type to another.

Typically, it is a visual inspection of propeller blades, hubs, controls, and accessories for security, safety, and general condition. Visual inspection of the blades does not mean a careless or casual observation. The inspection should be meticulous enough to detect any flaw or defect that may exist.

Inspections performed at greater intervals of time (e.g., 25, 50, or 100 hours) usually include a visual check of:

1. Blades, spinners, and other external surfaces for excessive oil or grease deposits.

2. Weld and braze sections of blades and hubs for evidence of failure.

3. Blade, spinner, and hubs for nicks, scratches, or other flaws. Use a magnifying glass if necessary.

4. Spinner or dome shell attaching screws for tightness.

5. The lubricating requirements and oil levels, when applicable.

If a propeller is involved in an accident, and a possibility exists that internal damage may have occurred, or if a propeller has had a ground strike or sudden stoppage, the recommendations of the engine and propeller manufacturer's maintenance manual need to be adhered to. The propeller should be disassembled and inspected. Whenever a propeller is removed from a shaft, the hub cone seats, cones, and other contact parts should be examined to detect undue wear, galling, or corrosion.

It is also vitally important to keep up-to-date airworthiness directives (ADs) or service bulletins (SBs) for a propeller. Compliance with ADs is required to make the aircraft legally airworthy, but it is also important to follow the SBs. All work performed on the propeller, including AD and SB compliance, should be noted in the propeller logbook.

The propeller inspection requirements and maintenance procedures discussed in this section are representative of those in widespread use on most of the propellers described in this chapter. No attempt has been made to include detailed maintenance procedures for a particular propeller, and all pressures, figures, and sizes are solely for the purpose of illustration and do not have specific application. For maintenance information on a specific propeller, always refer to applicable manufacturer instructions.

Wood Propeller Inspection

Wood propellers should be inspected frequently to ensure airworthiness. Inspect for defects, such as cracks, dents, warpage, glue failure, delamination defects in the finish, and charring of the wood between the propeller and the flange due to loose propeller mounting bolts. Examine the wood close to the metal sleeve of wood blades for cracks extending outward on the blade. These cracks sometimes occur at the threaded ends of the lag screws and may be an indication of internal cracking of the wood. Check the tightness of the lag screws, which attach the metal sleeve to the wood blade, in accordance with the manufacturer's instructions. In-flight tip failures may be avoided by frequent inspections of the metal cap, leading edge strip, and surrounding areas. Inspect for such defects as looseness or slipping, separation of soldered joints, loose screws, loose rivets, breaks, cracks, eroded sections, and corrosion. Inspect for separation between the metal leading edge and the cap, which would indicate the cap is moving outward in the direction of centrifugal force. This condition is often accompanied by discoloration and loose rivets. Inspect the tip for cracks by grasping it with your hand and slightly twisting about the longitudinal blade centerline and by slightly bending the tip backward and forward. If the leading edge and the cap have separated, carefully inspect for cracks at this point. Cracks usually start at the leading edge of the blade. Inspect moisture holes to ensure that they are open. A fine line appearing in the fabric or plastic may indicate a crack in the wood. Check the trailing edge of the propeller blades for bonding, separation, or damage.

Metal Propeller Inspection

Metal propellers and blades are generally susceptible to fatigue failure resulting from the concentration of stresses at the bottoms of sharp nicks, cuts, and scratches. It is necessary, therefore, to frequently and carefully inspect them for such defects and make repairs promptly. The inspection of steel blades may be accomplished by either visual, fluorescent penetrant or magnetic particle inspection. The visual inspection is easier if the steel blades are covered with engine oil or rust-preventive compound. The full length of the leading edge (especially near the tip), the full length of the trailing edge, the grooves and shoulders on the shank, and all dents and scars should be examined with a magnifying glass to decide whether defects are scratches or cracks.

Tachometer inspection is a very important part of the overall propeller inspection. Operation with an inaccurate tachometer may result in restricted rpm operation and damaging high stresses. This could shorten blade life and could result in catastrophic failure. If the tachometer is inaccurate, then the propeller could be turning much faster than it is rated to turn, providing extra stress. Accuracy of the engine tachometer should be verified at 100-hour intervals or at annual inspection, whichever occurs first. Hartzell Propeller recommends using a tachometer that is accurate within ± 10 rpm and has an appropriate calibration schedule.

Aluminum Propeller Inspection

Carefully inspect aluminum propellers and blades for cracks and other flaws. A transverse crack or flaw of any size is cause for rejection. No repairs are permitted to the shanks (roots or hub ends) of aluminum-alloy, adjustable-pitch blades. The shanks must be within manufacturer's limits. Multiple deep nicks and gouges on the leading edge and face of the blade is cause for rejection. Use dye penetrant or fluorescent dye penetrant to confirm suspected cracks found in the propeller. Refer any unusual condition or appearance revealed by these inspections to the manufacturer.

Composite Propeller Inspection

Composite blades need to be visually inspected for nicks, gouges, loose material, erosion, cracks and debonds, and lightning strike. *[Figure 7-32]* Composite blades are inspected for delaminations and debonds by tapping the blade or cuff (if applicable) with a metal coin. If an audible change is apparent, sounding hollow or dead, a debond or delamination is likely. *[Figure 7-33]* Blades that incorporate a "cuff" have a different tone when coin tapped in the cuff area. To avoid confusing the sounds, coin tap the cuff area and the transition area between the cuff and the blade separately from the blade area. Additional nondestructive testing (NDT) techniques for composite materials, such as phased array inspections, and ultrasound inspections, are available for more detailed inspections.

Repairs to propellers are often limited to minor type repairs. Certified mechanics are not allowed to perform major repairs on propellers. Major repairs need to be accomplished by a certificated propeller repair station.

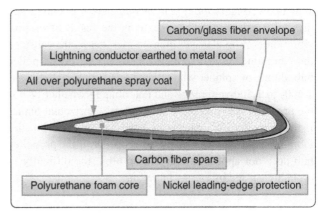

Figure 7-32. *Composite blade construction.*

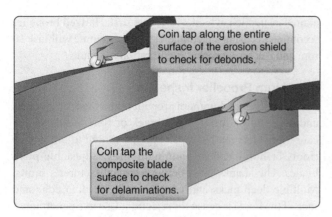

Coin tap along the entire surface of the erosion shield to check for debonds.

Coin tap the composite blade suface to check for delaminations.

Figure 7-33. *Coin-tap test to check for debonds and delaminations.*

Propeller Vibration

Although vibration can be caused by the propeller, there are numerous other possible sources of vibration that can make troubleshooting difficult. If a propeller vibrates, whether due to balance, angle, or track problems, it typically vibrates throughout the entire rpm range, although the intensity of the vibration may vary with the rpm. If a vibration occurs only at one particular rpm or within a limited rpm range (e.g., 2200–2350 rpm), the vibration is not normally a propeller problem but a problem of a poor engine-propeller match. If a propeller vibration is suspected but cannot be positively determined, the ideal troubleshooting method is to temporarily replace the propeller with one known to be airworthy and then test fly the aircraft if possible. Blade shake is not the source of vibration problems. Once the engine is running, centrifugal force holds the blades firmly (approximately 30,000–40,000 pounds) against blade bearings. Cabin vibration can sometimes be improved by reindexing the propeller to the crankshaft. The propeller can be removed, rotated 180°, and reinstalled. The propeller spinner can be a contributing factor to an out-of-balance condition. An indication of this would be a noticeable spinner wobble while the engine is running. This condition is usually caused by inadequate shimming of the spinner front support or a cracked or deformed spinner.

When powerplant vibration is encountered, it is sometimes difficult to determine whether it is the result of engine vibration or propeller vibration. In most cases, the cause of the vibration can be determined by observing the propeller hub, dome, or spinner while the engine is running within a 1,200- to 1,500-rpm range and determining whether or not the propeller hub rotates on an absolutely horizontal plane. If the propeller hub appears to swing in a slight orbit, the vibration is usually caused by the propeller. If the propeller hub does not appear to rotate in an orbit, the difficulty is probably caused by engine vibration.

When propeller vibration is the reason for excessive vibration, the difficulty is usually caused by propeller blade imbalance, propeller blades not tracking, or variation in propeller blade angle settings. Check the propeller blade tracking and then the low-pitch blade angle setting to determine if either is the cause of the vibration. If both propeller tracking and low blade angle setting are correct, the propeller is statically or dynamically unbalanced and should be replaced, or re-balanced if permitted by the manufacturer.

Blade Tracking

Blade tracking is the process of determining the positions of the tips of the propeller blades relative to each other (blades rotating in the same plane of rotation). Tracking shows only the relative position of the blades, not their actual path. The blades should all track one another as closely as possible. The difference in track at like points must not exceed the tolerance specified by the propeller manufacturer. The design and manufacture of propellers is such that the tips of the blades give a good indication of tracking. The following method for checking tracking is normally used:

1. Chock the aircraft so it cannot be moved.

2. Remove one spark plug from each cylinder. This makes the propeller easier and safer to turn.

3. Rotate one of the blades so it is pointing down.

4. Place a solid object (e.g., a heavy wooden block that is at least a couple of inches higher off the ground than the distance between the propeller tip and the ground) next to the propeller tip so that it just touches or attaches a pointer/indicator to the cowling itself. *[Figure 7-34]*

5. Rotate the propeller slowly to determine if the next blade tracks through the same point (touches the block/pointer). Each blade track should be within $\frac{1}{16}$ inch

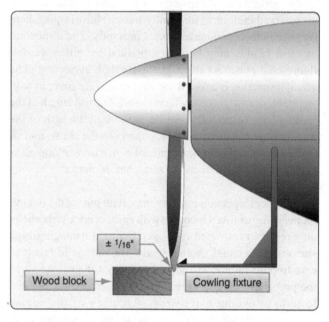

Figure 7-34. *Propeller blade tracking.*

(plus or minus) from the opposite blade's track.

6. An out-of-track propeller, may be due to one or more propeller blades being bent, a bent propeller flange, or propeller mounting bolts that are either over- or undertorqued. An out-of-track propeller causes vibration and stress to the airframe and engine and may cause premature propeller failure.

Checking & Adjusting Propeller Blade Angles

When you find an improper blade angle setting during installation or when indicated by engine performance, follow basic maintenance guidelines. From the applicable manufacturer's instructions, obtain the blade angle setting and the station at which the blade angle is checked. Do not use metal scribes or other sharply pointed instruments to mark the location of blade stations or make reference lines on propeller blades, since such surface scratches can induce failure (stress concentrator), eventually resulting in blade failure. Use a bench-top protractor if the propeller is removed from the aircraft. *[Figure 7-35]* Use a handheld protractor (a digital protractor provides an easy measurement) to check blade angle if the propeller is installed on the aircraft or is placed on the knife-edge balancing stand. *[Figure 7-36]*

Universal Propeller Protractor

The universal propeller protractor can be used to check propeller blade angles when the propeller is on a balancing stand or installed on the aircraft engine. *Figure 7-37* shows the parts and adjustments of a universal propeller protractor. The following instructions for using the protractor apply to a propeller installed on the engine. Turn the propeller until the first blade to be checked is horizontal with the leading edge up. Place the corner spirit level at right angles to the face of the protractor. Align degree and vernier scales by turning the disc adjuster before the disc is locked to the ring. The locking device is a pin that is held in the engaged position by a spring. The pin can be released by pulling it outward and turning it 90°.

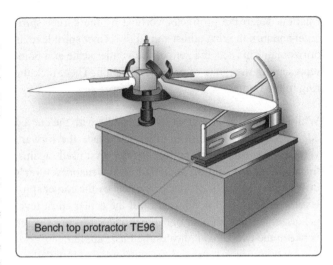

Figure 7-36. *Bench top protractor.*

Release the ring-to-frame lock (a right-hand screw with thumb nut) and turn the ring until both ring and disc zeros are at the top of the protractor.

Check the blade angle by determining how much the flat side of the block slants from the plane of rotation. First, locate a point to represent the plane of rotation by placing the protractor vertically against the end of the hub nut or any convenient surface known to lie in the plane of propeller

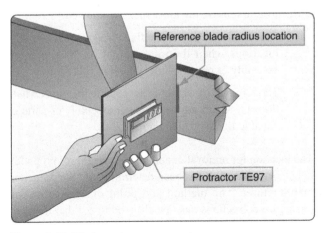

Figure 7-35. *Blade angle measurement.*

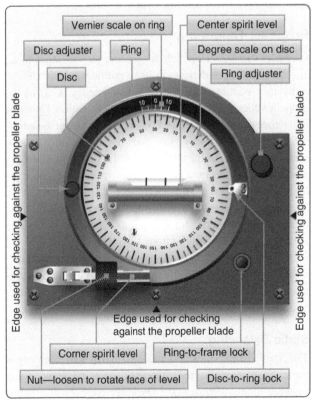

Figure 7-37. *Universal propeller protractor.*

rotation. Keep the protractor vertical by the corner spirit level and turn the ring adjuster until the center spirit level is horizontal. This sets the zero of the vernier scale at a point representing the plane of propeller rotation. Then, lock the ring to the frame.

While holding the protractor by the handle with the curved edge up, release the disc-to-ring lock. Place the forward vertical edge (the edge opposite the one first used) against the blade at the station specified in the manufacturer's instructions. Keep the protractor vertical by the corner spirit level and turn the disc adjuster until the center spirit level is horizontal. The number of degrees and tenths of a degree between the two zeros indicates the blade angle.

In determining the blade angle, remember that ten points on the vernier scale are equal to nine points on the degree scale. The graduations on the vernier scale represent tenths of a degree, but those of the degree scale represent whole degrees. The number of tenths of a degree in the blade angle is given by the number of vernier scale spaces between the zero of the vernier scale and the vernier scale graduation line nearest to perfect alignment with a degree scale graduation line. This reading should always be made on the vernier scale. The vernier scale increases in the same direction that the protractor scale increases. This is opposite to the direction of rotation of the moving element of the protractor. After making any necessary adjustment of the blade, lock it in position and repeat the same operations for the remaining blades of the propeller.

Propeller Balancing

Propeller unbalance, which is a source of vibration in an aircraft, may be either static or dynamic. Propeller static imbalance occurs when the center of gravity (CG) of the propeller does not coincide with the axis of rotation. Dynamic unbalance results when the CG of similar propeller elements, such as blades or flyweights, does not follow in the same plane of rotation. Since the length of the propeller assembly along the engine crankshaft is short in comparison to its diameter, and since the blades are secured to the hub so they lie in the same plane perpendicular to the running axis, the dynamic unbalance resulting from improper mass distribution is negligible, provided the track tolerance requirements are met. Another type of propeller unbalance, aerodynamic unbalance, results when the thrust (or pull) of the blades is unequal. This type of unbalance can be largely eliminated by checking blade contour and blade angle setting.

Static Balancing

The knife-edge test stand has two hardened steel edges mounted to allow the free rotation of an assembled propeller between them. [Figure 7-38] The knife-edge test stand must be located in a room or area that is free from any air motion,

and preferably removed from any source of heavy vibration.

The standard method of checking propeller assembly balance involves the following sequence of operations:

1. Insert a bushing in the engine shaft hole of the propeller.

2. Insert a mandrel or arbor through the bushing.

3. Place the propeller assembly so that the ends of the arbor are supported upon the balance stand knife-edges. The propeller must be free to rotate.

If the propeller is properly balanced statically, it remains at any position in which it is placed. Check two-bladed propeller assemblies for balance: first with the blades in a vertical position and then with the blades in a horizontal position. Repeat the vertical position check with the blade positions reversed; that is, with the blade that was checked in the downward position placed in the upward position.

Check a three-bladed propeller assembly with each blade placed in a downward vertical position. [Figure 7-39]

During a propeller static balance check, all blades must be at the same blade angle. Before conducting the balance check, inspect to see that each blade has been set at the same blade angle.

Unless otherwise specified by the manufacturer, an acceptable balance check requires that the propeller assembly have no tendency to rotate in any of the positions previously described. If the propeller balances perfectly in all described positions, it should also balance perfectly in all intermediate positions. When necessary, check for balance in intermediate positions to verify the check in the originally described positions. [Figure 7-40]

When a propeller assembly is checked for static balance and there is a definite tendency of the assembly to rotate, certain corrections to remove the unbalance are allowed.

1. The addition of permanent fixed weights at acceptable locations when the total weight of the propeller assembly or parts is under the allowable limit.

2. The removal of weight at acceptable locations when the total weight of the propeller assembly or parts is equal to the allowable limit.

The location for removal or addition of weight for propeller unbalance correction has been determined by the propeller manufacturer. The method and point of application of unbalance corrections must be checked to see that they are according to applicable drawings.

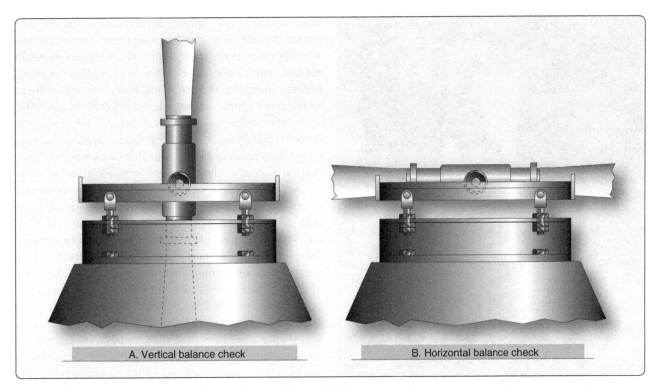

Figure 7-38. *Positions of two-bladed propeller during a balance check.*

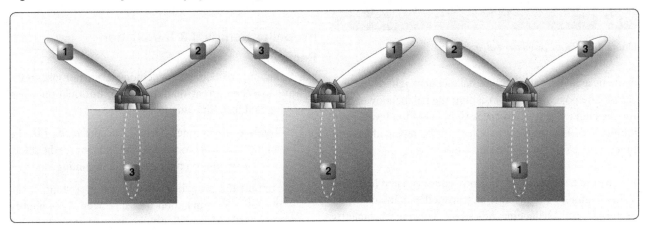

Figure 7-39. *Positions of three-bladed propeller during balance check.*

Dynamic Balancing

Propellers can also be dynamically balanced (spin balanced) with an analyzer kit to reduce the vibration levels of the propeller and spinner assembly. Some aircraft have the system hardwired in the aircraft and on other aircraft the sensors and cables need to be installed before the balancing run. Balancing the propulsion assembly can provide substantial reductions in transmitted vibration and noise to the cabin and also reduces excessive damage to other aircraft and engine components. The dynamic imbalance could be caused by mass imbalance or any aerodynamic imbalance. Dynamic balancing only improves the vibration caused by mass unbalance of the externally rotating components of the propulsion system. Balancing does not reduce the vibration

level if the engine or aircraft is in poor mechanical condition.

Defective, worn, or loose parts will make balancing impossible. Several manufacturers make dynamic propeller balancing equipment, and their equipment operation could differ. The typical dynamic balancing system consists of a vibration sensor that is attached to the engine close to the propeller, and an analyzer unit that calculates the weight and location of balancing weights.

Balancing Procedure

Face the aircraft directly into the wind (maximum 20 knots) and place chocks at the wheels. When you have installed the analyzing equipment, run the engine up at low cruise rpm; the

Figure 7-40. *Static propeller balancing.*

dynamic analyzer calculates the balancing weight required at each blade position. After installing the balancing weights, run the engine up again to check if the vibration levels have diminished. This process may have to be repeated several times before satisfactory results are achieved.

A dynamic balancing example procedure is listed here, but always refer to the aircraft and propeller manuals when performing any balancing procedure. Dynamic balance is accomplished by using an accurate means of measuring the amount and location of the dynamic imbalance. The number of balance weights installed must not exceed the limits specified by the propeller manufacturer. Follow the dynamic balance equipment manufacturer's instructions for dynamic balance in addition to the specifications of the propeller.

Most equipment use an optical pickup that senses reflective tape for rpm reading. Also, there is an accelerometer mounted to the engine that senses vibration in inches per second (ips).

Visually inspect the propeller assembly before dynamic balancing. The first runup of a new or overhauled propeller assembly may leave a small amount of grease on the blades and inner surface of the spinner dome. Use Stoddard solvent (or equivalent) to completely remove any grease on the blades

or inner surface of the spinner dome. Visually examine each propeller blade assembly for evidence of grease leakage. Visually examine the inner surface of the spinner dome for evidence of grease leakage. If there is no evidence of grease leakage, lubricate the propeller in accordance with the maintenance manual. If grease leakage is evident, determine the location of the leak and correct before relubricating the propeller and dynamic balancing. Before dynamic balance, record the number and location of all balance weights. Static balance is accomplished at a propeller overhaul facility when an overhaul or major repair is performed. Twelve equally spaced locations are recommended for weight attachment. Install the balancing weights using aircraft quality 10-32 or AN-3 type screws or bolts. Balance weight screws attached to the spinner bulkhead must protrude through the self-locking nuts or nut plates a minimum of one thread and a maximum of four threads. Unless otherwise specified by the engine or airframe manufacturer, Hartzell recommends that the propeller be dynamically balanced to a reading of 0.2 ips, or less. If reflective tape is used for dynamic balancing, remove the tape immediately after balancing is completed. Make a record in the propeller logbook of the number and location of dynamic balance weights, and static balance weights if they have been reconfigured.

Propeller Removal & Installation
Removal

The following procedure is for demonstration purposes only. Always use the current manufacturer's information when removing and installing any propeller.

1. Remove the spinner dome in accordance with the spinner removal procedures. Cut and remove the safety wire (if installed) on the propeller mounting studs.

2. Support the propeller assembly with a sling. If the propeller is reinstalled and has been dynamically balanced, make an identifying mark (with a felt-tipped pen only) on the propeller hub and a matching mark on the engine flange to make sure of proper orientation during reinstallation to prevent dynamic imbalance.

3. Unscrew the four mounting bolts from the engine bushings. Unscrew the two mounting nuts and the attached studs from the engine bushings. If the propeller is removed between overhaul intervals, mounting studs, nuts, and washers may be reused if they are not damaged or corroded.

    Caution: Remove the propeller from the mounting flange with care to prevent damaging the propeller mounting studs. Using the support sling, remove the propeller from the mounting flange.

4. Place the propeller on a cart for transport.

Installation

A flange propeller has six studs configured in a four-inch circle. Two special studs that also function as dowel pins are provided to transfer torque and index the propeller with respect to the engine crankshaft. The dowel pin locations used on a particular propeller installation are indicated in the propeller model stamped on the hub. Perform the applicable steps under Spinner Pre-Installation and clean the engine flange and propeller flange with quick dry Stoddard solvent or methyl ethyl ketone (MEK). Install the O-ring in the O-ring groove in the hub bore. **Note:** When the propeller is received from the factory, the O-ring has usually been installed. With a suitable support, such as a crane hoist or similar equipment, carefully move the propeller assembly to the aircraft engine mounting flange in preparation for installation.

Install the propeller on the engine flange. Make certain to align the dowel studs in the propeller flange with the corresponding holes in the engine mounting flange. The propeller may be installed on the engine flange in a given position, or 180° from that position. Check the engine and airframe manuals to determine if either manual specifies a propeller mounting position.

Caution: Mounting hardware must be clean and dry to prevent excessive preload of the mounting flange.

Caution: Tighten nuts evenly to avoid hub damage.

Install the propeller mounting nuts (dry) with spacers. Torque the propeller mounting nuts (dry) in accordance with the proper specifications and safety wire the studs in pairs (if required by the aircraft maintenance manual) at the rear of the propeller mounting flange.

Servicing Propellers

Propeller servicing includes cleaning, lubricating, and replenishing operating lubrication supplies.

Cleaning Propeller Blades

Aluminum and steel propeller blades and hubs are usually cleaned by washing the blades with a suitable cleaning solvent, using a brush or cloth but current manufacturer's information should always be used. Do not use acid or caustic materials. Power buffers, steel wool, steel brushes, or any other tool or substance that may scratch or mar the blade should be avoided. If a high polish is desired, a number of good grades of commercial metal polish are available. After completing the polishing operation, immediately remove all traces of polish. When the blades are clean, coat them with a clean film of engine oil or suitable equivalent.

To clean wooden propellers, use warm water and a mild soap, together with brushes or cloth. If a propeller has been subjected to salt water, flush it with fresh water until all traces of salt have been removed. This should be accomplished as soon as possible after the salt water has splashed on the propeller, regardless of whether the propeller parts are aluminum alloy, steel, or wood. After flushing, thoroughly dry all parts, and coat metal parts with clean engine oil or a suitable equivalent.

To remove grease or oil from propeller surfaces, apply Stoddard solvent or equivalent to a clean cloth and wipe the part clean. Using a noncorrosive soap solution, wash the propeller. Thoroughly rinse with water. Permit to dry. Aluminum and steel propeller blades and hubs usually are cleaned by washing the blades with a suitable cleaning solvent, using a brush or cloth. Do not use acid or caustic materials. Avoid power buffers, steel wool, steel brushes, or any other tool or substance that may scratch or mar the blade. If a high polish is desired, a number of good grades of commercial metal polish are available. After completing the polishing operation, immediately remove all traces of polish. When the blades are clean, coat them with a clean film of engine oil or suitable equivalent.

Charging the Propeller Air Dome

These instructions are general in nature and do not represent any aircraft procedure. Always check the correct manual before servicing any propeller system. Examine the propeller to make sure that it is positioned on the start locks and using the proper control, then charge the cylinder with dry air or nitrogen. The air charge valve is located on the cylinder as indicated in *Figure 7-41*. Nitrogen is the preferred charging medium. The correct charge pressure is identified by checking the correct table shown. The temperature is used to find the correct pressure to charge the hub air pressure.

Propeller Lubrication

Hydromatic propellers operated with engine oil and some sealed propellers do not require lubrication. Electric propellers require oils and greases for hub lubricants and pitch change drive mechanisms. Proper propeller lubrication procedures, with oil and grease specifications, are usually published in the manufacturer's instructions. Experience indicates that water sometimes gets into the propeller blade bearing assembly on some models of propellers. For this reason, the propeller manufacturer's greasing schedule must be followed to ensure proper lubrication of moving parts and protection from corrosion. Observe overhaul periods because most defects in propellers are not external but unseen internal corrosion. Dissimilar metals in the prop and hub create an environment ripe for corrosion, and the only way to properly inspect many of these areas is through a teardown. Extensive corrosion can dramatically reduce the strength of the blades or hub. Even

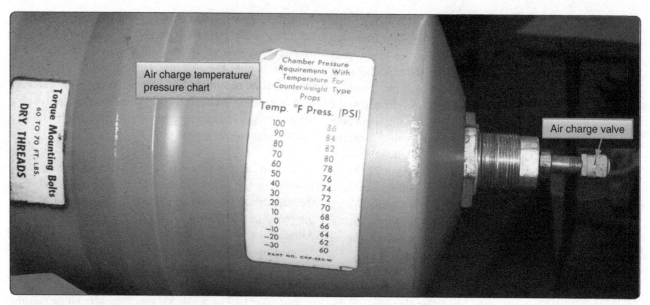

Figure 7-41. *Servicing air charge in propeller.*

seemingly minor corrosion may cause a blade or hub to fail an inspection. Because of the safety implications (blade loss), this is clearly an area in which close monitoring is needed.

One example of the lubrication requirements and procedures is detailed here for illustration purposes only. Lubrication intervals are important to adhere to because of corrosion implications. The propeller must be lubricated at intervals not to exceed 100 hours or at 12 calendar months, whichever occurs first. If annual operation is significantly less than 100 hours, calendar lubrication intervals should be reduced to 6 months. If the aircraft is operated or stored under adverse atmospheric conditions, such as high humidity, salt air, calendar lubrication intervals should be reduced to 6

months. Hartzell recommends that new or newly overhauled propellers be lubricated after the first 1 or 2 hours of operation because centrifugal loads pack and redistribute grease, which may result in a propeller imbalance. Redistribution of grease may also result in voids in the blade bearing area where moisture can collect. Remove the lubrication fitting from the cylinder-side hub half installed in the engine-side hub half. *[Figure 7-42]* Pump 1 fluid ounce (30 milliliters (ml)) grease into the fitting located nearest the leading edge of the blade on a tractor installation, or nearest the trailing edge on a pusher installation, until grease emerges from the hole where the fitting was removed, whichever occurs first.

Note: 1 fluid ounce (30 ml) is approximately six pumps with

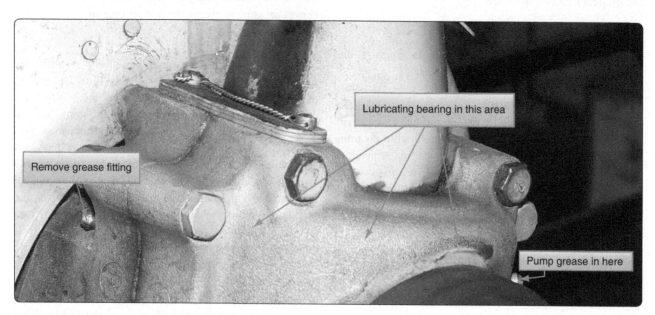

Figure 7-42. *Lubricating propeller bearings.*

a hand-operated grease gun. Reinstall the removed lubrication fittings. Tighten the fittings until snug. Make sure that the ball of each lubrication fitting is properly seated. Reinstall a lubrication fitting cap on each lubrication fitting. Perform grease replacement through attached pressure fittings (zerks) in accordance with the manufacturer's instructions.

Propeller Overhaul

Propeller overhaul should be accomplished at the maximum hours or calendar time limit, whichever occurs first. Upon receipt for overhaul, prepare a document that tracks the propeller components throughout the overhaul process. Research all applicable ADs, current specifications, and manufacturers' SBs for incorporation during the overhaul process. Double check the serial number and make notes on the work order regarding the general condition in which the propeller was received. As you disassemble and clean the unit, perform a preliminary inspection on all related parts. Record those revealing discrepancies requiring rework or replacement in the overhaul record by part number, along with the reason for the required action. Discard all threaded fasteners during disassembly and, with a few exceptions permitted by the manufacturer, replace with new components. Many specialized tools and fixtures are required in the disassembly and proper reassembly of propellers. These tools are generally model specific and range from massive 15-foot torque adapter bars and 100-ton presses down to tiny dowel pin alignment devices. Dimensionally inspect components that are subject to wear to the manufacturer's specifications. After passing inspection, anodize aluminum parts and cadmium plate steel parts for maximum protection against corrosion.

The Hub

Strip nonferrous hubs and components of paint and anodize and inspect for cracks using a liquid penetrant inspection (LPI) procedure. Etch, rinse, dry, and then immerse the parts in a fluorescent penetrant solution. After soaking in the penetrant, rinse them again and blow dry. Then, apply developer, which draws any penetrant caught in cracks or defects to the surface. Under an ultraviolet inspection lamp, the penetrant clearly identifies the flaw. Certain models of hubs are also eddy-current inspected around critical, high-stress areas. Eddy-current testing passes an electrical current through a conductive material that, when disturbed by a crack or other flaw, causes a fluctuation on a meter or CRT display. This method of inspection can detect flaws that are below the surface of the material and not exposed to the eye. Magnetic particle inspection (MPI) is used to locate flaws in steel parts. The steel parts of the propeller are magnetized by passing a strong electrical current through them. A suspension of fluorescent iron oxide powder and solvent is spread over the parts. While magnetized, the particles within the fluid

on the parts surface immediately align themselves with the discontinuity. When examined under black light, the crack or fault shows as a bright fluorescent line.

The first step in blade overhaul is the precise measurement of blade width, thickness, face alignment, blade angles, and length. Then, record the measurements on each blade's inspection record and check against the minimum acceptable overhaul specifications established by the manufacturer. Blade overhaul involves surface grinding and repitching, if necessary. Occasionally, blade straightening is also required. The manufacturer's specification dictates certain allowable limits within which a damaged blade may be cold straightened and returned to airworthy condition. Specialized tooling and precision measuring equipment permit pitch changes or corrections of less than one-tenth of one degree. To ensure accuracy, take frequent face alignment and angle measurements during the repair process. Precision hand grind the blade airfoil to remove all corrosion, scratches, and surface flaws. After completely removing all stress risers and faults, take final blade measurements and record on each blade's inspection record. Balance and match the propeller blades and anodize and paint them for long-term corrosion protection.

Prop Reassembly

When both the hubs and the blades have completed the overhaul process, the propeller is ready for final assembly. Recheck part numbers with the manufacturer's specifications. Lubricate and install the parts per each unit's particular overhaul manual. After final assembly, check both high- and low-pitch blade angles on constant-speed propellers for proper operation and leaks by cycling the propeller with air pressure through its blade angle range. Then, check the assembled propeller for static balance. If necessary, place weights on the hub areas of each "light" blade socket to bring about its proper balance. These weights should be considered part of the basic hub assembly and should not be moved during subsequent dynamic balancing to the engine.

As with most aircraft components, all of the hardware on the propeller assembly must be safety wired, unless secured by self-locking devices. Then, the final inspector fills out and signs maintenance release tags reflecting the work accomplished, applicable ADs, and all incorporated service documents. These documents certify that the major repairs and/or alterations that have been made meet established standards and that the propeller is approved for return to service. All minor repairs and minor alterations on propellers must be accomplished by a certified repair station, an airframe and powerplant technician (A&P), or a person working under the direct supervision of such a technician or an appropriately rated air carrier. Major repairs or alterations, including the overhaul of controllable pitch propellers, must be done by an appropriately rated repair station, manufacturer, or air carrier.

Troubleshooting Propellers

Some brief examples of troubleshooting problems and possible causes are provided in the following subsections. Always refer to the correct manual for actual information on troubleshooting.

Hunting & Surging

Hunting is characterized by a cyclic variation in engine speed above and below desired speed. Surging is characterized by a large increase/decrease in engine speed, followed by a return to set speed after one or two occurrences. If propeller is hunting, an appropriately licensed repair facility should check:

1. Governor,
2. Fuel control, and
3. Synchrophaser or synchronizer.

Engine Speed Varies with Flight Attitude (Airspeed)

Small variances in engine speed are normal and are no cause for concern. An increase in engine speed while descending or increasing airspeed with a nonfeathering propeller could be:

1. The governor not increasing oil volume in the propeller,
2. Engine transfer bearing leaking excessively, or
3. Excessive friction in blade bearings or pitch changing mechanism.

Failure to Feather or Feathers Slowly

Failure to feather or slow feathering of the propeller requires the FAA-certificated A&P technician to:

1. Refer to the air charge section in the maintenance manual if the air charge is lost or low.
2. Check for proper function and rigging of propeller/governor control linkage.
3. Check the governor drain function.
4. Check the propeller for misadjustment or internal corrosion (usually in blade bearings or pitch change mechanism) that results in excessive friction. This must be performed at an appropriately licensed propeller repair facility.

Figure 7-43. *Turboprop commuter.*

Figure 7-44. *Pratt & Whitney PT6 engine.*

Turboprop Engines & Propeller Control Systems

Turboprop engines are used for many single, twin, and commuter aircraft. *[Figure 7-43]* Smaller turboprop engines, such as the PT-6, are used on single and twin engine designs; the power ranges from 500 to 2,000 shaft horsepower. *[Figure 7-44]* Large commuter aircraft use turboprop engines, such as the P&W 150 and AE2100 that can deliver up to 5,000 shaft horsepower to power mid-sized to large turboprop aircraft. *[Figure 7-45]* The turboprop propeller is operated by a gas turbine engine through a reduction-gear assembly. It has proved to be an extremely efficient power source. The combination of propeller, reduction-gear assembly, and turbine engine is referred to as a turboprop powerplant.

The turbofan engine produces thrust directly; the turboprop engine produces thrust indirectly because the compressor and turbine assembly furnishes torque to a propeller, producing

Figure 7-45. *Pratt & Whitney 150 turboprop engine.*

the major portion of the propulsive force that drives the aircraft. The turboprop fuel control and the propeller governor are connected and operate in coordination with each other.

The power lever directs a signal from the flight deck to the fuel control for a specific amount of power from the engine. The fuel control and the propeller governor together establish the correct combination of rpm, fuel flow, and propeller blade angle to create sufficient propeller thrust to provide the desired power.

The propeller control system is divided into two types of control: one for flight and one for ground operation. For flight, the propeller blade angle and fuel flow for any given power lever setting are governed automatically according to a predetermined schedule. Below the "flight idle" power lever position, the coordinated rpm blade angle schedule becomes incapable of handling the engine efficiently. Here, the ground handling range, referred to as the beta range, is encountered. In the beta range of the throttle quadrant, the propeller blade angle is not governed by the propeller governor but is controlled by the power lever position. When the power lever is moved below the start position, the propeller pitch is reversed to provide reverse thrust for rapid deceleration of the aircraft after landing.

A characteristic of the turboprop is that changes in power are not related to engine speed, but to turbine inlet temperature. During flight, the propeller maintains a constant engine speed. This speed is known as the 100 percent rated speed of the engine, and it is the design speed at which most power and best overall efficiency can be obtained. Power changes are affected by changing the fuel flow. An increase in fuel flow causes an increase in turbine inlet temperature and a corresponding increase in energy available at the turbine. The turbine absorbs more energy and transmits it to the propeller in the form of torque. The propeller, in order to absorb the increased torque, increases blade angle, thus maintaining

Figure 7-46. *Reduction gearbox.*

constant engine rpm with added thrust.

Reduction Gear Assembly

The function of the reduction gear assembly is to reduce the high rpm from the engine to a propeller rpm that can be maintained without exceeding the maximum propeller tip speed (speed of sound). Most reduction gear assemblies use a planetary gear reduction. *[Figure 7-46]* Additional power takeoffs are available for propeller governor, oil pump, and other accessories. A propeller brake is often incorporated into the gearbox. The propeller brake is designed to prevent the propeller from windmilling when it is feathered in flight, and to decrease the time for the propeller to come to a complete stop after engine shutdown.

Turbo-Propeller Assembly

The turbo-propeller provides an efficient and flexible means of using the power of the engine at any condition in flight (alpha range). *[Figure 7-47]* For ground handling and reversing (beta range), the propeller can be operated to provide either zero or negative thrust. The major subassemblies of the propeller assembly are the barrel, dome, low-pitch stop assembly, overspeed governor, pitch control unit, auxiliary pump, feather and unfeather valves, torque motor, spinner, deice timer, beta feedback assembly, and propeller electronic control. Modern turboprop engines use dual Full Authority Digital Engine Control (FADEC) to control both engine and propeller. The spinner assembly is a cone-shaped configuration that mounts on the propeller and encloses the dome and barrel to reduce drag.

Propeller operation is controlled by a mechanical linkage from the flight deck-mounted power lever and the emergency engine shutdown handle (if the aircraft is provided with one) to the coordinator, which, in turn, is linked to the propeller control input lever. Newer designs use electronic throttle control that is linked to the FADEC controller.

Turbo-propeller control assemblies have a feathering system that feather the propeller when the engine is shut down in flight. The propeller can also be unfeathered during flight, if the engine needs to be started again. Propeller control systems for large turboprop engines differ from smaller engines because they are dual acting, which means that hydraulic pressure is used to increase and decrease propeller blade angle. *[Figure 7-48]*

Pratt & Whitney PT6 Hartzell Propeller System

The PT6 Hartzell propeller system incorporates three-, four-, or six-bladed propellers made of aluminum or composite materials. It is a constant-speed, feathering, reversing propeller system using a single-acting governor. Oil from the propeller governor feeds into the propeller shaft and to

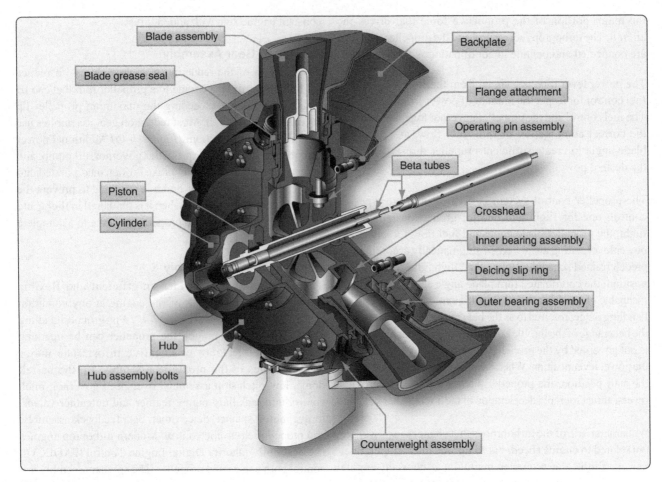

Figure 7-47. *Turboprop propeller.*

the servo piston via the oil transfer sleeve mounted on the propeller shaft. *[Figure 7-49]* As oil pressure increases, the servo piston is pushed forward, and the feather spring is compressed. Servo piston movement is transmitted to the propeller blade collars via a system of levers. When oil pressure is decreased, the return spring and flyweights force the oil out of the servo piston and change the blade pitch to a high pitch position. An increase in oil pressure drives the blades towards low pitch.

Engine oil is supplied to the governor from the engine oil supply. A gear pump, mounted at the base of the governor, increases the flow of oil going to the constant speed unit (CSU) relief valve. When the oil pressure reaches the desired level, the relief valve opens to maintain the governor oil pressure. When the speed selected by the pilot is reached, the flyweight force equals the spring tension of the speeder spring. The governor flyweights are then on speed. When the engine output power is increased, the power turbines tend to increase speed. The flyweights in the CSU sense this acceleration and the flyweights go into an overspeed condition because of the increase centrifugal force. This force causes the control valve to move up and restrict oil flow to the propeller dome. *[Figure 7-50]* The feathering spring increases the propeller

pitch to maintain the selected speed. Reducing power causes an under-speed of the flyweights, downward movement of the control valve, more oil in propeller dome, resulting in a lower pitch to control propeller speed. The propeller governor houses an electro-magnetic coil, which is used to match the rpm of both propellers during cruise. An aircraft supplied synchrophaser unit controls this function.

At low power, the propeller and governor flyweights do not turn fast enough to compress the speeder spring. *[Figure 7-51]* In this condition, the control valve moves down, and high pressure oil pushes the dome forward moving the blades towards low pitch. Any further movement pulls the beta rod and slip ring forward. The forward motion of the slip ring is transmitted to the beta valve via the beta lever and the carbon block. Forward movement of the beta valve stops the oil supply to the propeller. This prevents the blade angles from going any lower. This is the primary blade angle (PBA) and is the minimum blade angle allowed for flight operation. From this point, the propeller is in the beta mode. If the engine power is reduced when the propeller is at the primary blade angle, the propeller speed decreases since the blade angle does not change.

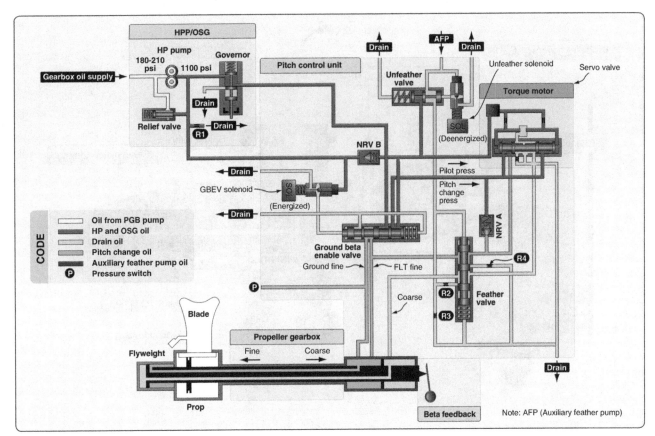

Figure 7-48. *Propeller control system schematic.*

The lock-pitch solenoid valve prevents the propeller from going into reverse or below the primary blade angle in the event of a beta system malfunction in flight. The solenoid is energized by a switch (airframe supplied) mechanically connected to the propeller slip-ring linkage via a second carbon block. As oil pressure leaks off around the propeller shaft oil transfer sleeve, the blade angle slowly drifts back toward high pitch. This deactivates the low pitch solenoid valve and restores the oil supply to the propeller servo. The low pitch solenoid valve cycles (close/open) as backup to the beta valve function. Moving the power lever backwards causes the reversing cam and cable to move the beta valve backward, allowing more oil to flow into the propeller dome, and causing the blades to go towards reverse pitch. *[Figure 7-52]*

As the blades move to reverse, the dome pulls the slip ring forward and moves the beta valve outward, restricting the oil flow. This stops the blade movement toward reverse. To obtain more reverse thrust, move the power lever back more to reset the beta valve inward, and repeat the process. Move the reset arm on the CSU rearward by the interconnecting rod at the same time the blade angle moves toward reverse. This causes the reset lever and reset post to move down in the CSU, bringing the reset lever closer to the speeder spring cup. As propeller speed increases due to the increase in engine

power, the governor flyweights begin to move outwards. Since the reset lever is closer to the speeder spring cup, the cup contacts the reset lever before the flyweights would normally reach the on-speed position (95 percent propeller speed instead of 100 percent). As the reset lever is pushed up by the flyweights/speeder spring cup, the Py air bleeds from the fuel control unit (FCU) which lowers the fuel flow, engine power, and thus propeller speed. In reverse, propeller speed remains 5 percent below the selected propeller speed so that the control valve remains fully open, and only the beta valve controls the oil flow to the propeller dome.

In this mode, the propeller speed is no longer controlled by changing the blade angle. It is now controlled by limiting engine power. Bringing the propeller lever to the feather position causes the speed selection lever on the CSU to push the feathering valve plunger and allows propeller servo oil to dump into the reduction gearbox sump. The pressure loss in the propeller hub causes the feathering spring and the propeller flyweights to feather the propeller. In the event of a propeller overspeed not controlled by the propeller overspeed governor (oil governor), the flyweights in the propeller governor move outward until the speeder spring cup contacts the reset lever. *[Figure 7-53]* The movement of the reset lever around its pivot point opens the Py air passage. Py bleeds into the reduction gearbox limiting the fuel supply to

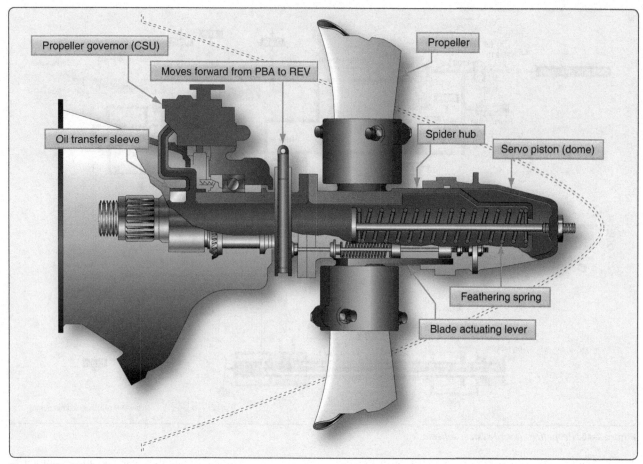

Figure 7-49. *Pitch change mechanism.*

the engine. This prevents the propeller/power turbines from accelerating beyond 106 percent rpm.

The oil overspeed governor houses a set of flyweights connected to a control valve that is driven by a beveled gear mounted on the propeller shaft. *[Figure 7-54]* The flyweight's centrifugal force is acting against two springs: a speeder spring and a reset spring. When the propeller speed reaches a specified limit (4 percent over maximum propeller speed), the governor flyweights lift the control valve and bleed off propeller servo oil into the reduction gearbox sump, causing the blade angle to increase. An increase in blade pitch puts more load on the engine and slows down the propeller. To test the unit, the speed reset solenoid is activated, and servo oil pressure pushes against the reset piston to cancel the effect of the reset spring. With less spring tension acting on the flyweights, the overspeed governor can be tested at speeds lower than maximum.

On twin installation, a second solenoid valve is mounted on the overspeed governor and is used in conjunction with the aircraft autofeather system. The system is switched on for takeoff and, in the event of an engine malfunction, energizes the solenoid valve to dump propeller servo oil into the reduction gearbox sump. The feathering spring and propeller

flyweights move the blade quickly to feather.

Hamilton Standard Hydromatic Propellers

Many of the hydromatic propellers are used with older type aircraft involved in cargo operations. A hydromatic propeller has a double-acting governor that uses oil pressure on both sides of the propeller piston. Many larger turboprop systems also use this type of system. The governors are similar in construction and principle of operation in normal constant-speed systems. The major difference is in the pitch-changing mechanism. In the hydromatic propeller, no flyweights are used, and the moving parts of the mechanism are completely enclosed. Oil pressure and the centrifugal twisting moment of the blades are used together to turn the blades to a lower angle. The main advantages of the hydromatic propeller are the large blade angle range and the feathering and reversing features.

This propeller system is a double-acting hydraulic propeller system in which the hydraulic pressure (engine oil pressure) on one piston dome is used against governor oil pressure on the other side of the piston. These two opposing hydraulic forces are used to control and change blade angle or pitch. Although hydromatic propeller systems are very old, some

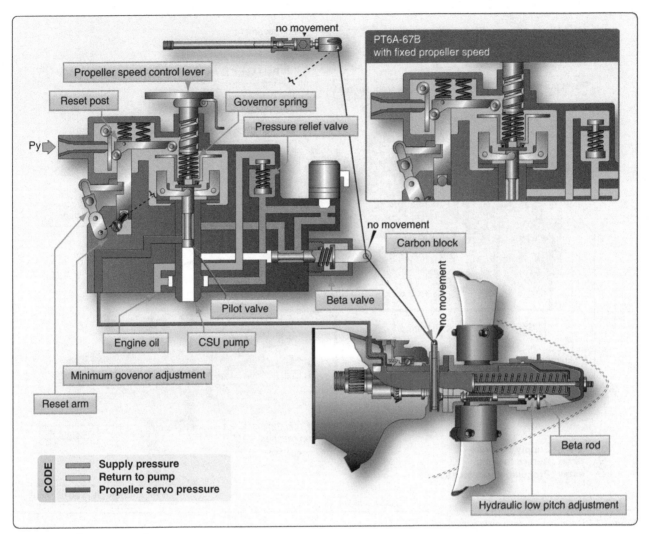

Figure 7-50. *Governing mode.*

are still used on radial engines. Larger new turboprop systems also use this opposing hydraulic force and double-acting governor systems.

The distributor valve or engine-shaft-extension assembly provides oil passages for governor or auxiliary oil to the inboard side of the piston and for engine oil to the outboard side. During unfeathering operation, the distributor shifts under auxiliary pressure and reverses these passages so that oil from the auxiliary pump flows to the outboard side of the piston and oil on the inboard side flows back to the engine. The engine-shaft-extension assembly is used with propellers that do not have feathering capabilities.

The hydromatic propeller *[Figure 7-55]* is composed of four major components:

1. The hub assembly,

2. The dome assembly,

3. The distributor valve assembly (for feathering on

single-acting propellers) or engine-shaft-extension assembly (for nonfeathering or double-acting propellers), and

4. The anti-icing assembly.

The hub assembly is the basic propeller mechanism. It contains both the blades and the mechanical means for holding them in position. The blades are supported by the spider and retained by the barrel. Each blade is free to turn about its axis under the control of the dome assembly.

The dome assembly contains the pitch-changing mechanism for the blades. Its major components are the:

1. rotating cam,

2. fixed cam,

3. piston, and

4. dome shell.

When the dome assembly is installed in the propeller hub,

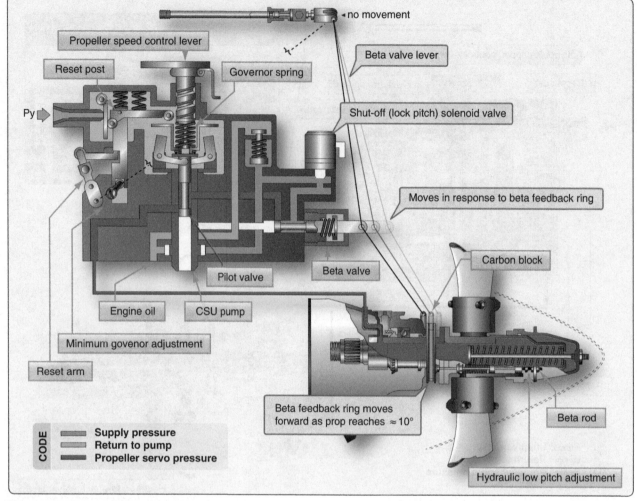

Figure 7-51. *Beta mode forward operation.*

the fixed cam remains stationary with respect to the hub. The rotating cam, which can turn inside the fixed cam, meshes with gear segments on the blades. The piston operates inside the dome shell and is the mechanism that converts engine and governor oil pressure into forces that act through the cams to turn propeller blades.

Principles of Operation

The pitch-changing mechanism of hydromatic propellers is a mechanical-hydraulic system in which hydraulic forces acting on a piston are transformed into mechanical twisting forces acting on the blades. Linear movement of the piston is converted to rotary motion by a cylindrical cam. A bevel gear on the base of the cam mates with bevel gear segments attached to the butt ends of the blades, thereby turning the blades. This blade pitch-changing action can be understood by studying the schematic in *Figure 7-56*.

The centrifugal force acting on a rotating blade includes a component force that tends to move the blade toward low pitch. As shown in *Figure 7-56*, a second force, engine oil

pressure, is supplied to the outboard side of the propeller piston to assist in moving the blade toward low pitch.

Propeller governor oil, taken from the engine oil supply and boosted in pressure by the engine-driven propeller governor, is directed against the inboard side of the propeller piston. It acts as the counterforce, which can move the blades toward higher pitch. By metering this high-pressure oil to, or draining it from, the inboard side of the propeller piston by means of the constant-speed control unit, the force toward high pitch can balance and control the two forces toward low pitch. In this way, the propeller blade angle is regulated to maintain a selected rpm.

The basic propeller control forces acting on the Hamilton Standard propeller are centrifugal twisting force and high pressure oil from the governor. The centrifugal force acting on each blade of a rotating propeller includes a component force that results in a twisting moment about the blade center line that tends, at all times, to move the blade toward low pitch. Governor pump output oil is directed by the governor

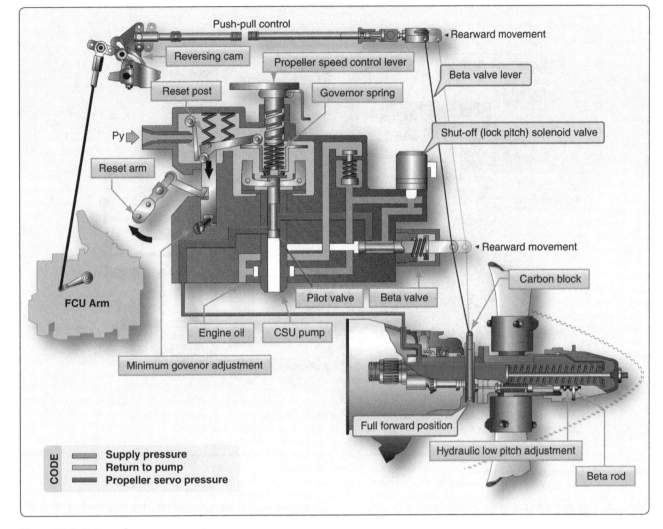

Figure 7-52. *Beta mode reverse operation.*

to either side of the propeller piston. The oil on the side of the piston opposite this high-pressure oil returns to the intake side of the governor pump and is used over again. Engine oil at engine supply pressure does not enter the propeller directly but is supplied only to the governor. During constant-speed operations, the double-acting governor mechanism sends oil to one side or the other of the piston as needed to keep the speed at a specified setting.

Feathering Operation

A typical hydromatic propeller feathering installation is shown in *Figure 7-57*. When the feathering push-button switch is depressed, the low current circuit is established from the battery through the push-button holding coil and from the battery through the solenoid relay. As long as the circuit remains closed, the holding coil keeps the push button in the depressed position. Closing the solenoid establishes the high current circuit from the battery to the feathering motor pump unit. The feathering pump picks up engine oil from the oil supply tank, boosts its pressure, if necessary,

to the relief valve setting of the pump, and supplies it to the governor high-pressure transfer valve connection. Auxiliary oil entering the high-pressure transfer valve connection shifts the governor transfer valve, which hydraulically disconnects the governor from the propeller and at the same time opens the propeller governor oil line to auxiliary oil. The oil flows through the engine transfer rings, through the propeller shaft governor oil passage, through the distributor valve port, between lands, and finally to the inboard piston end by way of the valve inboard outlet.

The distributor valve does not shift during the feathering operation. It merely provides an oil passageway to the inboard piston end for auxiliary oil and the outboard piston end for engine oil. The same conditions described for underspeed operation exist in the distributor valve, except that oil at auxiliary pressure replaces drain oil at the inboard end of the land and between lands. The distributor-valve spring is backed up by engine oil pressure, which means that at all times the pressure differential required to move the piston is

7-37

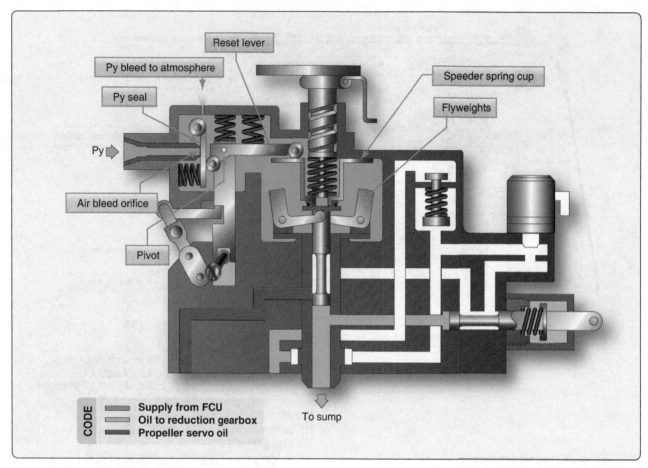

Figure 7-53. *Nf overspeed governor.*

identical with that applied to the distributor valve.

The propeller piston moves outboard under the auxiliary oil pressure at a speed proportional to the rate the oil is supplied. This piston motion is transmitted through the piston rollers operating in the oppositely inclined cam tracks of the fixed cam and the rotating cam and is converted by the bevel gears into the blade-twisting moment. Only during feathering or unfeathering is the low mechanical advantage portion of the cam tracks used. (The low mechanical advantage portion lies between the break and the outboard end of the track profile.) Oil at engine pressure, displaced from the outboard piston end, flows through the distributor valve outboard inlet, past the outboard end of the valve land, through the valve port, into the propeller shaft engine oil passage, and is finally delivered into the engine lubricating system. Thus, the blades move toward the full high-pitch (or feathered) angle.

Having reached the full-feathered position, further movement of the mechanism is prevented by contact between the high-angle stop ring in the base of the fixed cam and the stop lugs set in the teeth of the rotating cam. The pressure in the inboard piston end now increases rapidly, and upon reaching a set pressure, the electric cutout switch automatically opens.

This cutout pressure is less than that required to shift the distributor valve.

Opening the switch deenergizes the holding coil and releases the feathering push-button control switch. Release of this switch breaks the solenoid relay circuit, which shuts off the feathering pump motor. The pressures in both the inboard and outboard ends of the piston drop to zero, and, since all the forces are balanced, the propeller blades remain in the feathered position. Meanwhile, the governor high-pressure transfer valve has shifted to its normal position as soon as the pressure in the propeller governor line drops below that required to hold the valve open.

Unfeathering Operation

To unfeather a hydromatic propeller, depress and hold in the feathering switch push-button control switch. As in the case of feathering a propeller, the low-current control circuits from the battery through the holding coil and from the battery through the solenoid are completed when the solenoid closes. The high-current circuit from the battery starts the motor-pump unit, and oil is supplied at a high pressure to the governor transfer valve.

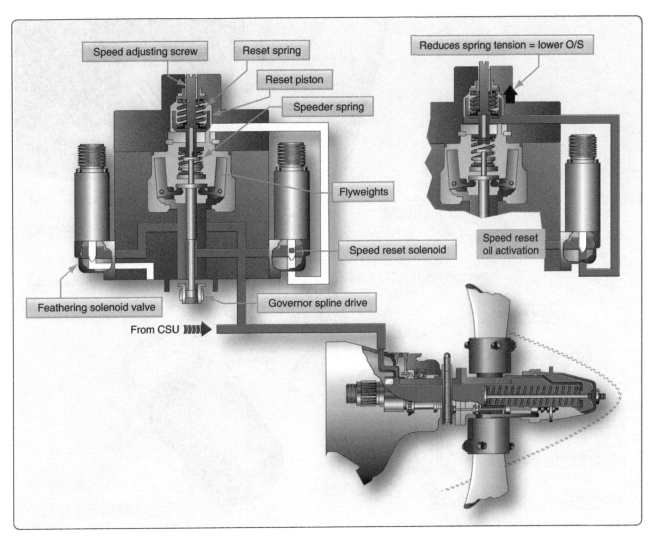

Figure 7-54. *Propeller overspeed governor.*

Auxiliary oil entering through the high-pressure transfer valve connection shifts the governor transfer valve and disconnects the governor from the propeller line; in the same operation, auxiliary oil is admitted. The oil flows through the engine oil transfer rings, through the propeller shaft governor oil passage, and into the distributor valve assembly.

When the unfeathering operation begins, the piston is in the extreme outboard position. The oil enters the inboard piston end of the cylinder by way of the distributor valve inboard outlet. As the pressure on the inboard end of the piston increases, the pressure against the distributor valve land builds up. When the pressure becomes greater than the combined opposing force of the distributor valve spring and the oil pressure behind this spring, the valve shifts. Once the valve shifts, the passages through the distributor valve assembly to the propeller are reversed. A passage is opened between lands and through a port to the outboard piston end by way of the distributor valve outlet. As the piston moves inboard under the auxiliary pump oil pressure, oil is displaced from the inboard

piston end through the inlet ports between the valve lands, into the propeller shaft engine oil lands, and into the propeller shaft engine oil passage where it is discharged into the engine lubricating system. At the same time, the pressure at the cutout switch increases and the switch opens. However, the circuit to the feathering pump and motor unit remains complete as long as the feathering switch is held in.

With the inboard end of the propeller piston connected to drain and auxiliary pressure flowing to the outboard end of the piston, the piston moves inboard, unfeathering the blades. As the blades are unfeathered, they begin to windmill and assist the unfeathering operation by the added force toward low pitch brought about by the centrifugal twisting moment. When the engine speed has increased to approximately 1,000 rpm, the operator shuts off the feathering pump motor. The pressure in the distributor valve and at the governor transfer valve decreases, allowing the distributor valve to shift under the action of the governor high-pressure transfer valve spring. This action reconnects the governor with the propeller and

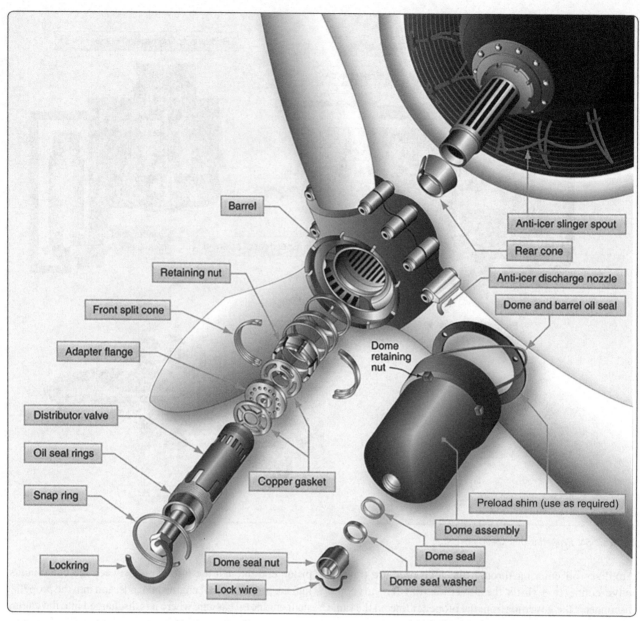

Figure 7-55. *Typical hydromatic propeller installation.*

establishes the same oil passages through the distributor valve that are used during constant-speed and feathering operations.

Setting the Propeller Governor

The propeller governor incorporates an adjustable stop that limits the maximum speed at which the engine can run. As soon as the takeoff rpm is reached, the propeller moves off the low-pitch stop. The larger propeller blade angle increases the load on the engine, thus maintaining the prescribed maximum engine speed. At the time of propeller, propeller governor, or engine installation, the following steps are normally taken to ensure that the powerplant obtains takeoff rpm. During ground runup, move the throttle to takeoff position and note the resultant rpm and manifold pressure. If the rpm obtained is higher or lower than the takeoff rpm prescribed in the

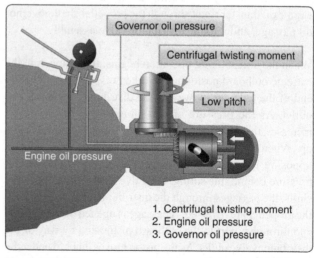

1. Centrifugal twisting moment
2. Engine oil pressure
3. Governor oil pressure

Figure 7-56. *Diagram of hydromatic propeller operational forces.*

manufacturer's instructions, reset the adjustable stop on the governor until the prescribed rpm is obtained.

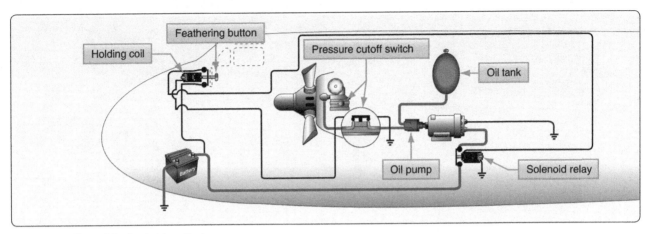

Figure 7-57. *Typical feathering installation.*

Chapter 8
Engine Removal & Replacement

Introduction

Procedures for removing or installing an aircraft engine usually vary widely with the type of aircraft and the type of engine. Thus, no single list of instructions can be provided as a guideline for all engines. Because of the many types of engine and aircraft installations and the large number of design variations within each type or category, representative examples have been selected to illustrate the most typical installation procedures for reciprocating, turboprop, and turbofan engines. There are some common tasks that must be accomplished when replacing an aircraft engine. Most engines require disconnecting and reconnecting electrical, hydraulic, fuel supply, intake and exhaust air path components, engine controls, and engine mounting connections to the airframe.

Reciprocating engines and gas turbine engines are used in this chapter to describe and represent general and typical procedures for engine buildup, removal, preservation, storage, and installation. Although these two types of engines have some common areas, each procedure has been included to ensure adequate coverage of the engines used in both heavy and light aircraft. It should be emphasized that while procedures for specific engines and aircraft are included in this chapter, many pertinent or mandatory references have been omitted because of their irrelevance to a general discussion. For this reason, always reference the applicable manufacturer's instructions before performing any phase of engine removal or installation.

Reasons for Removal of Reciprocating Engines

The following paragraphs outline the most common reasons for removing and replacing an engine. Information to aid in determining engine conditions that require removal is included; however, in every case, consult applicable manufacturer's instructions as the final authority in establishing the basis for engine replacement.

Engine or Component Lifespan Exceeded

Engine life is dependent upon such factors as operational use, the quality of manufacture or overhaul, the type of aircraft in which the engine is installed, the kind of operation being carried out, and the degree to which maintenance is accomplished. Thus, the manufacturer sets engine removal times. Based on service experience, it is possible to establish a maximum expected time before overhaul (TBO) or span of time within which an engine needs to be overhauled. Certain critical components of turbine engines such as turbine blades, turbine discs or combustion liners may have shorter life limits established by the manufacturer due to the stresses imposed on those parts during operation. The FAA requires that the manufacture identify and establish which parts have mandatory replacement times.

Engine Sudden Stoppage or Propeller Strike

Engine sudden stoppage causes a very rapid and complete engine stoppage. Propeller strikes can cause either reduction in speed or a complete engine stoppage. Either can be caused by engine seizure or by one or more of the propeller blades striking an object in such a way that revolutions per minute (rpm) goes to zero in less than one complete revolution of the propeller. Sudden stoppage may occur under such conditions as complete and rapid collapse of the landing gear, nosing over of the aircraft, or crash landing. Sudden stoppage can cause internal damage, such as cracked propeller gear teeth, gear train damage, crankshaft flyweights becoming detuned or misalignment, or damaged propeller bearings. When sudden stoppage occurs, the engine usually requires replacement or disassembly and inspection as per manufacturer's instructions.

Sudden Reduction in Speed

Sudden reduction in engine speed can occur when one or more of the propeller blades strike an object at a low engine rpm. After impact, the foreign object is cleared and the engine recovers rpm and continues to run unless stopped to prevent further damage. While taxiing an aircraft, sudden reduction in speed can occur when the propeller strikes a foreign object, such as a raised section in the runway, a tool box, or a portion of another airplane. When the accident occurs at high engine rpm, shocks are much more severe. When sudden reduction in rpm occurs, the following procedure can be used as a general rule, but you must comply with the manufacturer's information.

Make a thorough external inspection of the engine mount, crankcase, and nose section to determine whether any parts have been damaged. If damage is found that cannot be corrected by line maintenance, remove the engine. Internal components can be damaged, especially counter weights on the crankshaft.

Remove the engine oil screens or filters. Inspect them for

the presence of metal particles. Remove the engine sump plugs, drain the oil into a clean container, strain it through a clean cloth, and check the cloth and the strained oil for metal particles. Heavy metal particles in the oil indicate a definite engine failure, and the engine must be removed. However, if the metal particles present are similar to fine filings, continue the inspection of the engine to determine its serviceability. If there are no heavy metal particles in the engine oil, check again for metal in the oil system after operating the engine. Metal in the screens is a sign that the bearings have been compromised and are in the process of failing.

Remove the propeller and check the crankshaft, or the propeller drive shaft on reduction-gear engines, for misalignment. Clamp a test indicator to the nose section of the engine. Use the dial-indicator that has $\frac{1}{1,000}$-inch graduations. Remove the spark plugs from all the cylinders. Then, turn the crankshaft, and observe if the crankshaft, propeller shaft, or flange turns straight without any bending taking place. If there is an excessive runout (bend in the crankshaft or propeller flange) reading at the crankshaft or propeller-drive shaft at the front seat location, the engine should be removed. Consult the applicable manufacturer's instructions for permissible limits. If the crankshaft or propeller drive shaft runout does not exceed these limits, install a serviceable propeller. Make an additional check by tracking the propeller at the tip in the same plane, perpendicular to the axis of rotation, to assure that blade track tolerance is within the prescribed limits.

Start the engine to see if operation is smooth, without vibration, and the power output adequate. If the engine operates properly during this ground check, shut the engine down and repeat the inspection for metal particles in the oil system.

Metal Particles in the Oil

Metal particles in the engine oil screens or the magnetic chip detectors are generally an indication of partial internal failure of the engine. Carbon tends to break loose from the interior of the engine in rock-like pieces that have the appearance of metal. It is necessary to consider these possibilities when foreign particles are found on the engine oil screens or magnetic chip detectors.

Before removing an engine for suspected internal failure, as indicated by foreign material on the oil screens or oil sump plugs, determine if the foreign particles are ferrous metal by placing them close to a magnet to see if they are magnetic. If the material is not magnetic, it is not attracted by the magnet. Any ferrous metal in the oil screens is cause for concern. Very small amounts of nonferrous metal, especially after major engine maintenance, can sometimes be normal. If the particles are metal, determine the probable extent of internal damage. For example, if only small particles are found that

are similar in nature to filings, drain the oil system, and refill it. Then, ground-run the engine and reinspect the oil screens and magnetic chip detectors. If no further evidence of foreign material is found, continue the engine in service or per the manufacturer's instructions. However, engine performance should be closely observed for any indication of difficulty or internal failure.

Spectrometric Oil Analysis Engine Inspection Program

Spectrometric oil analysis program allows an oil sample to be analyzed and searched for the presence of minute metallic elements. Due to oil circulation throughout an aircraft engine, every lubricant that is in service contains microscopic particles of metallic elements called wear metals. As the engine operates over a certain amount of time, the oil picks up very small particles that stay suspended in the oil. Oil analysis programs identify and measure these particles in parts per million (PPM) by weight. The analyzed elements are grouped into categories, such as wear metals and additives, and their measurement in PPM provides the data that expert analysts can use as one of many tools to determine the engine's condition. If certain metals have an increase in PPM, it can be a signal of component wear or pending failure of the engine. The amount of wear metals is recorded and noted each time a sample is taken. If the amount of wear metals increases beyond a normal rate, then the operator can be notified quickly so repair, a recommended specific maintenance procedure, or inspection can be ordered.

The advantage of oil analysis is an increase in safety by noticing an engine problem before engine failure. It also saves money by finding engine problems before they become large problems or complete engine failure. This procedure can be used for both turbine and reciprocating engines. Oil analysis can be used to diagnose impending engine failure and would be a reason for removing the engine from the aircraft and sending it to overhaul.

Turbine Engine Condition Monitoring Programs

Many turbine engines are monitored by an engine condition program that helps determine the health of the engine in service. This can also be called trend analysis performance monitoring, but it consists mainly of monitoring certain engine parameters daily and watching for trend shifts or changes in the engine parameters. A shift in key parameters (change over time) could be a warning that the engine has serious internal deterioration and should be overhauled.

Engine Operational Problems

Engines are usually removed when there are consistent engine operational problems. Engine operational problems generally include, but are not limited to, one or more of the following conditions:

1. Excessive engine vibration; this is especially true with turbine engines.

2. Backfiring, or misfiring, either consistent or intermittent due to valve train or other mechanical defect in reciprocating engines.

3. Turbine engines that exceed normal operating parameters or life limited components exceeding maximum time in service or cycles.

4. Low power output, generally caused by low compression, with reciprocating engines and internal engine deterioration or damage with turbines.

General Procedures for Engine Removal & Installation

Preparation of Engines for Installation

After the decision has been made to remove an engine, the preparation of the replacement engine must be considered. The maintenance procedures and methods used vary widely. Commercial operators, whose maintenance operations require the most efficient and expeditious replacement of aircraft engines, usually rely on a system that utilizes the quick-engine-change assembly (QECA), also sometimes referred to as the engine power package. The QECA is essentially a powerplant and the necessary accessories installed in the engine.

Other operators of aircraft equipped with reciprocating engines sometimes use a different replacement method in these repair facilities because engine changes often occur at random intervals. Such replacement engines may be partially or wholly built up with the necessary accessories and subassemblies, or they may be stored as received from the manufacturer in packing boxes, cases, or cans and are uncrated and built up for installation only when needed to replace an engine.

QECA Buildup Method for Changing of Engines

Because the QECA system is most commonly used with large turbine engines used in the airlines, such engines are used to describe QECA buildup and installation procedures. Many of these procedures are applicable to all other methods of engine buildup and installation.

The following study of QECA buildup is not designed to outline procedures to follow in a practical application; always use those recommended by the manufacturer. The procedures included in this chapter provide a logical sequence in following a QECA and its components through the stages of a typical buildup to gain a better understanding of units and systems interconnection. The components of a

Figure 8-1. *Open cowling view of a typical power package.*

QECA are illustrated in *Figure 8-1.* As shown, the QECA consists of several units. On many aircraft, the engines are mounted in streamlined housings called nacelles that extend from the wings. These nacelles are divided into two main sections: wing nacelle and engine nacelle. The wing nacelle is that portion of the nacelle that is attached to the wing structure. The engine nacelle is that portion of the nacelle that is constructed separately from the wing. Also, the wing nacelles normally contain lines and units of the oil, fuel, and hydraulic systems, as well as linkages and other controls for the operation of the engine.

The firewall is usually the foremost bulkhead of the engine nacelle and differs from most other aircraft bulkheads in that it is constructed of stainless steel or some other fire-resistant material. *[Figure 8-2]* The primary purpose of the firewall is to

Figure 8-2. *Typical firewall with components mounted on it.*

confine any engine fire to the engine nacelle. It also provides a mounting surface for units within the engine nacelle and a point of disconnect for lines, linkages, and electrical wiring that are routed between the engine and the aircraft. Without this firewall, an engine fire would have ready access to the interior of the aircraft. Since the consequences of an engine fire are obvious, the necessity of sealing all unused openings in the firewall cannot be overstressed.

An aircraft engine and its accessories that have been in storage must undergo careful depreservation and inspection before they may be installed in an aircraft. This involves more than removing an engine from its container and bolting it to the aircraft. If the engine is stored in a pressurized metal container, the air valve should be opened to bleed off the air pressure. Depending upon the size of the valve, the air pressure should bleed off in somewhat less than 30 minutes.

Prepare the container for opening by removing the bolts that hold the two sections together. Then, attach a hoist to the "hoisting points" and lift the top section clear of the container and place it away from the work area. If the engine is installed in a wooden shipping case, it is necessary to carefully break the seal of the protective envelope and fold it down around the engine. Remove the dehydrating agent or desiccant bags and the humidity indicator from the outside of the engine. Also, remove and set safely aside any accessories that are not installed on the engine but are mounted on a special stand or otherwise installed inside the protective envelope with the engine.

Depreservation of an Engine

After the engine has been secured to an engine stand, all covers must be removed from the points where the engine was sealed or closed with ventilatory covers, such as the engine breathers, exhaust outlets, and accessory mounting-pad cover plates. As each cover is removed, inspect the uncovered part of the engine for signs of corrosion. Also, as the dehydrator plugs are removed from each cylinder, make a very careful check of the walls of any cylinder for which the dehydrator plug color indicates an unsafe condition. Care is emphasized in the inspection of the cylinders, even if it is necessary to remove a cylinder.

On radial engines, the inside of the lower cylinders and intake pipes should be carefully checked for the presence of excessive corrosion-preventive compound that has drained from throughout the interior of the engine and settled at these low points. This excessive compound could cause the engine to become damaged from a hydraulic lock (also referred to as liquid-lock) when a starting attempt is made.

The check for excessive corrosion-preventive compound in the cylinders of reciprocating engines can be made as the dehydrator plugs are removed from each cylinder. Much of the compound drains from the spark plug holes of the lower cylinders of a radial engine when the dehydrator plugs are removed. But some of the mixture remains in the cylinder head below the level of the spark plug hole and can be removed with a hand pump. *[Figure 8-3]* A more positive method is to remove the lower intake pipes and open the intake valve of the cylinder by rotating the crankshaft. This latter method allows the compound to drain from the cylinder through the open intake valve. If excessive compound is present in an upper cylinder, it can be removed with a hand pump.

The oil screens should be removed from the engine and thoroughly washed in an approved solvent to remove all accumulations that could restrict the oil circulation and cause engine failure. After the screens are cleaned, immerse them in clean oil and then reinstall them in the engine.

When the cover has been removed from the intake area, the silica gel desiccant bags (used to remove moisture from the engine in storage) must be removed from the engine area. If the engine uses a propeller, remove the protective covering from the propeller shaft and wash all corrosion-preventive compounds from both the inside and outside surfaces of the shaft. Then, coat the propeller shaft lightly with engine oil. Turbine engines require the removal of several covers on many external areas on the engine.

As a final check, see that the exterior of the engine is clean. Usually a quantity of compound runs out of the engine when the dehydrator plugs and oil screens are removed. To clean the engine, spray it with an approved commercial solvent.

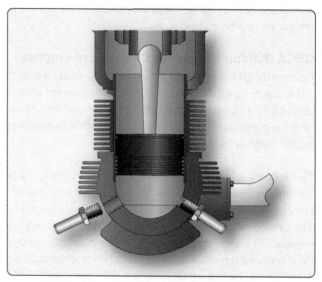

Figure 8-3. *Draining corrosion preventive compound.*

Inspection & Depreservation of Accessories

An engine's performance is no better than that of its accessories. Though the engine has been completely overhauled and is in top condition, any oversight or error in installing the accessories can result in improper engine operation or even irreparable damage to it.

Before depreserving any of the accessories enclosed with the engine, consult the storage data usually stenciled on the outside of the engine container or the records enclosed with the engine to determine how long the engine and accessories were in storage. Certain accessories that normally accompany an engine from overhaul are considered unsafe for use if their time in storage has exceeded a specified period. This time varies according to the limits prescribed by the manufacturer.

Any accessory that has been removed from an old engine that can be installed on the new one must be given a thorough inspection to determine its condition. This inspection includes a check for general condition, cleanliness, absence of corrosion, and absence of wear as evidenced by excessive play in the moving parts.

Some accessories must be replaced, regardless of their operating time, if the engine is being changed because of internal failure. Such accessories may have been contaminated by metal particles carried into their operating mechanisms by the engine oil that lubricates them.

Before installing any replacement accessory, check it visually for signs of corrosion and for freedom of operation. Always wipe the mounting pad, flange, and coupling clean before mounting the accessory, and install the proper gasket between the mounting pad and the accessory mounting flange. Lubricate the accessory drive shaft if so indicated in the manufacturer's instructions.

Inspection & Replacement of Powerplant External Units & Systems

The engine nacelle must be cleaned thoroughly before it is inspected. The design of an engine nacelle varies with different aircraft. Basically, it is a framework covered with removable cowling, in which the engine is mounted. This assembly is attached to the aircraft and incorporates an insulating firewall between the engine and the airframe. The interconnecting wiring, tubing, and linkages between the engine and its various systems and controls pass through the firewall.

Inspect the complete engine nacelle for condition of the framework and the sheet-metal cowling and riveted plates that cover the nacelle. The engine mounting frame assembly should be checked for any distortion of the steel tubing, such as bends, dents, flat spots, corrosion, or cracks. Use the dye penetrant inspection method to reveal a crack, porous area, or other defects.

The engine mounting bolts are usually checked for condition by magnetic particle inspection or other approved processes. While the bolts are removed, the bolt holes should be checked for elongation caused by the movement of an improperly tightened bolt.

Check the outer surface of all exposed electrical wiring for breaks, chafing, or other damage. Also, check the security of crimped or soldered cable ends. In addition, carefully inspect connector plugs for overall condition. Any item that is damaged must be repaired or replaced, depending on the extent of the damage.

Before installing an engine, inspect all tubing in the nacelle for dents, nicks, scratches, chafing, or corrosion. Check all tubing carefully for indications of fatigue or excessive flatness caused by improper or accidental bending. Thoroughly inspect all hoses used in various engine systems. Weather checking (a cracking of the outside covering of the hose) sometimes penetrates to the hose reinforcement. Replace any length of hose that shows indications of the cover peeling or flaking or has exposed fabric reinforcement. Replace a hose that shows indications of excessive cold flow. Cold flow is a term used to describe the deep and permanent impressions or cracks caused by hose clamp pressure.

Always replace a control rod if it is nicked or corroded deeply enough to affect its strength. If the corrosion cannot be removed by rubbing with steel wool, the pitting is too deep for safety.

On older aircraft, check the pulleys in the control system for freedom of movement. It is easy to spot a pulley that is not turning freely, for both it and the cable are worn from the cable sliding over the pulley instead of rolling free. The bearings of a pulley may be checked by inspecting the pulley for excessive play or wobble with the tension removed from the cable. The cable must also be inspected for corrosion and broken strands. Locate any broken strands by wiping the cable with a cloth.

Check bonding for fraying, loose attachment, and cleanness of terminal ends. The electrical resistance of the complete bond must not exceed the resistance values specified in the applicable manufacturer's instructions.

Inspect the exhaust stacks, collector ring, and tailpipe assembly for security, cracks, or excessive corrosion. Depending on the installation, these units, or parts of them, may be mounted on the engine before it is installed in the aircraft.

Check all air ducts for dents and for the condition of the fabric or rubber anti-chafing strips at the points where sections of duct are joined. The dents may be pounded out; the anti-chafing strips should be replaced if they are pulled loose from the duct or are worn to the point at which they no longer form a tight seal at the joint.

Thoroughly inspect the engine oil system and perform any required special maintenance upon it before installing a replacement engine. If an engine is being changed at the end of its normal time in service, it is usually necessary only to flush the oil system; however, if an engine has been removed for internal failure, usually some units of the oil system must be replaced and others thoroughly cleaned and inspected.

If the engine has been removed because of internal failure, the oil tank is generally removed to permit thorough cleaning. Also, the oil cooler and temperature regulator must be removed and sent to a repair facility for overhaul. The vacuum pump pressure line and the oil separator in the vacuum system must also be removed, cleaned, and inspected. Internal failure also requires that the propeller governor and feathering pump mechanism be replaced if these units are operated by engine oil pressure.

Preparing the Engine for Removal

Before starting to work on the aircraft or reciprocating engine, always be sure that the magneto switch is in the off position. Aircraft engines can be started accidentally by turning the propeller if the magneto switch is on.

Check to see that all fuel selectors or solenoid-operated fuel shutoff valves are closed. The fuel selector valves are either manually or solenoid operated. If solenoid-operated fuel shutoff valves are installed, it may be necessary to turn the battery switch on before the valves can be closed, since the solenoid depends on electricity for operation. These valves close the fuel line at the firewall between the engine and the aircraft. After ensuring that all fuel to the engine is shut off, disconnect the battery to eliminate the possibility of a hot wire starting a fire. If it is anticipated that the aircraft will be out of service for more than 6 days, the battery is usually removed and taken to the battery shop and placed on charge.

Also, a few other preparations should be made before starting to work on the engine removal. First, make sure that there are enough fire extinguishers near at hand to meet any possible emergency. Check the seals on these extinguishers to be sure the extinguishers have not been discharged. Then, check the wheel chocks. If these are not in place, the aircraft can, and probably will, inch forward or back during some crucial operation. Also, if the aircraft has a tricycle landing gear, be sure that the tail is supported so that the aircraft cannot tip back when the weight of the engine is removed from the forward end. It is not necessary to support the tail on some multiengine aircraft if only one engine is to be removed. In addition, the landing gear shock struts can be deflated to prevent them from extending as the engine weight is removed from the aircraft.

After taking these necessary precautions, begin removing the cowling from around the engine. As it is removed, clean it and check for cracks so that the necessary repairs can be made while the engine change is in progress. Place all cowling that does not need repair on a rack where it can be readily found when the time comes to reinstall it on the new engine. After removing the cowling, the propeller should be removed for inspection or repair.

Draining the Engine

Place a large metal pan (drip pan) on the floor under the engine to catch any spilled mixture or oil. Next, secure a clean container in which to drain the oil or corrosion-preventive mixture. Place the container beneath the engine, open the drain valve, and allow the oil to drain. *Figure 8-4* shows the points at which a typical aircraft engine oil system is drained. Other points at which the oil system is drained can typically include the oil cooler, oil return line, and engine sumps. All valves, drains, and lines must remain open until the oil system has been completely drained. After draining the oil, reinstall all drain plugs and close all drain valves. Then, wipe all excess oil from around the drain points.

Electrical Disconnects

Electrical disconnections are usually made at the engine firewall. When the basic engine is being removed, the electrical leads to such accessories as the starter and generators are disconnected at the units themselves. When

Figure 8-4. *Oil system drain points.*

disconnecting electrical leads, it is a good safety habit to disconnect the magnetos first and immediately ground them at some point on the engine or the assembly being removed. Most firewall disconnections of electrical conduit and cable are simplified by use of (Army/Navy) AN or (Military Standard) MS connectors. Each connector consists of two parts: a plug assembly and a receptacle assembly. To prevent accidental disconnection during airplane operation, the outlet is threaded to permit a knurled sleeve nut to be screwed to the outlet and then fastened with safety wire, if necessary.

A typical plug fitting assembly is shown in *Figure 8-5*. It also shows a typical junction box assembly, which is used as a disconnect on some aircraft engine installations. After the safety wire is broken, remove all of it from the sleeve nuts that hold the conduit to the junction boxes, as well as from the nuts on the connectors. Wrap moisture proof tape over the exposed ends of connectors to protect them from dirt and moisture. Also, do not leave long electrical cables or conduits hanging loose, since they may become entangled with some part of the aircraft while the engine is being hoisted. It is a good practice to coil all lengths of cable or flexible conduit neatly, and tie

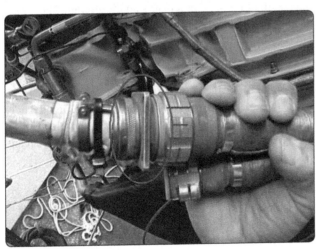

Figure 8-5. *Electrical connections.*

or tape them to some portion of the assembly being removed.

Disconnection of Engine Controls

The engine control rods and cables connect such units as the carburetor or fuel control throttle valve and the mixture control valve with their manually actuated control in the flight deck. The controls are sometimes disconnected by removing the turnbuckle that joins the cable ends. *[Figure 8-6]* A typical reciprocating engine control linkage consisting of a control rod attached to a bell crank is illustrated in *Figure 8-7*.

The control rod in the linkage shown has two rod-end assemblies, a clevis, and an eye screwed onto opposite ends. These rod-end assemblies determine the length of the control rod by the distance they are screwed onto it, and they are locked into position by check nuts. An antifriction bearing is usually mounted in the eye end of a rod. This eye is slipped over a bolt in the bell crank arm and is held in position by a castle nut safety with a cotter pin. The clevis rod end is slipped over the end of a bell crank arm, which also usually contains an antifriction bearing. A bolt is passed through the clevis and the bell crank eye, fastened with a castle nut, and safetied with a cotter pin. Most control rod ends are provided with a test hole in the shank for inspection. If safety wire can be inserted through the test hole, the terminal is not being held by the required number of threads.

Sometimes linkage assemblies do not include the antifriction bearings and are held in position only by a washer and cotter pin in the end of a clevis pin that passes through the bell crank and rod end. After the engine control linkages have been disconnected, the nuts and bolts should be replaced in the rod ends or bell crank arms to prevent their being lost. All control rods should be removed completely or tied back to prevent them from being bent or broken if they are struck by the replacement engine or QECA as it is being hoisted.

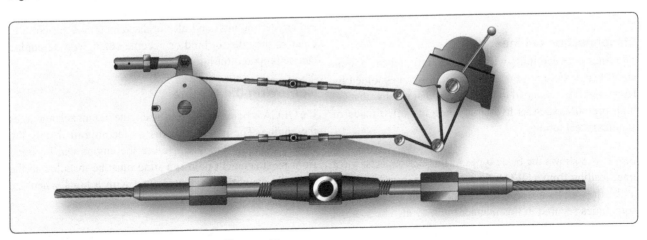

Figure 8-6. *Engine control cable and turnbuckle assembly.*

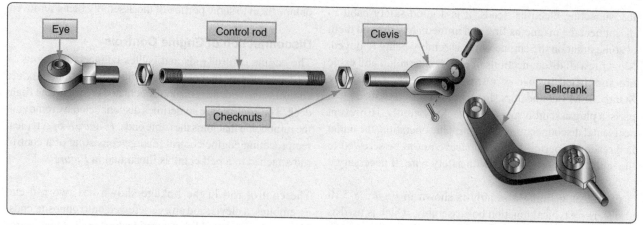

Figure 8-7. *Engine control linkage assembly.*

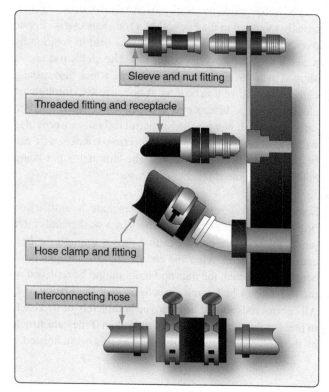

Figure 8-8. *Types of line disconnects.*

Disconnection of Lines

The lines between units within the aircraft and the engine are either flexible hose or aluminum-alloy tubes joined by lengths of hose clamped to them. Lines that must withstand high pressure, such as hydraulic lines, are often made of stainless steel tubing.

Figure 8-8 shows the basic types of line disconnects. Most lines leading from a QECA are secured to a threaded fitting at the firewall by a sleeve nut around the tubing. Hoses are sometimes secured in this manner but may also be secured by a threaded fitting on the unit to which they lead, or by a

hose clamp. The firewall fittings for some lines have a quick-disconnect fitting that contains a check valve to prevent the system from losing fluid when the line is disconnected. Metal tubing on some installations may also be disconnected at a point at which two lengths of it are joined together by a length of rubber hose. Such a disconnection is made by loosening the hose clamps and sliding the length of rubber hose over the length of tubing that remains on the aircraft. There may be some further variations in these types of disconnections, but they follow the same basic pattern.

Some type of a container should be used to collect any fuel, oil, or other fluid that may drain from the disconnected lines. After the lines have drained, they should be immediately plugged or covered with moisture-proof tape to prevent foreign matter from entering them, as well as to prevent any accumulated fluid from dripping out.

Other Disconnections

The points at which the various air ducts are disconnected depend upon the engine and the aircraft in which it is installed. Usually, the air intake ducts and the exhaust system must be disconnected so the basic engine or the QECA can be removed. After the engine connections are free (except the engine mounts) and all the disconnections are entirely clear so they do not bind or become entangled, the engine can be prepared for hoisting.

Removing the Engine

If a QECA is being removed, the engine mount accompanies the engine. The mount remains on the aircraft if only the engine is being removed. Before the engine can be freed from its attachment points, a sling must be installed so the engine's weight can be supported with a hoist when the mounting bolts are removed.

Aircraft engines, or QECAs, have marked points for attaching

Figure 8-9. *Hoisting sling attached to engine.*

a hoisting sling. The location of these attaching points varies according to the size and weight distribution of the engine. *Figure 8-9* shows a sling supporting an engine that has two attaching points. As a matter of safety, carefully inspect the sling for condition before installing it on the engine.

Before attaching the sling to the hoist, be sure that the hoist has sufficient capacity to lift the engine safely. The engine's center of gravity (CG) should also be taken into account as the engine is hoisted. A manually operated hoist mounted in a

Figure 8-10. *Hoist and frame assembly used for engine removal.*

portable frame is shown in *Figure 8-10*. This hoist assembly is specifically manufactured for the purpose of removing engines and other large assemblies from aircraft. Some frames are fitted with power-operated hoists. These should be used with care, since considerable damage can be done if an inexperienced operator allows a power-operated hoist to overrun. The hoist and frame should also be checked for condition before being used to lift the engine.

Hoisting the Engine

Before the hoist is hooked onto the engine sling, recheck the aircraft tail supports and the wheel chocks. Fasten lines to the engine, at points on the sides or rear, so that the engine can be controlled as it is being hoisted. Hook the hoist onto the sling and hoist the engine just enough to relieve the engine weight from the mount attachments. Remove the nuts from the mount attachments in the order recommended in the manufacturer's instructions for the aircraft. As the last nuts are being removed, pull back on the lines fastened to the engine (or force it back by other means if lines are not being used), thus steadying the engine. If bolts must be removed from the mount attachments, be sure the engine is under control before doing so. If the bolts are to remain in the mount attachments, the hoist can be gently maneuvered upward or downward as necessary after all the nuts have been removed. Meanwhile, gently relax the backward force on the engine just enough to allow the engine gradual forward movement when it is free from the mount attachments. When the hoist has removed all engine weight from the mount attachments, the engine should be eased gently forward, away from the aircraft. If the engine binds at any point, maneuver it with the hoist until it slips free.

The procedure just discussed applies to removal of most reciprocating and turbine aircraft engines. Any variation in details is outlined in the manufacturer's instructions. Before attempting any engine removal, always consult these instructions for the aircraft concerned. When the engine has been removed, it can be carefully lowered onto a stand. The engine should be fastened to the stand and prepared for the removal of accessories.

Hoisting & Mounting the Engine for Installation

When the new or overhauled engine is ready to be hoisted for installation, move the engine stand as close as possible to the nacelle in which the replacement is to be installed. Then, attach the sling to the engine and hook the hoist to the sling. Take up the slack until the hoist is supporting most of the engine weight. Next, remove the engine attaching bolts from the stand and hoist the engine clear.

The engine stand may be moved and the hoist frame positioned so the engine can be hoisted easily into the nacelle.

To prevent injury to the crew or damage to the aircraft or engine, be sure that the engine is steadied when moving the hoist frame.

Engine nacelles are rarely designed for the engine to be fitted and bolted into place as though it were being mounted on a bare wall. The engine must be guided into position and mated with its various connections, such as the mounting bolt holes and the exhaust tailpipe. This must be done despite such obstacles as the nacelle framework, ducts, or firewall connections and without leaving a trail of broken and bent parts, scratched paint, or crushed fingers.

When the engine has been aligned correctly in the nacelle, insert the mounting bolts into their holes and start all of the nuts on them. Always use the type of bolt and nut recommended by the manufacturer. Never use an unauthorized substitution of a different type or specification of nut and bolt than that prescribed.

The nuts on the engine mount bolts must be tightened to the torque recommended by the aircraft manufacturer. While the nuts are being tightened, the hoist should support the engine weight sufficiently to allow alignment of the mounting bolts. If the engine is permitted to exert upward or downward pressure on the bolts, it is necessary for the nuts to pull the engine into proper alignment. This results in nuts being tightened to the proper torque value without actually holding the engine securely to the aircraft.

The applicable manufacturer's instructions outline the sequence for tightening the mounting bolts to ensure security of fastening. After the nuts are safetied and the engine sling and hoist are removed, bonding strips should be connected across each engine mount to provide an electrical path from the mount to the airframe.

Mounting the engine in the nacelle is, of course, only the beginning. All the ducts, electrical leads, controls, tubes, and conduits must be connected before the engine can be operated.

Connections & Adjustments

There are no hard-and-fast rules that direct the order in which units or systems should be connected to the engine. Each maintenance organization normally supplies a worksheet or checklist to be followed during this procedure. This list is based upon past engine installations on each particular aircraft. If this is followed carefully, it serves as a guide for an efficient installation. The following instructions are not a sequence of procedures but a discussion of correct methods for completing an engine installation.

The system of ducts for routing air to the engine varies with all types of aircraft. In connecting them, the goal is to fit the ducts closely at all points of disconnect so that the air they route does not escape its intended path. The duct systems of some aircraft must be pressure checked for leaks. This is done by blocking the system at one end, supplying compressed air at a specified pressure at the other end, and then checking the rate of leakage.

The filters in the air induction system must be cleaned to ensure an unrestricted flow of clean air to the engine and its units. Because methods for cleaning air filters vary with the materials used in the filtering element, clean them in accordance with the technical instructions for the aircraft being serviced.

The exhaust system should also be carefully connected to prevent the escape of hot gases into the nacelle. When assembling the exhaust system, check all clamps, nuts, and bolts, and replace any in doubtful condition. During assembly, the nuts should be gradually and progressively tightened to the correct torque. The clamps should be tapped with a rawhide mallet as they are being tightened to prevent binding at any point. On some systems, a ball joint connects the stationary portion of the exhaust system to the portion that is attached to the engine. This ball joint absorbs the normal engine movement caused by the unbalanced forces of the engine operation. Ball joints must be installed with the specified clearance to prevent binding when expanded by hot exhaust gases.

Hoses used inside low-pressure systems are generally fastened into place with clamps. Before using a hose clamp, inspect it for security of welding or riveting and for smooth operation of the adjusting screw. A clamp that is badly distorted or materially defective should be rejected. Material defects include extremely brittle or soft areas that may easily break or stretch when the clamp is tightened. After a hose is installed in a system, it should be supported with rubber-lined supporting clamps at regular intervals.

Before installing metal tubing with threaded fittings, ensure the threads are clean and in good condition. Apply sealing compound, of the correct specification for the system, to the threads of the fittings before installing them. While connecting metal tubing, follow the same careful procedure for connecting hose fittings to prevent cross-threading and to ensure correct torque.

When connecting the starter, generator, or various other electrical units within the nacelle, make sure that all lead connections are clean and properly secured. On leads that are fastened to a threaded terminal with a nut, a lock washer is usually inserted under the nut to prevent the lead from working loose. When required, connector plugs can be safetied with steel wire to hold the knurled nut in the full-tight position.

Electrical leads within the engine nacelle are usually passed through either flexible or rigid conduit. The conduit must be anchored, as necessary, to provide a secure installation and bonded when required.

All engine controls must be accurately adjusted to ensure instantaneous response to the control setting. For flexibility, the engine controls are usually a combination of rods and cables. Since these controls are tailored to the model of aircraft in which they are installed, their adjustment must follow exactly the step-by-step procedure outlined in the manufacturer's instructions for each particular model of aircraft.

Figure 8-11 illustrates a simplified schematic drawing of a throttle control system for a reciprocating aircraft engine. Follow a general procedure for adjusting throttle controls. First, loosen the serrated throttle control arm at the carburetor and back off the throttle stop until the throttle valve is in

the fully closed position. After locking the cable drum into position with the locking pin, adjust the control rod to a specified length. Then, attach one end of the control rod to the locked cable drum, and reinstall the throttle control arm on the carburetor in the serrations that allow the other end of the control rod to be attached to it. This correctly connects the control arm to the cable drum.

Now, loosen the cable turnbuckles until the throttle control can be locked at the quadrant with the locking pin. Then, with both locking pins in place, adjust the cables to the correct tension as measured with a tensiometer. Remove the locking pins from the cable drum and quadrant.

Next, adjust the throttle control so that it has a slight cushion action at two positions on the throttle quadrant: one when the carburetor throttle valve is in the full-open position and the other when it is closed to the idle position (stop to stop).

Adjust the cushion by turning the cable turnbuckles equally in opposite directions until the throttle control cushion is correct at the full-open position of the throttle valve. Then, when the throttle arm stop is adjusted to the correct idle speed

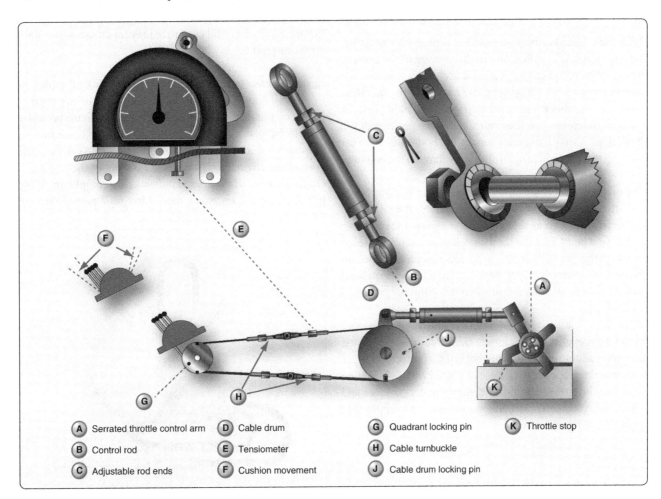

A	Serrated throttle control arm	D	Cable drum	G	Quadrant locking pin	K	Throttle stop
B	Control rod	E	Tensiometer	H	Cable turnbuckle		
C	Adjustable rod ends	F	Cushion movement	J	Cable drum locking pin		

Figure 8-11. *Schematic drawing of throttle control system.*

setting, the amount of cushion should be within tolerance at the idle speed position of the throttle valve. The presence of this cushion ensures that the travel of the throttle valve is not limited by the stops on the throttle control quadrant, but that they are opening fully and closing to the correct idle speed as determined by the throttle arm stop.

Adjustment of the engine controls is basically the same on all aircraft regarding the linkage adjustment to a predetermined length for a specific setting of the unit to be controlled. Then adjust cables, if used in the control system, to a specific tension with the control system locked. Finally, the full travel of the unit to be controlled is ensured by establishing the correct cushion in the controls. In general, the same basic procedure is used to connect the linkage of the remaining engine controls. After rigging the engine controls, safety the turnbuckles and castle nuts, and make certain the jam nuts on all control rods are tightened.

On multiengine aircraft, the amount of cushion of all engine controls on each quadrant must be equal so that all are aligned at any specific setting chosen. This eliminates the necessity of setting each control individually to synchronize engine operations.

After the engine has been installed, it is necessary to adjust the cowl flaps, if installed, so that the passage of the cooling air over the engine can be regulated accurately. Operate the system and recheck for opening and closing to the specified limits. Also, check the cowl flap position indicators, if installed, to ensure that they indicate the true position of the cowl flaps. Cowl flaps are doors at the bottom of the rear cowling that are used to control air flow through the cowling.

The oil cooler doors are adjusted in a manner similar to that used to adjust the cowl flaps. In some cases, the procedure is reversed in so far as the door is first adjusted to retract to a specified point, and the limit switch on the motor is set to cut out at this point. Then, the jackscrew is adjusted to permit the door to open only a specified distance, and the open limit switch is set to stop the motor when this point is reached.

After the engine has been completely installed and connected, install the propeller on the aircraft. Before doing so, the thrust bearing retaining nut should be checked for correct torque. If required, the propeller shaft must be coated with light engine oil before the propeller is installed; the propeller governor and anti-icing system must be connected according to applicable manufacturer's instructions.

Preparation of Engine for Ground & Flight Testing

Pre-Oiling

Before the new engine is flight tested, it must undergo a thorough ground check. Before this ground check can be made, several operations are usually performed on the engine.

To prevent failure of the engine bearings during the initial start, the engine should be pre-oiled. When an engine has been idle for an extended period of time, its internal bearing surfaces are likely to become dry at points where the corrosion-preventive mixture has dried out or drained away from the bearings. Hence, it is necessary to force oil throughout the entire engine oil system. If the bearings are dry when the engine is started, the friction at high rpm destroys the bearings before lubricating oil from the engine-driven oil pump can reach them.

There are several methods of pre-oiling an engine. The method selected should provide an expeditious and adequate pre-oiling service. Before using any pre-oiling method, remove one spark plug from each cylinder to allow the engine to be turned over more easily with the starter. Also, connect an external source of electrical power (auxiliary power unit) to the aircraft electrical system to prevent an excessive drain on the aircraft battery.

In using some types of pre-oilers, such as that shown in *Figure 8-12,* the oil line from the inlet side of the engine-driven oil pump must be disconnected to permit the pre-oiler tank to be connected at this point. Then, a line must be disconnected, or an opening made in the oil system at the nose of the engine, to allow oil to flow out of the engine. Oil flowing out of the engine indicates the completion of the pre-oiling operation, since the oil has now passed through the entire system.

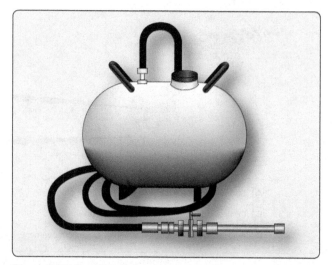

Figure 8-12. *Pre-oiler tank.*

In order to force oil from the pre-oiler tank through the engine, apply air pressure to the oil in the tank while the engine is being turned through with the starter. When this action has forced oil through the disconnection at the nose of the engine, stop cranking the engine and disconnect the pre-oiler tank. A motor-driven oil pump can also be used to pump oil through the engine during the pre-oiling operation.

When no external means of pre-oiling an engine are available, the engine oil pump may be used. Fill the engine oil tank, or crankcase, to the proper level. Then, with the mixture in the idle cutoff position (reciprocating engine), the fuel shutoff valve and ignition switches in the off position, and the throttles fully open, crank the engine with the starter until the oil pressure gauge mounted on the instrument panel indicates oil pressure.

After the engine has been pre-oiled, replace the spark plugs and connect the oil system. Generally, the engine should be operated within 4 hours of being pre-oiled; otherwise, the pre-oiling procedure normally must be repeated.

Fuel System Bleeding

To purge the fuel system of air locks, and to aid in flushing any traces of preservative oil from a pressure carburetor, fuel injector unit, or fuel control (turbine), remove the drain plug in the fuel unit chamber that is farthest from the fuel inlet to the fuel unit. In its place, screw a threaded fitting to a length of hose leading to a suitable container. Then, set the fuel control unit to flow fuel so that fuel is permitted to flow through the system. After ensuring the fuel shutoff and main fuel tank valves are open, turn on the fuel boost pump until there are no traces of preservative oil in the fuel being pumped through the system. The passage of air is indicated by the absence of air mixed in the fuel emerging from the end of the hose submerged in the container of fuel. Air trapped in the system should not be confused with the numerous small air bubbles that may appear as a result of the velocity of the fuel being ejected from the engine's fuel unit. Usually, after approximately a gallon of fuel has been bled off, the system can be considered safe for operation. After completing the bleeding operation, return all switches and controls to their normal, or off, position, and replace and safety all fuel unit connections disturbed.

Propeller Check

The propeller, if equipped, must be checked before, during, and after the engine has been ground operated. The propeller should be checked for proper torque on the mounting bolts, leaks, vibration, and for correct safety.

A propeller whose pitch-changing mechanism is electrically actuated may be checked before the engine is operated. Propellers whose pitch-changing mechanisms are oil actuated must be checked during engine operation after the normal operating oil temperature has been reached. In addition to checking the increase or decrease in rpm, the feathering cycle of the propeller should also be checked.

Checks & Adjustments After Engine Runup & Operation

After the engine has been ground operated, and again after flight test, operational factors must be adjusted, as necessary, and the entire installation given a thorough visual inspection. These adjustments often include fuel pressure and oil pressure, as well as rechecks of such factors as ignition timing, valve clearances, and idle speed and mixture. If these rechecks are indicated by the manner in which the engine performs.

After both the initial ground runup and the test flight, remove the oil sump plugs and screens and inspect for metal particles. Clean the screens before reinstalling them.

Check all lines for leakage and security of attachment. Especially, check the oil system hose clamps for security as evidenced by oil leakage at the hose connections. Also, inspect the cylinder holddown nuts or cap screws for security and safety. This check should also be performed after the flight immediately succeeding the test flight.

Rigging, Inspections, & Adjustments

The following instructions cover some of the basic inspections and procedures for rigging and adjusting fuel controls, fuel selectors, and fuel shutoff valves.

1. Inspect all bellcranks for looseness, cracks, or corrosion.

2. Inspect rod ends for damaged threads and the number of threads remaining after final adjustment.

3. Inspect cable drums for wear and cableguards for proper position and tension.

While rigging the fuel selector, power controls, and shutoff valve linkages, follow the manufacturer's step-by-step procedure for the particular aircraft model being rigged. The cables should be rigged with the proper tension with the rigging pins installed. The pins should be free to be removed without any binding; if they are hard to remove, the cables are not rigged properly and should be rechecked. The power lever should have the proper cushion at the idle and full-power positions. The pointers, or indicators, on the fuel control should be within limits. The fuel selectors must be rigged so that they have the proper travel and do not restrict the fuel flow to the engines.

Rigging Power Controls

Many older conventional turbofan engines use various power lever control systems. One of the common types is the cable and rod system. This system uses bellcranks, push-pull rods, drums, fairleads, flexible cables, and pulleys. All of these components make up the control system and must be adjusted or rigged from time to time. On single-engine aircraft, the rigging of the power lever controls is not very difficult. The basic requirement is to have the desired travel on the power lever and correct travel at the fuel control. On multiengine turbojet aircraft, the power levers must be rigged so that they are aligned at all power settings.

Most computer controlled engines have an electronic connection from the flight deck to the engine. This eliminates the need for any type of cable or linkages. In the computer controlled system, the computer sends electronic information through wires or buses to the fuel control to command it to follow pilot inputs from the flight deck.

On older style aircraft the power lever control cables and push-pull rods in the airframe system to the pylon and nacelle are not usually disturbed at engine change time and usually no rigging is required, except when some component has been changed. The control system from the pylon to the engine must be rigged after each engine change and fuel control change. *Figure 8-13* shows the control system from the bellcrank in the upper pylon to the fuel control.

Before adjusting the power controls at the engine, be sure that the power lever is free from binding and the controls have full throw on the console. If they do not have full throw or are binding, the airframe system should be checked, and the discrepancies repaired. After all adjustments have been

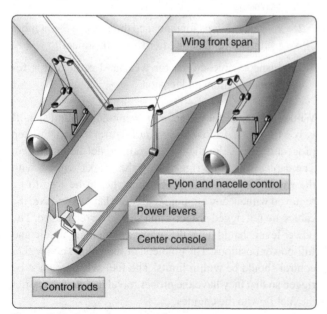

Figure 8-13. *Power lever control system.*

Wing front span

Pylon and nacelle control

Power levers

Center console

Control rods

made, move the power levers through their complete range, carefully inspecting for adequate clearance between the various push-pull rods and tubes. Secure all locknuts, cotter pins, and safety as required.

Adjusting the Fuel Control

The fuel control unit of the typical turbofan on older aircraft can be a hydromechanical device that schedules the quantity of fuel flowing to the engine so that the desired amount of thrust can be obtained. The amount of thrust is dictated by the position of the power lever in the flight deck and the particular operation of the engine. Thus, the thrust of the engine and the consequent rpm of its turbine are scheduled by fuel flow.

The fuel control unit of the engine is adjusted to trim the engine to obtain maximum thrust output of the engine when desired. The engine must be retrimmed after a fuel control unit is replaced, the engine does not develop maximum thrust, engine change, or excessive throttle stagger.

After trimming the engine, the idle rpm can be adjusted. The idle rpm is adjusted by turning the INC. IDLE screw an eighth of a turn at a time, allowing sufficient time for the rpm to stabilize between adjustments. Retard the power lever to idle and recheck the idle rpm.

If wind velocity is a factor, the aircraft should be headed into the wind while trimming or checking the trim on an engine. Since trimming accuracy decreases as windspeed and moisture content increase, the most accurate trimming is obtained under conditions of no wind and clear, moisture-free air. Do not trim when there is a tailwind because hot exhaust gases may be reingested. As a practical matter, the engine should never be trimmed when icing conditions exist because of the adverse effects on trimming accuracy. To obtain the most accurate results, the aircraft should always be headed into the wind while the engine is being trimmed.

With the aircraft headed into the wind, verify that the exhaust area is clear. Install an engine trim gauge to the T-fitting in the turbine discharge pressure line. Start the engine and allow it to stabilize for 5 minutes before attempting to adjust the fuel control. Refer to the applicable manufacturer's instructions for correct trim values. Compensate for temperature and pressure during the trimming process. If a hydromechanical fuel control is not within limits, turn the INC. MAX screw *[Figure 8-14]* about one-eighth turn in the appropriate direction. Repeat, if necessary, until the desired value is attained. If the aircraft is equipped with a pressure ratio gauge, set it to the correct value.

An example of a trim check using an electronic controlled fuel control must take into account temperature and pressure

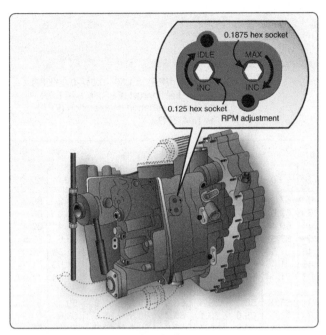

Figure 8-14. *Typical fuel control adjustments.*

for each parameter measured. The parameters checked can include:

1. Minimum idle (percent N_2).

2. Approach idle (percent N_2).

3. 2.5 bleed open (percent N_1).

4. 2.5 bleed closed (percent N_1).

5. Takeoff engine pressure ratio (EPR).

6. 95 percent takeoff thrust (EPR).

7. 90 percent thrust change decal (EPR).

The actual trim check would be done based on a temperature and pressure, such as the example in *Figure 8-15*. For these given temperature and pressures, the target parameter values can be derived from a chart in the manufacturer's manual. The engine is run up, and these values are checked against the tolerances given in the manual.

Turboprop Powerplant Removal & Installation

Since most turboprop powerplant removal and installation instructions are developed for QECA, the following procedures reflect those used for a typical QECA. The procedures for turboprop engine removal and installation are similar to those presented in the section of this chapter for turbojet engines, except for those systems related to the turboprop propeller.

Open the engine side panels and remove the nacelle access panels. Disconnect the engine thermocouple leads at the terminal board. Before disconnecting any lines, make sure that all fuel, oil, and hydraulic fluid valves are closed. Plug

all lines as they are disconnected to prevent entrance of foreign material.

Remove the clamps securing the bleed-air ducts at the firewall. Then, disconnect the electrical connector plugs, engine breather and vent lines, and fuel, oil, and hydraulic lines.

Disconnect the engine power lever and propeller control rods or cables. Remove the covers from the QECA lift points, attach the QECA sling, and remove slack from the cables using a suitable hoist. The sling must be adjusted to position the hoisting eye over the QECA CG. Failure to do so may result in engine damage.

Remove the engine mount bolts. The QECA is then ready to be removed. Recheck all of the disconnect points to make certain they are all disconnected prior to moving the engine. Move the engine forward, out of the nacelle structure, until it clears the aircraft. Lower the QECA into position on the QECA stand and secure it prior to removing the engine sling.

The installation procedures are essentially the reverse of the removal procedures. Move the QECA straight back into the nacelle structure and align the mount bolt holes and the firewall. Start all the bolts before torqueing. With all the bolts started, and using the correct torque wrench adapter, tighten the mount bolts to their proper torque. Remove the sling and install the access covers at the lift point. Using the reverse of the removal procedures, connect the various lines and connectors. New O-ring seals should be used. The manufacturer's instructions should be consulted for the proper torque limits for the various clamps and bolts.

After installation, an engine runup should be made. In general, the runup consists of checking proper operation of the powerplant and related systems. Several functional tests are performed to evaluate each phase of engine operation. The tests and procedures outlined by the engine or airframe manufacturers should be followed.

Reciprocating Helicopter Engine & QECA

The engine is installed facing aft with the propeller shaft approximately 39° above horizontal. The engine is supported by the engine mount, which is bolted to the fuselage structure. The installation of the engine provides for ease of maintenance by allowing easy access to all accessories and components when the engine access doors are opened. The QECA contains the engine, engine mount, engine accessories, engine controls, fuel system, lubrication system, ignition system, cooling system, and hydromechanical clutch and fan assembly.

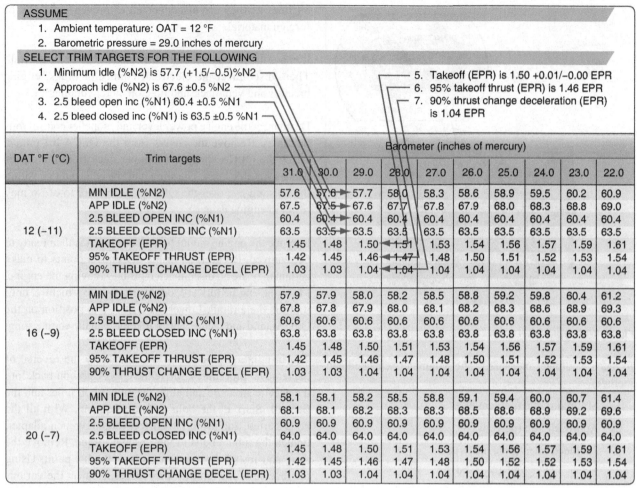

ASSUME
1. Ambient temperature: OAT = 12 °F
2. Barometric pressure = 29.0 inches of mercury

SELECT TRIM TARGETS FOR THE FOLLOWING
1. Minimum idle (%N2) is 57.7 (+1.5/–0.5)%N2
2. Approach idle (%N2) is 67.6 ±0.5 %N2
3. 2.5 bleed open inc (%N1) 60.4 ±0.5 %N1
4. 2.5 bleed closed inc (%N1) is 63.5 ±0.5 %N1
5. Takeoff (EPR) is 1.50 +0.01/–0.00 EPR
6. 95% takeoff thrust (EPR) is 1.46 EPR
7. 90% thrust change deceleration (EPR) is 1.04 EPR

DAT °F (°C)	Trim targets	Barometer (inches of mercury)									
		31.0	30.0	29.0	28.0	27.0	26.0	25.0	24.0	23.0	22.0
12 (–11)	MIN IDLE (%N2)	57.6	57.6	57.7	58.0	58.3	58.6	58.9	59.5	60.2	60.9
	APP IDLE (%N2)	67.5	67.5	67.6	67.7	67.8	67.9	68.0	68.3	68.8	69.0
	2.5 BLEED OPEN INC (%N1)	60.4	60.4	60.4	60.4	60.4	60.4	60.4	60.4	60.4	60.4
	2.5 BLEED CLOSED INC (%N1)	63.5	63.5	63.5	63.5	63.5	63.5	63.5	63.5	63.5	63.5
	TAKEOFF (EPR)	1.45	1.48	1.50	1.51	1.53	1.54	1.56	1.57	1.59	1.61
	95% TAKEOFF THRUST (EPR)	1.42	1.45	1.46	1.47	1.48	1.50	1.51	1.52	1.53	1.54
	90% THRUST CHANGE DECEL (EPR)	1.03	1.03	1.04	1.04	1.04	1.04	1.04	1.04	1.04	1.04
16 (–9)	MIN IDLE (%N2)	57.9	57.9	58.0	58.2	58.5	58.8	59.2	59.8	60.4	61.2
	APP IDLE (%N2)	67.8	67.8	67.9	68.0	68.1	68.2	68.3	68.6	68.9	69.3
	2.5 BLEED OPEN INC (%N1)	60.6	60.6	60.6	60.6	60.6	60.6	60.6	60.6	60.6	60.6
	2.5 BLEED CLOSED INC (%N1)	63.8	63.8	63.8	63.8	63.8	63.8	63.8	63.8	63.8	63.8
	TAKEOFF (EPR)	1.45	1.48	1.50	1.51	1.53	1.54	1.56	1.57	1.59	1.61
	95% TAKEOFF THRUST (EPR)	1.42	1.45	1.46	1.47	1.48	1.50	1.51	1.52	1.53	1.54
	90% THRUST CHANGE DECEL (EPR)	1.03	1.03	1.04	1.04	1.04	1.04	1.04	1.04	1.04	1.04
20 (–7)	MIN IDLE (%N2)	58.1	58.1	58.2	58.5	58.8	59.1	59.4	60.0	60.7	61.4
	APP IDLE (%N2)	68.1	68.1	68.2	68.3	68.3	68.5	68.6	68.9	69.2	69.6
	2.5 BLEED OPEN INC (%N1)	60.9	60.9	60.9	60.9	60.9	60.9	60.9	60.9	60.9	60.9
	2.5 BLEED CLOSED INC (%N1)	64.0	64.0	64.0	64.0	64.0	64.0	64.0	64.0	64.0	64.0
	TAKEOFF (EPR)	1.45	1.48	1.50	1.51	1.53	1.54	1.56	1.57	1.59	1.61
	95% TAKEOFF THRUST (EPR)	1.42	1.45	1.46	1.47	1.48	1.50	1.52	1.52	1.53	1.54
	90% THRUST CHANGE DECEL (EPR)	1.03	1.03	1.04	1.04	1.04	1.04	1.04	1.04	1.04	1.04

Figure 8-15. *Trim check data (Boeing).*

Removal of Helicopter QECA

Prior to removing the helicopter QECA, the engine should be preserved if it is possible to do so. Then, shut off the fuel supply to the engine and drain the oil. Make the disconnections necessary to remove the QECA, and then perform the following steps:

1. Attach the engine lifting sling to a hoist of at least a two-ton capacity.

2. Raise the hoist to apply a slight lift to the QECA. Loosen both engine mount lower attachment bolt nuts before leaving the upper attachment bolts.

3. Remove the bolts from the sway braces and remove both engine upper attachment bolts. Then, remove both engine mount lower attachment bolts and remove the QECA from the helicopter. Mount the power package in a suitable workstand and remove the sling.

Installation, Rigging, & Adjustment of Helicopter QECA

The installation of a new or an overhauled engine is in reverse of the removal procedure. The manufacturer's instructions

for the helicopter must be consulted to ascertain the correct interchange of parts from the old engine to the new engine. The applicable maintenance instructions should be followed. Refer to the Maintenance Instructions Manual and associated technical publications for detailed information concerning rigging the throttle, mixture control, cable tensions, and related data.

Testing the Engine Installation

Normal engine run-in procedures must be followed in accordance with the manufacturer's instructions. A flight test is usually performed after the engine has been installed and the engine controls have been adjusted.

Engine Mounts

Mounts for Reciprocating Engines

Most aircraft equipped with reciprocating engines use an engine mount structure made of welded steel tubing. The mount is constructed in one or more sections that incorporate the engine mount ring, bracing members (V-struts), and fittings for attaching the mount to the wing nacelle.

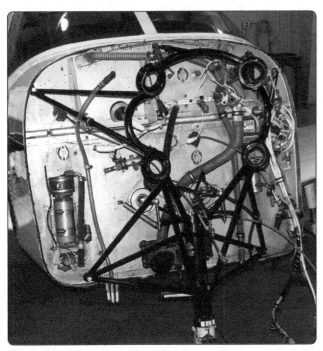

Figure 8-16. *Engine mounting ring.*

The engine mounts are usually secured to the aircraft by special heat-treated steel bolts. The importance of using only these special bolts can be readily appreciated, since they alone support the entire weight of, and withstand all, the stresses imposed by the engine and propeller in flight. The upper bolts support the weight of the engine while the aircraft is on the ground, but when the aircraft is airborne another stress is added. This stress is torsional and affects all bolts, not just the top bolts. A typical engine mount ring shown in *Figure 8-16* discloses fittings and attachment points located at four positions on the engine mount structure. Each fitting houses a dynamic engine mount.

The section of an engine mount where the engine is attached is known as the engine mount ring. It is usually constructed of steel tubing having a larger diameter than the rest of the mount structure. It is circular in shape so that it can surround the engine, which is near the point of balance for the engine. The engine is usually attached to the mount by dynafocal mounts, attached to the engine at the point of balance forward of the mount ring. Other types of mounting devices are also used to secure the different engines to their mount rings.

As aircraft engines became larger and produced more power, some method was needed to absorb their vibration. This demand led to the development of the rubber and steel engine-suspension units called shock mounts. This combination permits restricted engine movement in all directions. These vibration isolators are commonly known as flexible, or elastic, shock mounts. An interesting feature common to most shock mounts is that the rubber and metal parts are arranged so that,

under normal conditions, rubber alone supports the engine. Of course, if the engine is subjected to abnormal shocks or loads, the metal snubbers limit excessive movement of the engine. Dynafocal engine mounts, or vibration isolators, are units that give directional support to the engines. Dynafocal engine mounts have the mounting pad angled to point to the CG of the engines mass. *[Figure 8-16]*

Mounts for Turbofan Engines

The engine mounts on most turbofan engines perform the same basic functions of supporting the engine and transmitting the loads imposed by the engine to the aircraft structure. Most turbine engine mounts are made of stainless steel and are typically located as illustrated in *Figure 8-17*. Some engine mounting systems use two mounts to support the forward end of the engine and a single mount at the rear end.

Turbine Vibration Isolation Engine Mounts

The vibration isolator engine mounts support the power plants and isolate the airplane structure from adverse engine vibrations. Each power plant is generally supported by forward vibration isolator mounts and an aft vibration isolator mount.

The forward vibration isolator engine mounts carry vertical, side, and axial (thrust) loads and allow engine growth due to thermal expansion. The aft mounts take only vertical and side loads; however, they will also accommodate thermal expansion of the engine without applying axial loads to the engine flanges.

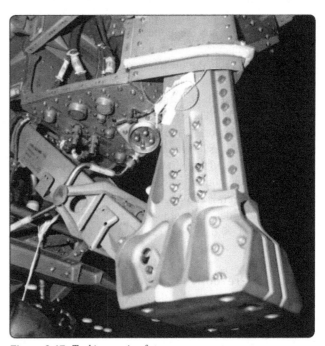

Figure 8-17. *Turbine engine front mount.*

The vibration isolators consist of a resilient material permanently enclosed in a metal case. As an engine vibrates, the resilient material deforms slightly, thereby dampening the vibrations before they reach the airplane structure. If complete failure or loss of the resilient material occurs, the isolators will continue to support the engine.

Preservation & Storage of Engines

An engine awaiting overhaul or return to service after overhaul must be given careful attention. It does not receive the daily care and attention necessary to detect and correct early stages of corrosion. For this reason, some definite action must be taken to prevent corrosion from affecting the engine. Engines that are not flown regularly may not achieve normal service life because of corrosion in and around the cylinders. The normal combustion process creates moisture and corrosive by-products that attack the unprotected surfaces of the cylinder walls, valves, and any other exposed areas that are unprotected. In engines that have accumulated 50 hours or more time in service in a short period, the cylinder walls have acquired a varnish that tends to protect them from corrosive action; engines under favorable atmospheric conditions can remain inactive for several weeks without evidence of damage by corrosion. This is the best-case scenario, but aircraft that operate close to oceans, lakes, rivers, and humid regions have a greater need for engine preservation than engines operated in dry low humid areas.

Corrosion-Preventive Materials

An engine in service is in a sense self-purging of moisture, since the heat of combustion evaporates the moisture in and around the engine, and the lubricating oil circulated through the engine temporarily forms a protective coating on the metal it contacts. If the operation of an engine in service is limited or suspended for a period of time, the engine is preserved to a varying extent, depending upon how long it is to be inoperative. There are three types of engine storage: active engine, temporary, and indefinite. An engine in active storage is defined as having at least one continuous hour of operation with an oil temperature of at least 165 °F to 200 °F and storage time not to exceed 30 days. Temporary storage describes an aircraft and engine that is not flown for 30 to 90 days, and indefinite storage is for an aircraft not to be flown for over 90 days or is removed from the aircraft for extended time.

Corrosion-Preventive Compounds

The preservation materials discussed are used for all types of engine storage. Corrosion-preventive compounds are petroleum-based products that form a wax-like film over the metal to which they are applied. Several types of corrosion-preventive compounds are manufactured according to

different specifications to fit the various aviation needs. The type mixed with engine oil to form a corrosion-preventive mixture is a relatively light compound that readily blends with engine oil when the mixture is heated to the proper temperature.

The light mixture is available in three forms: MIL-C-6529C type I, type II, or type III. Type I is a concentrate and must be blended with three parts of MIL-L-22851 or MIL-L-6082C (SAE J1966) grade 1100 oil to one part of concentrate. Type II is a ready-mixed material with MIL-L-22851 or grade 1100 oil and does not require dilution. Type III is a ready-mixed material with grade 1010 oil for use in turbine engines only. The light mixture is intended for use when a preserved engine is to remain inactive for less than 30 days. It is also used to spray cylinders and other designated areas.

The desired proportions of lubricating oil, and either heavy or light corrosion-preventive compound, must not be obtained by adding the compound to the oil already in the engine. The mixture must be prepared separately before applying to the engine or placing in an oil tank.

A heavy compound is used for the dip treating of metal parts and surfaces. It must be heated to a high temperature to be sufficiently liquid to effectively coat the objects to be preserved. A commercial solvent, or kerosene spray, is used to remove corrosion-preventive compounds from the engine or parts when they are being prepared for return to service.

Although corrosion-preventive compounds act as an insulator from moisture, in the presence of excessive moisture, they eventually break down and corrosion begins. Also, the compounds eventually dry because their oil base gradually evaporates. This allows moisture to contact the engine's metal and aids in corroding it. Therefore, when an engine is stored in a shipping case or container, some dehydrating (moisture removing) agent must be used to remove the moisture from the air in and around the engine.

Dehydrating Agents

There are a number of substances (referred to as desiccants) that can absorb moisture from the atmosphere in sufficient quantities to be useful as dehydrators. One of these is silica gel. This gel is an ideal dehydrating agent since it does not dissolve when saturated.

As a corrosion preventive, bags of silica gel are placed around and inside various accessible parts of a stored engine. It is also used in clear plastic plugs, called dehydrator plugs, that can be screwed into engine openings, such as the spark plug holes. Cobalt chloride is added to the silica gel used in dehydrator plugs. This additive makes it possible

Figure 8-18. *Dehydrator plug "pink" showing high humidity (Sacramento Sky Ranch).*

Figure 8-19. *Dehydrator plug "blue" showing low humidity (Sacramento Sky Ranch).*

for the plugs to indicate the moisture content, or relative humidity, of the air surrounding the engine. The cobalt-chloride-treated silica gel remains a bright blue color with low relative humidity; as the relative humidity increases, the shade of the blue becomes progressively lighter, becoming lavender at 30 percent relative humidity and fading through the various shades of pink *[Figure 8-18]*, until at 60 percent relative humidity it is a natural or white color. Some types

of dehydrator plugs can be dried by removing the silica gel and heating the gel to dry it out, returning it to its original blue color. *[Figure 8-19]* When the relative humidity is less than 30 percent, corrosion does not normally take place. Therefore, if the dehydrator plugs are bright blue, the air in the engine has so little moisture that internal corrosion is held to a minimum. This same cobalt-chloride-treated silica gel is used in humidity indicator envelopes. These envelopes can be fastened to the stored engine so that they can be inspected through a small window in the shipping case or metal engine container. All desiccants are sealed in containers to prevent their becoming saturated with moisture before they are used. Care should be taken never to leave the container open or improperly closed.

Engine Preservation & Return to Service

Before an engine is placed in temporary or indefinite storage, it should be operated and filled with a corrosion-preventive oil mixture added in the oil system to retard corrosion by coating the engine's internal parts. Drain the normal lubricating oil from the sump or system and replace with a preservative oil mixture according to the manufacturer's instructions. Operate the engine until normal operating temperatures are obtained for at least one hour.

Always take the appropriate precautions when turning or working around a propeller. After the flight, remove all the spark plug leads and the top spark plugs.

To prevent corrosion, spray each cylinder interior with corrosion-preventive mixture to prevent moisture and oxygen from contacting the deposits left by combustion. Spray the cylinders by inserting the nozzle of the spray gun into each spark plug hole and playing the gun to cover as much area as possible. Before spraying, each cylinder to be treated should be at the bottom center position and the oil at room temperature. This allows the entire inside of the cylinder to become coated with corrosion-preventive mixture. After spraying each engine cylinder at bottom center, respray each cylinder while the crankshaft is stationary with none of the cylinder's pistons at top dead center.

The crankshaft must not be moved after this final spraying, or the seal of corrosion-preventive mixture between the pistons and cylinder walls are broken. Air can then enter past the pistons into the engine. Also, the coating of corrosion-preventive mixture on the cylinder walls is scraped away, exposing the bare metal to possible corrosion. The engine should have a sign attached similar to the following: "DO NOT TURN CRANKSHAFT—ENGINE PRESERVED PRESERVATION DATE   ."

When preparing the engine for storage, dehydrator plugs are

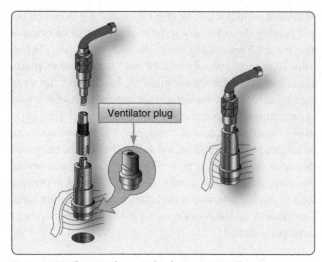

Figure 8-20. *Ignition harness lead support installation.*

screwed into the spark plug opening of each cylinder. If the engine is to be stored in a wooden shipping case, the ignition harness leads are attached to the dehydrator plugs with lead supports. *[Figure 8-20]* Special ventilatory plugs are installed in the spark plug holes of an engine stored horizontally in a storage container. Any engine being prepared for storage must receive thorough treatment around the exhaust ports. Because the residue of exhaust gases is potentially very corrosive, a corrosion-preventive mixture must be sprayed into each exhaust port, including the exhaust valve. After the exhaust ports have been thoroughly coated, a moisture-proof and oil-proof gasket backed by a metal or wooden plate should be secured over the exhaust ports using the exhaust stack mounting studs and nuts. These covers form a seal to prevent moisture from entering the interior of the engine through the exhaust ports. Engines stored in metal containers usually have special ventilatory covers. Another point at which the engine must be sealed is the intake manifold. If the carburetor is to remain on the engine during storage, the throttle valve should be wired open and a seal installed over the air inlet. But, if the carburetor is removed and stored separately, the seal is made at the carburetor mounting pad. The seal used in either instance can be an oil-proof and moisture-proof gasket, backed by a wooden or metal plate securely bolted into place. Silica gel should be placed in the intake manifold to absorb moisture. The silica gel bags are usually suspended from the cover plate. This eliminates the possibility of forgetting to remove the silica gel bags when the engine is eventually removed from storage. A ventilatory cover, without silica gel bags attached, can be used when the engine is stored in a metal container.

After the following details have been taken care of, the engine is ready to be packed into its container. If the engine has not been spray coated with corrosion-preventive mixture, the propeller shaft and propeller shaft thrust bearing must be coated with the compound. Then, a plastic sleeve, or moisture-proof paper, is secured around the shaft, and a threaded protector cap is screwed onto the propeller retaining nut threads.

All engine openings into which dehydrator plugs (or ventilatory plugs if the engine is stored in a metal container) have not been fitted must be sealed. At points where corrosion-preventive mixture can seep from the interior of the engine, such as the oil inlet and outlet, oil-proof and moisture-proof gasket material backed by a metal or wooden plate should be used. At other points moisture-proof tape can be used if it is carefully installed.

Before its installation in a shipping container, the engine should be carefully inspected to determine if the following accessories, which are not a part of the basic engine, have been removed: spark plugs and spark plug thermocouples, remote fuel pump adapters (if applicable), propeller hub attaching bolts (if applicable), starters, generators, vacuum pumps, hydraulic pumps, propeller governors, and engine-driven fuel pumps.

1. Remove seals and all desiccant bags.

2. Remove cylinder dehydrators and plugs or spark plugs from upper and lower spark plug holes.

3. Remove oil sump drain plug and drain the corrosion preventive mixture. Replace drain plug, torque and safety. Remove oil filter. Install new oil filter, torque and safety. Service the engine with oil in accordance with the manufacturer's instructions.

    Warning: To prevent serious bodily injury or death, accomplish the following before moving the propeller:

    a. Disconnect all spark plug leads.

    b. Verify that magneto switches are connected to magnetos and that they are in the off position and P-leads are grounded.

    c. Throttle position CLOSED.

    d. Mixture control IDLE-CUT-OFF.

    e. Set brakes and block aircraft wheels. Ensure that aircraft tiedowns are installed and verify that the cabin door latch is open.

    f. Do not stand within the arc of the propeller blades while turning the propeller.

4. Rotate propeller by hand several revolutions to remove preservative oil.

5. Service and install spark plugs and ignition leads in accordance with the manufacturer's instructions.

6. Service engine and aircraft in accordance with the manufacturer's instruction.

7. Thoroughly clean the aircraft and engine. Perform visual inspection.

8. Correct any discrepancies.

9. Conduct a normal engine start.

10. Perform operational test in accordance with operational inspection of the applicable Maintenance Manual.

11. Correct any discrepancies.

12 Perform a test flight in accordance with airframe manufacturer's instructions.

13. Correct any discrepancies prior to returning aircraft to service.

14. Change oil and filter after 25 hours of operation.

Engine Shipping Containers

For protection, engines are sealed in plastic or foil envelopes and can be packed in a wooden shipping case or in pressurized metal containers.

The engine is lowered into the shipping container so that the mounting plate can be bolted into position. The protective envelope is attached directly to the base of the shipping case. Then, the engine is lowered vertically onto the base and bolted directly to it. A carburetor not mounted on its reciprocating engine (or no provision is made to seal it in a small container to be placed inside the shipping case) can, in some cases, be fastened to a specially constructed platform bolted to the engine.

Before the protective envelope is sealed, silica gel should be placed around the engine to dehydrate the air sealed into the envelope. The amount of silica gel used is determined by the size of the engine. The protective envelope is then carefully gathered around the engine and partially sealed, leaving an opening at one end from which as much air as possible is exhausted. A vacuum applied to the container is very useful for this purpose and is also an aid in detecting any leaks in the envelope. The envelope is then completely sealed, usually by pressing the edges together and fusing them with heat.

Before lowering the shipping case cover over the engine, a quick inventory should be made. Be sure the humidity indicator card is placed so that it can be seen through the

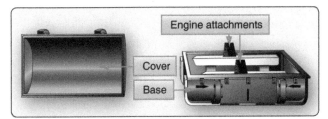

Figure 8-21. *Turbine engine shipping container.*

inspection window and that everything required is enclosed in the container. While lowering the wooden shipping case cover into position, be careful that it does not twist and tear the protective envelope. Secure the cover and stencil or mark the date of preservation on the case. Also, indicate whether the engine is repairable or serviceable.

There are several types of shipping containers in use. *[Figure 8-21]* Another type allows horizontal installation of an engine, thus eliminating the need for an extra hoist. The engine is simply lowered onto the base portion of the container and secured. Then, silica gel bags are packed into the container, usually in a special section. The amount of silica gel required in a metal container is generally greater than that needed in a wooden shipping case, since the volume of air in the metal container is much greater than that in the protective envelope installed around an engine in a wooden shipping case. Also, in the metal container the silica gel bags must dehydrate the interior of the engine, since ventilatory plugs are normally installed in the engine openings in place of dehydrator plugs. All records of the engine should be enclosed inside the shipping container or on the outside for accessibility. A humidity indicator should be fastened inside the containers with an inspection window provided. Then, the rubber seal between the base and the top of the container must be carefully inspected. This seal is usually suitable for re-use several times. After the top of the container has been lowered into position and fastened to the base of the container, dehydrated air at approximately 5 pounds per square inch (psi) pressure is forced into the container. The container should be checked for leaks by occasional rechecks of the air pressure, since radical changes in temperature affect the air pressure in the container.

Inspection of Stored Engines

Most maintenance shops provide a scheduled inspection system for engines in storage. Normally, the humidity indicators on engines stored in shipping cases are inspected every 30 days. When the protective envelope must be opened to inspect the humidity indicator, the inspection period may be extended to once every 90 days, if local conditions permit. The humidity indicator of a metal container is inspected every 180 days under normal conditions.

If the humidity indicator in a wooden shipping case shows by its color that more than 30 percent relative humidity is present in the air around the engine, all desiccants should be replaced. If more than half the dehydrator plugs installed in the spark plug holes indicate the presence of excessive moisture, the interior of the cylinders should be resprayed. If the humidity indicator in a metal container gives a safe blue indication, but air pressure has dropped below 1 psi, the container needs only to be brought to the proper pressure with dehydrated air. However, if the humidity indicator shows an unsafe (pink) condition, the engine should be represerved.

Preservation & Depreservation of Gas Turbine Engines

The procedures for preserving and depreserving gas turbine engines vary depending upon the length of inactivity, the type of preservative used, and whether or not the engine may be rotated during the inactive period. Much of the basic information on corrosion control presented in the section on reciprocating engines is applicable to gas turbine engines. However, the requirements for the types of preservatives and their use are normally different.

The lubrication system is usually drained and may or may not be flushed with preservative oil. The engine fuel system is generally filled with preservative oil, including the fuel control. Before the engine can be returned to service, the preservative oil must be completely flushed from the fuel system by motoring the engine and bleeding the fuel system. Always follow the manufacturer's instructions when performing any preservation or depreservation of gas turbine engines.

Chapter 9
Engine Fire Protection Systems

Introduction

Because fire is one of the most dangerous threats to an aircraft, the potential fire zones of all multiengine aircraft currently produced are protected by a fixed fire protection system. A "fire zone" is an area or region of an aircraft designated by the manufacturer to require fire detection and/or fire extinguishing equipment and a high degree of inherent fire resistance. The term "fixed" describes a permanently installed system in contrast to any type of portable fire extinguishing equipment, such as a hand-held fire extinguisher.

In accordance with Title 14 of the Code of Federal Regulations (14 CFR) parts 23 and 25, engine fire protection systems are mandatory on: multiengine turbine-powered aircraft, multiengine reciprocating engine-powered aircraft incorporating turbochargers, aircraft with engine(s) located where they are not readily visible from the flight deck, all commuter and transport category aircraft, and the auxiliary power unit (APU) compartment of any aircraft incorporating an APU. Fire protection systems are not mandatory for many single and twin reciprocating engine general aviation (GA) aircraft.

Several general failures or hazards can result in overheat conditions or fires peculiar to turbine engine aircraft because of their operating characteristics. The two major types of turbine failure can be classified as 1) thermodynamic and 2) mechanical.

Thermodynamic causes upset the proportion of air used to cool combustion temperatures to the levels that the turbine materials can tolerate. When the cooling cycle is upset, turbine blades can melt, causing a sudden loss of thrust. The rapid buildup of ice on inlet screens or inlet guide vanes can result in severe overheating, causing the turbine blades to melt or to be severed and thrown outward. Such failure can result in a severed tail cone and possible penetration of the aircraft structure, tanks, or equipment near the turbine wheel. In general, most thermodynamic failures are caused by ice, excess air bleed or leakage, or faulty controls that permit compressor stall or excess fuel.

Mechanical failures, such as fractured or thrown blades, can also lead to overheat conditions or fires. Thrown blades can puncture the tail cone, creating an overheat condition. Failure of forward stages of multi-stage turbines is usually

much more severe. Penetration of the turbine case by failed blades is a possible fire hazard, as is the penetration of lines and components containing flammable fluids.

A high flow of fuel through an improperly adjusted fuel nozzle can cause burn-through of the tail cone in some engines. Engine fires can be caused by burning fluid that occasionally runs out through the exhaust pipe.

Components

A complete fire protection system includes both a fire detection and a fire extinguishing system. To detect fires or overheat conditions, detectors are placed in the various zones to be monitored. Fires are detected in aircraft by using one or more of the following: overheat detectors, rate-of-temperature-rise detectors, and flame detectors. In addition to these methods, other types of detectors are used in aircraft fire protection systems but are not used to detect engine fires. For example, smoke detectors are better suited to monitor areas such as baggage compartments or lavatories, where materials burn slowly or smolder. Other types of detectors in this category include carbon monoxide detectors.

Fire protection systems on current-production aircraft do not rely on observation by crewmembers as a primary method of fire detection. An ideal fire detector system includes as many of the following features as possible:

1. A system that does not cause false warnings under any flight or ground condition.

2. Rapid indication of a fire and accurate location of the fire.

3. Accurate indication that a fire is out.

4. Indication that a fire has reignited.

5. Continuous indication for duration of a fire.

6. Means for electrically testing the detector system from the aircraft flight deck.

7. Detectors that resist damage from exposure to oil, water, vibration, extreme temperatures, or handling.

8. Detectors that are light in weight and easily adaptable to any mounting position.

9. Detector circuitry that operates directly from the aircraft power system without inverters.

10. Minimum electrical current requirements when not indicating a fire.

11. Each detector system should turn on a flight deck light, indicating the location of the fire, and have an audible alarm system.

12. A separate detector system for each engine.

Engine Fire Detection Systems

Several different types of fire detection systems are installed in aircraft to detect engine fires. Two common types used are spot detectors and continuously loop systems. Spot detector systems use individual sensors to monitor a fire zone. Examples of spot detector systems are the thermal switch system, the thermocouple system, the optical fire detection system, and the pneumatic-based thermal fire detection system. Continuous loop systems are typically installed on transport type aircraft and provide more complete fire detection coverage by using several loop-type sensors.

Thermal Switch System

A number of detectors or sensing devices are available. Many older model aircraft still operating have some type of thermal switch system or thermocouple system. A thermal switch system has one or more lights energized by the aircraft power system and thermal switches that control operation of the light(s). These thermal switches are heat-sensitive units that complete electrical circuits at a certain temperature. They are connected in parallel with each other, but in series with the indicator lights [Figure 9-1]. If the temperature rises above a set value in any one section of the circuit, the thermal switch closes, completing the light circuit to indicate a fire or overheat condition.

No set number of thermal switches is required; the exact number usually is determined by the aircraft manufacturer. On some installations, all the thermal detectors are connected to one light; others may have a separate thermal switch for each indicator light.

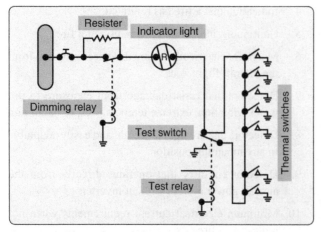

Figure 9-1. *Thermal switch fire circuit.*

Some warning lights are push-to-test lights. The bulb is tested by pushing it in to check an auxiliary test circuit. The circuit shown in *Figure 9-1* includes a test relay. With the relay contact in the position shown, there are two possible paths for current flow from the switches to the light. This is an additional safety feature. Energizing the test relay completes a series circuit and checks all the wiring and the light bulb.

Also included in the circuit shown in *Figure 9-1* is a dimming relay. By energizing the dimming relay, the circuit is altered to include a resistor in series with the light. In some installations, several circuits are wired through the dimming relay, and all the warning lights may be dimmed at the same time.

Thermocouple Systems

The thermocouple fire warning system operates on an entirely different principle than the thermal switch system. A thermocouple depends on the rate of temperature rise and does not give a warning when an engine slowly overheats or a short circuit develops. The system consists of a relay box, warning lights, and thermocouples. The wiring system of these units may be divided into the following circuits: (1) the detector circuit, (2) the alarm circuit, and (3) the test circuit. These circuits are shown in *Figure 9-2.*

The relay box contains two relays, the sensitive relay and the slave relay, and the thermal test unit. Such a box may contain from one to eight identical circuits, depending on the number of potential fire zones. The relays control the warning lights. In turn, the thermocouples control the operation of the relays. The circuit consists of several thermocouples in series with each other and with the sensitive relay.

Thermocouple leads are made from a variety of metals, depending on the maximum temperature to which they are exposed. Iron and constantan, or copper and constantan, are common for CHT measurement. Chromel and alumel are used for turbine EGT thermocouples. The point where these metals are joined and exposed to the heat of a fire is called a hot junction. There is also a reference junction enclosed in a dead air space between two insulation blocks.

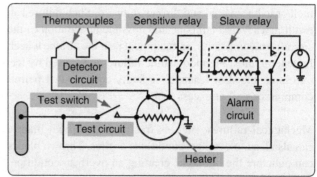

Figure 9-2. *Thermocouple fire warning circuit.*

A metal cage surrounds the thermocouple to give mechanical protection without hindering the free movement of air to the hot junction. Thermocouple leads are designed for a specific installation and may not be altered, if they are altered they will no longer be calibrated specific probe and instrument combination.

If the temperature rises rapidly, the thermocouple produces a voltage because of the temperature difference between the reference junction and the hot junction. If both junctions are heated at the same rate, no voltage results. In the engine compartment, there is a normal, gradual rise in temperature from engine operation; because it is gradual, both junctions heat at the same rate and no warning signal is given.

If there is a fire, however, the hot junction heats more rapidly than the reference junction. The reference junction is also commonly referred to as the cold junction. The ensuing voltage causes a current to flow within the detector circuit. Any time the current is greater than 4 milliamperes (0.004 ampere), the sensitive relay closes. This completes a circuit from the aircraft power system to the coil of the slave relay. The slave relay then closes and completes the circuit to the warning light to give a visual fire warning.

The total number of thermocouples used in individual detector circuits depends on the size of the fire zones and the total circuit resistance, which usually does not exceed 5 ohms. As shown in *Figure 9-2*, the circuit has two resistors. The resistor connected across the slave relay terminals absorbs the coil's self-induced voltage to prevent arcing across the points of the sensitive relay. The contacts of the sensitive relay are so fragile that they burn or weld if arcing is permitted.

When the sensitive relay opens, the circuit to the slave relay is interrupted and the magnetic field around its coil collapses. When this happens, the coil gets a voltage through self-induction, but with the resistor across the coil terminals, there is a path for any current flow as a result of this voltage. Thus, arcing at the sensitive relay contacts is eliminated.

Optical Fire Detection Systems

Optical sensors, often referred to as "flame detectors," are designed to alarm when they detect the presence of prominent, specific radiation emissions from hydrocarbon flames. The two types of optical sensors available are infrared (IR) and ultraviolet, based on the specific emission wave lengths they are designed to detect.

Infrared Optical Fire Protection

IR-based optical flame detectors are used primarily on light turboprop aircraft and helicopter engines. These sensors have proven to be very dependable and economical for the relatively benign environments of these applications.

Principle of Operation

Radiation emitted by the fire crosses the airspace between the fire and the detector and impinges on the detector front face and window. The window allows a broad spectrum of radiation to pass into the detector where it impinges on the face of the sensing device filter. The filter allows only radiation in a tight waveband centered around 4.3 micrometers in the IR to pass on to the radiation-sensitive surface of the sensing device. The radiation striking the sensing device minutely raises its temperature causing small thermoelectric voltages to be generated. These voltages are fed to an amplifier whose output is connected to various analytical electronic processing circuits. The processing electronics is tailored exactly to the time signature of all known hydrocarbon flame sources and ignores false alarm sources, such as incandescent lights and sunlight. Alarm sensitivity level is accurately controlled by a digital circuit. A typical warning system is illustrated in *Figure 9-3*.

Pneumatic Thermal Fire Detection

Pneumatic detectors are based on the principles of gas laws. The sensing element consists of a closed helium-filled tube connected at one end to a responder assembly. As the element is heated, the gas pressure inside the tube increases until the alarm threshold is reached. At this point, an internal switch closes and reports an alarm to the flight deck. The pneumatic detector integrity pressure switch opens and triggers the fault alarm if the pneumatic detector losses pressure, as in the case of a leak.

Continuous-Loop Detector Systems

Large commercial aircraft almost exclusively use continuous thermal sensing elements for powerplant protection, since these systems offer superior detection performance and coverage, and they have the proven ruggedness to survive in the harsh environment of modern turbofan engines.

A continuous-loop detector, or sensing system, permits more complete coverage of a fire hazard area than any of the spot-type temperature detectors. Continuous-loop systems are versions of the thermal switch system. They are overheat systems, heat-sensitive units that complete electrical circuits at a certain temperature. There is no rate-of-heat-rise sensitivity in a continuous-loop system. Two widely used types of continuous-loop systems are the Kidde and the Fenwal systems. This text briefly discusses the Fenwal system, while the Kidde system is discussed more in-depth.

Fenwal Continuous-Loop System

The Fenwal system uses a slender inconel tube packed with thermally sensitive eutectic salt and a nickel wire center

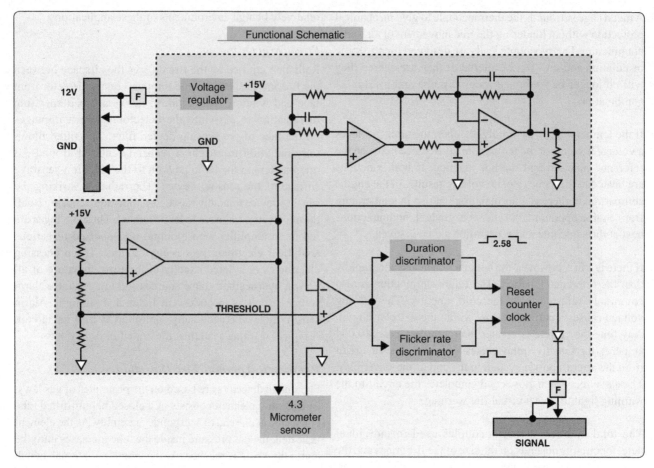

Figure 9-3. *Optical fire detection system circuit.*

conductor. *[Figure 9-4]* Lengths of these sensing elements are connected in series to a control unit. The elements may be of equal or varying length and of the same or different temperature settings. The control unit, operating directly from the power source, impresses a small voltage on the sensing elements. When an overheat condition occurs at any point along the element length, the resistance of the eutectic salt within the sensing element drops sharply, causing current to flow between the outer sheath and the center conductor. This current flow is sensed by the control unit, which produces a signal to actuate the output relay.

When the fire has been extinguished or the critical temperature lowered, the Fenwal system automatically returns to standby alert, ready to detect any subsequent fire or overheat condition. The Fenwal system may be wired to employ a "loop" circuit. In this case, should an open circuit occur, the system still signals fire or overheat. If multiple open circuits occur, only that section between breaks becomes inoperative.

Kidde Continuous-Loop System

In the Kidde continuous-loop system, two wires are imbedded in an inconel tube filled with a thermistor core material. *[Figure 9-5]* Two electrical conductors go through the length of the core. One conductor has a ground connection to the tube and the other conductor connects to the fire detection control unit.

As the temperature of the core increases, electrical resistance to ground decreases. The fire detection control unit monitors this resistance. If the resistance decreases to the overheat set point, an overheat indication occurs in the flight deck. Typically, a 10-second time delay is incorporated for the

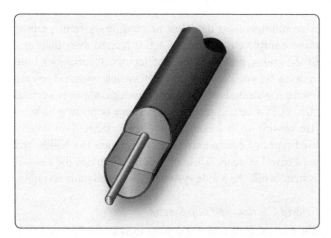

Figure 9-4. *Fenwal sensing element.*

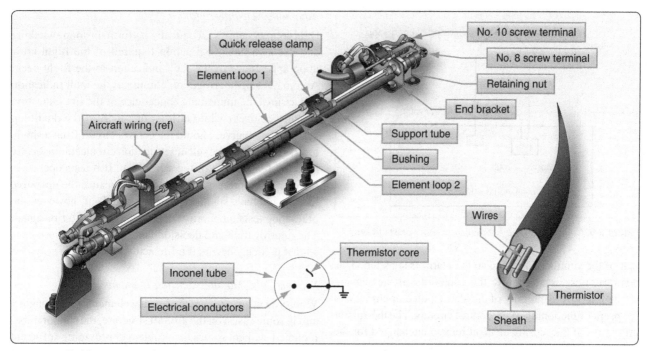

Figure 9-5. *Kidde continuous-loop system.*

overheat indication. If the resistance decreases more to the fire set point, a fire warning occurs. When the fire or overheat condition is gone, the resistance of the core material increases to the reset point and the flight deck indications go away.

The rate of change of resistance identifies an electrical short or a fire. The resistance decreases more quickly with an electrical short than with a fire. In addition to fire and overheat detection, the Kidde continuous-loop system can supply nacelle temperature data to the aircraft condition monitoring function of the Aircraft In-Flight Monitoring System (AIMS).

Sensing Element

The sensing element consists, essentially, of an infinite number of unit thermistors electrically in parallel along its length. The resistance of the sensing element is a function of the length heated, as well as the temperature-heating of less than the full length of element, which requires that portion to be heated to a higher temperature to achieve the same total resistance change. As a result, the system responds not to a fixed alarm temperature but to the sum of the resistances (in parallel) that reflects a nonarithmetic "average." The sensing element may be routed close to nonhazardous hot spots that may have a normal temperature well above the overall alarm temperature, without danger of causing a false alarm. This feature permits the alarm point to be set close to the maximum general ambient temperature, giving greater sensitivity to a general overheat or fire without being subject to false alarms from localized nonhazardous hot spots.

Combination Fire & Overheat Warning

The analog signal from the thermistor sensing element permits the control circuits to be arranged to give a two-level response from the same sensing element loop. The first is an overheat warning at a temperature level below the fire warning, indicating a general engine compartment temperature rise, which could be caused by leakage of hot bleed air or combustion gas into the engine compartment. It could be an early warning of fire and would alert the crew to appropriate action to reduce the engine compartment temperature. The second-level response would be at a level above that attainable by the leaking hot gas and would be the fire warning.

Temperature Trend Indication

The analog signal produced by the sensing element loop as its temperature changes can readily be converted to signals suitable for meter or cathode ray tube (CRT) display to indicate engine bay temperature increases from normal. A comparison of the readings from each loop system also provides a check on the condition of the fire detection system, because the two loops should normally read alike.

System Test

The integrity of the continuous-loop fire detection system may be tested by actuating a test switch in the flight deck, which switches one end of the sensing element loop from its control circuit to a test circuit, built into the control unit, that simulates the sensing element resistance change due to fire. *[Figure 9-6]* If the sensing element loop is unbroken, the resistance detected "seen" by the control circuit is now

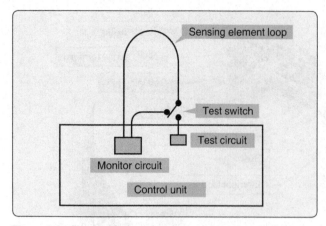

Figure 9-6. *Continuous-loop fire detection system test circuit.*

that of the simulated fire and so the alarm is signaled. This demonstrates, in addition to the continuity of the sensing element loop, the integrity of the alarm indicator circuit and the proper functioning of the control circuits. The thermistor properties of the sensing element remain unchanged for the life of the element (no chemical or physical changes take place on heating), so that it functions properly as long as it is electrically connected to the control unit.

Fault Indication

Provision can be made in the control unit to send a fault signal to activate a fault indicator whenever the short discriminator circuit detects a short in the sensing element loop. While this is a requirement in 14 CFR for transport category aircraft because such a short disables the fire detection system, it is offered as an option for other aircraft types in which it may not be a requirement.

Dual-Loop Systems

Dual-loop systems are, in essence, two complete basic fire detection systems with their output signals connected so that both must signal to result in a fire warning. This arrangement, called "AND" logic, results in greatly increased reliability against false fire warnings from any cause. Should one of the two loops be found inoperative at the preflight integrity test, a flight deck selector switch disconnects that loop and allows the signal from the other loop alone to activate the fire warning. Since the single operative loop meets all fire detector requirements, the aircraft can be safely dispatched, and maintenance deferred to a more convenient time. However, should one of the two loops become inoperative in flight and a fire subsequently occur, the fire signaling loop activates a flight deck fault signal that alerts the flight crew to select single-loop operation to confirm the possible occurrence of fire.

Automatic Self-Interrogation

Dual-loop systems automatically perform the loop switching and decision-making function required of the flight crew upon appearance of the fault indication in the flight deck. Automatic self-interrogation eliminates the fault indication and assures the immediate appearance of the fire indication should fire occur while at least one loop of the dual-loop system is operative. Should the control circuit from a single loop signal "fire," the self-interrogation circuit automatically tests the functioning of the other loop. If it tests operative, the circuit suppresses the fire signal (because the operative loop would have signaled if a fire existed). If, however, the other loop tests inoperative, the circuit outputs a fire signal. The interrogation and decision takes place in milliseconds, so that no delay occurs if a fire actually exists.

Support Tube-Mounted Sensing Elements

When you want to mount the sensing elements on the engine, and in some cases, on the aircraft structure, the support tube-mounted element solves the problem of providing sufficient element support points, and greatly facilitates the removal and reinstallation of the sensing elements for engine or system maintenance.

Most modern installations use the support tube concept of mounting sensing elements for better maintainability as well as increased reliability. The sensing element is attached to a prebent stainless steel tube by closely spaced clamps and bushings, where it is supported from vibration damage and protected from pinching and excessive bending. The support tube-mounted elements can be furnished with either single- or dual-sensing elements.

Being prebent to the designed configuration assures its installation in the aircraft precisely in its designed location, where it has the necessary clearance to be free from the possibility of the elements chafing against engine or aircraft structure. The assembly requires only a few attachment points, and removal for engine maintenance is quick and easy. Should the assembly require repair or maintenance, it is easily replaced with another assembly, leaving the repair for the shop. A damaged sensing element is easily replaced in the assembly. The assembly is rugged, easy to handle, and unlikely to suffer damage during handling for installation or removal.

Fire Detection Control Unit (Fire Detection Card)

The control unit for the simplest type of system typically contains the necessary electronic resistance monitoring and alarm output circuits, housed in a hermetically sealed aluminum case and filled with a mounting bracket and circular electrical connector. For more sophisticated systems, control modules may be employed that contain removable control cards having circuitry for individual hazard areas, and/ or unique functions. In the most advanced applications, the detection system circuitry controls all aircraft fire protection functions, including fire detection and extinguishing for engines, APUs, cargo bays, and bleed air systems.

Fire Zones

The powerplant installation has several designated fire zones: (1) the engine power section; (2) the engine accessory section; (3) except for reciprocating engines, any complete powerplant compartment in which no isolation is provided between the engine power section and the engine accessory section; (4) any APU compartment; (5) any fuel-burning heater and other combustion equipment installation; (6) the compressor and accessory sections of turbine engines; and (7) combustor, turbine, and tailpipe sections of turbine engine installations that contain lines or components carrying flammable fluids or gases. *Figure 9-7* shows fire protection for a large turbo fan engine.

In addition to the engine and nacelle area zones, other areas on multiengine aircraft are provided with fire detection and protection systems. These areas include baggage compartments, lavatories, APU, combustion heater installations, and other hazardous areas. Discussion of fire protection for these areas is not included in this section, which is limited to engine fire protection.

Engine Fire Extinguishing System

Commuter aircraft certificated under 14 CFR part 23 are required to have, at a minimum, a one-shot fire extinguishing system. All transport category aircraft certificated under 14 CFR part 25 are required to have two discharges, each of which produces adequate agent concentration. An individual one-shot system may be used for APUs, fuel burning heaters, and other combustion equipment. For each "other" designated fire zone, two discharges (two-shot system) must be provided, each of which produces adequate agent concentration. *[Figure 9-8]*

Fire Extinguishing Agents

The fixed fire extinguisher systems used in most engine fire protection systems are designed to dilute the atmosphere with an inert agent that does not support combustion. Many systems use perforated tubing or discharge nozzles to distribute the extinguishing agent. High rate of discharge (HRD) systems use open-end tubes to deliver a quantity of extinguishing agent in 1 to 2 seconds. The most common extinguishing agent still used today is Halon 1301 because of its effective firefighting capability and relatively low toxicity (U.L. classification Group 61). Noncorrosive, Halon 1301 does not affect the material it contacts and requires no clean-up when discharged. Halon 1301 is the current extinguishing agent for commercial aircraft, but a replacement is under development. Because Halon 1301 depletes the ozone layer only recycled Halon 1301 is currently available. Halon 1301 is used until a suitable replacement is developed. Some military aircraft use HCL-125, which the Federal Aviation Administration (FAA) is testing for use in commercial aircraft.

Carbon dioxide (CO_2) is an effective extinguishing agent. It is most often used in fire extinguishers that are available on the ramp to fight fires on the exterior of the aircraft, such as engine or APU fires. CO_2 has been used for many years to extinguish flammable fluid fires and fires involving electrical equipment. It is noncombustible and does not react with most substances. It provides its own pressure for discharge from the storage vessel, except in extremely cold climates where a booster charge of nitrogen may be added to winterize the system. Normally, CO_2 is a gas, but it is easily liquefied by compression and cooling. After liquefaction, CO_2 remains in a closed container as both liquid and gas. When CO_2 is then discharged to the atmosphere, most of the liquid expands to gas. Heat absorbed by the gas during vaporization cools the remaining liquid to −110 °F, and it becomes a finely divided white solid, dry ice snow.

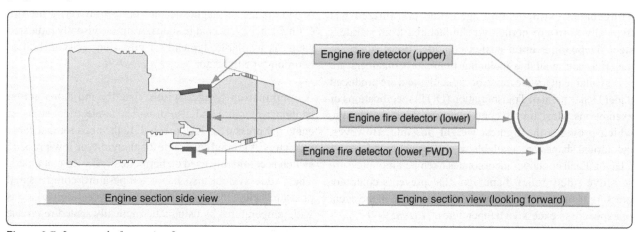

Figure 9-7. *Large turbofan engine fire zones.*

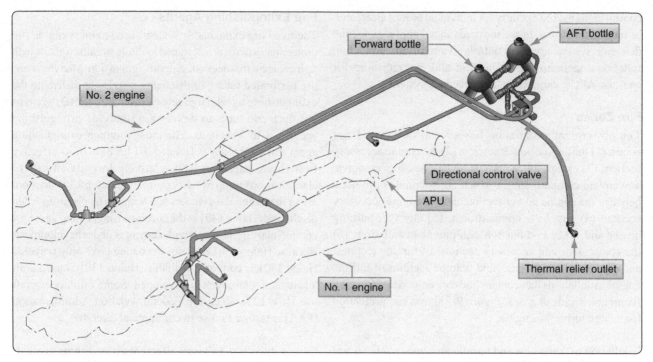

Figure 9-8. *Typical fire extinguishing system.*

Turbine Engine Ground Fire Protection

On many aircraft, means are usually provided for rapid access to the compressor, tailpipe, or burner compartments. Many aircraft systems are equipped with spring-loaded or pop-out access doors in the skin of the various compartments. Internal engine tailpipe fires that take place during engine shutdown or false starts can be blown out by motoring the engine with the starter. A running engine can be accelerated to rated speed to achieve the same result. If such a fire persists, a fire extinguishing agent can be directed into the tailpipe. It should be remembered that excessive use of CO_2, or other agents that have a cooling effect, can shrink the turbine housing on the turbine and cause the engine to disintegrate.

Containers

Fire extinguisher containers (HRD bottles) store a liquid halogenated extinguishing agent and pressurized gas (typically nitrogen) normally manufactured from stainless steel. Depending upon design considerations, alternate materials are available including titanium. Containers are also available in a wide range of capacities and are produced under Department of Transportation (DOT) specifications or exemptions. Most aircraft containers are spherical in design, which provides the lightest weight possible. However, cylindrical shapes are available where space limitations are a factor. Each container incorporates a temperature/pressure sensitive safety relief diaphragm that prevents container pressure from exceeding container test pressure in the event of exposure to excessive temperatures. *[Figure 9-9]*

Discharge Valves

Discharge valves are installed on the containers. A cartridge (squib) and frangible disc type valve are installed in the outlet of the discharge valve assembly. Special assemblies having solenoid-operated or manually-operated seat type valves are also available. Two types of cartridge disc-release techniques are used. Standard release type uses a slug driven by explosive energy to rupture a segmented closure disc. For high temperature or hermetically sealed units, a direct explosive impact type cartridge is used, which applies fragmentation impact to rupture a prestressed corrosion-resistant steel diaphragm. Most containers use conventional metallic gasket seals that facilitate refurbishment following discharge. *[Figure 9-10]*

Pressure Indication

A wide range of diagnostics are utilized to verify the fire extinguisher agent charge status. A simple visually indicated gauge is available, typically a vibration-resistant helical bourdon-type indicator. *[see Figure 9-9]*

A combination gauge switch visually indicates actual container pressure and also provides an electrical signal if container pressure is lost, precluding the need for discharge indicators. A ground checkable diaphragm-type low-pressure switch is commonly used on hermetically sealed containers. The Kidde system also has a temperature compensated pressure switch that tracks the container pressure variations with temperatures by using a hermetically sealed reference chamber.

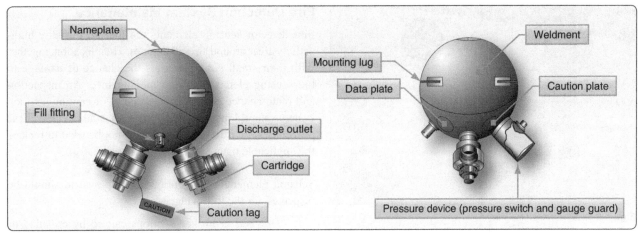

Figure 9-9. *Fire extinguisher containers (HRD bottles).*

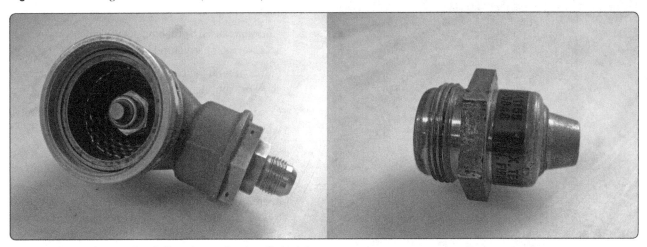

Figure 9-10. *Discharge valve (left) and cartridge (squib) (right).*

Two-Way Check Valve

A complete line of two-way check valves is available, manufactured from either lightweight aluminum or steel. These valves are required in a two-shot system to prevent the agent in a reserve container from backing up into the previous emptied main container. Valves are supplied with either MS-33514 or MS-33656 fitting configurations.

Discharge Indicators

Discharge indicators provide immediate visual evidence of container discharge on fire extinguishing systems. Two kinds of indicators can be furnished: thermal and discharge. Both types are designed for aircraft and skin mounting. *[Figure 9-11]*

Thermal Discharge Indicator (Red Disc)

The thermal discharge indicator is connected to the fire container relief fitting and ejects a red disc to show when container contents have dumped overboard due to excessive heat. The agent discharges through the opening created when the disc blows out. This gives the flight and maintenance crews an indication that the fire extinguisher container needs

to be replaced before the next flight.

Yellow Disc Discharge Indicator

If the flight crew activates the fire extinguisher system, a yellow disc is ejected from the skin of the aircraft fuselage. This is an indication for the maintenance crew that the fire extinguishing system was activated by the flight crew, and that the fire extinguishing container needs to be replaced before the next flight.

Fire Switch

Fire switches are typically installed on the center overhead panel or center console in the flight deck. *[Figure 9-12]* When the fire switch is activated, the following happens: the engine stops because the fuel control shuts off, the engine is isolated from the aircraft systems, and the fire extinguishing system is activated. Some aircraft use fire switches that need to be pulled and turned to activate the system, while others use a push-type switch with a guard. To prevent accidental activation of the fire switch, a lock is installed that releases the fire switch only when a fire has been detected. This lock can be manually released by the flight crew if the fire detection

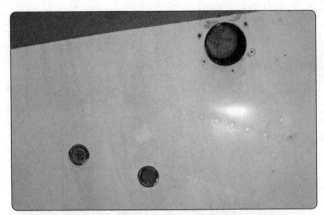

Figure 9-11. *Discharge indicators.*

Figure 9-12. *Engine fire switches.*

system malfunctions. *[Figure 9-13]*

Warning Systems

Visible and audible warning systems are installed in the flight deck to alert the flight crew. A horn sounds and one or several warning lights illuminate to alert the flight crew that an engine fire has been detected. These indications stop when the fire is extinguished.

Fire Detection System Maintenance

Fire detector sensing elements are located in many high-activity areas around aircraft engines. Their location, together with their small size, increases the chance of damage to the sensing elements during maintenance. An inspection and maintenance program for all types of continuous-loop systems should include the following visual checks. These procedures are examples and should not be used to replace the applicable manufacturer's instructions.

Sensing elements of a continuous-loop system should be inspected for the following:

1. Cracked or broken sections caused by crushing or squeezing between inspection plates, cowl panels, or engine components.

2. Abrasion caused by rubbing of the element on cowling, accessories, or structural members.

3. Pieces of safety wire or other metal particles that may short the spot-detector terminals.

4. Condition of rubber grommets in mounting clamps, which may be softened from exposure to oils or hardened from excessive heat.

5. Dents and kinks in sensing element sections. Limits on the element diameter, acceptable dents and kinks, and degree of smoothness of tubing contour are specified by the manufacturer. No attempt should be made to straighten any acceptable dent or kink, since stresses may be set up that could cause tubing failure. (See illustration of kinked tubing in *Figure 9-14*.)

6. Nuts at the end of the sensing elements *[Figure 9-15]* should be inspected for tightness and the presence of a safety wire. Loose nuts should be retorqued to the value specified by the manufacturer's instructions. Some types of sensing element connection joints require the use of copper crush gaskets, which should be replaced any time a connection is separated.

7. If shielded flexible leads are used, they should be inspected for fraying of the outer braid. The braided sheath is composed of many fine metal strands woven into a protective covering surrounding the inner insulated wire. Continuous bending of the cable or rough treatment can break these fine wires, especially those near the connectors.

8. Sensing element routing and clamping should be inspected carefully. *[Figure 9-14]* Long, unsupported sections may permit excessive vibration that can cause breakage. The distance between clamps on straight runs, usually about 8–10 inches, is specified by the manufacturer. At end connectors, the first support clamp is usually located about four to six inches from

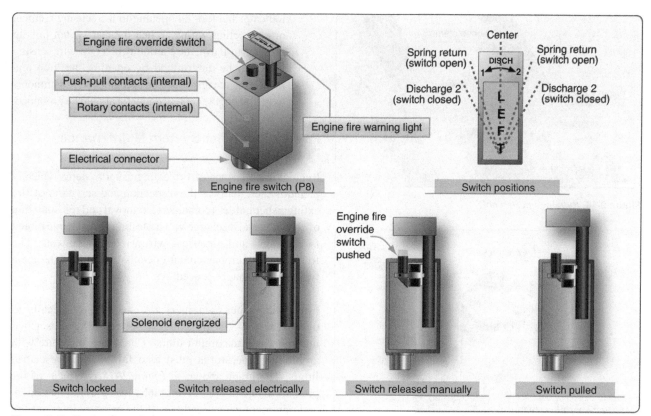

Figure 9-13. *Engine fire switch operation.*

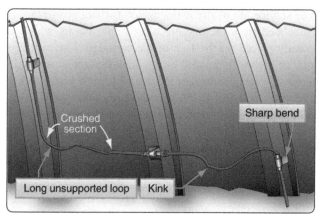

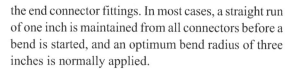

Figure 9-14. *Sensing element defects.*

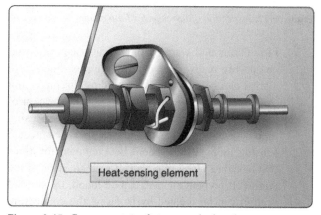

Figure 9-15. *Connector joint fitting attached to the structure.*

the end connector fittings. In most cases, a straight run of one inch is maintained from all connectors before a bend is started, and an optimum bend radius of three inches is normally applied.

9. Interference between a cowl brace and a sensing element can cause rubbing. *[Figure 9-16]* This interference may cause wear and short the sensing element.

10. Grommets should be installed on the sensing element so that both ends are centered on its clamp. The split end of the grommet should face the outside of the nearest bend. Clamps and grommets should fit the element snugly. *[Figure 9-17]*

Fire Detection System Troubleshooting

The following troubleshooting procedures represent the most common difficulties encountered in engine fire detection systems:

1. Intermittent alarms are most often caused by an intermittent short in the detector system wiring. Such shorts may be caused by a loose wire that occasionally touches a nearby terminal, a frayed wire brushing against a structure, or a sensing element rubbing against a structural member long enough to wear through the insulation. Intermittent faults often can be located by moving wires to re-create the short.

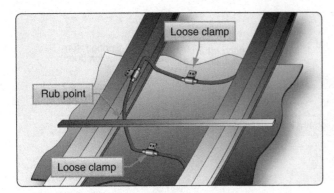

Figure 9-16. *Rubbing interference.*

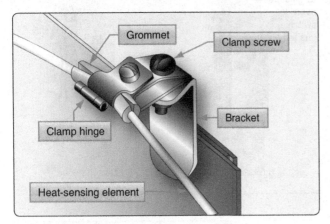

Figure 9-17. *Inspection of fire detector loop clamp.*

2. Fire alarms and warning lights can occur when no engine fire or overheat condition exists. Such false alarms can be most easily located by disconnecting the engine sensing loop connections from the control unit. If the false alarm ceases when the engine sensing loop is disconnected, the fault is in the disconnected sensing loop, which should be examined for areas that have been bent into contact with hot parts of the engine. If no bent element can be found, the shorted section can be located by isolating the connecting elements consecutively around the entire loop.

3. Kinks and sharp bends in the sensing element can cause an internal wire to short intermittently to the outer tubing. The fault can be located by checking the sensing element with an ohm meter while tapping the element in the suspected areas to produce the short.

4. Moisture in the detection system seldom causes a false fire alarm. If, however, moisture does cause an alarm, the warning persists until the contamination is removed or boils away, and the resistance of the loop returns to its normal value.

5. Failure to obtain an alarm signal when the test switch is actuated may be caused by a defective test switch or control unit, the lack of electrical power, inoperative

indicator light, or an opening in the sensing element or connecting wiring. When the test switch fails to provide an alarm, the continuity of a two-wire sensing loop can be determined by opening the loop and measuring the resistance. In a single-wire, continuous-loop system, the center conductor should be grounded.

Fire Extinguisher System Maintenance Practices

Regular maintenance of fire extinguisher systems typically includes such items as the inspection and servicing of fire extinguisher bottles (containers), removal and reinstallation of cartridge and discharge valves, testing of discharge tubing for leakage, and electrical wiring continuity tests. The following paragraphs contain details of some of the most typical maintenance procedures.

Fire extinguisher containers are checked periodically to determine that the pressure is between the prescribed minimum and maximum limits. Changes of pressure with ambient temperatures must also fall within prescribed limits. The graph shown in *Figure 9-18* is typical of the pressure temperature curve graphs that provide maximum and minimum gauge readings. If the pressure does not fall within the graph limits, the extinguisher container is replaced. The service life of fire extinguisher discharge cartridges is calculated from the manufacturer's date stamp, which is usually placed on the face of the cartridge. The cartridge service life recommended by the manufacturer is usually in terms of years. Cartridges are available with a service life of 5 years or more. To determine the unexpired service life of a discharge cartridge, it is usually necessary to remove the electrical leads and discharge line from the plug body, which can then be removed from the extinguisher container.

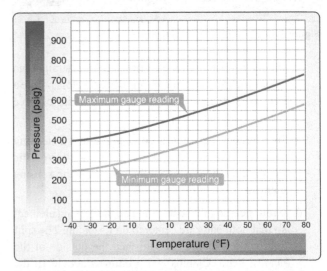

Figure 9-18. *Fire extinguisher container pressure-temperature chart.*

Be careful when replacing cartridge and discharge valves. Most new extinguisher containers are supplied with their cartridge and discharge valve disassembled. Before installation on the aircraft, properly assemble the cartridge in the discharge valve and connect the valve to the container, usually by means of a swivel nut that tightens against a packing ring gasket. [Figure 9-19]

If a cartridge is removed from a discharge valve for any reason, it should not be used in another discharge valve assembly, since the distance the contact point protrudes may vary with each unit. Thus, continuity might not exist if a used plug that had been indented with a long contact point were installed in a discharge valve with a shorter contact point.

The preceding material in this chapter is general in nature, addressing the principles involved and general procedures to be followed. When actually performing maintenance, always refer to the applicable maintenance manuals and other related publications pertaining to a particular aircraft.

Boeing 777 Aircraft Fire Detection & Extinguishing System

The following section discusses the fire detection and extinguishing system of the B777 aircraft. The information is included only for familiarization purposes.

Each engine has two fire detection loops: loop 1 and loop 2. A fire detection card in the system card file monitors the loops for fires, overheat conditions, and faults. There is a fire detection card for each engine.

Overheat Detection

If the fire detection loops detect an overheat condition, the fire detection card sends a signal to the AIMS and to the warning electronics unit. The following indications occur in the flight deck:

- The master caution lights come on.
- The caution aural operates.
- An engine overheat caution message shows.

Fire Detection

If an engine fire occurs, the fire detection card sends a signal to the AIMS and to the warning electronics unit, and a warning message illuminates. The following indications occur in the flight deck:

- The master warning lights come on.
- The fire warning aural operates.
- An engine fire warning message shows.
- The engine fire warning light comes on.

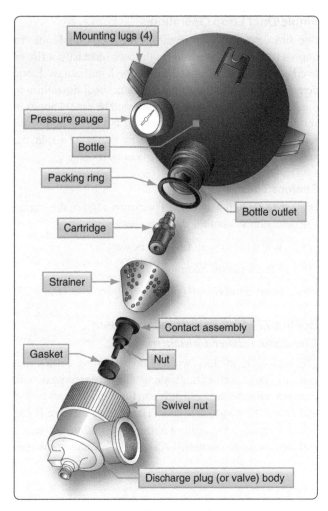

Figure 9-19. *Components of fire extinguisher container.*

- The fuel control switch fire warning light comes on.

Nacelle Temperature Recording

The fire detection card measures the average temperature of the loops. This data goes to the AIMS through the system's ARINC 629 buses and is recorded by the aircraft condition monitoring function.

Continuous Fault Monitoring

The fire detection card monitors the two loops and their wiring for defects. In normal (dual loop) operation, both loops must have a fire or overheat condition to cause the flight deck indications.

If a failure occurs in a loop, the fire detection card sends the data to the AIMS. A status message shows, and the system changes to single-loop operation. In this mode, fire/overheat indications occur when one loop is defective and the other has a fire or overheat condition.

Single/Dual Loop Operation

The fire detection card monitors the loops for faults. In normal (dual loop) operation, both loops must have a fire or overheat condition to cause the flight deck indications. If one detection loop fails, the card sends data about the failure to the AIMS, and a status message shows. The card changes to single-loop operation, if necessary. If both detection loops fail, an advisory message and status messages are displayed, and the fire detection system does not operate.

System Test

Built-in test equipment (BITE) performs a test of the engine fire detection system for these conditions:

- When the system first gets power.
- After a power interrupt.
- Every 5 minutes of operation. *[Figures 9-20 and 9-21]*

Boeing 777 Fire Extinguisher System
Fire Extinguisher Containers

The B777 aircraft has two fire extinguishing bottles that contain Halon fire extinguishing agent pressurized with nitrogen. The engine fire switches in the flight deck are pulled and rotated to release the Halon. Halon from each bottle can be discharged to the right or left engine. Engine indicating and crew alerting system (EICAS) messages, status messages,

and indicator lights show when the bottle pressure is low. The two engine fire extinguishing bottles are located behind the right sidewall lining of the forward cargo compartment, aft of the cargo door. *[Figure 9-22]*

The two engine fire extinguishing bottles are identical. Each bottle has these following components:

- A safety relief and fill port,
- A handle for removal and installation,
- A pressure switch,
- Two discharge assemblies,
- An identification plate, and
- Four mounting lugs. *[Figures 9-23, 9-24, 9-25]*

The bottles contain Halon fire extinguishing agent pressurized with nitrogen. If the pressure in the bottle becomes too high, the safety relief and fill port opens so that the bottle does not explode. The discharge assembly has an explosive squib. An electric current from the fire extinguishing circuit fires the squib. This releases the Halon through the discharge port. The pressure switch gives flight deck indications when bottle pressure decreases. The switch monitors the pressure inside the bottle and is normally open. When the pressure decreases because of a leak or bottle discharge, the switch closes an indicating circuit.

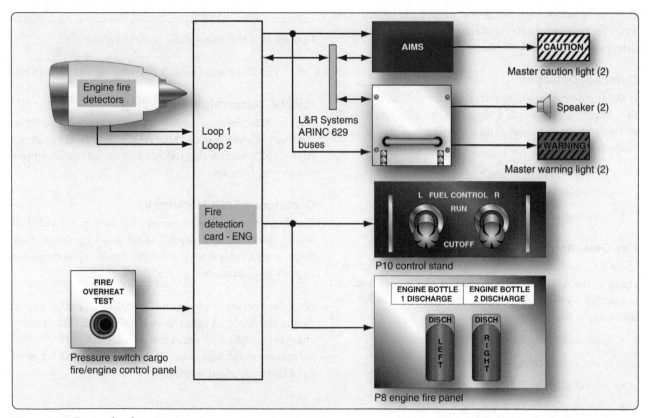

Figure 9-20. *Engine fire detection system.*

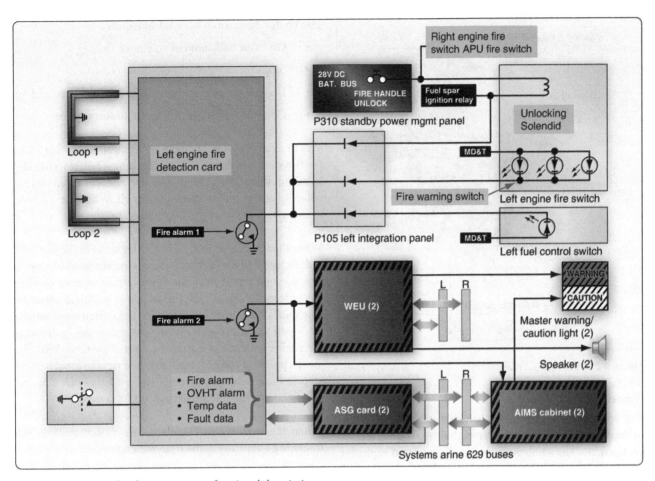

Figure 9-21. *Engine fire detection system functional description.*

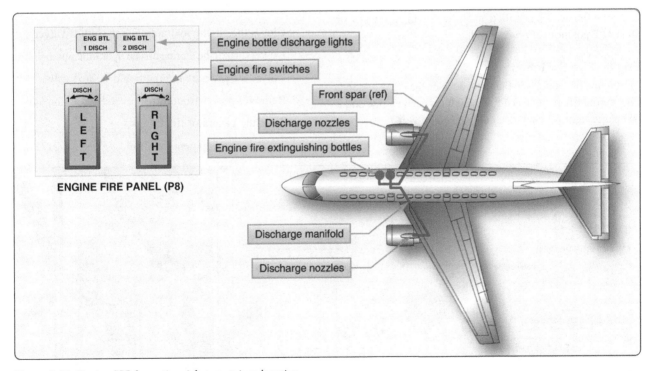

Figure 9-22. *Boeing 777 fire extinguisher container location.*

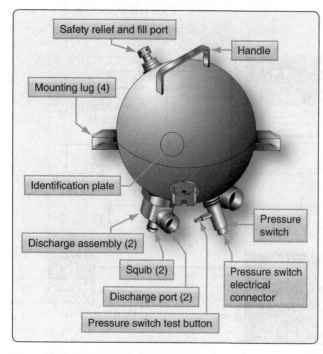

Figure 9-23. *Fire extinguishing bottle.*

Squib

The squib is installed in the discharge assembly at the bottom of the fire container. A fire container has two squibs, one for each engine. The squib is an electrically operated explosive device. When the squib is activated, it fires a slug through the breakable disc, and nitrogen pressure inside the bottle pushes the Halon through the discharge port. The squib fires when the fire switch is pulled and turned to the DISCH 1 or DISCH 2 position. *[Figure 9-22]*

Engine Fire Switches

The engine fire panel is in the flight deck on the P8 aisle stand. The engine fire panel has a fire switch for each engine and a discharge light for each fire bottle. *[Figure 9-26]*

The engine fire switch has four functions:

- Gives an indication of an engine fire,
- Stops the engine,
- Isolates the engine from the aircraft systems, and
- Controls the engine fire extinguishing system.

The fire switch assembly incorporates a solenoid that locks the fire switch so that the flight crew cannot pull it accidently. If an engine has a fire, the fire warning light comes on and the solenoid energizes to release the switch. When the solenoid is energized, the fire switch can be pulled.

When the fire detection system malfunctions or the solenoid is defective, and the flight crew wants to extinguish an engine fire, someone must push the fire override switch. The fire override switch allows the fire switch to be pulled when the solenoid is not energized. When the fire switch is pulled, the push-pull switch contacts operate electrical circuits that stop the engine and isolate it from the aircraft systems. With the switch pulled, it can be rotated to left or right to a mechanical stop at the discharge position. The rotary switch contacts close and operate the fire extinguishing system.

When the fire switch is pulled, the switch isolates the following aircraft systems from the engine:

- Closes the fuel spar valve.
- Deenergizes the engine fuel metering unit (FMU) cutoff solenoid.
- Closes the engine hydraulic pump shutoff valve.
- Depressurizes the engine driven hydraulic pump valve.
- Closes the pressure regulator and shutoff valve.
- Removes power from thrust reverser isolation valve.
- Trips the generator field.
- Trips the backup generator field. *[Figure 9-27]*

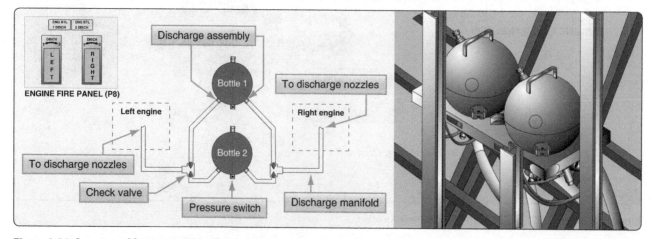

Figure 9-24. *Location of fire extinguishing bottles.*

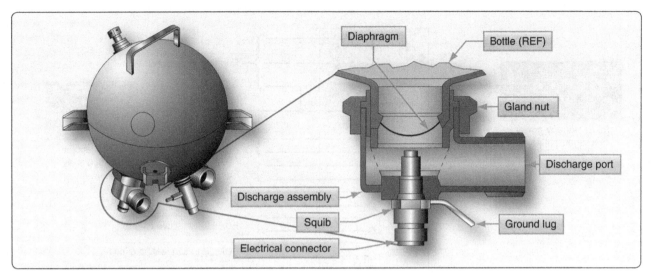

Figure 9-25. *Squib or cartridge.*

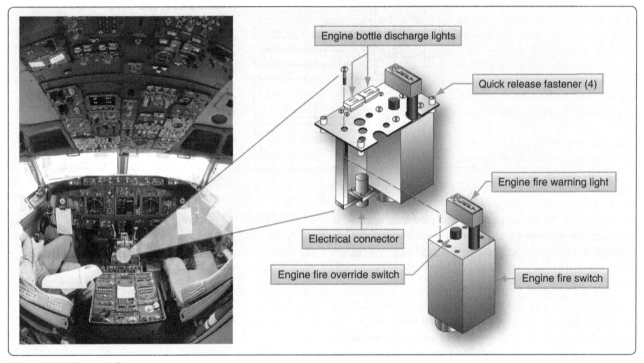

Figure 9-26. *Fire switch.*

Engine Fire Operation

If an engine has a fire, the engine fire detection system gives a fire warning in the flight deck. The engine fire warning lights come on to identify which fire switch to use to extinguish the fire. The solenoid in the fire switch energizes and releases the switch so that the fire switch can be pulled. If the solenoid does not energize, push the fire override switch to release the fire switch manually. When the fire switch is pulled, it stops the engine, and the fire switch isolates the engine from the aircraft systems.

If the fire warnings do not go away when the switch is pulled, position the switch to the DISCH 1 or DISCH 2 position, and hold the switch against the stop for one second. This fires the squib in the fire extinguisher container and releases the fire extinguishing agent into the engine nacelle. Ensure that the engine bottle discharge light comes on. If the first bottle does not extinguish the fire, the switch must be placed to the other DISCH position. This fires the squib for the other bottle.

APU Fire Detection & Extinguishing System

The APU fire protection system is similar in design to engine fire protection systems, but there are some differences. The APU is often operated with no personnel in the flight

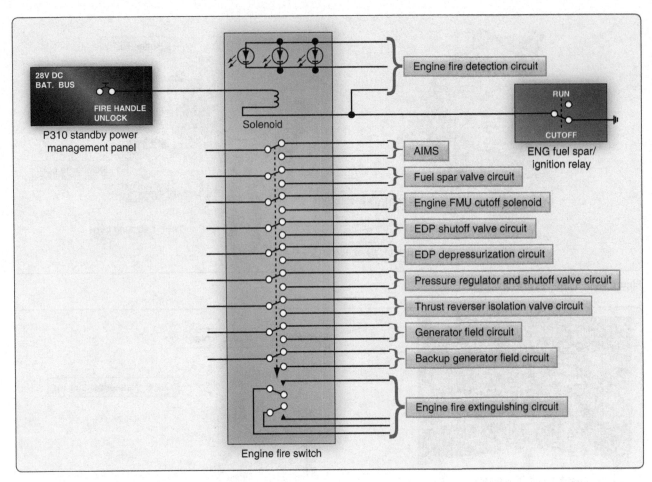

Figure 9-27. *Engine fire switch circuit.*

deck and; the APU fire protection system can operate in an unattended mode on the ground with the engines not running. If there is an APU fire in the unattended mode, the fire extinguisher discharges automatically. The APU operates in the attended mode when at least one engine is running. If there is an APU fire in this mode, the crew discharges the bottle manually. Fire switches are located on the cargo fire/ engine control panel and the service and APU shutdown panel located outside the aircraft on the nose landing gear. *[Figure 9-28]*

APU Fire Warning

If there is an APU fire, the APU fire detection system gives fire warnings and automatically stops the APU. The APU fire warning light comes on to identify the correct fire switch to use to extinguish the fire. The fire switch solenoid releases the switch so that it can be pulled up. If the APU is running, it stops when the fire switch is pulled. The fire switch isolates the APU from the aircraft systems.

Fire Bottle Discharge

If the fire warnings do not go away with the switch out, put the switch to the left or right DISCH position. Hold the

switch against the discharge stop for one second. This fires the bottle squib and releases the fire extinguishing agent into the APU compartment. Verify that the APU bottle discharge light comes on. *[Figure 9-29]*

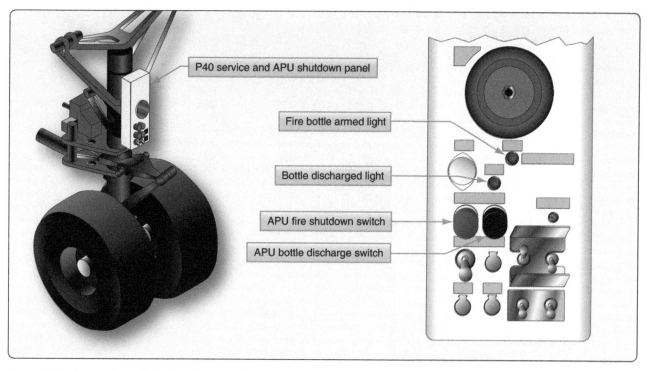

Figure 9-28. *P40 service and APU shutdown panel.*

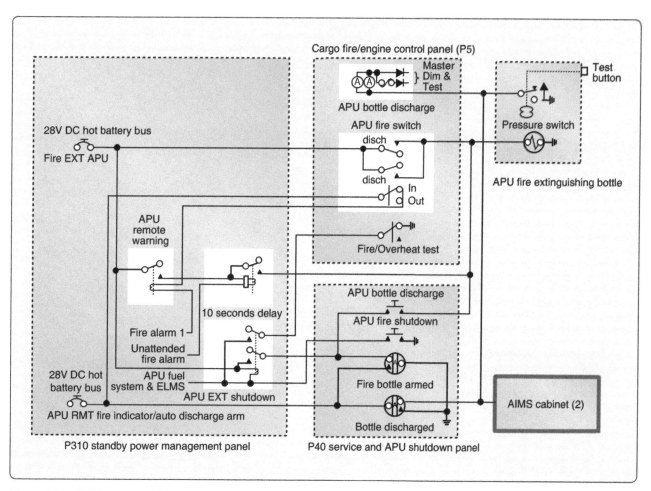

Figure 9-29. *APU fire extinguishing circuit.*

Chapter 10
Engine Maintenance & Operation

Reciprocating Engine Overhaul

Both maintenance and complete engine overhauls are performed normally at specified intervals. This interval is usually governed by the number of hours the powerplant has been in operation. The actual overhaul period for a specific engine is generally determined by the manufacturer's recommendations. Each engine manufacturer sets a total time in service when the engine should be removed from service and overhauled. Depending upon how the engine is used in service, the overhaul time can be mandatory. The overhaul time is listed in hours and is referred to as time before overhaul (TBO). For example, if an engine had a life of 2,000 hours and had operated 500 hours, it would have a TBO of 1,500 hours. Tests and experience have shown that operation beyond this period of time could result in certain parts being worn beyond their safe limits. For an overhauled engine to be as airworthy as a new one, worn parts, as well as damaged parts, must be detected and replaced during overhaul. The only way to detect all unairworthy parts is to perform a thorough and complete overhaul process while the engine is disassembled. The major purpose of overhaul is to inspect, repair, and replace worn engine parts.

A complete overhaul process includes the following ten steps: receiving inspection; disassembly; visual inspection; cleaning; structural inspection; non-destructive testing (NDT) inspection; dimensional inspection; repair and replacement; reassembly; and testing and break in. The inspection phases are the most precise and the most important phases of the overhaul. Inspection cannot be slighted or performed in a careless or incomplete manner. It is always recommended that complete records be made of the inspection process and kept with the engine records.

Each engine manufacturer provides very specific tolerances to which the engine parts must conform and provides general instructions to aid in determining the airworthiness of the part. However, in many cases, the final determination must be made by the technician. Although the determination must be made if the part is serviceable, repairable, or should be rejected, the technician should follow the manufacturer's manuals and information. When dimensional tolerances are concerned, the manufacturer publishes a new minimum and serviceable dimension for all critical component parts. Knowledge of the operating principles, strength, and stresses applied to a part is essential in making decisions regarding visible wear. When the powerplant technician signs the release for the return to service for an overhauled engine, this certifies that the complete overhaul process has been performed using methods, techniques, and practices acceptable to the Federal Aviation Administration (FAA) Administrator.

Top Overhaul

Reciprocating piston aircraft engines can be repaired by a top overhaul. This means an overhaul of those parts on top of the crankcase, without completely dismantling the engine. It includes removal of the units (i.e., exhaust collectors, ignition harness, intake pipes) necessary to remove the cylinders. The actual top overhaul consists of reconditioning the engine's cylinders by replacing or reconditioning the piston and piston rings, and reconditioning or plating the cylinder wall and valve-operating mechanism, including valve guides if needed. A top overhaul is a little misleading, because it is really an engine repair procedure and not a real overhaul as described earlier. Usually at this time, the accessories require no attention other than that normally required during ordinary maintenance functions. This repair is generally due to valves or piston rings wearing prematurely. Many stress that if an engine requires this much dismantling, it should be completely disassembled and receive a major overhaul.

Major Overhaul & Major Repairs

Major overhaul consists of the complete reconditioning of the powerplant. A reciprocating engine would require that the crankcase be disassembled per the FAA; a major overhaul is not generally a major repair. A certified powerplant-rated technician can perform or supervise a major overhaul of an engine if it is not equipped with an internal supercharger or has a propeller reduction system other than spur-type gears. At regular intervals, an engine should be completely dismantled, thoroughly cleaned, and inspected. Each part should be overhauled in accordance with the manufacturer's instructions and tolerances for the engine involved. At this time all accessories are removed, overhauled, and tested. Again, instructions from the manufacturer of the accessory concerned should be followed.

General Overhaul Procedures

Because of the continued changes and the many different types of engines in use, it is not possible to treat the specific overhaul of each engine in this text. However, there are various overhaul practices and instructions of a nonspecific nature that apply to all makes and models of engines.

Any engine to be overhauled completely should receive a runout check of its crankshaft or propeller shaft as a first step. Any question concerning crankshaft or propeller shaft replacement is resolved at this time, since a shaft whose runout is beyond limits must be replaced.

Throughout the life of a product (whether type-certificated or not), manufacturing defects, changes in service, or design improvements often occur. When that happens, the OEM frequently uses a safety bulletin (SB) to distribute the information to the operator of the aircraft. SBs are good information and should be strongly considered by the owner for implementation to the aircraft. However, SBs are not required unless they are referred to in an airworthiness directive (AD) note or if compliance is required as a part of the authorized inspection program. Refer to 14 CFR part 39, section 39.27.

Receiving Inspection

The receiving inspection consists of determining the general condition of the total engine as received, along with an inventory of the engine's components. The accessory information should be recorded, such as model and serial numbers, and the accessories should be sent to overhaul if needed. The overhaul records should be organized, and the appropriate manuals obtained and reviewed along with a review of the engine's history (log books). The engine's service bulletins, airworthiness directives, and type certificate compliance should be checked. The exterior of the engine should be cleaned after mounting it on an overhaul stand. *[Figure 10-1]*

Disassembly

As visual inspection immediately follows disassembly, all individual parts should be laid out in an orderly manner on a workbench as they are removed. To guard against damage and to prevent loss, suitable containers should be available in which to place small parts (nuts, bolts, etc.) during the disassembly operation.

Other practices to observe during disassembly include:

1. Drain the engine oil sumps and remove the oil filter. Drain the oil into a suitable container; strain it through a clean cloth. Check the oil and the cloth for metal particles.

2. Dispose of all safety devices (safety wire, cotter pins, etc.) as they are removed. Never use them a second time. Always replace with new safety devices.

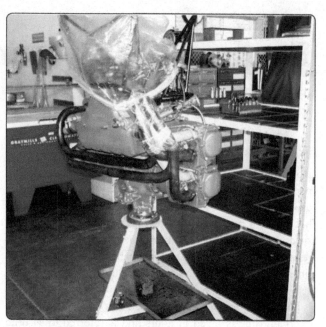

Figure 10-1. *Engine mounted on an overhaul stand.*

3. All loose studs, and loose or damaged fittings, should be carefully tagged to prevent being overlooked during inspection.

4. Always use the proper tool for the job. Use sockets and box end wrenches wherever possible. If special tools are required, use them rather than improvising.

Inspection Process

The inspection of engine parts during overhaul is divided into three categories:

1. Visual.

2. Structural.

3. Dimensional.

Many defects on the engine components can be detected visually, and a determination of airworthiness can be made at this time. If, by visual inspection, the component is determined to be unairworthy, the part is rejected, and no further inspection or repair is required. Structural failures can be determined by several different methods. Magnetic parts can readily be examined by the magnetic particle method. Other methods, such as dye penetrate, eddy current, ultra sound, and X-ray, can also be used. The first two methods are aimed at determining structural failures in the parts, while the last method deals with the size and shape of each part. By using very accurate measuring equipment, each engine component can be dimensionally evaluated and compared to service limits and standards (tolerances) set by the manufacturer.

Visual Inspection

Visual inspection should precede all other inspection procedures. Parts should not be cleaned before a preliminary visual inspection, since indications of a failure may often be detected from the residual deposits of metallic particles in some recesses in the engine.

Several terms are used to describe defects detected in engine parts during inspection. Some of the more common terms and definitions are:

1. Abrasion—an area of roughened scratches or marks usually caused by foreign matter between moving parts or surfaces.

2. Brinelling—one or more indentations on bearing races, usually caused by high static loads or application of force during installation or removal. Indentations are rounded or spherical due to the impression left by the contacting balls or rollers of the bearing.

3. Burning—surface damage due to excessive heat. It is usually caused by improper fit, defective lubrication, or over-temperature operation.

4. Burnishing—polishing of one surface by sliding contact with a smooth, harder surface. Usually no displacement nor removal of metal.

5. Burr—a sharp or roughened projection of metal usually resulting from machine processing.

6. Chafing—a condition caused by a rubbing action between two parts under light pressure that results in wear.

7. Chipping—breaking away of pieces of material, that is usually caused by excessive stress concentration or careless handling.

8. Corrosion—loss of metal by a chemical or electrochemical action. The corrosion products are easily removed by mechanical means. Iron rust is an example of corrosion.

9. Crack—a partial separation of material usually caused by vibration, overloading, internal stresses, defective assembly, or fatigue. Depth may be a few thousandths, to the full thickness of the piece.

10. Cut—loss of metal, usually to an appreciable depth over a relatively long and narrow area, by mechanical means, as would occur with the use of a saw blade, chisel, or sharp-edged stone striking a glancing blow.

11. Dent—a small, rounded depression in a surface usually caused by the part being struck with a rounded object.

12. Erosion—loss of metal from the surface by mechanical action of foreign objects, such as grit or fine sand. The eroded area is rough and may be lined in the direction that the foreign material moved relative to the surface.

13. Flaking—the breaking loose of small pieces of metal or coated surfaces, that is usually caused by defective plating or excessive loading.

14. Fretting—a condition of surface erosion caused by minute movement between two parts usually clamped together with considerable unit pressure.

15. Galling—a severe condition of chafing or fretting in which a transfer of metal from one part to another occurs. It is usually caused by a slight movement of mated parts having limited relative motion and under high loads.

16. Gouging—a furrowing condition in which a displacement of metal has occurred (a torn effect). It is usually caused by a piece of metal, or foreign material, between close moving parts.

17. Grooving—a recess, or channel, with rounded and smooth edges usually caused by faulty alignment of parts.

18. Inclusion—presence of foreign or extraneous material entirely within a portion of metal. Such material is introduced during the manufacture of rod, bar, or tubing by rolling or forging.

19. Nick—a sharp-sided gouge or depression with a V-shaped bottom, that is generally the result of careless handling of tools and parts.

20. Peening—a series of blunt depressions in a surface.

21. Pick up or scuffing—a buildup or rolling of metal from one area to another, that is usually caused by insufficient lubrication, clearances, or foreign matter.

22. Pitting—small hollows of irregular shape in the surface, usually caused by corrosion or minute mechanical chipping of surfaces.

23. Scoring—a series of deep scratches caused by foreign particles between moving parts or careless assembly or disassembly techniques.

24. Scratches—shallow, thin lines or marks, varying in degree of depth and width, caused by presence of fine foreign particles during operation or contact with other parts during handling.

25. Stain—a change in color, locally, causing a noticeably different appearance from the surrounding area.

26. Upsetting—a displacement of material beyond the normal contour or surface (a local bulge or bump). Usually indicates no metal loss.

Examine all gears for evidence of pitting or excessive wear.

These conditions are of particular importance when they occur on the teeth; deep pit marks in this area are sufficient cause to reject the gear. Bearing surfaces of all gears should be free from deep scratches. However, minor abrasions usually can be dressed out with a fine abrasive cloth.

All bearing surfaces should be examined for scores, galling, and wear. Considerable scratching and light scoring of aluminum bearing surfaces in the engine do no harm and should not be considered a reason for rejecting the part, provided it falls within the clearances set forth in the table of limits in the engine manufacturer's overhaul manual. Even though the part comes within the specific clearance limits, it is not satisfactory for re-assembly in the engine unless inspection shows the part to be free from other serious defects.

Ball bearings should be inspected visually and by feel for roughness, flat spots on balls, flaking or pitting of races, or scoring on the outside of races. All journals should be checked for galling, scores, misalignment, or out-of-round condition. Shafts, pins, etc., should be checked for straightness. This may be done, in most cases, by using V-blocks and a dial indicator.

Pitted surfaces in highly stressed areas, resulting from corrosion, can cause ultimate failure of the part. The following areas should be examined carefully for evidence of such corrosion:

1. Interior surfaces of piston pins.

2. The fillets at the edges of crankshaft main and crankpin journal surfaces.

3. Thrust bearing races.

If pitting exists on any of the surfaces mentioned, to the extent that it cannot be removed by polishing with crocus cloth or other mild abrasive, the part usually must be rejected.

Parts, such as threaded fasteners or plugs, should be inspected to determine the condition of the threads. Badly worn or mutilated threads cannot be tolerated; the parts should be rejected. However, small defects, such as slight nicks or burrs, may be dressed out with a small file, fine abrasive cloth, or stone. If the part appears to be distorted, badly galled, mutilated by overtightening, or from the use of improper tools, replace it with a new one.

Cylinder Head

Inspect the cylinder head for internal and external cracks. Use a bright light to inspect for cracks and investigate any suspicious areas with a magnifying glass or microscope. Carbon deposits must be cleaned from the inside of the head, and paint must be removed from the outside for this inspection.

Exterior cracks show up on the head fins where they have been damaged by tools or contact with other parts because of careless handling. Cracks near the edge of the fins are not dangerous, if the portion of the fin is removed and contoured properly. Cracks at the base of the fin are a reason for rejecting the cylinder. Cracks may also occur on the rocker box or in the rocker bosses.

Interior cracks almost always radiate from the valve seat bosses or the spark plug bushing boss. These cracks are usually caused by improper installation of the seats or bushings. They may extend completely from one boss to the other.

Inspect all the studs on the cylinder head for looseness, straightness, damaged threads, and proper length. Slightly damaged threads may be chased with the proper die. The length of the stud should be correct within $\pm\frac{1}{32}$ (0.03125) inch to allow for proper installation of safety devices.

Be sure the valve guides are clean before inspection. Often, carbon covers pits inside the guide. If a guide in this condition is put back in service, carbon again collects in the pits and valve sticking results. Besides pits, scores, and burned areas inside the valve guide, inspect them for wear or looseness.

Inspection of valve seat inserts before they are re-faced is mostly a matter of determining if there is enough of the seat left to correct any pitting, burning, scoring, or out-of-trueness.

Inspect the rocker shaft bosses for scoring, cracks, oversize, or out-of-roundness. Scoring is generally caused by the rocker shaft turning in the bosses, which means either the shaft was too loose in the bosses or a rocker arm was too tight on the shaft. Out-of-roundness is usually caused by a stuck valve. If a valve sticks, the rocker shaft tends to work up and down when the valve offers excessive resistance to opening. Inspect for out-of-roundness and oversize using a telescopic gauge and a micrometer.

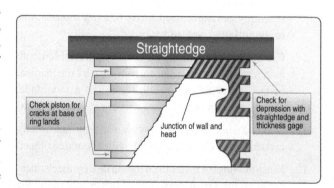

Figure 10-2. *Checking a piston head for flatness.*

Piston, Valve Train, & Piston Pin

When applicable, check for flatness of the piston head using a straightedge and thickness gauge. *[Figure 10-2]* If a depression is found, check for cracks on the inside of the piston. A depression in the top of the piston usually means that detonation has occurred within the cylinder.

Inspect the exterior of the piston for scores and scratches. Scores on the top ring land are not cause for rejection, unless they are excessively deep. Deep scores on the side of the piston are usually a reason for rejection.
Examine the piston for cracked skirts, broken ring lands, and scored piston-pin holes. Do not mistake casting marks or laps for a crack. During major overhaul, most pistons are generally replaced, as it requires more labor to clean and inspect the piston than it costs to replace it.

Crankshaft & Connecting Rods

Carefully inspect all surfaces of the crankshaft for cracks. Check the bearing surfaces for evidence of galling, scoring, or other damage. When a shaft is equipped with oil transfer tubes, check them for tightness.

Visual inspection of connecting rods should be done with the aid of a magnifying glass or bench microscope. A rod that is obviously bent or twisted should be rejected without further inspection. Inspect all surfaces of the connecting rods for cracks, corrosion, pitting, galling, or other damage. Galling is caused by a slight amount of movement between the surfaces of the bearing insert and the connecting rod during periods of high loading, such as that produced during over-speed or excessive manifold pressure operation. The visual evidence produced by galling appears as if particles from one contacting surface had welded to the other. Evidence of any galling is sufficient reason for rejecting the complete rod assembly. Galling is a distortion in the metal and is comparable to corrosion in the manner in which it weakens the metallic structure of the connecting rod.

Cleaning

After visually inspecting engine recesses for deposits of metal particles, it is important to clean all engine parts thoroughly to facilitate further inspection. Two processes for cleaning engine parts are:

1. Degreasing to remove dirt and sludge (soft carbon).

2. The removal of hard carbon deposits by decarbonizing, brushing or scraping, and grit-blasting.

Degreasing

Degreasing can be done by immersing or spraying the part in a suitable commercial solvent. *[Figure 10-3]* Extreme care must be used if any water-mixed degreasing solutions containing caustic compounds or soap are used. Such compounds, in addition to being potentially corrosive to aluminum and magnesium, may become impregnated in the pores of the metal and cause oil foaming when the engine is returned to service. Therefore, when using water-mixed solutions, it is imperative that the parts be rinsed thoroughly and completely in clear boiling water after degreasing. Regardless of the method and type of solution used, coat or spray all parts with lubricating oil immediately after cleaning to prevent corrosion.

Removing Hard Carbon

While the degreasing solution removes dirt, grease, and soft carbon, deposits of hard carbon almost invariably remain on many interior surfaces. To remove these deposits, they must first be loosened by immersion in a tank containing a decarbonizing solution (usually heated). A great variety of commercial decarbonizing agents are available. Decarbonizers, like the degreasing solutions previously mentioned, fall generally into two categories, water-soluble and hydrocarbons. The same caution concerning the use of water-soluble degreasers is applicable to water-soluble decarbonizers.

Caution: When using a decarbonizing solution on magnesium castings, avoid immersing steel and magnesium parts in the same decarbonizing tank, as this practice often results in damage to the magnesium parts from corrosion.

Decarbonizing will usually loosen most of the hard carbon deposits remaining after degreasing. However, the complete removal of all hard carbon generally requires brushing, scraping, or grit-blasting. In all of these operations, be careful to avoid damaging the machined surfaces. In particular, wire brushes and metal scrapers must never be used on any bearing or contact surface.

Follow the manufacturer's recommendations when grit-

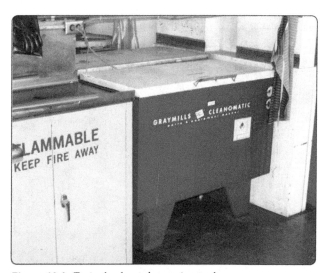

Figure 10-3. *Typical solvent degreasing tank.*

blasting parts for the abrasive material being used. Sand, rice, baked wheat, plastic pellets, glass beads, or crushed walnut shells are examples of abrasive substances that are used for grit-blasting parts. A grit-blasting machine is shown in *Figure 10-4*.

All machined surfaces must be masked properly and adequately, and all openings tightly plugged before blasting. The one exception to this is the valve seats, which may be left unprotected when blasting the cylinder head combustion chamber. It is often advantageous to grit-blast the seats, since this will cut the glaze which tends to form (particularly on the exhaust valve seat), thus facilitating subsequent valve seat reconditioning. Piston ring grooves may be grit-blasted if necessary; however, extreme caution must be used to avoid the removal of metal from the bottom and sides of the grooves. When grit-blasting housings, plug all drilled oil passages with rubber plugs or other suitable material to prevent the entrance of foreign matter.

The decarbonizing solution will generally remove most of the enamel on exterior surfaces. All remaining enamel should be removed by grit-blasting, particularly in the crevices between cylinder cooling fins.

At the conclusion of cleaning operations, rinse the part in petroleum solvent, dry and remove any loose particles of carbon or other foreign matter by air-blasting, and apply a liberal coating of preservative oil to all surfaces.

Magnesium parts should be cleaned thoroughly with a dichromate treatment prior to painting. This treatment consists of cleaning all traces of grease and oil from the part by using a neutral, noncorrosive degreasing medium followed by a rinse, after which the part is immersed for at least 45 minutes in a hot dichromate solution (three-fourths of a pound of sodium dichromate to 1 gallon of water at 180 °F

Figure 10-4. *Grit-blasting machine.*

to 200 °F). Then the part should be washed thoroughly in cold running water, dipped in hot water, and dried in an air blast. Immediately thereafter, the part should be painted with a prime coat and engine enamel in the same manner as that suggested for aluminum parts.

Some older engines used sludge chambers in the crankshafts, which were manufactured with hollow crankpins that serve as sludge removers. The sludge chambers require inspection and cleaning at overhaul. Sludge chambers are formed by means of spool-shaped tubes pressed into the hollow crankpins, or by plugs pressed into each end of the crankpin. If an engine has a sludge chamber or tubes, they must be removed for cleaning at overhaul. If these are not removed, accumulated sludge loosened during cleaning may clog the crankshaft oil passages and cause subsequent bearing failures. If the sludge chambers are formed by means of tubes pressed into the hollow crankpins, make certain they are re-installed correctly to avoid covering the ends of the oil passages. Due to improved oils, sludge chambers are no longer used with modern engines.

Structural Inspection

One of the best methods to double check your visual inspection findings is to supplement them with one of the forms of nondestructive testing, such as magnetic particle, dye penetrate, eddy current, ultrasound, and x-ray inspections. Defects in nonmagnetic parts (aluminum parts) can be found by all these methods except for magnetic particle inspection, which is used for magnetic or ferrous materials (steel).

Dye Penetrant Inspection

Dye penetrant inspection is a nondestructive test for defects open to the surface in parts made of any nonporous material. It is used with equal success on such metals as aluminum, magnesium, brass, copper, cast iron, stainless steel, and titanium. Dye penetrant inspection uses a penetrating liquid that enters a surface opening and remains there, making it clearly visible to the inspector. It calls for visual examination of the part after it has been processed, increasing the visibility of the defect so that it can be detected. Visibility of the penetrating material is increased by the addition of one of two types of dye: visible or fluorescent. When using a fluorescent dye, the inspection is accomplished using an ultraviolet (UV) light source (black light).

The steps for performing a dye penetrant inspection are:

1. Thorough cleaning of the metal surface.

2. Applying penetrant.

3. Removing penetrant with remover emulsifier or cleaner.

4. Drying the part.

5. Applying the developer.

6. Inspecting and interpreting results.

Eddy Current Inspection

Eddy currents are composed of free electrons under the influence of an induced electromagnetic field, that are made to drift through metal. Different meter readings are seen when the same metal is in different hardness states. Readings in the affected area are compared with identical materials in known unaffected areas for comparison. A difference in readings indicates a difference in the hardness state of the affected area. Eddy current inspection can frequently be performed without removing the surface coatings, such as primer, paint, and anodized films. It can be effective in detecting surface and subsurface corrosion, pots, and heat treat condition.

Ultrasonic Inspection

Ultrasonic detection equipment makes it possible to locate defects in all types of materials. There are three basic ultrasonic inspection methods:

1. Pulse-echo.

2. Through transmission.

3. Resonance.

Pulse-Echo

Flaws are detected by measuring the amplitude of signals reflected and the time required for these signals to travel between specific surfaces and the discontinuity.

Through Transmission

Through transmission inspection uses two transducers, one to generate the pulse and another placed on the opposite surface to receive it. A disruption in the sound path indicates a flaw and is displayed on the instrument screen. Through transmission is less sensitive to small defects than the pulse-echo method.

Resonance

This system differs from the pulse-echo method, in that the frequency of transmission may be continuously varied. The resonance method is principally used for thickness measurements when the two sides of the material being tested are smooth and parallel, and the backside is inaccessible. The point at which the frequency matches the resonance point of the material being tested is the thickness determining factor.

Magnetic Particle Inspection

Magnetic particle inspection is a method of detecting invisible cracks and other defects in ferromagnetic materials, such as iron and steel. It is not applicable to nonmagnetic materials.

The inspection process consists of magnetizing the part, and then applying ferromagnetic particles to the surface area to be inspected. The ferromagnetic particles (indicating medium) may be held in suspension in a liquid that is flushed over the part; the part may be immersed in the suspension liquid; or the particles, in dry powder form, may be dusted over the surface of the part. The wet process is more commonly used in the inspection of aircraft parts.

If a discontinuity is present, the magnetic lines of force are disturbed, and opposite poles exist on either side of the discontinuity. The magnetized particles form a pattern in the magnetic field between the opposite poles. This pattern, known as an indication, assumes the approximate shape of the surface projection of the discontinuity. A discontinuity may be defined as an interruption in the normal physical structure or configuration of a part.

X-ray

X-rays can penetrate material and disclose discontinuities through the metal or non-metal components, making it an excellent inspection process when needed to determine the structural integrity of an engine component. The penetrating radiation is projected through the part to be inspected and produces an invisible or latent image in the film. When processed, the film becomes a radiograph, or shadow picture, of the object. This inspection medium, as a portable unit, provides a fast and reliable means for checking the integrity of engine components.

Additional and more thorough information on NDT inspection is covered in detail in the Aviation Maintenance Technician Handbook - General (FAA-H-8083-30, as amended).

Dimensional Inspection

The dimensional inspection is used to assure that the engine's component parts and clearances meet the manufacturer's specifications. These specs are listed in a table of limits, which lists serviceable limits and the manufacturer's new part maximum and minimum dimensions. Many measuring tools are used to perform the dimensional inspection of the engine. Some examples of these devices are discussed as the procedure for measuring the engine's components for dimensional inspection is explained in the following paragraphs.

Cylinder Barrel

Inspect the cylinder barrel for wear, using a cylinder bore gauge *[Figure 10-5]*, a telescopic gauge, and micrometer or an inside micrometer. Dimensional inspection of the barrel consists of the following measurements:

1. Maximum taper of cylinder walls.

2. Maximum out-of-roundness.

Figure 10-5. *A cylinder bore gauge.*

3. Bore diameter.
4. Step.
5. Fit between piston skirt and cylinder.

All measurements involving cylinder barrel diameters must be taken at a minimum of two positions 90° apart in the particular plane being measured. It may be necessary to take more than two measurements to determine the maximum wear.

Taper of the cylinder walls is the difference between the diameter of the cylinder barrel at the bottom and the diameter at the top. The cylinder is usually worn larger at the top than at the bottom. This taper is caused by the natural wear pattern. At the top of the stroke, the piston is subjected to greater heat and pressure and more erosive environment than at the bottom of the stroke. Also, there is greater freedom of movement at the top of the stroke. Under these conditions, the piston wears the cylinder wall more at the top of the cylinder. In most cases, the taper ends with a ridge, that must be removed during overhaul. *[Figure 10-6]*

Where cylinders are built with an intentional choke, measurement of taper becomes more complicated. Cylinder choke is where the top of the cylinder has been made with the very top diameter of the cylinder smaller, to compensate for wear and expansion during operation. It is necessary to know exactly how the size indicates wear or taper. Taper can be measured in any cylinder by a cylinder dial gauge as long as there is not a sharp step. The dial gauge tends to ride up on the step and causes inaccurate readings at the top of the cylinder.

The measurement for out-of-roundness is usually taken at the top of the cylinder. However, a reading should also be taken at the skirt of the cylinder to detect dents or bends caused by careless handling. A step, or ridge, is formed in the cylinder by the wearing action of the piston rings. *[Figure 10-6]* The greatest wear is at the top of the ring travel limit. The ridge that results is likely to cause damage to the rings or piston. If the step exceeds tolerances, it should be removed by grinding the cylinder oversize, or it should be blended by hand-stoning to break the sharp edge.

A step also may be found where the bottom ring reaches the lowest travel. This step is rarely found to be excessive, but it should be checked.

Inspect the cylinder walls for rust, pitting, or scores. Mild damage of this sort can be removed when the cylinders are deglazed. With more extensive damage, the cylinder has to be reground or honed. If the damage is too deep to be removed by either of these methods, the cylinder usually will have to be rejected. Most engine manufacturers, or engine overhaul repair stations, have an exchange service on cylinders with damaged barrels.

Check the cylinder flange for warpage by placing the cylinder on a suitable jig. Check to see that the flange contacts the jig all the way around. The amount of warpage can be checked by using a thickness gauge. A cylinder whose flange is warped beyond the limits should be rejected.

Rocker Arms & Shafts

Inspect the valve rockers for cracks and worn, pitted, or scored tips. See that all oil passages are free from obstructions.

Inspect the shaft's diameter for correct size with a micrometer. Rocker shafts are often found to be scored and burned because of excessive turning in the cylinder head. Also, there may be some pickup on the shaft (bronze from the rocker bushing transferred to the steel shaft). Generally, this is caused by overheating and too little clearance between shaft and bushing. The clearance between the shaft and the bushing is most important.

Inspect the rocker arm bushing for correct size. Check for proper clearance between the shaft and the bushing. Very often the bushings are scored because of mishandling during disassembly. Check to see that the oil holes line up. .At least 50% of the hole in the bushing should align with the hole in the rocker arm.

On engines that use a bearing, rather than a bushing, inspect the bearing to make certain it has not been turning in the rocker arm boss. Also inspect the bearing to determine its serviceability.

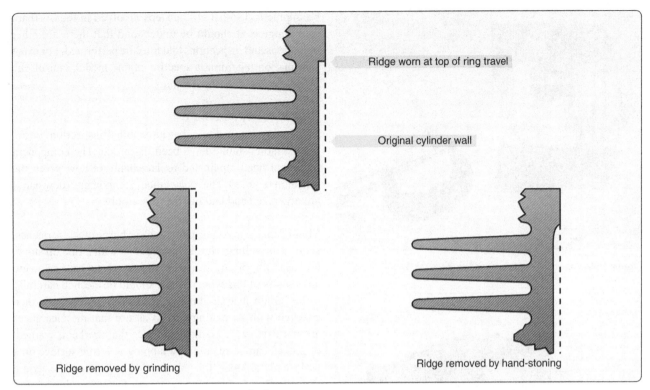

Figure 10-6. *Ridge or step formed in an engine cylinder.*

Crankshaft

Use extreme care in inspecting and checking the crankshaft for straightness. Place the crankshaft in V-blocks, supported at the locations specified in the applicable engine overhaul manual as in *Figure 10-7*. Using a surface plate and a dial indicator, measure the shaft runout. If the total indicator reading exceeds the dimensions given in the manufacturer's table of limits, the shaft must not be re-used. A bent crankshaft should not be straightened. Any attempt to do so results in rupture of the nitrided surface of the bearing journals, a condition that causes eventual failure of the crankshaft. Measure the outside diameter of the crankshaft main and rod-bearing journals using a micrometer. *[Figure 10-8]* Internal measurements can be made by using telescoping gauges, and then measuring the telescoping gauge with a micrometer. *[Figure 10-9]* Compare the resulting measurements with those in the table of limits.

Checking Alignment

Check bushings that have been replaced to determine if the bushing and rod bores are square and parallel to each other. The alignment of a connecting rod can be checked several ways. One method requires a push fit arbor for each end of the connecting rod, a surface plate, and two parallel blocks of equal height.

To measure rod squareness, or twist, insert the arbors into the rod bores. *[Figure 10-10]* Place the parallel blocks on

Figure 10-7. *Checking crankshaft runout.*

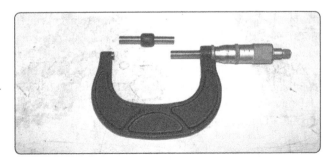

Figure 10-8. *A micrometer.*

Figure 10-9. *Telescoping gauges and micrometer combination.*

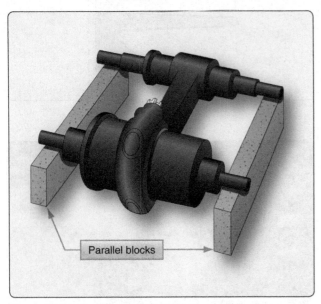

Parallel blocks

Figure 10-10. *Checking connecting rod squareness.*

a surface plate. Place the ends of the arbors on the parallel blocks. Using a thickness gauge, check the clearance at the points where the arbors rest on the blocks. This clearance, divided by the separation of the blocks in inches, gives the twist per inch of length.

To determine bushing or bearing parallelism (convergence), insert the arbors in the rod bores. Measure the distance between the arbors on each side of the connecting rod at points that are equidistant from the rod centerline. For exact parallelism, the distances checked on both sides should be the same. Consult the manufacturer's table of limits for the amount of misalignment permitted.

The preceding operations are typical of those used for most reciprocating engines and are included to introduce some of the operations involved in engine overhaul. It would be impractical to list all the steps involved in the overhaul of an engine. It should be understood that there are other operations and inspections that must be performed. For exact information regarding a specific engine model, consult the manufacturer's overhaul manual.

Repair & Replacement

The engine components that have failed inspection, or are unrepairable, should have been discarded. The component parts that need repair and replacement are now given the attention required. The replacement components (new parts) are organized and laid out for reassembly.

Minor damage to engine parts, such as burrs, nicks, scratches, scoring, or galling, should be removed with a fine oil stone, crocus cloth, or any similar abrasive substance. Following any repairs of this type, the part should be cleaned carefully to be certain that all abrasive has been removed, and then checked with its mating part to assure that the clearances are not excessive. Flanged surfaces that are bent, warped, or nicked can be repaired by lapping to a true surface on a surface plate. Again, the part should be cleaned to be certain that all abrasive has been removed. Defective threads can sometimes be repaired with a suitable die or tap. Small nicks can be satisfactorily removed with Swiss pattern files or small, edged stones. Pipe threads should not be tapped deeper to clean them, because this practice results in an oversized tapped hole. If galling or scratches are removed from a bearing surface of a journal, it should be buffed to a high polished finish.

In general, welding of highly-stressed engine parts can be accomplished only when approved by the manufacturer. However, welding may be accomplished using methods that are approved by the engine manufacturer, and if it can be reasonably expected that the welded repair will not adversely affect the airworthiness of the engine.

Many minor parts not subjected to high stresses may be safely repaired by welding. Mounting lugs, cowl lugs, cylinder fins, rocker box covers, and many parts originally fabricated by welding are in this category. The welded part should be suitably stress-relieved after welding. However, before welding any engine part, consult the manufacturer's instructions for the engine concerned to see if it is approved for repair by welding.

Parts requiring use of paint for protection or appearance should be repainted according to the engine manufacturer's recommendations. Aluminum alloy parts should have original, exterior painted surfaces rubbed smooth to provide a proper paint base. See that surfaces to be painted are thoroughly cleaned. Care must be taken to avoid painting mating surfaces. Exterior aluminum parts should be primed first with a thin coat of zinc chromate primer. After the primer is dry, parts should be painted with engine enamel, that should be air dried until hard, or baked for ½ hour at 82 °C (180 °F). Aluminum parts from which the paint has not been removed may be repainted without the use of a priming coat, provided no bare aluminum is exposed.

Any studs that are bent, broken, damaged, or loose must be replaced. After a stud has been removed, the tapped stud hole should be examined for size and condition of threads. If it is necessary to re-tap the stud hole, it also is necessary to use a suitable oversize stud. Studs that have been broken off flush with the case must be drilled and removed with suitable stud remover. Be careful not to damage any threads. When replacing studs, coat the coarse threads of the stud with an anti-seize compound.

Cylinder Assembly Reconditioning

Cylinder and piston assemblies are inspected according to the procedures contained in the engine manufacturer's manuals, charts, and service bulletins. A general procedure for inspecting and reconditioning cylinders is discussed in the following section to provide an understanding of the operations involved.

Visually inspect the head fins for other damage besides cracks. Dents or bends in the fins should be left alone unless there is danger of cracking. Where pieces of fin are missing, the sharp edges should be filed to a smooth contour. Fin breakage in a concentrated area causes dangerous local hot spots. Fin breakage near the spark plug bushings or on the exhaust side of the cylinder is obviously more dangerous than in other areas. When removing or re-profiling a cylinder fin, follow the instructions and the limits in the manufacturer's manual.

Inspect spark plug inserts for the condition of the threads and for looseness. Run a tap of the proper size through the bushing. Very often, the inside threads of the bushing are burned. If more than one thread is missing, the bushing should be rejected. Tighten a plug in the bushing to check for looseness.

Piston & Piston Pins

If the old piston is to be reused, or a new piston is to be used, measure the outside of the piston by means of a micrometer. Measurements must be taken in several directions and on the skirt, as well as on the lands section. Check these sizes against the cylinder size. Most engines use cam ground pistons to compensate for the greater expansion parallel to the pin during engine operation. The diameter of these pistons measures several thousandths of an inch larger at an angle to the piston pin hole, than parallel to the pin hole. Inspect the ring grooves for evidence of wear. The groove needs to be checked for side clearance with a feeler gauge to determine the amount of wear in the grooves. Examine the piston pin for scoring, cracks, excessive wear, and pitting. Check the clearance between the piston pin and the bore of the piston pin bosses using a telescopic gauge and a micrometer. Use the magnetic particle method to inspect the pin for cracks. Since the pins are often case hardened, cracks show up inside the pin more often than they do on the outside. Check the pin for bends using V-blocks and a dial indicator on a surface plate. [Figure 10-11] Measure the fit of the plugs in the pin. In many cases, the pistons and piston pins are routinely replaced at overhaul.

Valves & Valve Springs

The locations for checking runout and edge thickness of the valves are shown in Figure 10-12. Measure the edge thickness of valve heads. If, after re-facing, the edge thickness is less than the limit specified by the manufacturer, the valve must not be re-used. The edge thickness can be measured with sufficient accuracy by a dial indicator and a surface plate.

Using a magnifying glass, examine the valve in the stem area and the tip for evidence of cracks, nicks, or other indications of damage. This type of damage seriously weakens the valve, making it susceptible to failure. If superficial nicks and scratches on the valve indicate that it might be cracked, inspect it using a structural inspection method described later. Examine the valve springs for cracks, rust, broken ends, and compression. Cracks can be located by visual inspection or the magnetic particle method.

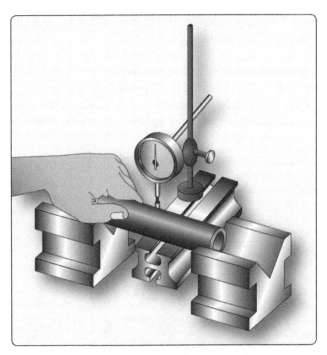

Figure 10-11. *Checking a piston pin for bends.*

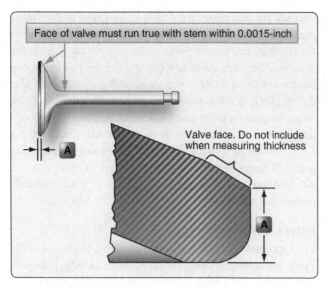
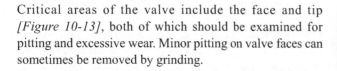

Figure 10-12. *Valve showing locations for checking runout and section for measuring edge thickness.*

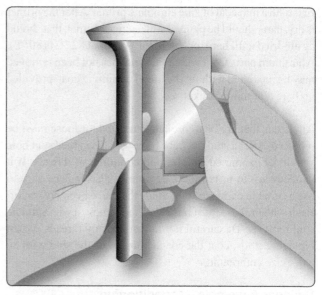

Figure 10-14. *Checking valve stretch with a manufacturer's gauge.*

Critical areas of the valve include the face and tip [*Figure 10-13*], both of which should be examined for pitting and excessive wear. Minor pitting on valve faces can sometimes be removed by grinding.

Inspect the valve for stretch and wear using a micrometer or a valve radius gauge. [*Figure 10-14*] If a micrometer is used, stretch is found as a smaller diameter of the valve stem near the neck of the valve. Measure the diameter of the valve stem and check the fit of the valve in its guide.

Examine the valve visually for physical damage and damage from burning or corrosion. Do not re-use valves that indicate damage of this nature.

Compression is tested with a valve spring compression tester. [*Figure 10-15*] The spring is compressed until its total height is that specified by the manufacturer. The dial on the tester should indicate the pressure, in pounds, required to compress

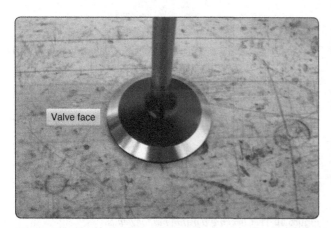

Figure 10-13. *Valve face surface.*

Figure 10-15. *Valve spring compression tester.*

the spring to the specified height. This must be within the pressure limits established by the manufacturer.

Refacing Valve Seats

The valve seat inserts of aircraft engine cylinders usually are in need of refacing at every overhaul. They are refaced to provide a true, clean, and correct size seat for the valve. When valve guides or valve seats are replaced in a cylinder, the seats must be made concentric with the valve guide.

Low power engines can use either bronze or steel seats. Bronze seats, although not widely used on current engines, are made of aluminum bronze or phosphor bronze alloys. Steel seats are commonly used for valve seats on higher powered engines and are made of heat-resistant steel with a layer of stellite steel alloy on the valve contact surface. Stellite seats can require a special stone to grind this very hard material.

Steel valve seats are refaced by grinding equipment. *[Figure 10-16]* Bronze seats are refaced preferably by the use of cutters or reamers, but they may be ground when this equipment is not available. The only disadvantage of using a stone on bronze is that the soft metal loads the stone to such an extent that much time is consumed in redressing the stone to keep it clean.

The equipment used on steel seats can be either wet or dry valve seat grinding equipment. The wet grinder uses a mixture of soluble oil and water to wash away the chips and to keep the stone and seat cool; this produces a smoother, more accurate job than the dry grinder. The stones may be either silicon carbide or aluminum oxide.

Before refacing the seat, make sure that the valve guide is in good condition, clean, and does not have to be replaced. Mount the cylinder firmly in the hold down fixture. An expanding

pilot is inserted in the valve guide from the inside of the cylinder, and an expander screw is inserted in the pilot from the top of the guide. *[Figure 10-17]* The pilot must be tight in the guide, because any movement can cause a poor grind. The fluid hose is inserted through one of the spark plug inserts.

The three grades of stones available for use are classified as rough, finishing, and polishing stones. The rough stone is designed to true and clean the seat. The finishing stone must follow the rough to remove grinding marks and produce a smooth finish. The polishing stone does just as the name implies and is used only where a highly polished seat is desired.

The stones are installed on special stone holders. The face of the stone is trued by a diamond dresser. The stone should be refaced whenever it is grooved or loaded, and when the stone is first installed on the stone holder. The diamond dresser also may be used to cut down the diameter of the stone. Dressing of the stone should be kept to a minimum as a matter of conservation; therefore, it is desirable to have sufficient stone holders for all the stones to be used on the job.

In the actual grinding job, considerable skill is required in handling the grinding gun. The gun must be centered accurately on the stone holder. If the gun is tilted off-center, chattering of the stone results, and a rough grind is produced. It is very important that the stone be rotated at a speed that permits grinding instead of rubbing. This speed is approximately 8,000 to 10,000 revolutions per minute (rpm). Excessive pressure on the stone can slow it down. It is not a good technique to let the stone grind at slow speed by putting pressure on the stone when starting or stopping the gun. The maximum pressure used on the stone at any time should be no more than that exerted by the weight of the gun.

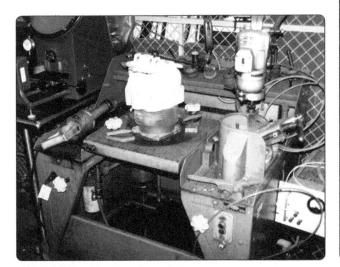

Figure 10-16. *Valve seat grinding equipment.*

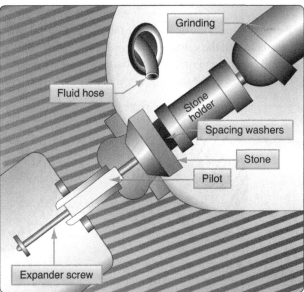

Figure 10-17. *Valve seat grinding setup.*

Another practice, conducive to good grinding, is to ease off on the stone every second or so to let the coolant wash away the chips on the seat. This rhythmic grinding action also helps keep the stone up to its correct speed. Since it is quite a job to replace a seat, remove as little material as possible during the grinding. Inspect the job frequently to prevent unnecessary grinding.

The rough stone is used until the seat is true to the valve guide and until all pits, scores, or burned areas are removed. *[Figure 10-18]* After refacing, the seat should be smooth and true. The finishing stone is used only until the seat has a smooth, polished appearance. Extreme caution should be used when grinding with the finishing stone to prevent chattering.

The size and trueness of the seat can be checked by several methods. Runout of the seat is checked with a special dial indicator and should not exceed 0.002 inch. The size of the seat may be determined by using Prussian blue. Prussian blue is used to check for contact transfer from one surface to the other. To check the fit of the seat, spread a thin coat of Prussian blue evenly on the seat. Press the valve onto the seat. The blue transferred to the valve indicates the contact surface. The contact surface should be one-third to two-thirds the width of the valve face and in the middle of the face. In some cases, a-go and no-go gauge is used in place of the valve when making the Prussian blue check. If Prussian blue is not used, the same check may be made by lapping the valve lightly to the seat. Lapping is accomplished by using a small amount of lapping compound placed between the valve face and seat. The valve is then moved in a rotary motion back and forth until the lapping compound grinds slightly into the surface. After cleaning the lapping contact compound off, a contact area can be seen. Examples of test results are shown in *Figure 10-19*.

If the seat contacts the upper third of the valve face, grind off the top corner of the valve seat. *[Figure 10-20]* Such grinding is called narrowing grinding. This permits the seat to contact the center third of the valve face without touching the upper portion of the valve face.

If the seat contacts the bottom third of the valve face, grind off the inner corner of the valve seat. *[Figure 10-21]* The seat is narrowed by a stone other than the standard angle. It is common practice to use a 15° angle and 45° angle cutting stone on a 30° angle valve seat, and a 30° angle and 75° angle stone on a 45° angle valve seat. *[Figure 10-22]*

If the valve seat has been cut or ground too much, the valve contacts the seat too far up into the cylinder head, and the valve clearance, spring tension, and the fit of the valve to the seat is affected. To check the height of a valve, insert the valve into the guide, and hold it against the seat. Check the height of the valve stem above the rocker box or some other fixed position.

Before refacing a valve seat, consult the overhaul manual for the particular model engine. Each manufacturer specifies the desired angle for grinding and narrowing the valve seat.

Valve Reconditioning

One of the most common jobs during engine overhaul is grinding the valves. The equipment used should preferably be a wet valve grinder. With this type of machine, a mixture of soluble oil and water is used to keep the valve cool and carry away the grinding chips.

Like many machine jobs, valve grinding is mostly a matter of setting up the machine. The following points should be checked or accomplished before starting a grind. True the stone by means of a diamond nib. The machine is turned on, and the diamond is drawn across the stone, cutting just deep enough to true and clean the stone. Determine the face angle of the valve being ground and set the movable head of the machine to correspond to this valve angle. Usually, valves are ground to the standard angles of 30° or 45°. However, in some instances, an interference fit of 0.5° or 1.5° less than the standard angle may be ground on the valve face.

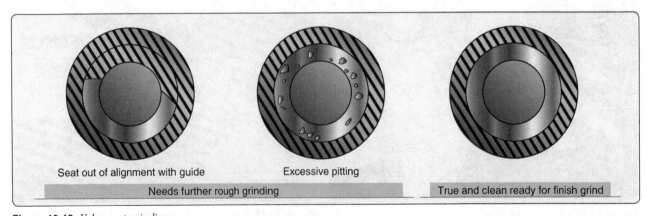

Seat out of alignment with guide   Excessive pitting

Needs further rough grinding   True and clean ready for finish grind

Figure 10-18. *Valve seat grinding.*

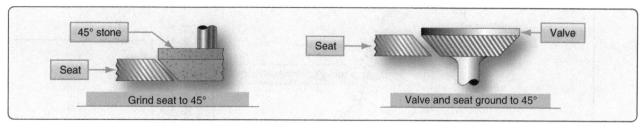

Figure 10-19. *Fitting the valve and seat.*

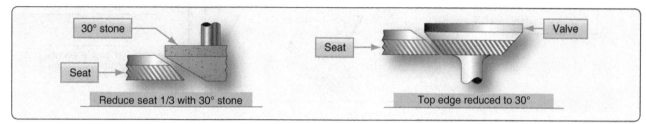

Figure 10-20. *Grinding top surface of the valve seat.*

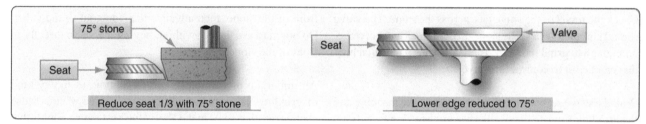

Figure 10-21. *Grinding the inner corner of the valve seat.*

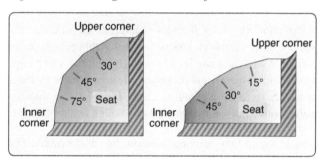

Figure 10-22. *Valve seat angles.*

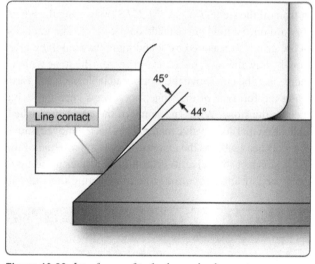

Figure 10-23. *Interference fit of valve and valve seat.*

The interference fit is used to obtain a more positive seal by means of a narrow contact surface. *[Figure 10-23]* Theoretically, there is a line contact between the valve and seat. With this line contact, the load that the valve exerts against the seat is concentrated in a very small area, thereby increasing the unit load at any one spot. The interference fit is especially beneficial during the first few hours of operation after an overhaul. The positive seal reduces the possibility of a burned valve or seat that a leaking valve might produce. After the first few hours of running, these angles tend to pound down and become identical.

Notice that the interference angle is ground into the valve, not the seat. It is easier to change the angle of the valve grinder work head than to change the angle of a valve seat grinder stone. Do not use an interference fit unless the manufacturer approves it.

Install the valve into the chuck and adjust the chuck so that the valve face is approximately 2 inches from the chuck. *[Figure 10-24]* If the valve is chucked any further out, there is danger of excessive wobble and also a possibility of grinding into the stem.

There are various types of valve grinding machines. In one type, the stone is moved across the valve face; in another, the valve is moved across the stone. Whichever type is used, the following procedures are typical of those performed when refacing a valve.

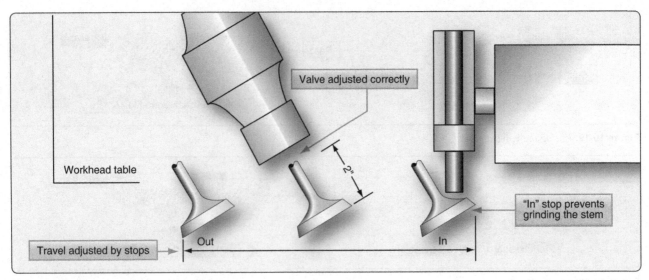

Figure 10-24. *Valve installed in grinding machine.*

Check the travel of the valve face across the stone. The valve should completely pass the stone on both sides, yet not travel far enough to grind the stem. There are stops on the machine that can be set to control this travel.

With the valve set correctly in place, turn on the machine and the grinding fluid so that it splashes on the valve face. Back the grinding wheel off all the way. Place the valve directly in front of the stone. *[Figure 10-25]* Slowly bring the wheel forward until a light cut is made on the valve. The intensity of the grind is measured by sound more than anything else. Slowly draw the valve back and forth across the stone without increasing the cut. Move the work head table back and forth using the full face of the stone, but always keep the valve face on the stone. When the sound of the grind diminishes, indicating that some valve material has been removed, move the workhead table to the extreme left to stop rotation of the valve. Inspect the valve to determine if further grinding is necessary. If another cut must be made, bring the valve in

front of the stone, then advance the stone out to the valve. Do not increase the cut without having the valve directly in front of the stone.

An important precaution in valve grinding, as in any kind of grinding, is to make light cuts only. Heavy cuts cause chattering, that may make the valve surface so rough that much time is lost in obtaining the desired finish.

After grinding, check the valve margin to be sure that the valve edge has not been ground too thin. A thin edge is called a feather edge and can lead to pre-ignition; the valve edge would burn away in a short period of time, and the cylinder would have to be overhauled again. *Figure 10-26* shows a valve with a normal margin and one with a feather edge.

The valve tip may be resurfaced on the valve grinder. The tip is ground to remove cupping or wear, and also to adjust valve clearances on some engines.

Figure 10-25. *Valve in chuck ready to grind.*

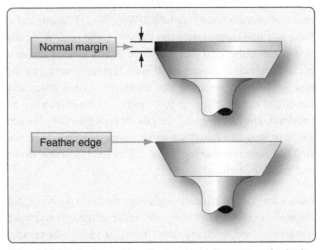

Figure 10-26. *Engine valves showing normal margin and a feather edge.*

The valve is held by a clamp on the side of the stone. *[Figure 10-27]* With the machine and grinding fluid turned on, the valve is pushed lightly against the stone and swung back and forth. Do not swing the valve stem off either edge of the stone. Because of the tendency for the valve to overheat during this grinding, be sure plenty of grinding fluid covers the tip.

Grinding of the valve tip may remove, or partially remove, the bevel on the edge of the valve. To restore this bevel, mount a V-way approximately 45° to the grinding stone. Hold the valve onto the V-way and twist the valve tip onto the stone. With a light touch, grind all the way around the tip. This bevel prevents scratching the valve guide when the valve is installed.

Valve Lapping & Leak Testing

After the grinding procedure is finished, it is sometimes necessary that the valve be lapped to the seat. This is done by applying a small amount of lapping compound to the valve face, inserting the valve into the guide, and rotating the valve with a lapping tool until a smooth, gray finish appears at the contact area. The appearance of a correctly lapped valve is shown in *Figure 10-28*.

After the lapping process is finished, be sure that all lapping compound is removed from the valve face, seat, and adjacent areas. The final step is to check the mating surface for leaks to see if it is sealing properly. This is done by installing the valve in the cylinder, holding the valve by the stem with the fingers, and pouring kerosene or solvent into the valve port. While holding finger pressure on the valve stem, check to see if the kerosene is leaking past the valve into the combustion chamber. If it is not, the valve re-seating operation is finished. If kerosene is leaking past the valve, continue the lapping operation until the leakage is stopped. The incorrect indications are of value in diagnosing improper valve and valve seat grinding. Incorrect indications, their cause, and remedy are shown in *Figure 10-29*.

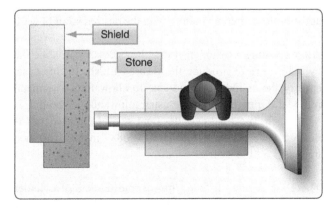

Figure 10-27. *Grinding a valve tip.*

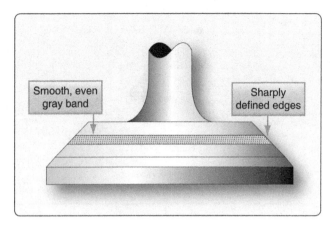

Figure 10-28. *A correctly lapped valve.*

Piston Repairs

Piston repairs are not required as often as cylinder repairs since most of the wear is between the piston ring and cylinder wall, valve stem and guide, and valve face and seat. A lesser amount of wear is encountered between the piston skirt and cylinder, ring and ring groove, or piston pin and bosses.

The most common repair is the removal of scores. Usually, these may be removed only on the piston skirt if they are very light. On engines where the entire rotating and reciprocating assembly is balanced, the pistons must weigh within one-fourth ounce of each other. When a new piston is installed, it must be within the same weight tolerance as the one removed. It is not enough to have the pistons matched alone; they must be matched to the crankshaft, connecting rods, piston pins, etc. To make weight adjustments on new pistons, the manufacturer provides a heavy section at the base of the skirt. To decrease weight, file metal evenly off the inside of this heavy section. The piston weight can be decreased easily, but welding, metalizing, or plating cannot be done to increase the piston weight.

If ring grooves are worn or stepped, the pistons are normally replaced. Small nicks on the edge of the piston pin boss may be sanded down. Deep scores inside the boss, or anywhere around the boss, are definite reasons for rejection. It has become more economical to replace pistons rather than reconditioning and reusing old ones, especially during overhaul.

Cylinder Grinding & Honing

If a cylinder has excessive taper, out-of-roundness, step, or its maximum size is beyond limits, it can be reground to the next allowable oversize. If the cylinder walls are lightly rusted, scored, or pitted, the damage may be removed by honing or lapping.

Regrinding a cylinder is a specialized job that the powerplant mechanic is not usually expected to be able to do. However,

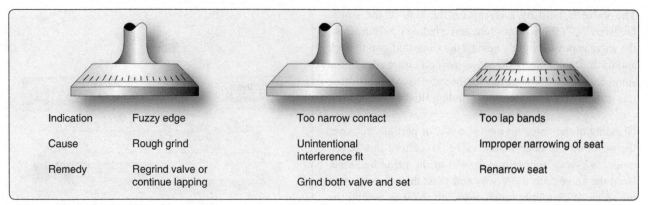

Figure 10-29. *Incorrectly lapped valves.*

Indication	Fuzzy edge	
Cause	Rough grind	
Remedy	Regrind valve or continue lapping	

Too narrow contact	
Unintentional interference fit	
Grind both valve and set	

Too lap bands	
Improper narrowing of seat	
Renarrow seat	

the mechanic must be able to recognize when a cylinder needs regrinding, and they must know what constitutes a good or bad job.

Generally, standard aircraft cylinder oversizes are 0.010 inch, 0.015 inches, 0.020 inch, or 0.030 inch. Aircraft cylinders have relatively thin walls and may have a nitrided surface, that must not be ground away. Nitriding is a surface hardening process that hardens the steel surface to a depth of several thousandths of an inch. Any one manufacturer usually does not allow all of the above oversizes. Some manufacturers do not allow regrinding to an oversize at all. The manufacturer's overhaul manual, or parts catalog, usually lists the oversizes allowed for a particular make and model engine.

To determine the regrind size, the standard bore size must be known. This usually can be determined from the manufacturer's specifications or manuals. The regrind size is figured from the standard bore. For example, a certain cylinder has a standard bore of 3.875 inches. To have a cylinder ground to 0.015 inches oversize, it is necessary to grind to a bore diameter of 3.890 inch (3.875 + 0.015). A tolerance of ±0.0005 inches is usually accepted for cylinder grinding.

Another factor to consider when determining the size to which a cylinder must be reground is the maximum wear that has occurred. If there are spots in the cylinder wall that are worn larger than the first oversize, then obviously it is necessary to grind to the next oversize to clean up the entire cylinder.

The type of finish desired in the cylinder is an important consideration when ordering a regrind. Some engine manufacturers specify a fairly rough finish on the cylinder walls, that allows the rings to seat even if they are not lapped to the cylinder. Other manufacturers desire a smooth finish to which a lapped ring seats without much change in ring or cylinder dimensions. The latter type of finish is more expensive to produce.

The standard used when measuring the finish of a cylinder wall is known as micro-inch root mean square (micro-inch RMS). In a finish where the depth of the grinding scratches are one-millionth (0.000001) of an inch deep, it is specified as 1 micro-inch RMS. Most aircraft cylinders are ground to a finish of 15 to 20 micro-inch RMS. Several low-powered engines have cylinders that are ground to a relatively rough 20- to 30-micro-inch RMS finish. On the other end of the scale, some manufacturers require a superfinish of approximately 4- to 6-micro-inch RMS.

Cylinder grinding is accomplished by a firmly mounted stone that revolves around the cylinder bore, as well as up and down the length of the cylinder barrel. *[Figure 10-30]* The cylinder, the stone, or both may move to get this relative movement. The size of the grind is determined by the distance the stone is set away from the centerline of the cylinder. Some cylinder bore grinding machines produce a perfectly straight bore, while others are designed to grind a choked bore. A choked bore grind refers to the manufacturing process in which the cylinder walls arc ground to produce a smaller internal diameter at the top than at the bottom. The purpose of this type grind or taper is to maintain a straight cylinder wall during operation. As a cylinder heats up during operation, the head and top of the cylinder are subjected to more heat than the bottom. This causes greater expansion at the top than at the bottom, thereby maintaining the desired straight wall.

After grinding a cylinder, it may be necessary to hone the cylinder bore to produce the desired finish. In this case, specify the cylinder regrind size to allow for some metal removal during honing. The usual allowance for honing is 0.001 inch. If a final cylinder bore size of 3.890 inches is desired, specify the regrind size of 3.889 inches, and then hone to 3.890 inches.

There are several different makes and models of cylinder hones. The burnishing hone is used only to produce the desired finish on the cylinder wall. The more elaborate

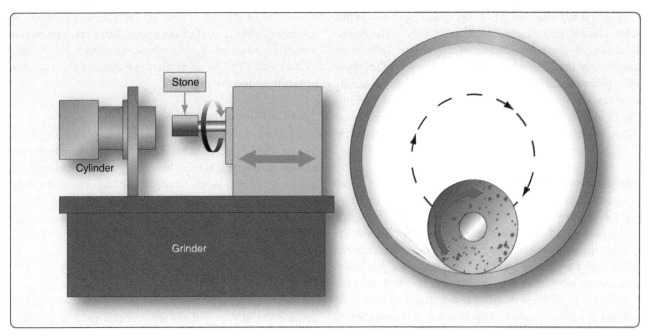

Figure 10-30. *Cylinder bore grinding.*

micromatic hone can also be used to straighten out the cylinder walls. A burnishing hone should not be used in an attempt to straighten cylinder walls. *[Figure 10-31]* Since the stones are only spring loaded, they follow the contour of the cylinder wall and may aggravate a tapered condition.

Deglazing the cylinder walls is accomplished with the use of a deglazing hone. A cross-hatch pattern must be placed on the cylinder wall to allow for piston ring break-in. This is accomplished by a deglazing hone turned by a drill being moved in and out of the cylinder rapidly. *[Figure 10-32]*

After the cylinders have been reground or deglazed, or both, check the size and wall finish, and check for evidence of overheating or grinding cracks before installing on an engine.

Reassembly

Before starting reassembly, all serviceable and new engine components need to be cleaned, organized, and laid out in the order they are to be assembled. A popular method of engine assembly is for the engine to be assembled at one work station with the same technicians completing the total assembly of the engine. It is also important to refer to the parts catalog to ensure that the correct hardware is used during the assembly of the engine. The engine overhaul manual should be referred to for information on the use of safety wire, self-locking nuts, and torque values. During assembly, the components should be pre-lubricated as the overhaul manual sets forth. It is important to follow the manufacturer's overhaul assembly procedures completely and perform all checks and procedures that are called for in the manual.

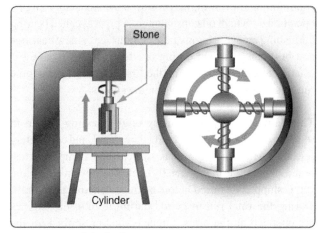

Figure 10-31. *Cylinder honing.*

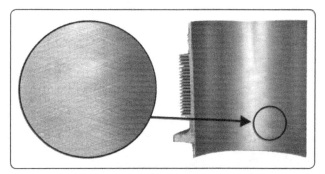

Figure 10-32. *Cross-hatch pattern on cylinder wall.*

Installation & Testing

Testing Reciprocating Engines

The procedures and equipment used in determining that an engine is ready for airworthy service and is in excellent mechanical condition, normally requires the use of a test stand, or test cell, although the aircraft can be used.

[Figure 10-33] The method of engine testing or run-in that takes place during overhaul prior to delivery of the engine is critical to the airworthiness of the engine. It must be emphasized that engine run-in is as vital as any other phase of engine overhaul, for it is the means by which the quality of a new or newly overhauled engine is checked, and it is the final step in the preparation of an engine for service. Thus, the reliability and potential service life of an engine is in question until it has satisfactorily passed the cell test.

The test serves a dual purpose. First, it accomplishes piston ring run-in and bearing burnishing. Second, it provides valuable information that it used to evaluate engine performance and determine engine condition. To provide proper oil flow to the upper portion of the cylinder barrel walls with a minimum loss of oil, it is important that piston rings be properly seated in the cylinder in which they are installed. The process is called piston ring run-in (break-in) and is accomplished chiefly by controlled operation of the engine in the high-speed range. Improper piston ring conditioning, or run-in, may result in unsatisfactory engine operation with high oil consumption. A process called bearing burnishing creates a highly polished surface on new bearings and bushings installed during overhaul. The burnishing is usually accomplished during the first periods of the engine run-in at comparatively slow engine speeds.

The failure of any part during engine testing or run-in requires that the engine be returned, repaired, and completely retested. After an engine has successfully completed test requirements, it is then specially treated to prevent corrosion, if it is shipped or stored before being installed in an aircraft. During the final run-in period during testing, the engines are operated on the proper grade of fuel prescribed for the particular kind of engine. The oil system is serviced with a mixture of corrosion-preventive compound and engine oil. The temperature of this mixture is maintained at 105 °C to 121 °C. Near the end of final run-in, corrosion-preventive mixture (CPM) is used as the engine lubricant. The engine induction passages and combustion chambers are also treated with CPM by an aspiration method. CPM is drawn or breathed into the engine.

Test Cell Requirements

The test cell requires an area to mount and hold the engine for testing. The cell needs to have the controls, instruments, and any special equipment to evaluate the total performance of the engine. A test club should be used for testing instead of a flight propeller. *[Figure 10-34]* A test club provides more cooling air flow and the correct amount of load. The operational tests and test procedures vary with individual engines, but the basic requirements are generally closely related.

Engine Instruments

The test cell control room contains the controls used to operate the engine and the instruments used to measure various temperatures and pressures, fuel flow, and other factors. These devices are necessary in providing an accurate check and an evaluation of the operating engine. The control room is separate from, but adjacent to, the space (test cell) that houses the engine being tested. The safe, economical, and reliable testing of modern aircraft engines depends largely upon the use of instruments. In engine run-in procedures, the same basic engine instruments are used as when the engine is installed in the aircraft, plus some additional connections to these instruments, and some indicating and measuring devices that cannot be practically installed in the aircraft. Instruments used in the testing procedures are inspected and calibrated periodically, as are instruments installed in the aircraft; thus, accurate information concerning engine operation is ensured.

Engine instruments can operate using different methods, some mechanically, some electrically, and some by sensing

Figure 10-33. *Test stand.*

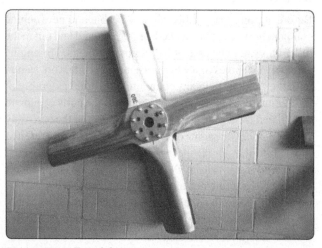

Figure 10-34. *Test club.*

the direct pressure of air or liquid. Some of the basic instruments are:

1. Carburetor air temperature gauge.
2. Fuel pressure gauge.
3. Fuel-flow meter.
4. Manifold pressure gauge.
5. Oil temperature gauge.
6. Oil pressure gauge.
7. Tachometer.
8. Exhaust gas temperature gauge.
9. Cylinder head temperature gauge.
10. Torquemeter.

Instrument markings, ranges of operation, minimum and maximum limits, and the interpretation of these markings are general to all the instruments. Generally, the instrument marking system consists of three colors: red, yellow, and green. A red line, or mark, indicates a point beyond which a dangerous operating condition exists. A red arc indicates a dangerous operating range due generally to an engine propeller vibration range. This arc can be passed through, but the engine cannot be operated in this area. Of the two, the red mark is used more commonly and is located radially on the cover glass or dial face. The yellow arc covers a given range of operation and is an indication of caution. Generally, the yellow arc is located on the outer circumference of the instrument cover glass or dial face. The green arc shows a normal and safe range of operation. When the markings appear on the cover glass, a white line is used as an index mark, often called a slippage mark. The white radial mark indicates any movement between the cover glass and the case, a condition that would cause mislocation of the other range and limit markings.

Carburetor Air Temperature (CAT) Indicator

Measured at the carburetor entrance, carburetor air temperature (CAT) is regarded by many as an indication of induction system ice formation. Although it serves this purpose, it also provides many other important items of information.

The powerplant is a heat machine, and the temperature of its components, or the fluids flowing through it, affects the combustion process either directly or indirectly. The temperature level of the induction air affects not only the charge density, but also the vaporization of the fuel. CAT is also useful for checking induction system condition. Backfiring is indicated as a momentary rise on the gauge, provided it is of sufficient severity for the heat to be sensed at the carburetor air-measuring point. A sustained induction system fire shows a continuous increase of CAT.

The CAT should be noted before starting and just after shutdown. The temperature before starting is the best indication of the temperature of the fuel in the carburetor body and tells whether vaporization is sufficient for the initial firing, or whether the mixture must be augmented by priming. If an engine has been shut down for only a short time, the residual heat in the carburetor may make it possible to rely on the vaporizing heat in the fuel and powerplant. Priming would then be unnecessary.

After shutdown, a high CAT is a warning that the fuel trapped in the carburetor will expand, producing high internal pressure. When a high temperature is present at this time, the fuel line and manifold valves should be open so that the pressure can be relieved by allowing fuel passage back to the tank. The CAT gauge indicates the temperature of the air before it enters the carburetor. The temperature reading is sensed by a bulb or electric sensor. In the test cell, the sensor is located in the air intake passage to the engine and, in an aircraft, it is located in the ram-air intake duct. The CAT gauge is calibrated in the centigrade scale. *[Figure 10-35]* This gauge, like many other multi-engine aircraft instruments, is a dual gauge; two gauges, each with a separate pointer and scale, are used in the same case.

Notice the range markings used. The yellow arc indicates a range from –10 °C to +15 °C, since the danger of icing occurs between these temperatures. The green range indicates the normal operating range from +15 °C to +40 °C. The red line indicates the maximum operating temperature of 40 °C; any operation at a temperature over this value places the engine in danger of detonation.

Fuel Pressure Indicator

The fuel pressure gauge is calibrated in pounds per square inch (psi) of pressure. It is used during the test run-in to measure engine fuel pressure at the carburetor inlet, the fuel

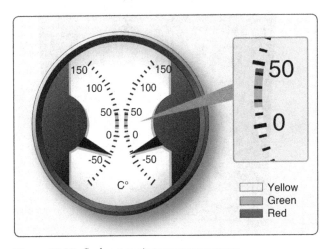

Figure 10-35. *Carburetor air temperature gauge.*

feed valve discharge nozzle, and the main fuel supply line. Fuel gauges are located in the operator's control room and are connected by flexible lines to the different points at which pressure readings are desired during the testing procedures.

In some aircraft installations, the fuel pressure is sensed at the carburetor or fuel injection unit inlet of each engine, and the pressure is indicated on individual gauges on the instrument panel. *[Figure 10-36]* The dial is calibrated in graduations and is extended and numbered. The numbers range from 0 to 10 in this example. The red line on the dial at the 2 pounds psi graduation shows the minimum fuel pressure allowed during flight. The green arc shows the desired range of operation, which is 2 to 9 psi. The red line at the 9 psi graduation indicates the maximum allowable fuel pressure. Fuel pressures vary with the type of fuel system installation and the size of the engine. When fuel injection systems are used, the fuel pressure range is much higher; the minimum allowable pressure is approximately 10 psi, and the maximum is generally 25 psi.

Oil Pressure Indicator

The main oil pressure reading is taken at the pressure side of the oil pump. Generally, there is only one oil pressure gauge for each aircraft engine. The oil pressure gauge dial does not show the pressure range or limits for all installations. *[Figure 10-36]* The actual markings for specific aircraft may be found in the aircraft specifications or Type Certificate Data Sheets. The lower red line at 25 psi indicates the minimum oil pressure permissible in flight. The green arc between 60 to 85 psi illustrates the desired operating oil pressure range. The red line at 100 psi indicates maximum permissible oil pressure.

Figure 10-36. *Engine instrument clusters.*

The oil pressure gauge indicates the pressure, in psi, that the oil of the lubricating system is being supplied to the moving parts of the engine. The engine should be shut down immediately if the gauge fails to register pressure when the engine is operating. Excessive oscillation of the gauge pointer indicates that there is air in the lines leading to the gauge, or that some unit of the oil system is functioning improperly.

Oil Temperature Indicator

During engine run-in in the test cell, engine oil temperature readings are taken at the oil inlet and outlet. From these readings, it can be determined if the engine heat transferred to the oil is low, normal, or excessive. This information is of extreme importance during the breaking-in process of large reciprocating engines. The oil temperature gauge line in the aircraft is connected at the oil inlet to the engine.

Three range markings are used on the oil temperature gauge. The green arc in *Figure 10-36*, on the dial, shows the minimum oil temperature permissible for ground operational checks or during flight. The green mark between 25 °F and below 245 °F shows the desired oil temperature for continuous engine operation. The red mark at 245 °F indicates the maximum permissible oil temperature.

Fuel-Flow Meter

The fuel-flow meter measures the amount of fuel delivered to the engine. During engine testing procedures, the fuel-flow to the engine can be measured by three different methods: a direct flow meter, a pressure-based flow meter, or a turbine sensor-based flow meter. The direct reading flow meter uses a series of calibrated tubes located in the control room. The tubes are of various sizes to indicate different volumes of fuel-flow. Each tube contains a float that can be seen by the operator, and as the fuel-flow through the tube varies, the float is either raised or lowered, indicating the amount of fuel-flow. From these indications, the operator can determine whether an engine is operating at the correct air-fuel mixture for a given power setting. Reciprocating engines on light aircraft usually use a fuel pressure gauge that is also used for the flow meter. This is because the fuel-flow is proportional to the fuel pressure in this system. Fuel-flow is normally measured in gallons per hour.

In most turbine aircraft installations, the fuel-flow indicating system consists of a transmitter and an indicator for each engine. The fuel-flow transmitter is conveniently mounted in the engine's accessory section and measures the fuel-flow between the engine-driven fuel pump and the fuel control device. The transmitter is an electrical device that contains a turbine that turns faster as the flow increases, which increases the electrical signal to the indicator. The fuel-flow transmitter is connected electrically to the indicator located on the aircraft flight deck, or on the test cell operator's panel. The reading on the indicator on turbine aircraft is calibrated to record the amount of fuel-flow in pounds of fuel per hour.

Manifold Pressure Indicator

The preferred type of instrument for measuring the manifold pressure on reciprocating engines is a gauge that records the pressure as an absolute pressure reading. Absolute pressure takes into account the atmospheric pressure plus the pressure in the intake manifold. To read the manifold pressure of the engines, a specially designed manifold pressure gauge that indicates absolute manifold pressure in inches of mercury ("Hg) is used. The red line indicates the maximum manifold pressure permissible during takeoff.

The manifold pressure gauge range markings and indications vary with different kinds of engines and installations. *Figure 10-37* illustrates the dial of a typical manifold pressure gauge and shows how the range markings are positioned. The green arc starts at 35 "Hg and continues to the 44 "Hg. The red line on the gauge, at 49 "Hg shows the manifold pressure recommended for takeoff. This pressure should not be exceeded.

Tachometer Indicator

The tachometer for reciprocating engines shows the engine crankshaft rpm. The system used for testing the engine is the same as the system in the aircraft installation. The tachometer, often referred to as TACH, is calibrated in hundreds with graduations at every 100-rpm interval. The dial shown in *Figure 10-38* starts at 0 rpm and goes to 35 (3,500 rpm). The green arc indicates the rpm range within operation that is permissible. The red line indicates the maximum rpm permissible during takeoff; any rpm beyond this value is an overspeed condition.

Turbine engines use percent rpm indicators due to the high rpm that the engines generally operate. Each rotating assembly in an engine has its own percent rpm indicator. The tachometer indicates speed of the compressor section. The 100 percent position on the indicator is the highest rpm at which the engine can operate. Red lines and green arcs operate the same as with reciprocating engines. Rotorcraft use two synchronous tachometers.

Cylinder Head Temperature Indicator

During the engine test procedures, the cylinder head temperatures of various cylinders on the reciprocating engine are normally tested. Thermocouples are connected to several cylinders and, by a selector switch, any cylinder head temperature can be indicated on the indicators. When installed in the aircraft, there is sometimes only one

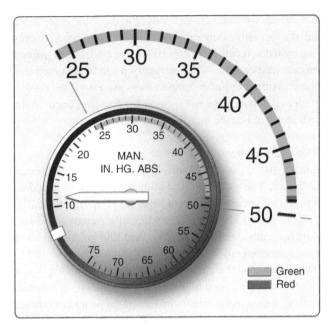

Figure 10-37. *Manifold pressure gauge.*

Figure 10-38. *Tachometer.*

thermocouple lead and indicator for each engine installed in an aircraft.

Cylinder head temperatures are indicated by a gauge connected to a thermocouple attached to the cylinder, that tests show to be the hottest on an engine in a particular installation. The thermocouple may be placed in a special gasket located under a rear spark plug, or in a special well in the top or rear of the cylinder head.

The temperature recorded at either of these points is merely a reference or control temperature; but as long as it is kept within the prescribed limits, the temperatures inside the cylinder dome, exhaust valve, and piston is within a satisfactory range. Since the thermocouple is attached to only one cylinder, it can do no more than give evidence of general engine temperature. While normally it can be assumed that the remaining cylinder temperatures are lower, conditions such as detonation are not indicated unless they occur in the cylinder that has the thermocouple attached.

The cylinder head temperature gauge range marking is similar to that of the manifold pressure and tachometer indicator. The cylinder head temperature gauge is a dual gauge that incorporates two separate temperature scales. *[Figure 10-39]* The scales are calibrated in increments of 10°, with numerals at the 0°, 100°, 200°, and 300° graduations. The space between any two graduation marks represents 10 °C. The blue arc on the gauge indicates the range within which operation is permitted in auto-lean. The bottom of this arc, 100 °C, indicates the minimum desired temperature to ensure efficient engine operation during flight. The top of the blue arc, 230 °C, indicates the temperature at which the mixture control must be moved to the "auto-rich" position. The green arc describes the range within which operation must be in auto-rich. The top of this arc, 248 °C, indicates maximum continuous power; all operation above this temperature is limited in time (usually 5 to 15 minutes). The red line indicates maximum permissible temperature, 260 °C.

Torquemeter

Most torque systems use an oil pressure output from a torque valve to indicate actual engine power output at various power settings. The torquemeter indicates the amount of torque being produced at the propeller shaft. A helical gear moves back and forth as the torque on the propeller shaft varies. This gear, acting on a piston, positions a valve that meters the oil pressure proportionally to the torque being produced. A change in pressure from the valve that is connected to a transducer is then converted to an electrical signal and is transmitted to the flight deck. The torquemeter can read out in foot-pounds

of torque, percent of horsepower, or horsepower. The earlier systems read out in psi, and the flight engineer converted this to the correct power setting. *[Figure 10-40]* Some systems use strain gauges to attach to the ring gear to provide an electrical signal directly to the readout.

Warning Systems

Many of the miscellaneous gauges and devices indicate only that a system is functioning or has failed to function. On some aircraft, a warning light illuminates when the fuel pressure or oil pressure is low.

Reciprocating Engine Operation

The operation of the powerplant is controlled from the flight deck. Some installations have numerous control handles and levers connected to the engine by rods, cables, bellcranks, pulleys, etc. In most cases, the control handles are conveniently mounted on quadrants in the flight deck. Placards, or markings, are placed on the quadrant to indicate the functions and positions of the levers. In some installations, friction clutches are installed to hold the controls in place.

Engine Instruments

The term engine instruments usually includes all instruments required to measure and indicate the functioning of the powerplant. The engine instruments are generally installed on the instrument panel so that all of them can easily be observed at one time. Manifold pressure, rpm, engine temperature, oil temperature, CAT, and the air-fuel ratio can be controlled by manipulating the flight deck controls. Coordinating the movement of the controls with the instrument readings protects against exceeding operating limits.

Engine operation is usually limited by specified operating ranges of the following:

1. Crankshaft speed (rpm).

2. Manifold pressure.

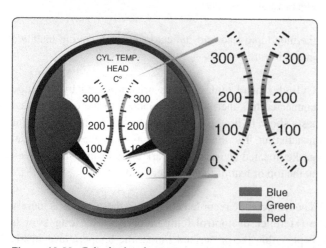

Figure 10-39. *Cylinder head temperature gauge.*

Figure 10-40. *Torquemeter readout.*

3. Cylinder head temperature.

4. CAT.

5. Oil temperature.

6. Oil pressure.

7. Fuel pressure.

8. Fuel-flow meter.

9. Air-fuel mixture setting.

The procedures, pressures, temperatures, and rpm used throughout this section are solely for the purpose of illustration and do not have general application. The operating procedures and limits used on individual makes and models of aircraft engines vary considerably from the values shown here. For exact information regarding a specific engine model, consult the applicable instructions.

Engine Starting

Before starting the engine, observe the manifold pressure gauge that should read approximate atmospheric (barometric) pressure when the engine is not running. At sea level, this is approximately 30 "Hg, and at fields above sea level, the atmospheric pressure is less, depending on the height above sea level. Also, observe all engine gauges for the correct reading for engine off settings.

Correct starting technique is an important part of engine operation. Improper procedures often are used, because some of the basic principles involved in engine operation are misunderstood. Read more about typical procedures for starting reciprocating engines in the Aviation Maintenance Technician Handbook - General.

Pre-Oiling

Engines that have undergone overhaul or major maintenance can have air trapped in some of the oil passages that must be removed before the first start. This is done by pre-oiling the engine by cranking, with the spark plugs removed, the engine with the starter or by hand (turning) until oil pressure is indicated. A second method is to pump oil under pressure through the oil system using an external pump until oil comes out of the oil outlet of the engine.

Hydraulic Lock

Whenever a radial engine remains shut down for any length of time beyond a few minutes, oil or fuel may drain into the combustion chambers of the lower cylinders or accumulate in the lower intake pipes ready to be drawn into the cylinders when the engine starts. *[Figure 10-41]* As the piston approaches top center of the compression stroke (both valves closed), this liquid being incompressible, stops piston movement. If the crankshaft continues to rotate, something

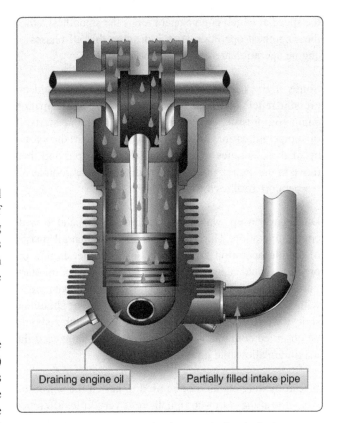

Draining engine oil | Partially filled intake pipe

Figure 10-41. *Initial step in developing a hydraulic lock.*

must give. Therefore, starting or attempting to start an engine with a hydraulic lock of this nature may cause the affected cylinder to blow out or, more likely, may result in a bent or broken connecting rod.

To eliminate a lock, remove either the front or rear spark plug of the lower cylinders and pull the propeller through in the direction of rotation. The piston expels any liquid that may be present. If the hydraulic lock occurs as a result of over-priming prior to initial engine start, eliminate the lock in the same manner (i.e., remove one of the spark plugs from the cylinder and rotate the crankshaft through two turns). Never attempt to clear the hydraulic lock by pulling the propeller through in the direction opposite to normal rotation. This tends to inject the liquid from the cylinder into the intake pipe with the possibility of a complete or partial lock occurring on the subsequent start.

Engine Warm-Up

Proper engine warm-up is important, particularly when the condition of the engine is unknown. Improperly adjusted idle mixture, intermittently firing spark plugs, and improperly adjusted engine valves all have an overlapping effect on engine stability. Therefore, the warm-up should be made at the engine speed where maximum engine stability is obtained. Experience has shown that the optimum warm-up speed is from 1,000 to 1,600 rpm. The actual speed selected should be the speed at which engine operation is the smoothest, since the smoothest operation is an indication that all phases of engine operation are the most stable.

Some engines incorporate temperature-compensated oil pressure relief valves. This type of relief valve results in high engine oil pressures immediately after the engine starts, if oil temperatures are very low. Consequently, start the warm-up of these engines at approximately 1,000 rpm and then move to the higher, more stable engine speed as soon as oil temperature reaches a warmer level.

During warm-up, watch the instruments associated with engine operation. This aids in making sure that all phases of engine operation are normal. For example, engine oil pressure should be indicated within 30 seconds after the start. Furthermore, if the oil pressure is not up to or above normal within 1 minute after the engine starts, the engine should be shut down. Cylinder head or coolant temperatures should be observed continually to see that they do not exceed the maximum allowable limit.

A lean mixture should not be used to hasten the warm-up. Actually, at the warm-up rpm, there is very little difference in the mixture supplied to the engine, whether the mixture is in a rich or lean position, since metering in this power range is governed by throttle position.

Carburetor heat can be used as required under conditions leading to ice formation. For engines equipped with a float-type carburetor, it is desirable to raise the CAT during warm-up to prevent ice formation and to ensure smooth operation.

The magneto safety check can be performed during warm-up. Its purpose is to ensure that all ignition connections are secure, and that the ignition system permits operation at the higher power settings used during later phases of the ground check. The time required for proper warm-up gives ample opportunity to perform this simple check, which may disclose a condition that would make it inadvisable to continue operation until after corrections have been made.

The magneto safety check is conducted with the propeller in the high rpm (low pitch) position, at approximately 1,000 rpm. Move the ignition switch from "both" to "right" and return to "both;" from "both" to "left" and return to "both;" from "both" to "off" momentarily and return to "both."

While switching from "both" to a single magneto position, a slight but noticeable drop in rpm should occur. This indicates that the opposite magneto has been properly grounded out. Complete cutting out of the engine when switching from "both" to "off" indicates that both magnetos are grounded

properly. While in the single magneto position, failure to obtain any rpm drop, or failure of the engine to cut out while switching to off, indicates that one or both ground connections are faulty. This indicates a safety problem; the magnetos are not secured at shut down and may fire if the propeller is turned.

Ground Check

The ground check is performed to evaluate the functioning of the engine by comparing power input, as measured by manifold pressure, with power output, as measured by rpm or torque.

The engine may be capable of producing a prescribed power, even rated takeoff, and not be functioning properly. Only by comparing the manifold pressure required during the check against a known standard is an unsuitable condition disclosed. The magneto check can also fail to show shortcomings, since the allowable rpm dropoff is only a measure of an improperly functioning ignition system and is not necessarily affected by other factors. Conversely, it is possible for the magneto check to prove satisfactory when an unsatisfactory condition is present elsewhere in the engine.

The ground check is made after the engine is thoroughly warm. It consists of checking the operation of the powerplant and accessory equipment by ear, by visual inspection, and by proper interpretation of instrument readings, control movements, and switch reactions. During the ground check, the aircraft should be headed into the wind, if possible, to take advantage of the cooling airflow. A ground check procedure is outlined below:

1. Control position check.

2. Cowl flaps (if equipped)—open.

3. Mixture—rich.

4. Propeller—high rpm.

5. Carburetor heat—cold.

6. Check propeller according to propeller manufacturer's instruction.

7. Open throttle to the run-up rpm setting as per manufacturer's instructions (specified RPM and manifold pressure).

8. Ignition system operational check.

In performing the ignition system operational check (magneto check), the power-absorbing characteristics of the propeller in the low fixed-pitch position are utilized. In switching to individual magnetos, cutting out the opposite plugs results in a slower rate of combustion, which gives the same effect as retarding the spark advance. The drop in engine speed is a measure of the power loss at this slower combustion rate.

When the magneto check is performed, a drop in torquemeter pressure indication is a good supplement to the variation in rpm. In cases where the tachometer scale is graduated coarsely, the torquemeter variation may give more positive evidence of the power change when switching to the individual magneto condition. A loss in torquemeter pressure not to exceed 10 percent can be expected when operating on a single magneto. By comparing the rpm drop with a known standard, the following are determined:

1. Proper timing of each magneto.

2. General engine performance as evidenced by smooth operation.

3. Additional check of the proper connection of the ignition leads.

Any unusual roughness on either magneto is an indication of faulty ignition caused by plug fouling or by malfunctioning of the ignition system. The operator should be very sensitive to engine roughness during this check. Lack of dropoff in rpm may be an indication of faulty grounding of one side of the ignition system. Complete cutting out when switching to one magneto is definite evidence that its side of the ignition system is not functioning. Excessive difference in rpm drop off between the left and right switch positions can indicate a difference in time between the left and right magnetos.

Sufficient time should be given to the check on each single switch position to permit complete stabilization of engine speed and manifold pressure. There is a tendency to perform this check too rapidly with resultant wrong indications. Operation as long as 1 minute on a single ignition system is not excessive.

Another point that must be emphasized is the danger of sticking tachometer. The tachometer should he tapped lightly to make sure the indicator needle moves freely. In some cases using older mechanical tachometers, sticking has caused errors in indication to the extent of 100 rpm. Under such conditions, the ignition system could have had as much as a 200 rpm drop with only a 100 rpm drop indicated on the instrument. In most cases, tapping the instrument eliminates the sticking and results in accurate readings.

In recording the results of time ignition system check, record the amount of the total rpm drop that occurs rapidly and the amount that occurs slowly. This breakdown in rpm drop provides a means of pinpointing certain troubles in the ignition system. This can reduce unnecessary work by confining maintenance to the specific part of the ignition system that is responsible for the trouble.

Fast rpm drop is usually the result of either faulty spark plugs or faulty ignition harness. This is true because faulty plugs or leads, take effect at once. The cylinder goes dead or starts firing intermittently the instant the switch is moved from "both" to the "right" or "left" position.

Slow rpm drop usually is caused by incorrect ignition timing or faulty valve adjustment. With late ignition timing, the charge is fired too late (in relation to piston travel) for the combustion pressures to build up to the maximum at the proper time. The result is a power loss greater than normal for single ignition because of the lower peak pressures obtained in the cylinder. However, this power loss does not occur as rapidly as that which accompanies a dead spark plug. This explains the slow rpm drop as compared to the instantaneous drop with a dead plug or defective lead. Incorrect valve clearances, through their effect on valve overlap, can cause the mixture to be too rich or too lean. The too rich or too lean mixture may affect one plug more than another, because of the plug location and show up as a slow rpm drop on the ignition check. Switch from "both" to "right" and return to "both." Switch from "both" to "left" and return to "both." Observe the rpm drop while operating on the right and left positions. The maximum drop should not exceed that specified by the engine manufacturer.

Fuel Pressure & Oil Pressure Check

Fuel pressure and oil pressure must be within the established tolerance (green arc) for the engine.

Propeller Pitch Check

The propeller is checked to ensure proper operation of the pitch control and the pitch-change mechanism. The operation of a controllable pitch propeller is checked by the indications of the tachometer and manifold pressure gauge when the propeller governor control is moved from one position to another. Because each type of propeller requires a different procedure, the applicable manufacturer's instructions should be followed.

Power Check

Specific rpm and manifold pressure relationship should be checked during each ground check. This can be done at the time the engine is run-up to make the magneto check. The purpose of this check is to measure the performance of the engine against an established standard. Calibration tests have determined that the engine is capable of delivering a given power at a given rpm and manifold pressure. The original calibration, or measurement of power, is made by means of a dynamometer in a test cell. During the ground check, power is measured with the propeller. With constant conditions of air density, the propeller, at any fixed-pitch position, always requires the same rpm to absorb the same horsepower from the engine. This characteristic is used in determining the condition of the engine.

With the governor control set for full low pitch, the propeller operates as a fixed-pitch propeller, because the engine is static. Under these conditions, the manifold pressure for any specific engine, with the mixture control in rich, indicates whether all the cylinders are operating properly. With one or more dead or intermittently firing cylinders, the operating cylinders must provide more power for a given rpm. Consequently, the carburetor throttle must be opened further, resulting in higher manifold pressure. Different engines of the same model using the same propeller installation, and at the same barometer and temperature readings, should require the same manifold pressure to within 1 "Hg. A higher than normal manifold pressure usually indicates a dead cylinder or late ignition timing. An excessively low manifold pressure for a particular rpm usually indicates that the ignition timing is early. Early ignition can cause detonation and loss of power at takeoff power settings.

The accuracy of the power check may be affected by the following variables:

1. Wind—any appreciable air movement (5 mph or more) changes the air load on the propeller blade when it is in the fixed-pitch position. A head wind increases the rpm obtainable with a given manifold pressure. A tail wind decreases the rpm.

2. Atmospheric temperatures—the effects of variations in atmospheric temperature tend to cancel each other. Higher carburetor intake and cylinder temperatures tend to lower the rpm, but the propeller load is lightened because of the less dense air.

3. Engine and induction system temperature—if the cylinder and carburetor temperatures are high because of factors other than atmospheric temperature, a low rpm results since the power is lowered without a compensating lowering of the propeller load.

4. Oil temperature—cold oil tends to hold down the rpm, since the higher viscosity results in increased friction horsepower losses.

Idle Speed & Idle Mixture Checks

Plug fouling difficulty is the inevitable result of failure to provide a proper idle mixture setting. The tendency seems to be to adjust the idle mixture on the extremely rich side and to compensate for this by adjusting the throttle stop to a relatively high rpm for minimum idling. With a properly adjusted idle mixture setting, it is possible to run the engine at idle rpm for long periods. Such a setting results in a minimum of plug fouling and exhaust smoking, and it pays dividends from the savings on the aircraft brakes after landing and while taxiing.

If the wind is not too strong, the idle mixture setting can be checked easily during the ground check as follows:

1. Close throttle.

2. Move the mixture control to the idle cutoff position and observe the change in rpm. Return the mixture control back to the rich position before engine cutoff.

As the mixture control lever is moved into idle cutoff, and before normal dropoff, one of two things may occur momentarily:

1. The engine speed may increase. An increase in rpm, but less than that recommended by the manufacturer (usually 20 rpm), indicates proper mixture strength. A greater increase indicates that the mixture is too rich.

2. The engine speed may not increase or may drop immediately. This indicates that the idle mixture is too lean. The idle mixture should be set to give a mixture slightly richer than best power, resulting in a 10- to 20-rpm rise after idle cutoff.

Engine Stopping

With each type of engine installation, specific procedures are used in stopping the engine. The general procedure, outlined in the following paragraphs, reduces the time required for stopping, minimizes backfiring tendencies, and prevents overheating of tightly baffled air-cooled engine during operation on the ground.

In stopping any aircraft engine, the controls are set as follows, irrespective of the type or fuel system installation.

1. Cowl flaps and any other shutters or doors are always placed in the full open position to avoid overheating the engine and are left in that position after the engine is stopped to prevent engine residual heat from deteriorating the ignition system.

2. Carburetor air-heater control is left in the cold position to prevent damage that may occur from backfire.

3. Constant speed propeller is usually stopped with the control set in the high pitch (decrease rpm) position to prevent exposure and corrosion of the pitch changing mechanism.

No mention is made of the throttle, mixture control, fuel selector valve, and ignition switches in the preceding set of directions because the operation of these controls varies with the type of carburetor used with the engine. An engine equipped with a carburetor incorporating an idle cutoff mixture control is stopped as follows:

1. Idle the engine by setting the throttle for 800 to 1,000 rpm.

2. Move the mixture control to the idle cutoff position. In a float-type carburetor, it equalizes the pressure in the float chamber and at the discharge nozzle.

3. After the propeller has stopped rotating, place the ignition switch in the off position.

In addition to the operations outlined previously, check the functioning of various items of aircraft equipment, such as generator systems, hydraulic systems, etc.

Basic Engine Operating Principles

Combustion Process

Normal combustion occurs when the air-fuel mixture ignites in the cylinder and burns progressively at a fairly uniform rate across the combustion chamber. When ignition is properly timed, maximum pressure is built up just after the piston has passed top dead center at the end of the compression stroke.

The flame fronts start at each spark plug and burn in more or less wavelike forms. [Figure 10-42] The velocity of the flame travel is influenced by the type of fuel, the ratio of the air-fuel mixture, and the pressure and temperature of the fuel mixture. With normal combustion, the flame travel is about 100 feet/second. The temperature and pressure within the cylinder rises at a normal rate as the air-fuel mixture burns.

Detonation

There is a limit, however, to the amount of compression and the degree of temperature rise that can be tolerated within an engine cylinder and still permit normal combustion. All fuels have critical limits of temperature and compression. Beyond this limit, they ignite spontaneously and burn with explosive violence. This instantaneous and explosive burning of the air-fuel mixture or, more accurately, of the latter portion of the charge is called detonation.

Detonation is the spontaneous combustion of the unburned charge ahead of the flame fronts after ignition of the charge. [Figure 10-43] During normal combustion, the flame fronts progress from the point of ignition across the cylinder. These flame fronts compress the gases ahead of them. At the same time, the gases are being compressed by the upward movement of the piston. If the total compression on the remaining unburned gases exceeds the critical point, detonation occurs.

The explosive burning during detonation results in an extremely rapid pressure rise. This rapid pressure rise and the high instantaneous temperature, combined with the high turbulence generated, cause a scrubbing action on the cylinder and the piston. This can burn a hole completely through the piston.

The critical point of detonation varies with the ratio of fuel to air in the mixture. Therefore, the detonation characteristic of the mixture can be controlled by varying the air-fuel ratio. At high power output, combustion pressures and temperatures are higher than they are at low or medium power. Therefore, at high power, the air-fuel ratio is made richer than is needed for good combustion at medium or low power output. This is done because, in general, a rich mixture does not detonate as readily as a lean mixture.

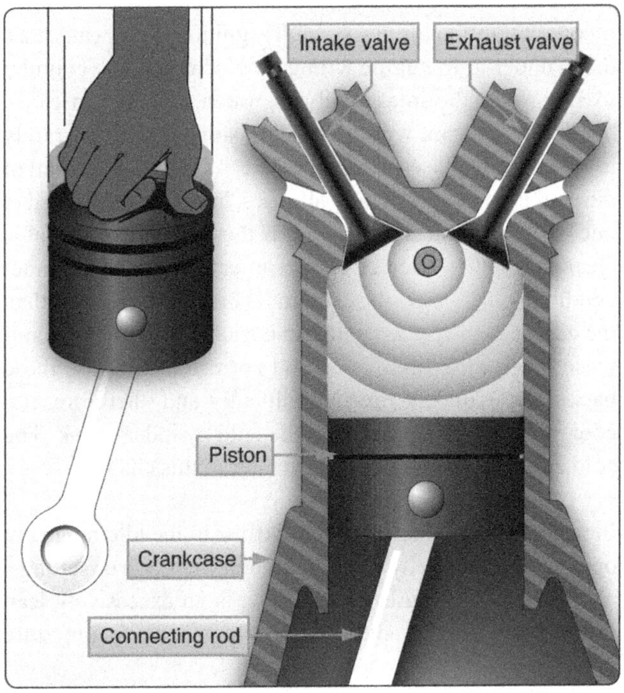

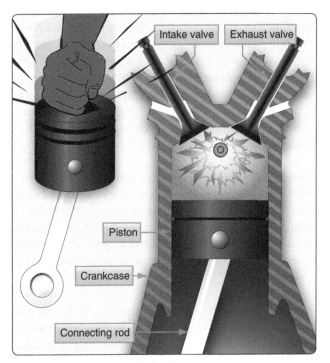

Figure 10-42. *Normal combustion within a cylinder.*

Figure 10-43. *Detonation within a cylinder.*

Unless detonation is heavy, there is no flight deck evidence of its presence. Light to medium detonation does not cause noticeable roughness, temperature increase, or loss of power. As a result, it can be present during takeoff and high-power climb without being known to the flight crew.

In fact, the effects of detonation are often not discovered until after teardown of the engine. When the engine is overhauled, however, the presence of severe detonation during its operation is indicated by dished piston heads, collapsed valve heads, broken ring lands, or eroded portions of valves, pistons, or cylinder heads.

The basic protection from detonation is provided in the design of the engine carburetor setting, which automatically supplies the rich mixtures required for detonation suppression at high power; the rating limitations, which include the maximum operating temperatures; and selection of the correct grade of fuel. The design factors, cylinder cooling, magneto timing, mixture distribution, degree of supercharging, and carburetor setting are taken care of in the design and development of the engine and its method of installation in the aircraft.

The remaining responsibility for prevention of detonation rests squarely in the hands of the ground and flight crews. They are responsible for observance of rpm and manifold pressure limits. Proper use of supercharger and fuel mixture, and maintenance of suitable cylinder head and carburetor-air-temperature (CAT) must be adhered to.

Pre-Ignition

Pre-ignition, as the name implies, means that combustion takes place within the cylinder before the timed spark jumps across the spark plug terminals. This condition can often be traced to excessive carbon or other deposits that cause local hot spots. Detonation often leads to pre-ignition. However, pre-ignition may also be caused by high-power operation on excessively lean mixtures. Pre-ignition is usually indicated in the flight deck by engine roughness, backfiring, and by a sudden increase in cylinder head temperature.

Any area within the combustion chamber that becomes incandescent serves as an igniter in advance of normal timed ignition and causes combustion earlier than desired. Pre-ignition may be caused by an area roughened and heated by detonation erosion. A cracked valve or piston, or a broken spark plug insulator, may furnish a hot point, that serves as a glow plug.

The hot spot can be caused by deposits on the chamber surfaces resulting from the use of leaded fuels. Normal carbon deposits can also cause pre-ignition. Specifically, pre-ignition is a condition similar to early timing of the spark. The charge in the cylinder is ignited before the required time for normal engine firing. However, do not confuse pre-ignition with the spark that occurs too early in the cycle. Pre-ignition is caused by a hot spot in the combustion chamber, not by incorrect ignition timing. The hot spot may be due to either an overheated cylinder or a defect within the cylinder.

The most obvious method of correcting pre-ignition is to reduce the cylinder temperature. The immediate step is to retard the throttle. This reduces the amount of fuel charge and the amount of heat generated. If a supercharger is in use, reduce manifold pressure as much as possible to reduce the charge temperature. Following this, the mixture should be enriched, if possible, to lower combustion temperature. If the engine is at high power when pre-ignition occurs, retarding the throttle for a few seconds may provide enough cooling to chip off some of the lead, or other deposit, within the combustion chamber. These chipped-off particles pass out through the exhaust.

Backfiring

When an air-fuel mixture does not contain enough fuel to consume all the oxygen, it is called a lean mixture. Conversely, a charge that contains more fuel than required is called a rich mixture. An extremely lean mixture either does not burn at all or burns so slowly that combustion is not complete at the end of the exhaust stroke. The flame lingers in the cylinder and then ignites the contents in the intake manifold or the induction system when the intake valve opens. This causes an explosion known as backfiring, which can damage the carburetor and other parts of the induction system.

Incorrect ignition timing, or faulty ignition wires, can cause the cylinder to fire at the wrong time, allowing the cylinder to fire when the intake valve is open, which can cause backfiring. A point worth stressing is that backfiring rarely involves the whole engine. Therefore, it is seldom the fault of the carburetor. In practically all cases, backfiring is limited to one or two cylinders. Usually, it is the result of faulty valve clearance setting, defective fuel injector nozzles, or other conditions that cause these cylinders to operate leaner than the engine as a whole. There can be no permanent cure until these defects are discovered and corrected. Because these backfiring cylinders fire intermittently and, therefore, run cool, they can be detected by the cold cylinder check. The cold cylinder check is discussed later in this chapter.

In some instances, an engine backfires in the idle range but operates satisfactorily at medium and high power settings. The most likely cause, in this case, is an excessively lean idle mixture. Proper adjustment of the idle air-fuel mixture usually corrects this problem.

Afterfiring

Afterfiring, sometimes called afterburning, often results when the air-fuel mixture is too rich. Overly rich mixtures are also slow burning; therefore, charges of unburned fuel are present in the exhausted gases. Air from outside the exhaust stacks mixes with this unburned fuel that ignites. This causes an explosion in the exhaust system. Afterfiring is perhaps more common where long exhaust ducting retains greater amounts of unburned charges. As in the case of backfiring, the correction for afterfiring is the proper adjustment of the air-fuel mixture.

Afterfiring can also be caused by cylinders that are not firing because of faulty spark plugs, defective fuel-injection nozzles. or incorrect valve clearance. The unburned mixture from these dead cylinders passes into the exhaust system, where it ignites and burns. Unfortunately, the resultant torching or afterburning can easily be mistaken for evidence of a rich carburetor. Cylinders that are firing intermittently can cause a similar effect. Again, the malfunction can be remedied only by discovering the real cause and correcting the defect. Dead or intermittent cylinders can be located by the cold cylinder check.

Factors Affecting Engine Operation

Compression

To prevent loss of power, all openings to the cylinder must close and seal completely on the compression and power strokes. In this respect, there are three items in the proper operation of the cylinder that must be operating correctly for maximum efficiency. First, the piston rings must be in good condition to provide maximum sealing during the stroke of the piston. There must be no leakage between the piston and the walls of the combustion chamber. Second, the intake and exhaust valves must close tightly so that there is no loss of compression at these points. Third, and very important, the timing of the valves (opening and closing) must be such that highest efficiency is obtained when the engine is operating at its normal rated rpm. A failure at any of these points results in greatly reduced engine efficiency.

Fuel Metering

The induction system is the distribution and fuel-metering part of the engine. Obviously, any defect in the induction system seriously affects engine operation. For best operation, each cylinder of the engine must be provided with the proper air-fuel mixture, usually metered by the carburetor. On some fuel-injection engines, fuel is metered by the fuel injector flow divider and fuel-injection nozzles.

The relation between air-fuel ratio and power is illustrated in *Figure 10-44*. As the fuel mixture is varied from lean to rich, the power output of the engine increases until it reaches

a maximum. Beyond this point, the power output falls off as the mixture is further enriched. This is because the fuel mixture is now too rich to provide perfect combustion. Note that maximum engine power can be obtained by setting the carburetor for one point on the curve.

In establishing the carburetor settings for an aircraft engine, the design engineers run a series of curves similar to the one shown. A curve is run for each of several engine speeds. If, for example, the idle speed is 600 rpm, the first curve might be run at this speed. Another curve might be run at 700 rpm, another at 800 rpm, and so on, in 100-rpm increments, up to takeoff rpm. The points of maximum power on the curves are then joined to obtain the best power curve of the engine for all speeds. This best power curve establishes the rich setting of the carburetor.

In establishing the detailed engine requirements regarding carburetor setting, the fact that the cylinder head temperature varies with air-fuel ratio must be considered. This variation is illustrated in the curve shown in *Figure 10-45*. Note that the cylinder head temperature is lower with the auto-lean setting than it is with the auto-rich mixture. This is exactly opposite common belief, but it is true. Furthermore, knowledge of this fact can be used to advantage by flight crews. If, during cruise, it becomes difficult to keep the cylinder head temperature within limits, the air-fuel mixture may be leaned out to get cooler operation. The desired cooling can then be obtained without going to auto-rich with its costly waste of fuel. The curve shows only the variation in cylinder head temperature. For a given rpm, the power output of the engine is less with the best-economy setting (auto-lean) than with the best-power mixture.

The decrease in cylinder head temperature with a leaner mixture holds true only through the normal cruise range. At higher power settings, cylinder temperatures are higher

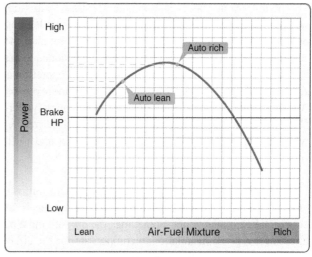

Figure 10-44. *Power versus air-fuel mixture curve.*

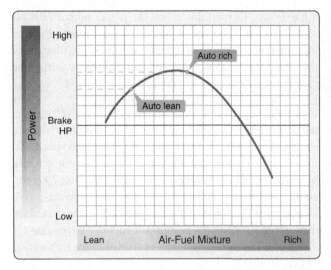

Figure 10-45. *Variation in head temperature with air-fuel mixture (cruise power).*

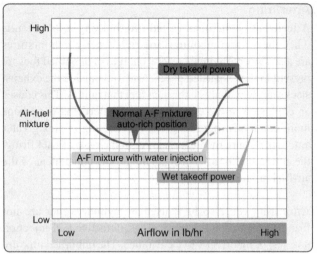

Figure 10-46. *Air-fuel curve for a water-injection engine.*

with the leaner mixtures. The reason for this reversal hinges on the cooling ability of the engine. As higher powers are approached, a point is reached where the airflow around the cylinders do not provide sufficient cooling. At this point, a secondary cooling method must be used. This secondary cooling is done by enriching the air-fuel mixture beyond the best-power point. Although enriching the mixture to this extent results in a power loss, both power and economy must be sacrificed for engine cooling purposes.

Many older, large, high-powered radial engines were influenced by the cooling requirements on air-fuel mixture, by effects of water injection. *Figure 10-46* shows an air-fuel curve for a water-injection engine. The dotted portion of the curve shows how the air-fuel mixture is leaned out during water injection. This leaning is possible because water, rather than extra fuel, is used as a cylinder coolant. These types of systems are not used on modern aircraft.

This permits leaning out to approximately best-power mixture without danger of overheating or detonation. This leaning out gives an increase in power. The water does not alter the combustion characteristics of the mixture. Fuel added to the auto-rich mixture in the power range during dry operation is solely for cooling. A leaner mixture would give more power. Actually, water or, more accurately, the antidetonant (water-alcohol) mixture is a better coolant than extra fuel. Therefore, water injection permits higher manifold pressures and a still further increase in power.

In establishing the final curve for engine operation, the engine's ability to cool itself at various power settings is, of course, taken into account. Sometimes the mixture must be altered for a given installation to compensate for the effect of cowl design, cooling airflow, or other factors on engine

cooling. The final air-fuel mixture curves take into account economy, power, engine cooling, idling characteristics, and all other factors that affect combustion.

Figure 10-47 shows a typical final curve for a float-type carburetor. Note that the air-fuel mixture at idle is the same in rich and in manual lean. The mixture remains the same until the low cruise range is reached. At this point, the curves separate and then remain parallel through the cruise and power ranges.

Note the spread between the rich and lean setting in the cruise range of both curves. Because of this spread, there is a decrease in power when the mixture control is moved from auto-rich to auto-lean with the engine operating in the cruise range. This is true because the auto-rich setting in the cruise range is very near the best power mixture ratio.

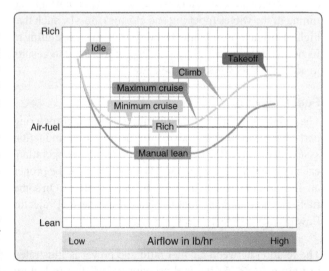

Figure 10-47. *Typical air-fuel mixture curve for a float-type carburetor.*

Therefore, any leaning out gives a mixture that is leaner than best power.

Idle Mixture

The idle mixture curve shows how the mixture changes when the idle mixture adjustment is changed. *[Figure 10-48]* Note that the greatest effect is at idling speeds. However, there is some effect on the mixture at airflows above idling. The airflow at which the idle adjustment effect cancels out varies from minimum cruise to maximum cruise. The exact point depends on the type of carburetor and the carburetor setting. In general, the idle adjustment affects the air-fuel mixture up to low cruise on engines equipped with float-type carburetors. This means that incorrect idle mixture adjustments can easily give faulty cruise performance, as well as poor idling.

There are variations in mixture requirements between one engine and another because of the fuel distribution within the engine and the ability of the engine to cool. Remember, a carburetor setting must be rich enough to supply a combustible mixture for the leanest cylinder. If fuel distribution is poor, the overall mixture must be richer than would be required for the same engine if distribution were good. The engine's ability to cool depends on such factors as cylinder design (including the design of the cooling fins), compression ratio, accessories on the front of the engine that cause individual cylinders to run hot, and the design of the baffling used to deflect airflow around the cylinder. At takeoff power, the mixture must be rich enough to supply sufficient fuel to keep the hottest cylinder cool.

Induction Manifold

The induction manifold provides the means of distributing air, or the air-fuel mixture, to the cylinders. Whether the manifold handles an air-fuel mixture or air alone depends on the type of fuel metering system used. On an engine equipped with

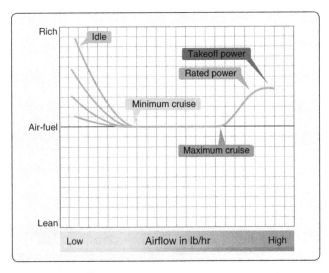

Figure 10-48. *Idle mixture curve.*

a carburetor, the induction manifold distributes an air-fuel mixture from the carburetor to the cylinders. On a fuel-injection engine, the fuel is delivered to injection nozzles, one in each cylinder, that provide the proper spray pattern for efficient burning. Thus, the mixing of fuel and air takes place at the inlet port to the cylinder. On a fuel-injection engine the induction manifold handles only air.

The induction manifold is an important item because of the effect it can have on the air-fuel mixture that finally reaches the cylinder. Fuel is introduced into the airstream by the carburetor in a liquid form. To become combustible, the fuel must be vaporized in the air. This vaporization takes place in the induction manifold, which includes the internal supercharger, if one is used. Any fuel that does not vaporize clings to the walls of the intake pipes. Obviously, this affects the effective air-fuel ratio of the mixture that finally reaches the cylinder in vapor form. This explains the reason for the apparently rich mixture required to start a cold engine. In a cold engine, some of the fuel in the airstream condenses out and clings to the walls of the manifold. This is in addition to that fuel that never vaporized in the first place. As the engine warms up, less fuel is required because less fuel is condensed out of the airstream and more of the fuel is vaporized, thus giving the cylinder the required air-fuel mixture for normal combustion.

Any leak in the induction system has an effect on the mixture reaching the cylinders. This is particularly true of a leak at the cylinder end of an intake pipe. At manifold pressures below atmospheric pressure, such a leak leans out the mixture. This occurs because additional air is drawn in from the atmosphere at the leaky point. The affected cylinder may overheat, fire intermittently, or even cut out altogether.

Operational Effect of Valve Clearance

While considering the operational effect of valve clearance, keep in mind that all aircraft reciprocating engines of current design use valve overlap. Valve overlap is when the intake and exhaust valves are open at the same time. This takes advantage of the momentum of the entering and exiting gases to improve the efficiency of getting air-fuel in and exhaust gases out. *Figure 10-49* shows the pressures at the intake and exhaust ports under two different sets of operating conditions. In one case, the engine is operating at a manifold pressure of 35 "Hg. Barometric pressure (exhaust back pressure) is 29 "Hg. This gives a pressure acting in the direction indicated by the arrow of differential of 6 "Hg (3 psi).

During the valve overlap period, this pressure differential forces the air-fuel mixture across the combustion chamber toward the open exhaust. This flow of air-fuel mixture forces ahead of it the exhaust gases remaining in the cylinder, resulting in complete scavenging of the combustion chamber.

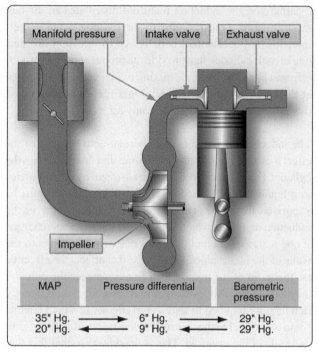

MAP	Pressure differential	Barometric pressure
35" Hg. →	6" Hg. →	29" Hg.
20" Hg. ←	9" Hg. ←	29" Hg.

Figure 10-49. *Effect of valve overlap.*

This, in turn, permits complete filling of the cylinder with a fresh charge on the following intake event. This is the situation in which valve overlap gives increased power.

There is a pressure differential in the opposite direction of 9 "Hg (4.5 psi) when the manifold pressure is below atmospheric pressure, for example, 20 "Hg. These cause air or exhaust gases to be drawn into the cylinder through the exhaust port during valve overlap.

In engines with collector rings, this inflow through the exhaust port at low power settings consists of burned exhaust gases. These gases are pulled back into the cylinder and mix with the incoming air-fuel mixture. However, these exhaust gases are inert; they do not contain oxygen. Therefore, the air-fuel mixture ratio is not affected much. With open exhaust stacks, the situation is entirely different. Here, fresh air containing oxygen is pulled into the cylinders through the exhaust. This leans out the mixture. Therefore, the carburetor must be set to deliver an excessively rich idle mixture so that, when this mixture is combined with the fresh air drawn in through the exhaust port, the effective mixture in the cylinder will be at the desired ratio.

At first thought, it does not appear possible that the effect of valve overlap on air-fuel mixture is sufficient to cause concern. However, the effect of valve overlap becomes apparent when considering idle air-fuel mixtures. These mixtures must be enriched 20 to 30 percent when open stacks, instead of collector rings (radial engines) are used on the same engine. *[Figure 10-50]* Note the spread at idle between

an open stack and an exhaust collector ring installation for engines that are otherwise identical. The mixture variation decreases as the engine speed or airflow is increased from idle into the cruise range.

Engine, airplane, and equipment manufacturers provide a powerplant installation that gives satisfactory performance. Cams are designed to give best valve operation and correct overlap. But valve operation is correct only if valve clearances are set and remain at the value recommended by the engine manufacturer. If valve clearances are set wrong, the valve overlap period is longer or shorter than the manufacturer intended. The same is true if clearances get out of adjustment during operation.

Where there is too much valve clearance, the valves do not open as wide or remain open as long as they should. This reduces the overlap period. At idling speed, it affects the air-fuel mixture, since a less-than-normal amount of air or exhaust gases is drawn back into the cylinder during the shortened overlap period. As a result, the idle mixture tends to be too rich.

When valve clearance is less than it should be, the valve overlap period is lengthened. A greater than normal amount of air, or exhaust gases, is drawn back into the cylinder at idling speeds. As a result, the idle mixture is leaned out at the cylinder. The carburetor is adjusted with the expectation that a certain amount of air or exhaust gases is drawn back into the cylinder at idling. If more or less air, or exhaust gases, are drawn into the cylinder during the valve overlap period, the mixture is too lean or too rich.

When valve clearances are wrong, it is unlikely that they are all wrong in the same direction. Instead, there is too much clearance on some cylinders and too little on others. Naturally,

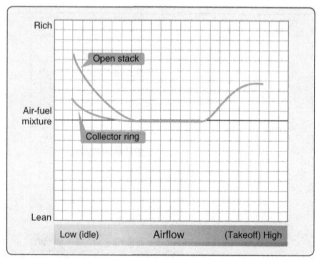

Figure 10-50. *Comparison of air-fuel mixture curves for open stack and collector ring installations.*

this gives a variation in valve overlap between cylinders. This results in a variation in air-fuel ratio at idling and lower-power settings, since the carburetor delivers the same mixture to all cylinders. The carburetor cannot tailor the mixture to each cylinder to compensate for variation in valve overlap. The effect of variation in valve clearance and valve overlap on the air-fuel mixture between cylinders is illustrated in *Figure 10-51*. Note how the cylinders with too little clearance run rich, and those with too much clearance run lean. Note also the extreme mixture variation between cylinders.

Valve clearance also effects volumetric efficiency. Any variations in air-fuel into, and exhaust gases out of, the cylinder affects the volumetric efficiency of the cylinder. With the use of hydraulic valve lifters that set the valve clearance automatically, engine operation has been greatly improved. Hydraulic lifters do have a limited range in which they can control the valve clearance, or they can become stuck in one position that can cause them to be a source of engine trouble. Normally engines equipped with hydraulic lifters require little to no maintenance.

Engine Troubleshooting

Troubleshooting is a systematic analysis of the symptoms that indicate engine malfunction. It would be impractical to list all the malfunctions that could occur in a reciprocating engine, so only the most common malfunctions are discussed. A thorough knowledge of the engine systems, applied with logical reasoning, solves most problems that may occur.

Figure 10-52 lists general conditions or troubles that may be encountered on reciprocating engines, such as engine fails to start. They are further divided into the probable causes contributing to such conditions. Corrective actions are indicated in the remedy column. The items are presented with consideration given to frequency of occurrence, ease of accessibility, and complexity of the corrective action indicated.

The need for troubleshooting normally is dictated by poor operation of the complete powerplant. Power settings for the type of operation at which any difficulty is encountered, in many cases, indicate that part of the powerplant that is the basic cause of difficulty.

The cylinders of an engine, along with any type of supercharging, form an air pump. Furthermore, the power developed in the cylinders varies directly with the rate that air can be consumed by the engine. Therefore, a measure of air consumption or airflow into the engine is a measure of power input. Ignoring for the moment such factors as humidity and exhaust back pressure, the manifold pressure gauge and the engine tachometer provide a measure of engine air consumption. Thus, for a given rpm, any change in power input is reflected by a corresponding change in manifold pressure.

The power output of an engine is the power absorbed by the propeller. Therefore, propeller load is a measure of power output. Propeller load, in turn, depends on the propeller rpm, blade angle, and air density. For a given angle and air

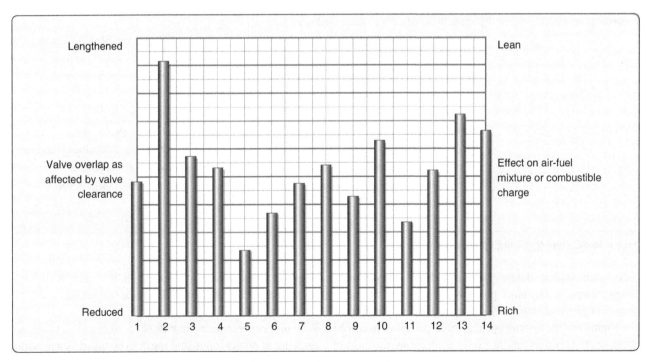

Figure 10-51. *Effect of variation in valve overlap on air-fuel mixture between cylinders.*

Trouble	Probable causes	Remedy
Engine fails to start.	• Lack of fuel.	• Check fuel system for leaks fill fuel tank.
		• Clean dirty lines, strainers, or fuel valves.
	• Underpriming.	• Use correct priming procedure.
	• Overpriming.	• Open throttle and "unload" engine by rotating the propeller.
	• Incorrect throttle setting.	• Open throttle to one-tenth of its range.
	• Defective spark plugs.	• Clean and re-gap or replace spark plugs.
	• Defective ignition wire.	• Test and replace any defective wires.
	• Defective or weak battery.	• Replace with charged battery.
	• Improper operation of magneto or breaker points.	• Check internal timing of magnetos.
	• Water in carburetor.	• Drain carburetor and fuel lines.
	• Internal failure.	• Check oil sump strainer for metal particles.
	• Magnetized impulse coupling, if installed.	• Demagnetize impulse coupling.
	• Frozen spark plug electrodes.	• Replace spark plugs or dry out plugs.
	• Mixture control in idle cutoff.	• Open mixture control.
	• Shorted ignition switch or loose ground.	• Check and replace or obtain correct idle.
Engine fails to idle properly.	• Incorrect carburetor idle speed adjustment.	• Adjust throttle stop to obtain correct idle.
	• Incorrect idle mixture.	• Adjust mixture. (Refer to engine manufacturer's handbook for proper procedure.)
	• Leak in the induction system.	• Tighten all connections in the induction system. Replace any defective parts.
	• Low cylinder compression.	• Check cylinder compression.
	• Faulty ignition system.	• Check entire ignition system.
	• Open or leaking primer.	• Lock or repair primer.
	• Improper spark plug setting for altitude.	• Check spark plug gap.
	• Dirty air filter.	• Clean or replace.
Low power and engine running uneven.	• Mixture too rich; indicated by sluggish engine operation, red exhaust flame, and black smoke.	• Check primer. Re-adjust carburetor mixture.
	• Mixture too lean; indicated by overheating or backfiring.	• Check fuel lines for dirt or other restrictions. Check fuel supply.
	• Leaks in induction system.	• Tighten all connections. Replace defective parts.
	• Defective spark plugs.	• Clean or replace spark plugs.
	• Improper grade of fuel.	• Fill tank with recommended grade.
	• Magneto breaker points not working properly.	• Clean points. Check internal timing of magneto.
	• Defective ignition wire.	• Test and replace any defective wires.
	• Defective spark plug terminal connectors.	• Replace connectors on spark plug wire.
	• Incorrect valve clearance.	• Adjust valve clearance.
		• Check and replace or repair.

Figure 10-52. *Troubleshooting opposed engines.*

density, propeller load (power output) is directly proportional to engine speed. The basic power of an engine is related to manifold pressure, fuel flow, and rpm. Because the rpm of the engine and the throttle opening directly control manifold pressure, the primary engine power controls are the throttle and the rpm control. An engine equipped with a fixed-pitch propeller has only a throttle control. In this case, the throttle setting controls both manifold pressure and engine rpm.

With proper precautions, manifold pressure can be taken as a measure of power input, and rpm can be taken as a measure of power output. However, the following factors must be

Trouble	Probable causes	Remedy
Low power and engine running uneven.	• Restriction in exhaust system. • Improper ignition timing.	• Remove restriction. • Check magnetos for timing and synchronization.
Engine fails to develop full power.	• Throttle lever out of adjustment. • Leak in induction system. • Restriction in carburetor airscoop. • Improper fuel. • Propeller governor out of adjustment. • Faulty ignition.	• Adjust throttle lever. • Tighten all connections and replace defective parts. • Examine airscoop and remove restriction. • Fill tank with recommended fuel. • Adjust governor. • Tighten all connections. Check system. Check ignition timing.
Rough running engine.	• Cracked engine mount(s). • Unbalanced propeller. • Defective mounting bushings. • Lead deposit on spark plugs. • Primer unlocked.	• Repair or replace engine mount(s). • Remove propeller and have it checked for balance. • Install new mounting bushings. • Clean or replace plugs. • Lock primer.
Low oil pressure.	• Insufficient oil. • Dirty oil strainers. • Defective pressure gauge. • Air lock or dirt in relief valve. • Leak in suction line or pressure line. • High oil temperature. • Stoppage in oil pump intake passage. • Worn or scored bearings.	• Check oil supply. • Remove and clean oil strainers. • Replace gauge. • Remove and clean oil pressure relief valve. • Check gasket between accessory housing crankcase. • See "high oil temperature" in trouble column. • Check line for obstruction. Clean suction strainer. • Overhaul engine.
High oil temperature.	• Insufficient air cooling. • Insufficient oil supply. • Clogged oil lines or strainers. • Failing or failed bearings. • Defective thermostats. • Defective temperature gauge. • Excessive blow-by.	• Check air inlet and outlet for deformation or obstruction. • Fill oil tank to proper level. • Remove and clean oil line or strainers. • Examine sump for metal particles and, if found, overhaul engine. • Replace thermostats. • Usually caused by weak or stuck rings. • Overhaul engine.
Excessive oil consumption.	• Failing or failed bearing. • Worn or broken piston rings. • Incorrect installation of piston rings. • External oil leakage. • Leakage through engine fuel pump vent. • Engine breather or vacuum pump breather.	• Check sump for metal particles and if found, an overhaul of engine is indicated. • Install new rings. • Install new rings. • Check engine carefully for leaking gaskets or O-rings. • Replace fuel pump seal. • Check engine, and overhaul or replace vacuum pump.

Figure 10-52. *Troubleshooting opposed engines (continued).*

considered:

1. Atmospheric pressure and air temperature must be considered, since they affect air density.

2. These measures of power input and power output should be used only for comparing the performance of an engine with its previous performance, or for comparing identical powerplants.

3. With a controllable propeller, the blades must be against their low-pitch stops, since this is the only blade position in which the blade angle is known and does not vary. Once the blades are off their low-pitch stops, the propeller governor takes over and maintains a constant rpm, regardless of power input or engine condition. This precaution means that the propeller control must be set to maximum or takeoff rpm, and the checks made at engine speeds below this setting.

Having relative measures of power input and power output, the condition of an engine can be determined by comparing input and output. This is done by comparing the manifold pressure required to produce a given rpm with the manifold pressure required to produce the same rpm at a time when the engine (or an identical powerplant) was known to be in top operating condition.

An example shows the practical application of this method of determining engine condition. With the propeller control set for takeoff rpm (full low blade angle), an engine may require 32 inches of manifold pressure to turn 2,200 rpm for the ignition check. On previous checks, this engine required only 30 inches of manifold pressure to turn 2,200 rpm at the same station (altitude) and under similar atmospheric conditions. Obviously, something is wrong; a higher power input (manifold pressure) is now required for the same power output (rpm). There is a good chance that one cylinder has a malfunction.

There are several standards against which engine performance can be compared. The performance of a particular engine can be compared with its past performance, provided adequate records are kept. Engine performance can be compared with that of other engines on the same aircraft or aircraft having identical installations.

If a fault does exist, it may be assumed that the trouble lies in one of the following systems:

1. Ignition system.
2. Fuel-metering system.
3. Induction system.
4. Power section (valves, cylinders, etc.).
5. Instrumentation.

If a logical approach to the problem is taken and the instrument readings properly utilized, the malfunctioning system can be pinpointed, and the specific problem in the defective system can be singled out.

The more information available about any particular problem, the better the opportunity for a rapid repair. Information that is of value in locating a malfunction includes:

1. Was any roughness noted? Under what conditions of operation?

2. What is the time on the engine and spark plugs? How long since last inspection?

3. Was the ignition system operational check and power check normal?

4. When did the trouble first appear?

5. Was backfiring or afterfiring present?

6. Was the full throttle performance normal?

From a different point of view, the powerplant is, in reality, a number of small engines turning a common crankshaft and being operated by two common phases: fuel metering and ignition. When backfiring, low power output or other powerplant difficulty is encountered, first find out which system, fuel metering or ignition, is involved and then determine whether the entire engine or only one cylinder is at fault. For example, backfiring normally is caused by:

1. Valves holding open or sticking open in one or more of the cylinders.

2. Lean mixture.

3. Intake pipe leakage.

4. An error in valve adjustment that causes individual cylinders to receive too small a charge or one too large, even though the mixture to the cylinders has the same air-fuel ratio.

Ignition system reasons for backfiring might be a cracked distributor block or a high-tension leak between two ignition leads. Either of these conditions could cause the charge in the cylinder to be ignited during the intake stroke. Ignition system troubles involving backfiring normally are not centered in the basic magneto, since a failure of the basic magneto would result in the engine not running, or it would run well at low speeds but cut out at high speeds. On the other hand, replacement of the magneto would correct a difficulty caused by a cracked distributor where the distributor is a part of the magneto.

If the fuel system, ignition system, and induction system are functioning properly, the engine should produce the correct bhp unless some fault exists in the basic power section.

Valve Blow-By

Valve blow-by is indicated by a hissing or whistle when pulling the propeller through prior to starting the engine, when turning the engine with the starter, or when running and blow-by past the intake valve is audible through the carburetor.

Correct valve blow-by immediately to prevent valve failure and possible engine failure by taking the following steps:

1. Perform a cylinder compression test to locate the faulty cylinder.

2. Check the valve clearance on the affected cylinder. If the valve clearance is incorrect, the valve may be sticking in the valve guide. To release the sticking valve, place a fiber drift on the rocker arm immediately over the valve stem and strike the drift several times with a mallet. Sufficient hand pressure should be exerted on the fiber drift to remove any space between the rocker arm and the valve stem prior to hitting the drift.

3. If the valve is not sticking and the valve clearance is incorrect, adjust it as necessary.

4. Determine whether blow-by has been eliminated by again pulling the engine through by hand or turning it with the starter. If blow-by is still present, it may be necessary to replace the cylinder.

Cylinder Compression Tests

The cylinder compression test determines if the valves, piston rings, and pistons are adequately sealing the combustion chamber. If pressure leakage is excessive, the cylinder cannot develop its full power. The purpose of testing cylinder compression is to determine whether cylinder replacement is necessary. The detection and replacement of defective cylinders prevents a complete engine change because of cylinder failure. It is essential that cylinder compression tests be made periodically. Low compression, for the most part, can be traced to leaky valves.

Conditions that affect engine compression are:

1. Incorrect valve clearances.

2. Worn, scuffed, or damaged piston.

3. Excessive wear of piston rings and cylinder walls.

4. Burned or warped valves.

5. Carbon particles between the face and the seat of the valve or valves.

6. Early or late valve timing.

Perform a compression test as soon as possible after the engine is shut down so that piston rings, cylinder walls, and other parts are still freshly lubricated. However, it is not necessary to operate the engine prior to accomplishing compression tests during engine buildup or on individually replaced cylinders. In such cases, before making the test, spray a small quantity of lubricating oil into the cylinder(s), and turn the engine over several times to seal the piston and rings in the cylinder barrel.

Be sure that the ignition switch is in the OFF position so that there is no accidental firing of the engine. Remove necessary cowling and the most accessible spark plug from each cylinder. When removing the spark plugs, identify them to coincide with the cylinder. Close examination of the plugs aid in diagnosing problems within the cylinder. Review the maintenance records of the engine being tested. Records of previous compression checks help in determining progressive wear conditions and in establishing the necessary maintenance actions.

Differential Pressure Tester

The differential pressure tester checks the compression of aircraft engines by measuring the leakage through the cylinders. The design of this compression tester is such that minute valve leakages can be detected, making possible the replacement of cylinders where valve burning is starting. The operation of the compression tester is based on the principle that, for any given airflow through a fixed orifice, a constant pressure drop across the orifice results.

As the airflow and pressure changes, pressure varies accordingly in the same direction. If air is supplied under pressure to the cylinder with both intake and exhaust valves closed, the amount of air that leaks by the valves or piston rings indicates their condition; the perfect cylinder would have no leakage. The differential pressure tester requires the application of air pressure to the cylinder being tested with the piston at top-center compression stroke. *[Figure 10-53]*

Guidelines for performing a differential compression test are:

1. Perform the compression test as soon as possible after engine shutdown to provide uniform lubrication of cylinder walls and rings.

2. Remove the most accessible spark plug from the cylinder, or cylinders, and install a spark plug adapter in the spark plug insert.

3. Connect the compression tester assembly to a 100 to 150 psi compressed air supply. *[Figure 10-54]* With the shutoff valve on the compression tester closed,

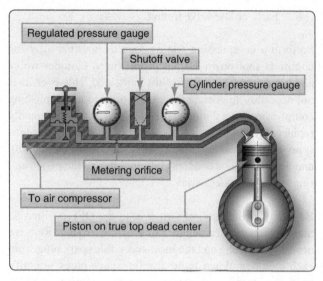

Figure 10-53. *Differential compression tester diagrams.*

adjust the regulator of the regulated pressure gauge compression tester to obtain 80 psi.

4. Open the shutoff valve and attach the air hose quick-connect fitting to the spark plug adapter. The shutoff valve, when open, automatically maintains a pressure in the cylinder of 15 to 20 psi when both the intake and exhaust valves are closed.

5. By hand, turn the engine over in the direction of rotation until the piston in the cylinder being tested comes up on the compression stroke against the 15 psi. Continue turning the propeller slowly in the direction of rotation until the piston reaches top dead center. Top dead center can be detected by a decrease in force required to move the propeller. If the engine is rotated past top dead center, the 15 to 20 psi tends to move the propeller in the direction of rotation. If this occurs, back the propeller up at least one blade prior to turning the propeller again in the direction of rotation. This backing up is necessary to eliminate the effect of backlash in the valve-operating

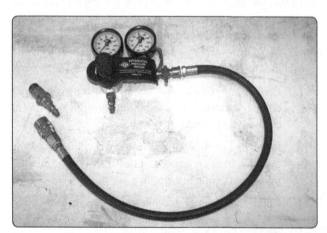

Figure 10-54. *Compression tester and adapter.*

mechanism and to keep the piston rings seated on the lower ring lands.

6. Close the shutoff valve in the compression tester and re-check the regulated pressure to see that it is 80 psi with air flowing into the cylinder. If the regulated pressure is more or less than 80 psi, readjust the regulator in the test unit to obtain 80 psi. When closing the shutoff valve, make sure that the propeller path is clear of all objects. There is sufficient air pressure in the combustion chamber to rotate the propeller if the piston is not on top dead center.

7. With regulated pressure adjusted to 80 psi, if the cylinder pressure reading indicated on the cylinder pressure gauge is below the minimum specified for the engine being tested, move the propeller in the direction of rotation to seat the piston rings in the grooves. Check all the cylinders and record the readings.

If low compression is obtained on any cylinder, turn the engine through with the starter, or re-start, and run the engine to takeoff power and re-check the cylinder, or cylinders, having low compression.

If the low compression is not corrected, remove the rocker-box cover and check the valve clearance to determine if the difficulty is caused by inadequate valve clearance. If the low compression is not caused by inadequate valve clearance, place a fiber drift on the rocker arm immediately over the valve stem and tap the drift several times with a 1 to 2 pound hammer to dislodge any foreign material that may be lodged between the valve and valve seat.

After staking the valve in this manner, rotate the engine with the starter and re-check the compression. Do not make a compression check after staking a valve until the crankshaft has been rotated either with the starter or by hand to re-seat the valve in normal manner. The higher seating velocity obtained when staking the valve will indicate valve seating, even though valve seats are slightly egged or eccentric. This procedure should only be performed if approved by the manufacturer.

Cylinders having compression below the minimum specified should be further checked to determine whether leakage is past the exhaust valve, intake valve, or piston. Excessive leakage can be detected (during the compression check):

1. At the exhaust valve by listening for air leakage at the exhaust outlet;

2. At the intake valve by escaping air at the air intake; and

3. Past the piston rings by escaping air at the engine

breather outlets.

Next to valve blow-by, the most frequent cause of compression leakage is excessive leakage past the piston. This leakage may occur because of lack of oil. To check this possibility, apply engine oil into the cylinder and around the piston. Then, re-check the compression. If this procedure raises compression to or above the minimum required, continue the cylinder in service. If the cylinder pressure readings still do not meet the minimum requirement, replace the cylinder. When it is necessary to replace a cylinder as a result of low compression, record the cylinder number and the compression value of the newly installed cylinder on the compression check sheet.

Cylinder Replacement

Reciprocating engine cylinders are designed to operate for a specified time before normal wear requires their overhaul. If the engine is operated as recommended and proficient maintenance is performed, the cylinders normally last until the engine has reached its TBO. It is known from experience that materials fail, and engines are abused through incorrect operation; this has a serious effect on cylinder life. Another reason for premature cylinder change is poor maintenance. Therefore, exert special care to ensure that all the correct maintenance procedures are adhered to when working on the engine. Some of the reasons for cylinder replacement are:

1. Low compression.
2. High oil consumption in one or more cylinders.
3. Excessive valve guide clearance.
4. Loose intake pipe flanges.
5. Loose or defective spark plug inserts.
6. External damage, such as cracks.

The cylinder is always replaced as a complete assembly, which includes piston, rings, valves, and valve springs. Obtain the cylinder by ordering the cylinder assembly under the part number specified in the engine parts catalog. Parts, such as valve springs, rocker arms, and rocker box covers, may be replaced individually.

Normally, all the cylinders in an engine are similar, all are standard size or all a certain oversize, and all are steel bore or all are chrome-plated. The size of the cylinder is indicated by a color code around the barrel between the attaching flange and the lower barrel cooling fin. In some instances, air-cooled engines are equipped with chrome-plated cylinders. Chrome-plated cylinders are usually identified by a paint band around the barrel between the attaching flange and the lower barrel cooling fin. This color band is usually international orange.

When installing a chrome-plated cylinder, do not use chrome-plated piston rings. The matched assembly includes the correct piston rings. However, if a piston ring is broken during cylinder installation, check the cylinder marking to determine what ring, chrome-plated or otherwise, is correct for replacement. Similar precautions must be taken to be sure that the correct size rings are installed.

Correct procedures and care are important when replacing cylinders. Careless work or the use of incorrect tools can damage the replacement cylinder or its parts. Incorrect procedures in installing rocker-box covers may result in troublesome oil leaks. Improper torque on cylinder hold down nuts or cap-screws can easily result in a cylinder malfunction and subsequent engine failure.

Cylinder Removal

Since these instructions are meant to cover all air-cooled engines, they are of a very general nature. The applicable manufacturer's maintenance manual should be consulted for torque values and special precautions applying to a particular aircraft and engine. However, always practice neatness and cleanliness, and always protect openings so that nuts, washers, tools, and miscellaneous items do not enter the engine's internal sections.

Assuming that all obstructing cowling and brackets have been removed, first remove the intake pipe and exhaust pipes. Plug or cover openings in the intake or diffuser section. Then, remove cylinder deflectors and any attaching brackets that would obstruct cylinder removal. Loosen the spark plugs and remove the spark plug lead clamps. Do not remove the spark plugs until ready to pull the cylinder off. Remove the rocker box covers. First, remove the nuts and then tap the cover lightly with a rawhide mallet or plastic hammer. Never pry the cover off with a screwdriver or similar tool.

Loosen the pushrod packing gland nuts or hose clamps, top and bottom. Pushrods are removed by depressing the rocker arms with a special tool, or by removing the rocker arm. Before removing the pushrods, turn the crankshaft until the piston is at top dead center on the compression stroke. This relieves the pressure on both intake and exhaust rocker arms. It is also wise to back off the adjusting nut as far as possible, because this allows maximum clearance for pushrod removal when the rocker arms are depressed.

On some model engines, or if the engine is rotated, tappets and springs of lower cylinders can fall out. Provision must be made to catch them as the pushrod and housing are removed. After removing the pushrods, examine them for markings or mark them so that they may be replaced in the same location as they were before removal. The ball ends are usually worn to fit the sockets in which they have been operating. Furthermore, on some engines, pushrods are not all of the same length. A good procedure is to mark the pushrods near the valve tappet ends No. 1 IN, No. 1 EX, No. 2 IN, No. 2 EX., etc. On fuel injection engines, disconnect the fuel injection line and any line clamps that interfere with cylinder removal.

The next step in removing the cylinder is to cut the lock wire or remove the cotter pin, and pry off the locking device from the cylinder-attaching cap-screws or nuts. Remove all the screws or nuts except two located 180° apart. Use the wrench specified for this purpose in the special tools section of the applicable manual.

Finally, while supporting the cylinder, remove the two remaining nuts and gently pull the cylinder away from the crankcase. Two technicians working together during this step, as well as during the remaining procedure for cylinder replacement, helps prevent damage or dropping of the cylinder. After the cylinder skirt has cleared the crankcase, but before the piston protrudes from the skirt, provide some means (usually a shop cloth) for preventing pieces of broken rings from falling into the crankcase. After the piston has been removed, remove the cloths and carefully check that all pieces were prevented from falling into the crankcase.

Place a support on the cylinder mounting pad and secure it with two cap-screws or nuts. Then, remove the piston and ring assembly from the connecting rod. A pin pusher or puller tool can be used when varnish makes it hard to remove the pin. If the special tool is not available and a drift is used to remove the piston pin, the connecting rod should be supported so that it does not have to take the shock of the blows. If this is not done, the rod may be damaged.

After the removal of a cylinder and piston, the connecting rod must be supported to prevent damage to the rod and crankcase. This can be done by supporting each connecting rod with the removed cylinder base oil seal ring looped around the rod and cylinder base studs.

Using a wire brush, clean the studs or cap-screws and examine them for cracks, damaged threads, or any other visible defects. If one cap-screw is found loose or broken at the time of cylinder removal, all the cap-screws for the cylinder should be discarded, since the remaining cap-screws may have been seriously weakened. A cylinder hold down stud failure places the adjacent studs under a greater operating pressure, and they are likely to be stretched beyond their elastic limit. The engine manufacturer's instruction must be followed for the number of studs that have to be replaced after a stud failure. When removing a broken stud, take proper precautions to prevent metal chips from entering the engine crankcase section. In all cases, both faces of the washers and the seating faces of stud nuts or cap-screws must be cleaned

and any roughness or burrs removed.

Cylinder Installation

See that all preservative oil accumulation on the cylinder and piston assembly is washed off with solvent and thoroughly dried with compressed air. Install the piston and ring assembly on the connecting rod. Be sure that the piston faces in the right direction. The piston number stamped on the bottom of the piston head should face toward the front of the engine. Lubricate the piston pin before inserting it. It should fit with a push fit. If a drift must be used, follow the same precaution that was taken during pin removal.

Oil the exterior of the piston assembly generously, forcing oil around the piston rings and in the space between the rings and grooves. Stagger the ring gaps around the piston and check to see that rings are in the correct grooves, and whether they are positioned correctly, as some are used as oil scrapers, others as pumper rings. The number, type, and arrangement of the compression and oil-control rings vary with the make and model of engine.

Perform any and all visual, structural, and dimensional inspection checks before installing the cylinder. Check the flange to see that the mating surface is smooth and clean. Coat the inside of the cylinder barrel generously with oil. Be sure that the cylinder oil-seal ring is in place and that only one seal ring is used.

Using a ring compressor, compress the rings to a diameter equal to that of the piston. With the piston at TDC, start the cylinder assembly down over the piston, making certain that the cylinder and piston plane remain the same. Ease the cylinder over the piston with a straight, even movement that moves the ring compressor as the cylinder slips on. Do not rock the cylinder while slipping it on the piston, since any rocking is apt to release a piston ring or a part of a ring from the ring compressor prior to the ring's entrance into the cylinder bore. A ring released in this manner expands and prevents the piston from entering the cylinder. Any attempt to force the cylinder onto the piston is apt to cause cracking or chipping of the ring or damage to the ring lands.

After the cylinder has slipped on the piston, so that all piston rings are in the cylinder bore, remove the ring compressor and the connecting rod guide. Then, slide the cylinder into place on the mounting pad. If cap-screws are used, rotate the cylinder to align the holes. While still supporting the cylinder, install two cap-screws or stud nuts 180° apart.

Install the remaining nuts or cap-screws and tighten them until they are snug. The hold down nuts, or cap-screws, must now be torqued to the value specified in the table of torque values in the engine manufacturer's service or overhaul manual. Apply the torque with a slow, steady motion until the prescribed value is reached. Hold the tension on the wrench for a sufficient length of time to ensure that the nut or cap-screw tightens no more at the prescribed torque value. In many cases, additional turning of the cap-screw, or nut, as much as one-quarter turn can be done by maintaining the prescribed torque on the nut for a short period of time. After the stud nuts, or cap-screws, have been torqued to the prescribed value, safety them in the manner recommended in the engine manufacturer's service manual.

Reinstall the push rods, push rod housings, rocker arms, barrel deflectors, intake pipes, ignition harness lead clamps and brackets, fuel injection line clamps and fuel injection nozzles (if removed), exhaust stack, cylinder head deflectors, and spark plugs. Remember that the push rods must be installed in their original locations and must not be turned end to end. Make sure that the push rod ball end seats properly in the tappet. If it rests on the edge or shoulder of the tappet during valve clearance adjustment and later drops into place, valve clearance is off.

Furthermore, rotating the crankshaft with the push rod resting on the edge of the tappet may bend the push rod. After installing the push rods and rocker arms, set the valve clearance. Before installing the rocker-box covers, lubricate the rocker arm bearings and valve stems. Check the rocker-box covers for flatness; re-surface them if necessary. After installing the gaskets and covers, tighten the rocker-box cover nuts to the specified torque. Always follow the recommended safety procedures.

Cold Cylinder Check

The cold cylinder check determines the operating characteristics of each cylinder of an air-cooled engine. The tendency for any cylinder, or cylinders, to be cold, or to be only slightly warm, indicates lack of combustion or incomplete combustion within the cylinder. This must be corrected if best operation and power conditions are to be obtained. The cold cylinder check is made with a cold cylinder indicator.

Engine difficulties that can be analyzed by use of the cold cylinder indicator are *[Figure 10-55]*:

1. Rough engine operation.

2. Excessive rpm drop during the ignition system check.

3. High manifold pressure for a given engine rpm during the ground check when the propeller is in the full low-pitch position.

4. Faulty mixture ratios caused by improper valve clearance.

In preparation for the cold cylinder check, head the aircraft into the wind to minimize irregular cooling of the individual cylinders and to ensure even propeller loading during engine operation.

Operate the engine on its roughest magneto at a speed between 1,200 and 1,600 rpm until the cylinder head temperature reading is stabilized. If engine roughness is encountered at more than one speed, or if there is an indication that a cylinder ceases operating at idle or higher speeds, run the engine at each of these speeds, and perform a cold cylinder check to pick out all the dead or intermittently operating cylinders. When low power output or engine vibration is encountered at speeds above 1,600 rpm when operating with the ignition switch on both, run the engine at the speed where the difficulty is encountered until the cylinder head temperatures have stabilized.

When cylinder head temperatures have reached the stabilized values, stop the engine by moving the mixture control to

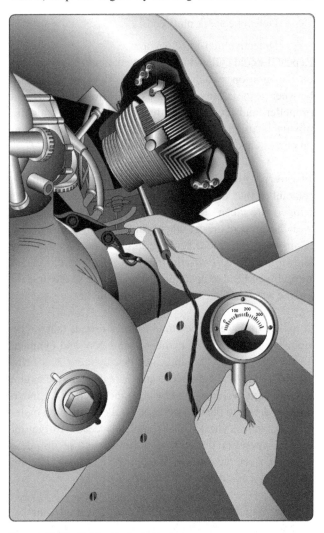

Figure 10-55. *Cold cylinder indicator.*

the idle cutoff or full lean position. When the engine ceases firing, turn off both ignition and master switches. Record the cylinder head temperature reading registered on the flight deck gauge. As soon as the propeller has ceased rotating, apply the instrument to each cylinder head, and record the relative temperature of each cylinder. Start with number one and proceed in numerical order around the engine, as rapidly as possible. To obtain comparative temperature values, a firm contact must be made at the same relative location on each cylinder. Note any outstandingly low (cold) values. Compare the temperature readings to determine which cylinders are dead (cold cylinders) or are operating intermittently.

Difficulties that may cause a cylinder to be inoperative (dead) when isolated to one magneto, either the right or left positions, are:

1. Defective spark plugs.

2. Incorrect valve clearances.

3. Leaking intake pipes.

4. Lack of compression.

5. Defective spark plug lead.

6. Defective fuel-injection nozzle.

Repeat the cold cylinder test for the other magneto positions on the ignition switch, if necessary. Cooling the engine between tests is unnecessary. The airflow created by the propeller, and the cooling effect of the incoming air-fuel mixture is sufficient to cool any cylinders that are functioning on one test and not functioning on the next.

In interpreting the results of a cold cylinder check, remember that the temperatures are relative. A cylinder temperature taken alone means little, but when compared with the temperatures of other cylinders on the same engine, it provides valuable diagnostic information. The readings shown in *Figure 10-56* illustrate this point. On this check, the cylinder head temperature gauge reading at the time the engine was shut down was 160 °C on both tests.

A review of these temperature readings reveals that, on the right magneto, cylinder number 3 runs cool and cylinders 5 and 6 run cold. This indicates that cylinder 3 is firing intermittently, and cylinders 5 and 6 are dead during engine operation on the plugs fired by the right magneto. Cylinders 4 and 6 are dead during operation on the plugs fired by the left magneto. Cylinder 6 is completely dead. An ignition system operational check would not disclose this dead cylinder, since the cylinder is inoperative on both right and left switch positions.

A dead cylinder can be detected during run-up, since an engine with a dead cylinder requires a higher than normal

Cylinder No.	Temperature readings	
	Right magneto	Left magneto
1	180	170
2	170	175
3	100	170
4	145	60
5	70	155
6	60	45

Figure 10-56. *Readings taken during a cold cylinder check.*

manifold pressure to produce any given rpm below the cut-in speed of the propeller governor. A dead cylinder could also be detected by comparing power input and power output with the aid of a torquemeter.

Defects within the ignition system that can cause a cylinder to go completely dead are:

1. Both spark plugs inoperative.

2. Both ignition leads grounded, leaking, or open.

3. A combination of inoperative spark plugs and defective ignition leads.

4. Faulty fuel-injection nozzles, incorrect valve clearances, and other defects outside the ignition system.

In interpreting the readings obtained on a cold cylinder check, the amount the engine cools during the check must be considered. To determine the extent to which this factor should be considered in evaluating the readings, re-check some of the first cylinders tested, and compare the final readings with those made at the start of the check. Another factor to be considered is the normal variation in temperature between cylinders and between rows. This variation results from those design features that affect the airflow past the cylinders.

Turbine Engine Maintenance

Turbine powerplant maintenance procedures vary widely according to the design and construction of the particular engine being serviced. The detailed procedures recommended by the engine manufacturer should be followed when performing inspections or maintenance. Maintenance information presented in this section is not intended to specify the exact manner in which maintenance operations are to be performed but is included to convey a general idea of the procedures involved. For inspection purposes, the turbine engine is divided into two main sections: the cold and hot.

Compressor Section

Maintenance of the compressor, or cold section, is one of concern because damage to blades can cause engine failure. Much of the damage to the blades arises from foreign matter being drawn into the turbine engine air intakes. The atmosphere near the ground is filled with tiny particles of dirt, oil, soot, and other foreign matter. A large volume of air is introduced into the compressor, and centrifugal force throws the dirt particles outward so that they build up to form a coating on the casing, the vanes, and the compressor blades. Accumulation of dirt on the compressor blades reduces the aerodynamic efficiency of the blades with resultant deterioration in engine performance. The efficiency of the blades is impaired by dirt deposits in a manner similar to that of an aircraft wing under icing conditions. Unsatisfactory acceleration and high exhaust gas temperature can result from foreign deposits on compressor components.

An end result of foreign particles, if allowed to accumulate in sufficient quantity, would be inefficiency. The condition can be remedied by periodic inspection, cleaning, and repair of compressor components.

Inspection & Cleaning

Minor damage to axial-flow engine compressor blades may be repaired if the damage can be removed without exceeding the allowable limits established by the manufacturer. Typical compressor blade repair limits are shown in *Figure 10-57*. Well-rounded damage to leading and trailing edges that is evident on the opposite side of the blade is usually acceptable without re-work, provided the damage is in the outer half of the blade only, and the indentation does not exceed values specified in the engine manufacturer's service and overhaul instruction manuals. When working on the inner half of the blade, damage must be treated with extreme caution. Repaired compressor blades are inspected by either magnetic particle or fluorescent penetrant inspection methods to ensure that all traces of the damage have been removed. All repairs must be well blended so that surfaces are smooth. *[Figure 10-58]* No cracks of any extent are tolerated in any area.

Whenever possible, stoning and local re-work of the blade should be performed parallel to the length of the blade. Re-work must be accomplished by hand, using stones, files, or emery cloth. Do not use a power tool to buff the entire area of the blade. The surface finish in the repaired area must be comparable to that of a new blade. On centrifugal flow engines, it is difficult to inspect the compressor inducers without first removing the air-inlet screen. After removing the screen, clean the compressor inducer and inspect it with a strong light. Check each vane for cracks by slowly turning the compressor. Look for cracks in the leading edges. A crack is usually cause for component rejection. The compressor inducers are normally the parts that are damaged by the impingement of foreign material during engine operation.

Compressor inducers are repaired by stoning out and blending the nicks and dents in the critical band (1½ to 2½ inches from the outside edge), if the depth of such nicks or dents does not exceed that specified in the engine manufacturer's service or overhaul instruction manuals. Repair nicks by stoning out material beyond the depth of damage to remove the resulting cold-worked metal. A generous radius must be applied at the edges of the blend. After blending the nick, it should be smoothed over with a crocus cloth. Pitting nicks or corrosion found on the sides of the inducer vanes are similarly removed by blending.

Causes of Blade Damage

Loose objects often enter an engine either accidentally or through carelessness. Foreign object damage (FOD), such as pencils, tools, and flashlights, are often drawn into the engine and can cause damage to the fan blades. *[Figure 10-59]* Do not carry any objects in pockets when working around operational turbine engines.

A compressor rotor can be damaged beyond repair by tools that are left in the air intake, where they are drawn into the engine on subsequent starts. A simple solution to the problem is to check the tools against a tool checklist. Prior to starting a turbine engine, make a minute inspection of engine inlet ducts to assure that items, such as nuts, bolts, lock wire, or tools, were not left there after work had been performed.

Figure 10-60 shows some examples of blade damage to an axial-flow engine. The descriptions and possible causes of blade damage are given in *Figure 10-61*. Corrosion pitting is not considered serious on the compressor stator vanes of axial-flow engines if the pitting is within the allowed tolerance. Do not attempt to repair any vane by straightening, brazing, welding, or soldering. Crocus cloth, fine files, and stones are used to blend out damage by removing a minimum of material and leaving a surface finish comparable to that of a new part. The purpose of this blending is to minimize stresses that concentrate at dents, scratches, or cracks.

The inspection and repair of air intake guide vanes, swirl vanes, and screens on centrifugal-flow engines necessitates the use of a strong light. Inspect screen assemblies for breaks, rips, or holes. Screens may be tin-dipped to tighten the wire mesh, provided the wires are not worn too thin. If the frame strip or lugs have separated from the screen frames, re-brazing may be necessary.

Inspect the guide and swirl vanes for looseness. Inspect the outer edges of the guide vanes, paying particular attention to

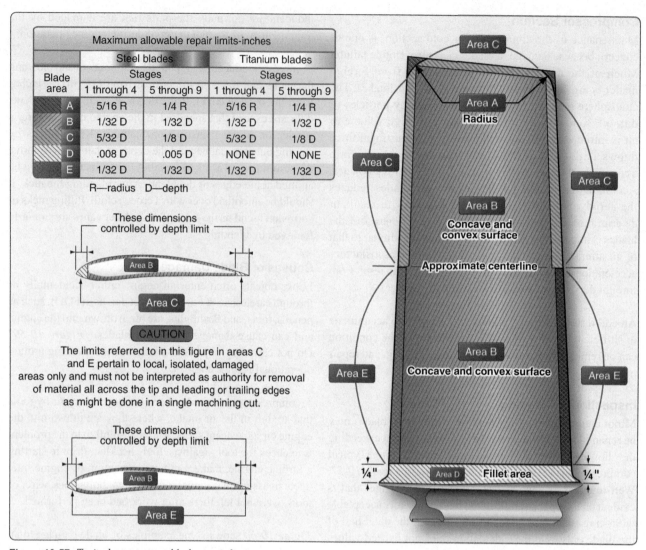

Maximum allowable repair limits-inches				
	Steel blades		Titanium blades	
Blade area	Stages		Stages	
	1 through 4	5 through 9	1 through 4	5 through 9
A	5/16 R	1/4 R	5/16 R	1/4 R
B	1/32 D	1/32 D	1/32 D	1/32 D
C	5/32 D	1/8 D	5/32 D	1/8 D
D	.008 D	.005 D	NONE	NONE
E	1/32 D	1/32 D	1/32 D	1/32 D

R—radius   D—depth

These dimensions controlled by depth limit

Area B

Area C

CAUTION

The limits referred to in this figure in areas C and E pertain to local, isolated, damaged areas only and must not be interpreted as authority for removal of material all across the tip and leading or trailing edges as might be done in a single machining cut.

These dimensions controlled by depth limit

Area B

Area E

Area C
Area A
Radius
Area C
Area C
Area B
Concave and convex surface
Approximate centerline
Area B
Concave and convex surface
Area E
Area E
¼"
Area D   Fillet area
¼"

Figure 10-57. *Typical compressor blade repair limits.*

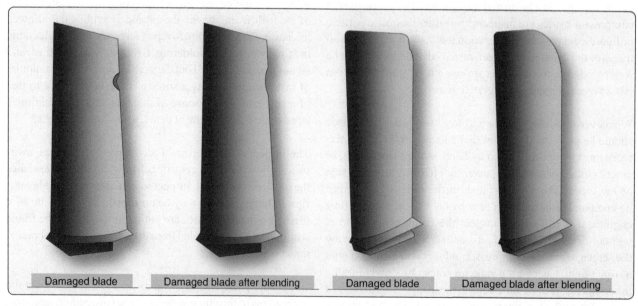

Damaged blade   Damaged blade after blending   Damaged blade   Damaged blade after blending

Figure 10-58. *Examples of repairs to damaged blades.*

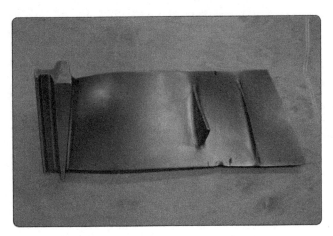

Figure 10-59. *Fan blade damage.*

the point of contact between the guides and swirl vanes for cracks and dents due to the impingement of foreign particles. Inspect the edges of the swirl vanes. Inspect the downstream edge of the guide vanes very closely, because cracks are generally more prevalent in this area. Cracks that branch or fork out so that a piece of metal could break free and fall into the compressor are cause for vane rejection.

Blending & Replacement

Because of the thin-sheet construction of hollow vanes, blending on the concave and convex surfaces, including the leading edge, is limited. Small, shallow dents are acceptable if the damage is of a rounded or gradual contour type and not a sharp or V-type, and if no cracking or tearing of vane material is evident in the damaged area.

Trailing edge damage may be blended, if one-third of the weld seam remains after repair. *[Figure 10-62]* Concave surfaces of rubber-filled vanes may have allowable cracks extending inward from the outer airfoil, provided there is no suggestion of pieces breaking away. Using a light and mirror, inspect each guide vane trailing edge and vane body for cracks or damage caused by foreign objects.

Any inspection and repair of turbine compressor section components require that the technician always use the specific manufacturer's current information for evaluation and limits of repairs.

Combustion Section Inspection

One of the controlling factors in the service life of the turbine engine is the inspection and cleaning of the hot section. Emphasis must be placed on the importance of careful inspection and repair of this section.

The following are general procedures for performing a hot section (turbine and combustion section) inspection. It is not intended to imply that these procedures are to be followed when performing repairs or inspections on turbine engines. However, the various practices are typical of those used on many turbine engines. Where a clearance or tolerance is shown, it is for illustrative purposes only. Always follow the instructions contained in the applicable manufacturer's maintenance and overhaul manuals.

The entire external combustion case should be inspected for evidence of hot spots, exhaust leaks, and distortions before the case is opened. After the combustion case has been opened, the combustion chambers can be inspected for localized overheating, cracks, or excessive wear. *[Figure 10-63]* Inspect the first stage turbine blades and nozzle guide vanes for cracks, warping, or FOD. Also inspect the combustion chamber outlet ducts and turbine nozzle for cracks and for evidence of FOD.

One of the most frequent discrepancies that are detected while inspecting the hot section of a turbine engine is cracking. These cracks may occur in many forms, and the only way to determine that they are within acceptable limits or if they are allowed at all, is to refer to the applicable engine manufacturer's service and overhaul manuals.

Cleaning the hot section is not usually necessary for a repair in the field, but in areas of high salt water or other chemicals a turbine rinse should be accomplished.

Engine parts can be degreased by using the emulsion-type cleaners or chlorinated solvents. The emulsion-type cleaners are safe for all metals, since they are neutral and noncorrosive. Cleaning parts by the chlorinated solvent method leaves the parts absolutely dry. If they are not to be subjected to further cleaning operations, they should be sprayed with a corrosion-preventive solution to protect them against rust or corrosion.

The hot section, which generally includes the combustion section and turbine sections, normally require inspections at regular intervals. The extent of disassembly of the engine to accomplish this inspection varies from different engine types. Most engines require that the combustion case be open for the inspection of the hot section. However, in performing this disassembly, numerous associated parts are readily accessible for inspection. The importance of properly supporting the engine and the parts being removed cannot be overstressed.

The alignment of components being removed and installed is also of the utmost importance. After all the inspections and repairs are made, the manufacturer's detailed assembly instructions should be followed. These instructions are important in efficient engine maintenance, and the ultimate life and performance of the engine. Extreme care must be

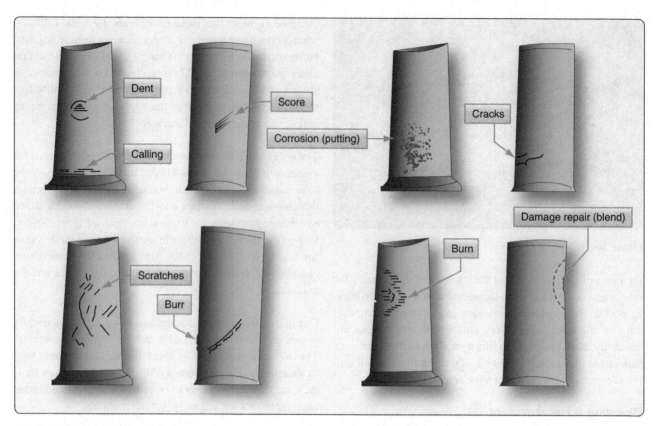

Figure 10-60. *Compressor blade damage.*

Term	Appearance	Usual Causes
• Blend	• Smooth repair of ragged edge or surface into the contour of surrounding area.	
• Bow	• Bent blade.	• Foreign objects.
• Burning	• Damage to surfaces evidenced by discoloration or, in severe cases, by flow of material.	• Excessive heat.
• Burr	• A ragged or turned out edge.	• Grinding or cutting operation.
• Corrosion (pits)	• Breakdown of the surface; pitted appearance.	• Corrosive agents—moisture, etc.
• Cracks	• A partial fracture (separation).	• Excessive stress due to shock, overloading, or faulty processing; defective materials; overheating.
• Dent	• Small, smoothly rounded hollow.	• Striking of a part with a dull object.
• Gall	• A transfer of metal from one surface to another.	• Severe rubbing.
• Gouging	• Displacement of material from a surface; a cutting or tearing effect.	• Presence of a comparatively large foreign body between moving parts.
• Growth	• Elongation of blade.	• Continued and/or excessive heat and centrifugal force.
• Pit	• (See corrosion).	
• Profile	• Contour of a blade or surface.	
• Score	• Deep scratches.	• Presence of chips between surfaces.
• Scratch	• Narrow shallow marks.	• Sand or fine foreign particles; careless handling.

Figure 10-61. *Blade maintenance terms.*

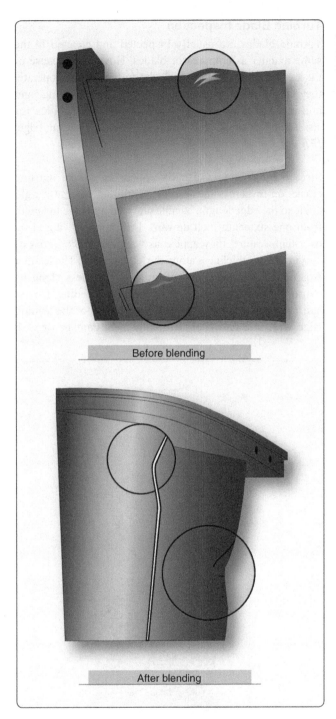

Before blending

After blending

Figure 10-62. *Guide vane trailing edge damage.*

taken during assembly to prevent dirt, dust, cotter pins, lock wire, nuts, washers, or other foreign material from entering the engine.

Marking Materials for Combustion Section Parts

Certain materials may be used for temporary marking during assembly and disassembly. Always refer to manufacturer's information for marking parts. Layout dye (lightly applied), a felt tip marker, or chalk may be used to mark parts that are

Figure 10-63. *Combustion case inspection.*

directly exposed to the engine's gas path, such as turbine blades and discs, turbine vanes, and combustion chamber liners. A wax marking pencil may be used for parts that are not directly exposed to the gas path. Do not use a wax marking pencil on a liner surface or a turbine rotor. The use of carbon alloy or metallic pencils is not recommended because of the possibility of causing intergranular corrosion attack, that could result in a reduction in material strength and cracking.

Inspection & Repair of Combustion Chambers

Inspect the combustion chambers and covers for cracks by using visible dye or fluorescent penetrant inspection method. Any cracks, nicks, or dents are usually cause for rejecting the component. Inspect the covers, noting particularly the area around the fuel drain bosses for any pits or corrosion. When repairing the combustion chamber liner, the procedures given in the appropriate engine manufacturer's overhaul instruction manual should be followed. If there is doubt that the liner is serviceable, it should be replaced.

Combustion chambers should be replaced or repaired if two cracks are progressing from a free edge so that their meeting is imminent and could allow a piece of metal that could cause turbine damage to break loose. Separate cracks in the baffle are acceptable. Cracks in the cone are rare but, at any location on this component, is cause for rejection of the liner. Cracks in the swirl vanes are cause for rejection of the liner. Loose swirl vanes may be repaired by silver brazing. Cracks in the front liner emanating from the air holes are acceptable, provided they do not exceed allowable limits. If such cracks fork or link with others, the liner must be repaired. If two cracks originating from the same air hole are diametrically opposite, the liner is acceptable. Radial cracks extending from the interconnector and spark igniter boss are acceptable, if they do not exceed allowable limits and if such cracks do not fork or link with others. Circumferential cracks around the boss pads should be repaired prior to re-use of the liner. Baffle cracks connecting more than two holes should be repaired.

After long periods of engine operation, the external surfaces of the combustion chamber liner location pads often show signs of fretting. This is acceptable, provided no resultant cracks or perforation of the metal is apparent. Any cover or chamber inadvertently dropped on a hard surface or mishandled should be thoroughly inspected for minute cracks that may elongate over a period of time and then open, creating a hazard.

Parts may be found where localized areas have been heated to an extent to buckle small portions of the chamber. Such parts are considered acceptable if the burning of the part has not progressed into an adjacent welded area, or to such an extent as to weaken the structure of the liner weldment. Buckling of the combustion chamber liner can be corrected by straightening the liner. Moderate buckling and associated cracks are acceptable in the row of cooling holes. More severe buckling that produces a pronounced shortening or tilting of the liner is cause for rejection. Upon completion of the repairs by welding, the liner should be restored as closely as possible to its original shape.

Fuel Nozzle & Support Assemblies
Clean all carbon deposits from the nozzles by washing with a cleaning fluid approved by the engine manufacturer and remove the softened deposits with a soft bristle brush. It is desirable to have filtered air passing through the nozzle during the cleaning operation to carry away deposits as they are loosened. Make sure all parts are clean. Dry the assemblies with clean, filtered air. Because the spray characteristics of the nozzle may become impaired, no attempt should be made to clean the nozzles by scraping with a hard implement or by rubbing with a wire brush. Inspect each component part of the fuel nozzle assembly for nicks and burrs. Many fuel nozzles can be checked by flowing fluid through the nozzle under pressure and closely checking the flow pattern coming for the nozzle.

Turbine Disc Inspection
The inspection for cracks is very important because cracks are not normally allowed. Crack detection, when dealing with the turbine disc and blades, is mostly visual, although structural inspection techniques can be used, such as penetrant methods and others to aid in the inspection. Cracks on the disc necessitate the rejection of the disc and replacement of the turbine rotor. Slight pitting caused by the impingement of foreign matter may be blended by stoning and polishing.

Turbine Blade Inspection
Turbine blades are usually inspected and cleaned in the same manner as compressor blades. However, because of the extreme heat under which the turbine blades operate, they are more susceptible to damage. Using a strong light and a magnifying glass, inspect the turbine blades for stress rupture cracks and deformation of the leading edge. *[Figures 10-64 and 10-65]*

Stress rupture cracks usually appear as minute hairline cracks on or across the leading or trailing edge at a right angle to the edge length. Visible cracks may range in length from one-sixteenth inch upward. Deformation, caused by over-temperature, may appear as waviness and/or areas of varying airfoil thickness along the leading edge. The leading edge must be straight and of uniform thickness along its entire length, except for areas repaired by blending. Do not confuse stress rupture cracks or deformation of the leading edge with foreign material impingement damage or with

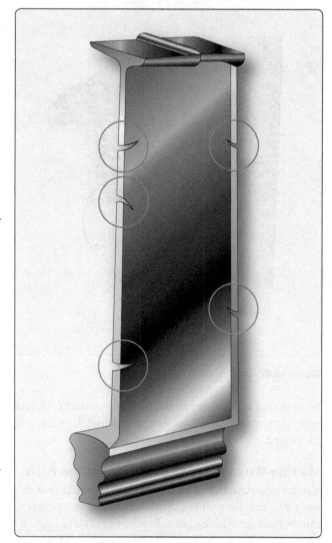

Figure 10-64. *Stress rupture cracks.*

blending repairs to the blade. When any stress rupture cracks or deformation of the leading edges of the first-stage turbine blades are found, an over-temperature condition must be suspected. Check the individual blades for stretch and the turbine disc for hardness and stretch. Blades removed for a detailed inspection or for a check of turbine disc stretch must be re-installed in the same slots from which they were removed. Number the blades prior to removal.

The turbine blade outer shroud should be inspected for air seal wear. If shroud wear is found, measure the thickness of the shroud at the worn area. Use a micrometer or another suitable and accurate measuring device that ensures a good reading in the bottom of the comparatively narrow wear groove. If the remaining radial thickness of the shroud is less than that specified, the stretched blade must be replaced. Typical blade inspection requirements are indicated in *Figure 10-66*. Blade tip curling within a one-half inch square area on the leading edge of the blade tip is usually acceptable if the curling is not sharp. Curling is acceptable on the trailing edge if it does not extend beyond the allowable area. Any sharp bends that may result in cracking or a piece breaking out of the turbine blade is cause for rejection, even though the curl may be within the allowable limits. Each turbine blade should be inspected for cracks.

Turbine Blade Replacement Procedure
Turbine blades are generally replaceable, subject to moment-weight limitations. These limitations are contained in the

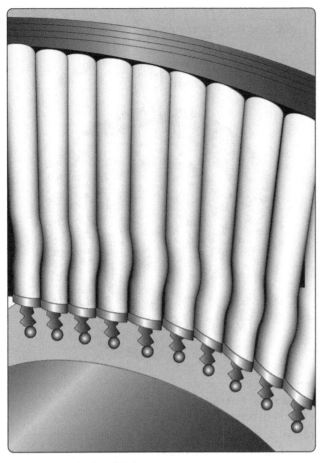

Figure 10-65. *Turbine blade waviness.*

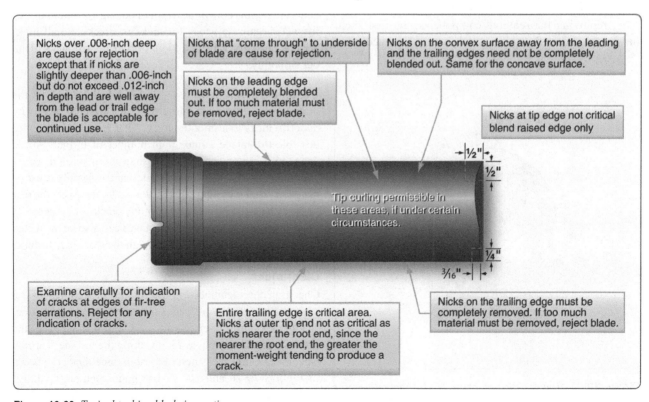

Nicks over .008-inch deep are cause for rejection except that if nicks are slightly deeper than .006-inch but do not exceed .012-inch in depth and are well away from the lead or trail edge the blade is acceptable for continued use.

Nicks that "come through" to underside of blade are cause for rejection.

Nicks on the leading edge must be completely blended out. If too much material must be removed, reject blade.

Nicks on the convex surface away from the leading and the trailing edges need not be completely blended out. Same for the concave surface.

Nicks at tip edge not critical blend raised edge only

Tip curling permissible in these areas, if under certain circumstances.

Examine carefully for indication of cracks at edges of fir-tree serrations. Reject for any indication of cracks.

Entire trailing edge is critical area. Nicks at outer tip end not as critical as nicks nearer the root end, since the nearer the root end, the greater the moment-weight tending to produce a crack.

Nicks on the trailing edge must be completely removed. If too much material must be removed, reject blade.

½"
½"
¼"
³⁄₁₆"

Figure 10-66. *Typical turbine blade inspection.*

engine manufacturer's applicable technical instructions. If visual inspection of the turbine assembly discloses several broken, cracked, or eroded blades, replacing the entire turbine assembly may be more economical than replacing the damaged blades. *[Figure 10-67]*

In the initial buildup of the turbine, a complete set of 54 blades made in coded pairs (two blades having the same code letters) is laid out on a bench in the order of diminishing moment-weight. The code letters, indicating the moment-weight balance in ounces, are marked on the rear face of the fir-tree section of the blade (viewing the blade as installed at final assembly of the engine). The pair of blades having the heaviest moment-weight is numbered 1 and 28; the next heaviest pair of blades is numbered 2 and 29; the third heaviest pair is numbered 3 and 30. This is continued until all the blades have been numbered. Mark a number 1 on the face of the hub on the turbine disc. The number 1 blade is then installed adjacent to the number 1 on the disc. *[Figure 10-68]*

The remaining blades are then installed consecutively in a clockwise direction, viewed from the rear face of the turbine disc. If there are several pairs of blades having the same code letters, they are installed consecutively before going to the next code letters. If a blade requires replacement, the diametrically opposite blade must also be replaced. Computer programs generally determine the location for turbine blades for turbine wheels on modern engines.

Turbine Nozzle Inlet Guide Vane Inspection

After removing the required components, the first stage turbine blades and turbine nozzle vanes are accessible

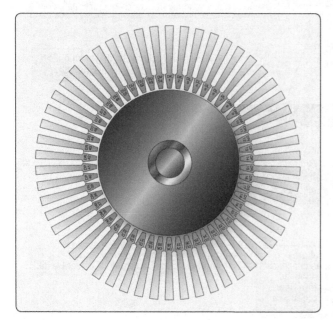

Figure 10-67. *Typical turbine rotor blade moment-weight distribution.*

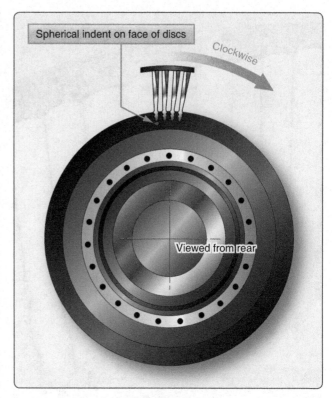

Figure 10-68. *Turbine blades.*

for inspection. The blade limits specified in the engine manufacturer's overhaul and service instruction manual should he adhered to. *Figure 10-69* shows where cracks usually occur on a turbine nozzle assembly. Slight nicks and dents are permissible if the depth of damage is within limits. Inspect the nozzle vanes for nicks or cracks. Small nicks are not cause for vane rejection, provided such nicks blend out smoothly.

Inspect the nozzle vane supports for defects caused by the impingement of foreign particles. Use a stone to blend any doubtful nicks to a smooth radius. Like turbine blades, it is possible to replace a maximum number of turbine nozzle vanes in some engines. If more than the maximum vanes are damaged, a new turbine nozzle vane assembly must be installed. With the tailpipe (exhaust nozzle) removed, the rear turbine stage can be inspected for any cracks or evidence of blade stretch. Additional nozzle stages can also be inspected with a strong light by looking through the rear-stage turbine.

Clearances

Checking the clearances is one of the procedures in the maintenance of the turbine section of a turbine engine. The manufacturer's service and overhaul manual gives the procedures and tolerances for checking the turbine. Turbine clearances being measured at various locations are shown in *Figures 10-70* and *10-71*. To obtain accurate readings, special tools provided by each manufacturer must be used

Turbine nozzle assembly

Turbine nozzle assembly at junction of combustion chamber outlet duct and turbine nozzle outer case

Cracked area along spot weld line on inner duct

Spot weld cracks on inner duct

Figure 10-69. *Turbine nozzle assembly defects.*

as described in the service instructions for specific engines.

Exhaust Section

The exhaust section of the turbine engine is susceptible to heat cracking. This section must be thoroughly inspected along with the inspection the combustion section and turbine section of the engine. Inspect the exhaust cone and exhaust nozzle for cracks, warping, buckling, or hot spots. Hot spots on the tail cone are a good indication of a malfunctioning fuel nozzle or combustion chamber.

The inspection and repair procedures for the hot section of any one gas turbine engine share similarities to those of other gas turbine engines. One usual difference is the nomenclature applied to the various parts of the hot section by the different manufacturers. Other differences include the manner of disassembly, the tooling necessary, and the repair methods and limits.

Engine Ratings

The flat rating of a turbine engine is the thrust performance that is guaranteed by the manufacturer for a new engine under specific operating conditions, such as takeoff, maximum continuous climb, and cruise power settings. The turbine inlet temperature is proportional to the energy available to turn the turbine. This means that the hotter the gases are that are entering the turbine section of the engine, the more power is available to turn the turbine wheel. The exhaust temperature is proportional to the turbine inlet temperature. Regardless of how or where the exhaust temperature is taken on the engine for the flight deck reading, this temperature is proportional to the temperature of the exhaust gases entering the first stage of inlet guide vanes. A higher EGT corresponds to a larger amount of energy to the turbine so it can turn the compressor faster. This works fine until the temperature reaches a point when the turbine inlet guide vanes start to be damaged. EGT must be held constant or lowered as the result of a prolonged hot section life and, at the same time, provide the thrust to meet the certification requirements.

Before high bypass turbofan engines, some older types of engines used water injection to increase thrust for takeoff (wet). This is the maximum allowable thrust for takeoff. The rating is obtained by actuating the water-injection system and setting the computed wet thrust with the throttle, in terms of a predetermined turbine discharge pressure or engine pressure ratio for the prevailing ambient conditions. The rating is restricted to takeoff, is time-limited, and has an altitude limitation. Water injection is not used very much on turbine engines any more.

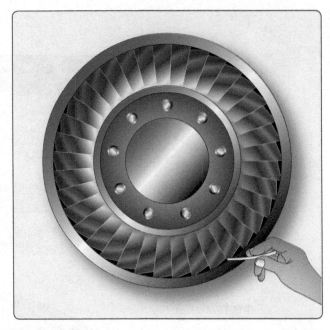

Figure 10-70. *Measuring the turbine blades to shroud (tip) clearances.*

Figure 10-71. *Measuring turbine wheel to exhaust cone clearance.*

Turbine Engine Instruments

Engine Pressure Ratio Indicator

Engine pressure ratio (EPR) is an indication of the thrust being developed by a turbofan engine and is used to set power for takeoff on many types of aircraft. It is instrumented by total pressure pickups in the engine inlet (Pt2) and in the turbine exhaust (Pt7). The reading is displayed in the flight deck by the EPR gauge, which is used in making engine power settings. *[Figure 10-72]*

Torquemeter (Turboprop Engines)

Only 10 to 15 percent of the thrust produced by a turboprop engine is from propulsive force derived from the jet thrust exiting the exhaust. Engine pressure ratio is not used as an indicator of the power produced by a turboprop engine. Turboprops are usually fitted with a torquemeter that measures torque applied to a shaft turned by the gas generator and power turbines of the turbine engine. The torquemeter can be operated by engine oil pressure metered through a valve that is controlled by a helical ring gear that moves in response to the applied torque. *[Figure 10-73]* This gear moves against a piston that controls the opening of a valve, which controls the oil pressure flow. This action makes the oil pressure proportional to torque being applied at the propeller shaft. Generally, transducer is used to transfer the oil pressure into an electrical signal to be read by the flight deck instrument. The read out in the flight deck is normally in lb/ft of torque, or percent horsepower. The torquemeter is very important as it is used to set power settings. This instrument must be calibrated at intervals to assure its accuracy.

Tachometer

Gas turbine engine speeds are measured by the engines rpm, which are also the compressor/turbine combination rpm of each rotating spool. Most turbofan engines have two or more spools, compressor, and turbine sections that turn independently at different speeds. Tachometers are usually calibrated in percent rpm so that various types of engines can be operated on the same basis of comparison. *[Figure 10-73]* Also, turbine speeds are generally very high, and the large numbers of rpm would make it very confusing. Turbofan engines with two spools or separate shafts, high

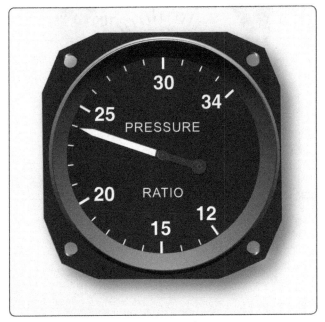

Figure 10-72. *Engine pressure ratio indications.*

pressure and low pressure spools, are generally referred to as N1 and N2, with each having their own indicator. The main purpose of the tachometer is to be able to monitor rpm under normal conditions, during an engine start, and to indicate an overspeed condition, if one occurs.

Exhaust Gas Temperature Indicator (EGT)

Exhaust gas temperature (EGT), turbine inlet temperature, (TIT), turbine gas temperature (TGT), interstage turbine temperature (ITT), and turbine outlet temperature (TOT) are all relative temperatures used to monitor the temperature of the exhaust gases entering the first stage turbine inlet guide vanes. Even though these temperatures are taken at different locations on the engine (each engine having one location), they are all relative to the temperature of the gases entering the first stage turbine inlet guide vanes.

Temperature is an engine operating limit and is used to monitor the mechanical integrity of the turbines, as well as to check engine operating conditions. Actually, the temperature of the gases entering the first stage turbine inlet guide vanes is the important consideration, since it is the most critical of all the engine variables. However, it is impractical to measure turbine inlet temperature in most engines, especially large engines. Consequently, temperature thermocouples are inserted at the turbine discharge, where the temperature provides a relative indication of that at the inlet. Although the temperature at this point is much lower than at the inlet, it provides surveillance over the engine's internal operating conditions. Several thermocouples are usually used, that are spaced at intervals around the perimeter of the engine exhaust duct near the turbine exit. The EGT indicator in the flight deck shows the average temperature measured by the individual thermocouples. *[Figure 10-73]*

Fuel-Flow Indicator

Fuel-flow instruments indicate the fuel flow in pounds per hour (lb/hr) from the engine fuel control. Fuel flow in turbine aircraft is measured in lb/hr instead of gallons, because the fuel weight is a major factor in the aerodynamics of large turbine aircraft. Fuel flow is of interest in monitoring fuel consumption and checking engine performance. *[Figure 10-73]*

Engine Oil Pressure Indicator

To guard against engine failure resulting from inadequate lubrication and cooling of the various engine parts, the oil supply to critical areas must be monitored. The oil pressure indicator usually shows the engine oil pump discharge pressure.

Engine Oil Temperature Indicator

The ability of the engine oil to lubricate and cool depends on the temperature of the oil, as well as the amount of

Figure 10-73. *Typical turbine engine instruments.*

oil supplied to the critical areas. An oil inlet temperature indicator frequently is provided to show the temperature of the oil as it enters the oil pressure pump. Oil inlet temperature is also an indication of proper operation of the engine oil cooler.

Turbine Engine Operation

The engine operating procedures presented here apply generally to turbofan, turboprop, turboshaft, and auxiliary power units (APU). The procedures, pressures, temperatures, and rpm that follow are intended primarily to serve as a guide. It should be understood that they do not have general application. The manufacturer's operating instructions should be consulted before attempting to start and operate any turbine engine.

A turbofan engine has only one power control lever. Adjusting the power lever, or throttle lever, sets up a thrust condition for which the fuel control meters fuel to the engine. Engines equipped with thrust reversers go into reverse thrust at throttle positions below idle. A separate fuel shutoff lever is usually provided on engines equipped with thrust reversers.

Prior to start, particular attention should be paid to the engine air inlet, the visual condition and free movement of the compressor and turbine assembly, and the parking ramp area fore and aft of the aircraft. The engine is started by using an external air power source, APU, or an already operating engine. Starter types and the engine starting cycle have been discussed previously. On multi-engine aircraft, the engines are usually started by an onboard APU that supplies the air pressure for a pneumatic starter on each engine. Air bled from the APU is used as a source of power for starting the engines.

During the start, it is necessary to monitor the tachometer, the oil pressure, and the exhaust gas temperature. The normal starting sequence is:

1. Rotate the compressor with the starter;

2. Turn the ignition on; and

3. Open the engine fuel valve, either by moving the throttle to idle or by moving a fuel shutoff lever or turning a switch.

Adherence to the procedure prescribed for a particular engine is necessary as a safety measure and to avoid a hot or hung start. A successful start is noted first by a rise in exhaust gas temperature. If the engine does not light up, meaning that fuel starts to burn inside of the engine within a prescribed period of time, or if the exhaust gas starting temperature limit is exceeded, a hot start, the starting procedure should be aborted. Hot starts are not common, but when they do occur, they can usually be stopped in time to avoid excessive temperature by observing the exhaust gas temperature constantly during the start. When necessary, the engine is cleared of trapped fuel or gases by continuing to rotate the compressor with the starter, but with the ignition and fuel turned off. If the engine did not light off during start after the allotted time, about 10 seconds although this time varies from engine to engine, the fuel must be shut off as the engine is being filled with unburned fuel. A hung start is when the engine lights off, but the engine will not accelerate to idle rpm.

Ground Operation Engine Fire

Move the fuel shutoff lever to the off position if an engine fire occurs, or if the fire warning light is illuminated during the starting cycle. Continue cranking or motoring the engine until the fire has been expelled from the engine. If the fire persists, CO_2 can be discharged into the inlet duct while it is being cranked. Do not discharge CO_2 directly into the engine exhaust, because it may damage the engine. If the fire cannot be extinguished, secure all switches and leave the aircraft. If the fire is on the ground under the engine overboard drain, discharge the CO_2 on the ground rather than on the engine. This also is true if the fire is at the tailpipe and the fuel is dripping to the ground and burning.

Engine Checks

Checking turbofan engines for proper operation consists primarily of simply reading the engine instruments and then comparing the observed values with those known to be correct for any given engine operating condition. After the engine has started, idle rpm has been attained, and the instrument readings have stabilized, the engine should be checked for satisfactory operation at idling speed. The oil pressure indicator, tachometer, and the exhaust gas temperature readings should be compared with the allowable ranges.

Checking Takeoff Thrust

Takeoff thrust is checked by adjusting the throttle to obtain a single, predicted reading on the engine pressure ratio indicator in the aircraft. The value for engine pressure ratio, which represents takeoff thrust for the prevailing ambient atmospheric conditions, is calculated from a takeoff thrust setting curve or, on newer aircraft, is a function of the onboard computer. This curve has been computed for static conditions. [Figure 10-74] Therefore, for all precise thrust checking, the aircraft should be stationary, and stable engine operation should be established. If it is needed for calculating thrust during an engine trim check, turbine discharge pressure (Pt7) is also shown on these curves. Appropriate manuals should be consulted for the charts for a specific make and model engine. Engine trimming procedure is also covered in Chapter 2, Engine Fuel & Fuel Metering Systems. The engine pressure ratio computed from the thrust setting curve represents thrust or a lower thrust call part power thrust used for testing. The aircraft throttle is advanced to obtain this predicted reading on the engine pressure ratio indicator, or the part power stop is engaged in the aircraft. If an engine develops the predicted thrust and if all the other engine instruments are reading within their proper ranges, engine operation is considered satisfactory. Full authority digital engine controls (FADEC) (computer controls) also have means of checking the engine with the results displayed on the flight deck.

Ambient Conditions

The sensitivity of gas turbine engines to compressor inlet air temperature and pressure necessitates that considerable care be taken to obtain correct values for the prevailing ambient air conditions when computing takeoff thrust. Some things to remember are:

1. The engine senses the air temperature and pressure at the compressor inlet. This is the actual air temperature just above the runway surface. When the aircraft is stationary, the pressure at the compressor inlet is the static field or true barometric pressure, and not the barometric pressure corrected to sea level that is normally reported by airport control towers as the altimeter setting. On FADEC engines, the computer reads this information and sends it to the engine controls.

2. Temperature sensed is the total air temperature (TAT) that is used by several onboard computers. The engine controls set the engine computers according to the TAT.

3. Relative humidity, which affects reciprocating engine power appreciably, has a negligible effect on turbine engine thrust, fuel flow, and rpm. Therefore, relative humidity is not usually considered when computing thrust for takeoff or determining fuel flow and rpm for routine operation.

Engine Shutdown

On turbine engines that have a thrust reverser, retarding the aircraft throttle to idle or power lever to OFF cuts the fuel supply to the engine and shuts down the engine. On engines equipped with thrust reversers, this is accomplished by means of a separate fuel shutoff lever or switch. When an engine has been operated at high power levels for extended periods of time, a cool down time should be allowed before shutting down. It is recommended the engine be operated at below a low power setting, preferably at idle for a period of 5 minutes to prevent possible seizure of the rotors. This applies, in particular, to prolonged operation at high rpm on the ground, such as during engine trimming. The turbine case and the turbine wheels operate at approximately the same temperature

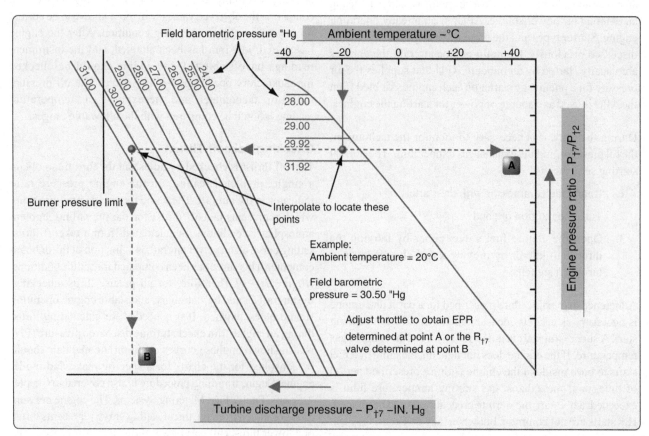

Figure 10-74. *Typical takeoff thrust setting curve for static conditions.*

when the engine is running. However, the turbine wheels are relatively massive, compared with the case, and are not cooled so readily. The turbine case is exposed to cooling air from both inside and outside the engine. Consequently, the case and the wheels lose their residual heat at different rates after the engine has been shut down. The case, cooling faster, tends to shrink upon the wheels, that are still rotating. Under extreme conditions, the turbine blades may squeal or seize; thus, a cooling period is required if the engine has been operating at prolonged high speed. Should the turbine wheels seize, no harm normally results, provided no attempt is made to turn the engine over until it has cooled sufficiently to free the wheels. In spite of this, every effort should be made to avoid seizure.

To ensure that fuel remains in the lines and that the engine-driven fuel pumps are not starved for fuel that lubricates the pumps, the aircraft fuel boost pump must be turned off after, not before, the throttle or the fuel shutoff lever is placed in the OFF position.

Generally, an engine should not be shut down by the fuel shutoff lever until after the aircraft throttle has been retarded to idle. Because the fuel shutoff valve is located on the fuel control discharge, a shutdown from high thrust settings results in high fuel pressures within the control that can harm the fuel system parts.

When an accurate reading of the oil level in the oil tank is needed following an engine shutdown, the engine should be operated and shut down with the oil check taking place within not more than 30 minutes after shutdown. Check the engine manuals for the specific procedure.

Troubleshooting Turbine Engines

Included in this section are typical guidelines for locating engine malfunctions on most turbine engines. Since it would be impractical to list all the malfunctions that could occur, only the most common malfunctions are covered. A thorough knowledge of the engine systems, applied with logical reasoning, solves most problems that may occur.

Figure 10-75 enumerates some malfunctions that may be encountered. Possible causes and suggested actions are given in the adjacent columns. The malfunctions presented herein are solely for the purpose of illustration and should not be construed to have general application. For exact information about a specific engine model, consult the applicable manufacturer's instructions.

Turboprop Operation

Turboprop engine operation is quite similar to that of a turbojet engine, except for the added feature of a propeller.

The starting procedure and the various operational features are very much alike. The turboprop chiefly requires attention to engine operating limits, the throttle or power lever setting, and the torquemeter pressure gauge. Although torquemeters indicate only the power being supplied to the propeller and not the equivalent shaft horsepower, torquemeter pressure is approximately proportional to the total power output and, thus, is used as a measure of engine performance. The torquemeter pressure gauge reading during the takeoff engine check is an important value. It is usually necessary to compute the takeoff power in the same manner as is done for a turbojet engine. This computation is to determine the maximum allowable exhaust gas temperature and the torquemeter pressure that a normally functioning engine should produce for the outside, or ambient, air temperature and barometric pressure prevailing at the time.

Troubleshooting Procedures for Turboprop Engines

All test run-ups, inspections, and troubleshooting should be performed in accordance with the applicable engine manufacturer's instructions. In *Figure 10-76*, the troubleshooting procedure for the turboprop reduction gear, torquemeter, and power section are combined because of their inter-relationships. The table includes the principal troubles, together with their probable causes and remedies.

Turbine Engine Calibration & Testing

Some of the most important factors affecting turbine engine life are EGT, engine cycles (a cycle is generally a takeoff and landing) and engine speed. Excess EGT of a few degrees reduces turbine component life. Low EGT materially reduces turbine engine efficiency and thrust. So, to make the engine highly efficient, the exhaust temperatures need to be as high as possible, while maintaining an EGT operating temperature that does not damage the turbine section of the engine. If the engine is operated at excess exhaust temperatures, engine deterioration occurs. Since the EGT temperature is set by the EGT temperature gauge, it is imperative that it is accurate. Excessive engine speed can cause premature engine wear and, if extreme, can cause engine failure.

One older type of calibration test unit used to analyze the turbine engine is the jetcal analyzer. *[Figure 10-77]* A jetcal analyzer is a portable instrument made of aluminum, stainless steel, and plastic. The major components of the analyzer are the thermocouple, rpm, EGT indicator, resistance, and insulation check circuits, as well as the potentiometer, temperature regulators, meters, switches, and all the necessary cables, probes, and adapters for performing all tests.

Turbine Engine Analyzer Uses

Many different types of analyzers are used each with its own function, including onboard systems that use computers to

Indicated Malfunction	Probable Causes	Suggested Action
Engine has low rpm, exhaust gas temperature, and fuel flow when set to expected engine pressure ratio.	• Engine pressure ratio indication has high reading error.	• Check inlet pressure line from probe to transmitter for leaks. • Check engine pressure ratio transmitter and indicator for accuracy.

Engine has high rpm, exhaust gas temperature, and fuel flow when set to expect engine pressure ratio.	• Engine pressure ratio indication has low reading error due to:
	  - Misaligned or cracked turbine discharge probe.
	  - Leak in turbine discharge pressure line from probe to transmitter.
	  - Inaccurate engine pressure ratio transmitter or indicator.
	  - Carbon particles collected in turbine discharge pressure line or restrictor orifices.
		• Check probe condition.
		• Pressure-test turbine discharge pressure line for leaks.
		• Check engine pressure ratio transmitter and indicator for accuracy.

Engine has high exhaust gas temperature, low rpm, and high fuel flow at all engine pressure ratio settings.	• Possible turbine damage and/or loss of turbine efficiency.	• Confirm indication of turbine damage by:
		  - Checking engine coast-down for abnormal noise and reduced time.
		  - Visually inspect turbine area with strong light.

Note: Engines with damage in turbine section may have tendency to hang up during starting.	• If only exhaust gas temperature is high, other parameters normal, the problem may be thermocouple leads or instrument.	• Re-calibrate exhaust gas temperature instrumentation.
Engine vibrates throughout rpm range, but indicated amplitude reduces as rpm is reduced.	• Turbine damage.	• Check turbine as outlined in preceding item.
Engine vibrates at high rpm and fuel flow when compared to constant engine pressure ratio.	• Damage in compressor section.	• Check compressor section for damage.
Engine vibrates throughout rpm range, but is more pronounced in cruise or idle rpm range.	• Engine-mounted accessory such as constant-speed drive, generator, hydraulic pump, etc.	• Check each component in turn.
No change in power setting parameters, but oil temperature high.	• Engine main bearings.	• Check scavenge oil filters and magnetic plugs.
Engine has higher than normal exhaust gas temperature during takeoff, climb, and cruise. Rpm and fuel flow higher than normal.	• Engine bleed-air valve malfunction. • Turbine discharge pressure probe or line to transmitter leaking.	• Check operation of bleed valve. • Check condition of probe and pressure line to transmitter.
Engine has high exhaust gas temperature at target engine pressure ratio for takeoff.	• Engine out of trim.	• Check engine with jetcal. Re-trim as desired.

Figure 10-75. *Troubleshooting turbojet engines.*

Indicated Malfunction	Probable Causes	Suggested Action
Engine rumbles during starting and at low power cruise conditions.	• Pressurizing and drain valve malfunction. • Cracked air duct. • Fuel control malfunction.	• Replace pressurizing and drain valves. • Repair or replace duct. • Replace fuel control.
Engine rpm hangs up during starting.	• Subzero ambient temperatures. • Compressor section damage. • Turbine section damage.	• If hang-up is due to low ambient temperature, engine usually can be started by turning on fuel booster pump or by positioning start lever to run earlier in the starting cycle. • Check compressor for damage. • Inspect turbine for damage.
High oil temperature.	• Scavenge pump failure. • Fuel heater malfunction.	• Check lubricating system and scavenge pumps. • Replace fuel heater.
High oil consumption.	• Scavenge pump failure. • High sump pressure. • Gearbox seal leakage.	• Check scavenge pumps. • Check sump pressure as outlined in manufacturer's maintenance manual. • Check gearbox seal by pressurizing overboard vent.
Overboard oil loss.	• Can be caused by high airflow through the tank, foaming oil, or unusual amounts of oil returned to the tank through the vent system.	• Check oil for foaming. • Vacuum-check sumps. • Check scavenge pumps.

Figure 10-75. *Troubleshooting turbojet engines (continued).*

test aircraft systems. Depending upon the specific analyzer used, procedures vary somewhat, but the basic test are outlined here. Always refer to the specific instructions associated with the analyzer being used.

Most analyzers may be used to:

1. Functionally check the aircraft EGT system for error, without running the engine or disconnecting the wiring.

2. Check individual thermocouples before placement in a parallel harness.

3. Check each engine thermocouple in a parallel harness for continuity.

4. Check the thermocouples and parallel harness for accuracy.

5. Check the resistance of the EGT circuit.

6. Check the insulation of the EGT circuit for shorts to ground, or for shorts between leads.

7. Check EGT indicators , either in or out of the aircraft, for error.

8. Determine engine rpm accuracy during engine testing. Added to this is the checking and troubleshooting of the aircraft tachometer system.

9. Establish the proper relationship between the EGT and engine rpm during engine run-up.

Analyzer Safety Precautions

Observe the following safety precautions while operating the engine analyzer or other types of test equipment:

1. Never use a voltammeter to check the potentiometer for continuity. If a voltammeter is used, damage to the galvanometer and standard battery cell results.

2. Check the thermocouple harness before engine run-up. This must be done because the circuit must be correct before the thermocouples can be used for true EGT pickup.

3. For safety, ground the jetcal analyzer when using an AC power supply. Any electrical equipment operated on AC power and utilizing wire-wound coils, such as the probes with the jetcal analyzer, has an induced voltage on the case that can be discharged if the equipment is not grounded. This condition is not apparent during dry weather, but on damp days the operator can be shocked slightly. Therefore, for the operator's protection, the jetcal analyzer should be grounded using the pigtail lead in the power inlet cable.

4. Use heater probes designed for use on the engine

Trouble	Probable causes	Remedy
Power unit fails to turn over during attempted start.	• No air to starter. • Propeller brake locked.	• Check started air valve solenoid and air supply. • Unlock brake by turning propeller by hand in direction of normal rotation.
Power unit fails to start.	• Starter speed low because of inadequate air supply to starter. • If fuel is not observed leaving the exhaust pipe during start, fuel selector valve may be inoperative because of low power supply or may be locked in "OFF." • Fuel pump inoperative. • Aircraft fuel filter dirty. • Fuel control cutoff valve closed.	• Check starter air valve solenoid and air supply. • Check power supply or electrically operated valves. Replace valves if defective. • Check pump for sheared drives or internal damage. Check for air leaks at outlet. • Clean filter and replace filtering elements if necessary. • Check electrical circuit to ensure that actuator is being energized. Replace actuator or control.
Engine fires, but will not accelerate to correct speed.	• Insufficient fuel supply to control unit. • Fuel control main metering valve sticking. • Fuel control bypass valve sticking open. • Drain valve stuck open. Starting fuel enrichment pressure switch setting too high.	• Check fuel system to ensure all valves are open and pumps are operative. • Flush system. Replace control. • Flush system. Replace control. • Replace drain valve. Replace pressure switch.
Acceleration temperature too high during starting.	• Fuel control bypass valve sticking closed. • Fuel control acceleration cam incorrectly adjusted. • Defective fuel nozzle. • Fuel control thermostat failure.	• Flush system. Replace control. • Replace control. • Replace nozzle with a known satisfactory unit. • Replace control.
Acceleration temperature during starting too low.	• Acceleration cam of fuel control incorrectly adjusted.	• Replace control.
Engine speed cycles after start.	• Unstable fuel control governor operation.	• Continue engine operation to allow control to condition itself.
Power unit oil pressure drops off severely.	• Oil supply low. • Oil pressure transmitter or indicator giving false indication.	• Check oil supply and refill as necessary. • Check transmitter or indicator and repair or replace if necessary.
Oil leakage at accessory drive seals.	• Seal failure.	• Replace seal or seals.
Engine unable to reach maximum controlled speed of 100 percent.	• Faulty propeller governor. • Faulty fuel control or air sensing tip.	• Replace propeller control assembly. • Replace faulty control. If dirty, use air pressure in reverse direction of normal flow through internal engine passage and sensing tip.
Vibration indication high.	• Vibration pickup or vibration meter malfunction.	• Calibrate vibration meter. • Start engine and increase power gradually. • Observe vibration indicator. If indications prove pickup to be at fault, replace it. If high vibration remains as originally observed, remove power unit for overhaul.

Figure 10-76. *Troubleshooting turboprop engines.*

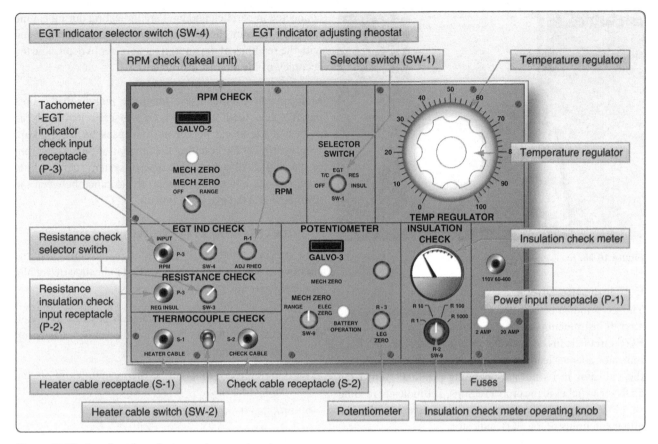

Figure 10-77. *Jetcal analyzer instrument compartment.*

thermocouples to be tested. Temperature gradients are very critical in the design of heater probes. Each type of aircraft thermocouple has its own specially designed probe. Never attempt to modify heater probes to test other types of thermocouples.

5. Do not leave heater probe assemblies in the exhaust nozzle during engine run-up.

6. Never allow the heater probes to go over 900 °C (1,652 °F). Exceeding these temperatures results in damage to the jetcal analyzer and heater probe assemblies.

Continuity Check of Aircraft EGT Circuit

To eliminate any error caused by one or more inoperative aircraft thermocouples, a continuity check is performed. The check is made by heating one heater probe to between 500 and 700 °C and placing the hot probe over each of the aircraft thermocouples, one at a time. The EGT indicator must show a temperature rise as each thermocouple is checked. When large numbers (eight or more) of thermocouples are used in the harness, it is difficult to see a rise on the aircraft instrument because of the electrical characteristics of a parallel circuit. Therefore, the temperature indication of the aircraft thermocouples is read on the potentiometer of the analyzer by using the check cable and necessary adapter.

Functional Check of Aircraft EGT Circuit

During the EGT system functional test and the thermocouple harness checks, the analyzer has a specific degree of accuracy at the test temperature, which is usually the maximum operating temperature of the turbine engine. *[Figure 10-78]* Each engine has its own maximum operating temperature, that can be found in applicable technical instructions.

The test is made by heating the engine thermocouples in the exhaust nozzle or turbine section to the engine test temperature. The heat is supplied by heater probes through the necessary cables. With the engine thermocouples hot, their temperature is registered on the aircraft EGT indicator. At the same time, the thermocouples embedded in the heater probes, which are completely isolated from the aircraft system, are picking up and registering the same temperature on the test analyzer.

The temperature registered on the aircraft EGT indicator should be within the specified tolerance of the aircraft system and the temperature reading on the temperature analyzer. When the temperature difference exceeds the allowable tolerance, troubleshoot the aircraft system.

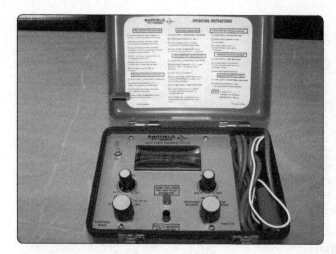

Figure 10-78. *EGT analyzer.*

EGT Indicator Check

The EGT indicator is tested after being removed from the aircraft instrument panel and disconnected from the aircraft EGT circuit leads. Attach the instrument cable and EGT indicator adapter leads to the indicator terminals and place the indicator in its normal operating position. Adjust the analyzer switches to the proper settings. The indicator reading should correspond to the readings of the analyzer within the allowable limits of the EGT indicator.

Correction for ambient temperature is not required for this test, as both the EGT indicator and analyzer are temperature compensated. The temperature registered on the aircraft EGT indicator should be within the specified tolerance of the aircraft system and the temperature reading on the analyzer readout. When the temperature difference exceeds the allowable tolerance, troubleshoot the aircraft system.

Resistance & Insulation Check

The thermocouple harness continuity is checked while the EGT system is being checked functionally. The resistance of the thermocouple harness is held to very close tolerances, since a change in resistance changes the amount of current flow in the circuit. A change of resistance gives erroneous temperature readings. The resistance and insulation check circuits make it possible to analyze and isolate any error in the aircraft system. How the resistance and insulation circuits are used is discussed with troubleshooting procedures.

Tachometer Check

To read engine speed with an accuracy of ±0.1 percent during engine run, the frequency of the tachometer-generator (older style) is measured by the rpm check analyzer. The scale of the rpm check circuit is calibrated in percent rpm to correspond to the aircraft tachometer indicator, which also reads in percent rpm. The aircraft tachometer and the rpm check circuit are

connected in parallel, and both are indicating during engine run-up. The rpm check circuit readings can be compared with the readings of the aircraft tachometer to determine the accuracy of the aircraft instrument.

Many newer engines use a magnetic pickup that counts passing gear teeth edges, which are seen electrically as pulses of electrical power as they pass by the pickup. *[Figure 10-79]* By counting the amount of pulses, the rpm of the shaft is obtained. This type of system requires little maintenance, other than setting the clearance between the gear teeth and the magnetic pickup.

Troubleshooting EGT System

An appropriate analyzer is used to test and troubleshoot the aircraft thermocouple system at the first indication of trouble, or during periodic maintenance checks.

The test circuits of the analyzer make it possible to isolate the troubles listed below. Following the list is a discussion of each trouble mentioned.

1. One or more inoperative thermocouples in engine parallel harness.

2. Engine thermocouples out of calibration.

3. EGT indicator error.

4. Resistance of circuit out of tolerance.

5. Shorts to ground.

6. Shorts between leads.

One or More Inoperative Thermocouples in Engine Parallel Harness

This error is found in the regular testing of aircraft thermocouples with a hot heater probe and is a broken lead wire in the parallel harness, or a short to ground in the harness. In the latter case, the current from the grounded thermocouple can leak off and never be shown on the indicator. However, this grounded condition can be found by using the insulation resistance check.

Engine Thermocouples Out of Calibration

When thermocouples are subjected for a period of time to oxidizing atmospheres, such as encountered in turbine engines, they drift appreciably from their original calibration. On engine parallel harnesses, when individual thermocouples can be removed, these thermocouples can be bench-checked, using one heater probe. The temperature reading obtained from the thermocouples should be within manufacturer's tolerances.

located and corrected.

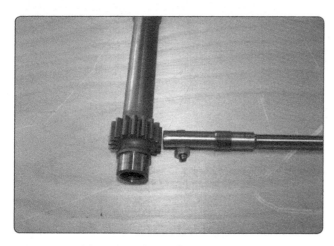

Figure 10-79. *Magnetic pickup and gear.*

EGT Circuit Error

This error is found by using the EGT and comparing the reading of the aircraft EGT indicator with the analyzer temperature reading. *[Figure 10-78]* The analyzer and aircraft temperature readings are then compared.

Resistance of Circuit Out of Tolerance

The engine thermocouple circuit resistance is a very important adjustment since a high-resistance condition gives a low indication on the aircraft EGT indicator. This condition is dangerous, because the engine is operating with excess temperature, but the high resistance makes the indicator read low. It is important to check and correct this condition.

Shorts to Ground/Shorts Between Leads

These errors are found by doing the insulation check using an ohmmeter. Resistance values from zero to 550,000 ohms can be read on the insulation check ohmmeter by selecting the proper range.

Troubleshooting Aircraft Tachometer System

A function of the rpm check is troubleshooting the aircraft tachometer system. The rpm check circuit in the analyzer is used to read engine speed during engine run-up with an accuracy of ±0.1 percent. The connections for the rpm check are the instrument cable and aircraft tachometer system lead to the tachometer indicator. After the connections have been made between the analyzer rpm check circuit and the aircraft tachometer circuit, the two circuits, now classed as one, are a parallel circuit. The engine is then run-up as prescribed in applicable technical instructions. Both systems can be read simultaneously.

If the difference between the readings of the aircraft tachometer indicator and the analyzer rpm check circuit exceeds the tolerance prescribed in applicable technical instructions, the engine must be stopped, and the trouble

Chapter 11
Light-Sport Aircraft Engines

Engine General Requirements

Engines used for light-sport aircraft and other types of aircraft, such as some experimental aircraft, ultralight aircraft, and powered parachutes, must be very light for the power they develop. Each aircraft requires thrust to provide enough forward speed for the wings to provide lift to overcome the weight of the aircraft. An aircraft that meets the requirements of the light-sport categories must meet the following requirements.

Note: All of the following requirements and regulations are subject to change. Always refer to the latest Federal Aviation Regulations for current information.

A light-sport aircraft means an aircraft, other than a rotorcraft or powered-lift, since its original certification, has continued to meet the following:

1. A maximum takeoff weight of not more than 1,320 pounds (lb) (600 kilograms (kg)) for aircraft not intended for operation on water; or 1,430 lb (650 kg) for an aircraft intended for operation on water.

2. A maximum airspeed in level flight with maximum continuous power (V_H) of not more than 120 knots calibrated airspeed (CAS) under standard atmospheric conditions at sea level.

3. A maximum never-exceed speed (V_{NE}) of not more than 120 knots CAS for a glider.

4. A maximum stalling speed or minimum steady flight speed without the use of lift-enhancing devices (V_{S1}) of not more than 45 knots CAS at the aircraft's maximum certificated takeoff weight and most critical center of gravity.

5. A maximum seating capacity of no more than two persons, including the pilot.

6. A single, reciprocating engine, if powered.

7. A fixed or ground-adjustable propeller, if a powered aircraft other than a powered glider.

8. A fixed or auto-feathering propeller system, if a powered glider.

9. A fixed-pitch, semirigid, teetering, two-blade rotor system, if a gyroplane.

10. A non-pressurized cabin, if equipped with a cabin.

11. Fixed landing gear, except for an aircraft intended for operation on water or a glider.

12. Fixed or retractable landing gear, or a hull, for an aircraft intended for operation on water.

13. Fixed or retractable landing gear for a glider.

Powered parachute means a powered aircraft comprised of a flexible or semirigid wing connected to a fuselage so that the wing is not in position for flight until the aircraft is in motion. The fuselage of a powered parachute contains the aircraft engine, a seat for each occupant, and is attached to the aircraft's landing gear.

Weight shift control aircraft means a powered aircraft with a framed pivoting wing and a fuselage controllable only in pitch and roll by the pilot's ability to change the aircraft's center of gravity with respect to the wing. Flight control of the aircraft depends on the wing's ability to flexibly deform rather than the use of control surfaces.

As the weight of an engine is decreased, the useful load that an aircraft can carry and the performance of the aircraft are obviously increased. Every excess pound of weight carried by an aircraft engine reduces its performance. Since light-sport aircraft have a narrow margin of useful load, engine weight is a very important concern with all of the light, low airspeed aircraft. Tremendous gains in reducing the weight of the aircraft engine through improvements in design, operating cycles, and metallurgy have resulted in engines with a much improved power to weight ratio.

A light-sport aircraft engine is reliable when it can perform at the specified ratings in widely varying flight attitudes and in extreme weather conditions. The engine manufacturer ensures the reliability and durability of the product by design, research, and testing. Although most of these engines are not certificated by the Federal Aviation Administration (FAA), close control of manufacturing and assembly procedures is generally maintained, and normally each engine is tested before it leaves the factory and meets certain American Society for Testing and Materials (ASTM) standards. Some engines used on light-sport aircraft are certificated by the FAA and these engines are maintained as per the manufacturer's instructions and Title 14 of the Code of Federal Regulations (14 CFR).

Most light-sport engines require a definite time interval

between overhauls. This is specified or implied by the engine manufacturer. The time between overhauls (TBO) varies with the type of engine (cycle), operating conditions, such as engine temperatures, amount of time the engine is operated at high-power settings, and the maintenance received. After reaching the time limit, the engine has to be overhauled. Sometimes this requires the engine to be shipped to an authorized manufacturer's overhaul facility. *[Figure 11-1]*

One consideration when selecting a light-sport engine is the shape, size, and number of cylinders of the engine. Since these engines range from single cylinder to multicylinder engines, the mounting in the airframe is important to maintain the view of the pilot, aircraft center of gravity, and to reduce aircraft drag.

Personnel Authorized to Perform Inspection & Maintenance on Light-Sport Engines

Given they meet all applicable regulations, the holder of a powerplant certificate can perform maintenance and inspections on light-sport engines. The holder of a sport pilot certificate may perform preventive maintenance on an aircraft owned or operated by that pilot and issued a special airworthiness certificate in the light-sport category under the provisions of 14 CFR part 43, section 43.3 (g). All maintenance must be performed in accordance with 14 CFR part 65, section 65.81, which describes specific experience requirements and current instructions for performing maintenance.

The following is used to determine eligibility for a repairman certificate (light-sport aircraft) and appropriate rating. To be eligible for a repairman certificate (light-sport aircraft), you must:

- Be at least 18 years old.

- Be able to read, speak, write, and understand English. If for medical reasons you cannot meet one of these

requirements, the FAA may place limits on the repairman certificate necessary to safely perform the actions authorized by the certificate and rating.

- Demonstrate the requisite skill to determine whether a light-sport aircraft is in a condition for safe operation.

- Be a citizen of the United States, or a citizen of a foreign country who has been lawfully admitted for permanent residence in the United States.

- To be eligible for a repairman certificate (light-sport aircraft) with an inspection rating, the applicant must:

  o Meet the requirements stated above for a repairman's certificate.

  o Complete a 16-hour training course acceptable to the FAA on inspecting the particular class of experimental light-sport aircraft for which these privileges are intended to be exercised.

- To be eligible for a repairman certificate (light-sport aircraft) with a maintenance rating, the applicant must:

  o Meet the requirements stated above for a repairman's certificate.

  o Complete a training course acceptable to the FAA on maintaining the particular class of light-sport aircraft upon which the privileges are intended to be exercised. The training course must, at a minimum provide the following number of hours of instruction:

  - For airplane class privileges: 120 hours.

  - Weight-shift control aircraft class privileges: 104 hours.

  - Powered parachute class privileges: 104 hours.

  - Lighter-than-air class privileges: 80 hours.

  - Glider class privileges: 80 hours.

Designation of Engine Type	For Engine S/N	Time Between Overhaul (TBO)	SB To Be Carried Out To Increase TBO
914 F	to 4,420.313	1,000 hours or 10 years, whichever comes first	SB-914-027 1,000 hours to 1,200 hours or 12 years, whichever comes first
914 F	from 4,420.314	1,200 hours or 12 years, whichever comes first	None
914 UL	to 4,418.103	1,000 hours or 10 years, whichever comes first	SB-914-027 1,000 hours to 1,200 hours or 12 years, whichever comes first
914 UL	from 4,418.104	1,200 hours or 12 years, whichever comes first	None

Figure 11-1. *Examples of TBO and calendar life for engines.*

The holder of a repairman certificate (light-sport aircraft) with an inspection rating may perform the annual condition inspection on a light-sport aircraft that is owned by the holder, has been issued an experimental certificate for operating a light-sport aircraft under 14 CFR part 21, section 21.191(i), and is in the same class of light-sport aircraft for which the holder has completed the training specified in the above paragraphs.

The holder of a repairman certificate (light-sport aircraft) with a maintenance rating may approve and return to service an aircraft that has been issued a special Airworthiness Certificate in the light-sport category under 14 CFR part 21, section 21.190, or any part thereof, after performing or inspecting maintenance (to include the annual condition inspection and the 100-hour inspection required by 14 CFR part 91, section 91.327), preventive maintenance, or an alteration (excluding a major repair or a major alteration on a product produced under an FAA approval). They may perform the annual condition inspection on a light-sport aircraft that has been issued an experimental certificate for operating a light-sport aircraft under 14 CFR part 21, section 21.191(i). However, they may only perform maintenance, preventive maintenance, and an alteration on a light-sport aircraft for which the holder has completed the training specified in the preceding paragraphs. Before performing a major repair, the holder must complete additional training acceptable to the FAA and appropriate to the repair performed.

The holder of a repairman certificate (light-sport aircraft) with a maintenance rating may not approve for return to service any aircraft or part thereof unless that person has previously performed the work concerned satisfactorily. If that person has not previously performed that work, the person may show the ability to do the work by performing it under the direct supervision of a certificated and appropriately rated mechanic, or a certificated repairman who has had previous experience in the specific operation concerned. The repairman may not exercise the privileges of the certificate unless the repairman understands the current instructions of the manufacturer and the maintenance manuals for the specific operation concerned.

Authorized Personnel That Meet FAA Regulations

All applicable aviation regulatory authority regarding maintenance procedures must be met. Maintenance organizations and personnel are encouraged to contact the manufacturer for more information and guidance on any of the maintenance procedures.

It is a requirement that every individual or maintenance provider possess the required special tooling, training, or experience to perform all tasks outlined. Maintenance providers that meet the following conditions outlined below may perform engine maintenance providing they meet all of the following FAA requirements:

- Knowledge of the specific task as a result of receiving authorized training from a training provider.

- Previous experience in performing the task and formal instruction from a manufacturer's authorized training facility or "on-the-job" instruction by a manufacturer's representative.

- A suitable work environment to prevent contamination or damage to engine parts or modules is needed.

- Suitable tools and fixtures as outlined in the manufacturers' Maintenance Manual should be used while performing maintenance requiring such tooling.

- Reasonable and prudent maintenance practices should be utilized.

Types of Light-Sport & Experimental Engines

Note: All information in this text is for educational illustrational purposes and is not to be used for actual aircraft maintenance. This information is not revised at the same rate as the maintenance manual; always refer to the current maintenance information when performing maintenance on any engine.

Light-Sport Aircraft Engines

Light-sport/ultralight aircraft engines can be classified by several methods, such as by operating cycles, cylinder arrangement, and air- or water-cooled. An inline engine generally has two cylinders, is two-cycle, and is available in several horsepower ranges. These engines may be either liquid-cooled, air-cooled, or a combination of both. They have only one crankshaft that drives the reduction gearbox or propeller directly. Most of the other cylinder configurations used are horizontally opposed, ranging from two to six cylinders from several manufacturers. These engines are either gear reduction or direct drive.

Two-Cycle, Two Cylinder Rotax Engine
Rotax 447 UL Single Capacitor Discharge Ignition (SCDI) & Rotax 503 UL Dual Capacitor Discharge Ignition (DCDI)
The Rotax inline cylinder arrangement has a small frontal area and provides improved streamlining. *[Figure 11-2]* The two cylinder, inline two-stroke engine, which is piston ported with air-cooled cylinder heads and cylinders, is available in a fan or free air-cooled version. Being a two-stroke cycle engine, the oil and fuel must be mixed in the fuel tank on some models. Other models use a lubrication system, such as the 503 oil injection lubrication system. This system does not mix the fuel and oil as the oil is stored in a separate tank.

Figure 11-2. *Rotax inline cylinder arrangement.*

Figure 11-3. *Rotax 582 engine.*

As the engine needs lubrication, the oil is injected directly from this tank. The typical ignition system is a breakerless ignition system with a dual ignition system used on the 503, and a single ignition system used on the 447 engine series. Both systems are of a magneto capacitor discharge design.

The engine is equipped with a carburetion system with one or two piston-type carburetors. One pneumatic driven fuel pump delivers the fuel to the carburetors. The propeller is driven via a flange connected gearbox with an incorporated shock absorber. The exhaust system collects the exhaust gases and directs them overboard. These engines come with an integrated alternating current (AC) generator (12V 170W) with external rectifier-regulator as an optional extra.

Rotax 582 UL DCDI

The Rotax 582 is a two-stroke engine, two cylinder inline with rotary valve inlet, has liquid-cooled cylinder heads, and cylinders that use an integrated water pump. *[Figure 11-3]* The lubrication system can be a fuel-oil mixture or oil injection lubrication. The ignition system is a dual ignition using a breakerless magneto capacitor discharge design. Dual piston type carburetors and a pneumatic fuel pump deliver the fuel to the cylinders. The propeller is driven via the prop flange connected gearbox with an incorporated torsional vibration shock absorber. This engine also uses a standard version exhaust system with an electric starter or manual rewind starter.

Description of Systems for Two-Stroke Engines
Cooling System of Rotax 447 UL SCDI & Rotax 503 UL DCDI

Two versions of air-cooling are available for these engines. The first method is free air-cooling, which is a process of engine cooling by an air-stream generated by aircraft speed and propeller. The second is fan cooling, which is cooling by an air-stream generated by a fan permanently driven from the crankshaft via a V-belt.

Cooling System of the Rotax 582 UL DCDI

Engine cooling for the Rotax 582 is accomplished by liquid cooled cylinders and cylinder heads. *[Figure 11-4]* The cooling system is in a two circuit arrangement. The cooling liquid is supplied by an integrated pump in the engine through the cylinders and the cylinder head to the radiator. The cooling system has to be installed, so that vapor coming from the cylinders and the cylinder head can escape to the top via a hose, either into the water tank of the radiator or to an expansion chamber. The expansion tank is closed by a pressure cap (with excess pressure valve and return valve). As the temperature of the coolant rises, the excess pressure valve opens, and the coolant flows via a hose at atmospheric pressure to the transparent overflow bottle. When cooling down, the coolant is sucked back into the cooling circuit.

Lubrication Systems
Oil Injection Lubrication of Rotax 503 UL DCDE & 582 UL DCDI

Generally, the smaller two cycle engines are designed to run on a mixture of gasoline and 2 percent oil that is premixed in the fuel tank. The engines are planned to run on an oil-gasoline mixture of 1:50. Other engines use oil injection systems that use an oil pump driven by the crankshaft via the pump gear that feeds the engine with the correct amount of fresh oil. The oil pump is a piston type pump with a metering system. Diffuser jets in the intake inject pump supplied two-stroke oil with the exact proportioned quantity needed. The oil quantity is defined by the engine rotations per minute and the oil pump lever position. This lever is actuated via a cable connected to the throttle cable. The oil comes to the pump

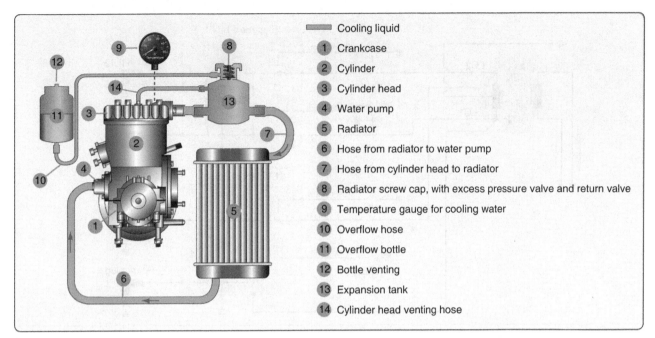

Cooling liquid	
1	Crankcase
2	Cylinder
3	Cylinder head
4	Water pump
5	Radiator
6	Hose from radiator to water pump
7	Hose from cylinder head to radiator
8	Radiator screw cap, with excess pressure valve and return valve
9	Temperature gauge for cooling water
10	Overflow hose
11	Overflow bottle
12	Bottle venting
13	Expansion tank
14	Cylinder head venting hose

Figure 11-4. *Rotax 582 cooling system.*

from an oil tank by gravity.

Note: In engines that use oil injection, the carburetors are fed with pure gasoline (no oil-gasoline mixture). The oil quantity in the oil tank must be checked before putting the engine into service as the oil is consumed during operation and needs to be replenished.

Electric System

The 503 UL DCDI, 582 UL DCDI engine types are equipped with a breakerless, single capacitor discharge ignition unit with an integrated generator. *[Figure 11-5]* The 447 UL SCDI engine is equipped with a breakerless, single capacitor discharge ignition unit with integrated generator. The ignition unit is completely free of maintenance and needs no external power supply. Two charging coils fitted on the generator stator, independent from each other, each feed one ignition circuit. The energy supplied is stored in the ignition capacitor. At the moment of ignition, the external triggers supply an impulse to the control circuits and the ignition capacitors are discharged via the primary winding of the ignition coil. The secondary winding supplies the high voltage for the ignition spark.

Fuel System

Due to higher lead content in aviation gas (AVGAS), operation can cause wear and deposits in the combustion chamber to increase. Therefore, AVGAS should only be used if problems are encountered with vapor lock or if the other fuel types are not available. Caution must be exercised to use only fuel suitable for the relevant climatic conditions, such as using winter fuel for summer operation.

Fuel-Oil Mixing Procedure

The following describes the process for fuel-oil mixing. Use a clean approved container of known volume. To help predilute the oil, pour a small amount of fuel into the container. Fill known amount of oil (two-stroke oil ASTM/Coordinating European Council (CEC) standards, API-TC classification (e.g., Castrol TTS) mixing ratio 1:50 (2 percent)), into container. Oil must be approved for air-cooled engines at 50:1 mixing ratio. Agitate slightly to dilute oil with gasoline. Add gasoline to obtain desired mixture ratio; use fine mesh screen. Replace the container cap and shake the container thoroughly. Then, using a funnel with a fine mesh screen to prevent the entry of water and foreign particles, transfer mixture from container into the fuel tank.

Warning: To avoid electrostatic charging at refueling, use only metal containers and ground the aircraft in accordance with the grounding specifications.

Opposed Light-Sport, Experimental, & Certificated Engines

Many certificated engines are used with light-sport and experimental aircraft. Generally, cost is a big factor when considering this type of powerplant. The certificated engines tend to be much more costly than the non-certificated engines, and are not ASTM approved.

Rotax 912/914

Figure 11-6 shows a typical four cylinder, four-stroke Rotax horizontally opposed engine. The opposed-type engine has

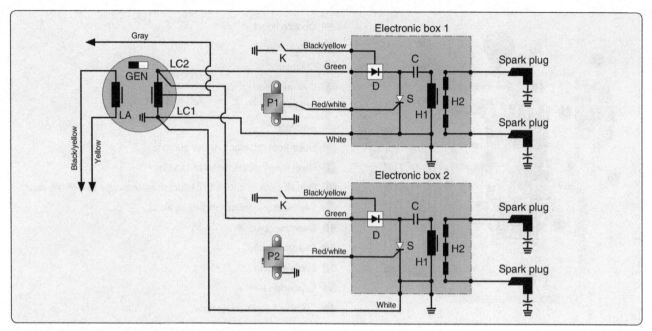

Figure 11-5 *Rotax 503 and 582 electrical system.*

Figure 11-6. *Typical four-cylinder, four-stroke horizontally opposed engine.*

two banks of cylinders directly opposite each other with a crankshaft in the center. The pistons of both cylinder banks are connected to the single crankshaft. The engine cylinder heads are both liquid-cooled and air-cooled; the air-cooling is mostly used on the cylinder. It is generally mounted with the cylinders in a horizontal position. The opposed-type engine has a low weight to horsepower ratio, and its narrow silhouette makes it ideal for horizontal installation on the aircraft wings (twin-engine applications). Another advantage is its low vibration characteristics. It is an ideal replacement for the Rotax 582 two-cylinder, two-stroke engine, which powers many of the existing light aircraft, as it is the same weight as the Rotax 582. These engines are ASTM approved for installation into light-sport category aircraft, with some models being FAA-certificated engines.

Description of Systems
Cooling System

The cooling system of the Rotax 914, shown in *Figure 11-7*, is designed for liquid cooling of the cylinder heads and ram-air cooling of the cylinders. The cooling system of the cylinder heads is a closed circuit with an expansion tank. *[Figure 11-8]* The coolant flow is forced by a water pump driven from the camshaft, from the radiator, to the cylinder heads. From the top of the cylinder heads, the coolant passes on to the expansion tank (1). Since the standard location of the radiator (2) is below engine level, the expansion tank located on top of the engine allows for coolant expansion.

The expansion tank is closed by a pressure cap (3) (with excess pressure valve and return valve). As the temperature of the coolant rises, the excess pressure valve opens and the coolant flows via a hose at atmospheric pressure to the transparent overflow bottle (4). When cooling down, the coolant is sucked back into the cooling circuit. Coolant temperatures are measured by means of temperature probes installed in the cylinder heads 2 and 3. The readings are taken on measuring the hottest point of cylinder head depending on engine installation. *[Figure 11-7]*

Fuel System

The fuel flows from the tank (1) via a coarse filter-water trap (2) to the two electric fuel pumps (3) connected in series. *[Figure 11-9]* From the pumps, fuel passes on via the fuel pressure control (4) to the two carburetors (5). Parallel to each fuel pump is a separate check valve (6) installed via the return line (7) that allows surplus fuel to flow back to the fuel tank. Inspection for possible constriction of diameter or

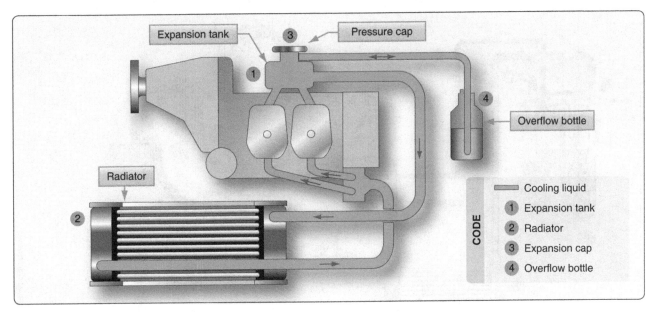

Figure 11-7. *Rotax 914 cooling system.*

Figure 11-8. *Water-cooled heads.*

obstruction must be accomplished to avoid overflowing of fuel from the carburetors. The return line must not have any resistance to flow. The fuel pressure control ensures that the fuel pressure is always maintained approximately 0.25 bar (3.63 pounds per square inch (psi)) above the variable boost pressure in the airbox and thus, ensures proper operation of the carburetors.

Lubrication System

The Rotax 914 engine is provided with a dry, sump-forced lubrication system with a main oil pump with integrated pressure regulator and an additional suction pump. *[Figure 11-10]* The oil pumps are driven by the camshaft. The main oil pump draws oil from the oil tank (1) via the oil cooler (2) and forces it through the oil filter to the points of lubrication. It also lubricates the plain bearings of the

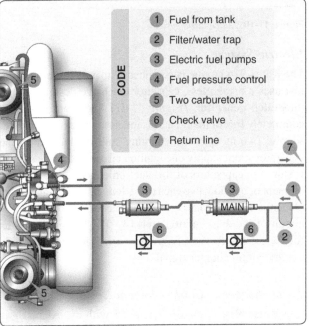

Figure 11-9. *Fuel system components.*

turbocharger and the propeller governor. The surplus oil emerging from the points of lubrication accumulates on the bottom of crankcase and is forced back to the oil tank by the blow-by gases. The turbocharger is lubricated via a separate oil line (from the main oil pump). The oil emerging from the lower placed turbocharger collects in the oil sump by a separate pump and is pumped back to the oil tank via the oil line (3). The oil circuit is vented via bore (5) in the oil tank. There is an oil temperature sensor in the oil pump flange for reading of the oil inlet temperature.

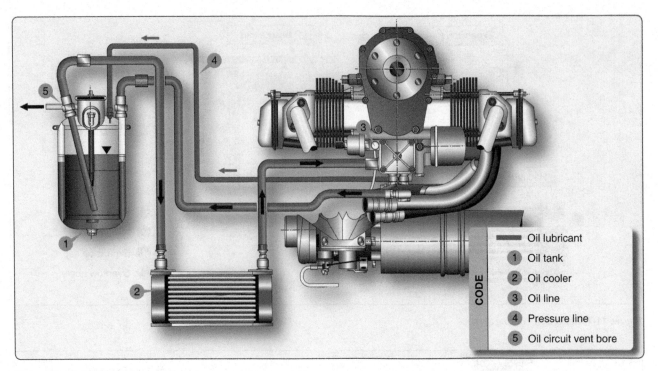

Figure 11-10. *Lubrication system.*

Electric System

The Rotax 914 engine is equipped with a dual ignition unit that uses a breakerless, capacitor discharge design with an integrated generator. *[Figure 11-11]* The ignition unit is completely free of maintenance and needs no external power supply. Two independent charging coils (1) located on the generator stator supply one ignition circuit each. The energy is stored in capacitors of the electronic modules (2). At the moment of ignition, two each of the four external trigger coils (3) actuate the discharge of the capacitors via the primary circuit of the dual ignition coils (4). The firing order is as follows: 1-4-2-3. The fifth trigger coil (5) is used to provide the revolution counter signal.

Turbocharger & Control System

The Rotax 914 engine is equipped with an exhaust gas turbocharger making use of the energy in the exhaust gas for compression of the intake air or for providing boost pressure to the induction system. The boost pressure in the induction system (airbox) is controlled by means of an electronically controlled valve (wastegate) in the exhaust gas turbine. The wastegate regulates the speed of the turbocharger and consequently the boost pressure in the induction system. The required nominal boost pressure in the induction system is determined by the throttle position sensor mounted on the carburetor 2/4. The sensor's transmitted position is linear from 0 to 115 percent, corresponding to a throttle position from idle to full power. *[Figure 11-12]* For correlation between throttle position and nominal boost pressure in the induction, refer to *Figure 11-13*. As shown in the diagram, with the throttle position at 108–110 percent results in a rapid rise of nominal boost pressure.

To avoid unstable boost, the throttle should be moved smoothly through this area either to full power (115 percent) or at a reduced power setting to maximum continuous power. In this range (108–110 percent throttle position), small changes in throttle position have a big effect on engine performance and speed. These changes are not apparent to the pilot from the throttle lever position. The exact setting for a specific performance is virtually impossible in this range and has to be prevented, as it might cause control fluctuations or surging. Besides the throttle position, overspeeding of the engine and too high intake air temperature have an effect on the nominal boost pressure. If one of the stated factors exceeds the specified limits, the boost pressure is automatically reduced, thus protecting the engine against over boost and detonation.

The turbo control unit (TCU) is furnished with output connections for an external red boost lamp and an orange caution lamp for indications of the functioning of the TCU. When switching on the voltage supply, the two lamps are automatically subject to a function test. Both lamps illuminate for one to two seconds, then they extinguish. If they do not, a check per the engine maintenance manual is necessary. If the orange caution lamp is not illuminated, then this signals that TCU is ready for operation. If the lamp is blinking, this indicates a malfunction of the TCU or its periphery systems. Exceeding the admissible boost pressure activates

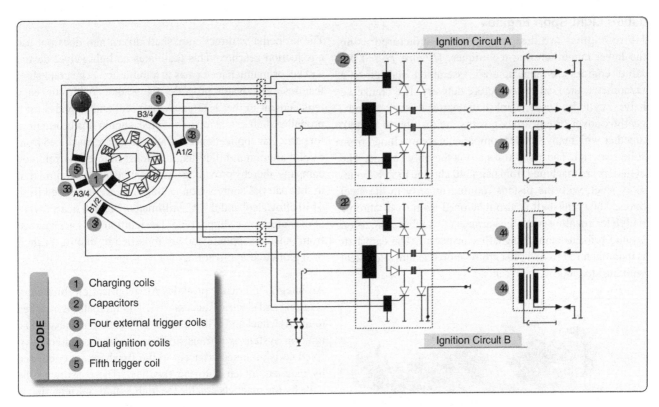

Figure 11-11. *Electric system.*

CODE
1 Charging coils
2 Capacitors
3 Four external trigger coils
4 Dual ignition coils
5 Fifth trigger coil

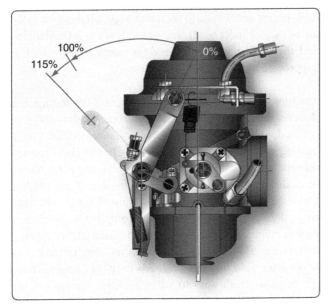

Figure 11-12. *Turbocharger control system throttle range and position.*

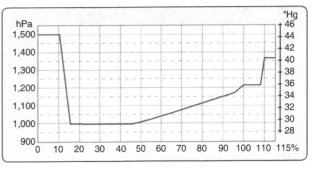

Figure 11-13. *Correlation between throttle position and nominal boost pressure.*

HKS 700T Engine

The HKS 700T engine is a four-stroke, two cylinder turbocharged engine equipped with an intercooler. *[Figure 11-14]* The horizontally opposed cylinders house four valves per cylinder, with a piston displacement of 709 cc. It uses an electronic control fuel injection system. A reduction gearbox is used to drive the propeller flange at a speed reduction ratio of 2.13 to 1. The engine is rated at 77 horsepower continuous and 80 horsepower takeoff (3 minutes) at 4,900 rpm and 5,300 rpm, respectively. A total engine weight of 126 pounds provides a good power to weight ratio. The 700T has a TBO of 500 hours.

and illuminates the red boost lamp continuously. The TCU registers the time of full throttle operation (boost pressure). Full throttle operation for longer than 5 minutes, with the red boost light illuminated, makes the red boost lamp start blinking. The red boost lamp helps the pilot to avoid full power operation for longer than 5 minutes or the engine could be subject to thermal and mechanical overstress.

Jabiru Light-Sport Engines

Jabiru engines are designed to be manufactured using the latest manufacturing techniques. [*Figure 11-15*] All Jabiru engines are manufactured, assembled, and ran on a Dynometer, then calibrated before delivery. The crankcase halves, cylinder heads, crankshaft, starter motor housings, gearbox cover (the gearbox powers the distributor rotors), together with many smaller components are machined from solid material. The sump (oil pan) is the only casting. The cylinders are machined from bar 4140 chrome molybdenum alloy steel, with the pistons running directly in the steel bores. The crankshaft is also machined from 4140 chrome molybdenum alloy steel, the journals of which are precision ground prior to being Magnaflux inspected. The camshaft is manufactured from 4140 chrome molybdenum alloy steel with nitrided journals and cams.

Figure 11-14. *HKS 700T engine.*

Figure 11-15. *Jabiru engines.*

The propeller is direct crankshaft driven and does not use a reduction gearbox. This facilitates its lightweight design and keeps maintenance costs to a minimum. The crankshaft features a removable propeller flange that enables the easy replacement of the front crankshaft seal and provides for a propeller shaft extension to be fitted, should this be required for particular applications. Cylinder heads are machined from a solid aluminum billet that is purchased directly from one company, thereby providing a substantive quality control trail to the material source. Connecting rods are machined from 4140 alloy steel and the 45 millimeters big end bearings are of the automotive slipper type. The ignition coils are sourced from outside suppliers and are modified by Jabiru for their own particular application.

An integral alternator provides AC rectification for battery charging and electrical accessories. The alternator is attached to the flywheel and is driven directly by the crankshaft. The ignition system is a transistorized electronic system; two fixed coils mounted adjacent to the flywheel are energized by magnets attached to the flywheel. The passing of the coils by the magnets creates the high voltage current, that is transmitted by high tension leads to the center post of two automotive type distributors, which are simply rotors and caps, before distribution to automotive spark plugs (two in the top of each cylinder head). The ignition system is fixed timing and, therefore, removes the need for timing adjustment. It is suppressed to prevent radio interference.

The ignition system is fully redundant, self-generating, and does not depend on battery power. The crankshaft is designed with a double bearing at the propeller flange end and a main bearing between each big end. Thrust bearings are located fore and aft of the front double bearing, allowing either tractor or pusher installation. Pistons are remachined to include a piston pin, circlip, and groove. They are all fitted with three rings, the top rings being cast iron to complement the chrome molybdenum cylinder bores. Valves are 7mm (stem diameter) and are manufactured specifically for the Jabiru engine. The valve drive train includes pushrods from the camshaft from the camshaft followers to valve rockers. The valves are Computer Numerical Control (CNC) machined from steel billet, induction hardened, polished on contact surfaces, and mounted on a shaft through Teflon coated bronze-steel bush. Valve guides are manufactured from aluminum/bronze. Replaceable valve seats are of nickel steel and are shrunk into the aluminum cylinder heads. The valve train is lubricated from the oil gallery. Engines use hydraulic lifters that automatically adjust valve clearance. An internal gear pump is driven directly by the camshaft and provides engine lubrication via an oil circuit that includes an automotive spin-on filter, oil cooler and built-in relief valve.

The standard engines are supplied with two ram-air cooling ducts, that have been developed by Jabiru to facilitate the cooling of the engine by directing air from the propeller to the critical areas of the engine, particularly the cylinder heads and barrels. The use of these ducts remove the need to design and manufacture baffles and the establishment of a plenum chamber, which is the traditional method of cooling air-cooled, aircraft engines. The fact that these baffles and plenum chamber are not required also ensures a cleaner engine installation, which in turn facilitates maintenance and inspection of the engine and engine components.

The engine is fitted with a 1.5 kilowatt starter motor that is also manufactured by Jabiru and provides very effective starting. The engine has very low vibration level; however, it is also supported by four large rubber shock mounts attached to the engine mounts at the rear of the engine. The fuel induction system uses a pressure compensating carburetor. Following the carburetor, the air-fuel mixture is drawn through a swept plenum chamber bolted to the sump casting, in which the mixture is warmed prior to entering short induction tubes attached to the cylinder heads.

An effective stainless steel exhaust and muffler system is fitted as standard equipment ensuring very quiet operations. For owners wanting to fit vacuum instruments to their aircraft, the Jabiru engines are designed with a vacuum pump drive direct mounted through a coupling on the rear of the crankshaft.

Jabiru 2200 Aircraft Engine

The Jabiru 2,200 cc aircraft engine is a four-cylinder, four-stroke horizontally opposed air-cooled engine. At 132 pounds (60 kgs) installed weight, it is one of the lightest four-cylinder, four-stroke aircraft engines. Small overall dimensions give it a small frontal area width (23.46 in, 596 mm) that makes it a good engine for tractor applications. The Jabiru engine is designed for either tractor or pusher installation. The Jabiru engine specifications are listed in *Figure 11-16*.

The Jabiru 3300 (120 hp) engine features *[Figure 11-17]*:

- 4-stroke,
- 3,300 cc engine (200 cubic inches),
- 6-cylinder horizontally opposed,
- 1 central camshaft,
- Fully machined aluminum alloy crankcase,
- Overhead valves (OHV) - push rod operated,
- Ram air-cooled,
- Wet sump lubrication - 4 liter capacity,

- Direct propeller drive,
- Dual transistorized magneto ignition,
- Integrated AC generator,
- Electric starter,
- Mechanical fuel pump, and
- Naturally aspirated - 1 pressure compensation carburetor.

Aeromax Aviation 100 (IFB) Aircraft Engine

Aeromax Aviation produces a version of a 100 hp engine called the Integral Front Bearing. The engine features a special made integral front bearing. *[Figure 11-18]* The engine uses an integral permanent magnet 35 amp alternator, lightweight starter, and dual ignition. The compact alternator and starter allow for a streamlined and aerodynamic cowl which improves the fuel efficiency of an experimental aircraft. The Aeromax aircraft engine is an opposed six-cylinder, air-cooled, and direct drive. Being a six-cylinder engine, it has smooth operation. The Aeromax engines are known for their heat dissipation qualities, provided the proper amount of cooling air is provided.

It features a crank extension supported by a massive integral front bearing (IFB) and bearing housing. These engines start out as a GM Corvair automobile core engine. These basic core engines are disassembled and each component that is reused is refurbished and remanufactured. The crankshaft in the Areomax 100 IFB aircraft engine is thoroughly inspected, including a magnaflux inspection. After ensuring the crank is free of any defects, it is extended by mounting the crank extension hub on its front. Then, the crank is ground true, with all five bearings' surfaces (four original and the new extended crank's front bearing), being true to each other and perpendicular to the crank's prop flange. *[Figure 11-19]*

All radiuses are smooth with no sharp corners where stress could concentrate. Every crankshaft is nitrated, which is a heat-chemical process that hardens the crank surfaces. The crank reinforcement coupled with the IFB is required to counter the additional dynamic and bending loads introduced on the crank in an aircraft application. The engine case is totally refurbished and checked for wear. Any studs or bolts that show wear are replaced. The engine heads are machined to proper specifications and all new valves, guides, and valve train components are installed. A three-angle valve grind and lapping ensure a good valve seal.

Once the engine is assembled, it is installed on a test stand, pre-lubricated, and inspected. The engine is, then, run several times for a total of two hours. The engine is carefully inspected after each run to ensure it is in excellent

Specifications: Jabiru 2200cc 85 HP Aircraft Engine	
Engine Features	Four-stroke
	Four-cylinder horizontally opposed
Opposed	One central camshaft
	Push rods
	Overhead valves (OHV)
(OHV)	Ram air-cooled
	Wet sump lubrication
	Direct propeller drive
	Dual transistorized magneto ignition
Magneto Ignition	Integrated AC generator 20 amp
Generator 20 amp	Electric starter
	Mechanical fuel pump
	Naturally aspirated - 1 pressure compensating carburetor
Pressure Compensating Carburetor	Six bearing crankshaft
Displacement	2,200 cc (134 cu. in.)
Bore	97.5 mm
Stroke	74 mm
Compression Ratio	8:1
Directional Rotation of Prop Shaft	Clockwise - pilot's view tractor applications
Ramp Weight	132 lb complete including exhaust, carburetor, starter motor, alternator, and ignition system
Ignition Timing	25° BTDC
Firing Order	1–3–2–4
Power Rating	85 hp @ 3,300 rpm
Fuel Consumption at 75% power	4 US gal/hr
Fuel	AVGAS 100 LL or auto gas 91 octane minimum
Oil	Aeroshell W100 or equivalent
Oil Capacity	2.3 quarts
Spark Plugs	NGK D9EA - automotive

Figure 11-16. *Jabiru 2200cc specifications.*

operating condition. At the end of test running the engine, the oil filter is removed and cut for inspection. Its internal condition is recorded. This process is documented and kept on file for each individual engine. Once the engine's proper performance is assured, it is removed and packaged in a custom built crate for shipping. Each engine is shipped with its engine service and operations manual. This manual contains information pertaining to installation, break–in, testing, tune-up, troubleshooting, repair, and inspection

procedures. The specifications for the Aeromax 100 engine are outlined in *Figure 11-20*.

Direct Drive VW Engines

Revmaster R-2300 Engine

The Revmaster R-2300 engine maintains Revmaster's systems and parts, including its RM-049 heads that feature large fins and a hemispherical combustion chamber. *[Figure 11-21]* It maintains the earlier R-2200 engine's top

Jabiru 3300cc Aircraft Engine	
Displacement	3,300 cc (202 cu.in.)
Bore	97.5 mm (3.838")
Stroke	74 mm (2.913")
Aircraft Engine	Jabiru 3,300 cc 120 hp
Compression Ratio	8:1
Directional Rotation of Prop Shaft	Clockwise - Pilot's view tractor applications
Ramp Weight	178 lbs (81 kg) complete including exhaust, carburetor, starter motor, alternator and ignition system
Ignition Timing	25° BTDC fixed timing
Firing order	1–4–5–2–3–6
Power Rating	120 hp @ 3,300 rpm
Fuel Consumption at 75% power	26 l/hr (6.87 US gal/hr)
Fuel	AVGAS 100 LL or auto gas 91 octane minimum
Oil	Aeroshell W100 or equivalent
Oil Capacity	3.5l (3.69 quarts)
Spark Plugs	NGK D9EA - automotive

Figure 11-17. *Jabiru 3300cc aircraft engine.*

Figure 11-18. *Aeromax direct drive, air-cooled, six-cylinder engine.*

Figure 11-19. *Front-end bearing on the 1000 IFB engine.*

horsepower (82) at 2,950 rpm continuous. *[Figure 11-22]* Takeoff power is rated at 85 at 3,350 rpm. The additional power comes from a bore of 94 mm plus lengthening of the R-2200's connecting rods, plus increasing the stroke from 78 to 84 mm. The longer stroke results in more displacement, and longer connecting rods yield better vibration and power characteristics. The lower cruise rpm allows the use of longer propellers, and the higher peak horsepower can be felt in shorter takeoffs and steeper climbs.

The Revmaster's four main bearing crankshaft runs on a 60 mm center main bearing, is forged from 4340 steel, and uses nitrided journals. Thrust is handled by the 55 mm #3 bearing at the propeller end of the crank. Fully utilizing its robust #4 main bearing, the Revmaster crank has built in oil-controlled propeller capability, a feature unique in this horsepower range; non-wood props are usable with these engines.

Moving from the crankcase and main bearings, the cylinders are made by using centrifugally cast chilled iron. The pistons are forged out of high quality aluminum alloy, machined and balanced in a set of four. There are two sizes of pistons, 92 mm and 94 mm, designed to be compatible with a 78 mm to 82 mm stroke crankshafts. The cylinder set also contains

Aeromax 100 Engine Specifications	
Power Output: 100 hp continuous at 3,200 rpm	Air-cooled
Displacement: 2.7 L	Six cylinders
Compression: 9:1	Dual ignition–single plug
Weight: 210 lb	Normally aspirated
Direct Drive	CHT max: 475° F
Rear Light weight. Starter and 45 amp alternator	New forged pistons
Counterclockwise rotation	Balanced and nitrated crank shaft
Harmonic balancer	New hydraulic lifters
Remanufactured case	New main/rod bearings
Remanufactured heads with new guides, valves, valve train, intake	New all replaceable parts
Remanufactured cylinders	New spark plug wiring harness
New light weight aluminum cylinder - optional	Remanufactured dual ignition distributor with new points set and electronic module
New high torque cam	New oil pump
New CNC prop hub and safety shaft	New oil pan
New Aeromax top cover and data plate	Engine service manual

Figure 11-20. *Aeromax 100 engine specifications.*

Figure 11-21. *Revmaster R-2300 engine.*

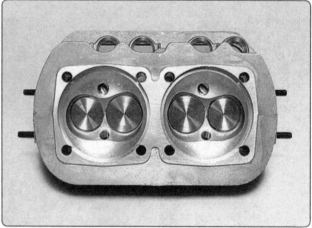

Figure 11-22. *Hemispherical combustion chamber within the Revmaster R-2300 Heads.*

piston rings, wrist pins, and locks. The direct-drive R-2300 uses a dual CDI ignition with eight coil spark to eight spark plugs, dual 20-amp alternators, oil cooler, and its proprietary Rev-Flo carburetor, while introducing the longer cylinders that do not require spacers. The automotive-based bearings, valves, valve springs, and piston rings (among others) make rebuilds easy and inexpensive.

Great Plains Aircraft Volkswagen (VW) Conversions

Great Plains Aircraft is one company that offers several configurations of the Volkswagen (VW) aircraft engine conversion. One very popular model is the front drive long block kits that offer a four-cycle, four-cylinder opposed engine with horsepower ranges from approximately 60-100. *[Figure 11-23]* The long block engine kits, which are the complete engine kits that are assembled, in the field or can be shipped completely assembled, are available from 1,600 cc up through 2,276 cc. All the engine kits are built from

Figure 11-23. *Great Plain's Volkswagen conversion.*

proven time tested components and are shipped with a Type One VW Engine Assembly Manual. This manual was written by the manufacturer, specifically for the assembly of their engine kits. Also included are how to determine service and maintenance procedures and many tips on how to set up and operate the engine correctly. The crankshaft used in the 2,180 cc to 2,276 cc engines is a 82 mm crankshaft made from a forged billet of E4340 steel, machined and magnafluxed twice. The end of the crankshaft features a ½-inch fine thread versus a 20 mm thread found on the standard automotive crank.

Teledyne Continental 0-200 Engine

The 0-200 Series engine has become a popular engine for use in light-sport aircraft. The 0-200-A/B is a four-cylinder, carbureted engine producing 100 brake hp and has a crankshaft speed of 2,750 rpm. *[Figure 11-24]* The engine has horizontally opposed air-cooled cylinders. The engine cylinders have an overhead valve design with updraft intake inlets and downdraft exhaust outlets mounted on the bottom of the cylinder. The 0-200-A/B engines have a 201 cubic inch displacement achieved by using a cylinder design with a 4.06-inch diameter bore and a 3.88-inch stroke. The dry weight of the engine is 170.18 pounds without accessories. The weight of the engine with installed accessories is approximately 215 pounds. Developed specifically for light aircraft, the 0-200-D engine has a dry weight with installed accessories of approximately 199 pounds. The engine is provided with four integral rear engine mounts. A crankcase breather port is located on the 1-3 side of the crankcase forward of the number 3 cylinder.

The engine lubrication system is a wet sump, high-pressure oil system. The engine lubrication system includes the internal engine-driven pressure oil pump, oil pressure relief valve, pressure oil screen mounted on the rear of the accessory case, and pressure instrumentation. A fitting is provided at the 1-3 side of the crankcase for oil pressure measurement. The oil sump capacity is six quarts maximum. The 0-200-A/B induction system consists of an updraft intake manifold with the air intake and throttle mounted below the engine. Engine

Figure 11-24. *0-200 Continental Engine.*

manifold pressure is measured at a port located on the 2-4 side of the intake air manifold. The 0-200-A/B is equipped with a carburetor that meters fuel flow as the flight deck throttle and mixture controls are changed.

Lycoming 0-233 Series Light-Sport Aircraft Engine

Lycoming Engines, a Textron Inc. company, produces an experimental non-certificated version of its 233 series light-sport aircraft engine. *[Figure 11-25]* The engine is light and capable of running on unleaded automotive fuels, as well as AVGAS. The engine features dual CDI spark ignition, an optimized oil sump, a streamlined accessory housing, hydraulically adjusted tappets, a lightweight starter, and a lightweight alternator with integral voltage regulator. It has a dry weight of 213 pounds (including the fuel pump) and offers continuous power ratings up to 115 hp at 2,800 rpm. In addition to its multi-gasoline fuel capability, it has proven to be very reliable with a TBO of 2,400 hours. The initial standard version of the engine is carbureted, but fuel injected configurations of the engine are also available.

Figure 11-25. *Lycoming 0-233 engine.*

General Maintenance Practices on Light-Sport Rotax Engines

Some specific maintenance practices that differ from conventional certificated engines is covered for background and educational acquaintance purposes only. Always refer to the current manufacturer's information when performing maintenance on any engine.

Safety regulations must be adhered to ensure maintenance personnel safety when performing maintenance and service work on any engine installation. The following information should be followed while performing maintenance.

The ignition should be off and the ignition system grounded with the battery disconnected. Secure the engine against unintentional operation. During maintenance work that requires ignition on and battery connected, secure the propeller against unintentional turning by hand, and secure and observe a propeller safety zone. This precautionary measure serves to avoid any injuries in case of an unintentional start of the engine, which can result in injuries or death. Remember, as long as the ground-cable (plead) is not properly connected to ground, the ignition is switched ON (hot).

Prevent contamination, such as metal chips, foreign material, and/or dirt, from entering the cooling, lubricating, and fuel system during maintenance. Severe burns and scalds may result if the engine is not allowed to cool down to outside air temperature before starting any work. Before reusing disassembled parts, clean with a suitable cleaning agent, check, and refit per instructions. Before every re-assembly, check for missing components. Only use adhesives, lubricants, cleaning agents, and solvents listed for use in the maintenance instructions. Observe the tightening torques for screws and nuts; overtorque or too loose connection could cause serious engine damage or failure.

The following are some general maintenance practices that provide for safety and good technique:

- Work only in a non-smoking area and not close to sparks or open flames.

- Always use the specified tools.

- During disassembling/reassembling the removal of any safety items (e.g., safety wiring, self-locking fastener) each part must be followed with the replacement of a new one.

- Once loosened, always replace self-securing (locking) nuts.

- Use clean screws and nuts only and inspect face of nuts and thread for damage.

- Check the contact faces and threads for damage and

replace if any damaged is detected.

- At reassembly of the engine, replace all sealing rings, gaskets, securing elements, O-rings, and oil seals.

- At disassembly of the engine, mark the engine's components as necessary to provide for locating the original position of the part.

- Parts should be replaced in the same position upon reassembly.

- Any used components have wear patterns that should be replaced or matched if reused. Ensure that these marks are not erased or washed off.

To perform maintenance, the technician must follow the manufacturer's instructions. Obtain, read, and understand the information pertaining to servicing of the light-sport or experimental engine.

Maintenance Schedule Procedures & Maintenance Checklist

All stated checks are visual inspections for damage and wear, unless otherwise stated. All listed work must be carried out within the specified period. For the intervals between maintenance work, a tolerance of + or – 0 hour is permissible, but these tolerances must not be exceeded. This means that if a 100 hour check is actually carried out at 110 hour, the next check is due at 200 hour + or – 10 hour and not at 210 hour + or – 10 hour. If maintenance is performed before the prescribed interval, the next maintenance check is to be done at the same interval (e.g., if first 100-hour check is done after 87 hours of operation, the next 100-hour check must be carried out after 187 hours of operation).

Checks are carried out per the maintenance checklists, where type and volume of maintenance work is outlined in key words. The lists must be photocopied and filled out for each maintenance check. The respective check (e.g., 100-hour check) must be noted on the top of each page of the maintenance checklist. All the maintenance work carried out must be initialed in the signature area by the aircraft mechanic performing the task. After maintenance, the completed checklists must be entered in the maintenance records. The maintenance must be confirmed in the log book. All discrepancies and remedial action must be recorded in a report of findings to be generated and maintained by the company authorized to carry out maintenance work. It is the responsibility of the aircraft operator to store and keep the records. Replacement of equipment (e.g., carburetor, fuel pump, governor) and execution of Service Bulletins must be entered in the log book, stating required information.

Carburetor Synchronization

For smooth idling, synchronization of the throttle valves is necessary. When synchronizing, slacken both Bowden cables, and detach the resonator hose (3) of the compensating tube (2) to separate the two air intake systems. *[Figure 11-26]* In this condition, no significant difference in the engine running should be noticeable. If adjustment is needed for synchronous basic throttle adjustments (mechanical synchronization), proceed as follows. *[Figures 11-27 and 11-28]*

Adjust the two Bowden cables for simultaneous opening of the throttle valves. Remove the cable fixation (4) on the throttle lever (1). Next, release the return spring (5) from its attachment on the throttle lever (1), and return the throttle lever (1) to its idle stop position (3) by hand. There should be no resistance during this procedure. Unscrew the idle speed adjustment screw (2) until it is free of the stop. Insert a 0.1 mm (0.004 in) feeler gauge (gap X) between the idle speed adjustment screw (2) and the carburetor idle stop (3), then gently turn the idle screw clockwise until contact is made with the 0.1 mm (0.004 in) feeler gauge. Pull out the feeler gauge and turn each idle speed adjustment screw (2) 1.5 turns in clockwise direction. Gently turn each idle mixture screw (6) clockwise until it is fully inserted and, then, open by 1½ turns counterclockwise. Hook the return spring (5) back up to the throttle lever (1) in its original position. Check that the throttle valve opens fully, automatically. Carry out the above procedure on both carburetors.

Note: The mechanical carburetor synchronization is sufficiently exact.

At this point, place the throttle lever in the flight deck to the idle stop position. Ensure that the throttle lever remains in this position during the next steps of the synchronization process. With the throttle lever in the idle stop position, move

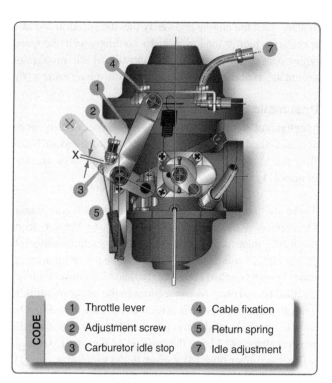

Figure 11-27. *Carburetor throttle lever.*

CODE			
1	Throttle lever	4	Cable fixation
2	Adjustment screw	5	Return spring
3	Carburetor idle stop	7	Idle adjustment

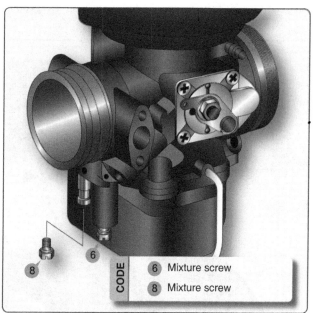

Figure 11-28. *Idle mixture screw.*

CODE		
6	Mixture screw
8	Mixture screw

the throttle lever (1) to the carburetor idle stop position, using the cable fixation (4), and secure the Bowden cable accordingly. As soon as the two carburetor Bowden cables are installed (throttle lever idle position), check that the idle speed adjustment screw (2) rests fully on the idle stop (3) without pressure.

Caution: An idle speed that is too low results in gearbox damage, and if an idle speed is too high, the engine is harder

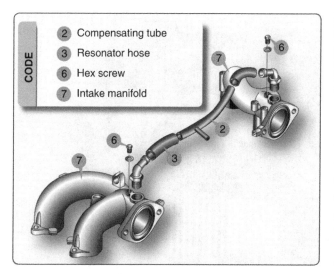

CODE		
2	Compensating tube
3	Resonator hose
6	Hex screw
7	Intake manifold

Figure 11-26. *Resonator hose and compensating tube.*

to start. Start the engine and verify the idle speed. If the idle speed is too high or too low, adjust accordingly with idle speed adjustment screw (2). Check the operational idle mixture of the engine. If necessary, adjust with the idle mixture screw (6).

Pneumatic Synchronization

Mechanical synchronization should have already been accomplished. The two carburetors are adjusted to equal flow rate at idling by use of a suitable flow meter or vacuum gauges (1).

There are two possible methods to connect test equipment. One option is to remove the hex screw (6) M6 x 6 from the intake manifold (7) and connect the vacuum gauge(s). *[Figure 11-26* and *Figure 11- 29]* Remove the compensating tube (2) with attached hoses (12) (connection between intake manifolds) and plug the connections in the intake manifolds. The other hook up option is to remove the compensating tube hose (2) from the push-on connection (5) after removing the tension clamp (4). Using the push-on connection (5), install a flexible rubber hose (8) leading to the vacuum gauge (1), using the balance tube (4). Install the other flexible rubber hose leading to the vacuum gauge. *[Figure 11-29]* Before proceeding any further, secure the aircraft on the ground using wheel chocks and ropes.

Warning: Secure and observe the propeller zone during engine operation.

Start the engine, verify the idle speed, and make any necessary corrections. If a setting correction of more than ½ turn is required, repeat mechanical synchronization to prevent too high a load on the idle stops. If the idle speed is too high, the maximum the idle screw can be unscrewed is a complete turn. If no satisfactory result can be achieved, inspect the idle jets for contamination and clean if necessary.

Caution: Also check for translucent, jelly-like contamination. Inspect for free flow.

Once the proper idling speed has been established, it is necessary to check the operating range above the idle speed. First, establish that the engine is developing full takeoff performance or takeoff rpm when selected in the flight deck. Then, the setting of the operating range (idle to full throttle) can be checked or adjusted.

Start and warm up engine as per the operator's manual.

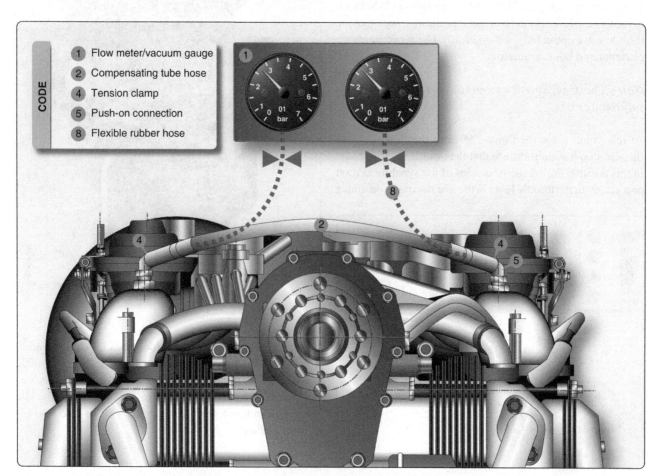

Figure 11-29. *Gauges attached to the engine.*

CODE
1 Flow meter/vacuum gauge
2 Compensating tube hose
4 Tension clamp
5 Push-on connection
8 Flexible rubber hose

Select full power and check that both pressure gauges are registering the same readings. If the same reading is not made on both gauges, shut down the engine and check that carburetor actuation has full travel and that the chokes are in the full off-position. If necessary, fit/modify the carburetor actuation as required to achieve full power on both carburetors. Once full power has been established on both carburetors, retard the throttle and observe the pressure gauge settings. The pressure gauges should show the same reading for both carburetors. Discrepancies must be compensated for by adjusting the off idle adjustment (7). *[Figure 11-27]* The carburetor with the lower indication must be advanced to match the higher one. This is done by shutting down the engine and loosening the locknut on the Bowden cable and screwing the off idle adjustment in by ½ turn, then tightening the locknut and retesting the engine. Final idle speed adjustment may be required by resetting the idle speed adjustment screws (2). *[Figure 11-27]* Equal adjustment must be made on both carburetors.

Any major adjustments require retesting to verify all parameters mentioned in this procedure are within limits. Install compensation tube assembly on the engine in reverse sequence of removal. Any minor differences in balance at idle speed is compensated for. Always follow the instructions of the instrument manufacturer.

Idle Speed Adjustment

If satisfactory idle speed adjustment cannot be achieved, inspection of the idle jet or additional pneumatic synchronization is necessary. Always carry out idle speed adjustment when the engine is warm. Basic adjustment of the idle speed is first accomplished by using the idle speed adjustment screw (2) of the throttle valve. *[Figure 11-27]*

Optimizing Engine Running

Optimizing the engine run is necessary only if not accomplished at carburetor synchronization. Close the idle mixture screw (6) by turning clockwise to screw in fully and, then, opening again by 1½ turns counterclockwise. *[Figures 11-27 and 11-28]* Starting from this basic adjustment, the idle mixture screw (6) is turned until the highest motor speed is reached. The optimum setting is the middle between the two positions, at which an rpm drop is noticed. Readjustment of the idle speed is carried out using the idle speed adjustment screw (2) and, if necessary, by slightly turning the idle mixture screw again. Turning the idle mixture control screw in a clockwise direction results in a leaner mixture and turning counterclockwise in a richer mixture.

Checking the Carburetor Actuation

The Bowden cables should be routed in such a way that carburetor actuation is not influenced by any movement of the engine or airframe, thus possibly falsifying idle speed setting and synchronization. *[Figure 11-30]* Each carburetor is actuated by two Bowden cables. At position 1, connection for throttle valve and at position 2, make the connection for the choke actuator. The Bowden cables must be adjusted so that the throttle valve and the choke actuation of the starting carburetor can be fully opened and closed. Bowden cables and lever must operate freely and not jam.

Warning: With carburetor actuation not connected, the throttle valve is fully open. The initial position of the carburetor is full throttle. Never start the engine with the actuation disconnected. Inspect Bowden cables and levers for free movement. Cables must allow for full travel of lever from stop to stop. Adjust throttle cables to a clearance of 1 mm (0.04 in). Inspect and lubricate linkage on carburetor and carburetor joints with engine oil. Inspect return springs (3) and engagement holes for wear.

Lubrication System

Oil Level Check

Always allow engine to cool down to ambient temperature before starting any work on the lubrication system. Severe burns and scalds may result from hot oil coming into contact with the skin. Switch off ignition and remove ignition key. To assure that the engine does not turn by the starter, disconnect the negative terminal of aircraft battery. Before checking the oil level, make sure that there is not excess residue oil in the crankcase. Prior to oil level check, turn the propeller

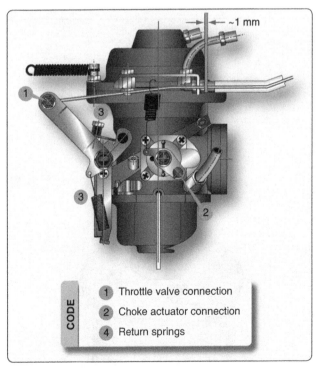

Figure 11-30. *Bowden cable routing.*

several times by hand in the direction of engine rotation to pump all the oil from the engine to the oil tank. This process is completed when air flows back to the oil tank. This air flow can be perceived as a gurgling noise when the cap of the oil tank is removed. The oil level in the oil tank should be between the two marks (maximum/minimum) on the oil dipstick, but must never fall below the minimum mark. *[Figure 11-31]* Replenish oil as required, but for longer flights, replenish oil to maximum mark to provide for more of an oil reserve. During standard engine operation, the oil level should be mid-way between the maximum and minimum marks a higher oil level (over servicing). Oil can escape through the venting (breather) passage.

Oil Change

It is advisable to check the oil level prior to an oil change, as it provides information about oil consumption. Run engine to warm the oil before beginning the procedure. Taking proper precautions, crank the engine by hand to transfer the oil from the crankcase. Remove the safety wire and oil drain screw (1) from the oil tank, drain the used oil, and dispose of as per environmental regulations. *[Figure 11-32]* Remove and replace oil filter at each oil change. It is not necessary to remove oil lines and other oil connections. Draining the suction lines, oil cooler, and return line is not necessary and must be avoided, as it results in air entering the oil system. Replacement of the oil filter and the oil change should be accomplished quickly and without interruption to prevent a draining of the oil system and the hydraulic tappets. Compressed air must not be used to blow through

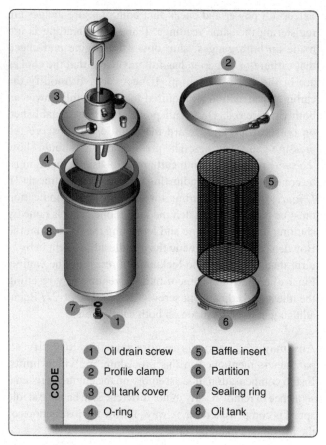

CODE			
1	Oil drain screw	5	Baffle insert
2	Profile clamp	6	Partition
3	Oil tank cover	7	Sealing ring
4	O-ring	8	Oil tank

Figure 11-32. *Oil tank.*

the oil system (or oil lines, oil pump housing, oil bores in the housing). Replace the oil drain screw torque and safety wire. Only use the appropriate oil in accordance with the latest operator's manual and service instruction. The engine must not be cranked when the oil system is open. After the oil change is accomplished, the engine should be cranked by hand in the direction of engine rotation (approximately 20 turns) to completely refill the entire oil circuit.

Cleaning the Oil Tank

Cleaning the oil is optional and requires venting of the oil system. It is only necessary to clean the oil tank and the inner parts if there is heavy oil contamination. The procedure for cleaning the oil tank is shown in *Figure 11-32.* Detach the profile clamp (2) and remove the oil tank cover (3), together with the O-ring (4) and the oil lines. Remove the inner parts of the oil tank, such as the baffle insert (5) and the partition (6). Clean oil tank (8) and inner parts (5, 6), and check for damage. Be aware that incorrect assembly of the oil tank components can cause engine faults or engine damage. Replace the drain screw with a new sealing ring (7) and tighten to 25 Newton meters (Nm) (18.5 ft/lb) and safety wire. Reassemble the oil tank by following the same steps in reverse order.

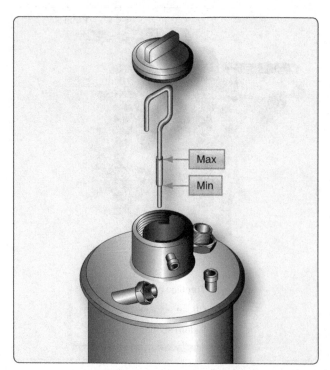

Figure 11-31. *Oil dipstick minimum and maximum marks.*

Inspecting the Magnetic Plug

Remove the magnetic plug and inspect it for accumulation of chips. *[Figure 11-33]* The magnetic plug (torx screw) is located on the crankcase between cylinder 2 and the gearbox. This inspection is important because it allows conclusions to be drawn on the internal condition of the gearbox and engine, and reveals information about possible damage. If a significant amount of metal chips are detected, the engine must be inspected, repaired, or overhauled. Steel chips in low numbers can be tolerated if the accumulation is below 3 mm (0.125 in). *[Figure 11-33]* In the case of unclear findings, flush the oil circuit and fit a new oil filter. Afterwards, conduct an engine test run and inspect the oil filter once more. If there are larger accumulations of metal chips on the magnetic plug, the engine must be repaired or overhauled in accordance with the manufacturer's instructions for continued airworthiness. A detailed inspection of affected engine components must be performed. If the oil circuit is contaminated, replace the oil cooler and flush the oil circuit, then trace the cause and remedy the situation. If the magnetic chip is found to have no metal, then clean and reinstall. Tighten the plug to a torque of 25 Nm (18.5 ft/lb). Safety wire the plug and inspect all systems for correct function.

Checking the Propeller Gearbox

The following free rotation check and friction torque check are necessary only on certificated engines and on engines with the overload clutch as an optional extra. Engines without the overload clutch (slipper clutch) still incorporate the torsional shock absorption. This design is similar to the system with overload clutch, but without free rotation. For this reason, the friction torque method cannot be applied on engines without the overload clutch.

Checking the Friction Torque in Free Rotation

Fit the crankshaft with a locking pin. *[Figure 11-34]* With the crankshaft locked, the propeller can be turned by hand 15 or 30 degrees, depending on the profile of the dog gears installed. This is the maximum amount of movement allowed

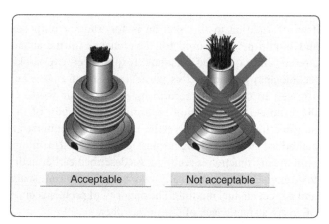

Figure 11-33. *Inspecting the magnetic plug.*

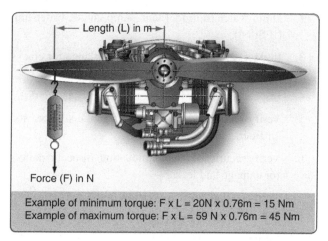

Example of minimum torque: F x L = 20N x 0.76m = 15 Nm
Example of maximum torque: F x L = 59 N x 0.76m = 45 Nm

Figure 11-34. *Checking propeller gearbox.*

by the dog gears in the torsional shock absorption unit.

Warning: Ignition OFF and system grounded. Disconnect negative terminal of aircraft battery.

Turn the propeller by hand back and forth between ramps, taking into consideration the friction torque. No odd noises or irregular resistance must be noticeable during this movement. Attach a calibrated spring scale to the propeller at a certain distance (L) from the center of the propeller. Measure the force required to pull the propeller through the 15 or 30 degree range of free rotation. Calculate friction torque Nm by multiplying the force Newton's (N) or pounds (lb) obtained on the spring scale by the distance the scale is attached from the center of the propeller (L). The distance measurement and torque measurement must be in the same units either standard or metric and cannot be mixed up. The friction torque must be between a minimum of 25 Nm and maximum of 60 Nm (18.5 to 44.3 ft/lb). A calculation example is as follows:

Friction Torque (FT) = Length (meters) x Torque (Newtons)
FT = 0.5 meters x 60 Newtons
FT = 30 Nm

Remove crankshaft locking pin and reinstall plug with new gasket. Reconnect negative terminal of aircraft battery. If the above mentioned friction torque is not achieved, inspect, repair, or overhaul the gearbox in accordance with the manufacturer's instructions for continued airworthiness. Testing the propeller flange is not normal maintenance but can be carried out if defects or cracks are suspected.

Daily Maintenance Checks

The following checklist should be used for daily maintenance checks. Repair, as necessary, all discrepancies before flight.

1. Verify ignition OFF.

2. Drain water from fuel tank sump and/or water trap (if fitted).

3. Inspect carburetor rubber socket or flange for cracks and verify secure attachment.

4. Inspect carburetor float chamber for water and dirt.

5. Verify security and condition of intake silencer and air filter.

6. Verify security of radiator mounting. Inspect radiators for damage and leaks.

7. Verify coolant level in overflow bottle and security of cap.

8. Verify coolant hoses for security, and inspect for leaks and chafing.

9. Inspect engine for coolant leaks (cylinder head, cylinder base, and water pump).

10. Verify oil content for rotary valve gear lubrication and security of oil cap.

11. Verify oil hoses for security, and inspect for leaks and chafing (rotary valve gear lubrication system and oil injection system).

12. Verify ignition coils/electronic boxes for secure mounting, and check ignition leads and all electrical wiring for secure connections and chafing.

13. Verify electric starter for secure mounting, and inspect cover for cracks.

14. Verify engine to airframe mounting for security and inspect cracks.

15. Verify fuel pump mounting for security, and inspect all fuel hose connections (filters, primer bulbs, and taps for security, leakage, chafing and kinks).

16. Verify fuel pump impulse hose for secure connections, and inspect for chafing and kinks.

17. Verify safety wiring of gearbox drain and level plugs.

18. Inspect rubber coupling for damage and aging (C type gearbox only).

19. Rotate engine by hand and listen for unusual noises (first, double verify ignition OFF).

20. Check propeller shaft bearing for clearance by rocking propeller.

21. Inspect throttle choke and oil pump lever cables for damage (end fittings, outer casing, and kinks).

Pre-flight Checks

The following checklist should be performed for all pre-flight checks. Repair, as necessary, all discrepancies and shortcoming before flight.

1. Verify ignition OFF.

2. Check fuel content.

3. Inspect for coolant leaks.

4. Verify oil tank content (oil injection engines).

5. Verify spark plug connectors for security.

6. Inspect engine and gearbox for oil leaks.

7. Inspect engine and gearbox for loose or missing nuts, bolts, and screws, and verify security of gearbox to engine mounting.

8. Inspect propeller for splits and chips. If any damage, repair and/or rebalance before use.

9. Verify security of propeller mounting.

10. Check throttle, oil injection pump, and choke actuation for free and full movement.

11. Verify that cooling fan turns when engine is rotated (air-cooled engines).

12. Inspect exhaust for cracks, security of mounting, springs, and hooks for breakage and wear, and verify safety wiring of springs.

13. Start engine after assuring that area is clear of bystanders.

14. Single ignition engines: check operation of ignition switch (flick ignition off and on again at idling).

15. Dual ignition engines: check operation of both ignition circuits.

16. Check operation of all engine instruments during warm up.

17. If possible, visually check engine and exhaust for excessive vibration during warm up (indicates propeller out of balance).

18. Verify that engine reaches full power rpm during takeoff roll.

Troubleshooting & Abnormal Operation

The information in this section is for training purposes and should never be used for maintenance on the actual aircraft. Only qualified personnel (experienced two-stroke technicians) trained on this particular type of engine are allowed to carry out maintenance and repair work. If the following information regarding the remedy of the malfunction does not solve the malfunction, contact an authorized facility. The engine must not be returned to service until the malfunction is rectified. As described earlier in the text, engines require basically two essentials to run: spark and correct air-fuel mixture. The majority of problems quite often are a simple lack of one or the other.

Troubleshooting

Follow an organized method of troubleshooting. This facilitates the identification of discrepancies or malfunctions.

- Fuel—start by checking the supply (tank), fittings (loose), filter (plugged), and float chamber (fouled).

- Spark—check for spark at the spark plugs.

Problems of a more complex nature are best left to an engine technician. The following are examples of engine troubles and potential fixes.

Engine Keeps Running With Ignition OFF

Possible cause: Overheating of engine.
Remedy: Let engine cool down at idling at approximately 2,000 engine rpm.

Knocking Under Load

Possible cause: Octane rating of fuel too low.
Remedy: Use fuel with higher octane rating.
Possible cause: Fuel starvation, lean mixture.
Remedy: Check fuel supply.

Abnormal Operation

Exceeding the Maximum Admissible Engine Speed

Reduce engine speed. Any overage of the maximum admissible engine speed must be entered by the pilot into the logbook, stating duration and extent of over-speed.

Exceeding Maximum Admissible Cylinder Head Temperature

Reduce engine power, setting to the minimum necessary, and carry out precautionary landing. Any exceeding of the maximum admissible cylinder head temperature must be entered by the pilot into the logbook, stating duration and extent of excess-temperature condition.

Exceeding Maximum Admissible Exhaust Gas Temperature

Reduce engine power, setting to the minimum necessary, and carry out precautionary landing. Any exceedence of the maximum admissible exhaust gas temperature must be entered by the pilot into the logbook, stating duration and extent of excess-temperature condition.

Engine Preservation

If the engine is not going to be used for an extended period of time, certain measures must be taken to protect engine against heat, direct sun light, corrosion, and formation of residues. In particular, the water bonded by the alcohol in the fuel causes increased corrosion problems during storage.

After each flight, activate choke for a moment before stopping engine. Close all engine openings like exhaust pipe, venting tube, and air filter to prevent entry of contamination and humidity. For engine storage of one to four weeks, proceed with preservation prior to engine stop or on the engine at operating temperature. Let the engine run at increased idle speed. Shut the engine down and secure against inadvertent engine start. Remove air filters and inject approximately 3 cubic cm of preservation oil or equivalent oil into the air intake of each carburetor. Restart the engine and run at increased idle speed for 10–15 seconds. Shut engine down and secure against inadvertent engine start. Close all engine openings, such as exhaust pipe, venting tube, and air filter, to prevent entry of contamination and humidity.

For engine storage of engine for longer than four weeks and up to one year, proceed with preservation prior to engine stop and on the engine at operating temperature. Let the engine run at increased idle speed. Remove air filters and inject approximately 6 cubic cm of preservative oil or equivalent oil into the air intake of each carburetor. Stop the engine. Remove spark plugs and inject approximately 6 cubic cm preservation oil or equivalent oil into each cylinder and slowly turn crankshaft 2 to 3 turns by hand to lubricate top end parts. Replace and re-torque the spark plugs. Drain gasoline from float chambers, fuel tank, and fuel lines. Drain coolant on liquid cooled engines to prevent any damage by freezing. Lubricate all carburetor linkages using the proper lubricates. Close all openings of the engine, such as exhaust pipe openings, venting tube, and air intake, to prevent entry of any foreign material and humidity. Protect all external steel parts by spraying with engine oil.

General Maintenance Practices for the Light-Sport Jabiru Engines

Note: Some specific maintenance practices that differ from conventional certificated engines is covered for background and educational acquaintance purposes only. Always refer to the current manufacturer's information when performing maintenance on any engine.

Engine & Engine Compartment Inspection

Check for oil, fuel exhaust, and induction leaks and clean the entire engine and compartment before inspection. Check flywheel screw tensions to 24 foot pounds. Check the carburetor air filter and clean it by removing it from the intake housing and blowing compressed air against the direction of the intake flow. For operation in heavy dust conditions, clean air filter at shorter intervals than recommended for normal conditions. A clogged filter reduces engine performance, as well as promotes premature engine wear. The engine baffles and air ducts should be checked for condition and functionality.

Two methods can be used to check the cylinders compression. The compression gauge method is used to measure compression using a compression tracer. Readings are taken with a fully open throttle valve at engine oil temperature between 30 °C and 70 °C (90 °F to 160 °F). If readings are below 6 bar (90 psi) a check of the pistons, cylinders, valves, and cylinder heads must be undertaken.

The second method uses the pressure differential test. Check cylinder compression for a maximum allowable pressure loss of 25 percent. As an alternative to a compression test, a pressure differential test (leak down) can be accomplished. This is a much better test of the condition of rings, bore, head sealing, and valves. This is the normal test used in aviation and requires specific equipment. The test is carried out with the engine in warm to hot condition. Input pressure is best set at 80 psi; a second gauge reads the differential. This is done with the piston on TDC on the firing stroke.

Note: The propeller needs to be restrained. A differential of lower than 80/60 (generally a 25 percent loss) indicates a problem.

Problems can be better identified by observing where air is escaping from the cylinder, blow-by. Some examples are as follows:

1. Blow-by through the crankcase vent indicates worn rings or bore.

2. Leaking from carburetor indicates a poor intake valve seal.

3. Leaking from exhaust indicates a poor exhaust valve seal.

4. Head leak indicates poor head to cylinder seal.

With the problem identified, the malfunction can then be corrected. Poor compression can be an indication of a serious problem. For example, continued operation with poor compression due to a poorly sealing valve can lead to eventual valve failure and heavy damage to the piston, connecting-rod, barrel, and head.

Lubrication System

The oil should be changed as required by the manufacturer. When changed, the oil filter should also be changed. Change the oil filter at every 50 hourly inspection. Drain the oil while engine is still warm and visually check for leaks. Fill the engine with oil (approximately 2.3 liters) and check oil level. Never exceed the maximum mark. Use only registered brand oils meeting the correct specifications. Do not drain the oil cooler during a normal oil change. The cooler holds only a small amount of old oil that has negligible effect on the new oil. Taking the hoses on and off the cooler can prematurely age the oil lines and lead to hoses slipping off the cooler.

Carburetor Adjustment & Checks

To adjust the engine's idle speed, adjust the idle stop screw (7 mm screw) against throttle lever. Standard idle mixture screw position is 1¼ turns out from the seated position. Fine adjustment may be necessary to give a smooth idle.

The mixture is set by selecting jet sizes. As supplied, the engine has jets to suit a majority of installations; however, the mixture may be affected by operation with a propeller that does not meet the requirements listed in the installation manual or by ambient temperature extremes. If an engine is to be used in these situations, an exhaust gas temperature (EGT) gauge should be fitted and monitored against the limits specified above. Do not change carburetor settings if EGT readings fall outside the range given without consulting with Jabiru Aircraft or the local authorized representative. The carburetor automatically adjusts the mixture to account for altitude. Visual inspection should include checks for carburetor joint degradation and carburetor linkage for full and free movement, correct positioning of stops and security.

Spark Plugs

When plugs are removed from a warm engine, the inspection of the tip of the spark plug can be used to indicate the health of the engine. If the tip of the plug is a light brown color, the plug is operating correctly. A black velvet, sooty looking plug tip generally is an indication of an overly rich mixture (check the choke, air filter, and intake). If the firing end tip is covered with oil, it is an indication of too much oil in the combustion chamber (check for worn piston rings and cylinder walls). When servicing the spark plugs, do not use steel or brass brushes for cleaning, and never sandblast plugs. Clean the spark plugs with a plastic brush in a solvent. Check electrode gap and, if necessary, adjust to 0.55–0.6mm (0.022 in–0.024 in) by carefully bending the electrode. Use the recommended Plugs (NGK D9EA) and place a suitable anti-seize compound on threads of the plug before installing them in the engine. Tighten spark plugs when the engine is cold and adjust engine to the correct torque value. Reconnect the ignition lead.

Exhaust System

Visually check the exhaust system for security of mounting, damage, rubbing, leaks, and general condition. Check nuts and bolts for tightness and condition; re-torque and replace if necessary.

Head Bolts

Check the head bolt torque after five hours of operation, and again after ten hours of operation. The bolts should, thereafter, be checked annually. Head bolts torque when cold to 20 ft/lb.

Tachometer & Sender

Many apparent engine problems can be caused through inaccurate tachometers. Where engine performance is observed to be outside limits, the tachometer should be checked against a calibrated instrument. Tachometer sender gap is 0.4mm (0.016 inches). The sender must have at least 60 percent covered by the tags fitted to the gearbox side of the flywheel. Ensure both tags are equal distance from sender.

Engine Inspection Charts

Note: Read all inspection requirement paragraphs prior to using these charts. *[Figure 11-35]*

Propeller	Engine and Engine Compartment
Spinner * *	Check flywheel screw tensions to 24 foot pounds*
Spinner flange * *	Carburetor air filter * *
Spinner screws * *	Engine baffles and air ducts *
Propeller * *	Cylinders *
Propeller bolts/nuts - Tension *	Crankcase & front crankcase seal *
Spinner/prop tracking * *	Hoses, lines and fittings * *
	Intake and exhaust systems *
	Ignition harness, distributor caps & rotors *
	NOTE: Check for oil, fuel exhaust and induction leaks, then clean entire engine and compartment before inspection.
Annual Inspection** **Each 100 Hours***	

Figure 11-35. *Engine inspection charts.*

Glossary

A

Abradable strip. A strip of material in the compressor housing of some axial-flow gas turbine engines. The tip of the compressor blade touches the abradable strip, and wears, or abrades a groove in it. This groove ensures the minimum tip clearance.

Abradable tip (compressor blade tip). The tip of some axial-flow compressor blades constructed so that it will abrade, or wear away, upon contact with the compressor housing, which ensures the minimum tip clearance between the blade and the housing.

Absolute pressure. Pressure referenced from zero pressure or a vacuum.

AC. Alternating current. Electrical current in which the electrons continually change their rate of flow and periodically reverse their direction.

ACC. Active clearance control. A system for controlling the clearance between tips of the compressor and turbine blades and the case of high-performance turbofan engines. When the engine is operating at maximum power, the blade tip clearance should be minimum, and the ACC system sprays cool fan discharge air over the outside of the engine case. This causes the case to shrink enough to decrease the tip clearance. For flight conditions that do not require such close clearance, the cooling air is turned off, and the case expands to its normal dimensions. The control of the ACC system is done by the FADEC, or full-authority digital electronic control.

Acceleration. The amount the velocity of an object is increased by a force during each second it is acted upon by that force. Acceleration is usually measured and expressed in terms of feet per second, per second (fps²).

Accessory end. The end of a reciprocating engine on which many of the accessories are mounted. Also, called the anti-propeller end.

Accumulator. A hydraulic component that stores a non-compressible fluid, such as oil, under pressure. An accumulator has two compartments separated by a flexible or movable partition with one compartment containing compressed air. When oil is pumped into the other compartment, the partition moves over, further compressing the air which holds pressure on the oil.

AD (ashless dispersant) oil. A mineral-based lubricating oil used in reciprocating engines. This oil does not contain any metallic ash-forming additives, but has additives that disperse the contaminants and hold them in suspension until they can be removed by filters.

ADC. Air data computer. An electronic computer in an aircraft that senses pitot pressure, static pressure, and total air temperature. It produces an indication of altitude, indicated airspeed, true airspeed, and Mach number. The output of ADC is usable by any of the engine or flight control computers.

ADI (antidetonation injection) system. A system used with some large reciprocating engines in which a mixture of water and alcohol is sprayed into the engine with the fuel when operating at extremely high power. The air/fuel mixture is leaned to allow the engine to develop its maximum power, and the ADI fluid absorbs excessive heat when it vaporizes.

Adiabatic change. A physical change that takes place within a material in which heat energy is neither added to the material, nor taken away. If a container of gas is compressed, with no heat energy added to or taken from it, the gas will become hotter; its temperature will rise.

Aeromatic propeller. A patented variable-pitch propeller that has flyweights around the blade shanks and the blades angled back from the hub to increase the effects of aerodynamic and centrifugal twisting forces. This propeller automatically maintains a relatively constant rpm for any throttle setting.

Aft-fan engine. A turbofan engine with the fan mounted behind the compressor section. The blades of an aft-fan are normally extensions of the free turbine blades.

Afterburner. A component in the exhaust system of a turbojet or turbofan engine used to increase the thrust for takeoff and for special flight conditions. Since much of the air passing through a gas turbine engine is used only for cooling, it still contains a great deal of oxygen. Fuel is sprayed into the hot, oxygen-rich exhaust in the afterburner, where it is ignited and burned to produce additional thrust.

Air bleed (carburetor component). A small hole in the fuel passage between the float bowl and the discharge nozzle of a float carburetor. Air drawn into the liquid fuel through the air bleed breaks the fuel up into an emulsion, making it easy to atomize and vaporize.

Air cooling. The removal of unwanted heat from an aircraft engine by transferring the heat directly into the air flowing over the engine components.

Air/fuel mixture ratio. The ratio of the weight of the air to that of the fuel in the mixture fed into the cylinders of an engine.

Air impingement starter. A turbine engine starter that basically consists of a nozzle that blows a stream of compressed air against the turbine blades to rotate the compressor for starting the engine.

Air-oil separator. A component in a turbine engine lubrication system that removes the air from the scavenged oil before it is returned to the oil tank.

Airworthiness Directive. Airworthiness Directives (ADs) are legally enforceable regulations issued by the FAA in accordance with 14 CFR part 39 to correct an unsafe condition in a product. Part 39 defines a product as an aircraft, engine, propeller, or appliance.

All-weather spark plug. A shielded spark plug designed for high altitude operation. The ceramic insulator is recessed into the shell to allow a resilient grommet on the ignition harness to provide a watertight seal. All weather spark plugs, also called high-altitude spark plugs, are identified by their 3/4-20 shielding threads.

Alpha control range (alpha mode). The flight operating mode from takeoff through landing for a turbo-prop engine. Alpha mode includes operations from 95% to 100% of the engine's rated rpm.

Altitude engine. An aircraft reciprocating engine equipped with a supercharger that allows it to maintain its rated sea-level horsepower to an established higher altitude.

Amateur-built aircraft. Aircraft built by individuals as a hobby rather than by factories as commercial products. Amateur-built or home-built aircraft do not fall under the stringent requirements imposed by the FAA on commercially built aircraft.

Ambient air pressure. The pressure of the air that surrounds an object.

Analog indicator. An indicator that shows the value of the parameter being measured by a number marked on a graduated dial aligned with a movable pointer.

Angle of attack. The acute angle between the chordline of a propeller blade and the relative wind. The angle of attack is affected by both the engine rpm and the forward speed of the aircraft.

Annual inspection. A complete inspection of the airframe and powerplant required for FAA-certificated aircraft operating under 14 CFR part 91 General Operating and Flight Rules, and not on one of the authorized special inspection programs. An annual inspection must be conducted every 12 calendar months, and it must be conducted by an aviation maintenance technician who holds an Airframe and Powerplant rating and an Inspection Authorization. The scope of an annual inspection is the same as that of a 100-hour inspection.

Annular duct. A duct, or passage, that surrounds an object. The annular fan-discharge duct surrounds the core engine.

Annular orifice. A ring-shaped orifice, normally one that surrounds another orifice.

Annulus. A ring or groove around the outside of a circular body or shaft, or around the inside of a cylindrical hole.

Annunciator panel. A panel of warning lights visible to the flight crew. The lights are identified by the name of the system they represent and are often covered with colored lenses. Red lights indicate a dangerous condition and green indicate a safe condition.

Anodizing. A hard, airtight, unbroken oxide film electrolytically deposited on an aluminum alloy surface to protect it from corrosion.

Anti-icing. Prevention of the formation of ice on a surface.

Anti-propeller end. The end of a reciprocating engine that does not attach to the propeller. Also called the accessory end.

APC. Absolute pressure controller.

APU. Auxiliary power unit. A small turbine- or reciprocating-engine-powered generator, hydraulic pump, and air pump. APUs are installed in the aircraft and are used to supply electrical power, air, and hydraulic pressure when the main engines are not running.

Aramid fiber. Fiber made from an organic compound of

carbon, hydrogen, oxygen, and nitrogen. It has high strength and low density. It is flexible under load and is able to withstand impact, shock, and vibration. Kevlar is a well-known aramid fiber.

Aromatic compound. A chemical compound such as toluene, xylene, and benzene that is blended with gasoline to improve its anti-detonation characteristics.

Articulating rod. See link rod.

Aspect ratio. The ratio of the length of an airfoil, such as a compressor blade, to its width.

Asymmetrical loading. The loading of a propeller disc that causes one side to produce more thrust than the other side.

ATF. Aerodynamic twisting force. The aerodynamic force that acts on a rotating propeller blade to increase its blade angle. The axis of rotation of a blade is near the center of its chordline, and the center of pressure is between the axis and the leading edge. Aerodynamic lift acting through the center of pressure tries to rotate the blade to a higher pitch angle.

Atomize. The process of breaking a liquid down into tiny droplets or a fine spray. Atomized liquids vaporize easily.

Augmentor tube. A long, specially shaped stainless steel tube mounted around the exhaust tail pipe of a reciprocating engine. As exhaust gases flow through the augmentor tube, they produce a low pressure in the engine compartment that draws in cooling air through the cylinder fins.

Automatic intake valve. An intake valve opened by low pressure created inside the cylinder as the piston moves down. There is no mechanical means of opening it.

Automatic mixture control (AMC). The device in a fuel metering system, such as a carburetor or fuel injection system, that keeps the air/fuel mixture ratio constant as the density of air changes with altitude.

Autosyn system. The registered trade name of a remote indicating instrument system. An Autosyn system uses an electromagnet rotor, excited with 400-hertz AC, and a three-phase distributed-pole stator.

Axial bearing load. The load on a bearing parallel to the shaft on which the bearing is mounted. Thrust produces an axial load on a bearing.

Axial turbine. A turbine that is turned by a fluid flowing through it in a direction that is approximately parallel to the

shaft on which the turbine wheel is mounted.

Axial-flow compressor. A type of compressor used in gas turbine engines. Air passes through the compressor in essentially a straight line, parallel to the axis of the compressor. The compressor is made of a number of stages of rotating compressor blades between stages of stationary stator vanes.

Axis of rotation. The center line about which a propeller rotates.

B

Babbitt. A soft silvery metal used for main bearing inserts in aircraft reciprocating engines. Babbitt is made of tin with small amounts of copper and antimony.

Back (propeller nomenclature). The curved surface of a propeller blade. The back of a propeller blade corresponds to the upper surface of an airplane wing.

Back-suction mixture control. A type of mixture control used in some float carburetors that regulates the air/fuel mixture by varying the air pressure above the fuel in the float bowl.

Baffle. A thin sheet metal shroud or bulkhead used to direct the flow of cooling air between and around the cylinder fins of an air-cooled reciprocating engine.

Bayonet stack. An exhaust stack with an elongated and flattened end. The gases leave the stack through a slot perpendicular to its length. Bayonet stacks decrease both exhaust back pressure and noise.

BDC. Bottom dead center. The position of a piston in a reciprocating engine when the piston is at the bottom of its stroke, and the wrist pin, crankpin, and center of the crankshaft are all in line.

Bell mouth. The shape of the inlet of an augmentor tube that forms a smooth converging duct. The bell mouth shape allows the maximum amount of air to be drawn into the tube.

Bell mouth inlet duct. A form of convergent inlet-air duct used to direct air into the compressor of a gas turbine engine. It is extremely efficient, and is used where there is little ram pressure available to force air into the engine. Bell mouth ducts are used in engine test cells and on engines installed in helicopters.

Benzene. A colorless, volatile, flammable, aromatic

hydrocarbon liquid which has the chemical formula C_6H_6. Benzene, which is sometimes called benzoil, is used as a solvent, a cleaning fluid, and a fuel for some special types of reciprocating engines.

Bernoulli's principle. A physical principle that explains the relationship between kinetic and potential energy in a stream of moving fluid. When energy is neither added to nor taken from the fluid, any increase in its velocity (kinetic energy) will result in a corresponding decrease in its pressure (potential energy).

Beta control range (Beta mode). The range of operation of a turboprop powerplant used for in-flight approach and ground handling of the engine and aircraft. Typically, the Beta mode includes operations from 65% to 95% of the engine's rated rpm.

Beta tube. A tube in a Garrett TPE331 turboprop powerplant that extends into the propeller pitch control to act as a follow-up device. It provides movement of the propeller blades in proportion to movement of the power lever.

Bezel. The rim which holds the glass cover in the case of an aircraft instrument.

BHP. Brake horsepower. The actual horsepower delivered to the propeller shaft of a reciprocating or turboprop engine.

Bidirectional fibers. Fibers in a piece of composite material arranged to sustain loads in two directions.

Bimetallic hairspring. A flat, spiral-wound spring made of two strips of metal laid side-by-side and welded together. The two metals have different coefficients of expansion, and as the temperature changes, the spiral either tightens or loosens. A bimetallic hair spring is used in a thermocouple temperature changes at the reference junction.

Bimetallic strip. A metal strip made of two different types of metal fastened together side by side. When heated, the two metals expand different amounts and the strip warps or bends.

BITE. Built-in test equipment. A troubleshooting system installed in many modern electronic equipment. BITE equipment monitors engine and airframe systems, and when a fault is found, isolates it and provides maintenance personnel with a code that identifies the LRU (line replaceable unit) that contains the fault.

Blade. The component of a propeller that converts the rotation of the propeller shaft into thrust. The blade of a propeller corresponds to the wing of an airplane.

Blending. A method of repairing damaged compressor and turbine blades. The damage is removed and the area is cleaned out with a fine file to form a shallow depression with generous radii. The file marks are then removed with a fine abrasive stone so the surface of the repaired area will match the surface of the rest of the blade.

Blisk. A turbine wheel machined from a single slab of steel. The disc and blades are an integral unit.

Blow-in doors. Spring-loaded doors in the inlet duct of some turbojet or turbofan engine installations that are opened by differential air pressure when inlet air pressure drops below that of the ambient air. Air flowing through the doors adds to the normal inlet air passing through the engine and helps prevent compressor stall.

BMEP. Brake mean effective pressure. The average pressure inside the cylinder of a reciprocating engine during the power stroke. BMEP, measured in pounds per square inch, relates to the torque produced by the engine and can be calculated when you know the brake horsepower.

Boost. A term for manifold pressure that has been increased above the ambient atmospheric pressure by a supercharger.

Bootstrapping. An action that is self-initiating or self-sustaining. In a turbocharger system, bootstrapping describes a transient increase in engine power that causes the turbocharger to speed up, which in turn causes the engine to produce more power.

Bore. The diameter of a reciprocating engine cylinder.

Borescope. An inspection tool for viewing the inside of a turbine engine without disassembling it. The instrument consists of a light, mirror, and magnifying lens mounted inside a small-diameter tube that is inserted into a turbine engine through borescope inspection ports.

Boss. An enlarged area in a casting or machined part. A boss provides additional strength to the part where holes for mounting or attaching parts are drilled.

Bottom. (verb) A condition in the installation of a propeller on a splined shaft when either the front or rear cone contacts an obstruction that prevents the cone from properly seating inside the propeller hub.

Bourdon tube. The major component in a gage-pressure measuring instrument. It is a thin-wall metal tube that has an elliptical cross section and is formed into a curve. One end of the tube is sealed and connected to an arm that moves the pointer across the instrument dial, and the open end is anchored to the instrument case. The pressure to be measured is directed into the open end, which causes the elliptical cross section to become more circular. As the cross section changes, the curve straightens and moves the pointer over the dial by an amount proportional to the amount of pressure.

Brayton cycle. The constant-pressure cycle of energy transformation used by gas turbine engines. Fuel is sprayed into the air passing through the engine and burned. Heat from the burning air/fuel mixture expands the air and accelerates it as it moves through the engine. The Brayton cycle is an open cycle in that the intake, compression, combustion, expansion, and exhaust events all take place at the same time, but in different locations within the engine.

British thermal unit (Btu). The basic unit of heat energy in the English system. One Btu is the amount of heat energy needed to raise the temperature of one pound of pure water from 60 °F to 61°F.

BSFC. Brake specific fuel consumption. A measure of the amount of fuel used for a given amount of power produced by a heat engine. BSFC is expressed in pounds of fuel burned per hour for each brake horse-power the engine is producing.

Buckets. The portions of aft-fan blades that are in the exhaust of the core engine. Buckets drive the fan from energy received from hot gases leaving the core engine.

Bungee cord. An elastic cord made of small strips of rubber encased in a loosely braided cloth tube that holds and protects the rubber, yet allows it to stretch. The energy in a stretched bungee cord may be used to crank a large aircraft engine.

Burner. See combustor.

Burnish. To smooth the surface of a metal part that has been damaged by a deep scratch or gouge. Metal piled at the edge of the damage is pushed back into the damage with a smooth, hard steel burnishing tool.

Butterfly valve. A flat, disc-shaped valve used to control the flow of fluid in a round pipe or tube. When the butterfly valve is across the tube, the flow is shut off, and when it is parallel with the tube, the obstruction caused by the valve is minimum, and the flow is at its greatest. Butterfly-type throttle valves are used to control the airflow through the fuel metering system.

Bypass engine. Another name for a turbofan engine. See turbofan engine.

Bypass ratio. The ratio of the mass of air moved by the fan to the mass of air moved by the core engine.

C

Calendar month. The measurement of time used by the FAA for inspection and certification purposes. One calendar month from a given date extends from that date until midnight of the last day of that month.

Cam. An eccentric, or lobe, on a rotating shaft that changes rotary motion into linear motion. A cam is mounted on the magnet shaft in a magneto to push upward on the insulated breaker point to separate, or open, the points when the magnet is in a particular location.

Cam engine. A reciprocating engine with axial cylinders arranged around a central shaft. Rollers on the pistons in the cylinders press against a sinusoidal cam mounted on the shaft to produce rotation of the shaft.

Cam-ground piston. A reciprocating engine piston that is not round, but is ground so that its diameter parallel to the wrist pin is slightly smaller than its diameter perpendicular to the pin. The mass of metal used in the wrist pin boss, the enlarged area around the wrist pin hole, expands when heated, and when the piston is at its operating temperature, it is perfectly round.

Can-annular combustor. A type of combustor used in some large turbojet and turbofan engines. It consists of individual cans into which fuel is sprayed and ignited. These cans mount on an annular duct which collects the hot gases and directs them uniformly into the turbine.

Capacitance afterfiring. The continuation of the spark across the gap in a shielded spark plug after the air/fuel mixture in the cylinder is ignited. Afterfiring is caused by the return of electrical energy stored in the capacitance of the shielded ignition leads. Capacitance afterfiring is eliminated by the use of a resistor in the spark plug.

Capacitor. An electrical component, formerly called a condenser, that consists of two large-area conductors, called plates, separated by an insulator. Electrons stored on one of the plates produces an electrostatic pressure difference between the plates.

Capillary tube. A glass or metal tube with a tiny inside diameter. Capillary action causes the fluid to move within the tube.

Carbon pile voltage regulator. A voltage regulator for a high output DC generator that uses a stack of pure carbon discs for the variable resistance element. A spring holds pressure on the stack to reduce its resistance when the generator output voltage is low. This allows maximum field current to flow. The field from an electro-magnet, whose strength varies directly with the generator voltage, opposes the spring to loosen the stack and increase its resistance when the generator voltage needs to be decreased. The increased resistance decreases the field current and reduces the output voltage.

Carbon track. A trail of carbon deposited by an arc across a high-voltage component such as a distributor block. Carbon tracks have a relatively low resistance to the high voltage and can cause misfiring and loss of engine power.

Cartridge starter. A self-contained starter used on some military aircraft. A cartridge similar in size to a shotgun shell is ignited in the starter breech. The expanding gases drive a piston attached to a helical spline that converts the linear movement of the piston into rotary motion to rotate the crankshaft.

Cascade effect. The cumulative effect that occurs when the output of one series of components serves as the input to the next series.

Catalyst. A substance used to change the speed, or rate, of a chemical action without being chemically changed itself.

Cavitating. The creation of low pressure in an oil pump when the inlet system is not able to supply all of the oil the pump requires. Prolonged cavitation can damage pump components.

Center of pressure. The point on the chordline of an airfoil where all aerodynamic forces are concentrated.

Center-line thrust airplane. A twin-engine airplane with both engines mounted in the fuselage. One is installed as a tractor in the front of the cabin. The empennage is mounted on booms.

Centrifugal compressor. A type of compressor that uses a vaned plate like impeller. Air is taken into the center, or eye, of the impeller and slung outward by centrifugal force into a diffuser where its velocity is decreased and its pressure increased.

Ceramic. Any of several hard, brittle, heat-resistant, noncorrosive materials made by shaping and then firing a mineral, such as clay, at a high temperature.

Channel-chromed cylinders. Reciprocating engine cylinders with hard chromium-plated walls. The surface of this chrome plating forms a spider web of tiny stress cracks. Deplating current enlarges the cracks and forms channels that hold lubricating oil on the cylinder wall.

Cheek (crankshaft). The offset portion of a crankshaft that connects the crankpin to the main bearing journals.

Chip detector. A component in a lubrication system that attracts and holds ferrous metal chips circulating with the engine oil. Some chip detectors are part of an electrical circuit. When metal particles short across the two contacts in the detector, the circuit is completed, and an annunciator light is turned on to inform the flight crew that metal particles are loose in the lubrication system.

Choke of a cylinder. The difference in the bore diameter of a reciprocating engine cylinder in the area of the head and in the center of the barrel.

Choke-ground cylinder. A cylinder of a reciprocating engine that is ground so that its diameter at the top of the barrel is slightly smaller than the diameter in the center of the stroke. The large mass of metal in the cylinder head absorbs enough heat to cause the top end of the barrel to expand more than the rest of the barrel. At normal operating temperature, the diameter of a choke-ground cylinder is uniform throughout.

Choke nozzle. A nozzle in a gas turbine engine that limits the speed of gases flowing through it. The gases accelerate until they reach the speed of sound, and a normal shock wave forms that prevents further acceleration.

Chordline. An imaginary line, passing through a propeller blade, joining the leading and trailing edges.

Cigarette. A commonly used name for a spark plug terminal connector used with a shielded spark plug.

Circular magnetism. A method of magnetizing a part for magnetic particle inspection. Current is passed through the part, and the lines of magnetic flux surround it. Circular magnetism makes it possible to detect faults that extend lengthwise through the part.

Circumferential coil spring (garter spring). A coil spring formed into a ring. This type of spring is used to hold segmented ring-type carbon seals tightly against a rotating shaft.

Claret red. A dark purplish pink to a dark gray purplish red color.

Class A fire. A fire with solid combustible materials such as

throttle, applies the brakes, and replies "contact", and then turns the ignition switch to BOTH. The propeller is then pulled through to start the engine.

Constant-displacement pump. A fluid pump that moves a specific volume of fluid each time it rotates.

Constant-pressure cycle of energy release. The cycle of energy transformation of a gas turbine engine. See Brayton cycle.

Constant-volume cycle of energy release. The cycle of energy transformation of a reciprocating engine. See Otto cycle.

Continuous magnetic particle inspection. A method of magnetic particle inspection in which the part is inspected by flowing a fluid containing particles of iron oxide over the part while the magnetizing current is flowing.

Contrarotating. Rotating in opposite directions. Turbine rotors are contrarotating when the different stages have a common center, but turn in opposite directions.

Convergent-divergent duct. A duct that has a decreasing cross section in the direction of flow (convergent) until a minimum area is reached. After this point, the cross section increases (divergent). Convergent-divergent ducts are called CD ducts or con-di ducts.

Convergent duct. A duct that has a decreasing cross section in the direction of flow.

Core engine. The gas generator portion of a turboshaft, turboprop, or turbofan engine. The core engine consists of the portion of the compressor used to supply air for the engine operation, diffuser, combustors, and turbine(s) used to drive the compressor. The core engine provides the high-velocity gas to drive the fan and/or any free turbines that provide power for propellers, rotors, pumps, or generators.

Cowling. The removable cover that encloses an aircraft engine.

Crankcase. The housing that encloses the crankshaft, camshaft, and many of the accessory drive gears of a reciprocating engine. The cylinders are mounted on the crankcase, and the engine attaches to the airframe by the crankcase.

Crankshaft. The central component of a reciprocating engine. This high-strength alloy steel shaft has hardened and polished bearing surfaces that ride in bearings in the

crankcase. Offset throws, formed on the crankshaft, have ground and polished surfaces on which the connecting rods ride. The connecting rods change the in-and-out motion of the pistons into rotation of the crankshaft.

Creep. The deformation of a metal part that is continually exposed to high centrifugal loads and temperatures.

Critical altitude. The altitude above which a reciprocating engine will no longer produce its rated horsepower with its throttle wide open.

Critical engine. The engine of a twin-engine airplane whose loss would cause the greatest yawing effect.

Critical Match number. The flight match number at which there is the first indication of air flowing over any part of the structure at a speed of Mach one, the local speed of sound.

CRT. Cathode ray tube. An electronic display tube in which a stream of electrons is attracted to the charged inner surface of the tube face. Acceleration grids and inner surface of the tube face. Acceleration grids and focusing grids speed the movement of the electrons and shape the beam to a pin-point size. Electrostatic or electromagnetic forces caused by deflection plates or coils move the beam over the face of the tube. The inside of the tube face is treated with a phosphor material that emits light when the electrons strike it.

Cryogenic fluid. A liquid which boils at a temperature lower than about 110 °K (-163 °C) under normal atmospheric pressure.

CSD. Constant-speed drive. A component used with either aircraft gas turbine or reciprocating engines to drive AC generators. The speed of the output shaft of the CSD is held constant while the speed of its input shaft varies. The CSD holds the speed of the generator, and the frequency of the AC constant as the engine speed varies through its normal operating range.

CTF. Centrifugal twisting force. The force acting about the longitudinal axis of a propeller blade, and which tries to rotate the blade to a low-pitch angle. As the propeller rotates, centrifugal force tries to flatten the blade so all of its mass rotates in the same plane.

Curtiss Jenny (Curtiss JN4-D). A World War I training airplane powered by a Curtiss OX-5 engine. It was widely available after the war and helped introduce aviation to the general public.

Customer bleed air. Air that is tapped off a turbine engine compressor and used for such airframe functions as the operation of air conditioning and pressurization systems.

Cylinder. The component of a reciprocating engine which houses the piston, valves, and spark plugs and forms the combustion chamber.

D

Data. The input for computer processing in the form of numerical information that represents characters or analog quantities.

Dataplate specifications. Specification of each gas turbine engine determined in the manufacturer's test cell when the engine was calibrated. This data includes the engine serial number with the EPR that produced a specific RPM. The technician refers to this information when trimming the engine.

Dataplate performance. The performance specifications of a turbine engine observed and recorded by the engine manufacturer or overhauler and recorded on the engine dataplate. This data includes the engine speed at which a specified EPR is attained. When trimming the engine, the technician uses this data as the goal.

DC. Direct current. Electrical current in which the electrons always flow in the same direction.

Deaerator. A component in a turbine engine lubrication system that removes air from the scavenged oil before it is returned to the tank.

Deceleration. The amount the velocity of an object, measured in feet per second, is decreased by a force during each second it is acted upon by that force. Deceleration is usually expressed in terms of feet per second, per second (fps²).

DeHaviland DH-4. An English designed observation airplane built in large quantities in the united States during World War I. After the war, surplus DH-4s were used for carrying the U.S. Mail.

Deicing. The removal of ice that has formed on a surface.

Density altitude. The altitude in standard air at which the density is the same as that of the existing air.

Detergent oil. A type of mineral oil with metallic-ash-forming additives that protects the inside of an engine from sludge and varnish buildup. Used in automotive engines, it has proven unsuitable for use in aircraft engines.

Detonation. An uncontrolled explosion inside the cylinder of a reciprocating engine. Detonation occurs when the pressure and temperature of the fuel inside the cylinder exceeds the critical pressure and temperature of the fuel. Detonation may be caused by using fuel that has a lower octane rating or performance number than is specified for the engine.

Dewar bottle. A special container used to store liquid oxygen and liquid nitrogen. A Dewar bottle has an inner and an outer container, and the space between them forms a vacuum. The two surfaces within the vacuum are silvered to reflect heat away from the container walls.

Differential pressure. A single pressure that is the difference between two opposing pressures.

Diffuser. A component in a gas turbine engine that decreases the velocity of air flowing through it and increases its pressure.

Digitized image. A modified image picked up by the miniature TV camera in the end of a fiber-optic probe. This image is converted into a digital electronic signal that eliminates unwanted portions of the viewed area and allows the desired image to be enhanced for a clearer view of the inside of a turbine engine.

Dipstick. A gage, in the form of a thin metal rod, used to measure the level of liquid in a reservoir. The dipstick is pushed into the reservoir until it contacts a built-in stop; then it is removed and visually inspected. The level of liquid in the reservoir is indicated by the amount of the dipstick wet by the liquid.

Dirigible. A large, cigar shaped, lighter-than-air flying machine. Dirigibles differ from balloons in that they are powered and can be steered.

Distributed pole stator winding. Alternator stator windings wound in a series of slots in the stator frame. A distributed pole stator is distinguished from a salient pole stator whose coils are wound around separate pole shoes that project inward from the field frame toward the rotor.

Distributor. A high-voltage selector switch that is gear-driven from the shaft of the rotating magnet in a magneto. The distributor rotor picks up the high voltage from the secondary winding of the coil and directs it to high-voltage terminals. From here, it is carried by high-tension ignition leads to the spark plugs.

Divergent duct. A duct that has an increased cross-sectional area in the direction of flow.

Downdraft carburetor. A carburetor that mounts on the top of a reciprocating engine. Air entering the engine flows downward through the carburetor.

ΔP (delta P). Differential pressure.

Droop. A progressive decrease in RPOM with load in a gas turbine engine whose speed is governed with a fly-weight-type governor in the fuel control. As the load increases, the pilot valve drops down to meter more fuel. The lower position of the valve decreases the compression of the speeder spring and allows the flyweights to assume an on-speed position at a lower rpm.

Dry-sump engine. An engine that carries its lubricating oil supply in a tank external to the engine.

Dual ignition. An ignition system of an aircraft reciprocating engine that has two of every critical unit, including two spark plugs in each cylinder. Dual ignition provides safety in the event of one system malfunctioning, but more important, igniting the air/fuel mixture inside the cylinder at two locations provides more efficient combustion of the air/fuel mixture in the cylinder.

Dual-spool gas turbine engine. An axial-flow turbine engine that has two compressors, each driven by its own stage or stages of turbines.

Duct heater. A thrust augmentation system, similar to an afterburner, where fuel is added to the fan-discharge air and burned.

Duct losses. A decrease in pressure of the air flowing into a gas turbine engine caused by friction.

Durability. A measure of engine life. Durability is usually measured in TBO hours.

Duty cycle. A schedule that allows a device to operate for a given period of time, followed by a cooling down period before the device can be operated again.

Dwell chamber. A chamber in a turbine engine into which the scavenged oil is returned. Entrained air separates from the oil in the dwell chamber before it is picked up by the pressure pump.

Dynamometer. A device used to measure the amount of torque being produced by an engine. The drive shaft of the engine is loaded with either an electric generator or a fluid pump, and the output of the generator or pump is measured and converted into units of torque. Torque at a specific rpm can be converted into brake horsepower.

Dyne. The unit of force that imparts an acceleration of one centimeter per second, per second to a mass of one gram. One dyne is equal to 2.248 · 10-6 pounds.

E

Eddy current. Current induced into a conductor due to a mobbing or non-uniform magnetic field.

EEC. Electronic engine control. An electronic fuel control for a gas turbine engine. The EEC senses the power-lever angle (PLA), engine RPM, bleed valve, and variable stator vane position, and the various engine pressures and temperatures. It meters the correct amount of fuel to the nozzles for all flight conditions, to prevent turbine over-speed and over-temperature.

Effective pitch. The actual distance a propeller advances in one revolution through the air.

E-gap angle. The position of the rotating magnet in a magneto when the breaker points are timed to open. The E-gap (efficiency gap) angle is several degrees of magnet rotation beyond the magnet's neutral position. At this point, the magnetic field stress is the greatest, and the change in flux is the greatest, inducing the maximum voltage in the secondary winding.

EGT. Exhaust gas temperature. The temperature of the gases as they leave the cylinder of a reciprocating engine or the turbine of a gas turbine engine.

EICAS. Engine indicating and crew alerting system. An electronic instrumentation system that monitors airframe and engine parameters and displays the essential information on a video display on the instrument panel. Only vital information is continually displayed, but when any sensed parameters fall outside of their allowable range of operation, they are automatically displayed.

Elastic limit. The maximum amount of tensile load, in pounds per square inch, that a material is able to withstand without permanent deformation.

Electrical potential. The electrical force caused by a deficiency of electrons in one location and an excess of electrons in another. Electrical potential is measured in volts.

Electrical steel. A low-carbon iron alloy that contains some silicon It is used as the core for transformers, field frames for generators and alternators, and the magnetic circuit of magnetos.

Electromagnet. A magnet produced by an electrical current flowing through a coil of wire. The coil is normally wound around a core of soft iron which has an extremely low retentivity, allowing it to lose its magnetism as soon as the current stops flowing.

Electromagnetic radiation. A method of transmitting energy from one location to another. Current caused by high voltage in the secondary winding of a magneto produces electric and magnetic fields which oscillate back and forth at a high frequency and extend out into space in the form of waves. These waves of electromagnetic radiation are received as interference by the radio receivers in the aircraft.

Electromotive force. A force that causes electrons to move from one atom to another within an electrical circuit. An electromotive force, or EMF, is the difference in the electrical pressure, or potential, that exists between two points. An EMF may be produced by converting mechanical movement, pressure, chemical, light, or heat energy into electrical energy. The basic unit of EMF is the volt.

Emulsion. A suspension of small globules of one material in another when the two materials will not mix. Oil and water will not mix, but they can be formed into an emulsion. An emulsion will separate into its components when it is allowed to sit.

Engine trimming. A maintenance procedure in which the fuel control on a gas turbine engine is adjusted to cause the engine to produce the required EGT or EPR at a specified rpm.

Entrained water. Water suspended in jet fuel. The amount of entrained water that can be held in the fuel is determined by the temperature of the fuel. When the fuel becomes cold, the water precipitates out and forms ice crystals on the fuel filter element.

Epicyclic reduction gears. A gear train in which a series of small planetary gears rotate around a central gear. More commonly called a planetary gear train.

EPR. Engine pressure ratio. The ratio of the turbine discharge total pressure to the compressor inlet total pressure. EPR is normally used as the parameter to determine the amount of thrust an axial-flow turbojet or turbofan engine is producing.

ESHP. Equivalent shaft horsepower. A measure of the power produced by a turboprop engine. ESHP takes into consideration both the shaft horsepower delivered to the propeller and the thrust developed at the engine exhaust. Under static conditions, one shaft horsepower is approximately equal to 2.5 pounds of thrust.

Ethanol. Alcohol made from cereal grains such as corn.

Ether. A volatile, highly flammable liquid that may be used to prime the cylinders of an aircraft engine when starting under extremely cold conditions.

Ethylene dibromide. A colorless, poisonous liquid $BrCH_2CH_2Br$ that is blended with leaded gasoline to help scavenge lead oxides.

Ethylene glycol. A form of alcohol used as a coolant for liquid-cooled aircraft engines. It is also used in automobile engines as a permanent antifreeze.

Eutectic. An alloy or solution that has the lowest possible constant melting point.

Evaporative cooling. See steam cooling.

Exceedance condition. A condition in which a parameter sensed by the EICAS exceeds the limits for which it is programmed.

Exhaust cone. The fixed conical fairing centered in the turbine wheel. The exhaust cone straightens the flow and prevents the hot gases from circulating over the rear face of the turbine wheel.

Exhaust nozzle. The opening at the rear of the exhaust pipe.

Expansion wave. The change in pressure and velocity of supersonic air as it passes over a surface that drops away from the flow. As the surface drops away, the air tries to follow it, and in changing its direction, the air speeds up to a higher supersonic speed, and its static pressure decreases. There is no change in the total amount of energy as air passes through an expansion wave.

External-combustion engine. A form of heat engine in which the fuel releases its energy outside of the engine. This released heat expands air which is used to perform useful work. Steam engines are a popular type of external combustion engine.

Extreme pressure (EP) lubricant. A lubricant that reacts with iron to form iron chlorides, sulfides, or phosphides on the surface of a steel part. These compounds reduce wear and damage to surfaces in heavy rubbing contact. EP lubricants are specially suited for lubricating gear trains.

F

FAA Form 337. The *Major Repair and Alteration* form that must be completed when an FAA-certificated aircraft or engine has been given a major repair or major alteration.

Face (propeller nomenclature). The flat surface of a propeller that strikes the air as the propeller rotates. The face of a propeller corresponds to the bottom of an airplane wing.

FADEC. Full-authority digital electronic control. A digital electronic fuel control for a gas turbine engine that is functioning during all engine operations, hence full authority. It includes the EEC (see EEC) and functions with the flight management computer. FADEC schedules the fuel to the nozzles in such a way that prevents overshooting power changes and over-temperature conditions. FADEC furnishes information to the EICAS (engine indication and crew alerting system).

Fan pressure ratio. The ratio of the fan-discharge pressure to the fan inlet pressure.

Feathering propeller. A controllable-pitch propeller whose blades can be moved into a high pitch angle of approximately 90°. Feathering the propeller of an inoperative engine prevents it from wind-milling and greatly decreases drag.

Feeler gages. A type of measuring tool consisting of strips of precision-ground steel of accurately measured thickness. Feeler gages are used to measure the distance between close-fitting parts, such as the clearances of a mechanical system or the distance by which moving contacts are separated.

FHP. Friction horsepower. The amount of horsepower used to turn the crankshaft, pistons, gears, and accessories in a reciprocating engine and to compress the air inside the cylinders.

Fiber optics. The technique of transmitting light or images through long, thin, flexible fibers of plastic or glass. Bundles of fibers are used to transmit complete images.

Fire sleeve. A covering of fire-resistant fabric used to protect flexible fluid lines that are routed through areas subject to high temperature.

Flame tubes. Small-diameter metal tubes that connect can-type combustors in a turbine engine to carry the ignition flame to all of the combustion chambers. The British call combustion liners flame tubes.

Flameout. A condition of turbine engine operation when the fire unintentionally goes out. Improper air/fuel mixture or interruption of the air flow through the engine can cause a flameout.

Flash point. The temperature to which a liquid must be raised for it to ignite, but not continue to burn when a flame is passed above it.

Flashing the field. A maintenance procedure for a DC generator that restores residual magnetism to the field frame. A pulse of current from a battery is sent through the field coils in the direction in which current normally flows. The magnetic field produced by this current magnetizes the steel frame of the generator.

Flashover. An ignition system malfunction in which the high voltage in the magneto distributor jumps to the wrong terminal. Flashover causes the wrong spark plug to fire. This reduces the engine power and produces vibration and excessive heat.

Flat-rated engine. A turboprop engine whose allowable output power is less than the engine is physically capable of producing.

Float carburetor. A fuel metering device that uses a float-actuated needle valve to maintain fuel level slightly below the edge of the discharge nozzle.

Flock. Pulverized wood or cotton fibers mixed with an adhesive. Flock, attached to a wire screen, acts as an effective induction air filter for small reciprocating engines.

Flow divider (reciprocating engine). The valve in an RSA fuel injection system that divides the fuel from the fuel control unit and distributes it to all of the cylinders. It compares with the manifold valve in a Teledyne-Continental fuel injection system.

Flow divider (turbine engine). A component in a turbine engine fuel system that routes all of the fuel to the primary nozzles or primary orifices when starting the engine or when the rpm is low. When the engine speed builds up, the flow divider shifts and opens a passage to send the majority of the fuel to the secondary nozzles or orifices.

FMC. Flight management computer. An electronic flight instrumentation system that enables the flight crew to initiate and implement a given flight plan and monitor its execution.

FOD. Foreign object damage. Damage to components in the gas path of a turbine engine, caused by ingested objects. Debris from the runway or ramp cause FOD on the ground.

Ice and birds cause most in-flight FOD.

Four-stroke cycle. A constant-volume cycle of energy transformation that has separate strokes for intake, compression, power, and exhaust.

Fractional distillation. Procedure used for separating various components from a physical mixture of liquids. Crude oil is a mixture of many different types of hydrocarbon fuels which can be separated by carefully raising its temperature. The first products to be released, those having the lowest boiling points, are some of the gaseous fuels; next are gasoline, kerosene, diesel fuel, heavy fuel oils, lubricating oils, and finally, tar and asphalt.

Frangible. Capable of being broken.

Free-turbine engine. A gas turbine engine with a turbine stage on a shaft independent of the shaft used to drive the compressor. Free turbines are used to drive the propeller reduction gear in a turboprop engine and the rotor transmission in a helicopter.

Freezing point. The temperature at which solids, such as wax crystals, separate from a hydrocarbon fuel as it is cooled.

Full-register position. The position of a magnet in a magneto when its poles are aligned with the pole shoes and the maximum amount of magnetic flux is flowing through the magnetic circuit.

G

Gauge pressure. Pressure referenced from existing atmospheric pressure.

Gas generator. The basic gas turbine engine. It consists of the compressor, diffuser, combustor, and turbine. The gas generator is also called the core engine.

Gas turbine engine. An internal combustion engine that burns its fuel in a constant-pressure cycle and uses the expansion of the air to drive a turbine which, in turn, rotates a compressor. Energy beyond that needed to rotate the compressor is used to produce torque or thrust.

General Aviation Airworthiness Alerts. While these documents are no longer published, they are still available at www.faa.gov. These are used to alert technicians of problems that have been found in specific models of aircraft, and reported on Malfunction and Defect Reports. Airworthiness Alerts suggest corrective action, but compliance with the suggestion is not mandatory.

General aviation. A term used to describe the total field of aviation operation except the military and airlines.

Geometric pitch. The distance a propeller would advance in one revolution if it were rotating in a solid.

Geopotential of the tropopause. The point in the standard atmosphere where the temperature stops dropping and becomes constant. This is the tropopause, or the dividing line between the troposphere and the stratosphere.

Gerotor pump. A form of constant-displacement pump that uses an external-tooth drive gear that meshes with and drives an internal-tooth gear that has one more space for a tooth than there are teeth on the drive gear. Both gears turn inside a close-tolerance housing. As the gears rotate, fluid flows between the teeth that are beginning to un-mesh, and is carried around the pump as the space continues to open up. On the discharge side of the pump, the teeth becomes smaller, fluid is forced out of the pump.

Glass flight deck. An aircraft instrument system that uses a few color cathode-ray-tube displays to replace a large number of mechanically actuated instruments.

Governor. A control used to automatically change the pitch of a constant speed propeller to maintain a constant engine rpm as air loads vary in flight.

GPU. Ground power unit. A service component used to supply electrical power and compressed air to an aircraft when it is operating on the ground.

Gross thrust. The thrust produced by a turbojet or turbofan engine when the engine is static or not moving. The air is considered to have no inlet velocity, and the velocity of the gas leaving the engine is considered to be the acceleration factor.

Ground-boosted engine. An aircraft reciprocating engine with a built-in supercharger that boosts the sea-level rated horsepower of the engine.

Gudgeon pin. The British name for a wrist pin, or piston pin. See wrist pin.

H

Half-wave rectifier. An electrical rectifier circuit that converts AC into pulsating DC. Only one alternation of each cycle is present in the output.

Halogenated hydrocarbon. A hydrocarbon compound in which one or more hydrogen atoms have been replaced with atoms of one of the halogen elements such as fluorine, chlorine, or bromine.

Head of pressure. Pressure exerted by a column of fluid and created by the height of the column.

Heat engine. A mechanical device that converts the chemical energy in a fuel into heat energy. The heat energy is then converted into mechanical energy and useful work.

Heli-Coil insert. The registered trade name of a special helical insert used to restore threads stripped from a bolt hole, or to reinforce the threads in an aluminum casting. The damaged threads are drilled out and new threads are cut with a special oversize tap. A coil of stainless steel wire, with a cross section in the shape of a diamond, is screwed into the hole and serves as the new threads. Heli-Coil inserts are also used to provide durable threads in soft metal castings. Some spark plug holes in aluminum alloy cylinder heads are fitted with Heli-Coil inserts to minimize the wear caused by repeated removal and installation of the spark plugs.

Helical spline. A spline that twists, or winds, around the periphery of a shaft. Helical splines are used to change linear motion into rotary motion of the shaft on which the splines are cut.

Helical spring. A spring wound in the form of a helix, or coil.

Helix. A spiral.

Heptanes. An organic compound, $CH_3(CH_2)_5CH_3$, that is used as the low reference fuel for rating the antidetonation characteristics of aviation gasoline.

Hermetically sealed. A complete seal, especially against the escape or entry of air.

Hertz. A unit of frequency equal to one cycle per second.

High-bypass ratio engine. A turbofan engine whose bypass ratio is 4:1 or greater.

High-pressure compressor. The second-stage compressor in a dual-spool gas turbine engine. The high pressure compressor is called the N2 compressor and is the one that is rotated by the starter for starting, and the one whose rpm is controlled by the fuel control.

High unmetered fuel pressure. Pressure in a Teledyne-Continental fuel injector pump that is adjusted by the variable orifice.

Home-built aircraft. See amateur-built aircraft.

Honing (cylinder wall treatment). Scratching the surface of the cylinder wall with an abrasive to produce a series of grooves of microscopic depth and uniform pattern. The honed pattern holds oil to lubricate the cylinder walls.

Horsepower. The most commonly used unit of mechanical power. One horsepower is equal to 33,000 foot-pounds of work done in one second.

Hot section. The portion of a gas turbine engine that operates at a high temperature. The hot section includes the combustion, turbine, and exhaust sections.

Hot-tank lubricating system. A turbine engine lubricating system in which the oil cooler is located in the pressure subsystem. The oil is returned to the tank without being cooled.

HRD fire extinguisher. A fire extinguisher that carries the extinguishing agent in a sealed sphere or cylinder. When the agent-discharged switch is closed, an ignited powder charge drives a cutter through a frangible disc which releases the agent. The entire contents of the container is emptied in much less than a second.

Hub (propeller component). The high-strength component inside a propeller that attaches the blades to the engine propeller shaft.

Hybrid compressor engine. A gas turbine engine that has both centrifugal and axial-flow compressors.

Hybrid spark plug. A fine-wire spark plug that has a platinum center electrode and iridium ground electrodes.

Hydraulic lock. A condition in which oil drains into the lower cylinders of a reciprocating engine and leaks past the piston rings to fill the combustion chamber. If the oil is not removed before the engine is started, it can cause serious damage.

Hydromechanical. Any device that combines fluid pressures with mechanical actions to achieve a desired result. In a hydromechanical fuel control used for a turbine engine, hydraulic servos are used in conjunction with the mechanical linkages.

I

Ice bridging. A spark plug failure that occurs when starting a reciprocating engine in extremely cold weather. When a cylinder fires, the air/fuel mixture is converted into carbon dioxide and water vapor. The water vapor condenses on the spark plug electrodes and forms ice that bridges the electrode gap and prevents the plug firing until the ice is melted. This normally requires removing the spark plugs from the engine.

IDG. Integrated drive generator. An AC generator installed on turbine engines. An IDG incorporates a brushless, three-phase AC generator and a constant-speed drive in a single component.

Igniter. The component in a turbine-engine ignition system that provides a high-energy spark for igniting the air/fuel mixture in the combustion chamber for starting.

IHP. Indicated horsepower. The theoretical horse-power a reciprocating engine develops.

IMEP. Indicated mean effective pressure. The average pressure existing inside the cylinder of a reciprocating engine during its power stroke.

Impulse coupling. A spring-loaded coupling between a magneto shaft and the drive gear inside the engine. When the engine is rotated for starting, the impulse coupling locks the magnet so it cannot turn. The spring in the coupling winds up as the crankshaft continues to turn, and when the piston is near top center, the coupling releases and spins the magnet, producing a hot and retarded spark.

Inline engine. A reciprocating engine with all of the cylinders arranged in a straight line.

Incandescent. Glowing because of intense heat.

Inconel. The registered trade name for an alloy of chromium, iron, and nickel. Inconel is similar to stainless steel, but cannot be hardened by heat treatment.

Inductive reactance. An opposition to the flow of AC or changing DC caused by inductance in the circuit. Inductive reactance, whose symbol is XL, causes a voltage drop, but it does not use power nor produce heat.

Inertia. The tendency of a body to resist acceleration. A body at rest will remain at rest or a body in motion will stay in motion in a straight line unless acted on by an outside force.

Inertia starter. A starter for a large reciprocating engine that uses energy stored in a rapidly spinning flywheel to turn the crankshaft.

Inlet guide vanes. A set of stator vanes in front of the first stage of compression in a gas turbine engine. The inlet guide vanes deflect the air entering the compressor in the correct direction for optimum operation. Inlet guide vanes may be fixed, or their angle may be controlled hydraulically by fuel from the fuel control.

Integral fuel tank. An aircraft fuel tank made by sealing off part of the structure so fuel can be carried in the structure itself.

Intercooler. An air-to-air heat exchanger installed between a turbosupercharger and the carburetor. Intercoolers decrease the temperature of compressed air to prevent detonation.

Interference angle (poppet valve dimension). The difference between the valve seat and the valve face angles. Normally, the valve seats are ground with between 0.5° and 1° greater angle than the valve face. This allows the face to touch the seat with a line contact that provides the best sealing.

Interference fit. A type of fit used when assembling certain mechanical devices. The hole is made smaller than the part that fits into it. The material containing the hole is heated to expand the hole, and the part that fits into the hole is chilled to shrink it. The parts are assembled, and when they reach the same temperature their fit is so tight they will not loosen in service.

Internal-combustion engine. A form of heat engine in which the fuel and air mixture is burned inside the engine to heat and expand the air so it can perform useful work.

Internal timing. The adjustment of the breaker points of a magneto so they will begin to open at the time the magnet is in its E-gap position.

Interpole. A field pole in a compound-wound DC generator used to minimize armature reaction. Interpoles are located between each of the regular field poles, and their coils are in series with the armature winding so all of the armature current flows through them. The magnetic field produced by the interpole coils cancels the distortion caused by the armature field and allows the brushed to remain in the neutral plane where there is no potential difference between the commutator segments. Keeping the brushes in the neutral plane minimizes sparking.

Inverted engine. An inline or V-engine in which the cylinders are mounted below the crankshaft.

Iridium. A very hard, brittle, highly corrosion-resistant, whitish-yellow, metallic chemical element. Iridium is used for the fine-wire electrodes in spark plugs that must operate in engines using fuel with an exceptionally high lead content.

Iso-octane. An organic compound used as the high reference fuel for rating the antidetonation characteristics of aviation gasoline $(CH_3)_2CHCH_2C(CH_3)_3$.

Isothermal change. A physical change that takes place within a material in which heat energy is added to or taken from the material as needed to keep its temperature constant.

J

Jet fuel. Fuel designed and produced to be used in aircraft gas turbine engines.

Jet propulsion. A method of propulsion by accelerating a relatively small mass of air through a large change in velocity.

Jeweler's file. A small, fine-cut, metalworking file used by jewelry manufacturers.

Joule. A measure of energy. In terms of electrical energy, one joule is equal to one watt-second.

Journal (bearing). A hardened and polished surface on a rotating shaft that rides in a plain bearing.

K

Kerosene. A light, almost colorless, hydrocarbon liquid obtained from crude oil through the fractional distillation process. Kerosene is the base for turbine engine fuel.

Kevlar. The registered trade name by DuPont for a patented aramid fiber.

Kinematic viscosity. The ratio of the absolute viscosity of a fluid to its density. Kinematic viscosity is measured in centistokes.

L

Labyrinth seal. A type of air and/or seal used around the main –shaft bearings in a gas turbine engine. The seal consists of a series of rotating blades that almost contact the seal land. A small amount of air flows between the seal and the land to prevent oil flowing past the seal.

Land (piston) The portion of a piston between the ring grooves.

Land (splined shaft). The portion of a splined shaft between the grooves.

Laser tachometer. A highly accurate tachometer that shines a laser beam on a rotating element that has reflective tape or a contrasting mark. The reflected laser beam is converted into electrical pulses which are counted and displayed on a monitoring instrument.

Last-chance oil filter. A small filter installed in the oil line to the bearing jet in a gas turbine engine. This filter traps any contaminants that have passed the main filter and holds them until the engine is disassembled for overhaul.

LCD. Liquid crystal display. A digital display that consists of two sheets of glass separated by a sealed-in, normally transparent liquid crystal material. The outer surface of each glass sheet has a transparent conductive coating with the viewing side etched into character-forming segments with leads going to the edges of the display. A voltage applied between the front and back coatings disrupts the orderly arrangement of molecules and causes the liquid to darken so that light cannot pass through it. The segment to which the voltage is applied appears as black against a reflected background.

Leading edge. The thick edge at the front of a propeller blade.

Lean die-out. A condition in which the fire in a gas turbine engine goes out because the air/fuel mixture ratio is too lean to sustain combustion.

Lean mixture. A air/fuel mixture that contains more than 15 parts of air to 1 part of fuel, by weight.

Line boring. A method of assuring concentricity of bored holes. A boring bar extends through all of the holes and cuts the inside diameters so they all have the same center.

Link rod. The rod in a radial engine that connects one of the piston wrist pins to a knuckle pin on the master rod. Also called articulating rods.

Liquid cooling. The removal of unwanted heat from an aircraft engine by transferring the heat into a liquid and then passing the heated liquid through a liquid-to-air heat exchanger (radiator) to transfer the heat into the ambient air.

Longitudinal magnetism. A method of magnetizing through a solenoid, or coil, that encircles the part so the lines of magnetic flux pass lengthwise through the part. Longitudinal magnetism makes it possible to detect faults that extend across the part.

Low bypass ratio engine. A turbofan engine whose bypass ratio is less than 2:1.

Low-pressure compressor. The first-stage compressor in a dual-spool gas turbine engine. The low-pressure compressor is called the N1 compressor and its speed is not governed. It seeks its own best speed as the atmospheric conditions change so it can furnish a relatively constant mass of air to the inlet of the second-stage compressor.

Low unmetered fuel pressure. Pressure in a Teledyne-Continental fuel injector pump that is adjusted by the relief valve.

LRU. Line replaceable unit. Aircraft components designed to be replaced as a unit while the aircraft is on the flight line.

M

M&D (Malfunction and Defect) report. A small postcard-like form (FAA Form 8330) used by repair stations, maintenance shops, and technicians to report an unacceptable condition to the FAA. Information on these forms provides the basis for the General Airworthiness Alerts and subsequent Airworthiness Directives.

Mach number. The ratio of the speed of an object through the air to the speed of sound under the same atmospheric conditions. An object traveling at the speed of sound is traveling at Mach one (M1.0).

Magnesyn system. The registered trade name of a remote indicating instrument system. A Magnesyn system uses a permanent magnet as its rotor and a toroidal coil excited by 400-hertz AC as its stator. A small magnet in the center of the indicator coil follows the movement of a larger magnet in the transmitter coil.

Magnetic field. The invisible, but measurable, force surrounding a permanent magnet or current-carrying conductor. This field is produced when the orbital axes of the electrons of the atoms in the material are all in alignment.

Magnetic flux. Lines of magnetic force that are assumed to leave a magnet at its north end and return to its south end. Lines of flux tend to be as short as possible and cannot cross each other.

Magnetic particle inspection. A method of non-destructive inspection for ferrous metal components. The part being inspected is magnetized and then flooded with a solution of iron oxide suspended in a light oil, much like kerosene. Any flaw, either on the surface or just below the surface, forms a north and south pole, and the iron oxide attracted to these poles helps locate the flaw. The iron oxide is normally treated with a fluorescent dye, and the inspection is conducted in a darkened booth. When an ultraviolet light (black light) is shone on the part, the treated iron oxide shows up as a brilliant line.

Major alteration. An alteration not listed in the aircraft, aircraft engine, or propeller specifications that might appreciably affect weight, balance, structural strength, powerplant operation, flight characteristics, or other qualities affecting airworthiness; an alteration not done according to accepted practices, or one that cannot be done by elementary operations.

Major overhaul. The disassembly, cleaning, and inspection of an engine and the repair and replacement of all parts that do not meet the manufacturer's specification.

Major repair. A repair to a component that if improperly done might appreciably affect weight, balance, structural strength, performance, powerplant operation, flight characteristics, or other qualities affecting airworthiness; a repair not done according to accepted practices, or one that cannot be done by elementary operations.

Mandrel. A precision steel bar on which a propeller is mounted for balancing. The mandrel is placed across two perfectly level knife-edge plates, and the propeller is allowed to rotate until it stops with its heavy point at the bottom.

Manifold pressure. The absolute pressure of the air inside the induction system of a reciprocating engine.

Manifold valve. See flow divider (reciprocating engine).

MAP. Manifold absolute pressure. The absolute pressure that exists within the induction system of a reciprocating engine. It is the MAP that forces air into the cylinders of the engine. MAP is commonly called manifold pressure.

Mass. A measure of the amount of matter in an object. For the purpose of measuring the mass of air flowing through a turbine engine, the weight of the air, in pounds per second, is divided by the acceleration due to gravity (32.3 feet per second).

Matrix (advanced composites). The material that bonds the fibers together in an advanced composite structure. The matrix carries the stresses into the fibers.

Matter. Something that has mass, takes up space, and exists as a solid, liquid, or gas.

Medium-bypass ratio engine. A turbofan engine whose bypass ratio is between 2:1 and 4:1.

MEK. Methul ethyl ketone. A volatile, water soluble, organic chemical compound that is used as a solvent to remove oily contaminants from ignition system components.

Methanol. Alcohol made from wood.

MFD. Multifunction display. A liquid crystal or CRT display that shows a number of parameters and replaces several analog-type indicators.

Microinches rms. A measure used for cylinder wall surface roughness. Twenty microinches rms means that the highest and lowest deviation from the average surface is 20 millionths of an inch.

Micron. A measurement used to identify the size of particles trapped by filters. One micron is a micro meter, or one millionth of a meter. It is 0.000039 inch.

Microprocessor. A single silicon chip that contains the arithmetic and logic functions of a computer.

Milliammeter. An instrument that measures electrical current in units of thousandths of an ampere.

Millibar. A unit of pressure in the metric system. One bar is a pressure of 14.5 psi, or 29.52 in. Hg. One millibar is one thousandth of a bar, or 0.01469 psi, or 0.02952 in. Hg.

Minor alteration. Any alteration that does not fit the definition of a major repair. See major repair.

Module (modular engine construction). The method of construction for mast modern gas turbine engine. The engine is made of several modules, or units, that can be removed and replaced or serviced independent of the rest of the engine.

Momentum. A force caused by the inertia of a moving body as it tries to keep the object moving in the same direction, at the same speed.

Motor. (verb) The act of rotating a turbine engine using the starter, with the ignition system deactivated. An engine is motored to force air through it to purge fuel fumes.

Multiple-can combustor. A combustor used in a gas turbine engine that consists of a series of individual burner cans, each made of an inner liner and an outer case. The individual cans are arranged around the periphery of a centrifugal compressor. Hot gases flow directly from the cans into the turbine.

N

N1. A symbol representing the rotational speed of the low-pressure compressor in a dual-spool gas turbine engine.

N2. A symbol representing the rotational speed of the high-pressure compressor in a dual-spool gas turbine engine.

NACA. National Advisory Committee for Aeronautics. This organization, dedicated to the technical development of aviation, has been superseded by NASA.

NACA cowling. A long-chord cowling used over a radial engine. The forward portion of this cowling has an aerodynamic shape that produces a forward pull, and the rear portion extends back to fair in with the fuselage. There is a narrow peripheral gap between the rear of the cowling and the fuselage for the cooling air to escape. Some NACA cowlings have controllable flaps over this opening to control the amount of cooling air that flows through the engine.

Nacelle. An enclosed compartment, normally in the leading edge of the wing, in which an aircraft engine is mounted.

Naphtha. A volatile, flammable liquid distilled from petroleum. It is used as a cleaning agent and solvent, and is present in some blended turbine-engine fuels.

NASA. National Aeronautics and Space Administration.

Naturally aspirated engine. A reciprocating engine that depends upon atmospheric pressure to force the air/fuel mixture into the cylinders. Naturally aspirated engines are neither supercharged nor turbocharged.

Net thrust. The thrust produced by a turbojet or turbofan engine in which the acceleration factor is the difference between the velocity of the incoming air and the velocity of the exhaust gases leaving the engine.

Neutral position. The position of the magnet in a magneto when its poles are between the pole shoes and no lines of flux are flowing through the magnetic circuit.

Newton. The unit of force needed to accelerate a mass of one kilogram one meter per second per second. One newton is equal to 1000,000 dynes, or 2.248×10^{-1} pound.

Nichrome. The registered trade name for an alloy of nickel and chromium. Nichrome wire is used for making electrical heater elements and precision wire-wound resistors. Nichrome's resistance is approximately 65 times that of copper.

Nitriding. A method of case hardening steel. Steel is placed in a retort (a sealed, high-temperature furnace), and heated to a specified temperature while surrounded by ammonia gas (NH_3). The ammonia breaks down into nitrogen and hydrogen, and the nitrogen unites with some of the alloying elements in the steel to form an extremely hard surface. Nitriding hardens crankshaft bearing surfaces and cylinder walls in reciprocating engines. It takes place at a lower temperature than other forms of case hardening, and does not cause warping.

Normal category airplane. An aircraft that is certificated under 14 CFR part 23 that is not certificated under the acrobatic, utility, or commuter category.

Normal shock wave. A type of pressure wave that forms at right angles to a surface when air moves at the speed of sound.

Notch sensitivity. A measure of the loss of strength of a material caused by the presence of a notch, or a V-shaped cut.

Nozzle guide vanes. See turbine inlet guide vanes.

O

Oblique shock wave. A pressure wave that forms on a sharp-pointed object when air flows past it at a supersonic speed.

Octane rating. A system used to rate the antidetonation characteristics of a reciprocating engine fuel. Fuel with an octane rating of 80 performs in a laboratory test engine the same as the fuel made of a mixture of 80% iso-octane and 20% heptanes.

Odometer. The portion of an automobile speedometer that indicates the distance traveled.

Offset throw (crankshaft design). Crank arms on a reciprocating engine crankshaft. The arms, or throws, to which the connecting rods and pistons are attached are offset from the center of the crankshaft to move the pistons in and out of the cylinder. The amount of the offset determines the stroke of the engine.

Oil analysis. A method of measuring the contents in parts per million of various chemical elements in oil. A sample of the oil is burned in an electric arc, and the resulting light is analyzed with a spectroscope which identifies the chemical elements in the oil and gives an indication of the amount of each element. This type of oil analysis is called a spectrometric oil analysis program, or SOAP.

Oil dilution. A method of temporarily decreasing the viscosity of the lubricating oil to make it possible to start a reciprocating engine when the temperature is very low. Before shutting the engine down, enough gasoline from the fuel system is mixed with the lubricating oil in the engine to dilute it so the starter can turn the engine over when the oil is cold and viscous. When the engine starts and the oil warms up, the gasoline evaporates.

Oil-damped bearing. A type of roller bearing installation in a gas turbine engine in which the outer race is installed in an oil damper compartment whose inside diameter is a few thousandths of an inch larger than the outside diameter of the outer race. Oil under pressure fills the oil damper compartment and allows the bearing to compensate for sight misalignment and to absorb vibrations of the shaft.

On-condition maintenance. A maintenance program that closely monitors the operating condition of an engine and allows major repairs or replacements to be made when engine performance deteriorates to a specific level.

On-speed condition. The speed condition in which the engine is turning at the rpm for which the propeller governor is set.

One-hundred-hour inspection. An inspection required by 14 CFR part 91, section 91.409 for FAA-certificated aircraft operated for hire or used for flight instruction for hire. A 100-hour inspection is identical in content to an annual inspection, but can be conducted by an aviation maintenance technician who holds an Airframe and Powerplant rating, but does not have an Inspection Authorization. See 14 CFR part 43, Appendix D for list of the items that must be included in an annual or 100-hour inspection.

Operating cycle. One complete series of events in the operation of a turbine engine that consists of starting the engine, taking off, landing, and shutting the engine down.

Optoelectronic device. An electronic device that produces, modulates, or senses electromagnetic radiation in the ultraviolet, visible light, or infrared portions of the energy spectrum.

Otto cycle. The constant-volume cycle of energy transformation used by reciprocating engines. A mixture of fuel and air is drawn into the cylinder as the piston moves to the bottom of its stroke. The mixture is compressed as the piston moves upward in the cylinder, and when the piston is near the top of its stroke, the mixture is electrically ignited and burns. The burning mixture heats and expands the air inside the cylinder and forces the piston down, performing useful work. The piston then moves back up, forcing the burned gases out of the cylinder.

Overboost. A condition of excessive manifold pressure in a reciprocating engine. Overboosting occurs when the supercharger is operated at too high a speed.

Overrunning clutch. A type of clutch that couples an input shaft with an output shaft. When the input shaft is driven, the output shaft rotates with it. When the output shaft is driven, the output shaft rotates with it. But when the output shaft is driven, the input shaft does not turn.

Overspeed condition. A speed condition in which the engine is turning at an rpm higher than that for which the propeller governor is set.

P

P-lead. Primary lead. The wire that connects the primary winding of a magneto to the ignition switch. The magneto is turned off by grounding its P-lead.

Pascal. The unit of pressure produced when one newton of force acts uniformly over an area of one square meter. One pascal is equal to $14.503 \cdot 10\text{-}5$ (0.00014503) psi. The kilopascal (kPa) is easier to manipulate. 1 kPa = 1,000 Pa = 0.14503 psi.

PCB. Plenum chamber burning. A method of thrust augmentation used on engines with vectored nozzles. Fuel injected into the fan-discharge air is burned to increase thrust.

Peak voltage. The voltage of AC electricity that is measured from zero voltage to the peak of either alternation.

Penetrant dwell time. The length of time a part is left in the penetrant when preparing it for inspection by the fluorescent or dye penetrant method. The hotter the part and the longer the penetrant dwell time, the smaller the fault that will be detected.

Performance number. The rating of antidetonation characteristics of a reciprocating engine fuel that is better than the high rating reference fuel, iso-octane. Performance numbers are greater than 100.

Permanent magnet. A piece of hardened steel that has been exposed to a strong magnetizing force which has aligned the spin axes of the electrons surrounding its atoms. The high retentivity of the material causes the electrons to retain their magnetic orientation.

Permanent-mold casting. A casting made in a reusable metal mold. The walls of permanent-mold castings can be made thinner than similar walls made by sand casting.

Permeability. A measure of the ease with which lines of magnetic flux can pass through a material.

Phase sequence, or phase rotation. The sequence with which the output phases of a three-phase generator are connected to the load. Reversing the phase sequence of a generator from A-B-C to A-C-B prevents the generator from being synchronized with the others on the bus.

Pi (π) filter. An electronic filter used to prevent radio frequency energy produced in the ignition exciter from feeding back into the aircraft electrical system. The filter is made of an inductor with a capacitor on its input and output. The name is derived from the resemblance of the three components on a schematic diagram to the Greek letter pi (π).

Pinion. A small gear that meshes with and drives a larger gear.

Piston (reciprocating engine component). The movable plug inside the cylinder of a reciprocating engine. The piston moves in and out to compress the air/fuel mixture and to transmit the force from the expanding gas in the cylinder to the crankshaft.

Piston pin. See wrist pin.

Pitch angle. The angle between the chordline of a propeller blade and the plane of rotation. See blade angle.

Pitch distribution. The gradual change in pitch angle of a propeller blade from the root to the tip.

Plane of rotation. The plane in which a propeller blade rotates. The plane of rotation is perpendicular to the propeller shaft.

Planetary gears. A type of large-ratio reduction gearing. A series of small planetary gears are mounted on a spider attached to the output shaft. The planetary gears rotate between a fixed sun gear and a driven ring gear.

Plenum chamber. An enclosed chamber in which air can be held at a pressure slightly higher than that of the surrounding air. Plenum chambers are used to stabilize the pressure of the air before it enters a double entry centrifugal compressor.

POH. Pilot's Operating Handbook. A document published by the airframe manufacturer and approved by the FAA that lists the operating conditions for a particular model of aircraft. Engine operating parameters are included in the POH.

Pole shoe. Inward extensions from the field frame of a generator around which the field coils are wound.

Poppet valve. A T-shaped valve with a circular head. Poppet valves are used to cover the intake and exhaust openings in the cylinder head of a reciprocating engine. The valves are held closed by one or more coil springs and are opened by a cam lobe or a rocker arm pushing on the end of the valve stem.

Porcelain. A hard, white, translucent ceramic material that was used as the insulator in some of the early aircraft spark plugs.

Positive-displacement pump. A fluid pump that moves a specific volume of fluid each time it rotates. Spur-gear pumps, gerotor pumps, and vane pumps are all positive-displacement pumps.

Power. The time rate of doing work. Power is found by dividing the amount of work done, measured in floor-pounds, by the time in seconds or minutes used to do the work. Power may be expressed in foot-pounds of work per minute or in horsepower. One horsepower is 33,000 foot-pounds of work done in one minute, or 550 foot pounds of work done in one second.

Power-assurance check. A test run made of a gas turbine engine to determine how its performance compares with its precious performance as new or freshly overhauled.

Powerplant. The complete installation of an aircraft engine, propeller, and all accessories needed for its proper function.

Pre-ignition. Ignition of the air/fuel mixture inside the cylinder of an engine before the time for normal ignition. Pre-ignition is often caused by incandescent objects inside the cylinder.

Prepreg. Preimpregnated fabric. A type of composite material in which the reinforcing fibers are encapsulated in an uncured resin. Prepreg materials are cut to size and shape and laid up with the correct ply orientation, and the entire component is cured with heat and pressure.

Pressure. A measure of force applied uniformly over a given unit of surface area.

Pressure altitude. The altitude in standard atmosphere at which the pressure is the same as the existing pressure.

Pressure carburetor. A carburetor installed on some aircraft reciprocating engines that uses the pressure difference between air inside the venture and ram air entering the carburetor to produce a fuel-metering force. Pressure carburetors have generally been replaced with continuous-flow fuel injection systems.

Pressure cooling. A method of air cooling a reciprocating engine in which the cylinders are enclosed in tight-fitting shrouds. The cowling is divided into two compartments by baffles and seals, with half of each cylinder in each compartment. Ram air is directed into one compartment, and the pressure in the other is decreased by air flowing over a flared exit or adjustable cowl flaps. The pressure difference across the cylinders causes cooling air to be drawn through the fins to remove the unwanted heat.

Pressure-injection carburetor. A multibarrel pressure carburetor used on large radial and V-engines. Fuel is metered on the basis of air mass flowing into the engine and is sprayed under pressure into the eye, or center, of the internal supercharger impeller.

Prevailing torque. The torque required to turn a threaded fastener before it contacts the surface it is intended to hold.

Primary winding. The winding in a magneto or ignition coil that is between the source of voltage and the breaker points. The primary winding is normally made of comparatively large diameter wire, and has a small number of turns, typically about 200.

Profile tip (compressor blade tip). The tip of an axial-flow compressor bladed whose thickness is reduced to give it a higher resonant frequency so it will not be subject to the vibrations that would affect a blade with a squared tip. The profile tip also provides a more aerodynamically efficient shape for the high velocity air that is moved by the blade. Profile tips often touch the housing and make a squealing noise as the engine is shut down. For this reason profile tips are often called squealer tips.

Profilometer. A precision measuring instrument used to measure the depth of the hone marks in the surface of a cylinder wall.

Prony brake. An instrument used to measure the amount of horsepower an engine is delivering to its output shaft. The engine is operated at a specific rpm, and a brake is applied to its output shaft. The amount of torque applied to the brake is measured, and this, with the rpm, is converted into brake horsepower.

Propeller. A device for propelling an aircraft that has blades on an engine-driven shaft and that, when rotated, produces by its action on the air, a thrust approximately perpendicular to its plane of rotation. It includes control components normally supplied by its manufacturer, but does not include main and auxiliary rotors or rotating airfoils of engines.

Propeller end. The end of a reciprocating engine to which the propeller is attached.

PropFan engine. The registered trade name by Hamilton Standard of an ultra-high-bypass turbine engine. See UHB engine.

Propulsive efficiency. A measure of the effectiveness with which an aircraft engine converts the fuel it burns into useful thrust. It is the ratio of the thrust horsepower produced by a propeller to the torque horsepower of the shaft turning the propeller. The nearer the speed of the aircraft is to the speed of the exhaust jet or propeller wake, the less kinetic energy is lost in the jet or wake, and the higher the propulsive efficiency.

PRT. Power recovery turbine. A turbine driven by exhaust gases from several cylinders of a reciprocating engine. Energy extracted from exhaust gases by the turbine is coupled, through a fluid clutch, to the engine crankshaft.

Pulsating DC. Direct current whose voltage periodically changes, but whose electrons flow in the same direction all of the time.

Pulse-jet engine. A type of air-breathing reaction engine used during World War II to power jet-propelled missiles. Fuel is sprayed into the combustion chamber and ignited. As the heated air expands, it closes the one-way shutter valve in the front of the engine and exits the engine through the nozzle at the rear. As soon as the pressure inside the combustion chamber decreases, air enters through the shutter valve and more fuel is ignited. The thrust is produced in a series of pulses.

Push fit. A fit between pieces in a mechanical assembly that is close enough to require the parts to be pushed together. A push fit is looser than a press fit, but closer than a free fit.

Pusher engine. An engine installed with the propeller facing the rear of the aircraft. Thrust produced by the propeller mounted on a pusher engine pushes rather than pulls the aircraft.

Pusher propeller. A propeller installed on an aircraft engine so that it faces the rear of the aircraft. Thrust from the propeller pushes rather than pulls the aircraft.

PV diagram. A diagram showing the relationship between the volume of a cylinder and the pressure during a cycle of engine operation.

Q

Quill shaft. A type of shaft used to couple parts of an engine that are subject to torsional loads. A quill shaft is a long, hardened steel shaft with splines on each end. One end splines into the drive shaft and the other end splines into the device being driven. Torsional vibrations are absorbed by the quill shaft twisting.

R

Radial bearing load. The load on a bearing perpendicular to the shaft on which the bearing is mounted. Centrifugal loads are radial loads.

Radial engine (static radial). A form of reciprocating engine in which the cylinders radiate out from a small central crankcase. The pistons in the cylinders drive a central crankshaft which in turn drives the propeller.

Radial-inflow turbine. A turbine, similar in appearance to a centrifugal compressor rotor. Radial-inflow turbines are used to drive the compressor in reciprocating engine turbochargers and some of the smaller APU turbine engines. Hot gases flow into the turbine from its outside rim, then radially inward through the vanes and out of the turbine at its center.

Radiation. See electromagnetic radiation.

Ram air. Air whose pressure has been increased by the forward motion of the aircraft. Ram air pressure is the same as pitot pressure.

Ram drag. The loss of thrust produced by a turbojet or turbofan engine caused by the increase of velocity of air entering the engine. Ram drag is the difference between gross thrust and net thrust.

Ram pressure. Pressure produced when a moving fluid is stopped.

Ram-recovery speed. The speed of an aircraft at which the ram effect caused by the forward movement increases the air pressure at the compressor inlet so that it is the same as that of the ambient air.

Ramjet engine. The simplest type of air-breathing reaction engine. Air entering the front of the engine at a high velocity has fuel sprayed into it and ignited. A barrier formed by the incoming air forces the expanding gases to leave through the nozzle at the rear. The energy added by the burning fuel accelerates the air and produces a forward thrust. Ramjet engines are used in some military unmanned aircraft that are initially boosted to a speed high enough for the engine to function.

Ratiometer indicator. An analog temperature measuring instrument in which the pointer deflection is proportional to the ratio between the current flowing in an internal reference circuit and that flowing through the temperature-sensing probe.

Reach (spark plug specification). The length of the threads on the shell of a spark plug.

Reaction engine. A form of heat engine that produces thrust by heating a mass of air inside the engine and discharging it at a high velocity through a specially shaped nozzle. The amount of thrust is determined by the mass of the air and the amount it is accelerated.

Reactive power. Wattless power in an AC circuit. It is the power consumed in the inductive and capacitive reactances. Reactive power is expressed in volt-amps reactive (var) or in kilovolt-amps reactive (kvar).

Reamed fir. The fit of a shaft in a hole in which the hole is drilled undersize and cut with a reamer to the correct diameter. Reamed holes have smooth walls and a consistent diameter.

Rebuilt engine. A used engine that has been completely disassembled, inspected, repaired as necessary, and reassembled, tested, and approved in the same manner and to the same tolerances and limits as a new engine, using either new or used parts. However, all parts used must conform to all production drawings, tolerances, and limits for new parts, or be of approved oversize or undersize dimensions for a new engine. According to 14 CFR part 91, section 91.421, a rebuilt engine is considered to have no precious operating history and may be issued a zero-time logbook. Only the engine manufacturer can rebuild an engine and issue a zero-time record.

Reciprocating engine. A type of heat engine that changes the reciprocating (back-and-forth) motion of pistons inside the cylinders into rotary motion of a crank-shaft.

Rectifier. A device that allows electrons to flow in one direction while preventing their flow in the opposite direction. Rectifiers are used to change AC into DC.

Reheat system. The British name for an afterburner. See afterburner.

Reid vapor pressure. The amount of pressure that must be exerted on a liquid to keep it from vaporizing. Reid vapor pressure is measured at 100 °F.

Reliability. The ability of an aircraft engine to perform its designed functions under widely varying operating conditions.

Residual magnetic particle inspection. A form of magnetic particle inspection for small steel parts that have a high degree of retentivity. The part is magnetized, removed, and inspected away from the magnetizing machine.

Residual magnetism. The magnetism that remains in the field frame of a generator when no current is flowing in the field coils.

Residual voltage. The voltage produced in a generator armature when the armature is rotated in the residual magnetism.

Resistor spark plug. A shielded spark plug with a resistor between the ignition lead terminal and the center electrode. The resistor stops the flow of secondary current when its voltage drops to a specified value. The resistor prevents capacitive afterfiring.

Retarded sparks. The timing of the firing of the spark plugs used to start a reciprocating engine. The sparks for starting occur later in terms of crankshaft rotation than those used for normal operation. Retarding the sparks prevent the engine from kicking back when it is being started.

Retentivity. The ability of a magnetizable material to retain the alignment of the magnetic domains after the magnetizing force has been removed. Hard steel normally has a high retentivity, while soft iron and electrical steel both have very low retentivity.

Reverse-flow combustor. A type of combustor in which the air from the compressor enters the combustor outer case and reverses its direction as it flows into the inner liner. It again reverses its direction as it flows into the inner liner. It again reverses its direction before it flows through the turbine. Reverse-flow combustors are used where engine length is critical.

RF energy. Electromagnetic energy with a frequency high enough to radiate from any conductor through which it is flowing.

Rich blowout. A condition in which the fire in a gas turbine engine goes out because the air/fuel mixture ratio is too rich to sustain combustion.

Rich mixture. A air/fuel mixture that contains less than 15 parts of air to 1 part of fuel, by weight.

Riffle file. A hand file with its teeth formed on a curved surface that resembles a spoon.

Rms. Root mean square. A dimension that is the square root of the average of an infinite number of varying values. An rms dimension is used to indicate the allowable surface roughness of a reciprocating engine cylinder wall.

Rocker arm. A pivoted arm on the cylinder head of a reciprocating engine. The pushrod forces one end of the rocker arm up, and as the other end moves down, it forces the poppet valve off of its seat.

Rocker box. The enclosed part of a reciprocating engine cylinder that houses the rocker arm and valve mechanism.

Rocket engine. A form of reaction engine whose fuel and oxidizer contain all of the oxygen needed for the release of heat energy. The released heat expands the gases which are ejected at a high velocity from a nozzle at the rear of the rocket. Because rocket engines carry their own oxygen, they can operate in outer space where there is no atmosphere.

Rotary radial engine. A form of reciprocating engine used in some early aircraft. The crankshaft is rigidly attached to the airframe, and the propeller, crankcase, and cylinders all revolve as a unit.

Rotating combustion (RC) engine. A form of internal combustion engine in which a rounded, triangular-shaped rotor with sliding seals at the apexes forms the combustion space inside an hourglass-shaped chamber. Expanding gases from the burning air/fuel mixture push the rotor around and turn a geared drive shaft in its center. The RC engine was conceived in Germany by Felix Wankel in 1955.

RPM. Revolutions per minute. A measure of rotational speed. One rpm is one revolution made in one minute.

Run in. A time of controlled operation of a new or freshly overhauled engine that allows the moving parts to wear together.

Run up. A procedure in which an aircraft engine is operated on the ground to determine its condition and performance.

Runout. A measure of the amount a shaft, flange, or disc is bent or fails to run true. Runout is normally measured with a dial indicator.

S

SAE. Society of Automotive Engineers. A professional organization that has formulated standards for the automotive and aviation industries.

Safety gap. A location in a magneto that allows a spark to jump to ground from the secondary circuit before the voltage rises high enough to damage the secondary insulation.

Sand casting. A method of molding metal parts in a mold made of sand. A pattern that duplicates the part to be molded is made of wood and is covered with a special casting sand that contains a resin to bind it. The mold is separated along a special parting line, and the pattern is removed. The mold is put back together, and molten metal is poured into the cavity. When the metal cools, the sand is broken away from the molded part. Sand casting is less expensive than permanent-mold casting.

Saybolt Seconds Universal (SSU) viscosity. A measurement of viscosity (resistance to flow)of a lubricating oil. The number of seconds needed for 60 milliliters of oil at a specified temperature to flow through a calibrated orifice. The viscosity number used for commercial aviation engine lubricating oil relates closely to the SSU viscosity of the oil at 210 °F.

Scavenge subsystem. The subsystem in the lubrication system of a gas turbine engine that collects oil after it has lubricated the bearings and gears and returns it to the oil tank.

Scimitar shape. The shape of the blades of the propellers mounted on UHB engines. The name is derived from the shape of a curved Asian sword that has its edge on the convex side. See UHB engine.

Scramjet. Supersonic combustion ramjet. A special type of ramjet engine whose fuel can be ignited while the vehicle is mobbing at a supersonic speed.

Scuffing. Severe damage to moving parts caused when one metal part moves across another without sufficient lubricant between them. Enough heat is generated by friction to cause the high points of the surfaces to weld together; continued movement tears, or scuffs, the metal.

Sea-level boosted engine. A reciprocating engine that has had its sea-level rated horsepower increased by supercharging. This is the same as a ground-boosted engine.

Secondary winding. The winding in a magneto or ignition coil that connects to the distributor rotor. The secondary winding is normally made of very small diameter wire and has a large number of turns, typically about 20,000.

Self-accelerating speed. The speed attained by a gas turbine engine during start-up that allows it to accelerate to its normal idling speed without assistance from the starter.

Semiconductor transducer. A piezoelectric crystal that converts input energy of one form, such as pressure, into output energy of another, such as an electrical signal.

Series-wound motor. An electric motor with field coils connected in series with the armature.

Serviceable limits. Limits included in a reciprocating engine overhaul manual. If a part measures outside of the new-parts limits, but within the serviceable limits, it will not likely wear to the point of causing engine failure within the next TBO interval.

Servo system. A type of automatic control system in which part of the output is fed back into the input.

Shaft horsepower. The horsepower actually available at a rotating shaft.

Shielding. The electrically conductive covering placed around an electrical component to intercept and conduct to ground any electromagnetic energy radiated from the device.

Short circuit. A low-resistance connection between two points in an electric circuit.

Shower of Sparks ignition system. A patented ignition system for reciprocating engines. An induction vibrator sends pulsating DC into a set of retard breaker points on one of the magnetos. This provides a hot and retarded spark for starting the engine.

Single-shaft turbine engine. A turboprop engine in which the propeller reduction gears are driven by the same shaft that drives the compressor for the gas generator.

Single-spool gas-turbine engine. A type of axial-flow-compressor gas turbine engine that has only one rotating element.

Skin radiator. A type of radiator used on some early liquid-cooled racing airplanes. The radiator was made of two thin sheets of brass, slightly separated so the heated coolant could flow between them. Skin radiators were mounted on the surface of the wing, on the sides of the fuselage, or on the floats of seaplanes. Air flowing over the smooth surface of the radiator removed heat from the coolant.

Slip (propeller specification). The difference between the geometric and effective pitch of a propeller.

Slip ring. A smooth, continuous ring of brass or copper mounted on the rotor shaft of an electrical generator or alternator. Brushes riding on the smooth surface of the slip ring carry current into and out of the rotor coil.

Slow-blow fuse. A special type of electrical circuit protection device that allows a momentary flow of excess current, but opens the circuit if the excessive flow is sustained.

Sludge. A heavy contaminant that forms in an aircraft engine lubricating oil because of oxidation and chemical decomposition of the oil.

Sludge plugs. Spool-shaped sheet metal plugs installed in the hollow throws of some engine crankshafts.

Slug. The unit of mass equal to that which experiences an acceleration of one foot per second, per second when a force of one pound acts on it. It is equal to 32.174 pounds, or 14.5939 kilograms, of mass. Also called a G-pound.

SOAP. Spectrometric oil analysis program. An oil analysis program in which a sample of oil is burned in an electric arc and an analysis is made of the wavelength composition of the resulting light. Each chemical element in the oil, when burned, produces light containing a unique band of frequencies. A computer analyzes the amount of each band of frequencies and prints out the number of parts of the element per million parts of the entire sample. SOAP can predict engine problems by warning the engine operator of an uncharacteristic increase of any elements in the oil.

Sound suppressor. The airframe component that replaces the turbine engine tail pipe. It reduces the distance the sounds made by the exhaust gases propagate by converting low-frequency vibrations.

Specific gravity. The ratio of the density of a material to the density of pure water.

Specific weight. The ratio of the weight of an aircraft engine to the brake horsepower it develops.

Spline. Parallel slots cut in the periphery of a shaft, parallel to its length. Matching slots, cut into the hub or wheel that fits on the shaft, lock the shaft into the device to transmit torque.

Sprag clutch. A freewheeling, nonreversible clutch that allows torque to be applied to a driven unit in one direction only.

Springback. A condition in the rigging of an aircraft engine control in which the stop at the engine is reached before the stop in the flight deck. The flight deck control moves slightly after the stop in the engine is reached, and when it is released, it springs back slightly.

Spur-gear pump. A form of constant-displacement fluid pump that uses two meshing spur-gears mounted in a close fitting housing. Fluid is taken into the housing where it fills the space between the teeth of the gears and is carried around the housing as the gears rotate. On the discharge side of the pump, the teeth of the two gears mesh, and the fluid is forced out of the pump.

Squat switch. An electrical switch actuated by the landing gear scissors on the oleo strut. When no weight is on the landing gear, the oleo piston is extended and the switch is in one position; but when weight is on the gear, the oleo strut compresses and the switch changes its position.

Squealer tip (compressor blade tip). See profile tip.

Squeeze film bearings. Another name for oil-damped bearings. See oil-damped bearings.

Stage length. The distance between landing points in airline operation.

Stage of a compressor. One disc of rotor blades and the following set of stator vanes in an axial-flow compressor.

Staggered timing. Ignition timing that causes the spark plug nearest the exhaust valve to fire a few degrees of crankshaft rotation before the spark plug nearest the intake valve.

Standard day conditions. Conditions that have been decided upon by the ICAO for comparing all aircraft and engine performance. The most basic standard day conditions are: temperature, 15 °C or 59 °F; altitude, mean sea level; pressure, 29.92 inches of mercury.

Standard J-1. A World War I training airplane powered by a Curtiss OX-5 engine.

Standpipe. A pipe which protrudes upward from the base of an oil tank and through which oil used for normal engine lubrication is drawn. In the event of a catastrophic leak when all oil available to the engine-driven pump is lost overboard, enough oil is available from an outlet below the standpipe to feather the propeller.

Starter-generator. A single-component starter and generator used on many smaller gas-turbine engines. It is used to start the engine, and when the engine is running, its circuitry is shifted so that it acts as a generator.

Static pressure. The pressure of an unmoving fluid.

Static rpm. The number of revolutions per minute an aircraft engine can produce when the aircraft is not moving.

Steam cooling. A method of liquid cooling in which the coolant, normally water, is allowed to absorb enough heat that it boils. The steam gives up its heat when it condenses back into a liquid.

Stellite. A nonferrous alloy of cobalt, chromium, and tungsten. Stellite is hard, water resistant, and corrosion resistant, and it does not soften until its temperature is extremely high. Stellite is welded to the faces of many reciprocating engine exhaust valves that operate at very high temperatures.

Stepping motor. A precision electric motor whose output shaft position is changed in steps by pulses from the control device. Stepping motors can make high-torque changes in small angular increments to their output shaft.

Stoichiometric mixture. The air/fuel mixture ratio that, when burned, leaves no uncombined oxygen nor any free carbon. It releases the maximum amount of heat, and therefore produces the highest exhaust gas temperature. A stoichiometric mixture of gasoline and air contains 15 pounds of air for 1 pound of gasoline.

Straight-through combustor. A combustor in a gas turbine engine through which the air from the compressor to the turbine flows in an essentially straight line.

Stratosphere. The upper part of the Earth's atmosphere. The stratosphere extends upward from the tropopause, which is approximately 36,000 feet above the surface of the Earth, to approximately 85,000 feet. The temperature of the air in the stratosphere remains constant at -56.5 °C (-69.7 °F).

Stress. A force within an object that tries to prevent an outside force from changing its shape.

Stroboscopic tachometer. A tachometer used to measure the speed of any rotating device without physical contact. A highly accurate variable-frequency oscillator triggers a high-intensity strobe light. When the lamp is flashing at the same frequency the device is rotating, the device appears to stand still.

Stroke. The distance the piston moves inside the cylinder.

Sump (aircraft engine component). A low point in an aircraft engine in which lubricating oil collects and is stored or transferred to an external oil tank. A removable sump attached to the bottom of the crankcase of a reciprocating engine is often called an oil pan.

Sump (fuel tank component). A low point in an aircraft fuel tank in which water and other contaminants collect and are held until they can be drained out.

Supercharged engine. A reciprocating engine that uses a mechanically driven compressor to increase the air pressure before it enters the engine cylinders.

Supercharger. An air compressor used to increase the pressure of the air being taken into the cylinders of a reciprocating engine.

Surface roughness. The condition of the surface of a reciprocating engine cylinder wall that has been honed to make it hold lubricating oil. Surface roughness is measured in micro-inches rms.

Surge. A condition of unstable airflow, through the compressor of a gas turbine engine, in which the compressor blades have an excessive angle of attack. Surge usually affects an entire stage of compression.

Synthetic oil. Oil made by chemical synthesis of a mineral, animal, or vegetable base. Synthetic oils have appropriate additives that give them such characteristics as low volatility, low pour point, high viscosity index, good lubricating qualities, low coke and lacquer formation, and low foaming.

T

Tachometer. An instrument that measures the rotational speed of an object.

TAI. Thermal anti-ice. A system used to prevent the formation of ice on an aircraft by flowing heated air inside the structure.

Tail pipe. The portion of the exhaust system of a gas turbine engine through which the gases leave. The tail pipe is often called the exhaust duct, or exhaust pipe.

TBO. Time between overhauls. A time period specified by the manufacturer of an aircraft engine as the maximum length of time an engine should be operated between overhauls without normal wear causing parts of the engine to be worn beyond safe limits. TBO depends upon proper operation and maintenance in accordance with the engine manufacturer's recommendations. The overhaul of an engine when it reaches its TBO hours is not mandatory, except for certain commercial operators that have the requirement written into their operations manual.

TDC. Top dead center. The position of a piston in a reciprocating engine when the piston is at the top of its stroke and the wrist pin, crankpin, and center of the crankshaft are all in line.

TEL. Tetraethyl lead.

Test club. A wide-blade, short-diameter propeller used on a reciprocating engine when it is run in a test cell. A test club applies a specific load to the engine and forces the maximum amount of air through the engine cooling fins.

Thermal efficiency. The ratio of the amount of useful work produced by a heat engine, to the amount of work that could be done by all of the heat energy available in the fuel burned.

Thermal expansion coefficient. A number that relates to the change in the physical dimensions of a material as the temperature of the material changes. The thermal expansion coefficient of aluminum is approximately twice that of steel.

Thermal shock. The sudden change in engine operating temperature that occurs when engine power is suddenly reduced at the same time the airspeed, thus the cooling, is increased. Thermal shock occurs when an aircraft is required to rapidly descend to a lower altitude.

Thermistor. A semiconductor material whose electrical resistance varies with its temperature.

Thermocouple. A device used to generate an electrical current. A thermocouple is made of two dissimilar metal wires whose ends are welded together to form a loop. A voltage exists in the loop proportional to the difference in temperature of the junctions at which the wires are joined. The amount of current flowing in the loop is determined by the types of metals used for the wires, the temperature difference between the junctions, and the resistance of the wires.

Thermosetting resin. A plastic resin that, once it has been hardened by heat, cannot be softened by heating again.

Thermostatic valve. A temperature-sensitive valve that controls the temperature of oil in an aircraft engine. When the oil is cold, the valve shifts and directs the oil through the oil cooler.

Thermoswitch. An electrical switch that closes a circuit when it is exposed to a specified high temperature.

Three-dimensional cam. A drum-shaped cam in a hydromechanical fuel control whose outer surface is ground so that followers riding on the surface, as the cam is moved up and down and rotated, can move mechanical linkages to control the fuel according to a preprogrammed schedule.

Throttle. The control in an aircraft that regulates the power or thrust the pilot wants the engine to produce.

Throw (crankshaft design). See offset throw.

Thrust horsepower. The horsepower equivalent of the thrust produced by a turbojet engine. Thrust horsepower is found by multiplying the net thrust of the engine, measured in pounds, by the speed of the aircraft, measured in miles per hour, and then dividing this by 375.

Thrust. The aerodynamic force produced by a propeller or turbojet engine as it forces a mass of air to the rear, behind the aircraft. A propeller produces its thrust by accelerating a large mass of air by a relatively small amount. A turbojet engine produces its thrust by accelerating a smaller mass of air by a much larger amount.

Time-Rite indicator. A patented piston-position indicator used to find the position of the piston in the cylinder of a reciprocating engine. The body of the Time-Rite indicator screws into a spark plug hole, and as the piston moves outward in the cylinder, it contacts the arm of the indicator. A pointer contacted by the arm moves across a calibrated scale to show the location of the piston in degrees of crankshaft rotation before top center.

Timing light. An indicator light used when timing magnetos to an engine to indicate when the breaker points open. Some timing lights incorporate an oscillator or buzzer that changes its pitch when the points open.

TIT. Turbine inlet temperature. The temperature of the gases from the combustion section of a gas turbine engine as they enter the turbine inlet guide vanes or the first stage of the turbine.

Toggle. A T-shaped handle fitted onto the end of a cable used to engage a simple starter with an overrunning clutch.

Top overhaul. An overhaul of the cylinders of an aircraft engine. The valves, pistons, and cylinders are overhauled, but the crankcase is not opened.

Torque. A force that produces or tries to produce rotation.

Total pressure. The pressure a column of moving fluid would have if it were stopped from its motion. Total pressure is the sum of dynamic pressure and static pressure.

Total temperature. The temperature of moving fluid that has been stopped from its motion. Total temperature is the sum of static temperature and the temperature rise caused by the ram effect as the fluid was stopped.

Townend ring. A type of ring cowling used over a single-row radial engine. The cross section of the ring is in the form of an airfoil that produces enough forward thrust to compensate for the cooling drag of the engine. In the United States, townend rings are often called speed rings.

Track. The path followed by a blade segment of a propeller or helicopter rotor in one rotation.

Tractor engine. An engine installed with the propeller facing the front of the aircraft. Thrust produced by the propeller mounted on a tractor engine pulls the aircraft through the air. tractor propeller. A propeller mounted on an airplane in such a way that its thrust pulls the aircraft.

Trailing edge. The thin edge at the rear of a propeller blade.

Transducer. A device that changes energy from one form to another. Commonly used transducers change mechanical movement or pressures into electrical signals.

Transformer. An electrical component used to change the voltage and current in an AC circuit.

Transonic range. Flight at Mach numbers between 0.8 and 1.2. In this range, some air passing over the aircraft is subsonic, and some is supersonic.

Trend monitoring. A system for comparing engine performance parameters with a baseline of these same parameters established when the engine was new or newly overhauled. Parameters such as EGT, rpm, fuel flow, and oil consumption are monitored on every flight, and the baseline is plotted. Any deviation from a normal increase or decrease warns the technician of an impending problem.

Tricresyl phosphate (TCP). A colorless, combustible compound, $(CH_3C_6H_4O)_3PO$, that is used as a plasticizer in aircraft dope and an additive in gasoline and lubricating oil.

TCP aids in scavenging lead deposits left in the cylinders when leaded fuel is burned.

TSFC. Thrust specific fuel consumption. A measure of efficiency of a turbojet or turbofan engine. It is a measure of the number of pounds of fuel burned per hour for each pound of thrust produced.

Turbine. A wheel fitted with vanes, or buckets, radiating outward from its circumference. The reactive or aerodynamic force caused by the fluid flowing through the vanes is converted into mechanical power that spins the shaft on which the wheel is mounted.

Turbine engine. See gas turbine engine.

Turbine inlet guide vanes. A series of stator vanes immediately ahead of the first-stage turbine. The function of the inlet guide vanes is to divert the hot gases in the proper direction to enter the turbine, and to provide a series of convergent ducts which increase the velocity of the gases.

Turbine nozzle. Another name for turbine inlet guide vanes.

Turbocharger. An exhaust-driven air compressor used to increase the power of a reciprocating engine. A turbocharger uses a small radial inflow turbine in the exhaust system to drive a centrifugal-type air compressor on the turbine shaft. The compressed air is directed into the engine cylinders to increase power.

Turbo-compound engine. A reciprocating engine that has power recovery turbines in its exhaust system. The power extracted from the exhaust by these turbines is directed into the engine crankshaft through a fluid coupling.

Turbofan engine. A type of gas turbine engine that has a set of lengthened blades on the low-pressure compressor or low-pressure turbine. Air moved by these special blades bypasses the core engine and produces between 30% and 75% of the total thrust.

Turbojet engine. A gas turbine engine that produces thrust by accelerating the air flowing through it. A minimum of energy is extracted by the turbine, with the majority used to produce an exhaust velocity much greater than the inlet velocity. The amount of thrust produced by the engine is determined by the amount the air is accelerated as it flows through the engine.

Turboprop engine. A turbine engine in which several stages of turbines are used to extract as much energy as possible. The turbines drive reduction gears which in turn drive a propeller.

Turboshaft engine. A turbine engine in which several stages of turbines are used to extract as much energy as possible. The turbines drive shafts which are used to drive helicopter rotors, generators, or pumps.

Turbosupercharger. A centrifugal air compressor driven by exhaust gases flowing through a turbine. The compressed air is used to increase the power produced by a reciprocating engine at altitude.

Two-spool engine. See dual-spool gas turbine engine.

Two-stroke cycle. A constant-volume cycle of energy transformation that completes its operating cycle in two strikes of the piston, one up and one down. When the piston moves up, fuel is pulled into the crankcase, and at the same time the air/fuel mixture inside the cylinder is compressed. When the piston is near the top of its stroke, a spark plug ignites the compressed air/fuel mixture, and the burning and expanding gases force the piston down. Near the bottom of the stroke, the piston uncovers an exhaust port and the burned gases leave the cylinder. When the piston moves further down, it uncovers the intake port, and a fresh charge of fuel and air are forced from the crankcase into the cylinder.

U

UDF engine. Un-ducted Fan™. The trade name registered by General Electric for a type of ultra-high-bypass turbofan engine that drives one or more wide-blade propellers that have between eight and twelve blades. These blades, which are not enclosed in a duct or shroud, are very thin, have wide chords, and are highly swept back in a scimitar shape that enables them to power airplanes flying in the speed range near Mach 0.8.

UHB (ultra-high-bypass) engine. A turbine that drives a pair of ducted or un-ducted contrarotating propellers which have eight to 12 variable-pitch blades. These blades are very thin, have wide chords, and are swept back with a scimitar shape that allows them to power airplanes flying in the speed range of Mach 0.8. The blades are made of advanced composites for high strength and light weight. USH engines may be of either the tractor or pusher type, and have a bypass ratio in excess of 30:1.

Underspeed condition. A speed condition in which the engine is turning at an RPM lower than that for which the propeller governor is set.

Unidirectional fibers. Fibers in a piece of composite material arranged so that they sustain loads in only one direction.

Updraft carburetor. A carburetor that mounts on the bottom of a reciprocating engine. Air entering the engine flows upward through the carburetor.

Upper-deck pressure. The absolute pressure of air at the inlet to the fuel metering system of a turbocharged engine. Upper-deck pressure is the same as the turbocharger discharge pressure.

V

V-blocks. A fixture that allows a shaft to be centered and rotated to measure any out-of-round condition.

V-engine. A form of reciprocating engine in which the cylinders are arranged in two banks. The banks are separated by an angle of between 45° and 90°. Pistons in two cylinders, one in each bank, are connected to each throw of the crankshaft.

Valence electrons. Electrons in the outer shell, or ring, around the nucleus of an atom. It is the valence electrons that give an atom its electrical characteristics and are the electrons that may be pulled loose from an atom to cause electrical current.

Valve overlap. The portion of the operating cycle of a four-stroke-cycle reciprocating engine during which both the intake and exhaust valves are off of their seats at the same time.

Vapor lock. A condition of fuel starvation that can occur in a reciprocating engine fuel system. If the fuel in the line between the tank and carburetor is heated enough for the fuel to vaporize, a bubble will form in the line. If the vapor pressure of the bubble is high enough, it will block the fuel and keep it from flowing to the engine.

Vapor pressure. The amount of pressure needed above a liquid to prevent it from evaporating.

Vaporize. The changing of a liquid into a vapor.

Vectored-thrust engine. A turbojet or turbofan engine with the fan and/or exhaust nozzles mounted in such a way that they may be rotated in flight to produce forward, vertically upward, or rearward thrust.

Velocity. A vector quantity that expresses both the speed an object is moving and the direction in which it is moving.

Velocity turbine. A turbine driven by forces produced by the velocity, rather than the pressure, of gases flowing through the vanes.

Venture. A specially shaped restrictor in a tube designed to speed up the flow of fluid passing through it. According to Bernoulli's principal, any time the flow of fluid speeds up without losing or gaining any energy from the outside, the pressure of the fluid decreases.

Vernier coupling. A timing coupling used with base-mounted magnetos. The vernier coupling allows the timing to be adjusted in increments of considerably less than one degree.

Vertical tape instrument. A tall rectangular instrument that displays the quantity of the parameter being measured by a movable strip of colored tape. The presentation resembles a vertical bar graph.

Vibration loop. A loop in a rigid fluid line used to prevent vibration from concentrating stresses that could cause the line to break.

VIFF. Vectoring in forward flight. A method of enhancing the maneuverability of an airplane by vectoring the exhaust gases and/or fan-discharge air to produce thrust components not parallel to the longitudinal axis of the aircraft.

Viscosimeter. An instrument used to measure the viscosity of a liquid. The time required for a given volume of liquid at a specified temperature to flow through a calibrated orifice is used to indicate the viscosity of the liquid.

Viscosity. The resistance of a fluid to flow. Viscosity is the stiffness of the fluid, or its internal friction.

Viscosity index (VI). A measure of change in viscosity of an oil as it changes temperature. The higher the viscosity index, the less the viscosity changes.

Viscosity index improver. An additive used to produce a multi-viscosity lubricating oil. The polymer additive expands as temperature increases and contracts as temperature decreases. VI improvers cause viscosity to increase as oil heats and decrease as it cools.

Volatile memory. Computer memory that is lost when the power to the computer is turned off.

Volatility. The characteristic of a liquid that relates to its ability to vaporize or change into a gas.

Volumetric efficiency. The ratio of the volume of the charge of the fuel and air inside the cylinder of a reciprocating engine to the total physical volume of the cylinder.

Von Ohain, Dr. Hans Pabst. The designer and developer of the first turbojet engine to power an airplane. His HeS3b engine was built in Germany by the Heinkel Company and it flew in a Heinkel He178 airplane on August 27, 1939.

Vortex. A whirling mass of air that sucks everything near it toward its center.

Vortex dissipator. A high-velocity stream of compressor bleed air blown from a nozzle into an area where vortices are likely to form. Vortex dissipaters destroy the vortices that would otherwise suck debris from the ground into engines mounted in pods that are low to the ground.

W

Wake. The high-velocity stream of turbulent air behind an operating aircraft engine.

Wankel engine. See rotating combustion (RC) engine.

Waste gate. A controllable butterfly valve in the exhaust pipe of a reciprocating engine equipped with an exhaust-driven turbocharger. When the waste gate is open, exhaust gases leave the engine through the exhaust pipe, and when it is closed, they leave through the turbine.

Watt. The basic unit of power in the metric system. One watt is the amount of power needed to do one joule (0.7376 foot-pound of work) in one second. One watt is $\frac{1}{746}$ horsepower.

Wet-sump engine. An engine that carries its lubricating oil supply in a reservoir that is part of the engine itself.

Wet-sump lubrication system. A lubrication system in which the oil supply is carried within the engine itself. Return oil drains into the oil reservoir by gravity.

Whittle, Sir Frank. The British Royal Air Force flying officer who in 1929 filed a patent application for a turbojet engine. Whittle's engine first flew in a Gloster E.28 on May 15, 1941. The first jet flight in America was made on October 2, 1942, in a Bell XP-59A that was powered by two Whittle-type General Electric I-A engines.

Windmilling propeller. A propeller that is rotated by air flowing over the blades rather than powered by the engine.

Work. The product of a force times the distance the force is moved.

Worm gear. A helical gear mounted on a shaft. The worm meshes with a spur gear whose teeth are cut at an angle to its face. A worm gear is an irreversible mechanism. The rotation of the shaft, on which the worm gear locks the spur gear so its shaft cannot be rotated.

Wrist pin. The hardened steel pin that attaches a piston to the small end of a connecting rod.

Y

Yaw. Rotation of an aircraft about its vertical axis.

Z

Zero-lash valve lifter. A hydraulic valve lifter that maintains zero clearance in the valve actuating mechanism.

Made in the USA
Las Vegas, NV
13 February 2024

85731978R00273